SI Prefixes

Multiplication Factor	Prefix†	Symbol
$1\ 000\ 000\ 000\ 000 = 10^{12}$	tera	T
$1\ 000\ 000\ 000 = 10^{9}$	giga	G
$1\ 000\ 000 = 10^{6}$	mega	M
$1\ 000 = 10^{3}$	kilo	k
$100 = 10^{2}$	hecto‡	h
$10 = 10^{1}$	deka‡	da
$0.1 = 10^{-1}$	deci‡	d
$0.01 = 10^{-2}$	centi‡	c
$0.001 = 10^{-3}$	milli	m
$0.000\ 001 = 10^{-6}$	micro	μ
$0.000\ 000\ 001 = 10^{-9}$	nano	n
$0.000\ 000\ 000\ 001 = 10^{-12}$	pico	p
$0.000\ 000\ 000\ 000\ 001 = 10^{-15}$	femto	f
$0.000\ 000\ 000\ 000\ 000\ 001 = 10^{-18}$	atto	a

† The first syllable of every prefix is accented so that the prefix will retain its identity. Thus, the preferred pronunciation of kilometer places the accent on the first syllable, not the second.

‡ The use of these prefixes should be avoided, except for the measurement of areas and volumes and for the nontechnical use of centimeter, as for body and clothing measurements.

Principal SI Units Used in Mechanics

Quantity	Unit	Symbol	Formula
Acceleration	Meter per second squared	. . .	m/s^2
Angle	Radian	rad	†
Angular acceleration	Radian per second squared	. . .	rad/s^2
Angular velocity	Radian per second	. . .	rad/s
Area	Square meter	. . .	m^2
Density	Kilogram per cubic meter	. . .	kg/m^3
Energy	Joule	J	$N \cdot m$
Force	Newton	N	$kg \cdot m/s^2$
Frequency	Hertz	Hz	s^{-1}
Impulse	Newton-second	. . .	$kg \cdot m/s$
Length	Meter	m	‡
Mass	Kilogram	kg	‡
Moment of a force	Newton-meter	. . .	$N \cdot m$
Power	Watt	W	J/s
Pressure	Pascal	Pa	N/m^2
Stress	Pascal	Pa	N/m^2
Time	Second	s	‡
Velocity	Meter per second	. . .	m/s
Volume, solids	Cubic meter	. . .	m^3
Liquids	Liter	l	$10^{-3}\ m^3$
Work	Joule	J	$N \cdot m$

† Supplementary unit (1 revolution $= 2\pi$ rad $= 360°$).
‡ Base unit.

U.S. Customary Units and Their SI Equivalents

Quantity	U.S. Customary Unit	SI Equivalent
Acceleration	ft/s^2	0.3048 m/s^2
	$in./s^2$	0.0254 m/s^2
Area	ft^2	0.0929 m^2
	in^2	645.2 mm^2
Energy	$ft \cdot lb$	1.356 J
Force	kip	4.448 kN
	lb	4.448 N
	oz	0.2780 N
Impulse	$lb \cdot s$	$4.448 \text{ N} \cdot \text{s}$
Length	ft	0.3048 m
	in.	25.40 mm
	mi	1.609 km
Mass	oz mass	28.35 g
	lb mass	0.4536 kg
	slug	14.59 kg
	ton	907.2 kg
Moment of a force	$lb \cdot ft$	$1.356 \text{ N} \cdot \text{m}$
	$lb \cdot in.$	$0.1130 \text{ N} \cdot \text{m}$
Moment of inertia		
Of an area	in^4	$0.4162 \times 10^6 \text{ mm}^4$
Of a mass	$lb \cdot ft \cdot s^2$	$1.356 \text{ kg} \cdot \text{m}^2$
Power	$ft \cdot lb/s$	1.356 W
	hp	745.7 W
Pressure or stress	lb/ft^2	47.88 Pa
	lb/in^2 (psi)	6.895 kPa
Velocity	ft/s	0.3048 m/s
	in./s	0.0254 m/s
	mi/h (mph)	0.4470 m/s
	mi/h (mph)	1.609 km/h
Volume, solids	ft^3	0.02832 m^3
	in^3	16.39 cm^3
Liquids	gal	3.785 l
	qt	0.9464 l
Work	$ft \cdot lb$	1.356 J

MECHANICS OF MATERIALS

Software to accompany this text.

Computer diskettes to accompany this text are available separately.
Please contact McGraw-Hill office nearest you or
McGraw-Hill Book Co., 21 Neythal Road, Singapore 2262

When ordering the diskettes, please quote Part No: 0-07-112939-D

MECHANICS OF MATERIALS

SECOND EDITION IN SI UNITS

Ferdinand P. Beer
Lehigh University

E. Russell Johnston, Jr.
University of Connecticut

With the collaboration of
John T. DeWolf
University of Connecticut

Adapted for SI units by
David C. Barton and Brian B. Miatt
University of Leeds

McGRAW-HILL BOOK COMPANY

London · New York · St Louis · San Francisco · Auckland
Bogotá · Caracas · Hamburg · Lisbon · Madrid · Mexico
Milan · Montreal · New Delhi · Panama · Paris · San Juan
São Paulo · Singapore · Sydney · Tokyo · Toronto

Mechanics of Materials, 2/E in SI units.

Metric Edition 1992

Exclusive rights by McGraw-Hill Book Co. Singapore for manufacture and export. This book cannot be re-exported from the country to which it is consigned by McGraw-Hill

10 09 08 07 06
20 09 08 07 06 05 04 03 02
PMP BJE
Published by
McGRAW-HILL Book Company Europe
Shoppenhangers Road, Maidenhead, Berkshire, SL6 2QL, England
Telephone 0628 23432
Fax 0628 770224

British Library Cataloguing in Publication Data

Beer, Ferdinand P.
   Mechanics of Materials. – 2 Rev. ed S.I. ed
   I. Title   II. Johnston, E. Russell
   III. Barton, David C.
   IV. Miatt, Brian B.
   620.112

When ordering this title, use ISBN 0-07-112939-1

Printed in Singapore

ABOUT THE AUTHORS

"How did you happen to write your books together, with one of you at Lehigh and the other at UConn, and how do you manage to keep collaborating on their successive revisions?" These are the two questions most often asked of our two authors.

The answer to the first question is simple. Russ Johnston's first teaching appointment was in the Department of Civil Engineering and Mechanics at Lehigh University. There he met Ferd Beer, who had joined that department two years earlier and was in charge of the courses in mechanics. Born in France and educated in France and Switzerland (he holds an M.S. degree from the Sorbonne and a Sc. D. degree in the field of theoretical mechanics from the University of Geneva), Ferd had come to the United States after serving in the French army during the early part of World War II and had taught for four years at Williams College in the Williams-MIT joint arts and engineering program. Born in Philadelphia, Russ had obtained a B.S. degree in civil engineering from the University of Delaware and a Sc. D. degree in the field of structural engineering from MIT.

Ferd was delighted to discover that the young man who had been hired chiefly to teach graduate structural engineering courses was not only willing but eager to help him reorganize the mechanics courses. Both believed that these courses should be taught from a few basic principles and that the various concepts involved would be best understood and remembered by the students if they were presented to them in a graphic way. Together they wrote lecture notes in statics and dynamics, to which they later added problems they felt would appeal to future engineers, and soon they produced the manuscript of the first edition of *Mechanics for Engineers*.

The second edition of this text and the first edition of *Vector Mechanics for Engineers* found Russ Johnston at Worcester Polytechnic Institute and the next editions at the University of Connecticut. In the meantime, both Ferd and Russ had assumed administrative responsibilities in their departments, and both were involved in research, consulting, and supervising graduate students—Ferd in the area of stochastic processes and random vibrations, and Russ in the area of elastic stability and structural analysis and design. However, their interest in improving the teaching of the basic mechanics courses had not subsided and they both taught sections of these courses as they kept revising their texts and began writing the manuscript of the first edition of *Mechanics of Materials*.

This brings us to the second question: How did the authors manage to work together so effectively after Russ Johnston had left Lehigh? Part of the answer may be provided by their phone bills and the money they have spent on postage. As the publication date of a new edition approaches, they call each other daily and rush to the post office with express-mail packages. There are also visits between the two families. At one time there were even joint camping trips, with both families pitching their tents next to each other. Now, with the advent of the fax machine, they do not need to meet so frequently.

Their collaboration has spanned the years of the revolution in computing. The first editions of *Mechanics for Engineers* and of *Vector Mechanics for Engineers* included notes on the proper use of the slide rule. To guarantee the accuracy of the answers given in the back of the book, the authors themselves used oversize 20-inch slide rules, then mechanical desk calculators complemented by tables of trigonometric functions, and later four-function electronic calculators. With the advent of the pocket multifunction calculators, all these were relegated to their respective attics, and the notes in the text on the use of the slide rule were replaced by notes on the use of calculators. In this edition of *Mechanics of Materials*, students are encouraged to use computers when available, and a note on the programming of singularity functions has been added to the text.

A new collaborator, John T. DeWolf, Professor of Civil Engineering at the University of Connecticut, has joined the Beer and Johnston team for this new edition. John holds a B.S. degree in engineering from the University of Hawaii and M.E. and Ph.D. degrees from Cornell University. His research interests are in the area of elastic stability, experimental analysis, and structural analysis and design.

Preface   xiii

List of Symbols   xvii

1.1   Introduction   1
1.2   Forces and Stresses   2
1.3   Axial Loading; Normal Stress   5
1.4   Shearing Stress   7
1.5   Bearing Stress in Connections   9
1.6   Application to the Analysis of Simple Structures   9
1.7   Stress on an Oblique Plane under Axial Loading   19
1.8   Stress under General Loading Conditions;
          Components of Stress   20
1.9   Ultimate and Allowable Stress: Factor of Safety   24
          Review and Summary   32

2.1   Introduction   39
2.2   Normal Strain under Axial Loading   40
2.3   Stress-Strain Diagram   42
*2.4 True Stress and True Strain   46
2.5   Hooke's Law; Modulus of Elasticity   47
2.6   Elastic versus Plastic Behavior of a Material   48
2.7   Repeated Loadings; Fatigue   50
2.8   Deformations of Members under Axial Loading   51
2.9   Statically Indeterminate Problems   60
2.10   Problems Involving Temperature Changes   64
2.11   Poisson's Ratio   74
2.12   Multiaxial Loading; Generalized Hooke's Law   75

*2.13	Dilatation; Bulk Modulus	77
2.14	Shearing Strain	79
2.15	Further Discussion of the Deformations under Axial Loading: Relation among E, ν, and G	82
2.16	Stress and Strain Distribution under Axial Loading; Saint-Venant's Principle	90
2.17	Stress Concentrations	93
2.18	Plastic Deformations	95
*2.19	Residual Stresses	99
	Review and Summary	105

3
TORSION 114

3.1	Introduction	114
3.2	Preliminary Discussion of the Stresses in a Shaft	116
3.3	Deformations in a Circular Shaft	117
3.4	Stresses in the Elastic Range	120
3.5	Angle of Twist in the Elastic Range	129
3.6	Statically Indeterminate Shafts	133
3.7	Design of Transmission Shafts	143
3.8	Stress Concentrations in Circular Shafts	145
*3.9	Plastic Deformations in Circular Shafts	150
*3.10	Circular Shafts Made of an Elastoplastic Material	151
*3.11	Residual Stresses in Circular Shafts	154
*3.12	Torsion of Noncircular Members	163
*3.13	Thin-Walled Hollow Shafts	166
	Review and Summary	174

4
PURE BENDING 183

4.1	Introduction	183
4.2	Prismatic Members in Pure Bending	184
4.3	Preliminary Discussion of the Stresses in Pure Bending	186
4.4	Deformations in a Symmetric Member in Pure Bending	187
4.5	Stresses and Deformations in the Elastic Range	190
4.6	Deformations in a Transverse Cross Section	194
4.7	Bending of Members Made of Several Materials	204
4.8	Stress Concentrations	208
*4.9	Plastic Deformations	218
*4.10	Members Made of an Elastoplastic Material	220
*4.11	Plastic Deformations of Members with a Single Plane of Symmetry	224
*4.12	Residual Stresses	224

4.13 Eccentric Axial Loading in a Plane of Symmetry   233
4.14 Unsymmetric Bending   242
4.15 General Case of Eccentric Axial Loading   247
*4.16 Bending of Curved Members   257
Review and Summary   269

5
TRANSVERSE LOADING   277

5.1 Introduction   277
5.2 Transverse Loading of Prismatic Members   278
5.3 Basic Assumption Regarding the Distribution of the Normal Stresses   280
5.4 Determination of the Shear on a Horizontal Plane   281
5.5 Determination of the Shearing Stresses τ_{xy} in a Beam   284
5.6 Shearing Stresses τ_{xy} in Common Types of Beams   286
*5.7 Further Discussion of the Distribution of Stresses in a Narrow Rectangular Beam   287
5.8 Shear on an Arbitrary Longitudinal Cut   294
5.9 Shearing Stresses in Thin-Walled Members   296
*5.10 Plastic Deformations   298
5.11 Stresses under Combined Loadings   307
*5.12 Unsymmetric Loading of Thin-Walled Members: Shear Center   320
Review and Summary   333

6
TRANSFORMATIONS OF STRESS AND STRAIN   339

6.1 Introduction   339
6.2 Transformation of Plane Stress   342
6.3 Principal Stresses; Maximum Shearing Stress   344
6.4 Mohr's Circle for Plane Stress   353
6.5 General State of Stress   362
6.6 Application of Mohr's Circle to the Three-Dimensional Analysis of Stress   364
*6.7 Yield Criteria for Ductile Materials under Plane Stress   367
*6.8 Fracture Criteria for Brittle Materials under Plane Stress   369
6.9 Stresses in Thin-Walled Pressure Vessels   377
*6.10 Transformation of Plane Strain   384
*6.11 Mohr's Circle for Plane Strain   387
*6.12 Three-Dimensional Analysis of Strain   390
*6.13 Measurements of Strain; Strain Rosette   393
Review and Summary   400

7

DESIGN OF BEAMS AND SHAFTS FOR STRENGTH 409

7.1	Introduction	409
7.2	Basic Considerations for the Design of Prismatic Beams	410
7.3	Shear and Bending-Moment Diagrams	413
7.4	Relations among Load, Shear, and Bending Moment	422
*7.5	Using Singularity Functions to Determine Shear and Bending Moment in a Beam	432
7.6	Principal Stresses in a Beam	443
7.7	Design of Prismatic Beams	446
*7.8	Beams of Constant Strength	458
*7.9	Design of Transmission Shafts	460
*7.10	Stresses under Applied Loads	461
	Review and Summary	470

8

DEFLECTION OF BEAMS BY INTEGRATION 478

8.1	Introduction	478
8.2	Deformation of a Beam under Transverse Loading	480
8.3	Equation of the Elastic Curve	481
*8.4	Direct Determination of the Elastic Curve from the Load Distribution	486
8.5	Statically Indeterminate Beams	488
*8.6	Using Singularity Functions to Determine the Slope and Deflection of a Beam	499
8.7	Method of Superposition	511
8.8	Application of Superposition to Statically Indeterminate Beams	512
	Review and Summary	522

9

DEFLECTION OF BEAMS BY MOMENT-AREA METHOD 530

*9.1	Introduction	530
*9.2	Moment-Area Theorems	531
*9.3	Application to Cantilever Beams and Beams with Symmetric Loadings	533

*9.4	Bending-Moment Diagrams by Parts	535
*9.5	Beams with Unsymmetric Loadings	544
*9.6	Maximum Deflection	546
*9.7	Statically Indeterminate Beams	554
	Review and Summary	563

10
ENERGY METHODS   570

10.1	Introduction	570
10.2	Strain Energy	571
10.3	Strain-Energy Density	573
10.4	Elastic Strain Energy for Normal Stresses	574
10.5	Elastic Strain Energy for Shearing Stresses	577
*10.6	Strain Energy for a General State of Stress	580
10.7	Impact Loading	593
10.8	Design for Impact Loads	595
10.9	Work and Energy under a Single Load	596
10.10	Deflection under a Single Load by the Work-Energy Method	598
*10.11	Work and Energy under Several Loads	608
*10.12	Castigliano's Theorem	610
*10.13	Deflections by Castigliano's Theorem	612
*10.14	Statically Indeterminate Structures	615
	Review and Summary	624

11
COLUMNS   631

11.1	Introduction	631
11.2	Stability of Structures	632
11.3	Euler's Formula for Pin-Ended Columns	634
11.4	Extension of Euler's Formula to Columns with Other End Conditions	637
*11.5	Eccentric Loading; The Secant Formula	650
11.6	Design of Columns under a Centric Load	660
11.7	Design of Columns under an Eccentric Load	674
	Review and Summary	684

APPENDICES   689

A	Moments of Areas	690
B	Typical Properties of Selected Materials Used in Engineering	700

C   Properties of Rolled-Steel Shapes   704
D   Beam Deflections and Slopes   716

Index   717
Answers to Even-Numbered Problems   725

PREFACE

The main objective of a basic mechanics course should be to develop in the engineering student the ability to analyze a given problem in a simple and logical manner and to apply to its solution a few fundamental and well-understood principles. This text is designed for the first course in mechanics of materials—or strength of materials—offered to engineering students in the sophomore or junior year. The authors hope that it will help the instructor achieve this goal in that particular course in the same way that their other texts may have helped them in statics and dynamics.

In this text the study of the mechanics of materials is based on the understanding of a few basic concepts and on the use of simplified models. This approach makes it possible to develop all the necessary formulas in a rational and logical manner, and to clearly indicate the conditions under which they may be safely applied to the analysis and design of actual engineering structures and machine components.

Free-body diagrams are used extensively throughout the text to determine external or internal forces. The use of "picture equations" will also help the students understand the superposition of loadings and the resulting stresses and deformations.

It is expected that students using this text will have completed a course in statics. However, Chap. 1 is designed to provide them with an opportunity to review the concepts learned in that course, while shear and bending-moment diagrams are covered in detail in Secs. 7.3 and 7.4. The properties of moments and centroids of areas are described in Appendix A; this material may be used to reinforce the discussion of the determination of normal and shearing stresses in beams (Chaps. 4 and 5).

The first five chapters of the text are devoted to the analysis of the stresses and of the corresponding deformations in various structural members, considering successively axial loading, torsion, pure bending, and transverse loading. Each analysis is based on a few basic concepts, namely, the conditions of equilibrium of the forces exerted on the member, the relations existing between stress and strain in the material, and the conditions imposed by the supports and loading of the member. The study of each type of loading is complemented by a large number of examples, sample problems, and problems to be assigned, all designed to strengthen the students' understanding of the subject. And as each new loading is considered, students are presented with problems combining the new loading with those studied previously.

The concept of stress at a point is introduced in Chap. 1, where it is shown that an axial load may produce shearing stresses as well as normal stresses, depending upon the section considered. The fact that stresses depend upon the orientation of the surface on which they are computed is emphasized again in Chaps. 3, 4, and 5 in the case of torsion, pure bending, and transverse loading. However, the discussion of computational techniques—such as Mohr's circle—used for the transformation of stress at a point is delayed until Chap. 6, after students have had the opportunity to solve problems involving a combination of the basic loadings and have discovered for themselves the need for such techniques. For a similar reason, shear and bending-moment diagrams are introduced only in Chap. 7, where they may be applied immediately to the design of beams and shafts. This approach has the additional advantage of maintaining the unity of presentation of the analysis of stresses.[†]

Statically indeterminate problems are first discussed in Chap. 2 and considered throughout the text for the various loading conditions encountered. Thus, students are presented at an early stage with a method of solution which combines the analysis of deformations with the conventional analysis of forces used in statics. In this way, they will have become thoroughly familiar with this fundamental method by the end of the course. In addition, this approach helps the students realize that stresses are statically indeterminate and can be computed only by considering the corresponding distribution of strains.

The concept of plastic deformation is introduced in Chap. 2, where it is applied to the analysis of members under axial loading. Problems involving the plastic deformation of circular shafts and of prismatic beams are also considered in optional sections of Chaps. 3 and 4, respectively. While some of this material may be omitted at the choice of the instructor, its inclusion in the body of the text will help the students realize the limitations of the assumption of a linear stress-strain relation and serve to caution them against the inappropriate use of the elastic torsion and flexure formulas.

The design of a given structure requires more than the determination of the normal and shearing stresses on a given element of that structure. The designer of a beam, for instance, must make sure that the allowable stresses will not be exceeded at any point of the beam, and also verify that the maximum deflection will not exceed a certain given value. In Chap. 6, students will learn to determine the maximum values of the normal and shearing stresses at a given point of a structure subjected to

---

[†] However, instructors who wish to discuss the shear and bending-moment diagrams at an earlier stage may cover Secs. 7.3 and 7.4 either together with Chap. 1 or immediately before Chap. 4 and assign Probs. 7.1 to 7.22 and 7.37 to 7.54 at that time.

any combination of the loadings considered in the previous chapters, while in the first part of Chap. 7 they will determine the maximum values of the shear and bending moment in a beam and the corresponding values of the shearing and normal stresses. This will provide them with the necessary prerequisites for the design of beams and shafts for strength, a topic discussed in the remainder of Chap. 7. The determination of the deflection of beams is carried out in Chaps. 8 and 9, with Chap. 8 devoted to the integration method, and Chap. 9 to the moment-area method. Presenting these two methods in separate chapters allows the instructor to use only one of them, or to cover them both in sequence. Chapter 10 discusses energy methods and Chap. 11 the analysis and design of columns.

In the first edition of this text, singularity functions were introduced in an optional section of Chap. 8 and used as an alternative method for determining the slope and deflection of beams. Because of the increased availability of computers and the fact that singularity functions are easily programmable, these functions have now been introduced in Sec. 7.5 as an alternative approach to the computation of the shear and bending moment in a beam. This has the added advantage of letting students familiarize themselves with singularity functions at an earlier stage and, thus, be better prepared to use them in Sec. 8.6 for the determination of slopes and deflections. While Secs. 7.5 and 8.6 are both optional, Sec. 8.6 should be included only if Sec. 7.5 has been covered earlier.

Additional topics such as residual stresses, torsion of noncircular and thin-walled members, bending of curved beams, shearing stresses in non-symmetrical members, and failure criteria, have been included in optional sections for use in courses of varying emphases. To preserve the integrity of the subject, these topics are presented in the proper sequence, wherever they logically belong. Thus, even when not covered in the course, they are highly visible and may be easily referred to by the students if needed in a later course or in engineering practice. For convenience, all optional sections have been indicated by asterisks.

Each chapter begins with an introductory section setting the purpose and goals of the chapter and describing in simple terms the material to be covered and its application to the solution of engineering problems. The body of the text has been divided into units, each consisting of one or several theory sections followed by sample problems and a large number of problems to be assigned. Each unit corresponds to a well-defined topic and generally may be covered in one lesson. Each chapter ends with a review and summary of the material covered in the chapter. Marginal notes have been included to help the students organize their review work, and cross references provided to help them find the portions of material requiring their special attention.

The theory sections include many examples designed to illustrate the material being presented and facilitate its understanding. The sample problems are intended to show some of the applications of the theory to the solution of engineering problems. Since they have been set up in much the same form that students will use in solving the assigned problems, the sample problems serve the double purpose of amplifying the text and demonstrating the type of neat and orderly work that students should cultivate in their own solutions. Most of the problems are of a practical nature and should appeal to engineering students. They are primarily designed, however, to illustrate the material presented in the text and help the students understand the basic principles used in mechanics of materials. The problems have been grouped according to the portions of material they illustrate and have been arranged in order of increasing difficulty. Problems requiring special attention have been indicated by asterisks. Answers to all even-numbered problems are given at the end of the book.

The introduction in the engineering curriculum of instruction in computer programming and the increasing availability of personal computers or mainframe terminals on most campuses make it now possible for engineering students to solve a greater number of challenging problems. In this new edition of *Mechanics of Materials*, a group of four or more problems designed to be solved with a computer has been added to the review problems at the end of each chapter. Developing the algorithm required to solve a given problem will benefit the students in two different ways: (1) it will help them gain a better understanding of the mechanics principles involved; (2) it will provide them with an opportunity to apply the skills acquired in their computer programming course to the solution of a meaningful engineering problem.

The authors wish to acknowledge the collaboration of Professor John T. DeWolf to this second edition of *Mechanics of Materials* and thank him especially for his contribution to the sections relating to the design of columns. They also wish to thank Dr. Philippe Marchal of Intellipro, Inc., for programming the tutorial software which complements this text.

Finally, the authors wish to gratefully acknowledge the many helpful comments and suggestions offered by the users of the first edition of their text. Special thanks are due to Professor Leon Y. Bahar of Drexel University and Professor Paul C. Paris of Washington University.

Ferdinand P. Beer
E. Russell Johnston, Jr.

LIST OF SYMBOLS

a	Constant; distance		$\mathbf{Q}$	Force
$\mathbf{A, B, C}, \ldots$	Forces; reactions		Q	First moment of area
$A, B, C, \ldots$	Points		r	Radius; radius of gyration
$A, \mathfrak{C}$	Area		$\mathbf{R}$	Force; reaction
b	Distance; width		R	Radius; modulus of rupture
c	Constant; distance; radius		s	Length
C	Centroid		S	Elastic section modulus
$C_1, C_2, \ldots$	Constants of integration		t	Thickness; distance; tangential deviation
d	Distance; diameter; depth		$\mathbf{T}$	Torque
D	Diameter		T	Temperature
e	Distance; eccentricity; dilatation		u, v	Rectangular coordinates
E	Modulus of elasticity		u	Strain-energy density
f	Frequency; function		U	Strain energy; work
$\mathbf{F}$	Force		v	Velocity
$F.S.$	Factor of safety		$\mathbf{V}$	Shearing force
G	Modulus of rigidity; shear modulus		V	Volume; shear
h	Distance; height		w	Width; distance; load per unit length
$\mathbf{H}$	Force		$\mathbf{W}, W$	Weight, load
H, J, K	Points		x, y, z	Rectangular coordinates; distance; displacements; deflections
$I, I_x, \ldots$	Moment of inertia		$\bar{x}, \bar{y}, \bar{z}$	Coordinates of centroid
$I_{xy}, \ldots$	Product of inertia		Z	Plastic section modulus
J	Polar moment of inertia		α, β, γ	Angles
k	Spring constant; shape factor; bulk modulus; constant		α	Coefficient of thermal expansion; influence coefficient
K	Stress concentration factor; torsional spring constant		γ	Shearing strain; specific weight
l	Length; span		δ	Deformation; displacement
L	Length; span		ϵ	Normal strain
L_e	Effective length		θ	Angle; slope
m	Mass		λ	Direction cosine
$\mathbf{M}$	Couple		ν	Poisson's ratio
$M, M_x, \ldots$	Bending moment		ρ	Radius of curvature; distance; density
n	Number; ratio of moduli of elasticity; normal direction		σ	Normal stress
p	Pressure		τ	Shearing stress
$\mathbf{P}$	Force; concentrated load		ϕ	Angle; angle of twist
q	Shearing force per unit length; shear flow		ω	Angular velocity

C H A P T E R   O N E

INTRODUCTION— CONCEPT OF STRESS

1.1. INTRODUCTION

The main objective of the study of the mechanics of materials is to provide the engineer with the means of analyzing and designing various machines and load-bearing structures.

Both the analysis and the design of a given structure involve the determination of *stresses* and *deformations*. This first chapter is devoted to the concept of *stress*.

After a short introduction (Sec. 1.2) emphasizing the difference between *forces* and *stresses,* we shall consider successively the *normal stresses* in a member under axial loading (Sec. 1.3), the *shearing stresses* caused by the application of equal and opposite transverse forces (Sec. 1.4), and the *bearing stresses* created by bolts, pins, and rivets in the members they connect (Sec. 1.5). These various concepts are applied in Sec. 1.6 to the analysis of a simple structure consisting of pin-connected, two-force members.

In Sec. 1.7, where a two-force member under axial loading is considered again, we shall find that the stresses on an *oblique plane* include both *normal* and *shearing* stresses, while in Sec. 1.8 we shall see that *six stress components* are required to describe the state of stress at a point in a body under the most general loading condition.

Finally, we shall discuss in Sec. 1.9 the determination from test specimens of the *ultimate strength* of a given material and the use of a *factor of safety* in the computation of the *allowable load* for a structural component made of that material.

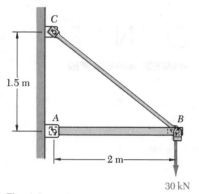

1.5 m

2 m

30 kN

Fig. 1.1

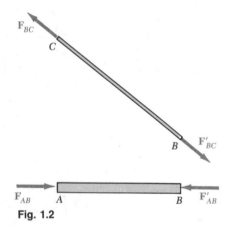

Fig. 1.2

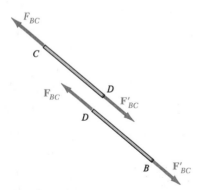

Fig. 1.4

1.2. FORCES AND STRESSES

Consider the structure shown in Fig. 1.1, consisting of a boom AB and a rod BC. We propose to determine whether a 30-kN load may safely be supported at B by this structure.

From our knowledge of statics, we recognize that AB and BC are two-force members acted upon at each end by equal and opposite axial forces: $\mathbf{F}_{AB}$ and $\mathbf{F}'_{AB}$ of magnitude F_{AB}, and $\mathbf{F}_{BC}$ and $\mathbf{F}'_{BC}$ of magnitude F_{BC} (Fig. 1.2). Drawing the free-body diagram of pin B and the corresponding force triangle (Fig. 1.3), we write from similar triangles

$$\frac{F_{AB}}{4 \text{ m}} = \frac{F_{BC}}{5 \text{ m}} = \frac{30 \text{ kN}}{3 \text{ m}}$$

and obtain

$$F_{AB} = 40 \text{ kN} \qquad F_{BC} = 50 \text{ kN}$$

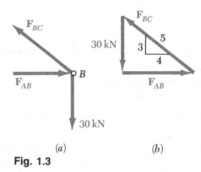

30 kN

30 kN

(a)

(b)

Fig. 1.3

Passing a section through rod BC at some arbitrary point D, we obtain two portions BD and CD (Fig. 1.4). Since 50-kN forces must be applied at D to both portions of the rod to keep them in equilibrium, we conclude that an internal force of 50 kN is produced in rod BC when a 30-kN load is applied at B, and further note that rod BC is in tension. A similar reasoning shows that the force in boom AB is 40 kN and that the boom is in compression.

While the results obtained represent a first and necessary step in the analysis of the structure, they do not tell us whether the given load may be safely supported. Whether rod BC, for example, will break or not under this loading depends not only upon the value found for the internal force F_{BC}, but also upon the cross-sectional area of the rod and the material of which the rod is made. Indeed, the internal force F_{BC} actually represents the resultant of elementary forces distributed over the entire

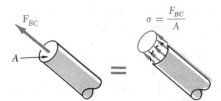

Fig. 1.5

area A of the cross section (Fig. 1.5) and the intensity of these distributed forces is equal to the force per unit area, F_{BC}/A, in the section. Whether or not the rod will break under the given loading clearly depends upon the ability of the material to withstand the corresponding value F_{BC}/A of the intensity of the distributed internal forces. It thus depends upon the force F_{BC}, the cross-sectional area A, and the material of the rod.

The force per unit area, or intensity of the forces distributed over a given section, is called the *stress* on that section and is denoted by the Greek letter σ (sigma). The stress in a member of cross-sectional area A subjected to an axial load **P** (Fig. 1.6) is therefore obtained by dividing the magnitude P of the load by the area A:

$$\sigma = \frac{P}{A} \tag{1.1}$$

A positive sign will be used to indicate a tensile stress (member in tension) and a negative sign to indicate a compressive stress (member in compression).

Since SI metric units are used in this discussion, with P expressed in newtons (N) and A in square meters (m^2), the stress σ will be expressed in N/m^2. This unit is called a *pascal* (Pa). However, one finds that the pascal is an exceedingly small quantity and that, in practice, multiples of this unit must be used, namely, the kilopascal (kPa), the megapascal (MPa), and the gigapascal (GPa). We have

$$1 \text{ kPa} = 10^3 \text{ Pa} = 10^3 \text{ N/m}^2$$

$$1 \text{ MPa} = 10^6 \text{ Pa} = 10^6 \text{ N/m}^2$$

$$1 \text{ GPa} = 10^9 \text{ Pa} = 10^9 \text{ N/m}^2$$

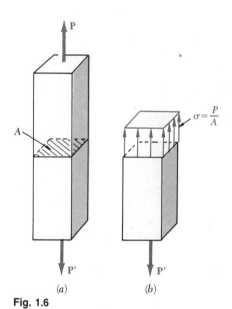

Fig. 1.6

Considering again rod *BC*, we shall assume that it is made of steel and has a diameter of 20 mm.   We have

$$P = F_{BC} = +50 \text{ kN} = +50 \times 10^3 \text{ N}$$

$$A = \pi r^2 = \pi \left(\frac{20 \text{ mm}}{2}\right)^2 = \pi (10 \times 10^{-3} \text{ m})^2 = 314 \times 10^{-6} \text{ m}^2$$

$$\sigma = \frac{P}{A} = \frac{+50 \times 10^3 \text{ N}}{314 \times 10^{-6} \text{ m}^2} = +159 \times 10^6 \text{ Pa} = +159 \text{ MPa}$$

To determine whether the rod *BC* may be used to support the given 30-kN load, we must compare the value obtained for σ under this loading with the maximum value of the stress which may be safely applied to steel.   From a table of properties of material we find that the maximum allowable stress in the type of steel to be used is $\sigma_{all} = 165$ MPa.   Since the actual value found for the stress in the rod is smaller, we conclude that rod *BC* can safely support the given load.   To be complete, our analysis should also include the determination of the compressive stress in the boom *AB*, as well as an investigation of the stresses produced in the pins and their bearings.   This will be discussed later in this chapter.   Finally, we should also determine whether the deformations produced by the given loading are acceptable.   The study of deformations under axial loads will be the subject of Chap. 2.

The engineer's role is not limited to the analysis of existing structures and machines subjected to given loading conditions; of equal if not greater importance is the *design* of new structures and machines, that is, the selection of the appropriate structural components to perform a given task.   As an example, let us return to the structure of Fig. 1.1, and assume that aluminum is to be used for rod *BC*. What should the diameter of the rod be if it is to safely carry the given load? From a table of properties of materials we first determine that the maximum allowable stress in the type of aluminum to be used is $\sigma_{all} = 100$ MPa.   Since the force in member *BC* will still be $P = F_{BC} = 50$ kN under the given loading, we must have, from Eq. (1.1),

$$\sigma_{all} = \frac{P}{A} \qquad A = \frac{P}{\sigma_{all}} = \frac{50 \times 10^3 \text{ N}}{100 \times 10^6 \text{ Pa}} = 500 \times 10^{-6} \text{ m}^2$$

and, since $A = \pi r^2$,

$$r = \sqrt{\frac{A}{\pi}} = \sqrt{\frac{500 \times 10^{-6} \text{ m}^2}{\pi}} = 12.62 \times 10^{-3} \text{ m} = 12.62 \text{ mm}$$

$$d = 2r = 25.2 \text{ mm}$$

We conclude that an aluminum rod 26 mm or more in diameter will be adequate.

1.3. AXIAL LOADING; NORMAL STRESS

As we have already indicated, rod *BC* of the example considered in the preceding section is a two-force member and, therefore, the forces $\mathbf{F}_{BC}$ and $\mathbf{F}'_{BC}$ acting on its ends *B* and *C* (Fig. 1.2) are directed along the axis of the rod. We say that the rod is under *axial loading*. The section we passed through the rod to determine the internal force in the rod and the corresponding stress was perpendicular to the axis of the rod; the internal force was therefore normal to the plane of the section (Fig. 1.5) and the corresponding stress is described as a *normal stress*. Thus, formula (1.1) gives us the *normal stress in a member under axial loading:*

$$\sigma = \frac{P}{A} \tag{1.1}$$

We should also note that, in formula (1.1), σ is obtained by dividing the magnitude *P* of the resultant of the internal forces distributed over the cross section by the area *A* of the cross section; it represents, therefore, the *average value* of the stress over the cross section, rather than the stress at a specific point of the cross section.

To define the stress at a given point *Q* of the cross section, we should consider a small area ΔA (Fig. 1.7). Dividing the magnitude of $\Delta \mathbf{F}$ by ΔA, we obtain the average value of the stress over ΔA. Letting ΔA approach zero, we obtain the stress at point Q:

$$\sigma = \lim_{\Delta A \to 0} \frac{\Delta F}{\Delta A} \tag{1.2}$$

In general, the value obtained for the stress σ at a given point *Q* of the section is different from the value of the average stress given by formula (1.1), and σ is found to vary across the section. In a slender rod subjected to equal and opposite concentrated loads **P** and **P**′ (Fig. 1.8*a*), this variation is small in a section away from the points of application of the concentrated loads (Fig. 1.8*c*), but it is quite noticeable in the neighborhood of these points (Fig. 1.8*b* and *d*).

It follows from Eq. (1.2) that the magnitude of the resultant of the distributed internal forces is

$$\int dF = \int_A \sigma \, dA$$

But the conditions of equilibrium of each of the portions of rod shown in Fig. 1.8 require that this magnitude be equal to the magnitude *P* of the concentrated loads. We have, therefore,

$$P = \int dF = \int_A \sigma \, dA \tag{1.3}$$

which means that the volume under each of the stress surfaces in Fig. 1.8 must be equal to the magnitude *P* of the loads. This, however, is the only

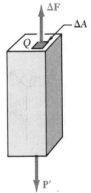

Fig. 1.7

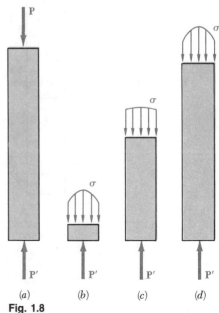

(a)　　(b)　　(c)　　(d)

Fig. 1.8

information that we can derive from our knowledge of statics, regarding the distribution of normal stresses in the various sections of the rod. The actual distribution of stresses in any given section is *statically indeterminate*. To learn more about this distribution, it is necessary to consider the deformations resulting from the particular mode of application of the loads at the ends of the rod. This will be discussed further in Chap. 2.

In practice, we shall assume that the distribution of normal stresses in an axially loaded member is uniform, except in the immediate vicinity of the points of application of the loads. The value σ of the stress is then equal to σ_{ave} and may be obtained from formula (1.1). However, we should realize that, when we assume a uniform distribution of stresses in the section, i.e., when we assume that the internal forces are uniformly distributed across the section, it follows from elementary statics† that the resultant **P** of the internal forces must be applied at the centroid C of the section (Fig. 1.9). This means that *a uniform distribution of stress is possible only if the line of action of the concentrated loads **P** and **P′** passes through the centroid of the section considered* (Fig. 1.10). This type of loading is called *centric loading* and will be assumed to take place in all straight two-force members found in trusses and pin-connected structures, such as the one considered in Fig. 1.1. However, if a two-force member is loaded axially, but *eccentrically* as shown in Fig. 1.11*a*, we find from the conditions of equilibrium of the portion of member shown in

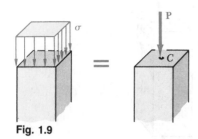

Fig. 1.9

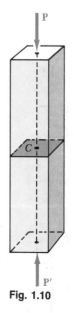

Fig. 1.10

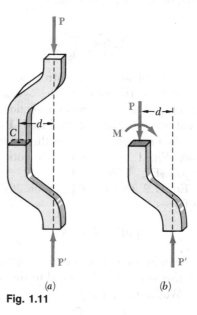

(a)   (b)

Fig. 1.11

† See Ferdinand P. Beer and E. Russell Johnston, Jr., *Mechanics for Engineers*, 4th ed., McGraw-Hill, New York, 1987, or *Vector Mechanics for Engineers*, 5th ed, McGraw Hill, New York, 1988, secs. 5.2 and 5.3.

Fig. 1.11*b* that the internal forces in a given section must be equivalent to a force **P** applied at the centroid of the section and a couple **M** of moment $M = Pd$. The distribution of forces—and, thus, the corresponding distribution of stresses—*cannot be uniform.* Nor can the distribution of stresses be symmetric as shown in Fig. 1.8. This point will be discussed in detail in Chap. 4.

1.4. SHEARING STRESS

The internal forces discussed in Secs. 1.2 and 1.3—and the corresponding stresses—were normal to the section considered. A very different type of stress is obtained when transverse forces **P** and **P′** are applied to a member *AB* (Fig. 1.12). Passing a section at *C* between the

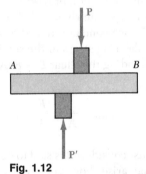

Fig. 1.12

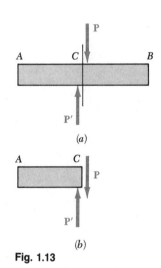

Fig. 1.13

points of application of the two forces (Fig. 1.13*a*), we obtain the diagram of portion *AC* shown in Fig. 1.13*b*. We conclude that internal forces must exist in the plane of the section, and that their resultant is equal to **P**. These elementary internal forces are called *shearing forces*, and the magnitude *P* of their resultant is the *shear* in the section. Dividing the shear *P* by the area *A* of the cross section, we obtain the *average shearing stress* in the section. Denoting the shearing stress by the Greek letter τ (tau), we write

$$\tau_{\text{ave}} = \frac{P}{A} \tag{1.4}$$

It should be emphasized that the value obtained is an average value of the shearing stress over the entire section. Contrary to what we said earlier for normal stresses, the distribution of shearing stresses across the section *cannot* be assumed uniform. As we shall see in Chap. 5, the actual value τ of the shearing stress varies from zero at the surface of the member to a maximum value τ_{max} which may be much larger than the average value τ_{ave}.

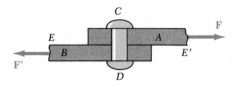

Fig. 1.14

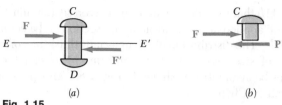

(a)   (b)

Fig. 1.15

Shearing stresses are commonly found in bolts, pins, and rivets used to connect various structural members and machine components. Consider, for example, the two plates A and B which are connected by a rivet CD (Fig. 1.14). If the plates are subjected to tension forces of magnitude F, stresses will develop in the section of rivet corresponding to the plane EE'. Drawing the diagrams of the rivet and of the portion located above the plane EE' (Fig. 1.15), we conclude that the shear P in the section is equal to F. The average shearing stress in the section is obtained, according to formula (1.4), by dividing the shear $P = F$ by the area A of the cross section:

$$\tau_{\text{ave}} = \frac{P}{A} = \frac{F}{A} \tag{1.5}$$

The rivet we have just considered is said to be in *single shear*. Different loading situations may arise, however. For example, if splice plates C and D are used to connect plates A and B (Fig. 1.16), shear will take place in rivet HJ in each of the two planes KK' and LL' (and similarly in rivet EG). The rivets are said to be in *double shear*. To determine the average shearing stress in each plane, we draw free-body diagrams of rivet HJ and of the portion of rivet located between the two planes (Fig. 1.17). Observing that the shear P in each of the sections is $P = F/2$, we conclude that the average shearing stress is

$$\tau_{\text{ave}} = \frac{P}{A} = \frac{F}{2A} \tag{1.6}$$

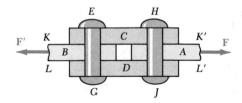

Fig. 1.16

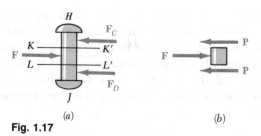

(a)   (b)

Fig. 1.17

1.5. BEARING STRESS IN CONNECTIONS

Bolts, pins, and rivets create stresses in the members they connect, along the *bearing surface*, or surface of contact. For example, consider again the two plates A and B connected by a rivet CD, that we have discussed in the preceding section (Fig. 1.14). The rivet exerts on plate A a force P equal and opposite to the force F exerted by the plate on the rivet (Fig. 1.18). The force P represents the resultant of elementary forces distributed on the inside surface of a half-cylinder of diameter d and of length t equal to the thickness of the plate. Since the distribution of these forces—and of the corresponding stresses—is quite complicated, one uses in practice an average nominal value σ_b of the stress, called the *bearing stress*, obtained by dividing the load P by the area of the rectangle representing the projection of the rivet on the plate section (Fig. 1.19). Since this area is equal to td, where t is the plate thickness and d the diameter of the rivet, we have

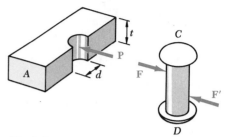

Fig. 1.18

$$\sigma_b = \frac{P}{A} = \frac{P}{td} \qquad (1.7)$$

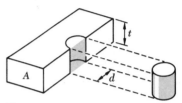

Fig. 1.19

1.6. APPLICATION TO THE ANALYSIS OF SIMPLE STRUCTURES

We are now in a position to determine the stresses in the members and connections of various simple two-dimensional structures.

a. Determination of the Normal Stress in Two-Force Members. The first step consists of determining the *force* in each of the members. In the case of the structure shown in Fig. 1.1 of Sec. 1.2, this could be done by considering the equilibrium of a single joint. In more involved problems, it will usually be necessary to first consider the free-body diagram of the entire structure and to determine the reactions at the various supports by writing the three equilibrium equations for a rigid body,

$$\Sigma F_x = 0 \qquad \Sigma F_y = 0 \qquad \Sigma M_A = 0 \qquad (1.8)$$

where A is a point in the plane of the structure. The forces in the two-force members may then be determined by considering successively the equilibrium of the various joints.† In some cases, it may also be advantageous to draw the free-body diagram of a portion of the structure and to write the equilibrium equations (1.8) for that portion.‡ If the structure

---

†See Beer and Johnston, *Vector Mechanics for Engineers*, secs. 4.1 to 4.4 and 6.4, or *Mechanics for Engineers*, secs. 3.13 to 3.15 and 6.4.

‡See Beer and Johnston, *Vector Mechanics for Engineers* or *Mechanics for Engineers*, sec. 6.7.

contains multiforce members, the same equations (1.8) may be written for each multiforce member.†

As we saw in Sec. 1.2 for a member subjected to a centric loading, the normal stress σ may be obtained from Eq. (1.1), that is, by dividing the force P in that member by the cross-sectional area of the member. If the cross section is not constant, the stress will be largest where the cross-sectional area A is smallest. For example, returning to the structure of Fig. 1.1 discussed in Sec. 1.2, we shall now specify that the 20-mm-diameter rod BC has flat ends of rectangular cross section 20 by 40 mm, and that boom AB has a uniform rectangular cross section 30 by 50 mm and is fitted with a clevis at end B (Fig. 1.20). Both members are connected at B by a pin from which the 30-kN load is suspended by means of a U-shaped bracket. Boom AB is supported at A by a pin fitted into a double bracket, while rod BC is connected by a pin at C to a single

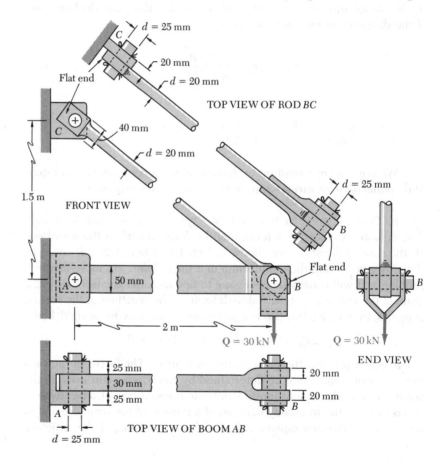

Fig. 1.20

†See Beer and Johnston, *Vector Mechanics for Engineers* or *Mechanics for Engineers*, secs. 6.9 to 6.11.

bracket. All pins are 25 mm in diameter. As we found in Sec. 1.2, the force in rod BC is $F_{BC} = 50$ kN (tension) and the area of its circular cross section is $A = 314 \times 10^{-6}$ m²; the corresponding average normal stress is $\sigma_{BC} = +159$ MPa. However, the flat parts of the rod are also under tension and at the narrowest section, where a hole is located, we have

$$A = (20 \text{ mm})(40 \text{ mm} - 25 \text{ mm}) = 300 \times 10^{-6} \text{ m}^2$$

The corresponding average value of the stress, therefore, is

$$(\sigma_{BC})_{\text{end}} = \frac{P}{A} = \frac{50 \times 10^3 \text{ N}}{300 \times 10^{-6} \text{ m}^2} = 167 \text{ MPa}$$

Note that this is an *average value;* close to the hole, the stress will actually reach a much larger value, as we shall see in Sec. 2.17. It is clear that, under an increasing load, the rod will fail near one of the holes rather than in its cylindrical portion; its design, therefore, could be improved by increasing the width or the thickness of the flat ends of the rod.

Turning now our attention to boom AB, we recall from Sec. 1.2 that the force in the boom is $F_{AB} = 40$ kN (compression). Since the area of the boom's rectangular cross section is $A = 30$ mm $\times$ 50 mm $= 1.5 \times 10^{-3}$ m², the average value of the normal stress in the main part of the rod, between pins A and B, is

$$\sigma_{AB} = -\frac{40 \times 10^3 \text{ N}}{1.5 \times 10^{-3} \text{ m}^2} = -26.7 \times 10^6 \text{ Pa} = -26.7 \text{ MPa}$$

Note that the sections of minimum area at A and B are not under stress, since the boom is in compression, and, therefore, *pushes* on the pins (instead of *pulling* on the pins as rod BC does).

b. Determination of the Shearing Stress in Various Connections. To determine the shearing stress in a connection such as a bolt, pin, or rivet, we first clearly show the forces exerted by the various members it connects. Thus, in the case of pin C of our example (Fig. 1.21a), we draw Fig. 1.21b, showing the 50-kN force exerted by member BC on the pin, and the equal and opposite force exerted by the bracket. Drawing now the diagram of the portion of the pin located below the plane DD' where shearing stresses occur (Fig. 1.21c), we conclude that the shear in that plane is $P = 50$ kN. Since the cross-sectional area of the pin is

$$A = \pi r^2 = \pi \left(\frac{25 \text{ mm}}{2}\right)^2 = \pi (12.5 \times 10^{-3} \text{ m})^2 = 491 \times 10^{-6} \text{ m}^2$$

we find that the average value of the shearing stress in the pin at C is

$$\tau_{\text{ave}} = \frac{P}{A} = \frac{50 \times 10^3 \text{ N}}{491 \times 10^{-6} \text{ m}^2} = 102 \text{ MPa}$$

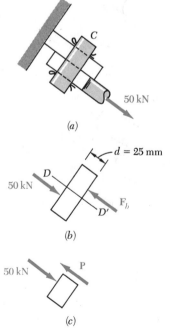

Fig. 1.21

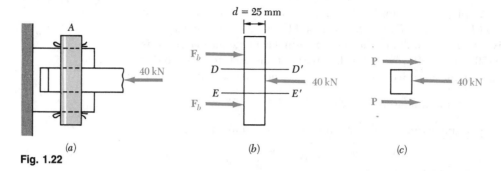

Fig. 1.22

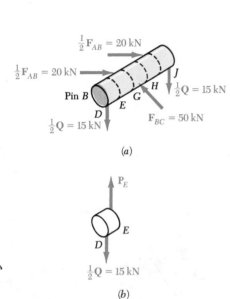

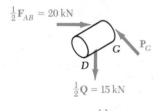

Fig. 1.23

Considering now the pin at A (Fig. 1.22), we note that it is in double shear. Drawing the free-body diagrams of the pin and of the portion of pin located between the planes DD' and EE' where shearing stresses occur, we conclude that $P = 20$ kN and that

$$\tau_{ave} = \frac{P}{A} = \frac{20 \text{ kN}}{491 \times 10^{-6} \text{ m}^2} = 40.7 \text{ MPa}$$

Considering the pin at B (Fig. 1.23a), we note that the pin may be divided into five portions which are acted upon by forces exerted by the boom, rod, and bracket. Considering successively the portions DE (Fig. 1.23b) and DG (Fig. 1.23c), we conclude that the shear in section E is $P_E = 15$ kN, while the shear in section G is $P_G = 25$ kN. Since the loading of the pin is symmetric, we conclude that the maximum value of the shear in pin B is $P_G = 25$ kN, and that the largest shearing stresses occur in sections G and H, where

$$\tau_{ave} = \frac{P_G}{A} = \frac{25 \text{ kN}}{491 \times 10^{-6} \text{ m}^2} = 50.9 \text{ MPa}$$

c. Determination of the Bearing Stresses. To determine the nominal bearing stress at A in member AB, we use formula (1.7) of Sec. 1.5. From Fig. 1.20, we have $t = 30$ mm and $d = 25$ mm. Recalling that $P = F_{AB} = 40$ kN, we have

$$\sigma_b = \frac{P}{td} = \frac{40 \text{ kN}}{(30 \text{ mm})(25 \text{ mm})} = 53.3 \text{ MPa}$$

To obtain the bearing stress in the bracket at A, we use $t = 2(25 \text{ mm}) = 50$ mm and $d = 25$ mm:

$$\sigma_b = \frac{P}{td} = \frac{40 \text{ kN}}{(50 \text{ mm})(25 \text{ mm})} = 32.0 \text{ MPa}$$

The bearing stresses at B in member AB, at B and C in member BC, and in the bracket at C are found in a similar way.

SAMPLE PROBLEM 1.1

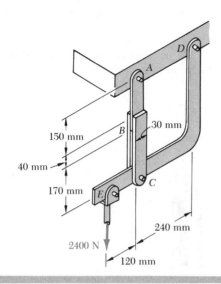

In the hanger shown the upper portion of link ABC is 9-mm thick and the lower portions are each 6-mm thick. Epoxy resin is used to bond the upper and lower portions together at B. The pin at A is of 9-mm diameter while a 6-mm-diameter pin is used at C. Determine (a) the shearing stress in pin A, (b) the shearing stress in pin C, (c) the largest normal stress in link ABC, (d) the average shearing stress on the bonded surfaces at B, (e) the bearing stress in the link at C.

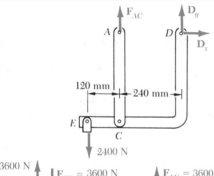

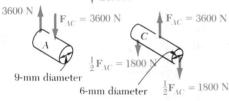

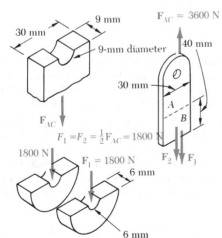

Free Body: Entire Hanger. Since the link ABC is a two-force member, the reaction at A is vertical; the reaction at D is represented by its components $\mathbf{D}_x$ and $\mathbf{D}_y$. We write

$$+\text{)}\Sigma M_D = 0: \qquad (2400\ \text{N})(360\ \text{mm}) - F_{AC}(240\ \text{mm}) = 0$$

$$F_{AC} = +3600\ \text{N} \qquad F_{AC} = 3600\ \text{N} \qquad \textit{tension}$$

a. Shearing Stress in Pin A. Since this 9-mm-diameter pin is in single shear, we write

$$\tau_A = \frac{F_{AC}}{A} = \frac{3600\ \text{N}}{\frac{1}{4}\pi(9\ \text{mm})^2} \qquad \tau_A = 56.6\ \text{MPa} \blacktriangleleft$$

b. Shearing Stress in Pin C. Since this 6-mm-diameter pin is in double shear, we write

$$\tau_C = \frac{\frac{1}{2}F_{AC}}{A} = \frac{1800\ \text{N}}{\frac{1}{4}\pi(6\ \text{mm})^2} \qquad \tau_C = 63.7\ \text{MPa} \blacktriangleleft$$

c. Largest Normal Stress in Link ABC. The largest stress is found where the area is smallest; this occurs at the cross section at A where the 9-mm hole is located. We have

$$\sigma_A = \frac{F_{AC}}{A_{\text{net}}} = \frac{3600\ \text{N}}{(9\ \text{mm})(30\ \text{mm} - 9\ \text{mm})} = \frac{3600\ \text{N}}{189\ \text{mm}^2} \qquad \sigma_A = 19.05\ \text{MPa} \blacktriangleleft$$

d. Average Shearing Stress at B. We note that bonding exists on both sides of the upper portion of the link and that the shear force on each side is $F_1 = (3600\ \text{N})/2 = 1800\ \text{N}$. The average shearing stress on each surface is thus

$$\tau_B = \frac{F_1}{A} = \frac{1800\ \text{N}}{(30\ \text{mm})(40\ \text{mm})} \qquad \tau_B = 1.500\ \text{MPa} \blacktriangleleft$$

e. Bearing Stress in Link at C. For each portion of the link, $F_1 = 1800\ \text{N}$ and the nominal bearing area is $(6\ \text{mm})(6\ \text{mm}) = 36\ \text{mm}^2$.

$$\sigma_b = \frac{F_1}{A} = \frac{1800\ \text{N}}{36\ \text{mm}^2} \qquad \sigma_b = 50.0\ \text{MPa} \blacktriangleleft$$

PROBLEMS

1.1 and 1.2 Two solid cylindrical rods are welded together at B as shown. Determine the normal stress at the midpoint of each rod.

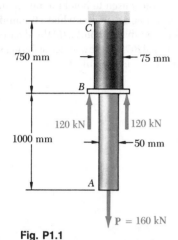

Fig. P1.1

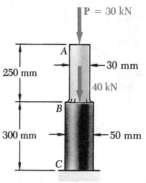

Fig. P1.2

1.3 In Prob. 1.2, determine the magnitude of the force **P** for which the same normal stress occurs in each rod.

1.4 In Prob. 1.1, determine the magnitude of the force **P** for which the tensile stress in rod AB has the same magnitude as the compressive stress in rod BC.

1.5 The axial load in the column supporting the timber beam shown is 100 kN.   Determine the length l of the bearing plate for which the average bearing stress on the timber is 3 MPa.

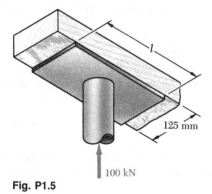

Fig. P1.5

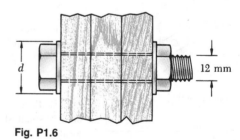

Fig. P1.6

1.6 Three wooden planks are fastened together by a series of bolts to form a column.   The diameter of each bolt is 12 mm, and the inside diameter of each washer is 15 mm, which is slightly larger than the diameter of the holes in the planks.   Knowing that the outside diameter of each washer is $d = 30$ mm and that the average bearing stress between the washers and the planks is not to exceed 5 MPa, determine the largest allowable value of the axial normal stress in each bolt.

1.7 For the structure of Prob. 1.6, determine the required outside diameter d of the washers, knowing that the axial normal stress in each bolt is 20 MPa and that the average bearing stress between the washers and the planks is not to exceed 5 MPa.

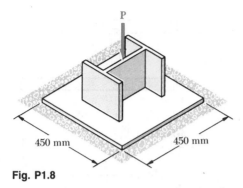

Fig. P1.8

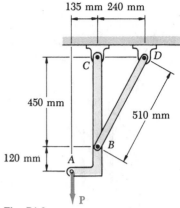

Fig. P1.9

1.8   An axial load **P** is supported by a short W250 × 67 column of cross-sectional area $A = 8580$ mm^2 and is distributed to a concrete foundation by an 450-mm-square plate as shown.   Knowing that the average stress in the column should not exceed 250 MPa and that the average bearing stress on the concrete foundation should not exceed 15 MPa, determine the largest allowable load P.

1.9 and 1.10   Knowing that the central portion of the link *BD* has a uniform cross-sectional area of 800 mm^2, determine the magnitude of the load **P** for which the normal stress in that portion of *BD* is 50 MPA.

1.11   Link *AC* has a uniform rectangular cross section 3 mm thick and 25 mm wide.   Determine the normal stress in the central portion of that link when $\alpha = 0$.

1.12   Solve Prob. 1.11, assuming that $\alpha = 90°$.

1.13   Each of the four vertical links connecting the two horizontal members has a uniform rectangular cross section of 10 × 40 mm, and each pin has a 14-mm diameter.   Determine the maximum value of the average normal stress caused by the 24-kN load in the links connecting (*a*) points *B* and *E*, (*b*) points *C* and *F*.

1.14   Solve Prob. 1.13, assuming that the 24-kN load is directed upward.

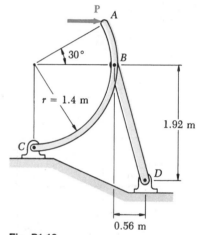

Fig. P1.10

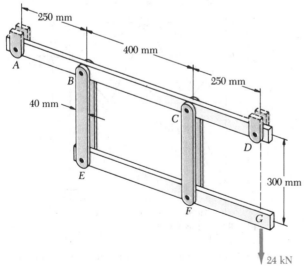

Fig. P1.13

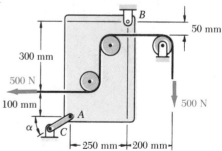

Fig. P1.11

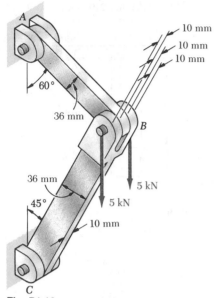

Fig. P1.16

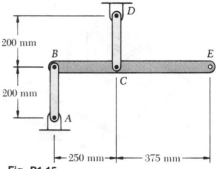

Fig. P1.15

1.15 Each of the links AB and CD has a uniform rectangular cross section of 6×25 mm and is connected to its support and to the horizontal member BCE by 10-mm-diameter pins. Knowing that the average normal stress in either link is not to exceed 175 MPa, determine the largest load which may be applied at point E if this load is directed (a) vertically downward, (b) vertically upward.

1.16 Two 5-kN vertical loads are applied to pin B of the assembly shown. Knowing that a 12-mm-diameter pin is used at each connection, determine the maximum value of the average normal stress in (a) link AB, (b) link BC.

1.17 For the truss and loading shown, determine the normal stress in member BD, knowing that the cross-sectional area of the member is 1250 mm^2.

1.18 Determine the smallest allowable cross-sectional area of member DE of the truss shown if the normal stress in that member is not to exceed 100 MPa for the given loading.

1.19 Determine the smallest allowable cross-sectional area of member DE of the truss shown if the normal stress in that member is not to exceed 200 MPa for the given loading.

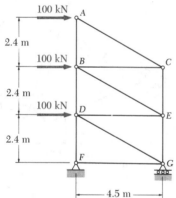

Fig. P1.17 and P1.18

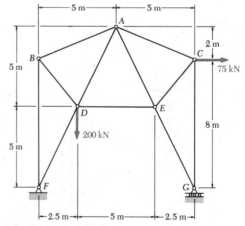

Fig. P1.19 and P1.20

1.20 For the truss and loading shown, determine the normal stress in member AD, knowing that the cross-sectional area of the member is 1200 mm^2.

1.21 Knowing that link *DE* is 25 mm wide and 3 mm thick, determine the normal stress in the central portion of that link when $\theta = 0$.

1.22 Knowing that link *DE* is 25 mm wide and 3 mm thick, determine the normal stress in the central portion of that link when $\theta = 90°$.

1.23 An aircraft tow bar is positioned by means of a single hydraulic cylinder connected by a 25-mm-diameter steel rod to two identical arm-and-wheel units *DEF*. The weight of the entire tow bar is 2 kN, and its center of gravity is located at *G*. For the position shown, determine the normal stress in the rod.

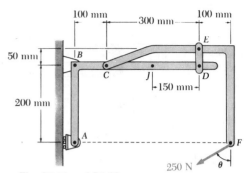

Fig. P1.21 and P1.22

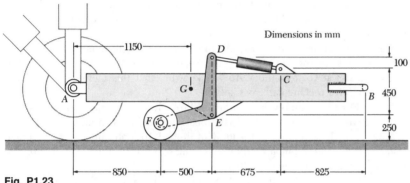

Fig. P1.23

1.24 In the marine crane shown, link *CD* is known to have a uniform cross section of 50 × 150 mm. For the loading shown, determine the normal stress in the central portion of that link.

1.25 Two wooden planks, each 12 mm thick and 225 mm wide, are joined by the dry mortise joint shown. Knowing that the wood used shears off along its grain when the average shearing stress reaches 8 MPa, determine the magnitude *P* of that axial load which will cause the joint to fail.

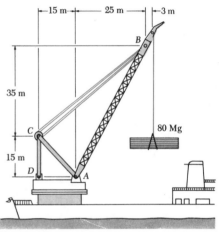

Fig. P1.24

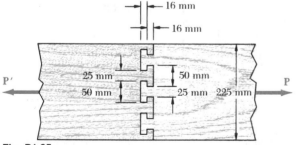

Fig. P1.25

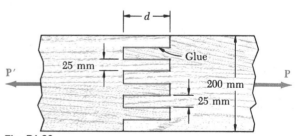

Fig. P1.26

1.26 Two wooden planks, each 15 mm thick and 200 mm wide, are joined by the glued mortise joint shown. Knowing that the joint will fail when the average shearing stress in the glue reaches 900 kPa, determine the required length *d* of the cuts if the joint is to withstand an axial load of magnitude $P = 4$ kN.

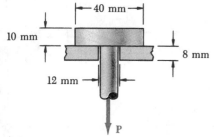

Fig. P1.27

1.27 A load **P** is applied to a steel rod supported as shown by an aluminum plate into which a 12-mm-diameter hole has been drilled. Knowing that the maximum allowable shearing stress is 180 MPa in the steel rod and 70 MPa in the aluminum plate, determine the largest load which may be applied to the rod.

1.28 Determine the diameter of the largest circular hole which may be punched into a sheet of polystyrene 6 mm thick, knowing that the force exerted by the punch is 45 kN and that an 55 MPa average shearing stress is required to cause the material to fail.

1.29 For the assembly and loading of Prob. 1.13, determine (*a*) the average shearing stress in the pin at *B*, (*b*) the bearing stress at *B* in link *BE*, (*c*) the bearing stress at *B* in member *ABCD*, knowing that this member has a 15 × 50-mm uniform rectangular cross section.

1.30 For the assembly and loading of Prob. 1.13, determine (*a*) the average shearing stress in the pin at *C*, (*b*) the bearing stress at *C* in link *CF*, (*c*) the bearing stress at *C* in member *ABCD*, knowing that this member has a 15 × 50-mm uniform rectangular cross section.

1.31 For the assembly and loading of Prob. 1.16, determine (*a*) the average shearing stress in the pin at *A*, (*b*) the bearing stress at *A* in member *AB*.

1.32 For the assembly and loading of Prob. 1.16, determine (*a*) the average shearing stress in the pin at *C*, (*b*) the bearing stress at *C* in member *BC*, (*c*) the bearing stress at *B* in member *BC*.

1.33 The hydraulic cylinder *CF*, which partially controls the position of rod *DE*, has been locked in the position shown. Member *BD* is 15 mm thick and is connected at *C* to the vertical rod by a 9-mm-diameter bolt. Knowing that *P* = 2 kN and $\theta = 75°$, determine (*a*) the average shearing stress in the bolt, (*b*) the bearing stress at *C* in member *BD*.

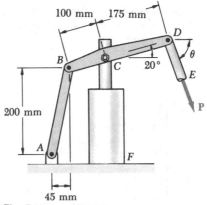

Fig. P1.33 and P1.34

1.34 The hydraulic cylinder *CF*, which partially controls the position of rod *DE*, has been locked in the position shown. Link *AB* has a uniform rectangular cross section of 12 × 25 mm and is connected at *B* to member *BD* by an 8-mm-diameter pin. Knowing that the maximum allowable average shearing stress in the pin is 140 MPa, determine (*a*) the largest force **P** which may be applied at *E* when $\theta = 60°$, (*b*) the corresponding bearing stress at *B* in link *AB*, (*c*) the corresponding maximum value of the normal stress in link *AB*.

1.7. STRESS ON AN OBLIQUE PLANE UNDER AXIAL LOADING

In the preceding sections, axial forces exerted on a two-force member (Fig. 1.24*a*) were found to cause normal stresses in that member (Fig. 1.24*b*), while transverse forces exerted on pins and rivets (Fig. 1.25*a*) were found to cause shearing stresses in those connections (Fig. 1.25*b*). The reason such a relation was observed between axial forces and normal stresses on one hand, and transverse forces and shearing stresses on the other, was because stresses were being determined only on planes perpendicular to the axis of the member or connection. As we shall see in this section, axial forces cause both normal and shearing stresses on planes which are not perpendicular to the axis of the member. Similarly, transverse forces exerted on a pin or rivet cause both normal and shearing stresses on planes which are not perpendicular to the axis of the pin or rivet.

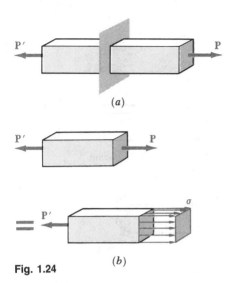

Fig. 1.24

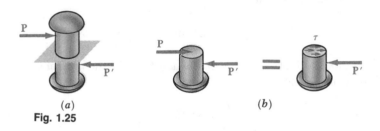

Fig. 1.25

Consider the two-force member of Fig. 1.24, which is subjected to axial forces **P** and **P′**. If we pass a section forming an angle θ with a normal plane (Fig. 1.26*a*) and draw the free-body diagram of the portion of member located to the left of that section (Fig. 1.26*b*), we find from the equilibrium conditions of the free body that the distributed forces acting on the section must be equivalent to the force **P.**

Resolving **P** into components **F** and **V,** respectively normal and tangential to the section (Fig. 1.26*c*), we have

$$F = P \cos \theta \qquad V = P \sin \theta \qquad (1.9)$$

The force **F** represents the resultant of normal forces distributed over the section, and the force **V** the resultant of shearing forces (Fig. 1.26*d*). The average values of the corresponding normal and shearing stresses are obtained by dividing, respectively, F and V by the area A_θ of the section:

$$\sigma = \frac{F}{A_\theta} \qquad \tau = \frac{V}{A_\theta} \qquad (1.10)$$

Substituting for F and V from (1.9) into (1.10), and observing from Fig. 1.26*c* that $A_0 = A_\theta \cos \theta$, or $A_\theta = A_0/\cos \theta$, where A_0 denotes the area of

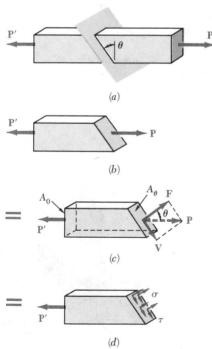

Fig. 1.26

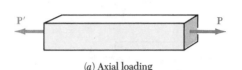

(a) Axial loading

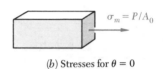

(b) Stresses for $\theta = 0$

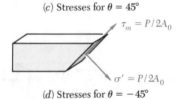

(c) Stresses for $\theta = 45°$

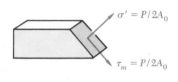

(d) Stresses for $\theta = -45°$

Fig. 1.27

a section perpendicular to the axis of the member, we obtain

$$\sigma = \frac{P\cos\theta}{A_0/\cos\theta} \qquad \tau = \frac{P\sin\theta}{A_0/\cos\theta}$$

or

$$\sigma = \frac{P}{A_0}\cos^2\theta \qquad \tau = \frac{P}{A_0}\sin\theta\cos\theta \tag{1.11}$$

We note from the first of Eqs. (1.11) that the normal stress σ is maximum when $\theta = 0$, i.e., when the plane of the section is perpendicular to the axis of the member, and that it approaches zero as θ approaches $90°$. We check that the value of σ when $\theta = 0$ is

$$\sigma_m = \frac{P}{A_0} \tag{1.12}$$

as we found earlier in Sec. 1.2. The second of Eqs. (1.11) shows that the shearing stress τ is zero for $\theta = 0$ and $\theta = 90°$, and that for $\theta = 45°$ it reaches its maximum value

$$\tau_m = \frac{P}{A_0}\sin 45°\cos 45° = \frac{P}{2A_0} \tag{1.13}$$

The first of Eqs. (1.11) indicates that, when $\theta = 45°$, the normal stress σ' is also equal to $P/2A_0$:

$$\sigma' = \frac{P}{A_0}\cos^2 45° = \frac{P}{2A_0} \tag{1.14}$$

The results obtained in Eqs. (1.12), (1.13), and (1.14) are shown graphically in Fig. 1.27. We note that the same loading may produce either a normal stress $\sigma_m = P/A_0$ and no shearing stress (Fig. 1.27b), or a normal and a shearing stress of the same magnitude $\sigma' = \tau_m = P/2A_0$ (Fig. 1.27c and d), depending upon the orientation of the section.

1.8. STRESS UNDER GENERAL LOADING CONDITIONS; COMPONENTS OF STRESS

The examples of the previous sections were limited to members under axial loading and connections under transverse loading. Most structural members and machine components are under more involved loading conditions.

Consider a body subjected to several loads $\mathbf{P}_1$, $\mathbf{P}_2$, etc. (Fig. 1.28). To understand the stress condition created by these loads at some point Q within the body, we shall first pass a section through Q, using a plane parallel to the yz plane. The portion of the body to the left of the section is subjected to some of the original loads, and to normal and shearing forces distributed over the section. We shall denote by $\Delta\mathbf{F}^x$ and $\Delta\mathbf{V}^x$, respectively, the normal and the shearing forces acting on a small area ΔA

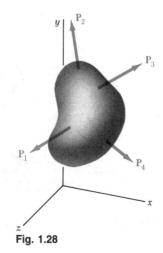

Fig. 1.28

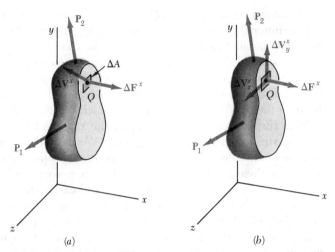

Fig. 1.29   (a)   (b)

surrounding point Q (Fig. 1.29a). Note that the superscript x is used to indicate that the forces $\Delta \mathbf{F}^x$ and $\Delta \mathbf{V}^x$ act on a surface perpendicular to the x axis. While the normal force $\Delta \mathbf{F}^x$ has a well-defined direction, the shearing force $\Delta \mathbf{V}^x$ may have any direction in the plane of the section. We shall, therefore, resolve $\Delta \mathbf{V}^x$ into two component forces, $\Delta \mathbf{V}_y^x$ and $\Delta \mathbf{V}_z^x$, in directions parallel to the y and z axes, respectively (Fig. 1.29b). Dividing now the magnitude of each force by the area ΔA, and letting ΔA approach zero, we define the three stress components shown in Fig. 1.30:

$$\sigma_x = \lim_{\Delta A \to 0} \frac{\Delta F^x}{\Delta A}$$

$$\tau_{xy} = \lim_{\Delta A \to 0} \frac{\Delta V_y^x}{\Delta A} \qquad \tau_{xz} = \lim_{\Delta A \to 0} \frac{\Delta V_z^x}{\Delta A} \tag{1.15}$$

We note that the first subscript in σ_x, τ_{xy}, and τ_{xz} is used to indicate that the stresses under consideration are exerted *on a surface perpendicular to the x axis*. The second subscript in τ_{xy} and τ_{xz} identifies *the direction of the component*. The normal stress σ_x is positive if the corresponding arrow points in the positive x direction, i.e., if the body is in tension, and negative otherwise. Similarly, the shearing stress components τ_{xy} and τ_{xz} are positive if the corresponding arrows point, respectively, in the positive y and z directions.

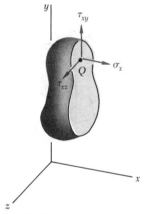

Fig. 1.30

The above analysis may also be carried out by considering the portion of body located to the right of the vertical plane through Q (Fig. 1.31). The same magnitudes, but opposite directions, are obtained for the normal and shearing forces $\Delta \mathbf{F}^x$, $\Delta \mathbf{V}_y^x$, and $\Delta \mathbf{V}_z^x$. Therefore, the same values are also obtained for the corresponding stress components, but since the section in Fig. 1.31 now faces the *negative x axis*, a positive sign for σ_x will indicate that the corresponding arrow points *in the negative x direction*. Similarly, positive signs for τ_{xy} and τ_{xz} will indicate that the corresponding arrows point, respectively, in the negative y and z directions, as shown in Fig. 1.31.

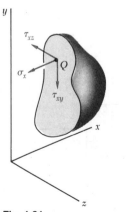

Fig. 1.31

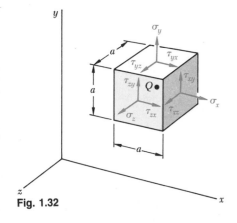

Fig. 1.32

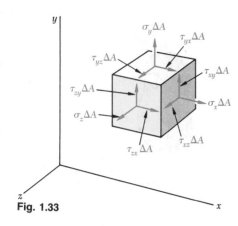

Fig. 1.33

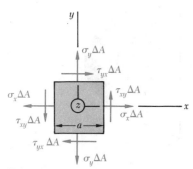

Fig. 1.34

Passing a section through Q parallel to the zx plane, we define in the same manner the stress components, σ_y, τ_{yz}, and τ_{yx}. Finally, a section through Q parallel to the xy plane yields the components σ_z, τ_{zx}, and τ_{zy}.

To facilitate the visualization of the stress condition at point Q, we shall consider a small cube of side a centered at Q and the stresses exerted on each of the six faces of the cube (Fig. 1.32). The stress components shown in the figure are σ_x, σ_y, and σ_z, which represent the normal stress on faces respectively perpendicular to the x, y, and z axes, and the six shearing stress components τ_{xy}, τ_{xz}, etc. We recall that, according to the definition of the shearing stress components, τ_{xy} represents the y component of the shearing stress exerted on the face perpendicular to the x axis, while τ_{yx} represents the x component of the shearing stress exerted on the face perpendicular to the y axis. Note that only three faces of the cube are actually visible in Fig. 1.32, and that equal and opposite stress components act on the hidden faces. While the stresses acting on the faces of the cube differ slightly from the stresses at Q, the error involved is small and vanishes as side a of the cube approaches zero.

Important relations between the shearing stress components will now be derived. Let us consider the free-body diagram of the small cube centered at point Q (Fig. 1.33). The normal and shearing forces acting on the various faces of the cube are obtained by multiplying the corresponding stress components by the area ΔA of each face. Selecting coordinate axes centered at Q, we write the six equilibrium equations:

$$\Sigma F_x = 0 \qquad \Sigma F_y = 0 \qquad \Sigma F_z = 0 \qquad (1.16)$$

$$\Sigma M_x = 0 \qquad \Sigma M_y = 0 \qquad \Sigma M_z = 0 \qquad (1.17)$$

Since forces equal and opposite to the forces actually shown in Fig. 1.33 are acting on the hidden faces of the cube, it is clear that Eqs. (1.16) are satisfied. Turning now to Eqs. (1.17), we first consider the last of these equations, $\Sigma M_z = 0$. Using a projection on the xy plane (Fig. 1.34), we note that the only forces with moments about the z axis different from zero are the shearing forces. These forces form two couples, one of counterclockwise (positive) moment $(\tau_{xy} \Delta A)a$, the other of clockwise (negative) moment $-(\tau_{yx} \Delta A)a$. We write therefore

$$+\curvearrowleft\Sigma M_z = 0: \qquad (\tau_{xy} \Delta A)a - (\tau_{yx} \Delta A)a = 0$$

from which we conclude that

$$\boxed{\tau_{xy} = \tau_{yx}} \qquad (1.18)$$

The relation obtained shows that the y component of the shearing stress exerted on a face perpendicular to the x axis is equal to the x component of the shearing stress exerted on a face perpendicular to the y axis. From the remaining two equations (1.17), we derive in a similar manner the relations

$$\tau_{yz} = \tau_{zy} \qquad \tau_{zx} = \tau_{xz} \qquad (1.19)$$

We conclude from Eqs. (1.18) and (1.19) that only six stress components are required to define the condition of stress at a given point Q, instead of nine as originally assumed. These six components are σ_x, σ_y, σ_z, τ_{xy}, τ_{yz}, and τ_{zx}. We also note that, at a given point, *shear cannot take place in one plane only;* an equal shearing stress must be exerted on another plane perpendicular to the first one. For example, considering again the rivet of Fig. 1.25 and a small cube at the center Q of the rivet (Fig. 1.35a), we find that shearing stresses of equal magnitude must be exerted on the two horizontal faces of the cube and on the two faces which are perpendicular to the forces **P** and **P′** (Fig. 1.35b).

Before concluding our discussion of stress components, let us consider again the case of a member under axial loading. If we consider a small cube with faces respectively parallel to the faces of the member and recall the results obtained in Sec. 1.7, we find that the conditions of stress in the member may be described as shown in Fig. 1.36a; the only stresses

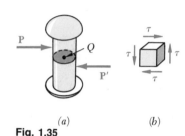

(a)   (b)

Fig. 1.35

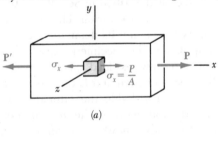

(a)

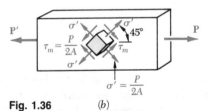

Fig. 1.36   (b)

are normal stresses σ_x exerted on the faces of the cube which are perpendicular to the x axis. However, if the small cube is rotated by 45° about the z axis so that its new orientation matches the orientation of the sections considered in Fig. 1.27c and d, we conclude that normal and shearing stresses of equal magnitude are exerted on four faces of the cube (Fig. 1.36b). We thus observe that the same loading condition may lead to different interpretations of the stress situation at a given point, depending upon the orientation of the element considered. More will be said about this in Chap. 6.

1.9. ULTIMATE AND ALLOWABLE STRESS; FACTOR OF SAFETY

In the preceding sections we learned to determine the stresses in rods and pins under simple loading conditions. In later chapters we shall learn to determine stresses in more complex situations. In engineering applications, however, the determination of stresses is seldom an end in itself. Rather, the knowledge of stresses is used by the engineer to assist in the performance of the following major tasks:

1. The *analysis* of existing or proposed structures and machines, in order to predict their behavior under specified loading conditions
2. The *design* of new structures and machines which will safely and economically perform a specified function

In order to perform either of the above tasks, we must know how the material to be used will act under known loading conditions. For a given material this is determined by performing specific tests on prepared samples of the material. For example, we may prepare a test specimen of steel and place it in a laboratory testing machine to subject it to a known centric axial tensile force, as described in Sec. 2.3. As the magnitude of the force is increased, we measure various changes in the specimen, for example, changes in its length and its diameter. Eventually the largest force which may be applied to the specimen is reached, and the specimen either breaks or begins to carry less load. This largest force is called the *ultimate load* for the test specimen and is denoted by P_U. Since the applied load is centric, we may divide the ultimate load by the original cross-sectional area of the rod to obtain the *ultimate normal stress* of the material used. This stress, also known as the *ultimate strength in tension* of the material, is

$$\sigma_U = \frac{P_U}{A} \qquad (1.20)$$

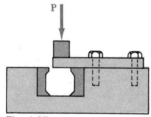

Fig. 1.37

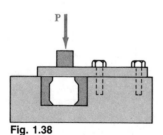

Fig. 1.38

Several test procedures are available to determine the *ultimate shearing stress*, or *ultimate strength in shear*, of a material. The one most commonly used involves the twisting of a circular tube (Sec. 3.9). A more direct, if less accurate, procedure consists in clamping a rectangular or round bar in a shear tool (Fig. 1.37) and applying an increasing load P until the ultimate load P_U for single shear is obtained. If the free end of the specimen rests on both of the hardened dies (Fig. 1.38), the ultimate load for double shear is obtained. In either case, the ultimate shearing stress τ_U is obtained by dividing the ultimate load by the total area over which shear has taken place. We recall that, in the case of single shear, this area is the cross-sectional area A of the specimen, while in double shear it is equal to twice the cross-sectional area.

A structural member or a machine component must be designed so that its ultimate load is considerably larger than the load the member or component will be allowed to carry under normal conditions of utilization. This smaller load is referred to as the *allowable load* and, sometimes, as the *working load* or *design load*. Thus, only a fraction of the ultimate load-carrying capacity of the member is utilized when the allowable load is applied. The remaining portion of the load-carrying capacity of the member is kept in reserve to assure its safe performance. The ratio of the ultimate load to the allowable load is defined as the *factor of safety*.† We write

$$\text{Factor of safety} = F.S. = \frac{\text{ultimate load}}{\text{allowable load}} \qquad (1.21)$$

In many applications a linear relationship exists between a load and the stress caused by the load. When this is the case, the factor of safety may also be expressed as

$$\text{Factor of safety} = F.S. = \frac{\text{ultimate stress}}{\text{allowable stress}} \qquad (1.22)$$

The determination of the factor of safety which should be used for various applications is one of the most important engineering tasks. On the one hand, if a factor of safety is chosen too small, the possibility of failure becomes unacceptably large; on the other hand, if a factor of safety is chosen unnecessarily large, the result is an uneconomical or nonfunctional design. The choice of the factor of safety which is appropriate for a given design application requires engineering judgment based on many considerations, such as the following:

1. *Variations which occur in material properties.* The composition, strength, and dimensions of materials are all subject to small variations during manufacture. In addition, material properties may be altered and residual stresses introduced through heating or deformation which may occur in the material during storage, transportation, or construction.
2. *The number of loadings which may be expected during the life of the structure or machine.* For most materials the ultimate stress decreases as the number of load applications is increased. This phenomenon is known as *fatigue* and, if ignored, may result in sudden failure (see Sec. 2.7).
3. *The type of loadings which are planned for in the design, or which may occur in the future.* Very few loadings are known with

---

† In some fields of engineering, notably aeronautical engineering, the *margin of safety* is used in place of the factor of safety. The margin of safety is defined as the factor of safety minus one; that is, margin of safety = $F.S. - 1.00$.

complete accuracy—most design loadings are engineering estimates. In addition, future alterations or changes in usage may introduce changes in the actual loading. Larger factors of safety are also required for dynamic, cyclic, or impulsive loadings.

4. *The type of failure which may occur.* Brittle materials fail suddenly, usually with no prior indication that collapse is imminent. On the other hand, ductile materials, such as structural steel, undergo a substantial deformation called *yielding* before failing, thus providing a warning that overloading exists. However, most buckling or stability failures are sudden, whether the material is brittle or not. When the possibility of sudden failure exists, a larger factor of safety should be used than when failure is preceded by obvious warning signs.

5. *Uncertainty due to methods of analysis.* All design methods are based on certain simplifying assumptions which result in calculated stresses being approximations of actual stresses.

6. *Deterioration which may occur in the future due to poor maintenance or due to unpreventable natural causes.* A larger factor of safety is necessary in locations where conditions such as rusting and decay are difficult to control or even to discover.

7. *The importance of a given member to the integrity of the whole structure.* Bracing and secondary members may in many cases be designed with a factor of safety lower than that used for primary members.

In addition to the above considerations, there is the additional consideration concerning the risk to life and property which a failure would produce. Where a failure would produce no risk to life and only minimal risk to property, the use of a smaller factor of safety may be considered. Finally, there is the practical consideration that, unless a careful design with a nonexcessive factor of safety is used, a structure or machine might not perform its design function. For example, high factors of safety may have an unacceptable effect on the weight of an aircraft.

For the majority of structural and machine applications, factors of safety are specified by design specifications or building codes written by committees of experienced engineers working with professional societies, with industries, or with federal, state, or city agencies. Examples of such design specifications and building codes are

1. *Steel:* American Institute of Steel Construction, Specifications for the Design and Erection of Structural Steel for Buildings

2. *Concrete:* American Concrete Institute, Building Code Requirement for Reinforced Concrete

3. *Timber:* National Forest Products Association, National Design Specifications for Stress-Grade Lumber and Its Fastenings

4. *Highway bridges:* American Association of State Highway Officials, Standard Specifications for Highway Bridges

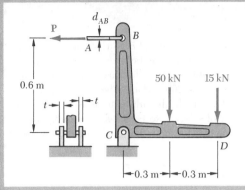

SAMPLE PROBLEM 1.2

Two forces are applied to the bracket BCD as shown. (a) Knowing that the control rod AB is to be made of a steel having an ultimate normal stress of 600 MPa, determine the diameter of the rod for which the factor of safety with respect to failure will be 3.3. (b) The pin at C is to be made of a steel having an ultimate shearing stress of 350 MPa. Determine the diameter of the pin C for which the factor of safety with respect to shear will also be 3.3. (c) Determine the required thickness of the bracket supports at C knowing that the allowable bearing stress of the steel used is 300 MPa.

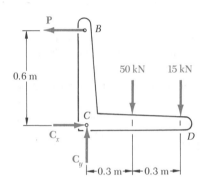

Free Body: Entire Bracket. The reaction at C is represented by its components C_x and C_y.

$$+\!\!\uparrow\!\Sigma M_C = 0: \qquad P(0.6\text{ m}) - (50\text{ kN})(0.3\text{ m}) - (15\text{ kN})(0.6\text{ m}) = 0 \qquad P = 40\text{ kN}$$

$$\Sigma F_x = 0: \qquad C_x = 40\text{ kN}$$
$$\Sigma F_y = 0: \qquad C_y = 65\text{ kN} \qquad C = \sqrt{C_x^2 + C_y^2} = 76.3\text{ kN}$$

a. **Control Rod AB.** Since the factor of safety is to be 3.3, the allowable stress is

$$\sigma_{\text{all}} = \frac{\sigma_U}{F.S.} = \frac{600\text{ MPa}}{3.3} = 181.8\text{ MPa}$$

For P = 40 kN the cross-sectional area required is

$$A_{\text{req}} = \frac{P}{\sigma_{\text{all}}} = \frac{40\text{ kN}}{181.8\text{ MPa}} = 220 \times 10^{-6}\text{ m}^2$$

$$A_{\text{req}} = \frac{\pi}{4}d_{AB}^2 = 220 \times 10^{-6}\text{ m}^2 \qquad d_{AB} = 16.74\text{ mm} \quad \blacktriangleleft$$

b. **Shear in Pin C.** For a factor of safety of 3.3, we have

$$\tau_{\text{all}} = \frac{\tau_U}{F.S.} = \frac{350\text{ MPa}}{3.3} = 106.1\text{ MPa}$$

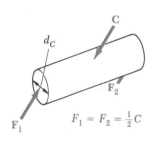

Since the pin is in double shear, we write

$$A_{\text{req}} = \frac{C/2}{\tau_{\text{all}}} = \frac{(76.3\text{ kN})/2}{106.1\text{ MPa}} = 360\text{ mm}^2$$

$$A_{\text{req}} = \frac{\pi}{4}d_C^2 = 360\text{ mm}^2$$

$$d_C = 21.4\text{ mm} \qquad \text{Use: } d_C = 22\text{ mm} \quad \blacktriangleleft$$

The next larger size pin available is of 22-mm diameter and should be used.

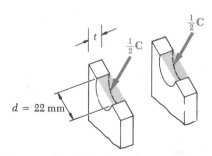

c. **Bearing at C.** Using d = 22 mm, the nominal bearing area of each bracket is 22t. Since the force carried by each bracket is C/2 and the allowable bearing stress is 300 MPa, we write

$$A_{\text{req}} = \frac{C/2}{\sigma_{\text{all}}} = \frac{(76.3\text{ kN})/2}{300\text{ MPa}} = 127.2\text{ mm}^2$$

Thus $22t = 127.2$ $\qquad t = 5.78\text{ mm}$ $\qquad\qquad$ Use: t = 6 mm $\quad \blacktriangleleft$

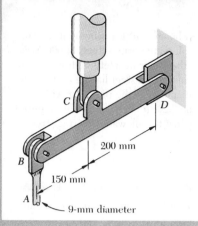

A

9-mm diameter

150 mm

B

200 mm

C

D

The rigid beam BCD is attached by bolts to a control rod at B, to a hydraulic cylinder at C, and to a fixed support at D. The diameters of the bolts used are: $d_B = d_D = 8$ mm, $d_C = 12$ mm. Each bolt acts in double shear and is made from a steel for which the ultimate shearing stress is $\tau_U = 300$ MPa. The 9-mm-diameter control rod AB is made of a steel for which the ultimate tensile stress is $\sigma_U = 450$ MPa. If the minimum factor of safety is to be 3.0 for the entire unit, determine the largest upward force which may be applied by the hydraulic cylinder at C.

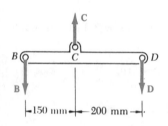

Solution. The factor of safety with respect to failure must be 3.0 or more in each of the three bolts and in the control rod. These four independent criteria will be considered separately.

Free Body: Beam BCD. We first determine the force at C in terms of the force at B and in terms of the force at D.

$+\!\!\uparrow\!\Sigma M_D = 0$: $\qquad B(350 \text{ mm}) - C(200 \text{ mm}) = 0 \qquad C = 1.750\ B$ (1)

$+\!\!\uparrow\!\Sigma M_B = 0$: $\qquad -D(350 \text{ mm}) + C(150 \text{ mm}) = 0 \qquad C = 2.33\ D$ (2)

Control Rod. For a factor of safety of 3.0 we have

$$\sigma_{\text{all}} = \frac{\sigma_U}{F.S.} = \frac{450 \text{ MPa}}{3.0} = 150 \text{ MPa}$$

The allowable force in the control rod is

$$B = \sigma_{\text{all}}(A) = (150 \text{ MPa})\tfrac{1}{4}\pi(9 \text{ mm})^2 = 9.54 \text{ kN}$$

Using Eq. (1) we find the largest permitted value of C:

$$C = 1.750\ B = 1.750(9.54 \text{ kN}) \qquad C = 16.70 \text{ kN} \quad \triangleleft$$

Bolt at B. $\tau_{\text{all}} = \tau_U/F.S. = (300 \text{ MPa})/3 = 100$ MPa. Since the bolt is in double shear, the allowable value of force B is

$$B = \tau_{\text{all}}(2A) = (100 \text{ MPa})\frac{2\pi}{4}(8 \text{ mm})^2 = 10.05 \text{ kN}$$

From Eq. (1): $\qquad C = 1.750\ B = 1.750(10.05 \text{ kN}) \qquad C = 17.59 \text{ kN} \quad \triangleleft$

Bolt at D. Since this bolt is the same as bolt B, the allowable force is $D = B = 10.05$ kN. Using Eq. (2):

$$C = 2.33\ D = 2.33(10.05 \text{ kN}) \qquad C = 23.4 \text{ kN} \quad \triangleleft$$

Bolt at C. We again have $\tau_{\text{all}} = 100$ MPa and write

$$C = \tau_{\text{all}}(2A) = (100 \text{ MPa})\frac{2\pi}{4}(12 \text{ mm})^2 \qquad C = 22.6 \text{ kN} \quad \triangleleft$$

Summary. We have found separately four maximum allowable values of the force C. In order to satisfy all these criteria we must choose the *smallest* value, namely:

$$C = 16.70 \text{ kN} \quad \blacktriangleleft$$

PROBLEMS

1.35 Two wooden members of 80 × 120-mm uniform rectangular cross section are joined by the simple glued scarf splice shown.   Knowing that $P = 12$ kN, determine the normal and shearing stresses in the glued splice.

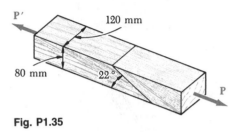

Fig. P1.35

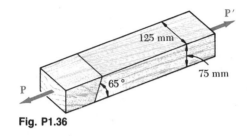

Fig. P1.36

1.36 Two wooden members of 75 × 125-mm uniform rectangular cross section are joined by the simple glued joint shown.   Knowing that $P = 3.5$ kN, determine the normal and shearing stresses in the glued joint.

1.37 Knowing that the maximum allowable tensile stress for the glued joint of Prob. 1.36 is 400 kPa, determine (a) the largest axial load P which may be applied, (b) the corresponding shearing stress in the joint.

1.38 Knowing that the maximum allowable shearing stress for the glued splice of Prob. 1.35 is 600 kPa, determine (a) the largest axial load P which may be applied, (b) the corresponding normal stress in the splice.

1.39 A steel pipe of 300-mm outer diameter is fabricated from 8-mm-thick plate by welding along a helix which forms an angle of 20° with a plane perpendicular to the axis of the pipe.   Knowing that a 250-kN axial force **P** is applied to the pipe, determine σ and τ in directions, respectively, normal and tangential to the weld.

1.40 For the pipe of Prob. 1.39, knowing that the maximum allowable stresses in directions normal and tangential to the weld are $\sigma = 150$ MPa and $\tau = 100$ MPa, respectively, determine the magnitude P of the largest axial force which may be applied to the pipe.

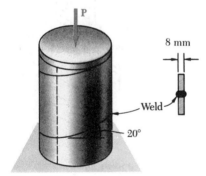

Fig. P1.39

1.41 Members AB and BE of the truss shown consist of rods made of the same metal alloy.   It is known that a 20-mm-diameter rod made of the same alloy was tested to failure and that an ultimate load of 150 kN was recorded.   Using a factor of safety of 3.2, determine the required diameter of (a) rod AB, (b) rod BE.

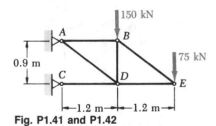

Fig. P1.41 and P1.42

1.42 Members AB and BE of the truss shown consist of rods made of the same metal alloy.   Knowing that AB has a 28-mm diameter and that the ultimate load for that rod is 300 kN, determine (a) the factor of safety for AB, (b) the required diameter of BE if both rods are to have the same factor of safety.

1.43 The horizontal link BC is 6 mm thick, has a width $w = 30$ mm, and is made of a steel with a 500-MPa ultimate strength in tension. What was the factor of safety used if the structure shown was designed to support a load $P = 45$ kN?

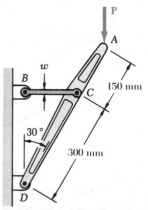

Fig. P1.43 and P1.44

1.44 The horizontal link BC is 6 mm thick and is made of a steel with a 450-MPa ultimate strength in tension. What should be the width w of the link if the structure shown was designed to support a load $P = 35$ kN with a factor of safety equal to 3?

1.45 The wooden members are joined by plywood splice plates which are fully glued on the surfaces in contact. Knowing that the clearance between the ends of the members is 6 mm and that the ultimate shearing stress in the glued joint is 2.5 MPa, determine the length L for which the factor of safety is 2.75 for the loading shown.

1.46 For the joint and loading of Prob. 1.45, determine the factor of safety when $L = 180$ mm.

1.47 For the joint and loading of Prob. 1.36, determine the factor of safety, knowing that the ultimate strength of the glue is 1 MPa in tension and 1.5 MPa in shear.

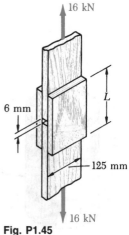

Fig. P1.45

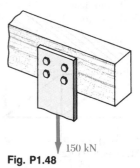

Fig. P1.48

1.48 Four steel bolts are used to attach the plate shown to a wooden beam. Knowing that the ultimate shearing stress in the steel used is 350 MPa and that a factor of safety of 3.3 is required, determine the smallest allowable diameter of the bolts to be used.

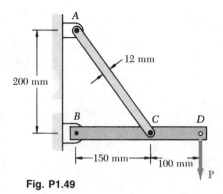

Fig. P1.49

1.49 Link *AC* is made of a steel with a 400-MPa ultimate normal stress and has a 6 × 12 mm uniform rectangular cross section.   It is connected to a support at *A* and to member *BCD* at *C* by 9-mm-diameter pins, while member *BCD* is connected to its support at *B* by a 7.5-mm-diameter pin; all of the pins are made of a steel with a 175 MPa ultimate shearing stress and are in single shear.   Knowing that a factor of safety of 3.25 is desired, determine the largest load **P** which may be applied at *D*.   Note that link *AC* is not reinforced around the pin holes.

1.50 In the steel structure shown, a 6-mm-diameter pin is used at *C* and 10-mm-diameter pins are used at *B* and *D*.   The ultimate shearing stress is 150 MPa at all connections, and the ultimate normal stress is 400 MPa in link *BD*.   Knowing that a factor of safety of 3 is desired, determine the largest load **P** which may be applied at *A*.   Note that link *BD* is not reinforced around the pin holes.

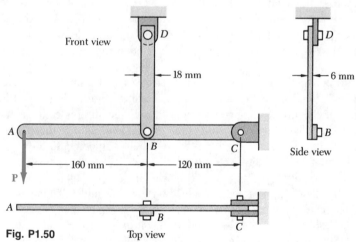

Fig. P1.50

1.51 Solve Prob. 1.50, assuming that the structure has been redesigned to use 12-mm-diameter pins at *B* and *D* and that no other change has been made.

1.52 Solve Prob. 1.49, assuming that the structure has been redesigned to use 7.5-mm-diameter pins at *A* and *C* as well as at *B* and that no other change has been made.

1.53 Solve Prob. 1.50, assuming that the structure has been redesigned to use 11.5-mm-diameter pins at *B* and *D* and that no other change has been made.

Axial loading. Normal stress

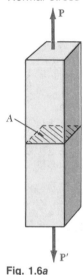

Fig. 1.6a

Transverse forces. Shearing stress

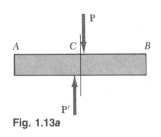

Fig. 1.13a

REVIEW AND SUMMARY

This chapter was devoted to the concept of *stress* and to an introduction to the methods used for the analysis and design of machines and load-bearing structures.

We first considered a straight two-force member under *axial loading* [Sec. 1.2]. The *normal stress* σ in that member was obtained by dividing the magnitude P of the load by the cross-sectional area A of the member (Fig. 1.6a). We wrote

$$\sigma = \frac{P}{A} \qquad (1.1)$$

As noted in Sec. 1.3, the value of σ obtained in this way represents the *average stress* over the section rather than the stress at a specific point Q of the section. Considering a small area ΔA surrounding Q and the magnitude ΔF of the force exerted on ΔA, we defined the stress at point Q as

$$\sigma = \lim_{\Delta A \to 0} \frac{\Delta F}{\Delta A} \qquad (1.2)$$

In general, the value obtained for the stress σ at point Q is different from the value of the average stress given by formula (1.1) and is found to vary across the section. However, this variation is small in any section away from the points of application of the loads. In practice, therefore, the distribution of the normal stresses in an axially loaded member is assumed to be *uniform*, except in the immediate vicinity of the points of application of the loads.
However, for the distribution of stresses to be uniform in a given section, it is necessary that the line of action of the loads **P** and **P′** pass through the centroid C of the section. Such a loading is called a *centric* axial loading. In the case of an *eccentric* axial loading, the distribution of stresses is *not* uniform. Stresses in members subjected to an eccentric axial loading will be discussed in Chap. 4.

When equal and opposite *transverse forces* **P** and **P′** of magnitude P are applied to a member AB (Fig. 1.13a), *shearing stresses* τ are created over any section located between the points of application of the two forces [Sec. 1.4]. These stresses vary greatly across the section and their distribution *cannot* be assumed uniform. However, dividing the magnitude P—referred to as the *shear* in the section—by the cross-sectional area A, we defined the *average shearing stress* over the section:

$$\tau_{\text{ave}} = \frac{P}{A} \qquad (1.4)$$

Shearing stresses are found in bolts, pins, or rivets connecting two structural members or machine components. For example, in the case of rivet CD (Fig. 1.14), which is in *single shear*, we wrote

$$\tau_{ave} = \frac{P}{A} = \frac{F}{A} \qquad (1.5)$$

while, in the case of rivets EG and HJ (Fig. 1.16), which are both in *double shear*, we had

$$\tau_{ave} = \frac{P}{A} = \frac{F}{2A} \qquad (1.6)$$

Bolts, pins, and rivets also create stresses in the members they connect, along the *bearing surface*, or surface of contact [Sec. 1.5]. The rivet CD of Fig. 1.14, for example, creates stresses on the semicylindrical surface of plate A with which it is in contact (Fig. 1.18). Since the distribution of these stresses is quite complicated, one uses in practice an average nominal value σ_b of the stress, called *bearing stress*, obtained by dividing the load P by the area of the rectangle representing the projection of the rivet on the plate section. Denoting by t the thickness of the plate and by d the diameter of the rivet, we wrote

$$\sigma_b = \frac{P}{A} = \frac{P}{td} \qquad (1.7)$$

In Sec. 1.6, we applied the concepts introduced in the previous sections to the analysis of a simple structure consisting of two pin-connected members supporting a given load. We determined successively the normal stresses in the two members, paying special attention to their narrowest sections, the shearing stresses in the various pins, and the bearing stress at each connection.

In Sec. 1.7, we considered the stresses created on an *oblique section* in a two-force member under axial loading. We found that both *normal* and *shearing* stresses occurred in such a situation. Denoting by θ the angle formed by the section with a normal plane (Fig. 1.26a) and by A_0 the area of a section perpendicular to the axis of the member, we derived the following expressions for the normal stress σ and the shearing stress τ on the oblique section:

$$\sigma = \frac{P}{A_0} \cos^2 \theta \qquad \tau = \frac{P}{A_0} \sin \theta \cos \theta \qquad (1.11)$$

We observed from these formulas that the normal stress is maximum and equal to $\sigma_m = P/A_0$ for $\theta = 0$, while the shearing stress is maximum and equal to $\tau_m = P/2A_0$ for $\theta = 45°$. We also noted that $\tau = 0$ when $\theta = 0$, while $\sigma = P/2A_0$ when $\theta = 45°$.

Single and double shear

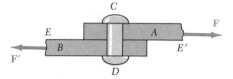

Fig. 1.14

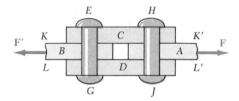

Fig. 1.16

Bearing stress

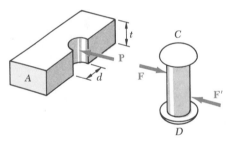

Fig. 1.18

Stresses on an oblique section

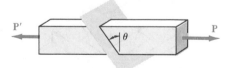

Fig. 1.26a

Stress under general loading

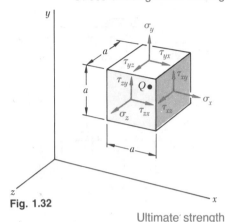

Fig. 1.32

Ultimate strength

Ultimate load and allowable load

Factor of safety

Next, we discussed the state of stress at a point Q in a body under the most general loading condition [Sec. 1.8]. Considering a small cube centered at Q (Fig. 1.32), we denoted by σ_x the normal stress exerted on a face of the cube perpendicular to the x axis, and by τ_{xy} and τ_{xz}, respectively, the y and z components of the shearing stress exerted on the same face of the cube. Repeating this procedure for the other two faces of the cube and observing that $\tau_{xy} = \tau_{yx}$, $\tau_{yz} = \tau_{zy}$, and $\tau_{zx} = \tau_{xz}$, we concluded that *six stress components* are required to define the state of stress at a given point Q, namely, σ_x, σ_y, σ_z, τ_{xy}, τ_{yz}, and τ_{zx}.

Finally, we discussed various concepts used in the analysis and design of engineering structures [Sec. 1.9]. The *ultimate load* of a given structural member or machine component is the load at which the member or component is expected to fail; it is computed from the *ultimate stress* or *ultimate strength* of the material used, as determined by a laboratory test on a specimen of that material. The ultimate load should be considerably larger than the *allowable load*, i.e., the load that the member or component will be allowed to carry under normal conditions. The ratio of the ultimate load to the allowable load is defined as the *factor of safety*:

$$\text{Factor of safety} = F.S. = \frac{\text{ultimate load}}{\text{allowable load}} \quad (1.21)$$

The determination of the factor of safety which should be used in the design of a given structure depends upon a number of considerations, some of which were listed in Sec. 1.9.

REVIEW PROBLEMS

1.54 A 50-kN axial load is applied to a short wooden post which is supported by a concrete footing resting on undisturbed soil. Determine (*a*) the maximum bearing stress on the concrete footing, (*b*) the size of the footing for which the average bearing stress on the soil is 150 kPa.

1.55 Two solid cylindrical rods AC and CD are welded together at C and are subjected to the loading shown. Determine the average normal stress in each of the portions AB, BC, and CD.

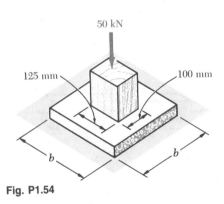

Fig. P1.54

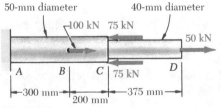

Fig. P1.55

1.56 Two wooden members of uniform rectangular cross section of sides $a =$ 100 mm and $b = 60$ mm are joined by a simple glued joint as shown. Knowing that the ultimate stresses for the joint are $\sigma_U = 1.26$ MPa in tension and $\tau_U = 1.50$ MPa in shear and that $P = 6$ kN, determine the factor of safety for the joint when (a) $\alpha = 20°$, (b) $\alpha = 35°$, (c) $\alpha = 45°$. For each of these values of α, also determine whether the joint will fail in tension or in shear if P is increased until rupture occurs.

1.57 Two wooden members of uniform rectangular cross section of sides a and b are joined by a simple glued joint as shown. Denoting by σ_U and τ_U, respectively, the ultimate strength of the joint in tension and in shear and by P the magnitude of the axial force exerted at each end of the member, (a) derive expressions for the factors of safety relative to failure of the joint in tension and in shear, respectively, (b) show that the overall factor of safety for the glued joint is equal to the former if $\tan \alpha > \sigma_U/\tau_U$ and to the latter otherwise.

1.58 Cable BD has a breaking strength of 100 kN, and the pin at A has a diameter of 10 mm and is made of steel with an ultimate shearing stress of 350 MPa. Determine the factor of safety for the loading shown.

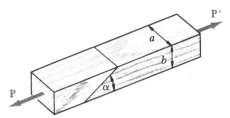

Fig. P1.56 and P1.57

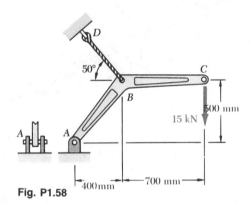

Fig. P1.58

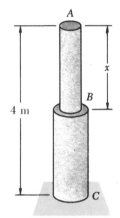

Fig. P1.59

1.59 A 4-m concrete column designed to stand under its own weight consists of two cylindrical elements AB and BC, 225 mm and 375 mm in diameter, respectively. (a) Determine the length of element AB for which the value of the maximum normal stress in the column is smallest. (b) Express this value in terms of the maximum stress σ_0 in a 4-m column of the same material but of uniform circular cross section.

1.60 A load **P** is supported as shown by a steel pin which has been inserted in a short wooden member hanging from the ceiling. The ultimate strength of the wood used is 60 MPa in tension and 7.5 MPa in shear, while the ultimate strength of the steel is 150 MPa in shear. Knowing that the diameter of the pin is $d = 15$ mm and that the magnitude of the load is $P = 16$ kN, determine (a) the factor of safety for the pin, (b) the average bearing stress on the wood, (c) the required values of b and c if the factor of safety for the wooden member is to be the same as that found in part a for the pin.

1.61 For the support of Prob. 1.60, knowing that $b = 40$ mm, $c = 55$ mm, and $d = 12$ mm, determine (a) the largest allowable load **P** if an overall factor of safety of 3 is desired, (b) the corresponding average bearing stress on the wood.

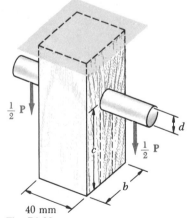

Fig. P1.60

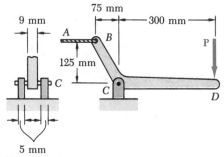

75 mm

9 mm

300 mm

A

B

125 mm

P

C

C

D

5 mm

Fig. P1.62 and P1.63

1.62 A 5-mm-diameter pin is used at connection C of the pedal shown. Knowing that $P = 600$ N, determine (a) the average shearing stress in the pin, (b) the nominal bearing stress in the pedal at C, (c) the nominal bearing stress in each of the support brackets at C.

1.63 Knowing that a force P of magnitude 800 N is applied to the pedal shown, determine (a) the diameter of the pin at C for which the average shearing stress in the pin is 35 MPa, (b) the corresponding bearing stress in each of the support brackets at C.

1.64 A steel plate 6 mm thick is embedded in a concrete wall to anchor a high-strength cable as shown. The diameter of the hole in the plate is 20 mm, the ultimate tensile stress of the steel used is 250 MPa, and the ultimate bonding stress between plate and concrete is 2 MPa. If a factor of safety of 3.60 is desired when $P = 10$ kN, determine (a) the required width a of the plate, (b) the minimum depth b to which the plate should be embedded in the wall. (Neglect the normal stresses between the concrete and the end of the plate.)

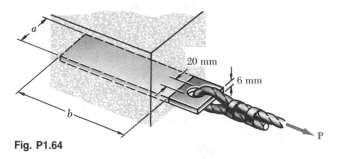

a

20 mm

6 mm

b

P

Fig. P1.64

1.65 Determine the factor of safety for the cable anchor of Prob. 1.64 when $P = 13$ kN, knowing that $a = 50$ mm and $b = 200$ mm.

The following problems are designed to be solved with a computer.

1.C1 A solid rod consisting of n cylindrical elements welded together is subjected to the loading shown. The diameter of element i is denoted by d_i and the load applied to its right end by $\mathbf{P}_i$, with the magnitude P_i of this load being assumed positive if $\mathbf{P}_i$ is directed as shown and negative otherwise. (a) Write a computer program which can be used with any consistent set of units to determine the average normal stress in each element of the rod. (b) Use this program to solve Probs. 1.1, 1.2, and 1.55.

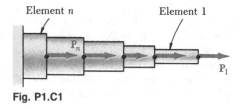

Element n

Element 1

P_n

P_1

Fig. P1.C1

1.C2 Each of the four vertical links connecting the two horizontal members has a uniform rectangular cross section of 10×40 mm and is made of a steel with an ultimate strength in tension of 400 MPa, while each of the pins at C and F is made of a steel with an ultimate strength in shear of 150 MPa. (*a*) Write a computer program which, for given values of a, b, and P and for values of the diameter d of the pins from 4 to 36 mm at 2-mm intervals, can be used to calculate (1) the maximum value of the average normal stress in the links CF, (2) the average shearing stress in the pins at C and F, (3) the factor of safety for the links CF, (4) the factor of safety for the pins at C and F, (5) the overall factor of safety F.S. for the links and the pins. (*b*) Use this program to solve Prob. 1.13*b* and Prob. 1.30*a*. (*c*) Determine for the values of a, b, and P used in these two problems what pin diameter d would yield the largest overall factor of safety F.S. and the corresponding value of F.S.

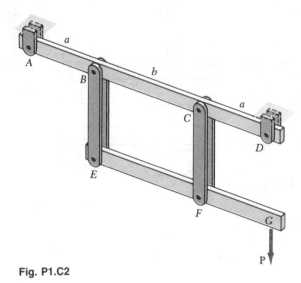

Fig. P1.C2

1.C3 Two wooden members of uniform rectangular cross section of sides a and b are joined by a simple glued joint as shown and subjected to an axial load of magnitude P. (*a*) Denoting by σ_U and τ_U, respectively, the ultimate strength of the joint in tension and in shear, write a computer program which, for given values of a, b, P, σ_U and τ_U, expressed in any consistent set of units, and for values of α from 5 to 85° at 5° intervals, can be used to calculate (1) the normal stress in the joint, (2) the shearing stress in the joint, (3) the factor of safety relative to failure in tension, (4) the factor of safety relative to failure in shear, (5) the overall factor of safety for the glued joint. (*b*) Apply this program, using the dimensions, loading, and material properties specified in Probs. 1.36 and 1.47 and in Prob. 1.56. (*c*) In each of these two cases check that the shearing stress is maximum for $\alpha = 45°$ and has the value defined by Eq. (1.13); in both cases also verify the property indicated in part *b* of Prob. 1.57.

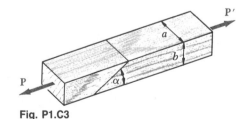

Fig. P1.C3

1.C4 Link *BD* is made of a steel with an ultimate strength in tension of 400 MPa, and pins *B*, *C*, and *D* of a steel with an ultimate strength in shear of 150 MPa. Note that pins *B* and *D* are in single shear while pin *C* is in double shear. Also note that link *BD* is not reinforced around the pin holes. (*a*) Write a computer program which, for given values of the dimensions *a*, *b*, *w*, *t*, of the diameter *d* of pins *B* and *D*, and of the diameter d_C of pin *C*, and for a given factor of safety, can be used to calculate the maximum allowable load *P* and also indicate which, among the tensile stress in link *BD*, the shearing stress in pins *B* and *D*, and the shearing stress in pin *C*, is critical. (*b*) Apply this program to the solution of Probs. 1.50, 1.51, and 1.53.

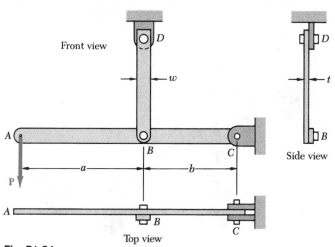

Fig. P1.C4

C H A P T E R   T W O

STRESS AND STRAIN —AXIAL LOADING

2.1. INTRODUCTION

In Chap. 1 we analyzed the stresses created in various members and connections by the loads applied to a structure or machine. We also learned to design simple members and connections so that they would not fail under specified loading conditions. Another important aspect of the analysis and design of structures relates to the *deformations* caused by the loads applied to a structure. Clearly, it is important to avoid deformations so large that they may prevent the structure from fulfilling the purpose for which it was intended. But the analysis of deformations may also help us in the determination of stresses. Indeed, it is not always possible to determine the forces in the members of a structure by applying only the principles of statics. This is because statics is based on the assumption of undeformable, rigid structures. By considering engineering structures as *deformable* and analyzing the deformations in their various members, it will be possible for us to compute forces which are *statically indeterminate*, i.e., indeterminate within the framework of statics. Also, as we indicated in Sec. 1.3, the distribution of stresses in a given member is statically indeterminate, even when the force in that member is known. To determine the actual distribution of stresses within a member, it is thus necessary to analyze the deformations which take place in that member. In this chapter, we shall discuss the deformations of a structural member such as a rod, bar, or plate under *axial loading*.

First, we shall define the *normal strain* ϵ in the member as the *deformation of the member per unit length*. Plotting the stress σ versus the strain ϵ as the load applied to the member increases, we shall obtain a *stress-strain diagram* for the material used. From such a diagram we shall be able to determine some important properties of the material, such as its *modulus of elasticity*, and whether the material is *ductile* or *brittle* (Secs. 2.2 to 2.5). From the stress-strain diagram, we shall also be able to determine whether the strains in the specimen will disappear after

39

the load has been removed—in which case the material is said to behave *elastically*—or whether a *permanent set* or *plastic deformation* will result (Sec. 2.6).

In Sec. 2.7, we shall discuss the phenomenon of *fatigue*, which causes structural or machine components to fail after a very large number of repeated loadings, even though the stresses remain in the elastic range.

The first part of the chapter ends with Sec. 2.8, which is devoted to the determination of the deformation of various types of members under various conditions of axial loading.

In Secs. 2.9 and 2.10, we shall consider *statically indeterminate problems*, i.e., problems in which the reactions and the internal forces *cannot* be determined from statics alone. The equilibrium equations derived from the free-body diagram of the member under consideration will be complemented by relations involving deformations and obtained from the geometry of the problem.

In Secs. 2.11 to 2.15, additional characteristic material constants will be introduced. They include *Poisson's ratio*, which relates lateral and axial strain, the *bulk modulus*, which characterizes the change in volume of a material under hydrostatic pressure, and the *modulus of rigidity*, which relates the components of the shearing stress and shearing strain.

In the text material described so far, stresses are assumed uniformly distributed in any given cross section; they are also assumed to remain within the elastic range. In Sec. 2.17 we shall consider *stress concentrations* near circular holes and fillets in flat bars, and in Secs. 2.18 and 2.19 we shall discuss stresses and deformations in members made of a ductile material when the yield point of the material is exceeded. As we shall see, permanent *plastic deformations* and *residual stresses* result from such loading conditions.

2.2. NORMAL STRAIN UNDER AXIAL LOADING

Let us consider a rod BC, of length L and uniform cross-sectional area A, which is suspended from B (Fig. 2.1a). If we apply a load $\mathbf{P}$ to end C, the rod elongates (Fig. 2.1b). Plotting the magnitude P of the load against the deformation δ (Greek letter delta), we obtain a certain load-deformation diagram (Fig. 2.2). While this diagram contains information useful to the analysis of the rod under consideration, it cannot be used directly to predict the deformation of a rod of the same material but of different dimensions. Indeed, we observe that, if a deformation δ is produced in rod BC by a load $\mathbf{P}$, a load $2\mathbf{P}$ is required to cause the same deformation in a rod $B'C'$ of the same length L, but of cross-sectional area $2A$ (Fig. 2.3). We note that, in both cases, the value of the stress is the same: $\sigma = P/A$. On the other hand, a load $\mathbf{P}$ applied to a rod $B''C''$, of the same cross-sectional area A, but of length $2L$, causes a deformation 2δ in that rod (Fig. 2.4), i.e., a deformation twice as large as the deformation δ

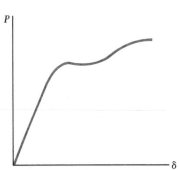

Fig. 2.1

Fig. 2.2 Load-deformation diagram.

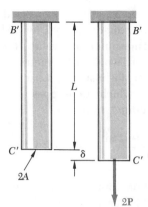

Fig. 2.3

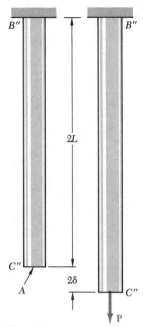

Fig. 2.4

it produces in rod BC. But in both cases the ratio of the deformation over the length of the rod is the same; it is equal to δ/L. This observation brings us to introduce the concept of *strain:* We define the *normal strain* in a rod under axial loading as the *deformation per unit length* of that rod. Denoting the normal strain by ϵ (Greek letter epsilon), we write

$$\epsilon = \frac{\delta}{L} \tag{2.1}$$

Plotting the stress $\sigma = P/A$ against the strain $\epsilon = \delta/L$, we obtain a curve which is characteristic of the properties of the material and does not depend upon the dimensions of the particular specimen used. This curve is called a *stress-strain diagram* and will be discussed in detail in Sec. 2.3.

Since the rod BC considered in the preceding discussion had a uniform cross section of area A, the normal stress σ could be assumed to have a constant value P/A throughout the rod. Thus, it was appropriate to define the strain ϵ as the ratio of the total deformation δ over the total length L of the rod. In the case of a member of variable cross-sectional area A, however, the normal stress $\sigma = P/A$ varies along the member, and it is necessary to define the strain at a given point Q by considering a small element of undeformed length Δx (Fig. 2.5). Denoting by $\Delta\delta$ the deformation of the element under the given loading, we define the *normal strain at point Q* as

$$\epsilon = \lim_{\Delta x \to 0} \frac{\Delta\delta}{\Delta x} = \frac{d\delta}{dx} \tag{2.2}$$

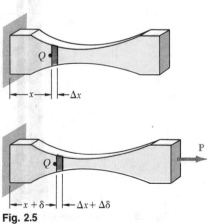

Fig. 2.5

Since deformation and length are expressed in the same units, the normal strain ϵ obtained by dividing δ by L (or $d\delta$ by dx) is a *dimensionless quantity*. Thus, the same numerical value is obtained for the normal strain in a given member, whether SI metric units or U.S. customary units are used. Consider, for instance, a bar of length $L = 0.600$ m and uniform cross section, which undergoes a deformation $\delta = 150 \times 10^{-6}$ m. The corresponding strain is

$$\epsilon = \frac{\delta}{L} = \frac{150 \times 10^{-6} \text{ m}}{0.600 \text{ m}} = 250 \times 10^{-6} \text{ m/m} = 250 \times 10^{-6}$$

Note that the deformation could have been expressed in micrometers: $\delta = 150 \ \mu$m. We would then have written

$$\epsilon = \frac{\delta}{L} = \frac{150 \ \mu\text{m}}{0.600 \text{ m}} = 250 \ \mu\text{m/m} = 250 \ \mu$$

and read the answer as "250 micros."

2.3. STRESS-STRAIN DIAGRAM

We saw in Sec. 2.2 that the diagram representing the relation between stress and strain in a given material is an important characteristic of the material. To obtain the stress-strain diagram of a material, one usually conducts a *tensile test* on a specimen of the material. One type of specimen commonly used is shown in Fig. 2.6. The cross-sectional area of the cylindrical central portion of the specimen has been accurately determined and two gage marks have been inscribed on that portion at a distance L_0 from each other. The distance L_0 is known as the *gage length* of the specimen.

The test specimen is then placed in a testing machine (Fig. 2.7), which is used to apply a centric load **P**. As the load **P** increases, the distance L between the two gage marks also increases (Fig. 2.8). The distance L is measured with a dial gage, and the elongation $\delta = L - L_0$ is recorded for each value of P. A second dial gage is often used simultaneously to measure and record the change in diameter of the specimen. From each pair of readings P and δ, the stress σ is computed by dividing P by the original cross-sectional area A_0 of the specimen, and the strain ϵ by dividing the elongation δ by the original distance L_0 between the two gage marks. The stress-strain diagram may then be obtained by plotting ϵ as an abscissa and σ as an ordinate.

Stress-strain diagrams of various materials vary widely, and different tensile tests conducted on the same material may yield different results,

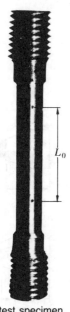

Fig. 2.6 Typical tensile-test specimen.

Fig. 2.7 Universal testing machine. (*Courtesy Detroit Testing Machine Co.*)

Fig. 2.8

depending upon the temperature of the specimen and the speed of loading. It is possible, however, to distinguish some common characteristics among the stress-strain diagrams of various groups of materials and to divide materials into two broad categories on the basis of these characteristics, namely, the *ductile* materials and the *brittle* materials.

Ductile materials, which comprise structural steel, as well as many alloys of other metals, are characterized by their ability to *yield* at normal temperatures. As the specimen is subjected to an increasing load, its length first increases linearly with the load and at a very slow rate. Thus, the initial portion of the stress-strain diagram is a straight line with a steep slope (Fig. 2.9). However, after a critical value σ_Y of the stress has been reached, the specimen undergoes a large deformation with a relatively small increase in the applied load. This deformation is caused by slippage of the material along oblique surfaces and is due, therefore, primarily to shearing stresses. As we may note from the stress-strain diagrams of two typical ductile materials (Fig. 2.9), the elongation of the specimen after it has started to yield may be 200 times as large as its deformation before yield. After a certain maximum value of the load has been

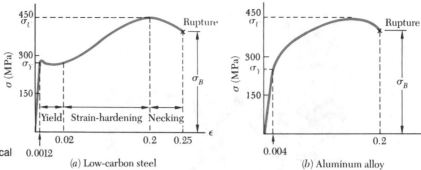

Fig. 2.9 Stress-strain diagrams of two typical ductile materials.

(a) Low-carbon steel　　　(b) Aluminum alloy

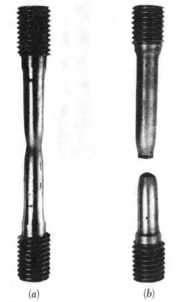

(a)　　　　　(b)

Fig. 2.10 Tested specimen of a ductile material.

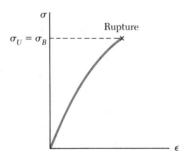

Fig. 2.11 Stress-strain diagram for a typical brittle material.

reached, the diameter of a portion of the specimen begins to decrease, due to local instability (Fig. 2.10a). This phenomenon is known as *necking*. After necking has begun, somewhat lower loads are sufficient to keep the specimen elongating further, until it finally ruptures (Fig. 2.10b). We note that rupture occurs along a cone-shaped surface which forms an angle of approximately 45° with the original surface of the specimen. This indicates that shear is primarily responsible for the failure of ductile materials, and confirms the fact that, under an axial load, shearing stresses are largest on surfaces forming an angle of 45° with the load (cf. Sec. 1.7). The stress σ_Y at which yield is initiated is called the *yield strength* of the material, the stress σ_U corresponding to the maximum load applied to the specimen is known as the *ultimate strength*, and the stress σ_B corresponding to rupture is called the *breaking strength*.

Brittle materials, which comprise cast iron, glass, and stone, are characterized by the fact that rupture occurs without any noticeable prior change in the rate of elongation (Fig. 2.11). Thus, for brittle materials, there is no difference between the ultimate strength and the breaking strength. Also, the strain at the time of rupture is much smaller for brittle than for ductile materials. From Fig. 2.12, we note the absence of any necking of the specimen in the case of a brittle material, and observe that rupture occurs along a surface perpendicular to the load. We conclude from this observation that normal stresses are primarily responsible for the failure of brittle materials.†

The stress-strain diagrams of Fig. 2.9 show that structural steel and aluminum, while both ductile, have different yield characteristics. In the case of structural steel (Fig. 2.9a), the stress remains constant over a large range of values of the strain after the onset of yield. Later the stress

---

†The tensile tests described in this section were assumed to be conducted at normal temperatures. However, a material which is ductile at normal temperatures may display the characteristics of a brittle material at very low temperatures, while a normally brittle material may behave in a ductile fashion at very high temperatures. At temperatures other than normal, therefore, one should refer to *a material in a ductile state* or to *a material in a brittle state*, rather than to a ductile or brittle material.

must be increased to keep elongating the specimen, until the maximum value σ_U has been reached. This is due to a property of the material known as strain-hardening. The yield strength of structural steel may be determined during the tensile test by watching the load dial. After increasing steadily, the load is observed to suddenly drop to a slightly lower value which is maintained for a certain period while the specimen keeps elongating. In a very carefully conducted test, one may be able to distinguish between the *upper yield point*, which corresponds to the load reached just before yield starts, and the *lower yield point*, which corresponds to the load required to maintain yield. Since the upper yield point is transient, the lower yield point should be used to determine the yield strength of the material.

In the case of aluminum (Fig. 2.9*b*) and of many other ductile materials, the onset of yield is not characterized by a horizontal portion of the stress-strain curve. Instead, the stress keeps increasing—although not linearly—until the ultimate strength is reached. Necking then begins, leading eventually to rupture. For such materials, one may define the yield strength σ_Y by the offset method. The yield strength at 0.2% offset, for example, is obtained by drawing through the point of the horizontal axis of abscissa $\epsilon = 0.2\%$ (or $\epsilon = 0.002$), a line parallel to the initial straight-line portion of the stress-strain diagram (Fig. 2.13). The stress σ_Y corresponding to the point Y obtained in this fashion is defined as the yield strength at 0.2% offset.

A standard measure of the ductility of a material is its *percent elongation*, which is defined as

$$\text{Percent elongation} = 100 \frac{L_B - L_0}{L_0}$$

where L_0 and L_B denote, respectively, the initial length of the tensile test specimen and its final length at rupture. For example, a minimum percent elongation of 20 percent for a 200-mm gage length is required for the most common grade of structural steel, known as A36. We note that this means that the average strain at rupture should be at least 0.20.

Another measure of ductility which is sometimes used is the *percent reduction in area*, defined as

$$\text{Percent reduction in area} = 100 \frac{A_0 - A_B}{A_0}$$

where A_0 and A_B denote, respectively, the initial cross-sectional area of the specimen and its minimum cross-sectional area at rupture. For structural steel, percent reductions in area of 60 to 70 percent are common.

Fig. 2.12 Tested specimen of a brittle material.

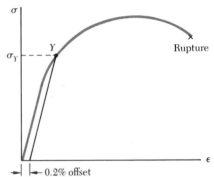

Fig. 2.13 Determination of yield strength by offset method.

Thus far, we have discussed only tensile tests. If a specimen made of a ductile material were loaded in compression instead of tension, the stress-strain curve obtained would be essentially the same through its initial straight-line portion and through the beginning of the portion corresponding to yield and strain-hardening. Particularly noteworthy is the fact that for a given steel, the yield strength is the same in both tension and compression. For larger values of the strain, the tension and compression stress-strain curves diverge, and it should be noted that necking cannot occur in compression.

For most brittle materials, one finds that the ultimate strength in compression is much larger than the ultimate strength in tension. This is due to the presence of flaws, such as microscopic cracks or cavities, which tend to weaken the material in tension, while not appreciably affecting its resistance to compressive failure.

*2.4. TRUE STRESS AND TRUE STRAIN

We recall that the stress plotted in the diagrams of Figs. 2.9 and 2.11 was obtained by dividing the load P by the cross-sectional area A_0 of the specimen measured before any deformation had taken place. Since the cross-sectional area of the specimen decreases as P increases, the stress plotted in our diagrams does not represent the actual stress in the specimen. The difference between the *engineering stress* $\sigma = P/A_0$ that we have computed and the *true stress* $\sigma_t = P/A$ obtained by dividing P by the cross-sectional area A of the deformed specimen becomes apparent in ductile materials after yield has started. While the engineering stress σ, which is directly proportional to the load P, decreases with P during the necking phase, the true stress σ_t, which is proportional to P but also inversely proportional to A, is observed to keep increasing until rupture of the specimen occurs.

Many scientists also use a definition of strain different from that of the *engineering strain* $\epsilon = \delta/L_0$. Instead of using the total elongation δ and the original value L_0 of the gage length, they use all the successive values of L that they have recorded. Dividing each increment ΔL of the distance between the gage marks, by the corresponding value of L, they obtain the elementary strain $\Delta\epsilon = \Delta L/L$. Adding the successive values of $\Delta\epsilon$, they define the *true strain* ϵ_t:

$$\epsilon_t = \Sigma \, \Delta\epsilon = \Sigma(\Delta L/L)$$

Replacing the summation by an integral, they can also express the true strain as follows:

$$\epsilon_t = \int_{L_0}^{L} \frac{dL}{L} = \ln\frac{L}{L_0} \tag{2.3}$$

The diagram obtained by plotting true stress versus true strain (Fig. 2.14) reflects more accurately the behavior of the material. As we have already noted, there is no decrease in true stress during the necking phase. Also, the results obtained from tensile and from compressive tests will yield essentially the same plot when true stress and true strain are used. This is not the case for large values of the strain when the engineering stress is plotted versus the engineering strain. However, engineers, whose responsibility is to determine whether a load P will produce an acceptable stress and an acceptable deformation in a given member, will want to use a diagram based on the engineering stress $\sigma = P/A_0$ and the engineering strain $\epsilon = \delta/L_0$, since these expressions involve data which are available to them, namely the cross-sectional area A_0 and the length L_0 of the member in its undeformed state.

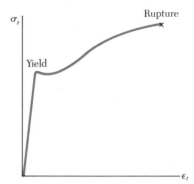

Fig. 2.14 True stress versus true strain for a typical ductile material.

2.5. HOOKE'S LAW; MODULUS OF ELASTICITY

Most engineering structures are designed to undergo relatively small deformations, involving only the straight-line portion of the corresponding stress-strain diagram. For that initial portion of the diagram (Fig. 2.9), the stress σ is directly proportional to the strain ϵ, and we may write

$$\sigma = E\,\epsilon \tag{2.4}$$

This relation is known as *Hooke's law*, after the English mathematician Robert Hooke (1635–1703). The coefficient E is called the *modulus of elasticity* of the material involved, or also *Young's modulus*, after the English scientist Thomas Young (1773–1829). Since the strain ϵ is a dimensionless quantity, the modulus E is expressed in the same units as the stress σ, namely in pascals or one of its multiples.

The largest value of the stress for which Hooke's law may be used for a given material is known as the *proportional limit* of that material. In the case of ductile materials possessing a well-defined yield point, as in Fig. 2.9a, the proportional limit almost coincides with the yield point. For other materials, the proportional limit cannot be defined as easily, since it is difficult to determine with accuracy the value of the stress σ for which the relation between σ and ϵ ceases to be linear. But from this very difficulty we may conclude for such materials that using Hooke's law for values of the stress slightly larger than the actual proportional limit will not result in any significant error.

Some of the physical properties of structural metals, such as strength, ductility, corrosion resistance, etc., may be greatly affected by alloying, heat treatment, and the manufacturing process used. For example, we note from the stress-strain diagrams of pure iron and of three different grades of steel (Fig. 2.15) that large variations in the yield

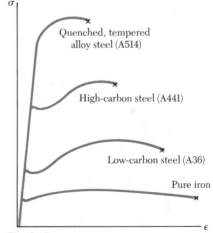

Fig. 2.15

strength, ultimate strength, and final strain (ductility) exist among these four metals. All of them, however, possess the same modulus of elasticity; in other words, their "stiffness" or ability to resist a deformation within the linear range is the same. Therefore, if a high-strength steel is substituted for a lower-strength steel in a given structure, and if all dimensions are kept the same, the structure will have an increased load-carrying capacity, but its stiffness will remain unchanged.

2.6. ELASTIC VERSUS PLASTIC BEHAVIOR OF A MATERIAL

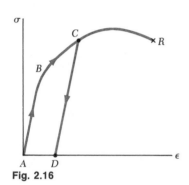

Fig. 2.16

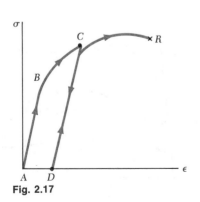

Fig. 2.17

If the strains caused in a test specimen by the application of a given load disappear when the load is removed, the material is said to behave *elastically*. The largest value of the stress for which the material behaves elastically is called the *elastic limit* of the material.

If the material has a well-defined yield point as in Fig. 2.9a, the elastic limit, the proportional limit (Sec. 2.5), and the yield point are essentially equal. In other words, the material behaves elastically and linearly as long as the stress is kept below the yield point. If the yield point is reached, however, yield takes place as described in Sec. 2.3 and, when the load is removed, the stress and strain decrease in a linear fashion, along a line CD parallel to the straight-line portion AB of the loading curve (Fig. 2.16). The fact that ϵ does not return to zero after the load has been removed indicates that a *permanent set* or *plastic deformation* of the material has taken place. For most materials, the plastic deformation depends, not only upon the maximum value reached by the stress, but also upon the time elapsed before the load is removed. The stress-dependent part of the plastic deformation is referred to as *slip*, and the time-dependent part—which is also influenced by the temperature—as *creep*.

When a material does not possess a well-defined yield point, the elastic limit cannot be determined with precision. However, assuming the elastic limit equal to the yield strength as defined by the offset method (Sec. 2.3) results only in a small error. Indeed, referring to Fig. 2.13, we note that the straight line used to determine point Y also represents the unloading curve after a maximum stress σ_Y has been reached. While the material does not behave truly elastically, the resulting plastic strain is as small as the selected offset.

If, after being loaded and unloaded (Fig. 2.17), the test specimen is loaded again, the new loading curve will closely follow the earlier unloading curve until it almost reaches point C; it will then bend to the right and connect with the curved portion of the original stress-strain diagram. We note that the straight-line portion of the new loading curve is longer than the corresponding portion of the initial one. Thus, the proportional limit and the elastic limit have increased as a result of the strain-hardening which occurred during the earlier loading of the specimen. However, since the point of rupture R remains unchanged, the ductility of the specimen, which should now be measured from point D, has decreased.

We have assumed in our discussion that the specimen was loaded twice in the same direction, i.e., that both loads were tensile loads. We shall now consider the case when the second load is applied in a direction opposite to that of the first one.

We shall assume that the material is mild steel, for which the yield strength is the same in tension and in compression. The initial load is tensile and is applied until point C has been reached on the stress-strain diagram (Fig. 2.18). After unloading (point D), a compressive load is

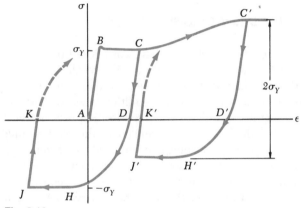

Fig. 2.18

applied, causing the material to reach point H, where the stress is equal to $-\sigma_Y$. We note that portion DH of the stress-strain diagram is curved and does not show any clearly defined yield point. This is referred to as the *Bauschinger effect*. As the compressive load is maintained, the material yields along line HJ.

If the load is removed after point J has been reached, the stress returns to zero along line JK, and we note that the slope of JK is equal to the modulus of elasticity E. The resulting permanent set AK may be positive, negative, or zero, depending upon the lengths of the segments BC and HJ. If a tensile load is applied again to the test specimen, the portion of the stress-strain diagram beginning at K (dashed line) will curve up and to the right until the yield stress σ_Y has been reached.

If the initial loading is large enough to cause strain-hardening of the material (point C'), unloading takes place along line $C'D'$. As the reverse load is applied, the stress becomes compressive, reaching its maximum value at H' and maintaining it as the material yields along line $H'J'$. We note that while the maximum value of the compressive stress is less than σ_Y, the total change in stress between C' and H' is still equal to $2\sigma_Y$.

If point K or K' coincides with the origin A of the diagram, the permanent set is equal to zero, and the specimen may appear to have returned to its original condition. However, internal changes will have taken place and, while the same loading sequence may be repeated, the specimen will rupture without any warning after relatively few repetitions. This indicates that the excessive plastic deformations to which the specimen was subjected have caused a radical change in the characteristics of the material. Reverse loadings into the plastic range, therefore, are seldom allowed, and only under carefully controlled conditions. Such situations occur in the straightening of damaged materiel and in the final alignment of a structure or machine.

2.7. REPEATED LOADINGS; FATIGUE

In the preceding sections we have considered the behavior of a test specimen subjected to an axial loading. We recall that, if the maximum stress in the specimen does not exceed the elastic limit of the material, the specimen returns to its initial condition when the load is removed. Indeed, we might conclude that a given loading may be repeated many times, provided that the stresses remain in the elastic range. Such a conclusion is correct for loadings repeated a few dozen or even a few hundred times. However, as we shall see, it is not correct when loadings are repeated thousands or millions of times. In such cases rupture will occur at a stress much lower than the static breaking strength; this phenomenon is known as *fatigue*. A fatigue failure is of a brittle nature, even for materials which are normally ductile.

Fatigue must be considered in the design of all structural and machine components which are subjected to repeated or to fluctuating loads. The number of loading cycles which may be expected during the useful life of a component varies greatly. For example, a beam supporting an industrial crane may be loaded as many as two million times in 25 years (about 300 loadings per working day), an automobile crankshaft will be loaded about half a billion times if the automobile is driven 200,000 miles, and an individual turbine blade may be loaded several hundred billion times during its lifetime.

Some loadings are of a fluctuating nature. For example, the passage of traffic over a bridge will cause stress levels which will fluctuate about the stress level due to the weight of the bridge. A more severe condition occurs when a complete reversal of the load occurs during the loading cycle. The stresses in the axle of a railroad car, for example, are completely reversed after each half-revolution of the wheel.

The number of loading cycles required to cause the failure of a specimen through repeated successive loadings and reverse loadings may be determined experimentally for any given maximum stress level. If a series of tests is conducted, using different maximum stress levels, the resulting data may be plotted as a *σ-n* curve. For each test, the maximum stress *σ* is plotted as an ordinate and the number of cycles *n* as an abscissa; because of the large number of cycles required for rupture, the cycles *n* are plotted on a logarithmic scale.

A typical *σ-n* curve for steel is shown in Fig. 2.19. We note that, if the applied maximum stress is high, relatively few cycles are required to cause rupture. As the magnitude of the maximum stress is reduced, the number of cycles required to cause rupture increases, until a stress, known as the *endurance limit*, is reached. The endurance limit is the stress for which failure does not occur, even for an indefinitely large number of loading cycles. For a low-carbon steel, such as structural steel, the endurance limit is about one-half of the ultimate strength of the steel.

For nonferrous metals, such as aluminum and copper, a typical *σ-n* curve (Fig. 2.19) shows that the stress at failure continues to decrease as the number of loading cycles is increased. For such metals, one defines the *fatigue limit* as the stress corresponding to failure after a specified number of loading cycles, such as 500 million.

Examination of test specimens, of shafts, of springs, and of other components which have failed in fatigue shows that the failure was initiated at a microscopic crack or at some similar imperfection. At each loading, the crack was very slightly enlarged. During successive loading cycles, the crack propagated through the material until the amount of undamaged material was insufficient to carry the maximum load, and an abrupt, brittle failure occurred. Because fatigue failure may be initiated at any crack or imperfection, the surface condition of a specimen has an important effect on the value of the endurance limit obtained in testing. The endurance limit for machined and polished specimens is higher than for rolled or forged components, or for components which are corroded. In applications in or near sea water, or in other applications where corrosion is expected, a reduction of up to 50% in the endurance limit may be expected.

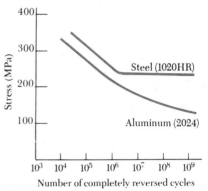

Fig. 2.19

2.8. DEFORMATIONS OF MEMBERS UNDER AXIAL LOADING

Consider a homogeneous rod *BC* of length *L* and uniform cross section of area *A* subjected to a centric axial load **P** (Fig. 2.20). If the resulting axial stress $\sigma = P/A$ does not exceed the proportional limit of the material, we may apply Hooke's law and write

$$\sigma = E \, \epsilon \qquad (2.4)$$

from which it follows that

$$\epsilon = \frac{\sigma}{E} = \frac{P}{AE} \qquad (2.5)$$

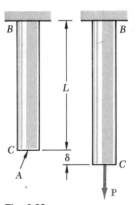

Fig. 2.20

Recalling that the strain ϵ was defined in Sec. 2.2 as $\epsilon = \delta/L$, we have

$$\delta = \epsilon L \qquad (2.6)$$

and, substituting for ϵ from (2.5) into (2.6):

$$\delta = \frac{PL}{AE} \qquad (2.7)$$

Equation (2.7) may be used only if the rod is homogeneous (constant E), has a uniform cross section of area A, and is loaded at its ends. If the rod is loaded at other points, or if it consists of several portions of various cross sections and possibly of different materials, we must divide it into component parts which satisfy individually the required conditions for the application of formula (2.7). Denoting respectively by P_i, L_i, A_i, and E_i the internal force, length, cross-sectional area, and modulus of elasticity corresponding to part i, we express the deformation of the entire rod as

$$\delta = \sum_i \frac{P_i L_i}{A_i E_i} \qquad (2.8)$$

We recall from Sec. 2.2 that, in the case of a rod of variable cross section (Fig. 2.5), the strain ϵ depends upon the position of the point Q where it is computed and is defined as $\epsilon = d\delta/dx$. Solving for $d\delta$ and substituting for ϵ from Eq. (2.5), we express the deformation of an element of length dx as

$$d\delta = \epsilon \, dx = \frac{P \, dx}{AE}$$

The total deformation δ of the rod is obtained by integrating this expression over the length L of the rod:

$$\delta = \int_0^L \frac{P \, dx}{AE} \qquad (2.9)$$

Formula (2.9) should be used in place of (2.7), not only when the cross-sectional area A is a function of x, but also when the internal force P depends upon x, as is the case for a rod hanging under its own weight.

Example 2.01

Determine the deformation of the steel rod shown in Fig. 2.21a under the given load ($E = 200$ GPa).

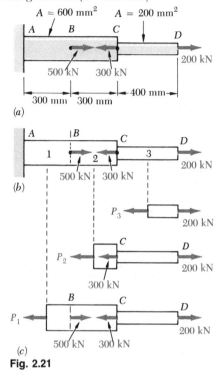

(a)

(b)

(c)

Fig. 2.21

We divide the rod into the three component parts shown in Fig. 2.21b and write

$$L_1 = L_2 = 0.300 \text{ m} \qquad L_3 = 0.400 \text{ m}$$
$$A_1 = A_2 = 600 \times 10^{-6} \text{ m}^2 \qquad A_3 = 200 \times 10^{-6} \text{ m}^2$$

To find the internal forces P_1, P_2, and P_3, we must pass sections through each of the component parts, drawing each time the free-body diagram of the portion of rod located to the right of the section (Fig. 2.21c).   Expressing that each of the free bodies is in equilibrium, we obtain successively

$$P_1 = 400 \text{ kN} = 400 \times 10^3 \text{ N}$$
$$P_2 = -100 \text{ kN} = -100 \times 10^3 \text{ N}$$
$$P_3 = 200 \text{ kN} = 200 \times 10^3 \text{ N}$$

Carrying the values obtained into Eq. (2.8), we have:

$$\delta = \sum_i \frac{P_i L_i}{A_i E_i} = \frac{1}{E}\left(\frac{P_1 L_1}{A_1} + \frac{P_2 L_2}{A_2} + \frac{P_3 L_3}{A_3} \right)$$

$$= \frac{1}{200 \times 10^9}\left[\frac{(400 \times 10^3)(0.300)}{600 \times 10^{-6}} \right.$$

$$\left. + \frac{(-100 \times 10^3)(0.300)}{600 \times 10^{-6}} + \frac{(200 \times 10^3)(0.400)}{200 \times 10^{-6}} \right]$$

$$\delta = 2.75 \times 10^{-3} \text{ m} = 2.75 \text{ mm}$$

The rod BC of Fig. 2.20, which was used to derive formula (2.7), and the rod AD of Fig. 2.21, which has just been discussed in Example 2.01, both had one end attached to a fixed support.   In each case, therefore, the deformation δ of the rod was equal to the displacement of its free end.   When both ends of a rod move, however, the deformation of the rod is measured by the *relative displacement* of one end of the rod with respect to the other.   Consider, for instance, the assembly shown in Fig. 2.22a, which consists of three elastic bars of length L connected by a rigid pin at A.   If a load **P** is applied at B (Fig. 2.22b), each of the three bars will deform.   Since the bars AC and AC' are attached to fixed supports at C and C', their common deformation is measured by the displacement δ_A of point A. On the other hand, since both ends of bar AB move, the deformation of AB is measured by the difference between the displacements δ_A and δ_B of points A and B, i.e., by the relative displacement of B with respect to A.   Denoting this relative displacement by $\delta_{B/A}$, we write

$$\delta_{B/A} = \delta_B - \delta_A = \frac{PL}{AE} \qquad (2.10)$$

where A is the cross-sectional area of AB and E its modulus of elasticity.

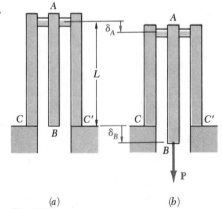

Fig. 2.22

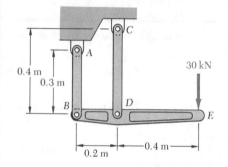

SAMPLE PROBLEM 2.1

The rigid bar *BDE* is supported by two links *AB* and *CD*. Link *AB* is made of aluminum ($E = 70$ GPa) and has a cross-sectional area of 500 mm²; link *CD* is made of steel ($E = 200$ GPa) and has a cross-sectional area of 600 mm². For the 30-kN force shown, determine the deflection (*a*) of *B*, (*b*) of *D*, (*c*) of *E*.

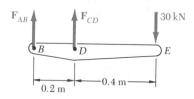

Free Body: Bar BDE

$+\uparrow\Sigma M_B = 0:$ $-(30 \text{ kN})(0.6 \text{ m}) + F_{CD}(0.2 \text{ m}) = 0$

$F_{CD} = +90 \text{ kN}$ $F_{CD} = 90 \text{ kN}$ *tension*

$+\uparrow\Sigma M_D = 0:$ $-(30 \text{ kN})(0.4 \text{ m}) - F_{AB}(0.2 \text{ m}) = 0$

$F_{AB} = -60 \text{ kN}$ $F_{AB} = 60 \text{ kN}$ *compression*

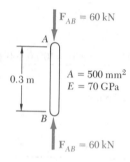

***a*. Deflection of *B*.** Since the internal force in link *AB* is compressive, we have $P = -60$ kN

$$\delta_B = \frac{PL}{AE} = \frac{(-60 \times 10^3 \text{ N})(0.3 \text{ m})}{(500 \times 10^{-6} \text{ m}^2)(70 \times 10^9 \text{ Pa})} = -514 \times 10^{-6} \text{ m}$$

The negative sign indicates a contraction of member *AB*, and, thus, an upward deflection of end *B*: $\delta_B = 0.514 \text{ mm} \uparrow$ ◄

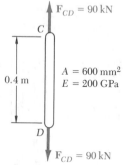

***b*. Deflection of *D*.** Since in rod *CD*, $P = 90$ kN, we write

$$\delta_D = \frac{PL}{AE} = \frac{(90 \times 10^3 \text{ N})(0.4 \text{ m})}{(600 \times 10^{-6} \text{ m}^2)(200 \times 10^9 \text{ Pa})}$$

$$= 300 \times 10^{-6} \text{ m} \qquad\qquad \delta_D = 0.300 \text{ mm} \downarrow \quad ◄$$

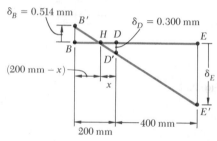

***c*. Deflection of *E*.** We denote by *B′* and *D′* the displaced positions of points *B* and *D*. Since the bar *BDE* is rigid, points *B′*, *D′*, and *E′* lie in a straight line and we write

$$\frac{BB'}{DD'} = \frac{BH}{HD} \qquad \frac{0.514 \text{ mm}}{0.300 \text{ mm}} = \frac{(200 \text{ mm}) - x}{x} \qquad x = 73.7 \text{ mm}$$

$$\frac{EE'}{DD'} = \frac{HE}{HD} \qquad \frac{\delta_E}{0.300 \text{ mm}} = \frac{(400 \text{ mm}) + (73.7 \text{ mm})}{73.7 \text{ mm}}$$

$$\delta_E = 1.928 \text{ mm} \downarrow \quad ◄$$

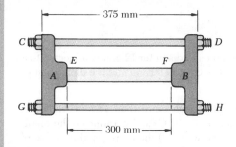

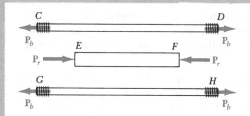

SAMPLE PROBLEM 2.2

The rigid castings A and B are connected by two 18 mm-diameter steel bolts CD and GH and are in contact with the ends of a 36 mm-diameter aluminum rod EF. Each bolt is single-threaded with a pitch of 2 mm, and after being snugly fitted, the nuts at D and H are both tightened one-quarter of a turn. Knowing that E is 200 GPa for steel and 70 GPa for aluminum, determine the normal stress in the rod.

Deformations

Bolts CD and GH. Tightening the nuts causes tension in the bolts. Because of symmetry, both are subjected to the same internal force P_b and undergo the same deformation δ_b. We have

$$\delta_b = +\frac{P_b L_b}{A_b E_b} = +\frac{P_b(0.375 \text{ m})}{\frac{1}{4}\pi(0.018 \text{ m})^2(200 \text{ GPa})} = +7.368 \times 10^{-9}\, P_b \qquad (1)$$

Rod EF. The rod is in compression. Denoting by P_r the magnitude of the force in the rod and by δ_r the deformation of the rod, we write

$$\delta_r = -\frac{P_r L_r}{A_r E_r} = -\frac{P_r(0.300 \text{ m})}{\frac{1}{4}\pi(0.036 \text{ m})^2(70 \text{ GPa})} = -4.210 \times 10^{-9}\, P_r \qquad (2)$$

Displacement of D Relative to B. Tightening the nuts one-quarter of a turn causes ends D and H of the bolts to undergo a displacement of $\frac{1}{4}(2 \text{ mm})$ *relative to* casting B. Considering end D, we write

$$\delta_{D/B} = \tfrac{1}{4}(0.002 \text{ m}) = 0.0005 \text{ m} \qquad (3)$$

But $\delta_{D/B} = \delta_D - \delta_B$, where δ_D and δ_B represent the displacements of D and B. If we assume that casting A is held in a fixed position while the nuts at D and H are being tightened, these displacements are equal to the deformations of the bolts and of the rod, respectively. We have, therefore,

$$\delta_{D/B} = \delta_b - \delta_r \qquad (4)$$

Substituting from (1), (2), and (3) into (4), we obtain

$$0.0005 \text{ m} = 7.368 \times 10^{-9}\, P_b + 4.210 \times 10^{-9}\, P_r \qquad (5)$$

Free Body: Casting B

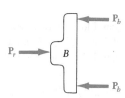

$$\xrightarrow{+}\Sigma F = 0: \qquad\qquad P_r - 2P_b = 0, \qquad P_r = 2P_b \qquad (6)$$

Forces in Bolts and Rod
Substituting for P_r from (6) into (5), we have

$$0.0005 \text{ m} = 7.368 \times 10^{-9}\, P_b + 4.210 \times 10^{-9}(2\, P_b)$$
$$P_b = 31.67 \text{ kN}$$
$$P_r = 2P_b = 2(31.67 \text{ kN}) = 63.34 \text{ kN}$$

Stress in Rod

$$\sigma_r = \frac{P_r}{A_r} = \frac{63.34 \text{ kN}}{\frac{1}{4}\pi(0.036 \text{ m})^2} \qquad \sigma_r = 62.2 \text{ MPa} \quad \blacktriangleleft$$

PROBLEMS

2.1 A polystyrene rod of length 300 mm and diameter 12 mm is subjected to a 3.5-kN tensile load. Knowing that $E = 3.1$ GPa, determine (a) the elongation of the rod, (b) the normal stress in the rod.

2.2 A 60-m-long steel wire must not stretch more than 48 mm when the tension in the wire is 6 kN. Knowing that $E = 200$ GPa, determine (a) the smallest diameter which may be selected for the wire, (b) the corresponding value of the normal stress.

2.3 An 80-m-long wire of 5-mm diameter is made of a steel with $E = 200$ GPa and an ultimate tensile strength of 400 MPa. If a factor of safety of 3.2 is desired, what is (a) the largest allowable tension in the wire, (b) the corresponding elongation of the wire?

2.4 A nylon thread is subjected to a 10-N tension. Knowing that $E = 2.8$ GPa and that the maximum allowable normal stress is 40 MPa determine (a) the required diameter of the thread, (b) the corresponding percent increase in the length of the thread.

2.5 A control rod made of yellow brass must stretch 3 mm when subjected to a 4-kN load. Knowing that $E = 105$ GPa and that the maximum allowable normal stress is 410 MPa, determine (a) the smallest diameter which may be selected for the rod, (b) the corresponding required length of the rod.

2.6 A 4-mm-diameter aluminum wire is observed to stretch 25 mm when the tension in the wire is 400 N. Knowing that $E = 70$ GPa and that the ultimate tensile strength of the aluminum used is 110 MPa, determine (a) the length of the wire, (b) the factor of safety.

2.7 A 1.5-m-long aluminum rod should not stretch more than 1 mm and the normal stress should not exceed 40 MPa when the rod is subjected to a 3-kN axial load. Knowing that $E = 70$ GPa, determine the required diameter of the rod.

2.8 A nylon thread is to be subjected to an 11-N tension. Knowing that $E = 3.1$ GPa, that the maximum allowable normal stress is 40 MPa and that the length of the thread should not increase by more than 1%, determine the required diameter of the thread.

2.9 Two solid cylindrical rods are joined at B and loaded as shown. Rod AB is made of steel ($E = 200$ GPa), and rod BC of brass ($E = 105$ GPa). Determine (a) the total deformation of the composite rod ABC, (b) the deflection of point B.

2.10 Two solid cylindrical rods are joined at B and loaded as shown. Rod AB is made of steel ($E = 200$ GPa), and rod BC of brass ($E = 105$ GPa). Determine (a) the total deformation of the composite rod ABC, (b) the deflection of point B.

2.11 For the composite rod of Prob. 2.10, determine (a) the load P for which the total deformation of the rod is −0.2 mm, (b) the corresponding deflection of point B.

2.12 For the composite rod of Prob. 2.9, determine (a) the load P for which the total deformation of the rod is zero, (b) the corresponding deflection of point B.

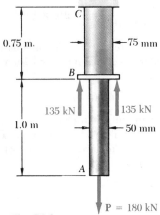

Fig. P2.9

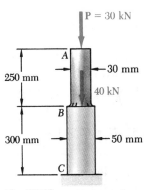

Fig. P2.10

2.13 Two solid cylindrical rods *AC* and *CD*, both of the same aluminum alloy ($E = 70$ GPa), are welded together at *C* and subjected to the loading shown. Determine (*a*) the total deformation of the composite rod *ACD*, (*b*) the deflection of point *C*.

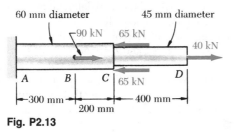

Fig. P2.13

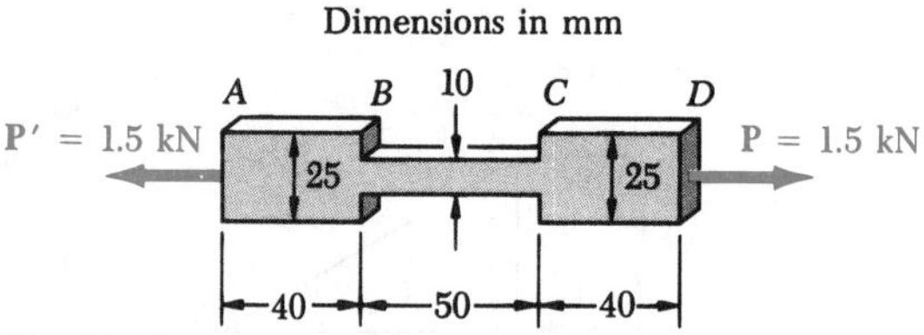

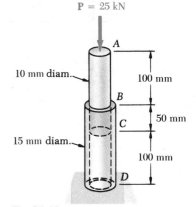

Fig. P2.14

2.14 The specimen shown has been cut from a 5-mm-thick sheet of vinyl ($E = 3.10$ GPa) and is subjected to a 1.5-kN tensile load. Determine (*a*) the total deformation of the specimen, (*b*) the deformation of its central portion *BC*.

2.15 A solid brass rod of length 150 mm and diameter 10 mm may fit exactly inside a brass tube of the same length, of outer diameter 15 mm and inner diameter 10 mm. A 50 mm portion of the rod is bonded to the tube, with the rest of the rod sticking out of the tube. The resulting member is to support a 25 kN load as shown. Knowing that $E = 105$ GPa, determine (*a*) the deflection of point *A*, (*b*) the maximum value of the average normal stress in the member.

2.16 A 250-mm-long aluminum tube ($E = 70$ GPa) of 36-mm outer diameter and 28-mm inner diameter may be closed at both ends by means of single-threaded screw-on covers of 1.5-mm pitch. With one cover screwed on tight, a solid brass rod ($E = 105$ GPa) of 25-mm diameter is placed inside the tube and the second cover is screwed on. Since the rod is slightly longer than the tube, it is observed that the cover must be forced against the rod by rotating it one-quarter of a turn before it can be tightly closed. Determine (*a*) the average normal stress in the tube and in the rod, (*b*) the deformation of the tube and of the rod.

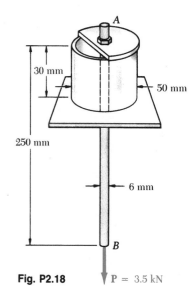

Fig. P2.15

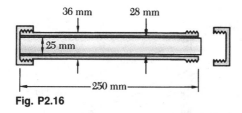

Fig. P2.16

2.17 Solve Prob. 2.16, assuming that the tube is made of brass ($E = 105$ GPa) and the rod of aluminum ($E = 70$ GPa).

2.18 A 3-mm-thick hollow polystyrene cylinder ($E = 3.1$ GPa) and a rigid circular plate (only part of which is shown) are used to support a 250-mm-long steel rod *AB* ($E = 200$ GPa) of 6-mm diameter. If a 3.5-kN load **P** is applied at *B*, determine (*a*) the elongation of rod *AB*, (*b*) the deflection of point *B*, (*c*) the average normal stress in rod *AB*.

2.19 For the composite rod of Prob. 2.9, determine the largest allowable load *P* if the total deformation of the rod is not to exceed 1 mm and the maximum normal stress is not to exceed 180 MPa.

Fig. P2.18

2.20 For the composite rod of Prob. 2.10, determine the largest allowable load P if the absolute values of the total deformation of the rod and of the maximum normal stress are not to exceed 0.2 mm and 75 MPa, respectively.

2.21 For the specimen of Prob. 2.14, determine the largest allowable load P if the deformation of portion AB and the total deformation of the specimen are not to exceed 0.2 mm and 1 mm, respectively.

2.22 For the rod and support of Prob. 2.18, determine the largest allowable load P if the deflections of points A and B are not to exceed 0.1 mm and 0.3 mm, respectively.

2.23 For the steel truss ($E = 200$ GPa) and loading shown, determine the deformations of members BD and DE, knowing that their cross-sectional areas are 1.30×10^{-3} m^2 and 1.95×10^{-3} m^2, respectively.

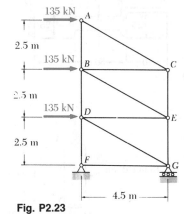

Fig. P2.23

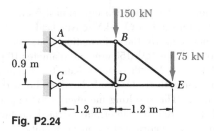

Fig. P2.24

2.24 Members AB and BE of the truss shown consist of 25-mm-diameter steel rods ($E = 200$ GPa). For the loading shown, determine the elongation of (a) rod AB, (b) rod BE.

2.25 Each of the four vertical links connecting the two horizontal members is made of aluminum ($E = 70$ GPa) and has a uniform rectangular cross section of 10×40 mm. For the loading shown, determine the deflection of (a) point E, (b) point F, (c) point G.

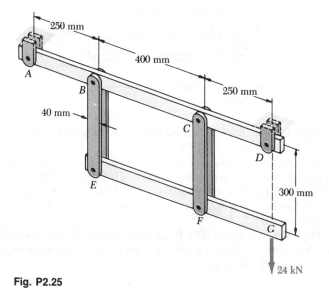

Fig. P2.25

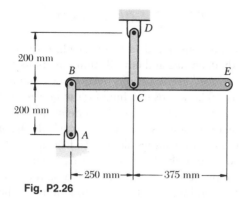

Fig. P2.26

2.26 Each of the links AB and CD is made of steel ($E = 200$ GPa) and has a uniform rectangular cross section of 6×25 mm. Determine the largest load which may be suspended from point E if the deflection of E is not to exceed 0.25 mm.

2.27 A homogeneous cable of length L and uniform cross section is suspended from one end. (*a*) Denoting by ρ the density (mass per unit volume) of the cable and by E its modulus of elasticity, determine the elongation of the cable due to its own weight. (*b*) Show that the same elongation would be obtained if the cable were horizontal and if a force equal to half of its weight were applied at each end.

2.28 Determine the deflection of the apex A of a homogeneous paraboloid of revolution of height h, density ρ, and modulus of elasticity E due to its own weight.

2.29 Determine the deflection of the apex A of a homogeneous circular cone of height h, density ρ, and modulus of elasticity E due to its own weight.

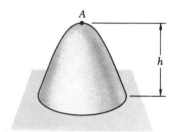

Fig. P2.28

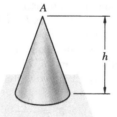

Fig. P2.29

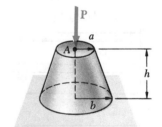

Fig. P2.30

2.30 A vertical load **P** is applied at the center A of the upper section of a homogeneous frustum of a circular cone of height h, minimum radius a, and maximum radius b. (*a*) Denoting by E the modulus of elasticity of the material and neglecting the effect of its weight, determine the deflection of point A. (*b*) Show that the same result would be obtained if the load **P** were applied at the center A of the upper section of a homogeneous cylinder of height h and of elliptic cross section with semiminor axis a and semimajor axis b.

2.31 The volume of a tensile specimen is essentially constant while plastic deformation occurs. If the initial diameter of the specimen is d_1, show that when the diameter is d, the true strain is $\epsilon_t = 2 \ln(d_1/d)$.

2.32 Denoting by ϵ the "engineering strain" in a tensile specimen, show that the true strain is $\epsilon_t = \ln(1 + \epsilon)$.

2.9. STATICALLY INDETERMINATE PROBLEMS

In the problems considered in the preceding section, we could always use free-body diagrams and equilibrium equations to determine the internal forces produced in the various portions of a member under given loading conditions. The values obtained for the internal forces were then entered into Eq. (2.8) or (2.9) to obtain the deformation of the member.

There are many problems, however, in which the internal forces cannot be determined from statics alone. In fact, in most of these problems the reactions themselves—which are external forces—cannot be determined by simply drawing a free-body diagram of the member and writing the corresponding equilibrium equations. The equilibrium equations must be complemented by relations involving deformations obtained by considering the geometry of the problem. Because statics is not sufficient to determine either the reactions or the internal forces, problems of this type are said to be *statically indeterminate*. The following examples will show how to handle this type of problem.

Example 2.02

A rod of length L, cross-sectional area A_1, and modulus of elasticity E_1, has been placed inside a tube of the same length L, but of cross-sectional area A_2 and modulus of elasticity E_2 (Fig. 2.23a). What is the deformation of the rod and tube when a force **P** is exerted on a rigid end plate as shown?

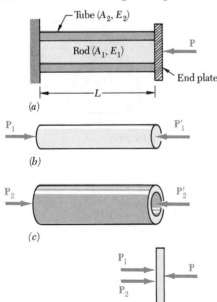

Fig. 2.23

Denoting by P_1 and P_2, respectively, the axial forces in the rod and in the tube, we draw free-body diagrams of all three elements (Fig. 2.23b, c, d). Only the last of the diagrams yields any significant information, namely:

$$P_1 + P_2 = P \qquad (2.11)$$

Clearly, one equation is not sufficient to determine the two unknown internal forces P_1 and P_2. The problem is statically indeterminate.

However, the geometry of the problem shows that the deformations δ_1 and δ_2 of the rod and tube must be equal. Recalling Eq. (2.7), we write

$$\delta_1 = \frac{P_1 L}{A_1 E_1} \qquad \delta_2 = \frac{P_2 L}{A_2 E_2} \qquad (2.12)$$

Equating the deformations δ_1 and δ_2, we obtain:

$$\frac{P_1}{A_1 E_1} = \frac{P_2}{A_2 E_2} \qquad (2.13)$$

Equations (2.11) and (2.13) may be solved simultaneously for P_1 and P_2:

$$P_1 = \frac{A_1 E_1 P}{A_1 E_1 + A_2 E_2} \qquad P_2 = \frac{A_2 E_2 P}{A_1 E_1 + A_2 E_2}$$

Either of Eqs. (2.12) may then be used to determine the common deformation of the rod and tube.

Example 2.03

A bar AB of length L and uniform cross section is attached to rigid supports at A and B before being loaded. What are the stresses in portions AC and BC due to the application of a load **P** at point C (Fig. 2.24a)?

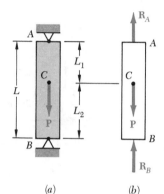

Fig. 2.24   (a)   (b)

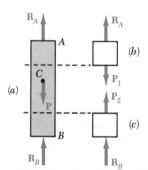

Fig. 2.25

Drawing the free-body diagram of the bar (Fig. 2.24b), we obtain the equilibrium equation

$$R_A + R_B = P \tag{2.14}$$

Since this equation is not sufficient to determine the two unknown reactions R_A and R_B, the problem is statically indeterminate.

However, the reactions may be determined if we observe from the geometry that the total elongation δ of the bar must be zero. Denoting by δ_1 and δ_2, respectively, the elongations of the portions AC and BC, we write

$$\delta = \delta_1 + \delta_2 = 0$$

or, expressing δ_1 and δ_2 in terms of the corresponding internal forces P_1 and P_2:

$$\delta = \frac{P_1 L_1}{AE} + \frac{P_2 L_2}{AE} = 0 \tag{2.15}$$

But we note from the free-body diagrams shown respectively in parts b and c of Fig. 2.25 that $P_1 = R_A$ and $P_2 = -R_B$. Carrying these values into (2.15), we write

$$R_A L_1 - R_B L_2 = 0 \tag{2.16}$$

Equations (2.14) and (2.16) may be solved simultaneously for R_A and R_B; we obtain $R_A = PL_2/L$ and $R_B = PL_1/L$. The desired stresses σ_1 in AC and σ_2 in BC are obtained by dividing, respectively, $P_1 = R_A$ and $P_2 = -R_B$ by the cross-sectional area of the bar:

$$\sigma_1 = \frac{PL_2}{AL} \qquad \sigma_2 = -\frac{PL_1}{AL}$$

Superposition Method. We may observe that a structure is statically indeterminate whenever it is held by more supports than are required to maintain its equilibrium. This results in more unknown reactions than available equilibrium equations. It is often found convenient to designate one of the reactions as *redundant* and to eliminate the corresponding support. Since the stated conditions of the problem cannot be arbitrarily changed, the redundant reaction must be maintained in the solution. But it will be treated as an *unknown load* which, together with the other loads, must produce deformations which are compatible with the original constraints. The actual solution of the problem is carried out by considering separately the deformations caused by the given loads and by the redundant reaction, and by adding—or *superposing*—the results obtained.†

---

† The general conditions under which the combined effect of several loads may be obtained in this way are discussed in Sec. 2.12.

Example 2.04

Determine the reactions at A and B for the steel bar and loading shown in Fig. 2.26, assuming a close fit at both supports before the loads are applied.

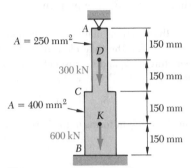

$A = 250 \text{ mm}^2$

300 kN

$A = 400 \text{ mm}^2$

600 kN

150 mm

150 mm

150 mm

150 mm

Fig. 2.26

We shall consider the reaction at B as redundant and release the bar from that support. The reaction $\mathbf{R}_B$ is now considered as an unknown load (Fig. 2.27a) and will be determined from the condition that the deformation δ of the rod must be equal to zero. The solution is carried out by considering separately the deformation δ_L caused by the given loads (Fig. 2.27b) and the deformation δ_R due to the redundant reaction $\mathbf{R}_B$ (Fig. 2.27c).

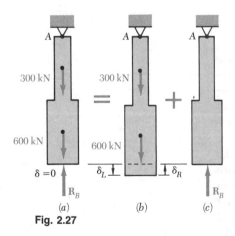

$(a) \quad (b) \quad (c)$

Fig. 2.27

The deformation δ_L is obtained from Eq. (2.8) after the bar has been divided into four portions, as shown in Fig. 2.28.

Following the same procedure as in Example 2.01, we write

$$P_1 = 0 \qquad P_2 = P_3 = 600 \times 10^3 \text{ N} \qquad P_4 = 900 \times 10^3 \text{ N}$$
$$A_1 = A_2 = 400 \times 10^{-6} \text{ m}^2 \qquad A_3 = A_4 = 250 \times 10^{-6} \text{ m}^2$$
$$L_1 = L_2 = L_3 = L_4 = 0.150 \text{ m}$$

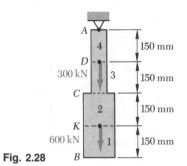

Fig. 2.28

Substituting these values into Eq. (2.8), we obtain

$$\delta_L = \sum_{i=1}^{4} \frac{P_i L_i}{A_i E} = \left(0 + \frac{600 \times 10^3 \text{ N}}{400 \times 10^{-6} \text{ m}^2} \right.$$
$$\left. + \frac{600 \times 10^3 \text{ N}}{250 \times 10^{-6} \text{ m}^2} + \frac{900 \times 10^3 \text{ N}}{250 \times 10^{-6} \text{ m}^2} \right) \frac{0.150 \text{ m}}{E}$$
$$\delta_L = \frac{1.125 \times 10^9}{E} \tag{2.17}$$

Considering now the deformation δ_R due to the redundant reaction R_B, we divide the bar into two portions, as shown in Fig. 2.29, and write

$$P_1 = P_2 = -R_B$$
$$A_1 = 400 \times 10^{-6} \text{ m}^2 \qquad A_2 = 250 \times 10^{-6} \text{ m}^2$$
$$L_i = L_2 = 0.300 \text{ m}$$

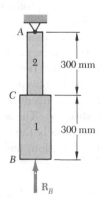

300 mm

300 mm

Fig. 2.29

Substituting these values into Eq. (2.8), we obtain

$$\delta_R = \frac{P_1 L_1}{A_1 E} + \frac{P_2 L_2}{A_2 E} = -\frac{(1.95 \times 10^3) R_B}{E} \qquad (2.18)$$

Expressing that the total deformation δ of the bar must be zero, we write

$$\delta = \delta_L + \delta_R = 0 \qquad (2.19)$$

and, substituting for δ_L and δ_R from (2.17) and (2.18) into (2.19),

$$\delta = \frac{1.125 \times 10^9}{E} - \frac{(1.95 \times 10^3) R_B}{E} = 0$$

Solving for R_B, we have

$$R_B = 577 \times 10^3 \text{ N} = 577 \text{ kN}$$

The reaction R_A at the upper support is obtained from the free-body diagram of the bar (Fig. 2.30). We write

$$+\uparrow \Sigma F_y = 0: \quad R_A - 300 \text{ kN} - 600 \text{ kN} + R_B = 0$$
$$R_A = 900 \text{ kN} - R_B = 900 \text{ kN} - 577 \text{ kN} = 323 \text{ kN}$$

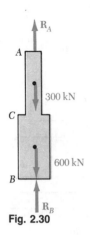

Fig. 2.30

Once the reactions have been determined, the stresses and strains in the bar may easily be obtained. It should be noted that, while the total deformation of the bar is zero, each of its component parts *does deform* under the given loading and restraining conditions.

Example 2.05

Determine the reactions at A and B for the steel bar and loading of Example 2.04, assuming now that a 4.50-mm clearance exists between the bar and the ground before the loads are applied (Fig. 2.31). Assume $E = 200$ GPa.

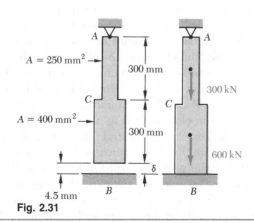

Fig. 2.31

We follow the same procedure as in Example 2.04. Considering the reaction at B as redundant, we compute the deformations δ_L and δ_R caused respectively by the given loads and by the redundant reaction $\mathbf{R}_B$. However, in this case the total deformation is not zero, but $\delta = 4.5$ mm. We write therefore

$$\delta = \delta_L + \delta_R = 4.5 \times 10^{-3} \text{ m} \qquad (2.20)$$

Substituting for δ_L and δ_R from (2.17) and (2.18) into (2.20), and recalling that $E = 200$ GPa $= 200 \times 10^9$ Pa, we have

$$\delta = \frac{1.125 \times 10^9}{200 \times 10^9} - \frac{(1.95 \times 10^3) R_B}{200 \times 10^9} = 4.5 \times 10^{-3}$$

Solving for R_B, we obtain

$$R_B = 115.4 \times 10^3 \text{ N} = 115.4 \text{ kN}$$

The reaction at A is obtained from the free-body diagram of the bar (Fig. 2.30):

$$+\uparrow \Sigma F_y = 0: \quad R_A - 300 \text{ kN} - 600 \text{ kN} + R_B = 0$$
$$R_A = 900 \text{ kN} - R_B = 900 \text{ kN} - 115.4 \text{ kN} = 785 \text{ kN}$$

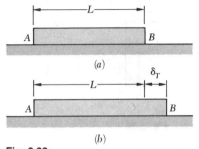

Fig. 2.32

2.10. PROBLEMS INVOLVING TEMPERATURE CHANGES

All of the members and structures that we have considered so far were assumed to remain at the same temperature while they were being loaded. We shall now consider various situations involving changes in temperature.

Let us first consider a homogeneous rod AB of uniform cross section, which rests freely on a smooth horizontal surface (Fig. 2.32a). If the temperature of the rod is raised by ΔT, we observe that the rod elongates by an amount δ_T which is proportional to both the temperature change ΔT and the length L of the rod (Fig. 2.32b). We have

$$\delta_T = \alpha(\Delta T)L \tag{2.21}$$

where α is a constant characteristic of the material, called the *coefficient of thermal expansion*. Since δ_T and L are both expressed in units of length, α represents a quantity *per degree C*.

With the deformation δ_T must be associated a strain $\varepsilon_T = \delta_T/L$. Recalling Eq. (2.21), we conclude that

$$\epsilon_T = \alpha \, \Delta T \tag{2.22}$$

The strain ϵ_T is referred to as a *thermal strain*, since it is caused by the change in temperature of the rod. In the case we are considering here, there is *no stress associated with the strain* ϵ_T.

Let us now assume that the same rod AB of length L is placed between two fixed supports at a distance L from each other (Fig. 2.33a). Again, there is neither stress nor strain in this initial condition. If we raise the temperature by ΔT, the rod cannot elongate because of the restraints imposed on its ends; the elongation δ_T of the rod is thus zero. Since the rod is homogeneous and of uniform cross section, the strain ϵ_T at any point is $\epsilon_T = \delta_T/L$ and, thus, also zero. However, the supports will exert equal and opposite forces $\mathbf{P}$ and $\mathbf{P'}$ on the rod after the temperature has been raised, to keep it from elongating (Fig. 2.33b). It thus follows that a state of stress (with no corresponding strain) is created in the rod.

As we prepare to determine the stress σ created by the temperature change ΔT, we observe that the problem we have to solve is statically indeterminate. Therefore, we should first compute the magnitude P of the reactions at the supports from the condition that the elongation of the rod is zero. Using the superposition method described in Sec. 2.9, we

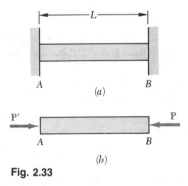

Fig. 2.33

detach the rod from its support B (Fig. 2.34a) and let it elongate freely as it undergoes the temperature change ΔT (Fig. 2.34b). According to formula (2.21), the corresponding elongation is

$$\delta_T = \alpha(\Delta T)L$$

Applying now to end B the force $\mathbf{P}$ representing the redundant reaction, and recalling formula (2.7), we obtain a second deformation (Fig. 2.34c)

$$\delta_P = \frac{PL}{AE}$$

Expressing that the total deformation δ must be zero, we have

$$\delta = \delta_T + \delta_P = \alpha(\Delta T)L + \frac{PL}{AE} = 0$$

from which we conclude that

$$P = -AE\alpha(\Delta T)$$

and that the stress in the rod due to the temperature change ΔT is

$$\sigma = \frac{P}{A} = -E\alpha(\Delta T) \qquad (2.23)$$

It should be kept in mind that the result we have obtained here and our earlier remark regarding the absence of any strain in the rod *apply only in the case of a homogeneous rod of uniform cross section.* Any other problem involving a restrained structure undergoing a change in temperature must be analyzed on its own merits. However, the same general approach may be used, i.e., we may consider separately the deformation due to the temperature change and the deformation due to the redundant reaction and superpose the solutions obtained.

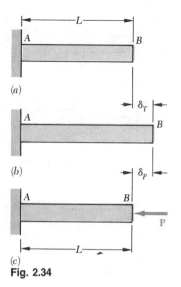

(a)

(b)

(c)

Fig. 2.34

Example 2.06

Determine the values of the stress in portions AC and CB of the steel bar shown (Fig. 2.35) when the temperature of the bar is $-50°C$, knowing that a close fit exists at both of the rigid supports when the temperature is $+25°C$. Use the values $E = 200$ GPa and $\alpha = 12 \times 10^{-6}/°C$ for steel.

We first determine the reactions at the supports. Since the problem is statically indeterminate, we detach the bar from its support at B and let it undergo the temperature change

$$\Delta T = (-50°C) - (25°C) = -75°C$$

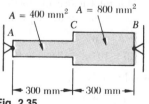

Fig. 2.35

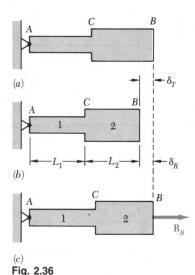

(a)

(b)

(c)

Fig. 2.36

The corresponding deformation (Fig. 2.36b) is

$$\delta_T = \alpha(\Delta T)L = (12 \times 10^{-6}/°C)(-75°C)(0.6 \text{ m})$$
$$= -540 \times 10^{-6} \text{ m}$$

Applying now the unknown force $\mathbf{R}_B$ at end B (Fig. 2.36c), we use Eq. (2.8) to express the corresponding deformation δ_R. Substituting

$$L_1 = L_2 = 0.3 \text{ m}$$
$$A_1 = 400 \times 10^{-6} \text{ m}^2 \qquad A_2 = 800 \times 10^{-6} \text{ m}^2$$
$$P_1 = P_2 = R_B \qquad E = 200 \times 10^9 \text{ Pa}$$

into Eq. (2.8), we write

$$\delta_R = \frac{P_1 L_1}{A_1 E} + \frac{P_2 L_2}{A_2 E}$$

$$= \frac{R_B}{200 \times 10^9 \text{ Pa}}\left(\frac{0.3 \text{ m}}{400 \times 10^{-6} \text{ m}^2} + \frac{0.3 \text{ m}}{800 \times 10^{-6} \text{ m}^2}\right)$$

$$= (5.625 \times 10^{-9} \text{ m/N})R_B$$

Expressing that the total deformation of the bar must be zero as a result of the imposed constraints, we write

$$\delta = \delta_T + \delta_R = 0$$
$$\delta = -540 \times 10^6 \text{ m} + (5.625 \times 10^{-9} \text{ m/N})R_B = 0$$

from which we obtain

$$R_B = 96.0 \times 10^3 \text{ N} = 96.0 \text{ kN}$$

The reaction at A is equal and opposite.

Noting that the forces in the two portions of the bar are $P_1 = P_2 = 96.0$ kN, we obtain the following values of the stress in portions AC and CB of the bar:

$$\sigma_1 = \frac{P_1}{A_1} = \frac{96.0 \times 10^3 \text{ N}}{400 \times 10^{-6} \text{ m}^2} = 240 \text{ MPa}$$

$$\sigma_2 = \frac{P_2}{A_2} = \frac{96.0 \times 10^3 \text{ N}}{800 \times 10^{-6} \text{ m}^2} = 120 \text{ MPa}$$

We cannot emphasize too strongly the fact that, while the *total deformation* of the bar must be zero, the deformations of the portions AC and CB are not zero. A solution of the problem based on the assumption that these deformations are zero would therefore be wrong. Neither can the values of the strain in AC or CB be assumed equal to zero. To amplify this point, we shall now determine the strain ϵ_{AC} in portion AC of the bar. The strain ϵ_{AC} may be divided into two component parts; one is the thermal strain ϵ_T produced in the unrestrained bar by the temperature change ΔT (Fig. 2.36b). From Eq. (2.22) we write

$$\epsilon_T = \alpha \Delta T = (12 \times 10^{-6}/°C)(-75°C)$$
$$= -900 \times 10^{-6} = -900 \, \mu$$

The other component of ϵ_{AC} is associated with the stress σ_1 due to the force $\mathbf{R}_B$ applied to the bar (Fig. 2.36c). From Hooke's law, we express this component of the strain as

$$\frac{\sigma_1}{E} = \frac{240 \text{ MPa}}{200 \text{ GPa}} = 1200 \times 10^{-6} = 1200 \, \mu$$

Adding the two components of the strain in AC, we obtain

$$\epsilon_{AC} = \epsilon_T + \frac{\sigma_1}{E} = -900 \, \mu + 1200 \, \mu$$
$$= +300 \, \mu$$

A similar computation yields the strain in portion CB of the bar:

$$\epsilon_{CB} = \epsilon_T + \frac{\sigma_2}{E} = -900 \, \mu + 600 \, \mu$$
$$= -300 \, \mu$$

The deformations δ_{AC} and δ_{CB} of the two portions of the bar are expressed respectively as

$$\delta_{AC} = \epsilon_{AC}(AC) = (+300 \, \mu)(0.3 \text{ mm}) = +90 \, \mu\text{m}$$
$$\delta_{CB} = \epsilon_{CB}(CB) = (-300 \, \mu)(0.3 \text{ m}) = -90 \, \mu\text{m}$$

We thus check that, while the sum $\delta = \delta_{AC} + \delta_{CB}$ of the two deformations is zero, neither of the deformations is zero.

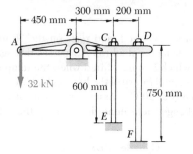

The 10-mm-diameter rod CE and the 15-mm-diameter rod DF are attached to the rigid bar $ABCD$ as shown. Knowing that the rods are made of aluminum and using $E = 70$ GPa, determine (a) the force in each rod caused by the loading shown, (b) the corresponding deflection at point A.

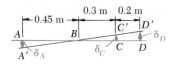

Statics. Considering the free body of bar $ABCD$, we note that the reaction at B and the forces exerted by the rods are indeterminate. However, using statics, we may write

$$+)\Sigma M_B = 0: \quad (32\text{ kN})(0.45\text{ m}) - F_{CE}(0.3\text{ m}) - F_{DF}(0.5\text{ m}) = 0$$
$$0.3F_{CE} + 0.5F_{DF} = 14.4 \times 10^3 \tag{I}$$

Geometry. After application of the 32-kN load, the position of the bar is $A'BC'D'$. From the similar triangles BAA', BCC', and BDD' we have

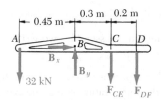

$$\frac{\delta_C}{0.3\text{ m}} = \frac{\delta_D}{0.5\text{ m}} \qquad \delta_C = 0.6\delta_D \tag{2}$$

and

$$\frac{\delta_A}{0.45\text{ m}} = \frac{\delta_D}{0.5\text{ m}} \qquad \delta_A = 0.9\delta_D \tag{3}$$

Deformations. Using Eq. (2.8), we have

$$\delta_C = \frac{F_{CE}L_{CE}}{A_{CE}E} \qquad \delta_D = \frac{F_{DF}L_{DF}}{A_{DF}E}$$

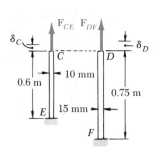

Substituting for δ_C and δ_D into (2), we write

$$\delta_C = 0.6\delta_D \qquad \frac{F_{CE}L_{CE}}{A_{CE}E} = 0.6\frac{F_{DF}L_{DF}}{A_{DF}E}$$

$$F_{CE} = 0.6\frac{L_{DF}}{L_{CE}}\frac{A_{CE}}{A_{DF}}F_{DF} = 0.6\left(\frac{0.75\text{ m}}{0.60\text{ m}}\right)\left[\frac{\frac{1}{4}\pi(0.010\text{ m})^2}{\frac{1}{4}\pi(0.015\text{ m})^2}\right]F_{DF} \qquad F_{CE} = 0.333F_{DF}$$

Force in Each Rod. Substituting for F_{CE} into (1), we have

$$0.3(0.333F_{DF}) + 0.5F_{DF} = 14.4 \times 10^3 \qquad F_{DF} = 24\text{ kN} \blacktriangleleft$$
$$F_{CE} = 0.333F_{DF} = 0.333(24\text{ kN}) \qquad F_{CE} = 8\text{ kN} \blacktriangleleft$$

Deflections. The deflection of point D is

$$\delta_D = \frac{F_{DF}L_{DF}}{A_{DF}E} = \frac{(24\text{ kN})(0.75\text{ m})}{\frac{1}{4}\pi(0.015\text{ m})^2(70\text{ GPa})} \qquad \delta_D = 1.455\text{ mm}$$

Using (3), we write

$$\delta_A = 0.9\delta_D = 0.9(1.455\text{ mm}) \qquad \delta_A = 1.310\text{ mm} \blacktriangleleft$$

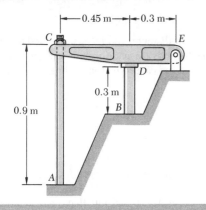

- 0.45 m
- 0.3 m
- 0.3 m
- 0.9 m

C E D B A

SAMPLE PROBLEM 2.4

The rigid bar CDE is attached to a pin support at E and rests on the 30-mm-diameter brass cylinder BD. A 22-mm-diameter steel rod AC passes through a hole in the bar and is secured by a nut which is snugly fitted when the temperature of the entire assembly is 20°C. The temperature of the brass cylinder is then raised to 50°C while the steel rod remains at 20°C. Assuming that no stresses were present before the temperature change, determine the stress in the cylinder.

Rod AC: Steel	Cylinder BD: Brass
$E = 200$ GPa	$E = 105$ GPa
$\alpha = 12 \times 10^{-6}/°C$	$\alpha = 18.8 \times 10^{-6}/°C$

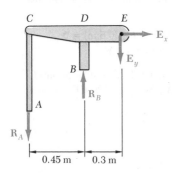

Statics. Considering the free body of the entire assembly, we write

$$+\uparrow\Sigma M_E = 0: \qquad R_A(0.75\text{ m}) - R_B(0.3\text{ m}) = 0 \qquad R_A = 0.4R_B \qquad (1)$$

Deformations. We shall use the method of superposition, considering $\mathbf{R}_B$ as redundant. With the support at B removed, the temperature rise of the cylinder causes point B to move down through δ_T. The reaction $\mathbf{R}_B$ must cause a deflection δ_1 of the same magnitude as δ_T so that the final deflection of point B will be zero (Fig. 3).

Deflection δ_T. Due to a temperature rise of $50° - 20° = 30°C$, the length of the brass cylinder increases by δ_T.

$$\delta_T = L(\Delta T)\alpha = (0.3\text{ m})(30°C)(18.8 \times 10^{-6}/°C) = 169.2 \times 10^{-6}\text{ m} \downarrow$$

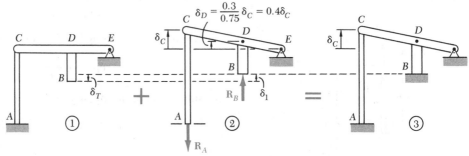

Deflection δ_1. From Fig. 2 we note that $\delta_D = 0.4\delta_C$ and that $\delta_1 = \delta_D + \delta_{B/D}$.

$$\delta_C = \frac{R_A L}{AE} = \frac{R_A(0.9\text{ m})}{\frac{1}{4}\pi(0.022\text{ m})^2(200\text{ GPa})} = 11.84 \times 10^{-9}R_A \uparrow$$

$$\delta_D = 0.40\delta_C = 0.4(11.84 \times 10^{-9}R_A) = 4.74 \times 10^{-9}R_A \uparrow$$

$$\delta_{B/D} = \frac{R_B L}{AE} = \frac{R_B(0.3\text{ m})}{\frac{1}{4}\pi(0.03\text{ m})^2(105\text{ GPa})} = 4.04 \times 10^{-9}R_B \uparrow$$

We recall from (1) that $R_A = 0.4R_B$ and write

$$\delta_1 = \delta_D + \delta_{B/D} = [4.74(0.4R_B) + 4.04R_B]10^{-9} = 5.94 \times 10^{-9}R_B \uparrow$$

But $\delta_T = \delta_1$: $\qquad 169.2 \times 10^{-6}\text{ m} = 5.94 \times 10^{-9}R_B \qquad R_B = 28.5\text{ kN}$

Stress in Cylinder $\qquad \sigma_B = \frac{R_B}{A} = \frac{28.5\text{ kN}}{\frac{1}{4}\pi(0.03)^2} \qquad \sigma_B = 40.3\text{ MPa} \blacktriangleleft$

PROBLEMS

2.33 A 1.2-m concrete post is reinforced by four steel bars, each of 19-mm diameter. Knowing that $E_s = 200$ GPa and $E_c = 25$ GPa, determine the normal stresses in the steel and concrete when a 700-kN centric force is applied to the post.

2.34 A 250-mm bar of 15×30-mm rectangular cross section consists of two aluminum layers, 5-mm thick, brazed to a center brass layer of the same thickness. If it is subjected to centric forces of magnitude $P = 30$ kN, and knowing that $E_a = 70$ GPa and $E_b = 105$ GPa, determine the normal stress (a) in the aluminum layers, (b) in the brass layer.

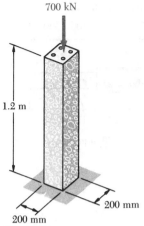

700 kN

1.2 m

200 mm

200 mm

Fig. P2.33

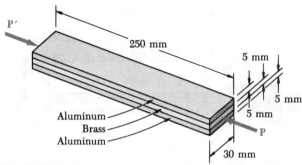

P′

250 mm

5 mm

5 mm

5 mm

Aluminum

Brass

Aluminum

P

30 mm

Fig. P2.34

2.35 Determine the deformation of the composite bar of Prob. 2.34 if it is subjected to centric forces of magnitude $P = 45$ kN.

2.36 Compressive centric forces of 180 kN are applied at both ends of the assembly shown by means of rigid end plates. Knowing that $E_s = 200$ GPa and $E_a = 70$ GPa, determine (a) the normal stresses in the steel core and the aluminum shell, (b) the deformation of the assembly.

2.37 A polystyrene rod consisting of two cylindrical portions AB and BC is restrained at both ends and supports two 26-kN loads as shown. Knowing that $E = 3.1$ GPa, determine (a) the reactions at A and C, (b) the normal stress in each portion of rod.

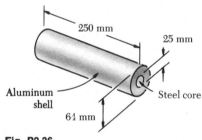

250 mm

25 mm

Aluminum
shell

Steel core

64 mm

Fig. P2.36

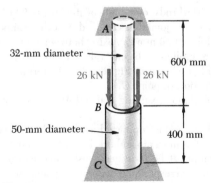

A

32-mm diameter

600 mm

26 kN

26 kN

B

50-mm diameter

400 mm

C

Fig. P2.37

Dimensions in mm

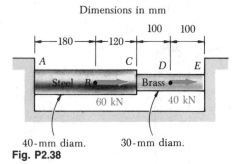

Fig. P2.38

40-mm diam.   30-mm diam.

2.38 Two cylindrical rods, one of steel and the other of brass, are joined at C and restrained by rigid supports at A and E. For the loading shown and knowing that $E_s = 200$ GPa and $E_b = 105$ GPa, determine (a) the reactions at A and E, (b) the deflection of point C.

2.39 Solve Prob. 2.38, assuming that rod AC is made of brass and rod CE of steel.

2.40 A 300-mm-long brass tube of 32-mm outer diameter and 3-mm thickness is placed in a vise which is adjusted so that its jaws just touch the ends of the tube without exerting any pressure on them. Two forces P and Q of magnitude $P = 190$ kN and $Q = 160$ kN are then applied to the tube as shown. Knowing that $E = 103$ GPa, determine (a) the forces exerted by the vise on the tube at A and D, (b) the elongation of portion BC of the tube.

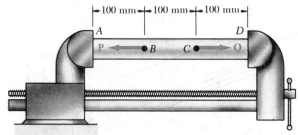

Fig. P2.40

2.41 Solve Prob. 2.40, assuming that after the forces P and Q have been applied, the vise is adjusted to increase the distance between its jaws by 0.25 mm.

2.42 Solve Prob. 2.40, assuming that after the forces P and Q have been applied, the vise is adjusted to decrease the distance between its jaws by 0.25 mm.

Dimensions in mm

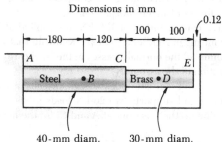

Fig. P2.43   40-mm diam.   30-mm diam.

2.43 Two cylindrical rods, one made of steel ($E_s = 200$ GPa) and the other of brass ($E_b = 105$ GPa), are joined at C. End A of the composite rod obtained in this way is fixed, while a 0.12-mm gap exists between end E and a vertical wall. A 60-kN force is then applied at B and a 40-kN force at D, both horizontal and directed to the right (as shown in Fig. P2.38). Determine (a) the reactions at A and E, (b) the deflection of point C.

2.44 Solve Prob. 2.43, assuming that rod AC is made of brass and rod CE of steel.

2.45 The rigid bar AD is supported as shown by two steel wires of 1.5-mm diameter ($E = 200$ GPa) and a pin and bracket at D. Knowing that the wires were initially taut, determine (a) the additional tension in each wire when a 500-N load P is applied at B, (b) the corresponding deflection of point B.

Dimensions in mm

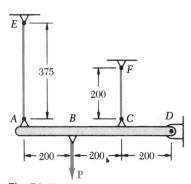

Fig. P2.45

2.46 A rigid bar is suspended from a fixed plate by means of four wires as shown. The wires attached to pegs A and D are of aluminum ($E_a = 70$ GPa) and have a diameter of 2.5 mm, while the wires attached to pegs B and C are of steel ($E_s = 200$ GPa) and have a diameter of 2 mm. Knowing that all the wires are initially taut, determine (a) the additional tension in each wire when a 2-kN load is applied to the middle of the bar, (b) the corresponding elongation of the wires.

2.47 Solve Prob. 2.46, assuming that the wires attached to pegs A and D are of steel ($E_s = 200$ GPa) and have a diameter of 2.5 mm, while the wires attached to pegs B and C are of aluminum ($E_a = 70$ GPa) and have a diameter of 2 mm.

2.48 Links BC and DE are both made of steel ($E = 200$ GPa) and are 12 mm wide and 6 mm thick. Determine (a) the force in each link when 2.5-kN force $\mathbf{P}$ is applied to the rigid member AF as shown, (b) the corresponding deflection of point A.

Dimensions in mm

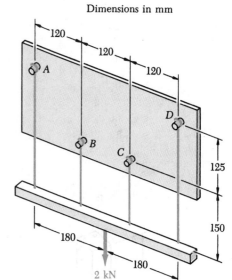

Dimensions in mm

Fig. P2.46

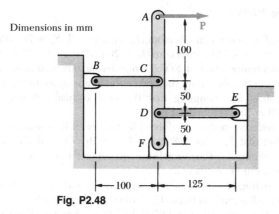

Fig. P2.48

2.49 Each of the steel rods AD and CE ($E = 200$ GPa) has an 8-mm diameter and is single-threaded at its upper end with a pitch of 2 mm. Knowing that after being snugly fitted the nut at A is tightened two full turns, determine (a) the tension in each rod, (b) the deflection of point A of the rigid member ABC.

2.50 Solve Prob. 2.49, assuming that after being snugly fitted the nut at C, rather than the nut at A, is tightened two full turns.

2.51 The rigid bar $ABCD$ is suspended from three identical wires as shown. Knowing that $a = b$, determine the tension in each wire caused by the load $\mathbf{P}$ applied at C.

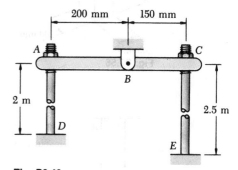

Fig. P2.49

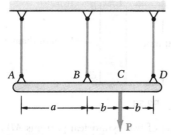

Fig. P2.51 and P2.52

2.52 The rigid bar $ABCD$ is suspended from three identical wires as shown. Knowing that $a = 2b$, determine the tension in each wire caused by the load $\mathbf{P}$ applied at C.

2.53   A steel railroad track ($E = 200$ GPa, $\alpha = 11.7 \times 10^{-6}/°C$) was laid out at a temperature of 0°C.   Determine the normal stress in a rail when the temperature reaches 50°C, assuming that the rails (*a*) are welded to form a continuous track, (*b*) are 12 m long with 6 mm gaps between them.

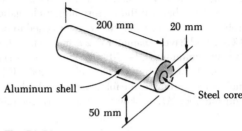

Fig. P2.54

2.54   The assembly shown consists of an aluminum shell ($E_a = 70$ GPa, $\alpha_a = 23.6 \times 10^{-6}/°C$) fully bonded to a steel core ($E_s = 200$ GPa, $\alpha_s = 11.7 \times 10^{-6}/°C$) and is unstressed at a temperature of 20°C.   Considering only axial deformations, determine the stress in the aluminum shell when the temperature reaches 180°C.

2.55   Solve Prob. 2.54, assuming that the core is made of brass ($E_b = 105$ GPa, $\alpha_b = 20.9 \times 10^{-6}/°C$) instead of steel.

2.56   A 1.2-m concrete post is reinforced by four steel bars, each of 19-mm diameter.   Knowing that $E_s = 200$ GPa, $\alpha_s = 11.7 \times 10^{-6}/°C$ and $E_c = 25$ GPa, $\alpha_c = 9.9 \times 10^{-6}/°C$, determine the normal stressed induced in the steel and in the concrete by a temperature rise of 45°C.

2.57   A rod consisting of two cylindrical portions *AB* and *BC* is restrained at both ends.   Portion *AB* is made of brass ($E_b = 105$ GPa, $\alpha_b = 20.9 \times 10^{-6}/°C$) and portion *BC* of steel ($E_s = 200$ GPa, $\alpha_s = 11.7 \times 10^{-6}/°C$).   Knowing that the rod is initially unstressed, determine (*a*) the normal stresses induced in portions *AB* and *BC* by a temperature rise of 50°C, (*b*) the corresponding deflection of point *B*.

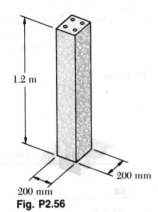

Fig. P2.56

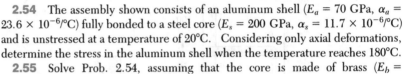

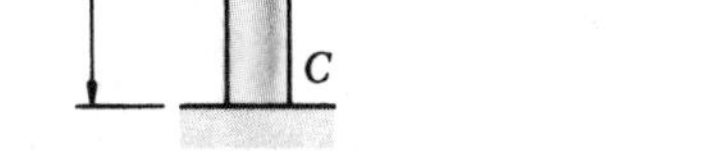

Fig. P2.57

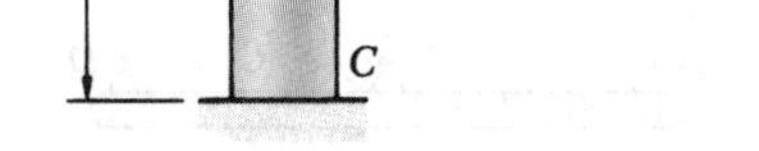

Fig. P2.58

2.58   A rod consisting of two cylindrical portions *AB* and *BC* is restrained at both ends.   Portion *AB* is made of steel ($E_s = 200$ GPa, $\alpha_s = 11.7 \times 10^{-6}/°C$) and portion *BC* of brass ($E_b = 105$ GPa, $\alpha_b = 20.9 \times 10^{-6}/°C$).   Knowing that the rod is initially unstressed, determine (*a*) the normal stresses induced in portions *AB* and *BC* by a temperature rise of 50°C, (*b*) the corresponding deflection of point *B*.

2.59   Solve Prob. 2.58, assuming that portion AB of the composite rod is made of brass and portion BC of steel.

2.60   Solve Prob. 2.57, assuming that portion AB of the composite rod is made of steel and portion BC of brass.

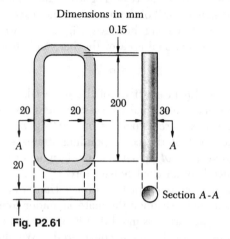

Dimensions in mm

Fig. P2.61

2.61   An aluminum rod ($E_a = 70$ GPa, $\alpha_a = 23.6 \times 10^{-6}$/°C) and a steel link ($E_s = 200$ GPa, $\alpha_s = 11.7 \times 10^{-6}$/°C) have the dimensions shown at a temperature of 20°C. The steel link is heated until the aluminum rod can be fitted freely into it. The temperature of the whole assembly is then raised to 150°C. Determine the final stress (a) in the rod, (b) in the link.

2.62   The temperature of the composite rod of Prob. 2.43 is raised by 80°C. Knowing that $E_s = 200$ GPa, $\alpha_s = 11.7 \times 10^{-6}$/°C, $E_b = 105$ GPa, $\alpha_b = 20.9 \times 10^{-6}$/°C, and that no force is applied at B or D, determine (a) the normal stresses in portions AC and CE, (b) the deformation of portion AC.

2.63   Rod AB is made of brass ($E_b = 105$ GPa, $\alpha_b = 20.9 \times 10^{-6}$/°C) and rod CD of aluminum ($E_a = 70$ GPa, $\alpha_a = 23.6 \times 10^{-6}$/°C). Knowing that at 15°C a 0.5-mm gap exists between the ends of the two rods, determine (a) the normal stress in each rod after the temperature has been raised to 85°C, (b) the deformation of rod AB at that time.

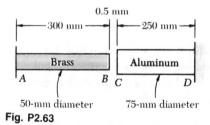

Fig. P2.63

2.64   For the rods of Prob. 2.63, determine (a) the temperature at which the stress in rod AB will be -140 MPa, (b) the corresponding deformation of rod AB.

2.65   For the rods of Prob. 2.49, we assume that after being snugly fitted each of the nuts at A and C is loosened half a turn. The temperature of both rods is then lowered by 60°C. Knowing that $E_s = 200$ GPa and $\alpha_s = 11.7 \times 10^{-6}$/°C, determine the final stress (a) in rod AD, (b) in rod CE.

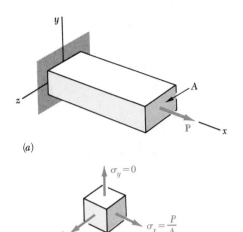

(a)

(b)

Fig. 2.37

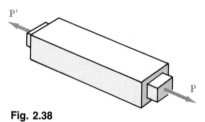

Fig. 2.38

2.11. POISSON'S RATIO

We saw in the earlier part of this chapter that, when a homogeneous slender bar is axially loaded, the resulting stress and strain satisfy Hooke's law, as long as the elastic limit of the material is not exceeded. Assuming that the load **P** is directed along the x axis (Fig. 2.37a), we have $\sigma_x = P/A$, where A is the cross-sectional area of the bar, and, from Hooke's law,

$$\epsilon_x = \sigma_x/E \qquad (2.24)$$

where E is the modulus of elasticity of the material.

We also note that the normal stresses on faces respectively perpendicular to the y and z axes are zero: $\sigma_y = \sigma_z = 0$ (Fig. 2.37b). It would be tempting to conclude that the corresponding strains ϵ_y and ϵ_z are also zero. This, however, *is not the case.* In all engineering materials, the elongation produced by an axial tensile force **P** in the direction of the force is accompanied by a contraction in any transverse direction (Fig. 2.38).† We have already assumed the material under consideration to be homogeneous, i.e., we have assumed that its various mechanical properties were independent of the point considered. We shall now further assume the material to be *isotropic*, i.e., we shall assume that its mechanical properties are also independent of the direction considered. Under this additional assumption, the strain must have the same value for any transverse direction: $\epsilon_y = \epsilon_z$. This value is referred to as the *lateral strain*, and the absolute value of the ratio of the lateral strain over the axial strain is called *Poisson's ratio*, after the French mathematician Siméon Denis Poisson (1781–1840), and is denoted by the Greek letter ν (nu). We have

$$\nu = \left| \frac{\text{lateral strain}}{\text{axial strain}} \right| \qquad (2.25)$$

or

$$\nu = -\frac{\epsilon_y}{\epsilon_x} = -\frac{\epsilon_z}{\epsilon_x} \qquad (2.26)$$

for the loading condition represented in Fig. 2.37. Solving Eq. (2.26) for ϵ_y and ϵ_z, and recalling (2.24), we write the following relations, which fully describe the condition of strain under an axial load applied in a direction parallel to the x axis:

$$\epsilon_x = \frac{\sigma_x}{E} \qquad \epsilon_y = \epsilon_z = -\frac{\nu\sigma_x}{E} \qquad (2.27)$$

---

†It would also be tempting, but equally wrong, to assume that the volume of the rod remains unchanged as a result of the combined effect of the axial elongation and transverse contraction (see Sec. 2.13).

Example 2.07

A 500-mm-long, 16-mm-diameter rod made of a homogeneous, isotropic material is observed to increase in length by 300 μm, and to decrease in diameter by 2.4 μm when subjected to an axial 12-kN load. Determine the modulus of elasticity and Poisson's ratio of the material.

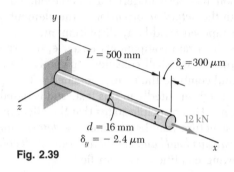

Fig. 2.39

The cross-sectional area of the rod is

$$A = \pi r^2 = \pi(8 \times 10^{-3} \text{ m})^2 = 201 \times 10^{-6} \text{ m}^2$$

Choosing the x axis along the axis of the rod (Fig. 2.39), we write

$$\sigma_x = \frac{P}{A} = \frac{12 \times 10^3 \text{ N}}{201 \times 10^{-6} \text{ m}^2} = 59.7 \text{ MPa}$$

$$\epsilon_x = \frac{\delta_x}{L} = \frac{300 \ \mu\text{m}}{500 \text{ mm}} = 600 \times 10^{-6}$$

$$\epsilon_y = \frac{\delta_y}{d} = \frac{-2.4 \ \mu\text{m}}{16 \text{ mm}} = -150 \times 10^{-6}$$

From Hooke's law, $\sigma_x = E\epsilon_x$, we obtain

$$E = \frac{\sigma_x}{\epsilon_x} = \frac{59.7 \text{ MPa}}{600 \times 10^{-6}} = 99.5 \text{ GPa}$$

and, from Eq. (2.26),

$$\nu = -\frac{\epsilon_y}{\epsilon_x} = -\frac{-150 \times 10^{-6}}{600 \times 10^{-6}} = 0.25$$

2.12. MULTIAXIAL LOADING; GENERALIZED HOOKE'S LAW

All the examples considered so far in this chapter have dealt with slender members subjected to axial loads, i.e., to forces directed along a single axis. Choosing this axis as the x axis, and denoting by P the internal force at a given location, the corresponding stress components were found to be $\sigma_x = P/A$, $\sigma_y = 0$, and $\sigma_z = 0$.

We shall now consider structural elements subjected to loads acting in the directions of the three coordinate axes and producing normal stresses σ_x, σ_y, and σ_z which are all different from zero (Fig. 2.40). This condition is referred to as a *multiaxial loading*. Note that this is not the general stress condition described in Sec. 1.8, since no shearing stresses are included among the stresses shown in Fig. 2.40.

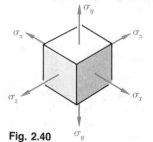

Fig. 2.40

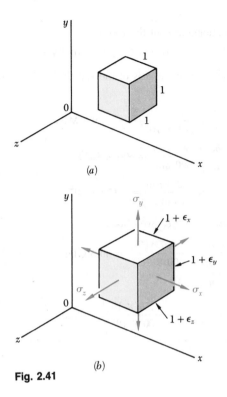

Fig. 2.41

Consider an element of material in the shape of a cube (Fig. 2.41*a*). We may assume the side of the cube to be equal to unity, since it is always possible to select the side of the cube as a unit of length. Under the given multiaxial loading, the element will deform into a *rectangular parallelepiped* of sides equal, respectively, to $1 + \epsilon_x$, $1 + \epsilon_y$, and $1 + \epsilon_z$, where ϵ_x, ϵ_y, and ϵ_z denote the values of the normal strain in the directions of the three coordinate axes (Fig. 2.41*b*). We should note that, as a result of the deformations of the other elements of the material, the element under consideration may also undergo a translation, but we are concerned here only with the *actual deformation* of the element, and not with any possible superimposed rigid-body displacement.

In order to express the strain components ϵ_x, ϵ_y, ϵ_z in terms of the stress components σ_x, σ_y, σ_z, we shall consider separately the effect of each stress component and combine the results obtained. The approach we propose here will be used repeatedly in this text, and is based on the *principle of superposition*. This principle states that the effect of a given combined loading on a structure may be obtained by *determining separately the effects of the various loads and combining the results obtained*, providing that the following conditions are satisfied:

1. Each effect is linearly related to the load which produces it.
2. The deformation resulting from any given load is small and does not affect the conditions of application of the other loads.

In the case of a multiaxial loading, the first condition will be satisfied if the stresses do not exceed the proportional limit of the material, and the second condition will also be satisfied if the stress on any given face does not cause deformations of the other faces, which are large enough to affect the computation of the stresses on those faces.

Considering first the effect of the stress component σ_x, we recall from Sec. 2.11 that σ_x causes a strain equal to σ_x/E in the *x* direction, and strains equal to $-\nu\sigma_x/E$ in each of the *y* and *z* directions. Similarly, the stress component σ_y, if applied separately, will cause a strain σ_y/E in the *y* direction and strains $-\nu\sigma_y/E$ in the other two directions. Finally, the stress component σ_z causes a strain σ_z/E in the *z* direction and strains $-\nu\sigma_z/E$ in the *x* and *y* directions. Combining the results obtained, we conclude that the components of strain corresponding to the given multiaxial loading are

$$
\begin{aligned}
\epsilon_x &= +\frac{\sigma_x}{E} - \frac{\nu\sigma_y}{E} - \frac{\nu\sigma_z}{E} \\[1mm]
\epsilon_y &= -\frac{\nu\sigma_x}{E} + \frac{\sigma_y}{E} - \frac{\nu\sigma_z}{E} \\[1mm]
\epsilon_z &= -\frac{\nu\sigma_x}{E} - \frac{\nu\sigma_y}{E} + \frac{\sigma_z}{E}
\end{aligned}
\qquad (2.28)
$$

The relations (2.28) are referred to as the *generalized Hooke's law for a multiaxial loading.* As we indicated earlier, the results obtained are valid only as long as the stresses do not exceed the proportional limit, and as long as the deformations involved remain small. We also recall that a positive value for a stress component signifies tension, and a negative value compression. Similarly, a positive value for a strain component indicates expansion in the corresponding direction, and a negative value contraction.

Example 2.08

The steel block shown (Fig. 2.42) is subjected to a uniform pressure on all its faces. Knowing that the change in length of edge AB is $-24\,\mu$m, determine (*a*) the change in length of the other two edges, (*b*) the pressure p applied to the faces of the block. Assume $E = 200$ GPa and $\nu = 0.29$

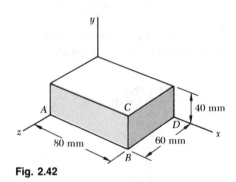

Fig. 2.42

(*a*) *Change in Length of Other Edges.* Substituting $\sigma_x = \sigma_y = \sigma_z = -p$ into the relations (2.28), we find that the three strain components have the common value

$$\epsilon_x = \epsilon_y = \epsilon_z = -\frac{p}{E}(1 - 2\nu) \qquad (2.29)$$

Since

$$\epsilon_x = \delta_x/AB = -24\,\mu\text{m}/80\text{ mm} = -300\,\mu$$

we obtain

$$\epsilon_y = \epsilon_z = \epsilon_x = -300\,\mu$$

from which it follows that

$$\delta_y = \epsilon_y(BC) = (-300\,\mu)(40\text{ mm}) = -12\,\mu\text{m}$$
$$\delta_z = \epsilon_z(BD) = (-300\,\mu)(60\text{ mm}) = -18\,\mu\text{m}$$

(*b*) *Pressure.* Solving Eq. (2.29) for p, we write

$$p = -\frac{E\epsilon_x}{1 - 2\nu} = -\frac{(200\text{ GPa})(-300\,\mu)}{1 - 0.58} = 142.9\text{ MPa}$$

---

*2.13. DILATATION; BULK MODULUS

In this section we shall examine the effect of the normal stresses σ_x, σ_y, and σ_z on the volume of an element of material. Let us consider the element shown in Fig. 2.41. In its unstressed state, it is in the shape of a cube of unit volume; and under the stresses σ_x, σ_y, σ_z, it deforms into a rectangular parallelepiped of volume

$$v = (1 + \epsilon_x)(1 + \epsilon_y)(1 + \epsilon_z)$$

Since the strains ϵ_x, ϵ_y, ϵ_z are much smaller than unity, their products will be even smaller and may be omitted in the expansion of the product. We have, therefore,

$$v = 1 + \epsilon_x + \epsilon_y + \epsilon_z$$

Denoting by e the change in volume of our element, we write

$$e = v - 1 = 1 + \epsilon_x + \epsilon_y + \epsilon_z - 1$$

or

$$e = \epsilon_x + \epsilon_y + \epsilon_z \tag{2.30}$$

Since the element had originally a unit volume, the quantity e represents *the change in volume per unit volume;* it is referred to as the *dilatation* of the material. Substituting for ϵ_x, ϵ_y, and ϵ_z from Eqs. (2.28) into (2.30), we write

$$e = \frac{\sigma_x + \sigma_y + \sigma_z}{E} - \frac{2\nu(\sigma_x + \sigma_y + \sigma_z)}{E}$$

$$e = \frac{1 - 2\nu}{E}(\sigma_x + \sigma_y + \sigma_z) \tag{2.31}\dagger$$

A case of special interest is that of a body subjected to a uniform hydrostatic pressure p. Each of the stress components is then equal to $-p$ and Eq. (2.31) yields

$$e = -\frac{3(1 - 2\nu)}{E}p \tag{2.32}$$

Introducing the constant

$$k = \frac{E}{3(1 - 2\nu)} \tag{2.33}$$

we write Eq. (2.32) in the form

$$e = -\frac{p}{k} \tag{2.34}$$

The constant k is known as the *bulk modulus* or *modulus of compression* of the material. It is expressed in the same units as the modulus of elasticity E, that is, in pascals.

Observation and common sense indicate that a stable material subjected to a hydrostatic pressure can only *decrease* in volume; thus the

---

† Since the dilatation e represents a change in volume, it must be independent of the orientation of the element considered. It then follows from Eqs. (2.30) and (2.31) that the quantities $\epsilon_x + \epsilon_y + \epsilon_z$ and $\sigma_x + \sigma_y + \sigma_z$ are also independent of the orientation of the element. This property will be verified in Chap. 6.

dilatation e in Eq. (2.34) is negative, from which it follows that the bulk modulus k is a positive quantity. Referring to Eq. (2.33), we conclude that $1 - 2\nu > 0$, or $\nu < \frac{1}{2}$. On the other hand, the very definition of Poisson's ratio (Sec. 2.11) requires it to be a positive quantity. We thus conclude that, for any engineering material,

$$0 < \nu < \tfrac{1}{2} \qquad (2.35)$$

We may note that an ideal material having a value of ν equal to zero could be stretched in one direction without any lateral contraction. On the other hand, an ideal material for which $\nu = \frac{1}{2}$, and thus $k = \infty$, would be perfectly incompressible ($e = 0$). Referring to Eq. (2.31) we also note that, since $\nu < \frac{1}{2}$ in the elastic range, stretching an engineering material in one direction, for example in the x direction ($\sigma_x > 0$, $\sigma_y = \sigma_z = 0$), will result in an increase of its volume ($e > 0$).†

Example 2.09

Determine the change in volume ΔV of the steel block shown in Fig. 2.42, when it is subjected to the hydrostatic pressure $p = 180$ MPa. Use $E = 200$ GPa and $\nu = 0.29$.

From Eq. (2.33), we determine the bulk modulus of steel,

$$k = \frac{E}{3(1 - 2\nu)} = \frac{200 \text{ GPa}}{3(1 - 0.58)} = 158.7 \text{ GPa}$$

and, from Eq. (2.34), the dilatation,

$$e = -\frac{p}{k} = -\frac{180 \text{ MPa}}{158.7 \text{ GPa}} = -1.134 \times 10^{-3}$$

Since the volume V of the block in its unstressed state is

$$V = (80 \text{ mm})(40 \text{ mm})(60 \text{ mm}) = 192 \times 10^3 \text{ mm}^3$$

and since e represents the change in volume per unit volume, $e = \Delta V/V$, we have

$$\Delta V = eV = (-1.134 \times 10^{-3})(192 \times 10^3 \text{ mm}^3)$$
$$\Delta V = -218 \text{ mm}^3$$

2.14. SHEARING STRAIN

When we derived in Sec. 2.12 the relations (2.28) between normal stresses and normal strains in a homogeneous isotropic material, we assumed that no shearing stresses were involved. In the more general stress situation represented in Fig. 2.43, shearing stresses τ_{xy}, τ_{yz}, and τ_{zx} will be present (as well, of course, as the corresponding shearing stresses τ_{yx}, τ_{zy}, and τ_{xz}). These stresses have no direct effect on the normal strains and, as long as all the deformations involved remain small, they will not affect the derivation nor the validity of the relations (2.28). The shearing stresses, however, will tend to deform a cubic element of material into an *oblique* parallelepiped.

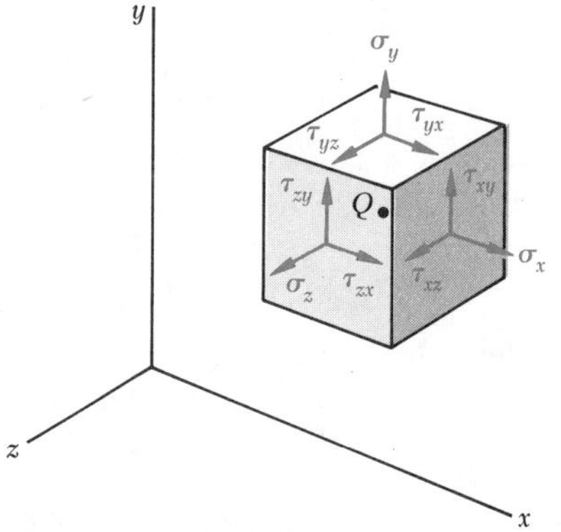

Fig. 2.43

---

† However, in the plastic range, the volume of the material remains nearly constant.

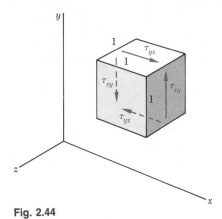

Fig. 2.44

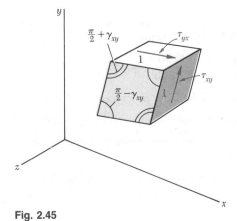

Fig. 2.45

Consider first a cubic element of side one (Fig. 2.44) subjected to no other stresses than the shearing stresses τ_{xy} and τ_{yx} applied to faces of the element respectively perpendicular to the x and y axes. (We recall from Sec. 1.8 that $\tau_{xy} = \tau_{yx}$.) The element is observed to deform into a rhomboid of sides equal to one (Fig. 2.45). Two of the angles formed by the four faces under stress are reduced from $\dfrac{\pi}{2}$ to $\dfrac{\pi}{2} - \gamma_{xy}$, while the other two are increased from $\dfrac{\pi}{2}$ to $\dfrac{\pi}{2} + \gamma_{xy}$. The small angle γ_{xy} (expressed in radians) defines the *shearing strain* corresponding to the x and y directions. When the deformation involves a *reduction* of the angle formed by the two faces oriented respectively toward the positive x and y axes (as shown in Fig. 2.45), the shearing strain γ_{xy} is said to be *positive*; otherwise, it is said to be negative.

We should note that, as a result of the deformations of the other elements of the material, the element under consideration may also undergo an overall rotation. However, as was the case in our study of normal strains, we are concerned here only with the *actual deformation* of the element, and not with any possible superimposed rigid-body displacement.†

Plotting successive values of τ_{xy} against the corresponding values of γ_{xy}, we obtain the shearing stress-strain diagram for the material under consideration. This may be accomplished by carrying out a torsion test, as we shall see in Chap. 3. The diagram obtained is similar to the normal stress-strain diagram obtained for the same material from the tensile test described earlier in this chapter. However, the values obtained for the yield strength, ultimate strength, etc., of a given material are only about half as large in shear as they are in tension. As was the case for normal stresses and strains, the initial portion of the shearing stress-strain diagram is a straight line. For values of the shearing stress which do not

---

†In defining the strain γ_{xy} some authors arbitrarily assume that the actual deformation of the element is accompanied by a rigid-body rotation such that the horizontal faces of the element do not rotate. The strain γ_{xy} is then represented by the angle through which the other two faces have rotated (Fig. 2.46). Others assume a rigid-body rotation such that the horizontal faces rotate through $\frac{1}{2}\gamma_{xy}$ counterclockwise and the vertical faces through $\frac{1}{2}\gamma_{xy}$ clockwise (Fig. 2.47). Since both assumptions are unnecessary and may lead to confusion, we prefer in this text to associate the shearing strain γ_{xy} with the *change in the angle* formed by the two faces, rather than with the *rotation of a given face* under restrictive conditions.

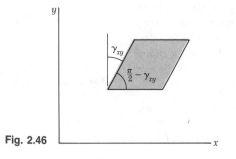

Fig. 2.46

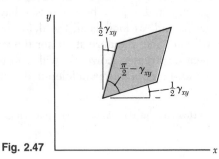

Fig. 2.47

exceed the proportional limit in shear, we may therefore write for any homogeneous isotropic material,

$$\tau_{xy} = G\gamma_{xy} \tag{2.36}$$

This relation is known as *Hooke's law for shearing stress and strain*, and the constant G is called the *modulus of rigidity* or *shear modulus* of the material. Since the strain γ_{xy} was defined as an angle in radians, it is dimensionless, and the modulus G is expressed in the same units as τ_{xy}, that is, in pascals. The modulus of rigidity G of any given material is less than one-half, but more than one-third of the modulus of elasticity E of that material.†

Considering now a small element of material subjected to shearing stresses τ_{yz} and τ_{zy} (Fig. 2.48a), we define the shearing strain γ_{yz} as the change in the angle formed by the faces under stress. The shearing strain γ_{zx} is defined in a similar way by considering an element subjected to shearing stresses τ_{zx} and τ_{xz} (Fig. 2.48b). For values of the stress which do not exceed the proportional limit, we may write the two additional relations

$$\tau_{yz} = G\gamma_{yz} \qquad \tau_{zx} = G\gamma_{zx} \tag{2.37}$$

where the constant G is the same as in Eq. (2.36).

For the general stress condition represented in Fig. 2.43, and as long as none of the stresses involved exceeds the corresponding proportional limit, we may apply the principle of superposition and combine the results obtained in this section and in Sec. 2.12. We obtain the following group of equations representing the generalized Hooke's law for a homogeneous isotropic material under the most general stress condition:

$$\epsilon_x = +\frac{\sigma_x}{E} - \frac{\nu\sigma_y}{E} - \frac{\nu\sigma_z}{E}$$

$$\epsilon_y = -\frac{\nu\sigma_x}{E} + \frac{\sigma_y}{E} - \frac{\nu\sigma_z}{E}$$

$$\epsilon_z = -\frac{\nu\sigma_x}{E} - \frac{\nu\sigma_y}{E} + \frac{\sigma_z}{E} \tag{2.38}$$

$$\gamma_{xy} = \frac{\tau_{xy}}{G} \qquad \gamma_{yz} = \frac{\tau_{yz}}{G} \qquad \gamma_{zx} = \frac{\tau_{zx}}{G}$$

An examination of Eqs. (2.38) might lead us to believe that three distinct constants, E, ν, and G, must first be determined experimentally, if we are to predict the deformations caused in a given material by an arbitrary combination of stresses. Actually, only two of these constants need be determined experimentally for any given material. As we shall see in the next section, the third constant may then be obtained through a very simple computation.

† See Prob. 2.90.

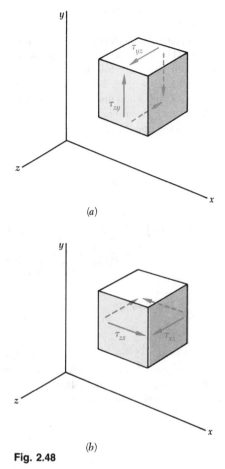

(a)

(b)

Fig. 2.48

Example 2.10

A rectangular block of a material with a modulus of rigidity $G = 600$ MPa is bonded to two rigid horizontal plates. The lower plate is fixed, while the upper plate is subjected to a horizontal force **P** (Fig. 2.49). Knowing that the upper plate moves through 0.8 mm under the action of the force, determine (*a*) the average shearing strain in the material, (*b*) the force **P** exerted on the upper plate.

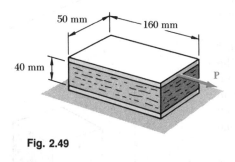

Fig. 2.49

(*a*) *Shearing Strain.* We select coordinate axes centered at the midpoint *C* of edge *AB* and directed as shown (Fig. 2.50). According to its definition, the shearing strain γ_{xy} is equal to the angle formed by the vertical and the line *CF* join-

ing the midpoints of edges *AB* and *DE*. Noting that this is a very small angle and recalling that it should be expressed in radians, we write

$$\gamma_{xy} \approx \tan \gamma_{xy} = \frac{0.8 \text{ mm}}{40 \text{ mm}} \qquad \gamma_{xy} = 0.020 \text{ rad}$$

(*b*) *Force Exerted on Upper Plate.* We first determine the shearing stress τ_{xy} in the material. Using Hooke's law for shearing stress and strain, we have

$$\tau_{xy} = G\gamma_{xy} = (600 \text{ MPa})(0.020) = 12 \text{ MPa}$$

The force exerted on the upper plate is thus

$$P = \tau_{xy} A = (12 \times 10^6 \text{ Pa})(0.160 \text{ m} \times 0.050 \text{ m})$$
$$P = 96 \text{ kN}$$

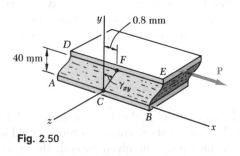

Fig. 2.50

2.15. FURTHER DISCUSSION OF DEFORMATIONS UNDER AXIAL LOADING; RELATION AMONG *E*, ν, AND *G*

We saw in Sec. 2.11 that a slender bar subjected to an axial tensile load **P** directed along the *x* axis will elongate in the *x* direction and contract in both of the transverse *y* and *z* directions. If ϵ_x denotes the axial strain, the lateral strain is expressed as $\epsilon_y = \epsilon_z = -\nu\epsilon_x$, where ν is Poisson's ratio. Thus, an element in the shape of a cube of side equal to one and oriented as shown in Fig. 2.51*a* will deform into a rectangular parallelepiped of sides $1 + \epsilon_x$, $1 - \nu\epsilon_x$, and $1 - \nu\epsilon_x$. (Note that only one face of the element is shown in the figure.) On the other hand, if the element is oriented at 45° to the axis of the load (Fig. 2.51*b*), the face shown in the figure is observed to deform into a rhombus. We conclude that the axial load **P** causes in this element a shearing strain γ' equal to the amount by which each of the angles shown in Fig. 2.51*b* increases or decreases.†

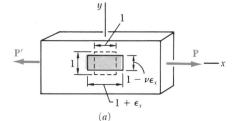

(*a*)

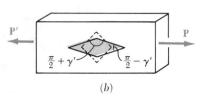

(*b*)

Fig. 2.51

† Note that the load **P** also produces normal strains in the element shown in Fig. 2.51*b* (see Prob. 2.74).

The fact that shearing strains, as well as normal strains, result from an axial loading should not come to us as a surprise, since we already observed at the end of Sec. 1.8 that an axial load **P** causes normal and shearing stresses of equal magnitude on four of the faces of an element oriented at 45° to the axis of the member. This was illustrated in Fig. 1.36 which, for convenience, has been repeated here. It was also shown in Sec. 1.7 that the shearing stress is maximum on a plane forming an angle of 45° with the axis of the load. It follows from Hooke's law for shearing stress and strain that the shearing strain γ' associated with the element of Fig. 2.51b is also maximum: $\gamma' = \gamma_m$.

While a more detailed study of the transformations of strain will be postponed until Chap. 6, we shall derive in this section a relation between the maximum shearing strain $\gamma' = \gamma_m$ associated with the element of Fig. 2.51b and the normal strain ϵ_x in the direction of the load. We shall consider for this purpose the prismatic element obtained by intersecting the cubic element of Fig. 2.51a by a diagonal plane (Fig. 2.52a and b). Referring to Fig. 2.51a, we conclude that this new element will deform into the element shown in Fig. 2.52c, which has horizontal and vertical sides respectively equal to $1 + \epsilon_x$ and $1 - \nu\epsilon_x$. But the angle formed by the oblique and horizontal faces of the element of Fig. 2.52b is precisely half of one of the right angles of the cubic element considered in Fig. 2.51b. The angle β into which this angle deforms must therefore be

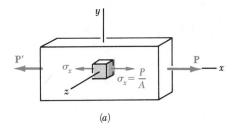

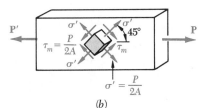

(a)

(b)

Fig. 1.36 (repeated)

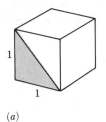

(a)

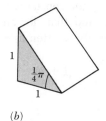

(b)

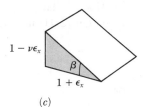

(c)

Fig. 2.52

equal to half of $\dfrac{\pi}{2} - \gamma_m$. We write

$$\beta = \frac{\pi}{4} - \frac{\gamma_m}{2}$$

Applying the formula for the tangent of the difference of two angles, we obtain

$$\tan \beta = \frac{\tan \dfrac{\pi}{4} - \tan \dfrac{\gamma_m}{2}}{1 + \tan \dfrac{\pi}{4} \tan \dfrac{\gamma_m}{2}} = \frac{1 - \tan \dfrac{\gamma_m}{2}}{1 + \tan \dfrac{\gamma_m}{2}}$$

or, since $\gamma_m/2$ is a very small angle,

$$\tan \beta = \frac{1 - \dfrac{\gamma_m}{2}}{1 + \dfrac{\gamma_m}{2}} \qquad (2.39)$$

But, from Fig. 2.52c, we observe that

$$\tan \beta = \frac{1 - \nu\epsilon_x}{1 + \epsilon_x} \qquad (2.40)$$

Equating the right-hand members of (2.39) and (2.40), and solving for γ_m, we write

$$\gamma_m = \frac{(1 + \nu)\epsilon_x}{1 + \dfrac{1 - \nu}{2}\epsilon_x}$$

Since $\epsilon_x \ll 1$, the denominator in the expression obtained may be assumed equal to one; we have, therefore,

$$\gamma_m = (1 + \nu)\epsilon_x \qquad (2.41)$$

which is the desired relation between the maximum shearing strain γ_m and the axial strain ϵ_x.

To obtain a relation among the constants E, ν, and G, we recall that, by Hooke's law, $\gamma_m = \tau_m/G$, and that, for an axial loading, $\epsilon_x = \sigma_x/E$. Equation (2.41) may therefore be written as

$$\frac{\tau_m}{G} = (1 + \nu)\frac{\sigma_x}{E}$$

or

$$\frac{E}{G} = (1 + \nu)\frac{\sigma_x}{\tau_m} \qquad (2.42)$$

We now recall from Fig. 1.36 that $\sigma_x = P/A$ and $\tau_m = P/2A$, where A is the cross-sectional area of the member. It thus follows that $\sigma_x/\tau_m = 2$. Substituting this value into (2.42) and dividing both members by 2, we obtain the relation

$$\frac{E}{2G} = 1 + \nu \qquad (2.43)$$

which may be used to determine one of the constants E, ν, or G from the other two. For example, solving Eq. (2.43) for G, we write

$$G = \frac{E}{2(1 + \nu)} \qquad (2.43')$$

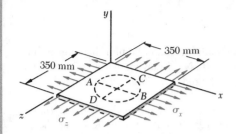

A circle of diameter $d = 200$ mm is scribed on an unstressed aluminum plate of thickness $t = 18$ mm. Forces acting in the plane of the plate later cause normal stresses $\sigma_x = 85$ MPa and $\sigma_z = 150$ MPa. For $E = 70$ GPa and $\nu = \frac{1}{3}$, determine the change in (a) the length of diameter AB, (b) the length of diameter CD, (c) the thickness of the plate, (d) the volume of the plate.

Hooke's Law. We note that $\sigma_y = 0$. Using Eqs. (2.28) we find the strain in each of the coordinate directions.

$$\epsilon_x = +\frac{\sigma_x}{E} - \frac{\nu\sigma_y}{E} - \frac{\nu\sigma_z}{E}$$

$$= \frac{1}{70\ \text{GPa}}\left[(85\ \text{MPa}) - 0 - \frac{1}{3}(150\ \text{MPa})\right] = +0.500 \times 10^{-3}$$

$$\epsilon_y = -\frac{\nu\sigma_x}{E} + \frac{\sigma_y}{E} - \frac{\nu\sigma_z}{E}$$

$$= \frac{1}{70\ \text{GPa}}\left[-\frac{1}{3}(85\ \text{MPa}) + 0 - \frac{1}{3}(150\ \text{MPa})\right] = -1.119 \times 10^{-3}$$

$$\epsilon_z = -\frac{\nu\sigma_x}{E} - \frac{\nu\sigma_y}{E} + \frac{\sigma_z}{E}$$

$$= \frac{1}{70\ \text{GPa}}\left[-\frac{1}{3}(85\ \text{MPa}) - 0 + (150\ \text{MPa})\right] = +1.738 \times 10^{-3}$$

a. **Diameter AB.** The change in length is $\delta_{B/A} = \epsilon_x d$.

$$\delta_{B/A} = \epsilon_x d = (+0.500 \times 10^{-3})(200\ \text{mm})$$

$$\delta_{B/A} = +100\ \mu\text{m} \quad \blacktriangleleft$$

b. **Diameter CD.**

$$\delta_{C/D} = \epsilon_z d = (+1.738 \times 10^{-3})(200\ \text{mm})$$

$$\delta_{C/D} = +348\ \mu\text{m} \quad \blacktriangleleft$$

c. **Thickness.** Recalling that $t = 18$ mm, we have

$$\delta_t = \epsilon_y t = (-1.119 \times 10^{-3})(18\ \text{mm})$$

$$\delta_t = -20.1\ \mu\text{m} \quad \blacktriangleleft$$

d. **Volume of the Plate.** Using Eq. (2.30), we write

$$e = \epsilon_x + \epsilon_y + \epsilon_z = (+0.500 - 1.119 + 1.738)10^{-3} = +1.119 \times 10^{-3}$$

$$\Delta V = eV = +1.119 \times 10^{-3}[(0.35\ \text{m})(0.35\ \text{m})(0.018\ \text{m})] = +2.47 \times 10^{-6}\ \text{m}^3$$

$$\Delta V = +2470\ \text{mm}^3 \quad \blacktriangleleft$$

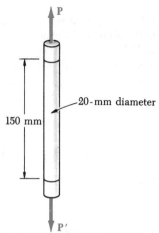

Fig. P2.66 and P2.67

PROBLEMS

2.66 In a standard tensile test an aluminum rod of 20-mm diameter is subjected to a tensile force of magnitude $P = 30$ kN. Knowing that $E = 70$ GPa and $\nu = 0.35$, determine (*a*) the elongation of the rod in a 150-mm gage length, (*b*) the change in diameter of the rod.

2.67 In a standard tensile test a 20-mm-diameter rod made of an experimental plastic is subjected to a tensile force of magnitude $P = 6$ kN. Knowing that an elongation of 14 mm and a decrease in diameter of 0.85 mm are observed in a 150-mm gage length, determine the modulus of elasticity, the modulus of rigidity, and Poisson's ratio for the material.

2.68 A line of slope 4:10 was scribed on a yellow-brass plate, 150 mm wide and 6 mm thick. Using the data available in Appendix B, determine the slope of the line when the plate is subjected as shown to a 200-kN centric axial load.

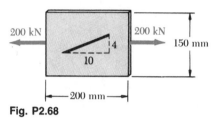

Fig. P2.68

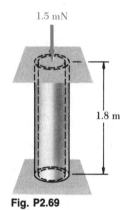

Fig. P2.69

2.69 A 1.8-m length of steel pipe of 300 mm outer diameter and 12-mm wall thickness is used as a short column to carry a 1.5-mN centric axial load. Using the data available in Appendix B for structural steel, determine (*a*) the change in length of the pipe, (*b*) the change in the outer diameter of the pipe, (*c*) the change in wall thickness of the pipe.

2.70 A 20-mm square was scribed on the side of a large steel pressure vessel. After pressurization, the biaxial stress condition of the square is as shown. Using the data available in Appendix B for structural steel, determine the change in length of (*a*) side AB, (*b*) side BC, (*c*) diagonal AC.

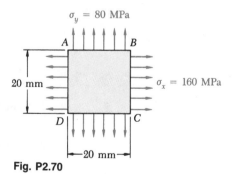

Fig. P2.70

2.71 For the square of Prob. 2.70, determine the percent change in the slope of diagonal DB due to the pressurization of the vessel.

2.72 The aluminum rod *AD* is fitted with a jacket which is used to apply a hydrostatic pressure of 40 MPa to the 300 mm portion *BC* of the rod. Knowing that *E* = 70 GPa and *v* = 0.36, determine (*a*) the change in the total length *AD* of the rod, (*b*) the change in the diameter at the middle of the rod.

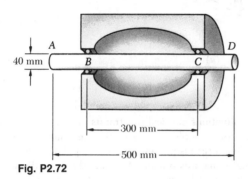

40 mm

A
B
C
D

300 mm

500 mm

Fig. P2.72

2.73 For the rod of Prob. 2.72, determine the forces which should be applied to the ends *A* and *D* of the rod (*a*) if the axial strain in portion *BC* of the rod is to remain zero as the hydrostatic pressure is applied, (*b*) if the total length *AD* of the rod is to remain unchanged.

2.74 For a member under axial loading, express the normal strain ϵ' in a direction forming an angle of 45° with the axis of the load in terms of the axial strain ϵ_x by (*a*) comparing the hypotenuses of the triangles shown in Fig. 2.52, which represent respectively an element before and after deformation, (*b*) using the values of the corresponding stresses σ' and σ_x shown in Fig. 1.36, and the generalized Hooke's law.

2.75 A line of slope *b*:*a*, where *a* and *b* are both different from zero, was scribed on a plate when it was unstressed. The plate is then subjected to the biaxial state of stress shown. (*a*) Show that within the elastic range and for any engineering material, the percent change in the slope of the line may be expressed as $100(\sigma_y - \sigma_x)/2G$. (*Hint:* Use an expansion in series, neglecting higher-order terms in σ_x/E and σ_y/E.) (*b*) Use the result obtained in part *a* to solve Probs. 2.68 and 2.71.

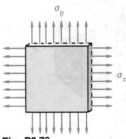

Fig. P2.75

2.76 In many situations it is known that the normal stress in a given direction is zero, for example, $\sigma_z = 0$ in the case of the thin plate shown. For this case, which is known as *plane stress*, show that if the strains ϵ_x and ϵ_y have been determined experimentally, we may express σ_x, σ_y, and ϵ_z as follows:

$$\sigma_x = E \frac{\epsilon_x + v\epsilon_y}{1 - v^2}$$

$$\sigma_y = E \frac{\epsilon_y + v\epsilon_x}{1 - v^2}$$

$$\epsilon_z = -\frac{v}{1 - v}(\epsilon_x + \epsilon_y)$$

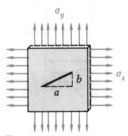

Fig. P2.76

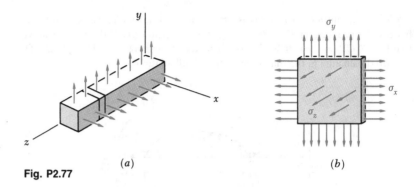

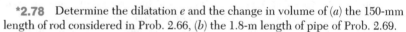

(a)　　　　　　　　　　　　　　　　(b)

Fig. P2.77

2.77 In many situations physical constraints prevent strain from occurring in a given direction, for example, $\epsilon_z = 0$ in the case shown, where longitudinal movement of the long prism is prevented at every point. Plane sections perpendicular to the longitudinal axis remain plane and the same distance apart. Show that, for this situation, which is known as *plane strain*, we may express σ_z, ϵ_x, and ϵ_y as follows:

$$\sigma_z = \nu\,(\sigma_x + \sigma_y)$$

$$\epsilon_x = \frac{1}{E}[(1 - \nu^2)\sigma_x - \nu\,(1 + \nu)\sigma_y]$$

$$\epsilon_y = \frac{1}{E}[(1 - \nu^2)\sigma_y - \nu\,(1 + \nu)\sigma_x]$$

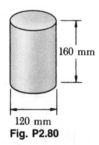

160 mm

120 mm

Fig. P2.80

*2.78 Determine the dilatation e and the change in volume of (a) the 150-mm length of rod considered in Prob. 2.66, (b) the 1.8-m length of pipe of Prob. 2.69.

*2.79 Determine the change in volume of the rod of Prob. 2.72 (a) by computing the dilatation of the material, (b) by subtracting the original volume of portion BC from its final volume.

*2.80 A cylindrical block of brass, 160 mm high and 120 mm in diameter, is lowered into the ocean to a depth where the pressure is 75 MPa (about 7500 m below the surface). Knowing that $E = 105$ GPa and $\nu = 0.35$, determine (a) the change in height of the block, (b) its change in diameter, (c) its change in volume.

*2.81 For the block of Prob. 2.80, determine the pressure which should be applied (a) to its top and bottom faces only, (b) to its cylindrical surface only, to cause the same change in volume as the hydrostatic pressure defined in Prob. 2.80. In each case, also find the corresponding changes in the height and diameter of the block.

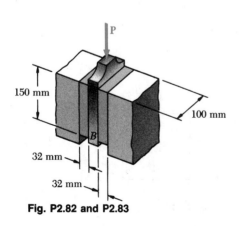

150 mm

100 mm

32 mm

32 mm

Fig. P2.82 and P2.83

2.82 A vibration isolation unit consists of two blocks of hard rubber bonded to a plate AB and to rigid supports as shown. Knowing that a force of magnitude $P = 25$ kN causes a deflection $\delta = 1.5$ mm of plate AB, determine the modulus of rigidity of the rubber used.

2.83 A vibration isolation unit consists of two blocks of hard rubber with a modulus of rigidity $G = 19$ MPa bonded to a plate AB and to rigid supports as shown. Denoting by P the magnitude of the force applied to the plate and by δ the corresponding deflection, determine the effective spring constant P/δ of the system.

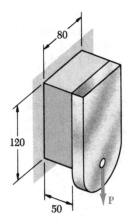

Dimensions in mm
Fig. P2.84

2.84 The plastic block shown is bonded to a rigid support and to a vertical plate to which a 240-kN load **P** is applied. Knowing that for the plastic used $G = 1050$ MPa, determine the deflection of the plate.

2.85 What load **P** should be applied to the plate of Prob. 2.84 to produce a 1.5-mm deflection?

2.86 Two blocks of rubber with a modulus of rigidity $G = 12$ MPa are bonded to rigid supports and to a plate AB. Knowing that $c = 100$ mm and $P = 45$ kN, determine the smallest allowable dimensions a and b of the blocks if the shearing stress in the rubber is not to exceed 1.4 MPa and the deflection of the plate is to be at least 5 mm.

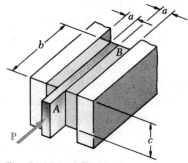

Fig. P2.86 and P2.87

2.87 Two blocks of rubber with a modulus of rigidity $G = 10$ MPa are bonded to rigid supports and to a plate AB. Knowing that $b = 200$ mm and $c = 125$ mm, determine the largest allowable load P and the smallest allowable thickness a of the blocks if the shearing stress in the rubber is not to exceed 1.5 MPa and the deflection of the plate is to be at least 6 mm.

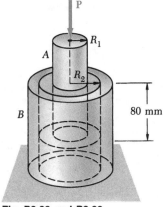

Fig. P2.88 and P2.89

*2.88 A vibration isolation support consists of a rod A of radius $R_1 = 10$ mm and a tube B of inner radius $R_2 = 25$ mm bonded to an 80-mm-long hollow rubber cylinder with a modulus of rigidity $G = 12$ MPa. Determine the largest allowable force **P** which may be applied to rod A if its deflection is not to exceed 2.50 mm.

*2.89 A vibration isolation support consists of a rod A of radius R_1 and a tube B of inner radius R_2 bonded to an 80-mm-long hollow rubber cylinder with a modulus of rigidity $G = 10.93$ MPa. Determine the required value of the ratio R_2/R_1 if a 10-kN force **P** is to cause a 2-mm deflection of rod A.

*2.90 Show that for any given material, the ratio G/E of the modulus of rigidity over the modulus of elasticity is always less than $\frac{1}{2}$ but more than $\frac{1}{3}$. [*Hint:* Refer to Eq. (2.43) and to Sec. 2.13.]

*2.91 The material constants E, G, k, and ν are related by Eqs. (2.33) and (2.43). Show that any one of these constants may be expressed in terms of any other two constants. For example, show that (*a*) $k = GE/(9G - 3E)$ and (*b*) $\nu = (3k - 2G)/(6k + 2G)$.

2.16. STRESS AND STRAIN DISTRIBUTION UNDER AXIAL LOADING; SAINT-VENANT'S PRINCIPLE

We have assumed so far that, in an axially loaded member, the normal stresses are uniformly distributed in any section perpendicular to the axis of the member. As we saw in Sec. 1.3, such an assumption may be quite in error in the immediate vicinity of the points of application of the loads. However, the determination of the actual stresses in a given section of the member requires the solution of a statically indeterminate problem.

In Sec. 2.9, we saw that statically indeterminate problems involving the determination of *forces* may be solved by considering the *deformations* caused by these forces. It is thus reasonable to conclude that the determination of the *stresses* in a member requires the analysis of the *strains* produced by the stresses in the member. This is essentially the approach found in advanced textbooks, where the mathematical theory of elasticity is used to determine the distribution of stresses corresponding to various modes of application of the loads at the ends of the member. Given the more limited mathematical means at our disposal, we shall be satisfied with determining the distribution of stresses in the particular case when two rigid plates are used to transmit the loads to the member (Fig. 2.53).

Fig. 2.53

If the loads are applied at the center of each plate,† the plates will move toward each other without rotating, causing the member to get shorter, while increasing in width and thickness. It is reasonable to assume that the member will remain straight, that plane sections will remain plane, and that all elements of the member will deform in the same way, since such an assumption is clearly compatible with the given end conditions. This is illustrated in Fig. 2.54, which shows a rubber model before and after loading.‡ Now, if all elements deform in the same way, the distribution of strains throughout the member must be uniform. In other words, the axial strain ϵ_y and the lateral strain $\epsilon_x = -\nu\epsilon_y$ are constant. But, if the stresses do not exceed the proportional limit, Hooke's law applies and we may write $\sigma_y = E\epsilon_y$, from which it follows that the normal stress σ_y is also constant. Thus, the distribution of stresses is uniform throughout the member and, at any point,

$$\sigma_y = (\sigma_y)_{\text{ave}} = \frac{P}{A}$$

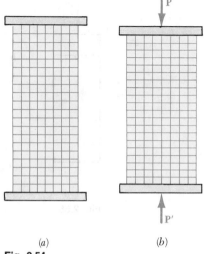

(a)   (b)

Fig. 2.54

On the other hand, if the loads are concentrated, as illustrated in Fig. 2.55, the elements in the immediate vicinity of the points of application of the loads are subjected to very large stresses, while other elements near the ends of the member are unaffected by the loading. This may be verified by observing that strong deformations, and thus large strains and large stresses, occur near the points of application of the loads, while no deformation takes place at the corners. As we consider elements farther and farther from the ends, however, we note a progressive equalization of the deformations involved, and thus a more nearly uniform distribution of the strains and stresses across a section of the member. This is further illustrated in Fig. 2.56, which shows the result of the calculation by advanced mathematical methods of the distribution of stresses across various sections of a thin rectangular plate subjected to concentrated loads. We note that at a distance b from either end, where b is the width of the plate, the stress distribution is nearly uniform across the section, and the value of the stress σ_y at any point of that section may be assumed equal to the average value P/A. Thus, at a distance equal to, or greater than, the width of the member, the distribution of stresses across a given section is the same, whether the member is loaded as shown in Fig. 2.53 or Fig. 2.55. In other words, except in the immediate vicinity of the points of

Fig. 2.55

---

† More precisely, the common line of action of the loads should pass through the centroid of the cross section (cf. Sec. 1.3).

‡ Note that for long, slender members, another configuration is possible, and indeed will prevail, if the load is sufficiently large; the member *buckles* and assumes a curved shape. This will be discussed in Chap. 11.

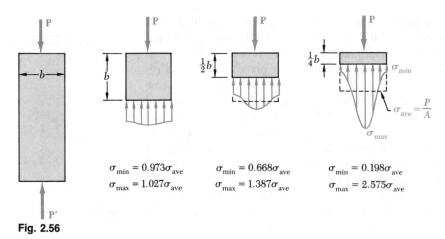

$$\sigma_{min} = 0.973\sigma_{ave}$$
$$\sigma_{max} = 1.027\sigma_{ave}$$

$$\sigma_{min} = 0.668\sigma_{ave}$$
$$\sigma_{max} = 1.387\sigma_{ave}$$

$$\sigma_{min} = 0.198\sigma_{ave}$$
$$\sigma_{max} = 2.575\sigma_{ave}$$

Fig. 2.56

application of the loads, the stress distribution may be assumed independent of the actual mode of application of the loads. This statement, which applies not only to axial loadings, but to practically any type of load, is known as *Saint-Venant's principle*, after the French mathematician and engineer Adhémar Barré de Saint-Venant (1797–1886).

While Saint-Venant's principle makes it possible to replace a given loading by a simpler one for the purpose of computing the stresses in a structural member, we should keep in mind two important points when applying this principle:

1. The actual loading and the loading used to compute the stresses must be *statically equivalent*.
2. Stresses cannot be computed in this manner in the immediate vicinity of the points of application of the loads. Advanced theoretical or experimental methods must be used to determine the distribution of stresses in these areas.

We should also observe that the plates used to obtain a uniform stress distribution in the member of Fig. 2.54 must allow the member to freely expand laterally. Thus, the plates cannot be rigidly attached to the member; we must assume them to be just in contact with the member, and smooth enough not to impede the lateral expansion of the member. While such end conditions can actually be achieved for a member in compression, they cannot be physically realized in the case of a member in tension. It does not matter, however, whether or not an actual fixture can be made and used to load a member so that the distribution of stresses in the member is uniform. The important thing for us is to be able to *imagine a model* which will allow such a distribution of stresses, and to keep this model in mind so that we may later compare it with the actual loading conditions encountered in practice.

2.17. STRESS CONCENTRATIONS

As we saw in the preceding section, the stresses near the points of application of concentrated loads may reach values much larger than the average value of the stress in the member. When a structural member contains a discontinuity, such as a hole or a sudden change in cross section, high localized stresses may also occur near the discontinuity. Figures 2.57 and 2.58 show the distribution of stresses in critical sections corresponding to two such situations. Figure 2.57 refers to a flat bar with a *circular hole* and shows the stress distribution in a section passing through the center of the hole. Figure 2.58 refers to a flat bar consisting of two portions of different widths connected by *fillets;* it shows the stress distribution in the narrowest part of the connection, where the highest stresses occur.

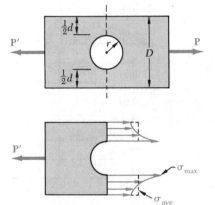

Fig. 2.57 Stress distribution near circular hole in flat bar under axial loading.

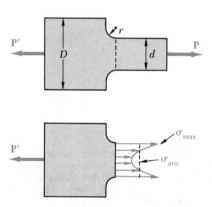

Fig. 2.58 Stress distribution near fillets in flat bar under axial loading.

These results were obtained experimentally through the use of a photoelastic method. Fortunately for the engineer who has to design a given member and cannot afford to carry out such an analysis, the results obtained are independent of the size of the member and of the material used; they depend only upon the ratios of the geometric parameters involved, i.e., upon the ratio r/d in the case of a circular hole, and upon the ratios r/d and D/d in the case of fillets. Furthermore, the designer is more interested in the *maximum value* of the stress in a given section, than in the actual distribution of stresses in that section, since his main concern is to determine *whether* the allowable stress will be exceeded under a given loading, and not *where* this value will be exceeded. For this reason, one defines the ratio

$$K = \frac{\sigma_{\max}}{\sigma_{\text{ave}}} \tag{2.44}$$

of the maximum stress over the average stress computed in the critical (narrowest) section of the discontinuity. This ratio is referred to as the stress-concentration *factor* of the given discontinuity. Stress-concentration factors may be computed once and for all in terms of the ratios of the geometric parameters involved, and the results obtained may be expressed in the form of tables or of graphs, as shown in Fig. 2.59. To determine the maximum stress occurring near a discontinuity in a given member subjected to a given axial load P, the designer needs only to compute the average stress $\sigma_{ave} = P/A$ in the critical section, and multiply the result obtained by the appropriate value of the stress-concentration factor K. We should note, however, that this procedure is valid only as long as σ_{max} does not exceed the proportional limit of the material, since the values of K plotted in Fig. 2.59 were obtained by assuming a linear relation between stress and strain.

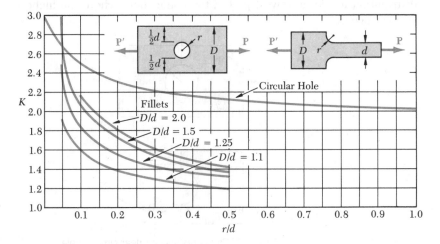

Fig. 2.59 Stress concentration factors for *flat bars* under axial loading.† Note that the average stress must be computed across the narrowest section: $\sigma_{ave} = P/td$, where t is the thickness of the plate.

Example 2.11

Determine the largest axial load **P** which may be safely supported by a flat steel bar consisting of two portions, both 10 mm thick, and respectively 40 and 60 mm wide, connected by fillets of radius $r = 8$ mm. Assume an allowable normal stress of 165 MPa.

We first compute the ratios

$$\frac{D}{d} = \frac{60 \text{ mm}}{40 \text{ mm}} = 1.50 \qquad \frac{r}{d} = \frac{8 \text{ mm}}{40 \text{ mm}} = 0.20$$

Using the curve in Fig. 2.59 corresponding to $D/d = 1.50$, we find that the value of the stress-concentration factor corresponding to $r/d = 0.20$ is

$$K = 1.72$$

Carrying this value into Eq. (2.44) and solving for σ_{ave}, we have

$$\sigma_{ave} = \frac{\sigma_{max}}{1.72}$$

But σ_{max} cannot exceed the allowable stress $\sigma_{all} = 165$ MPa. Substituting this value for σ_{max}, we find that the average stress in the narrower portion ($d = 40$ mm) of the bar should not exceed the value

$$\sigma_{ave} = \frac{165 \text{ MPa}}{1.72} = 96 \text{ MPa}$$

Recalling that $\sigma_{ave} = P/A$, we have

$$P = A\sigma_{ave} = (40 \text{ mm})(10 \text{ mm})(96 \text{ MPa}) = 38.4 \times 10^3 \text{ N}$$

$$P = 38.4 \text{ kN}$$

†M. M. Frocht, "Photoelastic Studies in Stress Concentration," *Mechanical Engineering*, August 1936, pp. 485–489.

2.18. PLASTIC DEFORMATIONS

The results obtained in the preceding sections were based on the assumption of a linear stress-strain relationship. In other words, we assumed that the proportional limit of the material was never exceeded. This is a reasonable assumption in the case of brittle materials, which rupture without yielding. In the case of ductile materials, however, this assumption implies that the yield strength of the material is not exceeded. The deformations will then remain within the elastic range and the structural member under consideration will regain its original shape after all loads have been removed. If, on the other hand, the stresses in any part of the member exceed the yield strength of the material, plastic deformations occur and most of the results obtained in earlier sections cease to be valid. A more involved analysis, based on a nonlinear stress-strain relationship must then be carried out.

While an analysis taking into account the actual stress-strain relationship is beyond the scope of this text, we shall be able to gain considerable insight into plastic behavior by considering an idealized *elastoplastic material* for which the stress-strain diagram consists of the two straight-line segments shown in Fig. 2.60. We may note that the stress-strain diagram for mild steel in the elastic and plastic ranges is similar to this idealization. As long as the stress σ is less than the yield strength σ_Y, the material behaves elastically and obeys Hooke's law, $\sigma = E\epsilon$. When σ reaches the value σ_Y, the material starts yielding and keeps deforming plastically under a constant load. If the load is removed, unloading takes place along a straight-line segment CD parallel to the initial portion AY of the loading curve. The segment AD of the horizontal axis represents the strain corresponding to the permanent set or plastic deformation resulting from the loading and unloading of the specimen. While no actual material behaves exactly as shown in Fig. 2.60, this stress-strain diagram will prove useful in discussing the plastic deformations of ductile materials such as mild steel.

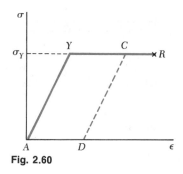

Fig. 2.60

Example 2.12

A rod of length $L = 500$ mm and cross-sectional area $A = 60$ mm^2 is made of an elastoplastic material having a modulus of elasticity $E = 200$ GPa in its elastic range and a yield point $\sigma_Y = 300$ MPa. The rod is subjected to an axial load until it is stretched 7 mm and the load is then removed. What is the resulting permanent set?

Referring to the diagram of Fig. 2.60, we find that the maximum strain, represented by the abscissa of point C, is

$$\epsilon_C = \frac{\delta_C}{L} = \frac{7 \text{ mm}}{500 \text{ mm}} = 14 \times 10^{-3}$$

On the other hand, the yield strain, represented by the abscissa of point Y, is

$$\epsilon_Y = \frac{\sigma_Y}{E} = \frac{300 \times 10^6 \text{ Pa}}{200 \times 10^9 \text{ Pa}} = 1.5 \times 10^{-3}$$

The strain after unloading is represented by the abscissa ϵ_D of point D. We note from Fig. 2.60 that

$$\epsilon_D = AD = YC = \epsilon_C - \epsilon_Y$$
$$= 14 \times 10^{-3} - 1.5 \times 10^{-3} = 12.5 \times 10^{-3}$$

The permanent set is the deformation δ_D corresponding to the strain ϵ_D. We have

$$\delta_D = \epsilon_D L = (12.5 \times 10^{-3})(500 \text{ mm}) = 6.25 \text{ mm}$$

Example 2.13

An 800-mm long cylindrical rod of cross-sectional area $A_r = 45$ mm^2 is placed inside a tube of the same length and of cross-sectional area $A_t = 60$ mm^2. The ends of the rod and tube are attached to a rigid support on one side, and to a rigid plate on the other, as shown in the longitudinal section of Fig. 2.61. The rod and tube are both assumed to be elastoplastic, with moduli of elasticity $E_r = 200$ GPa and $E_t = 100$ GPa, and yield strengths $(\sigma_r)_Y = 200$ MPa and $(\sigma_t)_Y = 250$ MPa. Draw the load-deflection diagram of the rod-tube assembly when a load **P** is applied to the plate as shown.

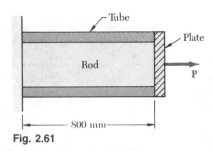

Fig. 2.61

We first determine the internal force and the elongation of the rod as it begins to yield:

$$(P_r)_Y = (\sigma_r)_Y A_r = (200 \text{ MPa})(45 \text{ mm}^2) = 9 \text{ kN}$$

$$(\delta_r)_Y = (\epsilon_r)_Y L = \frac{(\sigma_r)_Y}{E_r} L = \frac{200 \text{ MPa}}{200 \text{ GPa}} (800 \text{ mm}) = 0.8 \text{ mm}$$

Since the material is elastoplastic, the force-elongation diagram *of the rod alone* consists of an oblique straight line and of a horizontal straight line, as shown in Fig. 2.62a. Following the same procedure for the tube, we have

$$(P_t)_Y = (\sigma_t)_Y A_t = (250 \text{ MPa})(60 \text{ mm}^2) = 15 \text{ kN}$$

$$(\delta_t)_Y = (\epsilon_t)_Y L = \frac{(\sigma_t)_Y}{E_t} L = \frac{250 \text{ MPa}}{100 \text{ GPa}} (800 \text{ mm}) = 2 \text{ mm}$$

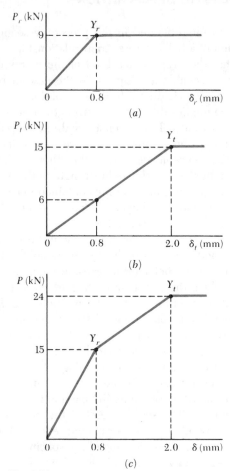

Fig. 2.62

The load-deflection diagram of the tube alone is shown in Fig. 2.62b. Observing that the load and deflection of the rod-tube combination are, respectively,

$$P = P_r + P_t \qquad \delta = \delta_r = \delta_t$$

we draw the required load-deflection diagram by adding the ordinates of the diagrams obtained for the rod and for the tube (Fig. 2.62c). Points Y_r and Y_t correspond to the onset of yield in the rod and in the tube, respectively.

Example 2.14

If the load **P** applied to the rod-tube assembly of Example 2.13 is increased from zero to 19.5 kN and decreased back to zero, determine (a) the maximum elongation of the assembly, (b) the permanent set after the load has been removed.

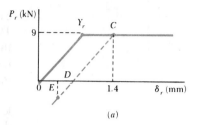

(a)

(a) *Maximum Elongation.* Referring to Fig. 2.62c, we observe that the load $P_{max} = 19.5$ kN corresponds to a point located on the segment $Y_r Y_t$ of the load-deflection diagram of the assembly. Thus, the rod has reached the plastic range, with $P_r = (P_r)_Y = 9$ kN and $\sigma_r = (\sigma_r)_Y = 200$ MPa, while the tube is still in the elastic range, with

$$P_t = P - P_r = 19.5 \text{ kN} - 9 \text{ kN} = 10.5 \text{ kN}$$

$$\sigma_t = \frac{P_t}{A_t} = \frac{10.5 \text{ kN}}{60 \text{ mm}^2} = 175 \text{ MPa}$$

$$\delta_t = \epsilon_t L = \frac{\sigma_t}{E_t} L = \frac{175 \text{ MPa}}{100 \text{ GPa}} (800 \text{ mm}) = 1.40 \text{ mm}$$

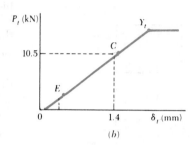

(b)

The maximum elongation of the assembly, therefore, is

$$\delta_{max} = \delta_t = 1.40 \text{ mm}$$

(b) *Permanent Set.* As the load **P** decreases from 19.5 kN to zero, the internal forces P_r and P_t both decrease along a straight line, as shown in Fig. 2.63a and b, respectively. The force P_r decreases along line CD parallel to the initial portion of the loading curve, while the force P_t decreases along the original loading curve, since the yield stress was not exceeded in the tube. Their sum P, therefore, will decrease along a line CE parallel to the portion $0Y_r$ of the load-deflection curve of the assembly (Fig. 2.63c). Referring to Fig. 2.62c, we find that the slope of $0Y_r$, and thus of CE, is

$$m = \frac{15 \text{ kN}}{0.8 \text{ mm}} = 18.75 \text{ kN/mm}$$

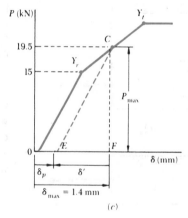

(c)

Fig. 2.63

The segment of line FE in Fig. 2.63c represents the deformation δ' of the assembly during the unloading phase, and the segment $0E$ the permanent set δ_p after the load **P** has been removed. From triangle CEF we have

$$\delta' = -\frac{P_{max}}{m} = -\frac{19.50 \text{ kN}}{18.75 \text{ kN/mm}} = -1.04 \text{ mm}$$

The permanent set is thus

$$\delta_p = \delta_{max} + \delta' = 1.40 - 1.04 = 0.36 \text{ mm}$$

We recall that the discussion of stress concentrations of Sec. 2.17 was carried out under the assumption of a linear stress-strain relationship. The stress distributions shown in Figs. 2.57 and 2.58, and the values of the stress-concentration factors plotted in Fig. 2.59 cannot be used, therefore, when plastic deformations take place, i.e., when the value of σ_{max} obtained from these figures exceeds the yield strength σ_Y.

Let us consider again the flat bar with a circular hole of Fig. 2.57, and let us assume that the material is elastoplastic, i.e., that its stress-strain diagram is as shown in Fig. 2.60. As long as no plastic deformation takes place, the distribution of stresses is as indicated in Sec. 2.17 (Fig. 2.64a). We observe that the area under the stress-distribution curve represents the integral $\int \sigma \, dA$, which is equal to the load P. Thus this area, and the value of σ_{max}, must increase as the load P increases. As long as $\sigma_{max} \leq \sigma_Y$, all the successive stress distributions obtained as P increases will have the shape shown in Fig. 2.57 and repeated in Fig. 2.64a. However, as P is increased beyond the value P_Y corresponding to $\sigma_{max} = \sigma_Y$ (Fig. 2.64b), the stress-distribution curve must flatten in the vicinity of the hole (Fig. 2.64c), since the stress in the material considered cannot exceed the value σ_Y. This indicates that the material is yielding in the vicinity of the hole. As the load P is further increased, the plastic zone where yield takes place keeps expanding, until it reaches the edges of the plate (Fig. 2.64d). At that point, the distribution of stresses across the plate is uniform, $\sigma = \sigma_Y$, and the corresponding value $P = P_U$ of the load is the largest which may be applied to the bar without causing rupture.

It is interesting to compare the maximum value P_Y of the load which may be applied if no permanent deformation is to be produced in the bar, with the value P_U which will cause rupture. Recalling the definition of the average stress, $\sigma_{ave} = P/A$, where A is the net cross-sectional area, and the definition of the stress concentration factor, $K = \sigma_{max}/\sigma_{ave}$, we write

$$P = \sigma_{ave}A = \frac{\sigma_{max}A}{K} \qquad (2.45)$$

for any value of σ_{max} which does not exceed σ_Y. When $\sigma_{max} = \sigma_Y$ (Fig. 2.64b), we have $P = P_Y$, and Eq. (2.45) yields

$$P_Y = \frac{\sigma_Y A}{K} \qquad (2.46)$$

On the other hand, when $P = P_U$ (Fig. 2.64d), we have $\sigma_{ave} = \sigma_Y$ and

$$P_U = \sigma_Y A \qquad (2.47)$$

Comparing Eqs. (2.46) and (2.47), we conclude that

$$P_Y = \frac{P_U}{K} \qquad (2.48)$$

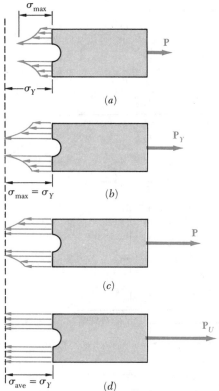

Fig. 2.64 Distribution of stresses in elastoplastic material under increasing load.

*2.19. RESIDUAL STRESSES

In Example 2.12 of the preceding section, we considered a rod which was stretched beyond the yield point. As the load was removed, the rod did not regain its original length; it had been permanently deformed. However, after the load was removed, all stresses disappeared. We should not assume that this will always be the case. Indeed, when only some of the parts of an indeterminate structure undergo plastic deformations, as in Example 2.14, or when different parts of the structure undergo different plastic deformations, the stresses in the various parts of the structure will not, in general, return to zero after the load has been removed. Stresses, called *residual stresses*, will remain in the various parts of the structure.

While the computation of the residual stresses in an actual structure may be quite involved, the following example will provide us with a general understanding of the method to be used for their determination.

Example 2.15

Determine the residual stresses in the rod and tube of Examples 2.13 and 2.14 after the load **P** has been increased from zero to 19.5 kN and decreased back to zero.

We observe from the diagrams of Fig. 2.65 that after the load **P** has returned to zero, the internal forces P_r and P_t are *not* equal to zero. Their values have been indicated by point E in parts a and b, respectively, of Fig. 2.65. It follows that the corresponding stresses are not equal to zero either after the assembly has been unloaded. To determine these residual stresses, we shall determine the reverse stresses σ'_r and σ'_t caused by the unloading and add them to the maximum stresses $\sigma_r = 200$ MPa and $\sigma_t = 175$ MPa found in part a of Example 2.14.

The strain caused by the unloading is the same in the rod and in the tube. It is equal to δ'/L, where δ' is the deformation of the assembly during unloading, which was found in Example 2.14. We have

$$\epsilon' = \frac{\delta'}{L} = \frac{-1.04 \text{ mm}}{800 \text{ mm}} = -1.30 \times 10^{-3}$$

The corresponding reverse stresses in the rod and tube are

$$\sigma'_r = \epsilon' E_r = (-1.30 \times 10^{-3})(200 \text{ GPa}) = -260 \text{ MPa}$$
$$\sigma'_t = \epsilon' E_t = (-1.30 \times 10^{-3})(100 \text{ GPa}) = -130 \text{ MPa}$$

The residual stresses are found by superposing the stresses due to loading and the reverse stresses due to unloading. We have

$$(\sigma_r)_{\text{res}} = \sigma_r + \sigma'_r = 200 \text{ MPa} - 260 \text{ MPa} = -60 \text{ MPa}$$
$$(\sigma_t)_{\text{res}} = \sigma_t + \sigma'_t = 175 \text{ MPa} - 130 \text{ MPa} = +45 \text{ MPa}$$

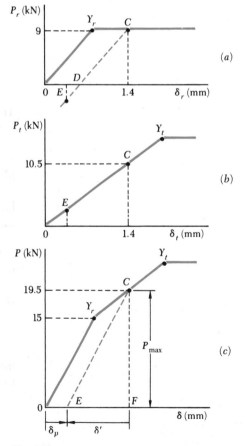

Fig. 2.65

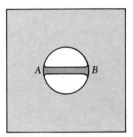

Fig. 2.66

Plastic deformations caused by temperature changes can also result in residual stresses. For example, consider a small plug which is to be welded to a large plate. For discussion purposes we shall consider the plug as a small rod AB which is to be welded across a small hole in the plate (Fig. 2.66). During the welding process the temperature of the rod will be raised to over 1000°C, at which temperature its modulus of elasticity, and hence its stiffness and stress, will be almost zero. Since the plate is large, its temperature will not be increased significantly above room temperature (20°C). Thus, when the welding is completed, we have rod AB at $T = 1000°C$, with no stress, attached to the plate which is at 20°C.

As the rod cools, its modulus of elasticity increases and, at about 500°C, will approach its normal value of about 200 GPa. As the temperature of the rod decreases further, we have a situation similar to that considered in Sec. 2.10 and illustrated in Fig. 2.33. Solving Eq. (2.23) for ΔT and making σ equal to the yield strength, $\sigma_Y = 300$ MPa, of average steel, and $\alpha = 12 \times 10^{-6}/°C$, we find the temperature change which will cause the rod to yield:

$$\Delta T = -\frac{\sigma}{E\alpha} = -\frac{300 \text{ MPa}}{(200 \text{ GPa})(12 \times 10^{-6}/°C)} = -125°C$$

This means that the rod will start yielding at about 375°C and will keep yielding at a fairly constant stress level, as it cools down to room temperature. As a result of the welding operation, a residual stress approximately equal to the yield strength of the steel used is thus created in the plug and in the weld.

Residual stresses also occur as a result of the cooling of metals which have been cast or hot rolled. In these cases, the outer layers cool more rapidly than the inner core. This causes the outer layers to reacquire their stiffness (E returns to its normal value) faster than the inner core. When the entire specimen has returned to room temperature, the inner core will have contracted more than the outer layers. The result is residual longitudinal tensile stresses in the inner core and residual compressive stresses in the outer layers.

Residual stresses due to welding, casting, and hot rolling may be quite large (of the order of magnitude of the yield strength). These stresses may be removed, when necessary, by reheating the entire specimen to about 600°C, and then allowing it to cool slowly over a period of 12 to 24 hours.

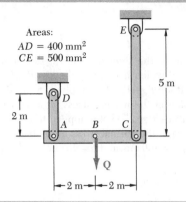

Areas:
$AD = 400 \text{ mm}^2$
$CE = 500 \text{ mm}^2$

SAMPLE PROBLEM 2.6

The rigid beam ABC is suspended from two steel rods as shown and is initially horizontal. The midpoint B of the beam is deflected 10 mm downward by the slow application of the force $\mathbf{Q}$, after which the force is slowly removed. Knowing that the steel used for the rods is elastoplastic with $E = 200$ GPa and $\sigma_Y = 300$ MPa, determine (a) the required maximum value of Q and the corresponding position of the beam, (b) the final position of the beam.

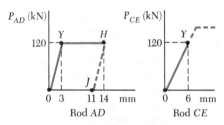

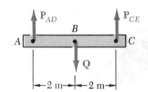

Load-deflection diagrams

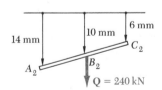

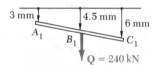

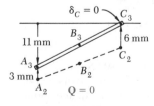

(a) Deflections for $\delta_B = 10$ mm ◄

(b) Final deflections ◄

Statics. Since $\mathbf{Q}$ is applied at the midpoint of the beam, we have

$$P_{AD} = P_{CE} \quad \text{and} \quad Q = 2P_{AD}$$

Elastic Action. The maximum value of Q and the maximum elastic deflection of point A occur when $\sigma = \sigma_Y$ in rod AD.

$$(P_{AD})_{max} = \sigma_Y A = (300 \text{ MPa})(400 \text{ mm}^2) = 120 \text{ kN}$$
$$Q_{max} = 2(P_{AD})_{max} = 2(120 \text{ kN}) \qquad Q_{max} = 240 \text{ kN} \quad ◄$$
$$\delta_{A_1} = \epsilon L = \frac{\sigma_Y}{E} L = \left(\frac{300 \text{ MPa}}{200 \text{ GPa}}\right)(2 \text{ m}) = 3 \text{ mm}$$

Since $P_{CE} = P_{AD} = 120$ kN, the stress in rod CE is

$$\sigma_{CE} = \frac{P_{CE}}{A} = \frac{120 \text{ kN}}{500 \text{ mm}^2} = 240 \text{ MPa}$$

The corresponding deflection of point C is

$$\delta_{C_1} = \epsilon L = \frac{\sigma_{CE}}{E} L = \left(\frac{240 \text{ MPa}}{200 \text{ GPa}}\right)(5 \text{ m}) = 6 \text{ mm}$$

The corresponding deflection of point B is

$$\delta_{B_1} = \tfrac{1}{2}(\delta_{A_1} + \delta_{C_1}) = \tfrac{1}{2}(3 \text{ mm} + 6 \text{ mm}) = 4.5 \text{ mm}$$

Since we must have $\delta_B = 10$ mm, we conclude that plastic deformation will occur.

Plastic Deformation. For $Q = 240$ kN, plastic deformation occurs in rod AD, where $\sigma_{AD} = \sigma_Y = 300$ MPa. Since the stress in rod CE is within the elastic range, δ_C remains equal to 6 mm. The deflection δ_A for which $\delta_B = 10$ mm is obtained by writing

$$\delta_{B_2} = 10 \text{ mm} = \tfrac{1}{2}(\delta_{A_2} + 6 \text{ mm}) \qquad \delta_{A_2} = 14 \text{ mm}$$

Unloading. As force $\mathbf{Q}$ is slowly removed, the force P_{AD} decreases along line HJ parallel to the initial portion of the load-deflection diagram of rod AD. The final deflection of point A is

$$\delta_{A_3} = 14 \text{ mm} - 3 \text{ mm} = 11 \text{ mm}$$

Since the stress in rod CE remained within the elastic range, we note that the final deflection of point C is zero.

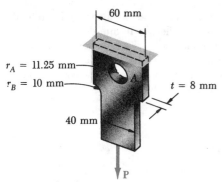

$r_A = 11.25$ mm

$r_B = 10$ mm

$t = 8$ mm

60 mm

40 mm

P

Fig. P2.92 and P2.93

PROBLEMS

2.92 Knowing that $\sigma_{all} = 150$ MPa for the steel bar shown, determine the maximum allowable centric axial force **P** which may be applied to the bar.

2.93 A centric axial force of magnitude $P = 40$ kN is applied to the steel bar shown. Determine the maximum value of the normal stress (a) at A, (b) at B.

2.94 A centric axial force of magnitude $P = 270$ kN is applied to the steel bar shown. Determine the maximum value of the normal stress in the bar, knowing that (a) $d = 120$ mm and $r = 15$ mm, (b) $d = 100$ mm and $r = 25$ mm, (c) $d = 75$ mm and $r = 37.5$ mm.

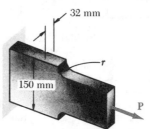

32 mm

r

150 mm

P

Fig. P2.94 and P2.95

2.95 The allowable normal stress is $\sigma_{all} = 125$ MPa for the steel bar shown. Determine the maximum allowable centric axial force **P** which may be applied to the bar, knowing that $d = 100$ mm and (a) $r = 15$ mm, (b) $r = 25$ mm.

2.96 The aluminum test specimen shown is subjected to two equal and opposite centric axial forces of magnitude P. (a) Knowing that $E = 70$ GPa and $\sigma_{all} = 200$ MPa, determine the maximum allowable value of P and the corresponding total elongation of the specimen. (b) Solve part a, assuming that the specimen has been replaced by an aluminum bar of the same length but of uniform 60 × 15-mm rectangular cross section.

2.97 For the test specimen of Prob. 2.96, determine the maximum value of the normal stress corresponding to a total elongation of 0.75 mm.

2.98 A centric axial force **P** is applied to the steel bar shown. Knowing that $\sigma_{all} = 165$ MPa and $b = 80$ mm, determine (a) the largest allowable value of P, (b) the smallest permissible value of r_f if σ_{all} is not to be exceeded at the fillet for the value of P found in part a.

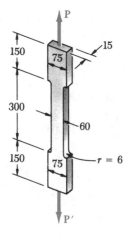

P

150

75

15

300

60

150

75

$r = 6$

P'

Dimensions in mm
Fig. P2.96

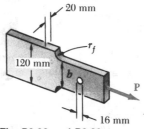

20 mm

r_f

120 mm

b

P

16 mm

Fig. P2.98 and P2.99

2.99 A centric axial force **P** of magnitude $P = 90$ kN is applied to the steel bar shown. Knowing that $b = 96$ mm, determine (a) the maximum normal stress at the hole, (b) the smallest permissible value of r_f if the maximum normal stress at the fillet is not to exceed the maximum normal stress at the hole.

2.100   The cylindrical rod AB has a length $L = 2$ m and a 32 mm diameter; it is made of a mild steel which is assumed to be elastoplastic with $E = 200$ GPa and $\sigma_Y = 250$ MPa.   A force **P** is applied to the rod until its end A has moved down by an amount δ_m.   Determine the maximum value of the force **P** and the permanent set of the rod after the force has been removed, knowing that (*a*) $\delta_m = 3$ mm, (*b*) $\delta_m = 6$ mm.

2.101   The cylindrical rod AB has a length $L = 1.5$ m and a 19 mm diameter; it is made of mild steel which is assumed to be elastoplastic with $E = 200$ GPa and $\sigma_Y = 250$ MPa.   A force **P** is applied to the rod and then removed to give it a permanent set δ_p.   Determine the maximum value of the force **P** and the maximum amount δ_m by which the rod should be stretched if the desired value of δ_p is (*a*) 2.5 mm, (*b*) 5.0 mm.

2.102   Rod ABC consists of two cylindrical portions AB and BC; it is made of a mild steel which is assumed to be elastoplastic with $E = 200$ GPa and $\sigma_Y = 250$ MPa.   A force **P** is applied to the rod and then removed to give it a permanent set $\delta_p = 2$ mm.   Determine the maximum value of the force **P** and the maximum amount δ_m by which the rod should be stretched to give it the desired permanent set.

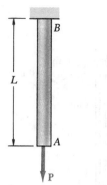

Fig. P2.100 and P2.101

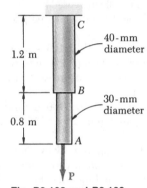

Fig. P2.102 and P2.103

2.103   Rod ABC consists of two cylindrical portions AB and BC; it is made of a mild steel which is assumed to be elastoplastic with $E = 200$ GPa and $\sigma_Y = 250$ MPa.   A force **P** is applied to the rod until its end A has moved down by an amount $\delta_m = 5$ mm.   Determine the maximum value of the force **P** and the permanent set of the rod after the force has been removed.

2.104   Rod AB consists of two portions, AC and CB, each 200 mm long and of 1.9×10^{-3} m^2 cross-sectional area.   Portion AC is made of a mild steel with $E = 200$ GPa and $\sigma_Y = 250$ MPa, and portion CB of a high-strength steel with $E = 200$ GPa and $\sigma_Y = 350$ MPa.   A load **P** is applied at C as shown.   (*a*) Assuming both steels to be elastoplastic, draw the load-deflection diagram for point C.   (*b*) If P is gradually increased from zero to 1.080 MN and then reduced back to zero, determine the maximum deflection of C, the maximum stress in each portion of rod, and the permanent deflection of C.

2.105   For the composite rod of Prob. 2.104, if P is gradually increased from zero until the deflection of point C reaches a maximum value δ_m and then decreased back to zero, determine the maximum value of P, the maximum stress in each portion of rod, and the permanent deflection of C when (*a*) $\delta_m = 0.28$ mm, (*b*) $\delta_m = 0.38$ mm.

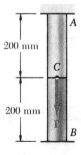

Fig. P2.104

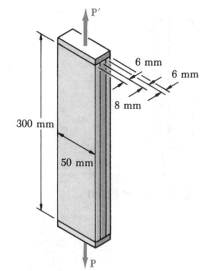

Fig. P2.106

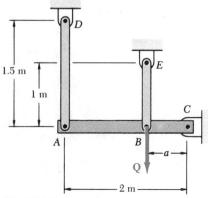

Fig. P2.112

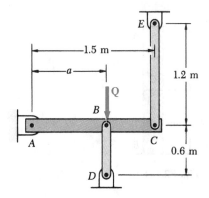

Fig. P2.114

2.106 A 300-mm bar of 20 × 50-mm rectangular cross section consists of two mild-steel layers, each 6 mm thick, bonded to an 8-mm-thick tempered-steel central layer. This composite bar is subjected as shown to a centric axial load of magnitude P. Assuming both steels to be elastoplastic with $E = 200$ GPa and with a yield strength equal to 250 and 690 MPa, respectively, for the mild and the tempered steel, draw the load-deflection diagram of the composite bar. Using this diagram, determine the maximum value of P, the maximum stress in each layer, and the permanent set of the bar when P is gradually increased from zero until the deformation of the bar reaches a maximum value δ_m and then decreased back to zero, assuming (a) $\delta_m = 0.621$ mm, (b) $\delta_m = 1.242$ mm.

2.107 For the composite bar of Prob. 2.106, if P is gradually increased from zero to 357 kN and then decreased back to zero, determine the maximum deformation of the bar, the maximum stress in each layer, and the permanent set of the bar.

*2.108 For the composite rod of Prob. 2.104, determine the residual stress in each portion of the rod if P is gradually increased from zero to 1.080 MN and then decreased back to zero.

*2.109 For the composite rod of Prob. 2.104, determine the residual stress in each portion of the rod if P is gradually increased from zero until the deflection of C reaches a maximum value δ_m and then decreased back to zero when (a) $\delta_m = 0.28$ mm, (b) $\delta_m = 0.38$ mm.

*2.110 For the composite bar of Prob. 2.106, determine the residual stress in each layer if P is gradually increased from zero until the deformation of the bar reaches a maximum value δ_m and then decreased back to zero when (a) $\delta_m = 0.621$ mm, (b) $\delta_m = 1.242$ mm.

*2.111 For the composite bar of Prob. 2.106, determine the residual stress in each layer if P is gradually increased from zero to 357 kN and then decreased back to zero.

*2.112 The rigid bar ABC is supported by two links, AD and BE, of uniform 38 × 6 mm rectangular cross section and made of a mild steel which is assumed to be elastoplastic with $E = 200$ GPa and $\sigma_Y = 250$ MPa. The magnitude Q of the force applied at B is gradually increased from zero to 225 kN and then decreased back to zero. Knowing that $a = 0.6$ m, determine (a) the maximum deflection of point B and the corresponding value of the normal stress in each link, (b) the final deflection of point B and the residual stress in each link. Assume that the links are braced so that they can carry compressive forces without buckling.

*2.113 Solve Prob. 2.112 if $a = 1.2$ m and if Q is gradually increased from zero to 147 kN and then decreased back to zero.

*2.114 The rigid bar ABC is supported by two links, BD and CE, of uniform 40 × 8-mm rectangular cross section and made of a mild steel which is assumed to be elastoplastic with $E = 200$ GPa and $\sigma_Y = 250$ MPa. The magnitude Q of the force applied at B is gradually increased from zero until the deflection of B reaches 0.840 mm and decreased back to zero. Knowing that $a = 0.9$ m, determine (a) the maximum value of Q and the corresponding value of the normal stress in each link, (b) the final deflection of point B and the residual stress in each link. Assume that the links are braced so that they can carry compressive forces without buckling.

*2.115 Solve Prob. 2.114 if $a = 0.6$ m and if Q is gradually increased from zero until the deflection of B reaches 0.720 mm and then decreased back to zero.

*2.116   A steel rod of length L and uniform cross section of area A is attached to rigid supports and is unstressed at a temperature of 20°C. The steel is assumed to be elastoplastic with $E = 200$ GPa and $\sigma_Y = 250$ MPa. Knowing that $\alpha = 11.7 \times 10^{-6}$/°C, determine (a) the stress in the rod after the temperature has been raised to 150°C, (b) the residual stress after the temperature has returned to 20°C.

*2.117   Solve Prob. 2.116, assuming that the temperature is being raised and lowered *for the second time*.

2.118   The steel rod ACB is attached to rigid supports and is unstressed at a temperature of 25°C. The steel is assumed to be elastoplastic with $E = 200$ GPa and $\sigma_Y = 250$ MPa. The temperature of both portions of the rod is then raised to 150°C. Knowing that $\alpha = 11.7 \times 10^{-6}$/°C, determine (a) the stress in both portions of the rod, (b) the deflection of point C.

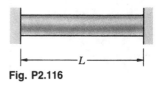
Fig. P2.116

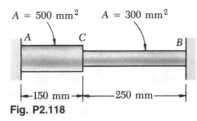

$A = 500$ mm^2 $A = 300$ mm^2

-150 mm-   -250 mm-
Fig. P2.118

*2.119   Solve Prob. 2.118, assuming that the temperature of the rod is raised to 150°C and is then returned to 25°C.

REVIEW AND SUMMARY

This chapter was devoted to the introduction of the concept of *strain*, to the discussion of the relationship between stress and strain in various types of materials, and to the determination of the deformations of structural components under axial loading.

Considering a rod of length L and uniform cross section and denoting by δ its deformation under an axial load $\mathbf{P}$ (Fig. 2.1), we defined the *normal strain* ϵ in the rod as the *deformation per unit length* [Sec. 2.2]:

$$\epsilon = \frac{\delta}{L} \qquad (2.1)$$

In the case of a rod of variable cross section, the normal strain was defined at any given point Q by considering a small element of rod at Q. Denoting by Δx the length of the element and by $\Delta \delta$ its deformation under the given load, we wrote

$$\epsilon = \lim_{\Delta x \to 0} \frac{\Delta \delta}{\Delta x} = \frac{d\delta}{dx} \qquad (2.2)$$

Normal strain

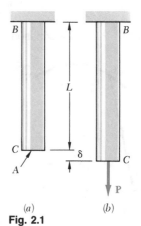

(a)   (b)
Fig. 2.1

Stress-strain diagram

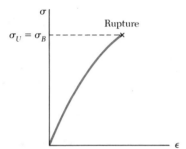

(a) Low-carbon steel   (b) Aluminum alloy

Fig. 2.9 Stress-strain diagrams of two typical ductile materials.

Fig. 2.11 Stress-strain diagram for a typical brittle material.

Hooke's law
Modulus of elasticity

Elastic limit. Plastic deformation

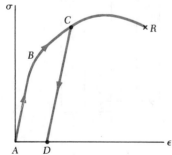

Fig. 2.16

Plotting the stress σ versus the strain ϵ as the load increased, we obtained a *stress-strain diagram* for the material used [Sec. 2.3]. From such a diagram, we were able to distinguish between *brittle* and *ductile* materials: A specimen made of a brittle material ruptures without any noticeable prior change in the rate of elongation (Fig. 2.11), while a specimen made of a ductile material *yields* after a critical stress σ_Y, called the *yield strength*, has been reached, i.e., the specimen undergoes a large deformation before rupturing, with a relatively small increase in the applied load (Fig. 2.9).

We noted in Sec. 2.5 that the initial portion of the stress-strain diagram is a straight line. This means that for small deformations, the stress is directly proportional to the strain:

$$\sigma = E\epsilon \tag{2.4}$$

This relation is known as *Hooke's law* and the coefficient E as the *modulus of elasticity* of the material. The largest stress for which Eq. (2.4) applies is the *proportional limit* of the material.

If the strains caused in a test specimen by the application of a given load disappear when the load is removed, the material is said to behave *elastically*, and the largest stress for which this occurs is called the *elastic limit* of the material [Sec. 2.6]. If the elastic limit is exceeded, the stress and strain decrease in a linear fashion when the load is removed and the strain does not return to zero (Fig. 2.16), indicating that a *permanent set* or *plastic deformation* of the material has taken place.

In Sec. 2.7, we discussed the phenomenon of *fatigue*, which causes the failure of structural or machine components after a very large number of repeated loadings, even though the stresses remain in the elastic range.

A standard fatigue test consists in determining the number n of successive loading-and-unloading cycles required to cause the failure of a specimen for any given maximum stress level σ, and plotting the resulting σ-n curve. The value of σ for which failure does not occur, even for an indefinitely large number of cycles, is known as the *endurance limit* of the material used in the test.

Fatigue. Endurance limit

Section 2.8 was devoted to the determination of the elastic deformations of various types of machine and structural components under various conditions of axial loading. We saw that if a rod of length L and uniform cross section of area A is subjected at its end to a centric axial load **P** (Fig. 2.20), the corresponding deformation is

Elastic deformation under axial loading

$$\delta = \frac{PL}{AE} \tag{2.7}$$

If the rod is loaded at several points or consists of several parts of various cross sections and possibly of different materials, the deformation δ of the rod must be expressed as the sum of the deformations of its component parts [Example 2.01]:

$$\delta = \sum_i \frac{P_i L_i}{A_i E_i} \tag{2.8}$$

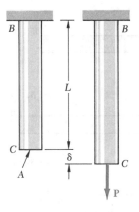

Fig. 2.20

Section 2.9 was devoted to the solution of *statically indeterminate problems*, i.e., problems in which the reactions and the internal forces *cannot* be determined from statics alone. The equilibrium equations derived from the free-body diagram of the member under consideration were complemented by relations involving deformations and obtained from the geometry of the problem. The forces in the rod and in the tube of Fig. 2.23a, for instance, were determined by observing, on one hand,

Statically indeterminate problems

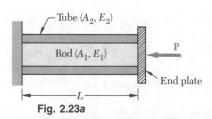

Tube (A_2, E_2)

Rod (A_1, E_1)

P

End plate

L

Fig. 2.23a

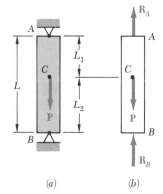

(a) (b) **Fig. 2.24**

that their sum is equal to P, and on the other, that they cause equal deformations in the rod and in the tube [Example 2.02]. Similarly, the reactions at the supports of the bar of Fig. 2.24 could not be obtained from the free-body diagram of the bar alone [Example 2.03]; but they could be determined by expressing that the total elongation of the bar must be equal to zero.

Problems with temperature changes

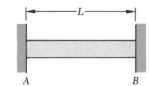

Fig. 2.33a

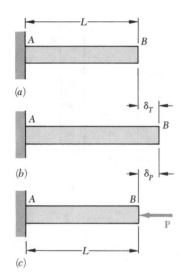

(a)

(b)

(c)

Fig. 2.34

In Sec. 2.10, we considered problems involving *temperature changes*. We first observed that if the temperature of an *unrestrained rod AB* of length L is increased by ΔT, its elongation is

$$\delta_T = \alpha(\Delta T)L \tag{2.21}$$

where α is the *coefficient of thermal expansion* of the material. We noted that the corresponding strain, called *thermal strain*, is

$$\epsilon_T = \alpha \Delta T \tag{2.22}$$

and that *no stress* is associated with this strain. However, if the rod AB is *restrained* by fixed supports (Fig. 2.33a), stresses develop in the rod as the temperature increases, due to the reactions at the supports. To determine the magnitude P of the reactions, we detached the rod from its support at B (Fig. 2.34) and considered separately the deformation δ_T of the rod as it expands freely due to the temperature change, and the deformation δ_P caused by the force $\mathbf{P}$ required to bring it back to its original length, so that it may be reattached to the support at B. Writing that the total deformation $\delta = \delta_T + \delta_P$ is equal to zero, we obtained an equation that could be solved for P. While the final strain in rod AB is clearly zero, this will generally not be the case for rods and bars consisting of elements of different cross sections or materials, since the deformations of the various elements will usually *not* be zero [Example 2.06].

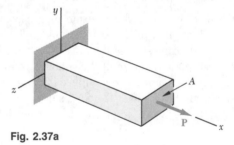

Fig. 2.37a

Lateral strain. Poisson's ratio

When an axial load $\mathbf{P}$ is applied to a homogeneous, slender bar (Fig. 2.37a), it causes a strain, not only along the axis of the bar but in any transverse direction as well [Sec. 2.11]. This strain is referred to as the *lateral strain*, and the ratio of the lateral strain over the axial strain is called *Poisson's ratio* and is denoted by ν (Greek letter nu). We wrote

$$\nu = \left| \frac{\text{lateral strain}}{\text{axial strain}} \right| \tag{2.25}$$

Recalling that the axial strain in the bar is $\epsilon_x = \sigma_x/E$, we expressed as follows the condition of strain under an axial loading in the x direction:

$$\epsilon_x = \frac{\sigma_x}{E} \qquad \epsilon_y = \epsilon_z = -\frac{\nu\sigma_x}{E} \tag{2.27}$$

This result was extended in Sec. 2.12 to the case of a *multiaxial loading* causing the state of stress shown in Fig. 2.40. The resulting strain condition was described by the following relations, referred to as the *generalized Hooke's law* for a multiaxial loading:

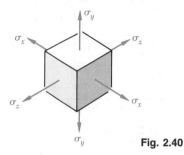

Fig. 2.40

$$\epsilon_x = + \frac{\sigma_x}{E} - \frac{\nu\sigma_y}{E} - \frac{\nu\sigma_z}{E}$$

$$\epsilon_y = - \frac{\nu\sigma_x}{E} + \frac{\sigma_y}{E} - \frac{\nu\sigma_z}{E} \qquad (2.28)$$

$$\epsilon_z = - \frac{\nu\sigma_x}{E} - \frac{\nu\sigma_y}{E} + \frac{\sigma_z}{E}$$

If an element of material is subjected to the stresses σ_x, σ_y, σ_z, it will deform and a certain change of volume will result [Sec. 2.13]. The *change in volume per unit volume* is referred to as the *dilatation* of the material and is denoted by e. We showed that

$$e = \frac{1 - 2\nu}{E}(\sigma_x + \sigma_y + \sigma_z) \qquad (2.31)$$

When a material is subjected to a hydrostatic pressure p, we have

$$e = -\frac{p}{k} \qquad (2.34)$$

where k is known as the *bulk modulus* of the material:

$$k = \frac{E}{3(1 - 2\nu)} \qquad (2.33)$$

As we saw in Chap. 1, the state of stress in a material under the most general loading condition involves shearing stresses, as well as normal stresses (Fig. 2.43). The shearing stresses tend to deform a cubic element of material into an oblique parallelepiped [Sec. 2.14]. Considering, for instance, the stresses τ_{xy} and τ_{yx} shown in Fig. 2.45 (which, we recall, are equal in magnitude), we noted that they cause the angles formed by the faces on which they act to either increase or decrease by a small angle γ_{xy}; this angle, expressed in radians, defines the *shearing strain* corresponding to the x and y directions. Defining in a similar way the shearing strains γ_{yz} and γ_{zx}, we wrote the relations

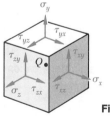

Fig. 2.43

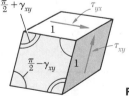

Fig. 2.45

$$\tau_{xy} = G\gamma_{xy} \qquad \tau_{yz} = G\gamma_{yz} \qquad \tau_{zx} = G\gamma_{zx} \qquad (2.36,37)$$

which are valid for any homogeneous isotropic material within its proportional limit in shear. The constant G is called the *modulus of rigidity* of the material and the relations obtained express *Hooke's law for shearing stress and strain*. Together with Eqs. (2.28), they form a group of equations representing the generalized Hooke's law for a homogeneous isotropic material under the most general stress condition.

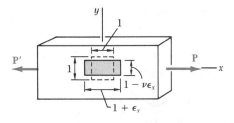

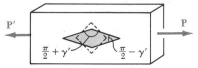

Fig. 2.51

We observed in Sec. 2.15 that while an axial load exerted on a slender bar produces only normal strains—both axial and transverse—on an element of material oriented along the axis of the bar, it will produce both normal and shearing strains on an element rotated through 45° (Fig. 2.51). We also noted that the three constants E, ν, and G are not independent; they satisfy the relation

$$\frac{E}{2G} = 1 + \nu \qquad (2.43)$$

which may be used to determine any of the three constants in terms of the other two.

Saint-Venant's principle

In Sec. 2.16, we discussed *Saint-Venant's principle*, which states that except in the immediate vicinity of the points of application of the loads, the distribution of stresses in a given member is independent of the actual mode of application of the loads. This principle makes it possible to assume a uniform distribution of stresses in a member subjected to concentrated axial loads, except close to the points of application of the loads, where stress concentrations will occur.

Stress concentrations

Stress concentrations will also occur in structural members near a discontinuity, such as a hole or a sudden change in cross section [Sec. 2.17]. The ratio of the maximum value of the stress occurring near the discontinuity over the average stress computed in the critical section is referred to as the *stress-concentration factor* of the discontinuity and is denoted by K:

$$K = \frac{\sigma_{\max}}{\sigma_{\text{ave}}} \qquad (2.44)$$

Values of K for circular holes and fillets in flat bars were given in Fig. 2.59 on page 94.

Plastic deformations

In Sec. 2.18, we discussed the *plastic deformations* which occur in structural members made of a ductile material when the stresses in some part of the member exceed the yield strength of the material. Our analysis was carried out for an idealized *elastoplastic material* characterized by the stress-strain diagram shown in Fig. 2.60 [Examples 2.12, 2.13, and 2.14]. Finally, in Sec. 2.19, we observed that when an indeterminate structure undergoes plastic deformations, the stresses do not, in general, return to zero after the load has been removed. The stresses remaining in the various parts of the structure are called *residual stresses* and may be determined by adding the maximum stresses reached during the loading phase and the reverse stresses corresponding to the unloading phase [Example 2.15].

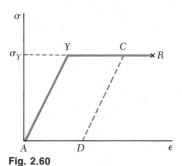

Fig. 2.60

REVIEW PROBLEMS

2.120 A brass rod AB of 10-mm diameter is attached to the base of a cylindrical brass vessel CD, the cross-sectional area of which is 325×10^{-6} m^2. Vessel CD is attached to a fixed support at C, and a plug E is attached to end A of the rod. Knowing that the modulus of elasticity of brass is 105 GPa, determine the magnitude of the load **P** for which the deflection of the plug is 1.25 mm downward.

2.121 The length of the 2-mm-diameter steel wire CD has been adjusted so that, with no load applied, a gap of 1.5 mm exists between the end B of the rigid beam ACB and a contact point E. Using $E = 200$ GPa, determine where a 225-N load should be applied to the beam to cause contact between B and E.

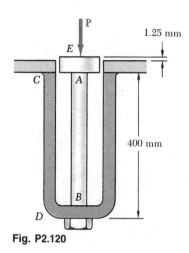

Fig. P2.120

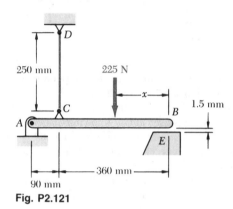

Fig. P2.121

2.122 The ends of the 15-mm-diameter steel rods BE and CD ($E = 200$ GPa) are single-threaded with a pitch of 2.25 mm. Knowing that after being snugly fitted, the nut at C is tightened one full turn, determine (*a*) the tension in rod CD, (*b*) the deflection of point C of the rigid member ABC.

2.123 Solve Prob. 2.122, assuming that after being snugly fitted, the nut at B is tightened one full turn.

2.124 Bar AB has a 1250-mm^2 cross-sectional area and is made of a mild steel which is assumed to be elastoplastic with $E = 200$ GPa and $\sigma_Y = 280$ MPa. Knowing that the force **F** gradually increases from 0 to 600 kN and then decreases back to zero, determine (*a*) the permanent deflection of point C, (*b*) the residual stress in the bar.

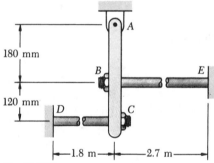

Fig. P2.122

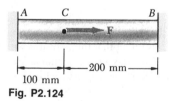

Fig. P2.124

2.125 The change in diameter of a large steel bolt is carefully measured as the nut is tightened. Using the data available for structural steel in Appendix B, determine the internal force in the bolt if the diameter is observed to decrease by 12 μm.

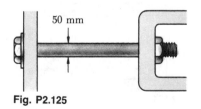

Fig. P2.125

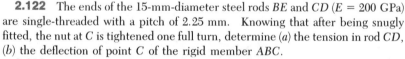

2.126 The steel rod ABC is attached to rigid supports and is unstressed at a temperature of 20°C. The steel is assumed to be elastoplastic with $E = 200$ GPa and $\sigma_Y = 250$ MPa. Knowing that $\alpha = 11.7 \times 10^{-6}/°C$, determine the stress in portion AC and the deflection of point C if the temperature of both portions of the rod is raised to (a) 87°C, (b) 120°C.

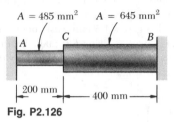

$A = 485$ mm² $A = 645$ mm²

200 mm   400 mm

Fig. P2.126

2.127 Solve Prob. 2.126, assuming that the temperature of the rod is returned to 20°C after being raised to (a) 87°C, (b) 120°C.

2.128 Steel wires of 3.25-mm diameter are used at A and B while an aluminum wire of 2-mm diameter is used at C. Knowing that each wire is initially taut, determine the additional tension in each wire when a 900-N force $\mathbf{P}$ is applied to the midpoint of the lower edge of the plate. Use $E_s = 200$ GPa for steel and $E_a = 70$ GPa for aluminum.

2.129 (a) Determine the radius r of the fillets for which the same maximum stress occurs at hole A and at the fillets. (b) If the allowable stress is 140 MPa, what is the corresponding allowable load $\mathbf{P}$?

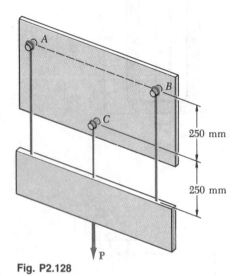

250 mm

250 mm

Fig. P2.128

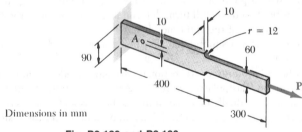

10   10

$r = 12$

90   A 60

400   300

Dimensions in mm

Fig. P2.129 and P2.130

2.130 (a) For $P = 25$ kN, determine the maximum stress in the plate shown. (b) Solve part a, assuming that the hole at A was not drilled.

2.131 The uniform rod BC has a cross-sectional area A and is made of a mild steel which may be assumed to be elastoplastic with a modulus of elasticity E and a yield strength σ_Y. Using the block-and-spring system shown, it is desired to simulate the deflection of end C of the rod as the axial force $\mathbf{P}$ is gradually applied and removed; that is, the deflection of points C and C' should be the same for all values of P. Denoting by μ the coefficient of friction between the block and the horizontal surface, derive an expression for (a) the required mass m of the block, (b) the required constant k of the spring.

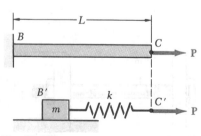

L

B C P

B' k C'

m P

Fig. P2.131

The following problems are designed to be solved with a computer. Write each program so that it can be used with any consistent set of units and in such a way that solid cylindrical elements may be defined by either their diameter or their cross-sectional area.

2.C1 A rod consisting of n elements, each of which is homogeneous and of uniform cross section, is subjected to the loading shown. The length of element i is denoted by L_i, its cross-sectional area by A_i, its modulus of elasticity by E_i, and the load applied to its right end by $\mathbf{P}_i$, the magnitude P_i of this load being assumed to be positive if $\mathbf{P}_i$ is directed to the right and negative otherwise. (*a*) Write a computer program which can be used to determine the average normal stress in each element, the deformation of each element, and the total deformation of the rod. (*b*) Use this program to solve Probs. 2.9, 2.10, 2.13, 2.14, and 2.15.

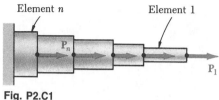

Fig. P2.C1

2.C2 Rod AB is horizontal with both ends fixed; it consists of n elements, each of which is homogeneous and of uniform cross section, and is subjected to the loading shown. The length of element i is denoted by L_i, its cross-sectional area by A_i, its modulus of elasticity by E_i, and the load applied to its right end by $\mathbf{P}_i$, the magnitude P_i of this load being assumed to be positive if $\mathbf{P}_i$ is directed to the right and negative otherwise. (Note that $P_1 = 0$.) (*a*) Write a computer program which can be used to determine the reactions at A and B, the average normal stress in each element, and the deformation of each element. (*b*) Use this program to solve Probs. 2.37, 2.38, 2.39, and 2.40.

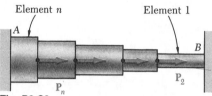

Fig. P2.C2

2.C3 Rod AB consists of n elements, each of which is homogeneous and of uniform cross section. End A is fixed, while initially there is a gap δ_0 between end B and the fixed vertical surface on the right. The length of element i is denoted by L_i, its cross-sectional area by A_i, its modulus of elasticity by E_i, and its coefficient of thermal expansion by α_i. After the temperature of the rod has been increased by ΔT, the gap at B is closed and the vertical surfaces exert equal and opposite forces on the rod. (*a*) Write a computer program which can be used to determine the magnitude of the reactions at A and B, the normal stress in each element, and the deformation of each element. (*b*) Use this program to solve Probs. 2.58, 2.60, 2.62, and 2.63.

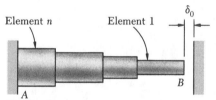

Fig. P2.C3

2.C4 Bar AB has a length L and is made of two different materials of given cross-sectional area, modulus of elasticity, and yield strength. The bar is subjected as shown to a load $\mathbf{P}$ which is gradually increased from zero until the deformation of the bar has reached a maximum value δ_m and then decreased back to zero. (*a*) Write a computer program which, for each of 25 values of δ_m equally spaced over a range extending from 0 to a value equal to 120% of the deformation causing both materials to yield, can be used to determine the maximum value P_m of the load, the maximum normal stress in each material, the permanent deformation δ_p of the bar, and the residual stress in each material. (*b*) Use this program to solve Probs. 2.104 and 2.108, 2.106 and 2.110, and 2.107 and 2.111.

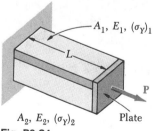

Fig. P2.C4

CHAPTER THREE

TORSION

3.1. INTRODUCTION

In the two preceding chapters, we discussed the stresses and strains in structural members subjected to axial loads, i.e., to forces directed along the axis of the member. In this chapter, we shall consider members which are in *torsion*. More specifically, we shall analyze the stresses and strains in members of circular cross section subjected to twisting couples, or *torques*, **T** and **T′** (Fig. 3.1). These couples have a common magnitude T, and opposite senses. They are vector quantities and may be represented either by curved arrows as in Fig. 3.1a, or by couple vectors as in Fig. 3.1b.

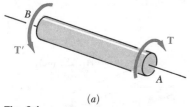
(a)

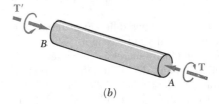
(b)

Fig. 3.1

Members in torsion are encountered in many engineering applications. The most common application is provided by *transmission shafts*, which are used to transmit power from one point to another, as from a steam turbine to an electric generator, or from a motor to a machine tool, or from the engine to the rear axle of an automobile. These shafts may either be solid as shown in Fig. 3.1, or they may be hollow. Considering, for example, the system consisting of the turbine A and the generator B connected by the transmission shaft AB (Fig. 3.2a), and breaking the system into its three component parts (Fig. 3.2b), we note that the tur-

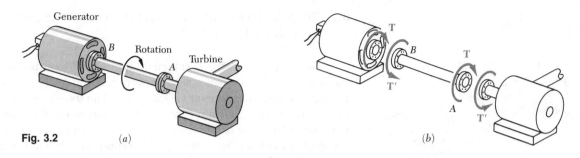

Fig. 3.2   (*a*)   (*b*)

bine exerts a twisting couple or torque **T** on the shaft, and that the shaft exerts an equal torque on the generator. The generator reacts by exerting the equal and opposite torque **T′** on the shaft, and the shaft by exerting the torque **T′** on the turbine.

We shall first analyze the stresses and deformations taking place in circular shafts. In Sec. 3.3, we shall demonstrate an important property of circular shafts: When a circular shaft is subjected to torsion, *every cross section remains plane and undistorted.* In other words, while the various cross sections along the shaft rotate through different angles, each cross section rotates as a solid rigid slab. This property will enable us to determine the *distribution of shearing strains* in a circular shaft and to conclude that *the shearing strain varies linearly with the distance from the axis of the shaft.*

Considering deformations in the *elastic range* and using Hooke's law for shearing stress and strain, we shall determine the *distribution of shearing stresses* in a circular shaft and derive the *elastic torsion formulas* [Sec. 3.4].

In Sec. 3.5, we shall determine the *angle of twist* of a circular shaft subjected to a given torque, assuming again elastic deformations. Section 3.6 will be devoted to the solution of problems involving *statically indeterminate shafts.*

In Sec. 3.7, we shall discuss the *design of transmission shafts.* We shall learn to determine the required physical characteristics of a shaft in terms of its speed of rotation and the power to be transmitted.

The torsion formulas cannot be used to determine stresses near sections where the loading couples are applied, nor near a section where an abrupt change in the diameter of the shaft occurs. Moreover, these formulas apply only within the elastic range of the material. In Sec. 3.8, we shall consider *stress concentrations* in circular shafts; and in Secs. 3.9 to 3.11, we shall discuss stresses and deformations in circular shafts made of a ductile material when the yield point of the material is exceeded. As we shall see, permanent *plastic deformations* and *residual stresses* result from such loading conditions.

Finally, we shall discuss the torsion of *noncircular members* [Sec. 3.12] and analyze the distribution of stresses in *thin-walled hollow noncircular shafts* [Sec. 3.13].

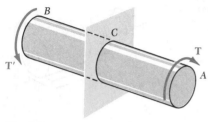

Fig. 3.3

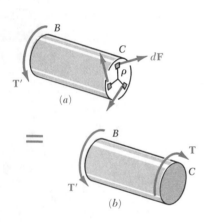

Fig. 3.4

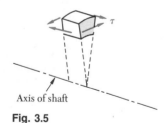

Fig. 3.5

3.2. PRELIMINARY DISCUSSION OF THE STRESSES IN A SHAFT

Considering a shaft AB subjected at A and B to equal and opposite torques $\mathbf{T}$ and $\mathbf{T}'$, we pass a section perpendicular to the axis of the shaft through some arbitrary point C (Fig. 3.3). The free-body diagram of the portion BC of the shaft must include the elementary shearing forces $d\mathbf{F}$ perpendicular to the radius of the shaft, which portion AC exerts on BC as the shaft is twisted (Fig. 3.4a). But the conditions of equilibrium for BC require that the system of these elementary forces be equivalent to an internal torque $\mathbf{T}$, equal and opposite to $\mathbf{T}'$ (Fig. 3.4b). Denoting by ρ the perpendicular distance from the force $d\mathbf{F}$ to the axis of the shaft, and expressing that the sum of the moments of the shearing forces $d\mathbf{F}$ about the axis of the shaft is equal in magnitude to the torque $\mathbf{T}$, we write

$$\int \rho\, dF = T$$

or, since $dF = \tau\, dA$, where τ is the shearing stress on the element of area dA,

$$\int \rho(\tau\, dA) = T \tag{3.1}$$

While the relation obtained expresses an important condition which must be satisfied by the shearing stresses in any given cross section of the shaft, it does *not* tell us how these stresses are distributed in the cross section. We thus observe, as we already did in Sec. 1.3, that the actual distribution of stresses under a given load is *statically indeterminate*, i.e., this distribution *cannot be determined by the methods of statics*. However, having assumed in Sec. 1.3 that the normal stresses produced by an axial centric load were uniformly distributed, we found later (Sec. 2.16) that this assumption was justified, except in the neighborhood of concentrated loads. A similar assumption with respect to the distribution of shearing stresses in an elastic shaft *would be wrong*. We must withhold any judgment regarding the distribution of stresses in a shaft until we have analyzed the *deformations* which are produced in the shaft. This will be done in the next section.

One more observation should be made at this point. As we have indicated in Sec. 1.8, shear cannot take place in one plane only. Consider the very small element of shaft shown in Fig. 3.5. We know that the torque applied to the shaft produces shearing stresses τ on the faces perpendicular to the axis of the shaft. But the conditions of equilibrium discussed in Sec. 1.8 require the existence of equal stresses on the faces formed by the two planes containing the axis of the shaft. That such shearing stresses actually occur in torsion may be demonstrated by considering a "shaft" made of separate slats pinned at both ends to disks as shown in Fig. 3.6a. If markings have been painted on two adjoining slats, it is observed that the slats slide with respect to each other when equal and opposite torques are applied to the ends of the "shaft" (Fig.

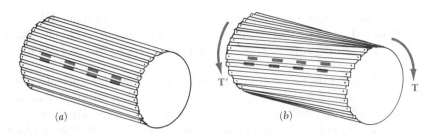

Fig. 3.6

3.6*b*). While sliding will not actually take place in a shaft made of a homogeneous and cohesive material, the tendency for sliding will exist, showing that stresses occur on longitudinal planes as well as on planes perpendicular to the axis of the shaft.†

3.3. DEFORMATIONS IN A CIRCULAR SHAFT

Consider a circular shaft which is attached to a fixed support at one end (Fig. 3.7*a*). If a torque **T** is applied to the other end, the shaft will twist, with its free end rotating through an angle ϕ called *the angle of twist* (Fig. 3.7*b*). Observation shows that, within a certain range of values of T, the angle of twist ϕ is proportional to T. It also shows that ϕ is proportional to the length L of the shaft. In other words, the angle of twist for a shaft of the same material and same cross section, but twice as long, will be twice as large under the same torque **T**. One purpose of our analysis will be to find the specific relation existing between ϕ, L, and T; another purpose will be to determine the distribution of shearing stresses in the shaft, which we were unable to obtain in the preceding section on the basis of statics alone.

At this point, we should note an important property of circular shafts: When a circular shaft is subjected to torsion, *every cross section remains plane and undistorted*. In other words, while the various cross sections along the shaft rotate through different amounts, each cross section rotates as a solid rigid slab. This is illustrated in Fig. 3.8*a*, which shows the deformations in a rubber model subjected to torsion. The property we are discussing is characteristic of circular shafts, whether solid or hollow; it is not enjoyed by members of noncircular cross section. For example, when a bar of square cross section is subjected to torsion, its various cross sections are warped and do not remain plane (Fig. 3.8*b*).

The fact that the cross sections of a circular shaft remain plane and undistorted is due to the fact that a circular shaft is *axisymmetric*, i.e., its appearance remains the same when it is viewed from a fixed position and rotated about its axis through an arbitrary angle. (Square bars, on the

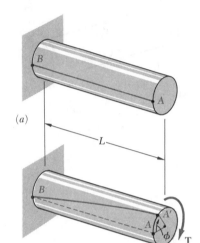

(a)

(b)
Fig. 3.7

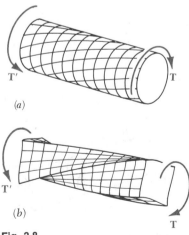

(a)

(b)
Fig. 3.8

---

†The twisting of a cardboard tube which has been slit lengthwise provides another demonstration of the existence of shearing stresses on longitudinal planes.

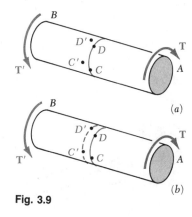

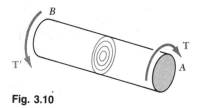

Fig. 3.9

Fig. 3.10

other hand, retain the same appearance only if they are rotated through 90° or 180°.) As we shall see presently, the axisymmetry of circular shafts may be used to prove theoretically that their cross sections remain plane and undistorted.

Consider the points C and D located on the circumference of a given cross section of the shaft, and let C' and D' be the positions they will occupy after the shaft has been twisted (Fig. 3.9a). The axisymmetry of the shaft and of the loading requires that the rotation which would have brought D into C should now bring D' into C'. Thus C' and D' must lie on the circumference of a circle, and the arc $C'D'$ must be equal to the arc CD (Fig. 3.9b). We shall now examine whether the circle on which C' and D' lie is different from the original circle. Let us assume that C' and D' do lie on a different circle and that the new circle is located to the left of the original circle, as shown in Fig. 3.9b. The same situation will prevail for any other cross section, since all the cross sections of the shaft are subjected to the same internal torque T, and an observer looking at the shaft from its end A will conclude that the loading causes any given circle drawn on the shaft to move *away* from him. But an observer located at B, to whom the given loading looks the same (a clockwise couple in the foreground and a counterclockwise couple in the background) will reach the opposite conclusion, i.e., he will conclude that the circle moves *toward* him. This contradiction proves that our assumption is wrong and that C' and D' lie on the same circle as C and D. Thus, as the shaft is twisted, the original circle just rotates in its own plane. Since the same reasoning may be applied to any smaller, concentric circle located in the cross section under consideration, we conclude that the entire cross section remains plane (Fig. 3.10).

The above argument does not preclude the possibility for the various concentric circles of Fig. 3.10 to rotate by different amounts when the shaft is twisted. But if that were so, a given diameter of the cross section would be distorted into a curve which might look as shown in Fig. 3.11a. An observer looking at this curve from A would conclude that the outer layers of the shaft get more twisted than the inner ones, while an observer looking from B would reach the opposite conclusion (Fig. 3.11b). This inconsistency leads us to conclude that any diameter of a given cross section remains straight (Fig. 3.11c) and, therefore, that any given cross section of a circular shaft remains plane and undistorted.

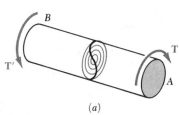

(a)

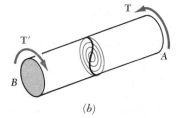

(b)

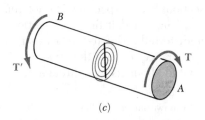

(c)

Fig. 3.11

Our discussion so far has ignored the mode of application of the twisting couples **T** and **T′**. If *all* sections of the shaft, from one end to the other, are to remain plane and undistorted, we must make sure that the couples are applied in such a way that the ends of the shaft themselves remain plane and undistorted. This may be accomplished by applying the couples **T** and **T′** to rigid plates, which are solidly attached to the ends of the shaft (Fig. 3.12*a*). We may then be sure that all sections will remain plane and undistorted when the loading is applied, and that the resulting deformations will occur in a uniform fashion throughout the entire length of the shaft. All of the equally spaced circles shown in Fig. 3.12*a* will rotate by the same amount relative to their neighbors, and each of the straight lines will be transformed into a curve (helix) intersecting the various circles at the same angle (Fig. 3.12*b*).

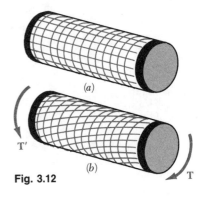

Fig. 3.12

The derivations given in this and the following sections will be based on the assumption of rigid end plates. Loading conditions encountered in practice may differ appreciably from those corresponding to the model of Fig. 3.12. The chief merit of this model is that it helps us define a torsion problem for which we can obtain an exact solution, just as the rigid-end-plates model of Sec. 2.16 made it possible for us to define an axial-load problem which could be easily and accurately solved. By virtue of Saint-Venant's principle, the results obtained for our idealized model may be extended to most engineering applications. However, we should keep these results associated in our mind with the specific model shown in Fig. 3.12.

We shall now determine the distribution of *shearing strains* in a circular shaft of length L and radius c which has been twisted through an angle ϕ (Fig. 3.13*a*). Detaching from the shaft a cylinder of radius ρ, we consider the small square element formed by two adjacent circles and two adjacent straight lines traced on the surface of the cylinder before any load is applied (Fig. 3.13*b*). As the shaft is subjected to a torsional load, the element deforms into a rhombus (Fig. 3.13*c*). We now recall from Sec. 2.14 that the shearing strain γ in a given element is measured by the change in the angles formed by the sides of that element. Since the circles defining two of the sides of the element considered here remain unchanged, the shearing strain γ must be equal to the angle between lines AB and $A'B$. (We recall that γ should be expressed in radians.)

We observe from Fig. 3.13*c* that, for small values of γ, we may express the arc length AA' as $AA' = L\gamma$. But, on the other hand, we have $AA' = \rho\phi$. It follows that $L\gamma = \rho\phi$, or

$$\gamma = \frac{\rho\phi}{L} \tag{3.2}$$

where γ and ϕ are both expressed in radians. The equation obtained shows, as we could have anticipated, that the shearing strain γ at a given point of a shaft in torsion is proportional to the angle of twist ϕ. It also

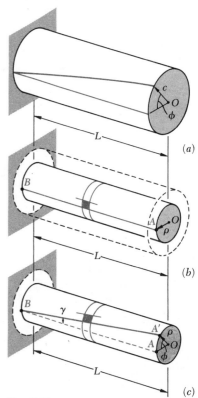

Fig. 3.13

shows that γ is proportional to the distance ρ from the axis of the shaft to the point under consideration. Thus, *the shearing strain in a circular shaft varies linearly with the distance from the axis of the shaft.*

It follows from Eq. (3.2) that the shearing strain is maximum on the surface of the shaft, where $\rho = c$. We have

$$\gamma_{max} = \frac{c\phi}{L} \qquad (3.3)$$

Eliminating ϕ from Eqs. (3.2) and (3.3) we may express the shearing strain γ at a distance ρ from the axis of the shaft as

$$\gamma = \frac{\rho}{c}\gamma_{max} \qquad (3.4)$$

3.4. STRESSES IN THE ELASTIC RANGE

No particular stress-strain relationship has been assumed so far in our discussion of circular shafts in torsion. We shall now consider the case when the torque **T** is such that all shearing stresses in the shaft remain below the yield strength τ_Y. We know from Chap. 2 that, for all practical purposes, this means that the stresses in the shaft will remain below the proportional limit and below the elastic limit as well. Thus, Hooke's law will apply and there will be no permanent deformation.

Recalling Hooke's law for shearing stress and strain from Sec. 2.14, we write

$$\tau = G\gamma \qquad (3.5)$$

where G is the modulus of rigidity or shear modulus of the material. Multiplying both members of Eq. (3.4) by G, we write

$$G\gamma = \frac{\rho}{c}G\gamma_{max}$$

or, making use of Eq. (3.5),

$$\tau = \frac{\rho}{c}\tau_{max} \qquad (3.6)$$

The equation obtained shows that, as long as the yield strength (or proportional limit) is not exceeded in any part of a circular shaft, *the shearing stress in the shaft varies linearly with the distance ρ from the axis of the shaft.* Figure 3.14a shows the stress distribution in a solid circular shaft of radius c, and Fig. 3.14b in a hollow circular shaft of inner radius c_1 and outer radius c_2. From Eq. (3.6), we find that, in the latter case,

$$\tau_{min} = \frac{c_1}{c_2}\tau_{max} \qquad (3.7)$$

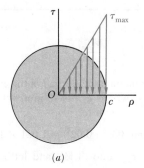 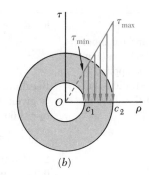

Fig. 3.14   (a)   (b)

We now recall from Sec. 3.2 that the sum of the moments of the elementary forces exerted on any cross section of the shaft must be equal to the magnitude T of the torque exerted on the shaft:

$$\int \rho(\tau \, dA) = T \tag{3.1}$$

Substituting for τ from (3.6) into (3.1), we write

$$T = \int \rho\tau \, dA = \frac{\tau_{max}}{c} \int \rho^2 \, dA$$

But the integral in the last member represents the polar moment of inertia J of the cross section with respect to its center O. We have therefore

$$T = \frac{\tau_{max}J}{c}$$

or, solving for τ_{max},

$$\tau_{max} = \frac{Tc}{J} \tag{3.8}$$

Substituting for τ_{max} from (3.8) into (3.6), we expess the shearing stress at any distance ρ from the axis of the shaft as

$$\tau = \frac{T\rho}{J} \tag{3.9}$$

Equations (3.8) and (3.9) are known as the *elastic torsion formulas*. We recall from statics that the polar moment of inertia of a circle of radius c is $J = \frac{1}{2}\pi c^4$. In the case of a hollow circular shaft of inner radius c_1 and outer radius c_2, the polar moment of inertia is

$$J = \tfrac{1}{2}\pi c_2^4 - \tfrac{1}{2}\pi c_1^4 = \tfrac{1}{2}\pi(c_2^4 - c_1^4) \tag{3.10}$$

We note that, in Eq. (3.8) or (3.9), T will be expressed in $N \cdot m$, c or ρ in meters, and J in m^4; we check that the resulting shearing stress will be expressed in N/m^2, that is, pascals (Pa).

Example 3.01

A hollow cylindrical steel shaft is 1.5 m long and has inner and outer diameters respectively equal to 40 and 60 mm (Fig. 3.15). (*a*) What is the largest torque which may be applied to the shaft if the shearing stress is not to exceed 120 MPa? (*b*) What is the corresponding minimum value of the shearing stress in the shaft?

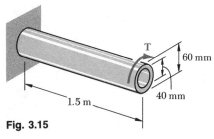

Fig. 3.15

(*a*) *Largest Permissible Torque.* The largest torque **T** which may be applied to the shaft is the torque for which $\tau_{max} = 120$ MPa. Since this value is less than the yield strength for steel, we may use Eq. (3.8). Solving this equation

for T, we have

$$T = \frac{J\tau_{max}}{c} \tag{3.11}$$

Recalling that the polar moment of inertia J of the cross section is given by Eq. (3.10), where $c_1 = \frac{1}{2}(40 \text{ mm}) = 0.02$ m and $c_2 = \frac{1}{2}(60 \text{ mm}) = 0.03$ m, we write

$$J = \tfrac{1}{2}\pi(c_2^4 - c_1^4) = \tfrac{1}{2}\pi(0.03^4 - 0.02^4) = 1.021 \times 10^{-6} \text{ m}^4$$

Substituting for J and τ_{max} into (3.11), and letting $c = c_2 = 0.03$ m, we have

$$T = \frac{J\tau_{max}}{c} = \frac{(1.021 \times 10^{-6} \text{ m}^4)(120 \times 10^6 \text{ Pa})}{0.03 \text{ m}} = 4.08 \text{ kN} \cdot \text{m}$$

(*b*) *Minimum Shearing Stress.* The minimum value of the shearing stress occurs on the inner surface of the shaft. It is obtained from Eq. (3.7), which expresses that τ_{min} and τ_{max} are respectively proportional to c_1 and c_2:

$$\tau_{min} = \frac{c_1}{c_2}\tau_{max} = \frac{0.02 \text{ m}}{0.03 \text{ m}}(120 \text{ MPa}) = 80 \text{ MPa}$$

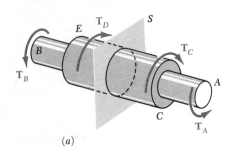

(*a*)

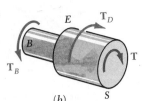

(*b*)

Fig. 3.16

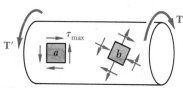

Fig. 3.17

The torsion formulas (3.8) and (3.9) were derived for a shaft of uniform circular cross section subjected to torques at its ends. However, they may also be used for a shaft of variable cross section or for a shaft subjected to torques at locations other than its ends (Fig. 3.16*a*). The distribution of shearing stress in a given cross section S of the shaft is obtained from Eq. (3.8), where J denotes the polar moment of inertia of that section, and where T represents the *internal torque* in that section. The value of T is obtained by drawing the free-body diagram of the portion of shaft located on one side of the section (Fig. 3.16*b*) and writing that the sum of the torques applied to that portion, including the internal torque **T,** is zero (see Sample Prob. 3.1).

Up to this point, our analysis of stresses in a shaft has been limited to shearing stresses. This is due to the fact that the element we had selected was oriented in such a way that its faces were either parallel or perpendicular to the axis of the shaft (Fig. 3.5). We know from earlier discussions (Secs. 1.7 and 1.8) that normal stresses, shearing stresses, or a combination of both may be found under the same loading condition, depending upon the orientation of the element which has been chosen. Consider the two elements *a* and *b* located on the surface of a circular shaft subjected to torsion (Fig. 3.17). Since the faces of element *a* are respectively parallel and perpendicular to the axis of the shaft, the only stresses on the element will be the shearing stresses defined by formula (3.8), namely $\tau_{max} = Tc/J$. On the other hand, the faces of element *b*, which form arbitrary angles with the axis of the shaft, will be subjected to a combination of normal and shearing stresses.

Let us consider the particular case of an element c (not shown) at 45° to the axis of the shaft. In order to determine the stresses on the faces of this element, we shall consider the two triangular elements shown in Fig. 3.18 and draw their free-body diagrams. In the case of the element of Fig. 3.18a, we know that the stresses exerted on the faces BC and BD are the shearing stresses $\tau_{max} = Tc/J$. The magnitude of the corresponding shearing forces is thus $\tau_{max}A_0$, where A_0 denotes the area of the face. Observing that the components along DC of the two shearing forces are equal and opposite, we conclude that the force **F** exerted on DC must be perpendicular to that face. It is a tensile force, and its magnitude is

$$F = 2(\tau_{max}A_0)\cos 45° = \tau_{max}A_0\sqrt{2} \tag{3.12}$$

The corresponding stress is obtained by dividing the force F by the area A of face DC. Observing that $A = A_0 \sqrt{2}$, we write

$$\sigma = \frac{F}{A} = \frac{\tau_{max}A_0\sqrt{2}}{A_0\sqrt{2}} = \tau_{max} \tag{3.13}$$

A similar analysis of the element of Fig. 3.18b shows that the stress on the face BE is $\sigma = -\tau_{max}$. Recalling Eq. (3.8), we conclude that the stresses exerted on the faces of an element c at 45° to the axis of the shaft (Fig. 3.19) are normal stresses equal to

$$\sigma_{45°} = \pm\frac{Tc}{J} \tag{3.14}$$

Thus, while the element a in Fig. 3.19 is in pure shear, the element c in the same figure is subjected to a tensile stress on two of its faces, and to a compressive stress on the other two. We also note that all the stresses involved have the same magnitude, Tc/J.†

As we saw in Sec. 2.3, ductile materials generally fail in shear. Therefore, when subjected to torsion, a specimen made of a ductile material breaks along a plane perpendicular to its longitudinal axis (Fig. 3.20a). On the other hand, brittle materials are weaker in tension than in shear. Thus, when subjected to torsion, a specimen made of a brittle material tends to break along surfaces which are perpendicular to the direction in which tension is maximum, i.e., along surfaces forming a 45° angle with the longitudinal axis of the specimen (Fig. 3.20b).

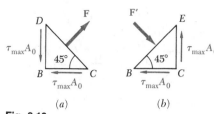

Fig. 3.18

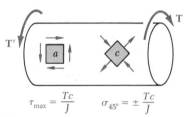

Fig. 3.19

Fig. 3.20   (a)   (b)

†Stresses on elements of arbitrary orientation, such as element b of Fig. 3.17, will be discussed in Chap. 6.

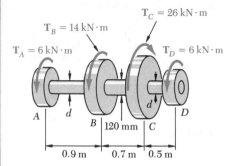

SAMPLE PROBLEM 3.1

Shaft BC is hollow with inner and outer diameters of 90 mm and 120 mm respectively. Shafts AB and CD are solid and of diameter d. For the loading shown, determine (a) the maximum and minimum shearing stress in shaft BC, (b) the required diameter d of shafts AB and CD if the allowable shearing stress in these shafts is 65 MPa.

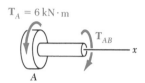

Equations of Statics. Denoting by T_{AB} the torque in shaft AB, we pass a section through shaft AB and, for the free body shown, we write

$$\Sigma M_x = 0: \qquad (6 \text{ kN} \cdot \text{m}) - T_{AB} = 0 \qquad T_{AB} = 6 \text{ kN} \cdot \text{m}$$

We now pass a section through shaft BC and, for the free body shown, we have

$$\Sigma M_x = 0: \quad (6 \text{ kN} \cdot \text{m}) + (14 \text{ kN} \cdot \text{m}) - T_{BC} = 0 \qquad T_{BC} = 20 \text{ kN} \cdot \text{m}$$

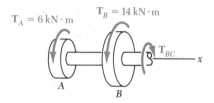

a. Shaft BC. For this hollow shaft we have

$$J = \frac{\pi}{2}(c_2^4 - c_1^4) = \frac{\pi}{2}[(0.060)^4 - (0.045)^4] = 13.92 \times 10^{-6} \text{ m}^4$$

Maximum Shearing Stress. On the outer surface, we have

$$\tau_{max} = \tau_2 = \frac{T_{BC}c_2}{J} = \frac{(20 \text{ kN} \cdot \text{m})(0.060 \text{ m})}{13.92 \times 10^{-6} \text{ m}^4} \qquad \tau_{max} = 86.2 \text{ MPa} \blacktriangleleft$$

Minimum Shearing Stress. We write that the stresses are proportional to the distance from the axis of the shaft.

$$\frac{\tau_{min}}{\tau_{max}} = \frac{c_1}{c_2} \qquad \frac{\tau_{min}}{86.2 \text{ MPa}} = \frac{45 \text{ mm}}{60 \text{ mm}} \qquad \tau_{min} = 64.7 \text{ MPa} \blacktriangleleft$$

b. Shafts AB and CD. We note that in both of these shafts the magnitude of the torque is $T = 6 \text{ kN} \cdot \text{m}$ and $\tau_{all} = 65 \text{ MPa}$. Denoting by c the radius of the shafts, we write

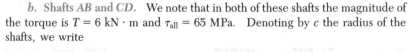

$$\tau = \frac{Tc}{J} \qquad 65 \text{ MPa} = \frac{(6 \text{ kN} \cdot \text{m})c}{\frac{\pi}{2}c^4}$$

$$c^3 = 58.8 \times 10^{-6} \text{ m}^3 \qquad c = 38.9 \times 10^{-3} \text{ m}$$

$$d = 2c = 2(38.9 \text{ mm}) \qquad d = 77.8 \text{ mm} \blacktriangleleft$$

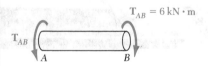

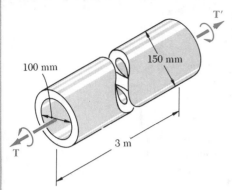

SAMPLE PROBLEM 3.2

The preliminary design of a large shaft connecting a motor to a generator calls for the use of a hollow shaft with inner and outer diameters of 100 mm and 150 mm respectively. Knowing that the allowable shearing stress is 85 MPa, determine the maximum torque which may be transmitted (*a*) by the shaft as designed, (*b*) by a solid shaft of the same weight, (*c*) by a hollow shaft of the same weight and of 200 mm outside diameter.

100 mm

150 mm

3 m

T

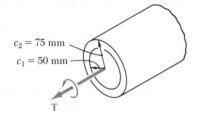

$c_2 = 75$ mm

$c_1 = 50$ mm

T

a. **Hollow Shaft as Designed.** For the hollow shaft we have

$$J = \frac{\pi}{2}(c_2^4 - c_1^4) = \frac{\pi}{2}[(0.075 \text{ m})^4 - (0.050 \text{ m})^4] = 39.9 \times 10^{-6} \text{ m}^4$$

Using Eq. (3.8), we write

$$\tau_{\max} = \frac{Tc_2}{J} \qquad 85 \text{ MPa} = \frac{T(0.075 \text{ m})}{39.9 \times 10^{-6} \text{ m}^4} \qquad T = 45.2 \text{ kN} \cdot \text{m} \quad \blacktriangleleft$$

b. **Solid Shaft of Equal Weight.** For the shaft as designed and this solid shaft to have the same weight and length, their cross-sectional areas must be equal.

$$A_{(a)} = A_{(b)}$$
$$\pi[(75 \text{ mm})^2 - (50 \text{ mm})^2] = \pi c_3^2 \qquad c_3 = 55.9 \text{ mm}$$

Since $\tau_{\text{all}} = 85$ MPa, we write

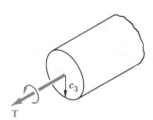

c_3

T

$$\tau_{\max} = \frac{Tc_3}{J} \qquad 85 \text{ MPa} = \frac{T(0.0559 \text{ m})}{\frac{\pi}{2}(0.0559 \text{ m})^4} \qquad T = 23.3 \text{ kN} \cdot \text{m} \quad \blacktriangleleft$$

c. **Hollow Shaft of 200-mm Diameter.** For equal weight, the cross-sectional areas again must be equal. We determine the inside diameter of the shaft by writing

$$A_{(a)} = A_{(c)}$$
$$\pi[(75 \text{ mm})^2 - (50 \text{ mm})^2] = \pi[(100 \text{ mm})^2 - c_5^2] \qquad c_5 = 82.92 \text{ mm}$$

For $c_5 = 82.92$ mm and $c_4 = 100$ mm,

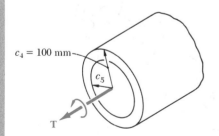

$c_4 = 100$ mm

c_5

T

$$J = \frac{\pi}{2}[(0.100 \text{ m})^4 - (0.08292 \text{ m})^4] = 82.82 \times 10^{-6} \text{ m}^4$$

With $\tau_{\text{all}} = 85$ MPa and $c_4 = 100$ mm,

$$\tau_{\max} = \frac{Tc_4}{J} \qquad 85 \text{ MPa} = \frac{T(0.100 \text{ m})}{82.82 \times 10^{-6} \text{ m}^4} \qquad T = 70.4 \text{ kN} \cdot \text{m} \quad \blacktriangleleft$$

PROBLEMS

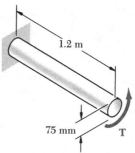

Fig. P3.1

Fig. P3.2

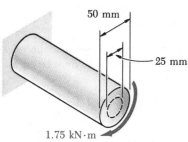

Fig. P3.3

3.1 (a) Determine the maximum shearing stress caused by a 5-kN · m torque **T** in the 75-mm-diameter solid aluminum shaft shown. (b) Solve part a, assuming that the solid shaft has been replaced by a hollow shaft of the same outer diameter and of 25-mm inner diameter.

3.2 (a) Determine the torque **T** which causes a maximum shearing stress of 45 MPa in the hollow cylindrical steel shaft shown. (b) Determine the maximum shearing stress caused by the same torque **T** in a solid cylindrical shaft of the same cross-sectional area.

3.3 A 1.75-kN · m torque is applied to the solid cylinder shown. Determine (a) the maximum shearing stress, (b) the percent of the torque carried by the inner 25-mm-diameter core.

3.4 (a) Determine the torque which may be applied to a solid shaft of 75-mm diameter without exceeding an allowable shearing stress of 80 MPa. (b) Solve part a, assuming that the solid shaft has been replaced by a hollow shaft of the same cross-sectional area and with an inner diameter equal to half its outer diameter.

3.5 The torques shown are exerted on pulleys A, B, and C. Knowing that both shafts are solid, determine the maximum shearing stress (a) in shaft AB, (b) in shaft BC.

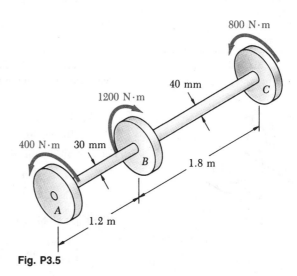

Fig. P3.5

3.6 Under normal operating conditions, the electric motor exerts a 1.2-kN · m torque at E.   Knowing that each shaft is solid, determine the maximum shearing stress (*a*) in shaft BC, (*b*) in shaft CD, (*c*) in shaft DE.

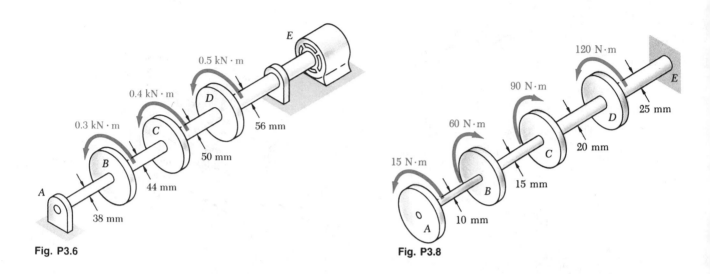

Fig. P3.6

Fig. P3.8

3.7 Solve Prob. 3.6, assuming that a 25-mm-diameter hole has been drilled into each shaft.

3.8 The torques shown are exerted on pulleys A, B, C, and D.   Knowing that each shaft is solid, determine (*a*) the shaft in which the maximum shearing stress occurs, (*b*) the magnitude of that stress.

3.9 Shaft AB is made of a steel with an allowable shearing stress of 90 MPa, and shaft BC is made of an aluminum alloy with an allowable shearing stress of 60 MPa.   Knowing that the diameter of shaft BC is 50 mm and neglecting the effect of stress concentration, determine (*a*) the largest torque **T** which may be applied at A if the allowable stress is not to be exceeded in shaft BC, (*b*) the corresponding required diameter of shaft AB.

3.10 Shaft AB has a 30-mm diameter and is made of a steel with an allowable shearing stress of 90 MPa, while shaft BC has a 50-mm diameter and is made of an aluminum alloy with an allowable shearing stress of 60 MPa.   Neglecting the effect of stress concentration, determine the largest torque **T** which may be applied at A.

3.11 The solid spindle AB has a diameter $d_s = 40$ mm and is made of a steel with an allowable shearing stress of 80 MPa, while sleeve CD is made of a brass with an allowable shearing stress of 50 MPa.   Determine the largest torque **T** which may be applied at A.

3.12 The solid spindle AB is made of a steel wih an allowable shearing stress of 80 MPa, and sleeve CD is made of a brass with an allowable shearing stress of 50 MPa.   Determine (*a*) the largest torque **T** which may be applied at A if the allowable stress is not to be exceeded in sleeve CD, (*b*) the corresponding required value of the diameter d_s of spindle AB.

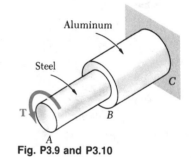

Fig. P3.9 and P3.10

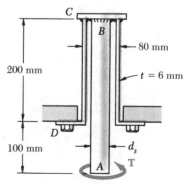

Fig. P3.11 and P3.12

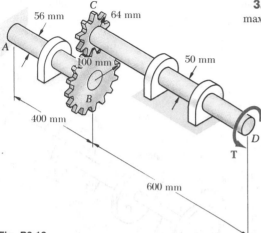

56 mm C 64 mm

A

100 mm

B

50 mm

400 mm

600 mm

D

T

Fig. P3.13

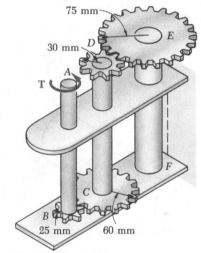

75 mm

30 mm D E

A

T

F

C

B

25 mm 60 mm

Fig. P3.17 and P3.18

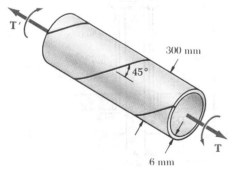

T'

300 mm

45°

T

6 mm

Fig. P3.19

3.13 Two solid steel shafts are connected by the gears shown. Determine the maximum shearing stress in each shaft when an 900-N · m torque **T** is applied at D.

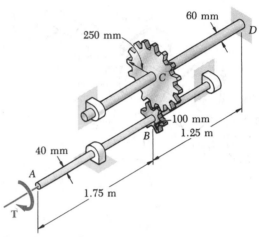

60 mm

250 mm

D

C

100 mm

B 1.25 m

40 mm

A

1.75 m

T

Fig. P3.14 and P3.15

3.14 Two solid steel shafts are connected by the gears shown. Determine the maximum shearing stress in each shaft when a 600-N · m torque **T** is applied at A.

3.15 Two solid steel shafts are connected by the gears shown. Determine the largest torque **T** which may be applied at A if the allowable shearing stress is 50 MPa in each shaft.

3.16 For the shafts of Prob. 3.13, determine the largest torque **T** which may be applied at D if the allowable shearing stress is 50 MPa in each shaft.

3.17 A torque of magnitude $T = 120$ N · m is applied to shaft AB of the gear train shown. Knowing that the allowable shearing stress is 75 MPa in each of the three solid shafts, determine the required diameter of (a) shaft AB, (b) shaft CD, (c) shaft EF.

3.18 A torque of magnitude $T = 100$ N · m is applied to shaft AB of the gear train shown. Knowing that the diameters of the three solid shafts are, respectively, $d_{AB} = 21$ mm, $d_{CD} = 30$ mm, and $d_{EF} = 40$ mm, determine the maximum shearing stress in (a) shaft AB, (b) shaft CD, (c) shaft EF.

3.19 A steel pipe of 300-mm outer diameter is fabricated from 6-mm-thick plate by welding along a helix which forms an angle of 45° with a plane perpendicular to the axis of the pipe. Knowing that the maximum allowable tensile stress in the weld is 80 MPa, determine the largest torque which may be applied to the pipe.

3.20 A cylindrical specimen 25 mm in diameter is subjected to a torque **T**. Determine the value of T for which the specimen will fail and the type of failure (tension or shear) if the specimen is made of (a) cold-rolled yellow brass ($\sigma_U = 510$ MPa, $\tau_U = 300$ MPa), (b) gray cast iron ($\sigma_U = 170$ MPa, $\tau_U = 240$ MPa).

3.21 While the exact distribution of the shearing stresses in a hollow cylindrical shaft is as shown in Fig. (a), an approximate value may be obtained for τ_{max} by assuming the stresses to be uniformly distributed over the area A of the cross section, as shown in Fig. (b), and then further assuming that all the elementary shearing forces act at a distance from O equal to the mean radius $r_m = \frac{1}{2}(c_1 + c_2)$ of the cross section. This approximate value is thus $\tau_0 = T/Ar_m$, where T is the applied torque. (a) Express the ratio τ_{max}/τ_0 of the true value of the maximum shearing stress and of its approximate value τ_0 in terms of the ratio c_1/c_2. (b) Determine the largest and smallest values of τ_{max}/τ_0 and the corresponding values of c_1/c_2.

3.22 (a) For the hollow shaft shown and a given allowable stress, determine the ratio T/w of the maximum allowable torque T and the weight per unit length w of the shaft. (b) Denoting by $(T/w)_0$ the value of this ratio computed for a solid shaft of the same radius c_2, express the ratio T/w for the hollow shaft in terms of $(T/w)_0$ and c_1/c_2.

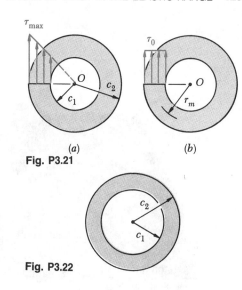

(a)   (b)

Fig. P3.21

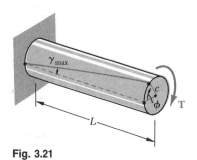

Fig. P3.22

3.5. ANGLE OF TWIST IN THE ELASTIC RANGE

In this section, we shall derive a relation between the angle of twist ϕ of a circular shaft and the torque $\mathbf{T}$ exerted on the shaft. The entire shaft will be assumed to remain elastic. Considering first the case of a shaft of length L and of uniform cross section of radius c subjected to a torque $\mathbf{T}$ at its free end (Fig. 3.21), we recall from Sec. 3.3 that the angle of twist ϕ and the maximum shearing strain γ_{max} are related as follows:

$$\gamma_{max} = \frac{c\phi}{L} \tag{3.3}$$

But, in the elastic range, the yield stress is not exceeded anywhere in the shaft, Hooke's law applies, and we have $\gamma_{max} = \tau_{max}/G$ or, recalling Eq. (3.8),

$$\gamma_{max} = \frac{\tau_{max}}{G} = \frac{Tc}{JG} \tag{3.15}$$

Equating the right-hand members of Eqs. (3.3) and (3.15), and solving for ϕ, we write

$$\phi = \frac{TL}{JG} \tag{3.16}$$

where ϕ is expressed in radians. The relation obtained shows that, within the elastic range, *the angle of twist ϕ is proportional to the torque T applied to the shaft.* This is in accordance with the experimental evidence cited at the beginning of Sec. 3.3.

Equation (3.16) provides us with a convenient method for determining the modulus of rigidity of a given material. A specimen of the mate-

Fig. 3.21

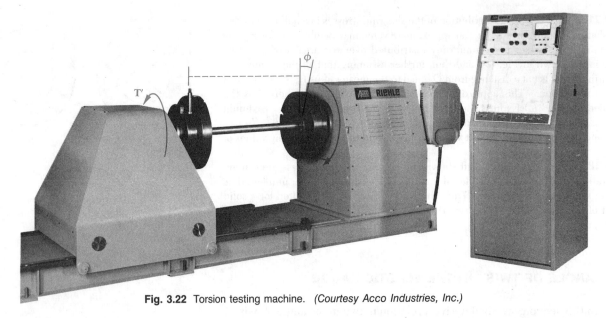

Fig. 3.22 Torsion testing machine. *(Courtesy Acco Industries, Inc.)*

rial, in the form of a cylindrical rod of known diameter and length, is placed in a *torsion testing machine* (Fig. 3.22). Torques of increasing magnitude T are applied to the specimen, and the corresponding values of the angle of twist ϕ in a length L of the specimen are recorded. As long as the yield stress of the material is not exceeded, the points obtained by plotting ϕ against T will fall on a straight line. The slope of this line represents the quantity JG/L, from which the modulus of rigidity G may be computed.

Example 3.02

What torque should be applied to the end of the shaft of Example 3.01 to produce a twist of 2°? Use the value $G = 80$ GPa for the modulus of rigidity of steel.

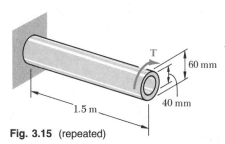

Fig. 3.15 (repeated)

Solving Eq. (3.16) for T, we write

$$T = \frac{JG}{L}\phi$$

Substituting the given values

$$G = 80 \times 10^9 \text{ Pa} \qquad L = 1.5 \text{ m}$$

$$\phi = 2°\left(\frac{2\pi \text{ rad}}{360°}\right) = 34.9 \times 10^{-3} \text{ rad}$$

and recalling from Example 3.01 that, for the given cross section,

$$J = 1.021 \times 10^{-6} \text{ m}^4$$

we have

$$T = \frac{JG}{L}\phi = \frac{(1.021 \times 10^{-6} \text{ m}^4)(80 \times 10^9 \text{ Pa})}{1.5 \text{ m}}(34.9 \times 10^{-3} \text{ rad})$$

$$T = 1.900 \times 10^3 \text{ N} \cdot \text{m} = 1.900 \text{ kN} \cdot \text{m}$$

Example 3.03

What angle of twist will create a shearing stress of 70 MPa on the inner surface of the hollow steel shaft of Examples 3.01 and 3.02?

The method of attack for solving this problem which first comes to mind is to use Eq. (3.9) to find the torque T corresponding to the given value of τ, and Eq. (3.16) to determine the angle of twist ϕ corresponding to the value of T just found.

A more direct solution, however, may be used. From Hooke's law, we first compute the shearing strain on the inner surface of the shaft:

$$\gamma_{min} = \frac{\tau_{min}}{G} = \frac{70 \times 10^6 \text{ Pa}}{80 \times 10^9 \text{ Pa}} = 875 \times 10^{-6}$$

Recalling Eq. (3.2), which was obtained by expressing the length of arc AA' in Fig. 3.13c in terms of both γ and ϕ, we have

$$\phi = \frac{L\gamma_{min}}{c_1} = \frac{1500 \text{ mm}}{20 \text{ mm}} (875 \times 10^{-6}) = 65.6 \times 10^{-3} \text{ rad}$$

To obtain the angle of twist in degrees, we write

$$\phi = (65.6 \times 10^{-3} \text{ rad})\left(\frac{360°}{2\pi \text{ rad}}\right) = 3.76°$$

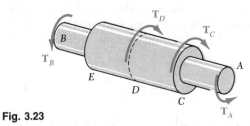

Fig. 3.23

Formula (3.16) for the angle of twist may be used only if the shaft is homogeneous (constant G), has a uniform cross section, and is loaded only at its ends. If the shaft is subjected to torques at locations other than its ends, or if it consists of several portions with various cross sections and possibly of different materials, we must divide it into component parts which satisfy individually the required conditions for the application of formula (3.16). In the case of the shaft AB shown in Fig. 3.23, for example, four different parts should be considered: AC, CD, DE, and EB. The total angle of twist of the shaft, i.e., the angle through which end A rotates with respect to end B, is obtained by adding *algebraically* the angles of twist of each component part. Denoting respectively by T_i, L_i, J_i, and G_i the internal torque, length, cross-sectional polar moment of inertia, and modulus of rigidity corresponding to part i, the total angle of twist of the shaft is expressed as

$$\phi = \sum_i \frac{T_i L_i}{J_i G_i} \tag{3.17}$$

The internal torque T_i in any given part of the shaft is obtained by passing a section through that part and drawing the free-body diagram of the portion of shaft located on one side of the section. This procedure, which has already been explained in Sec. 3.4 and illustrated in Fig. 3.16, is applied in Sample Prob. 3.3.

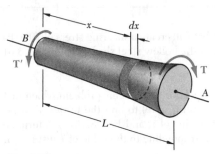

Fig. 3.24

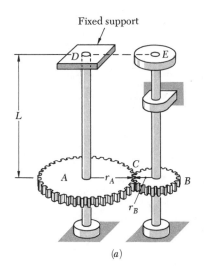

(a)

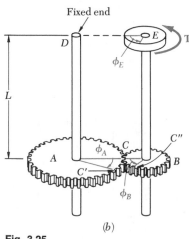

(b)

Fig. 3.25

In the case of a shaft with a variable circular cross section, as shown in Fig. 3.24, formula (3.16) may be applied to a disk of thickness dx. The angle by which one face of the disk rotates with respect to the other is thus

$$d\phi = \frac{T\,dx}{JG}$$

where J is a function of x which may be determined. Integrating in x from 0 to L, we obtain the total angle of twist of the shaft:

$$\phi = \int_0^L \frac{T\,dx}{JG} \tag{3.18}$$

The shaft shown in Fig. 3.21, which was used to derive formula (3.16), and the shaft of Fig. 3.15, which was discussed in Examples 3.02 and 3.03, both had one end attached to a fixed support. In each case, therefore, the angle of twist ϕ of the shaft was equal to the angle of rotation of its free end. When both ends of a shaft rotate, however, the angle of twist of the shaft is equal to the angle through which one end of the shaft rotates *with respect to the other*. Consider, for instance, the assembly shown in Fig. 3.25a, consisting of two elastic shafts AD and BE, each of length L, radius c, and modulus of rigidity G, which are attached to gears meshed at C. If a torque $\mathbf{T}$ is applied at E (Fig. 3.25b), both shafts will be twisted. Since the end D of shaft AD is fixed, the angle of twist of AD is measured by the angle of rotation ϕ_A of end A. On the other hand, since both ends of shaft BE rotate, the angle of twist of BE is equal to the difference between the angles of rotation ϕ_B and ϕ_E, i.e., the angle of twist is equal to the angle through which end E rotates with respect to end B. Denoting this relative angle of rotation by $\phi_{E/B}$, we write

$$\phi_{E/B} = \phi_E - \phi_B = \frac{TL}{JG}$$

Example 3.04

For the assembly of Fig. 3.25, and knowing that $r_A = 2r_B$, determine the angle of rotation of end E of shaft BE when the torque **T** is applied at E.

We first determine the torque $\mathbf{T}_{AD}$ exerted on shaft AD. Observing that equal and opposite forces **F** and **F'** are applied on the two gears at C (Fig. 3.26), and recalling that $r_A = 2r_B$, we conclude that the torque exerted on shaft AD is twice as large as the torque exerted on shaft BE; thus, $T_{AD} = 2T$.

Since the end D of shaft AD is fixed, the angle of rotation ϕ_A of gear A is equal to the angle of twist of the shaft and may be obtained by writing

$$\phi_A = \frac{T_{AD}L}{JG} = \frac{2TL}{JG}$$

Observing that the arcs CC' and CC'' in Fig. 3.25b must be equal, we write $r_A\phi_A = r_B\phi_B$ and obtain $\phi_B = (r_A/r_B)\phi_A = 2\phi_A$. We have, therefore,

$$\phi_B = 2\phi_A = \frac{4TL}{JG}$$

Considering now shaft BE, we recall that the angle of twist of the shaft is equal to the angle $\phi_{E/B}$ through which end E rotates with respect to end B. We have

$$\phi_{E/B} = \frac{T_{BE}L}{JG} = \frac{TL}{JG}$$

The angle of rotation of end E is obtained by writing

$$\phi_E = \phi_B + \phi_{E/B}$$

$$= \frac{4TL}{JG} + \frac{TL}{JG} = \frac{5TL}{JG}$$

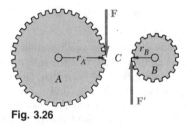

Fig. 3.26

3.6. STATICALLY INDETERMINATE SHAFTS

We saw in Sec. 3.4 that, in order to determine the stresses in a shaft, it was necessary to first calculate the internal torques in the various parts of the shaft. These torques were obtained from statics by drawing the free-body diagram of the portion of shaft located on one side of a given section and writing that the sum of the torques exerted on that portion was zero.

There are situations, however, where the internal torques cannot be determined from statics alone. In fact, in such cases the external torques themselves, i.e., the torques exerted on the shaft by the supports and connections cannot be determined from the free-body diagram of the entire shaft. The equilibrium equations must be complemented by relations involving the deformations of the shaft and obtained by considering the geometry of the problem. Because statics is not sufficient to determine the external and internal torques, the shafts are said to be *statically indeterminate*. The following example, as well as Sample Prob. 3.5, will show how to analyze statically indeterminate shafts.

Example 3.05

A circular shaft *AB* consists of a 250-mm-long, 20-mm-diameter steel cylinder, in which a 125-mm-long, 16-mm-diameter cavity has been drilled from end *B*. The shaft is attached to fixed supports at both ends, and a 120 N · m torque is applied at its mid-section (Fig. 3.27). Determine the torque exerted on the shaft by each of the supports.

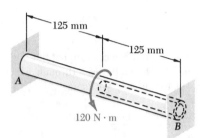

125 mm

125 mm

A

120 N · m

B

Fig. 3.27

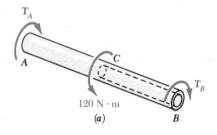

T_A

A

C

T_B

120 N · m

B

(a)

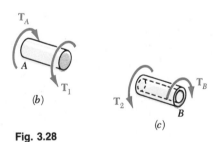

T_A

A

T_1

(b)

T_2

T_B

B

(c)

Fig. 3.28

Drawing the free-body diagram of the shaft and denoting by $\mathbf{T}_A$ and $\mathbf{T}_B$ the torques exerted by the supports (Fig. 3.28a), we obtain the equilibrium equation

$$T_A + T_B = 120 \text{ N} \cdot \text{m}$$

Since this equation is not sufficient to determine the two unknown torques $\mathbf{T}_A$ and $\mathbf{T}_B$, the shaft is statically indeterminate.

However, $\mathbf{T}_A$ and $\mathbf{T}_B$ may be determined if we observe that the total angle of twist of shaft *AB* must be zero, since both of its ends are restrained. Denoting by ϕ_1 and ϕ_2, respectively, the angles of twist of portions *AC* and *CB*, we write

$$\phi = \phi_1 + \phi_2 = 0$$

From the free-body diagram of a small portion of shaft including end *A* (Fig. 3.28b), we note that the internal torque T_1 in *AC* is equal to T_A; from the free-body diagram of a small portion of shaft including end *B* (Fig. 3.28c), we note that the internal torque T_2 in *CB* is equal to T_B. Recalling Eq. (3.16) and observing that portions *AC* and *CB* of the shaft are twisted in opposite senses, we write

$$\phi = \phi_1 + \phi_2 = \frac{T_A L_1}{J_1 G} - \frac{T_B L_2}{J_2 G} = 0$$

Solving for T_B, we have

$$T_B = \frac{L_1 J_2}{L_2 J_1} T_A$$

Substituting the numerical data

$$L_1 = L_2 = 125 \text{ mm}$$
$$J_1 = \tfrac{1}{2}\pi(10 \text{ mm})^4 = 15.71 \times 10^3 \text{ mm}^4$$
$$J_2 = \tfrac{1}{2}\pi[(10 \text{ mm})^4 - (8 \text{ mm})^4] = 9.27 \times 10^3 \text{ mm}^4$$

we obtain

$$T_B = 0.590 \, T_A$$

Substituting this expression into the original equilibrium equation, we write

$$1.590 \, T_A = 120 \text{ N} \cdot \text{m}$$
$$T_A = 75.5 \text{ N} \cdot \text{m} \qquad T_B = 44.5 \text{ N} \cdot \text{m}$$

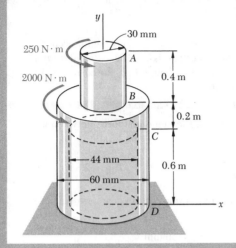

SAMPLE PROBLEM 3.3

The vertical shaft AD is attached to a fixed base at D and is subjected to the torques shown. A 44-mm-diameter hole has been drilled into portion CD of the shaft. Knowing that the entire shaft is made of steel for which $G = 80$ GPa, determine the angle of twist at end A.

Solution. Since the shaft consists of three portions AB, BC, and CD each of uniform cross section and each with a constant internal torque, Eq. (3.17) may be used.

Statics. Passing a section through the shaft between A and B and using the free body shown, we find

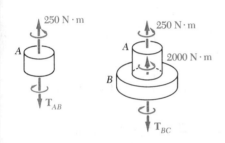

$$\Sigma M_y = 0: \qquad (250 \text{ N} \cdot \text{m}) - T_{AB} = 0 \qquad T_{AB} = 250 \text{ N} \cdot \text{m}$$

Passing now a section between B and C, we have

$$\Sigma M_y = 0: \quad (250 \text{ N} \cdot \text{m}) + (2000 \text{ N} \cdot \text{m}) - T_{BC} = 0 \qquad T_{BC} = 2250 \text{ N} \cdot \text{m}$$

Since no torque is applied at C,

$$T_{CD} = T_{BC} = 2250 \text{ N} \cdot \text{m}$$

Polar Moments of Inertia

$$J_{AB} = \frac{\pi}{2}c^4 = \frac{\pi}{2}(0.015 \text{ m})^4 = 0.0795 \times 10^{-6} \text{ m}^4$$

$$J_{BC} = \frac{\pi}{2}c^4 = \frac{\pi}{2}(0.030 \text{ m})^4 = 1.272 \times 10^{-6} \text{ m}^4$$

$$J_{CD} = \frac{\pi}{2}(c_2^4 - c_1^4) = \frac{\pi}{2}[(0.030 \text{ m})^4 - (0.022 \text{ m})^4] = 0.904 \times 10^{-6} \text{ m}^4$$

Angle of Twist. Using Eq. (3.17) and recalling that $G = 80$ GPa for the entire shaft, we have

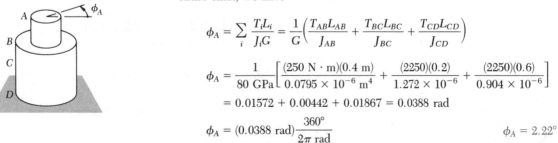

$$\phi_A = \sum_i \frac{T_i L_i}{J_i G} = \frac{1}{G}\left(\frac{T_{AB}L_{AB}}{J_{AB}} + \frac{T_{BC}L_{BC}}{J_{BC}} + \frac{T_{CD}L_{CD}}{J_{CD}}\right)$$

$$\phi_A = \frac{1}{80 \text{ GPa}}\left[\frac{(250 \text{ N} \cdot \text{m})(0.4 \text{ m})}{0.0795 \times 10^{-6} \text{ m}^4} + \frac{(2250)(0.2)}{1.272 \times 10^{-6}} + \frac{(2250)(0.6)}{0.904 \times 10^{-6}}\right]$$

$$= 0.01572 + 0.00442 + 0.01867 = 0.0388 \text{ rad}$$

$$\phi_A = (0.0388 \text{ rad})\frac{360°}{2\pi \text{ rad}} \qquad\qquad \phi_A = 2.22° \quad \blacktriangleleft$$

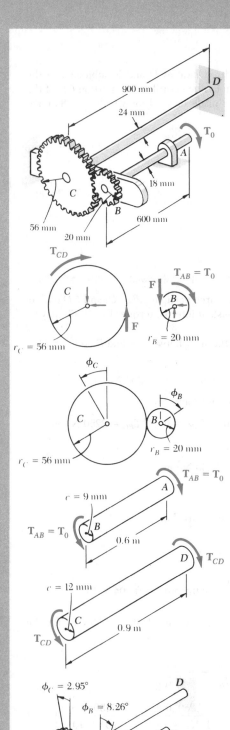

Two solid steel shafts are connected by the gears shown. Knowing for each shaft that $G = 80$ GPa and that the allowable shearing stress is 55 MPa, determine (a) the largest torque $\mathbf{T}_0$ which may be applied to end A of shaft AB, (b) the corresponding angle through which end A of shaft AB rotates.

Statics. Denoting by F the magnitude of the tangential force between gear teeth, we have

Gear B. $\Sigma M_B = 0$: $\qquad F(20 \text{ mm}) - T_0 = 0$

Gear C. $\Sigma M_C = 0$: $\qquad F(56 \text{ mm}) - T_{CD} = 0$ $\qquad\qquad T_{CD} = 2.8\,T_0 \qquad$ (1)

Kinematics. Noting that the peripheral motions of the gears are equal, we write

$$r_B\phi_B = r_C\phi_C \qquad \phi_B = \phi_C \frac{r_C}{r_B} = \phi_C \frac{56 \text{ mm}}{20 \text{ mm}} = 2.8\phi_C \qquad (2)$$

a. Torque T_0

Shaft AB. With $T_{AB} = T_0$ and $c = 0.009$ m, together with a maximum permissible shearing stress of 55 MPa, we write

$$\tau = \frac{T_{AB}c}{J} \qquad 55 \text{ MPa} = \frac{T_0(0.009 \text{ m})}{\dfrac{\pi}{2}(0.009 \text{ m})^4} \qquad T_0 = 63.0 \text{ N} \cdot \text{m}$$

Shaft CD. From (1) we have $T_{CD} = 2.8T_0$. With $c = 0.012$ m and $\tau_{\text{all}} = 55$ MPa, we write

$$\tau = \frac{T_{CD}c}{J} \qquad 55 \text{ MPa} = \frac{2.8T_0(0.012 \text{ m})}{\dfrac{\pi}{2}(0.012 \text{ m})^4} \qquad T_0 = 53.3 \text{ N} \cdot \text{m}$$

Maximum Permissible Torque. We choose the smaller value obtained for T_0

$$T_0 = 53.3 \text{ N} \cdot \text{m} \qquad \blacktriangleleft$$

b. Angle of Rotation at End A. We first compute the angle of twist for each shaft.

Shaft AB. For $T_{AB} = T_0 = 53.3$ N $\cdot$ m, we have

$$\phi_{A/B} = \frac{T_{AB}L}{GJ} = \frac{(53.3 \text{ N} \cdot \text{m})(0.6 \text{ m})}{(80 \text{ GPa})\dfrac{\pi}{2}(0.009 \text{ m})^4} = 0.0388 \text{ rad} = 2.22°$$

Shaft CD. $T_{CD} = 2.8T_0 = 2.8(53.3 \text{ N} \cdot \text{m})$

$$\phi_{C/D} = \frac{T_{CD}L}{GJ} = \frac{2.8(53.3 \text{ N} \cdot \text{m})(0.9 \text{ m})}{(80 \text{ GPa})\dfrac{\pi}{2}(0.012 \text{ m})^4} = 0.0515 \text{ rad} = 2.95°$$

Since end D of shaft CD is fixed, we have $\phi_C = \phi_{C/D} = 2.95°$. Using (2), we find the rotation of gear B is

$$\phi_B = 2.8\phi_C = 2.8(2.95°) = 8.26°$$

For end A of shaft AB, we have

$$\phi_A = \phi_B + \phi_{A/B} = 8.26° + 2.22° \qquad \phi_A = 10.48° \qquad \blacktriangleleft$$

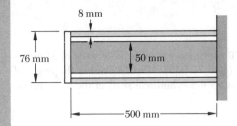

SAMPLE PROBLEM 3.5

A steel shaft and an aluminum tube are connected to a fixed support and to a rigid disk as shown in the cross section. Knowing that the initial stresses are zero, determine the maximum torque $\mathbf{T}_0$ which may be applied to the disk if the allowable stresses are 120 MPa in the steel shaft and 70 MPa in the aluminum tube. Use $G = 80$ GPa for steel and $G = 27$ GPa for aluminum.

Statics. *Free Body of Disk.* Denoting by $\mathbf{T}_1$ the torque exerted by the tube on the disk and by $\mathbf{T}_2$ the torque exerted by the shaft, we find

$$T_0 = T_1 + T_2 \tag{1}$$

Deformations. Since both the tube and the shaft are connected to the rigid disk, we have

$$\phi_1 = \phi_2 \qquad \frac{T_1 L_1}{J_1 G_1} = \frac{T_2 L_2}{J_2 G_2}$$

$$\frac{T_1(0.5 \text{ m})}{(2.003 \times 10^{-6} \text{ m}^4)(27 \text{ GPa})} = \frac{T_2(0.5 \text{ m})}{(0.614 \times 10^{-6} \text{ m}^4)(80 \text{ GPa})}$$

$$T_2 = 0.908 T_1 \tag{2}$$

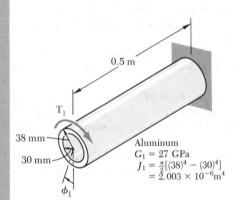

Aluminum
$G_1 = 27$ GPa
$J_1 = \frac{\pi}{2}[(38)^4 - (30)^4]$
$= 2.003 \times 10^{-6} \text{m}^4$

Shearing Stresses. We shall *assume* that the requirement $\tau_{\text{alum}} \leq 70$ MPa is critical. For the aluminum tube, we have

$$T_1 = \frac{\tau_{\text{alum}} J_1}{c_1} = \frac{(70 \text{ MPa})(2.003 \times 10^{-6} \text{ m}^4)}{0.038 \text{ m}} = 3690 \text{ N} \cdot \text{m}$$

Using Eq. (2), we compute the corresponding value T_2 and then find the maximum shearing stress in the steel shaft.

$$T_2 = 0.908 T_1 = 0.908(3690) = 3350 \text{ N} \cdot \text{m}$$

$$\tau_{\text{steel}} = \frac{T_2 c_2}{J_2} = \frac{(3350 \text{ N} \cdot \text{m})(0.025 \text{ m})}{0.614 \times 10^{-6} \text{ m}^4} = 136.4 \text{ MPa}$$

We note that the allowable steel stress of 120 MPa is exceeded; our assumption was *wrong.* Thus the maximum torque $\mathbf{T}_0$ will be obtained by making $\tau_{\text{steel}} = 120$ MPa. We first determine the torque $\mathbf{T}_2$.

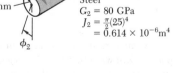

Steel
$G_2 = 80$ GPa
$J_2 = \frac{\pi}{2}(25)^4$
$= 0.614 \times 10^{-6} \text{m}^4$

$$T_2 = \frac{\tau_{\text{steel}} J_2}{c_2} = \frac{(120 \text{ MPa})(0.614 \times 10^{-6} \text{ m}^4)}{0.025 \text{ m}} = 2950 \text{ N} \cdot \text{m}$$

From Eq. (2), we have

$$2950 \text{ N} \cdot \text{m} = 0.908 T_1 \qquad T_1 = 3250 \text{ N} \cdot \text{m}$$

Using Eq. (1), we obtain the maximum permissible torque

$$T_0 = T_1 + T_2 = 3250 \text{ N} \cdot \text{m} + 2950 \text{ N} \cdot \text{m}$$

$$T_0 = 6.20 \text{ kN} \cdot \text{m} \quad \blacktriangleleft$$

PROBLEMS

3.23 (a) Determine the torque **T** which causes an angle of twist of 3° in the hollow cylindrical steel shaft shown (G = 77 GPa). (b) Determine the angle of twist caused by the same torque **T** in a solid cylindrical shaft of the same cross-sectional area.

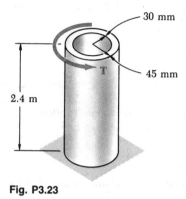

Fig. P3.23

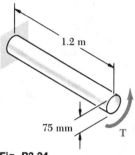

Fig. P3.24

3.24 (a) Determine the angle of twist caused by a 5-kN · m torque **T** in the 75-mm-diameter solid aluminum shaft shown (G = 26 GPa). (b) Solve part a, assuming that the solid shaft has been replaced by a hollow shaft of the same outer diameter and of 25-mm inner diameter.

3.25 Determine the largest allowable diameter of a 3-m-long steel rod (G = 77 GPa) if the rod is to be twisted through 90° without exceeding a shearing stress of 100 MPa.

3.26 While an oil well is being drilled at a depth of 2500 m, it is observed that the top of the 200-mm-diameter steel drill pipe (G = 77 GPa) rotates through 2.5 rev before the drilling bit starts to operate. Determine the maximum shearing stress caused in the pipe by torsion.

3.27 The torques shown are exerted on pulleys A, B, and C. Knowing that both shafts are solid and made of brass (G = 39 GPa), determine the angle of twist between (a) A and B, (b) A and C.

Fig. P3.27

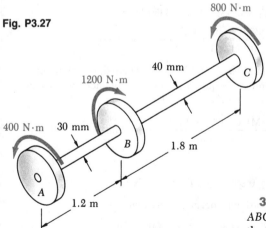

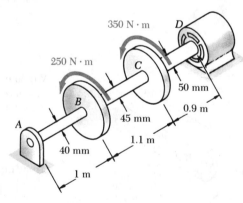

Fig. P3.28

3.28 The electric motor exerts a 600-N · m torque on the aluminum shaft ABCD when it is rotating at a constant speed. Knowing that G = 26 GPa and that the torques exerted on pulleys B and C are as shown, determine the angle of twist between (a) B and C, (b) B and D.

3.29 Solve Prob. 3.28, assuming that a 25-mm-diameter hole has been drilled into the entire shaft.

3.30 The torques shown are exerted on pulleys A, B, C, and D. Knowing that each shaft is solid, 120 mm long, and made of steel ($G = 77$ GPa), determine the angle of twist between (a) A and C, (b) A and E.

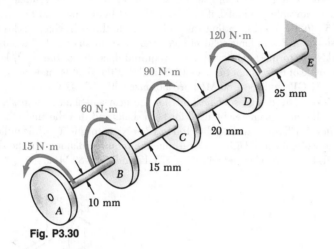

Fig. P3.30

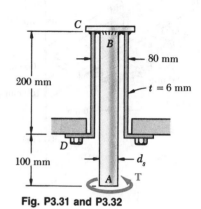

Fig. P3.31 and P3.32

3.31 The solid spindle AB has a diameter $d_s = 40$ mm and is made of a steel with $G = 77$ GPa and $\tau_{all} = 80$ MPa, while sleeve CD is made of a brass with $G = 38$ GPa and $\tau_{all} = 50$ MPa. Determine the largest angle through which end A may be rotated.

3.32 The solid spindle AB has a diameter $d_s = 45$ mm and is made of a steel with $G = 77$ GPa and $\tau_{all} = 80$ MPa, while sleeve CD is made of a brass with $G = 38$ GPa and $\tau_{all} = 50$ MPa. Determine (a) the largest torque **T** which may be applied at A if the given allowable stresses are not to be exceeded and if the angle of twist of sleeve CD is not to exceed 0.375°, (b) the corresponding angle through which end A will rotate.

3.33 Two solid steel shafts ($G = 77$ GPa) are connected by the gears shown. Determine the angle through which end A rotates when a 600-N $\cdot$ m torque **T** is applied at A.

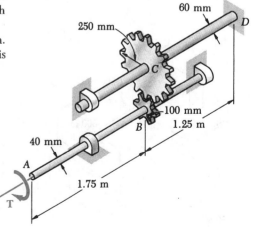

Fig. P3.33

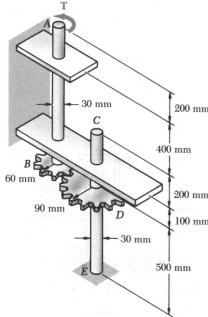

Fig. P3.34

3.34 Two solid steel shafts ($G = 77$ GPa) are connected by the gears shown. Determine the angle through which end A rotates when a 340-N · m torque **T** is applied at A.

3.35 For the gear-and-shaft system of Prob. 3.34, determine (a) the maximum allowable value of T if the angle of twist between D and E is not to exceed 2.5°, (b) the corresponding value of the angle through which end A will rotate.

3.36 For the gear-and-shaft system of Prob. 3.33, determine (a) the maximum allowable value of T if the angle of twist between C and D is not to exceed 0.75°, (b) the corresponding value of the angle through which end A will rotate.

3.37 A 1.5-m-long solid circular transmission shaft is to be subjected to a 12-kN · m torque. Determine the required diameter of the shaft, knowing that the angle of twist is not to exceed 1.5° and that the shaft is made of (a) a steel with $G = 77$ GPa and $\tau_{all} = 80$ MPa, (b) a bronze with $G = 42$ GPa and $\tau_{all} = 30$ MPa.

3.38 The design specifications for the gear-and-shaft system shown require that the same diameter be used for both shafts and that the angle through which pulley A will rotate when subjected to a 200-N · m torque $\mathbf{T}_A$ while pulley D is held fixed will not exceed 7.5°. Determine the required diameter of the shafts if both shafts are made of a steel with $G = 77$ GPa and $\tau_{all} = 80$ MPa.

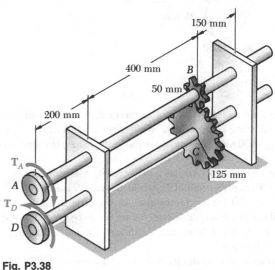

Fig. P3.38

3.39 Solve Prob. 3.39, assuming that both shafts are made of a brass with $G = 38$ GPa and $\tau_{all} = 55$ MPa.

3.40 A hole is punched at A in a plastic sheet by applying a 600-N force **P** to end D of lever CD, which is rigidly attached to the solid cylindrical shaft BC. Design specifications require that the displacement of D should not exceed 15 mm from the time the punch first touches the plastic sheet to the time it actually penetrates it. Determine the required diameter of shaft BC if the shaft is made of (a) a steel with $G = 77$ GPa and $\tau_{all} = 80$ MPa, (b) an aluminum with $G = 26$ GPa and $\tau_{all} = 70$ MPa.

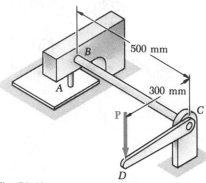

Fig. P3.40

3.41 The composite shaft shown consists of a 1.5-mm-thick steel jacket (G_s = 77 GPa) bonded to a 65-mm-diameter aluminum core (G_a = 26 GPa). Knowing that a 900-N · m torque **T** is applied at B, determine (a) the maximum shearing stress in the steel jacket, (b) the maximum shearing stress in the aluminum core, (c) the angle of twist at B.

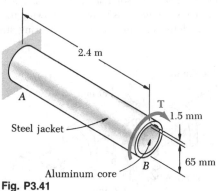

Steel jacket

Aluminum core

2.4 m

T

1.5 mm

65 mm

A

B

Fig. P3.41

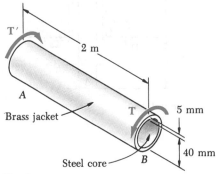

T'

Brass jacket

Steel core

2 m

T

5 mm

40 mm

A

B

Fig. P3.42

3.42 The composite shaft shown consists of a 5-mm-thick brass jacket (G_b = 39 GPa) bonded to a 40-mm-diameter steel core (G_s = 77 GPa). Knowing that the shaft is subjected to a 600-N · m torque, determine (a) the maximum shearing stress in the brass jacket, (b) the maximum shearing stress in the steel core, (c) the angle of twist of B relative to A.

3.43 For the composite shaft of Prob. 3.42, the allowable shearing stress is known to be 20 MPa in the brass jacket and 45 MPa in the steel core. Determine (a) the largest torque which may be applied to the shaft, (b) the corresponding angle of twist of B relative to A.

3.44 For the composite shaft of Prob 3.41, the allowable shearing stress is known to be 50 MPa in the steel jacket and 20 MPa in the aluminum core. Determine (a) the largest torque which may be applied to the shaft, (b) the corresponding angle of twist at B.

3.45 Two solid steel shafts (G = 77 GPa) are connected to a coupling disk B and to fixed supports at A and C. For the loading shown, determine (a) the reaction at each support, (b) the maximum shearing stress in shaft AB, (c) the maximum shearing stress in shaft BC.

3.46 Solve Prob. 3.45, assuming that shaft AB is replaced by a hollow shaft of the same outer diameter and of 25-mm inner diameter.

3.47 The solid cylinders AB and BC are bonded together at B and attached to fixed supports at A and C. Knowing that AB is made of aluminum (G = 26 GPa) and BC of brass (G = 39 GPa), determine for the loading shown (a) the reaction at each support, (b) the maximum shearing stress in AB, (c) the maximum shearing stress in BC.

3.48 Solve Prob. 3.47, assuming that AB is made of steel (G = 77 GPa) instead of aluminum.

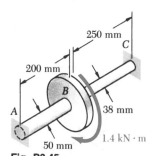

250 mm

C

200 mm

B

38 mm

A

50 mm

1.4 kN · m

Fig. P3.45

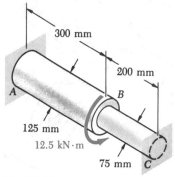

300 mm

200 mm

A

B

125 mm

12.5 kN·m

75 mm

C

Fig. P3.47

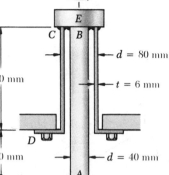

Fig. P3.49

3.49 The solid spindle AB and the sleeve CD are both bonded to the short cylinder E. Spindle AB is made of a steel with $G_s = 77$ GPa and $(\tau_{all})_s = 80$ MPa, while sleeve CD is made of a brass with $G_b = 38$ GPa and $(\tau_{all})_b = 50$ MPa. Determine the largest torque T which may be applied to cylinder E.

3.50 Knowing that a 2-kN · m torque T is applied to cylinder E of Prob. 3.49, determine (a) the maximum shearing stress in spindle AB, (b) the maximum shearing stress in sleeve CD.

3.51 Ends A and D of the two solid steel shafts AB and CD are fixed, while ends B and C are connected to gears as shown. Knowing that a 4-kN · m torque T is applied to gear B, determine the maximum shearing stress (a) in shaft AB, (b) in shaft CD.

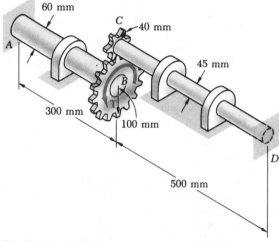

Fig. P3.51 and P3.52

3.52 Ends A and D of the two solid steel shafts AB and CD are fixed, while ends B and C are connected to gears as shown. Knowing that the allowable shearing stress is 50 MPa in each shaft, determine the largest torque T which may be applied to gear B.

3.53 An annular plate of thickness t and modulus of rigidity G is used to connect shaft AB of radius r_1 to tube CD of inner radius r_2. Knowing that a torque T is applied to end A of shaft AB and that end D of tube CD is fixed, (a) determine the magnitude and location of the maximum shearing stress in the annular plate, (b) show that the angle through which end B of the shaft rotates with respect to end C of the tube is

$$\phi_{B/C} = \frac{T}{4\pi G t}\left(\frac{1}{r_1^2} - \frac{1}{r_2^2}\right)$$

3.54 An annular brass plate ($G = 38$ GPa), of thickness $t = 6$ mm, is used to connect the brass shaft AB, of length $L_1 = 50$ mm and radius $r_1 = 30$ mm, to the brass tube CD, of length $L_2 = 125$ mm, inner radius $r_2 = 75$ mm, and 3-mm thickness. Knowing that a 2.8-kN · m torque T is applied to end A of shaft AB and that end D of tube CD is fixed, determine (a) the maximum shearing stress in the shaft-plate-tube system, (b) the angle through which end A rotates. (*Hint*: Use the formula derived in Prob. 3.53 to solve part b.)

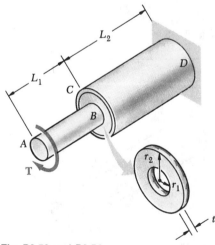

Fig. P3.53 and P3.54

3.55 A solid shaft and a hollow shaft are made of the same material and are of the same weight and length. Denoting by n the ratio c_1/c_2, show that the ratio T_s/T_h of the torque T_s in the solid shaft to the torque T_h in the hollow shaft is (a) $\sqrt{1 - n^2}/(1 + n^2)$ if the maximum stress is the same in both shafts, (b) $(1 - n^2)/(1 + n^2)$ if the angle of twist is the same for both shafts.

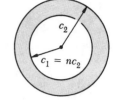

Fig. P3.55

3.56 A torque T is applied as shown to a solid tapered shaft AB. Show by integration that the angle of twist at A is

$$\phi_A = \frac{7TL}{12\pi Gc^4}$$

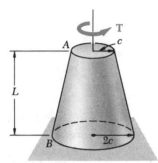

Fig. P3.56

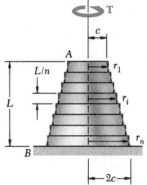

Fig. P3.57

3.57 Derive an approximate expression for the angle of twist of the tapered shaft of Prob. 3.56, replacing the tapered shaft by n cylindrical shafts of equal length and of radius $r_i = (n + i - \frac{1}{2})(c/n)$, where $i = 1, 2, \ldots, n$. What is the answer to Prob. 3.56 and the percentage error in that answer if the expression obtained is used with (a) $n = 4$, (b) $n = 8$, (c) $n = 20$, (d) $n = 100$. (*Note:* Solve parts c and d only if a computer or programmable calculator is available to you.)

3.7. DESIGN OF TRANSMISSION SHAFTS

The principal specifications to be met in the design of a transmission shaft are the *power* to be transmitted and the *speed of rotation* of the shaft. The role of the designer is to select the material and the dimensions of the cross section of the shaft, so that the maximum shearing stress allowable in the material will not be exceeded when the shaft is transmitting the required power at the specified speed.

To determine the torque exerted on the shaft, we recall from elementary dynamics that the power P associated with the rotation of a rigid body subjected to a torque $\mathbf{T}$ is

$$P = T\omega \qquad (3.19)$$

where ω is the angular velocity of the body expressed in radians per second. But $\omega = 2\pi f$, where f is the frequency of the rotation, i.e., the number of revolutions per second. The unit of frequency is thus 1 s^{-1} and is called a *hertz* (Hz). Substituting for ω into Eq. (3.19), we write

$$P = 2\pi f T \qquad (3.20)$$

We verify that, with f expressed in Hz and T in $\text{N} \cdot \text{m}$, the power will be expressed in $\text{N} \cdot \text{m/s}$, that is, in *watts* (W). Solving Eq. (3.20) for T, we obtain the torque exerted on a shaft transmitting the power P at a frequency of rotation f,

$$T = \frac{P}{2\pi f} \qquad (3.21)$$

where P, f, and T are expressed in the units indicated above.

After having determined the torque **T** which will be applied to the shaft and having selected the material to be used, the designer will carry the values of T and of the maximum allowable stress into the elastic torsion formula (3.8). Solving for J/c, we have

$$\frac{J}{c} = \frac{T}{\tau_{\text{max}}} \qquad (3.22)$$

and obtain in this way the minimum value allowable for the parameter J/c. We check that T will be expressed in $\text{N} \cdot \text{m}$, τ_{max} in Pa (or N/m^2), and J/c will be obtained in m^3. In the case of a solid circular shaft, $J = \frac{1}{2}\pi c^4$, and $J/c = \frac{1}{2}\pi c^3$; substituting this value for J/c into Eq. (3.22) and solving for c yields the minimum allowable value for the radius of the shaft. In the case of a hollow circular shaft, the critical parameter is J/c_2, where c_2 is the outer radius of the shaft; the value of this parameter may be computed from Eq. (3.10) of Sec. 3.4 to determine whether a given cross section will be acceptable.

Example 3.06

What size of shaft should be used for the rotor of a 7.5-kW motor operating at 3600 rpm if the shearing stress is not to exceed 60 MPa in the shaft?

We first express the frequency of the motor in cycles per second (or hertzes).

$$f = (3600 \text{ rpm}) \frac{1 \text{ Hz}}{60 \text{ rpm}} = 60 \text{ Hz} = 60 \text{ s}^{-1}$$

The torque exerted on the shaft is given by Eq. (3.21):

$$T = \frac{P}{2\pi f} = \frac{7.5 \text{ kW}}{2\pi (60 \text{ s}^{-1})} = 19.89 \text{ N} \cdot \text{m}$$

Substituting for T and τ_{max} into Eq. (3.22), we write

$$\frac{J}{c} = \frac{T}{\tau_{max}} = \frac{19.89 \text{ N} \cdot \text{m}}{60 \text{ MPa}} = 331.6 \times 10^{-9} \text{ m}^3$$

But $J/c = \frac{1}{2}\pi c^3$ for a solid shaft. We have therefore

$$\frac{1}{2}\pi c^3 = 331.6 \times 10^{-9} \text{ m}^3$$
$$c = 5.954 \times 10^{-3} \text{ m}$$
$$d = 2c = 11.91 \text{ mm}$$

A 12-mm shaft should be used.

Example 3.07

A shaft consisting of a steel tube of 50-mm outer diameter is to transmit 100 kW of power while rotating at a frequency of 20 Hz. Determine the tube thickness which should be used if the shearing stress is not to exceed 60 MPa.

The torque exerted on the shaft is given by Eq. (3.21):

$$T = \frac{P}{2\pi f} = \frac{100 \times 10^3 \text{ W}}{2\pi (20 \text{ Hz})} = 795.8 \text{ N} \cdot \text{m}$$

From Eq. (3.22) we conclude that the parameter J/c_2 must be at least equal to

$$\frac{J}{c_2} = \frac{T}{\tau_{max}} = \frac{795.8 \text{ N} \cdot \text{m}}{60 \times 10^6 \text{ N/m}^2} = 13.26 \times 10^{-6} \text{ m}^3 \quad (3.23)$$

But, from Eq. (3.10) we have

$$\frac{J}{c_2} = \frac{\pi}{2c_2} (c_2^4 - c_1^4) = \frac{\pi}{0.050} [(0.025)^4 - c_1^4] \quad (3.24)$$

Equating the right-hand members of Eqs. (3.23) and (3.24):

$$(0.025)^4 - c_1^4 = \frac{0.050}{\pi} (13.26 \times 10^{-6})$$
$$c_1^4 = 390.6 \times 10^{-9} - 211.0 \times 10^{-9} = 179.6 \times 10^{-9} \text{ m}^4$$
$$c_1 = 20.6 \times 10^{-3} \text{ m} = 20.6 \text{ mm}$$

The corresponding tube thickness is

$$c_2 - c_1 = 25 \text{ mm} - 20.6 \text{ mm} = 4.4 \text{ mm}$$

A tube thickness of 5 mm should be used.

3.8. STRESS CONCENTRATIONS IN CIRCULAR SHAFTS

The torsion formula $\tau_{max} = Tc/J$ was derived in Sec. 3.4 for a circular shaft of uniform cross section. Moreover, we had assumed earlier in Sec. 3.3 that the shaft was loaded at its ends through rigid end plates solidly attached to it. In practice, however, the torques are usually applied to the shaft through flange couplings (Fig. 3.29a) or through gears connected to the shaft by keys fitted into keyways (Fig. 3.29b). In both cases one should expect the distribution of stresses, in and near the section where the torques are applied, to be different from that given by the torsion formula. High concentrations of stresses, for example, will occur in the neighborhood of the keyway shown in Fig. 3.29b. The determination of these localized stresses may be carried out by experimental stress analysis methods or, in some cases, through the use of the mathematical theory of elasticity.

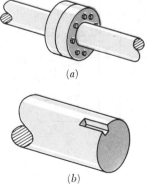

(a)

(b)

Fig. 3.29

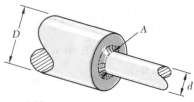

Fig. 3.30

As we indicated in Sec. 3.4, the torsion formula may also be used for a shaft of variable circular cross section. In the case of a shaft with an abrupt change in the diameter of its cross section, however, stress concentrations will occur near the discontinuity, with the highest stresses occurring at A (Fig. 3.30). These stresses may be reduced through the use of a fillet, and the maximum value of the shearing stress at the fillet may be expressed as

$$\tau_{\text{max}} = K\frac{Tc}{J} \tag{3.25}$$

where the stress Tc/J is the stress computed for the smaller-diameter shaft, and where K is a stress-concentration factor. Since the factor K depends only upon the ratio of the two diameters and the ratio of the radius of the fillet to the diameter of the smaller shaft, it may be computed once and for all and recorded in the form of a table or a graph, as shown in Fig. 3.31. We should note, however, that this procedure for determining localized shearing stresses is valid only as long as the value of τ_{max} given by Eq. (3.25) does not exceed the proportional limit of the material, since the values of K plotted in Fig. 3.31 were obtained under the assumption of a linear relation between shearing stress and shearing strain. If plastic deformations occur, they will result in values of the maximum stress lower than those indicated by Eq. (3.25).

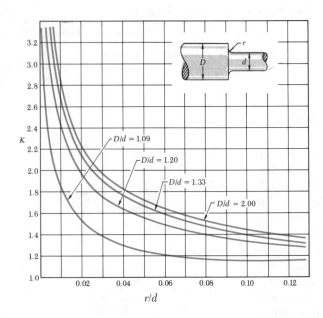

Fig. 3.31 Stress-concentration factors for fillets in circular shafts.†

†L. S. Jacobsen, "Torsional-Stress Concentrations in Shafts of Circular Cross Section and Variable Diameter," *Trans. A.S.M.E.*, vol. 47 (1925), pp. 619–638.

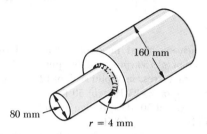

SAMPLE PROBLEM 3.6

The stepped shaft shown is to rotate at 900 rpm as it transmits power from a turbine to a generator. The grade of steel specified in the design has an allowable shearing stress of 55 MPa. (*a*) For the preliminary design shown, determine the maximum power that may be transmitted. (*b*) If in the final design the radius of the fillet is increased so that $r = 10$ mm, what will be the percent change, relative to the preliminary design, in the power which may be transmitted?

a. **Preliminary Design.** Using the notation of Fig. 3.31, we have: $D =$ 160 mm, $d = 80$ mm, $r = 4$ mm.

$$\frac{D}{d} = \frac{160 \text{ mm}}{80 \text{ mm}} = 2 \qquad \frac{r}{d} = \frac{4 \text{ mm}}{80 \text{ mm}} = 0.050$$

A stress concentration factor $K = 1.72$ is found from Fig. 3.31.

   Torque. Recalling Eq. (3.25), we write

$$\tau_{\max} = K \frac{Tc}{J} \qquad T = \frac{J}{c} \frac{\tau_{\max}}{K} \tag{1}$$

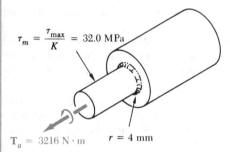

$\tau_m = \dfrac{\tau_{\max}}{K} = 32.0$ MPa

$\mathbf{T}_a = 3216$ N · m $r = 4$ mm

where J/c refers to the smaller-diameter shaft:

$$J/c = \tfrac{1}{2}\pi c^3 = \tfrac{1}{2}\pi (0.04 \text{ m})^3 = 100.5 \times 10^{-6} \text{ m}^3$$

and where

$$\frac{\tau_{\max}}{K} = \frac{55 \text{ MPa}}{1.72} = 32.0 \text{ MPa}$$

Substituting into Eq. (1), we find $T = (100.5 \times 10^{-6} \text{ m}^3)(32.0 \text{ MPa}) = 3216$ N · m.

   Power. Since $f = (900 \text{ rpm}) \dfrac{1 \text{ Hz}}{60 \text{ rpm}} = 15$ Hz, we write

$$P_a = 2\pi fT = 2\pi(15 \text{ Hz})(3216 \text{ N} \cdot \text{m}) = 303\,000 \text{ N} \cdot \text{m/s}$$

$$P_a = 303 \text{ kW} \blacktriangleleft$$

b. **Final Design.** For $r = 10$ mm,

$$\frac{D}{d} = 2 \qquad \frac{r}{d} = \frac{10 \text{ mm}}{80 \text{ mm}} = 0.125 \qquad K = 1.35$$

Following the procedure used above, we write

$$\frac{\tau_{\max}}{K} = \frac{55 \text{ MPa}}{1.35} = 40.7 \text{ MPa}$$

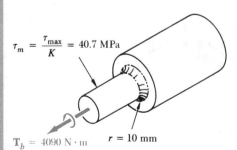

$\tau_m = \dfrac{\tau_{\max}}{K} = 40.7$ MPa

$\mathbf{T}_b = 4090$ N · m $r = 10$ mm

$$T = \frac{J}{c} \frac{\tau_{\max}}{K} = (100.5 \times 10^{-6} \text{ m}^3)(40.7 \text{ MPa}) = 4090 \text{ N} \cdot \text{m}$$

$$P_b = 2\pi fT = 2\pi(15 \text{ Hz})(4090 \text{ N} \cdot \text{m}) = 385\,000 \text{ N} \cdot \text{m/s}$$

$$P_b = 385 \text{ kW}$$

Percent Change in Power

$$\text{Percent change} = 100 \frac{P_b - P_a}{P_a} = 100 \frac{385 - 303}{303} = +27\% \blacktriangleleft$$

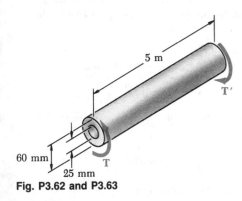

Fig. P3.62 and P3.63

PROBLEMS

3.58 Determine the maximum shearing stress in a solid shaft of 40-mm diameter as it transmits 60 kW at a speed of (a) 750 rpm, (b) 1500 rpm.

3.59 Determine the maximum shearing stress in a solid shaft of 12-mm diameter as it transmits 2.5 kW at a frequency of (a) 25 Hz, (b) 50 Hz.

3.60 Using an allowable shearing stress of 50 MPa, design a solid steel shaft to transmit 15 kW at a frequency of (a) 30 Hz, (b) 60 Hz.

3.61 Using an allowable shearing stress of 30 MPa, design a solid steel shaft to transmit 9 kW at a speed of (a) 1200 rpm, (b) 2400 rpm.

3.62 As the hollow steel shaft shown rotates at 180 rpm, a stroboscopic measurement indicates that the angle of twist of the shaft is 3°. Knowing that $G = 77$ GPa, determine (a) the power being transmitted, (b) the maximum shearing stress in the shaft.

3.63 The hollow steel shaft shown ($G = 77$ GPa, $\tau_{all} = 50$ MPa) rotates at 240 rpm. Determine (a) the maximum power which may be transmitted, (b) the corresponding angle of twist of the shaft.

3.64 A hollow steel drive shaft ($G = 77$ GPa) is 2.5 m long, and its outer and inner diameters are, respectively, 60 and 54 mm. Knowing that the shaft transmits 150 kW while rotating at 1500 rpm, determine (a) the maximum shearing stress, (b) the angle of twist of the shaft.

3.65 Knowing that $G = 77$ GPa and $\tau_{all} = 40$ MPa for the steel to be used, determine (a) the smallest permissible diameter of a solid shaft which must transmit 15 kW while rotating at 2400 rpm, (b) the corresponding angle of twist in a 1.8 m length of the shaft.

3.66 The two solid shafts and gears shown are used to transmit 12 kW from the motor at A, which rotates at a frequency of 20 Hz, to a machine tool at D. Knowing that each shaft has a diameter of 25 mm, determine the maximum shearing stress (a) in shaft AB, (b) in shaft CD.

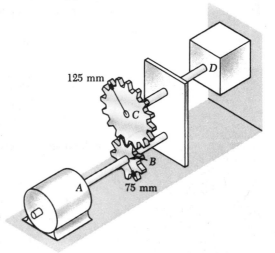

Fig. P3.66 and P3.67

3.67 The two solid shafts and gears shown are used to transmit 12 kW from the motor at A, which rotates at a frequency of 20 Hz, to a machine tool at D. Knowing that the maximum allowable shearing stress is 60 MPa for each shaft, determine the required diameter (a) of shaft AB, (b) of shaft CD.

3.68 A 2.5-m-long solid steel shaft ($G = 77$ GPa) is to transmit 10 kW at a speed of 1500 rpm. Determine the required diameter of the shaft, knowing that the allowable shearing stress is 30 MPa and that the angle of twist must not exceed 4°.

3.69 A 3-m-long solid steel shaft ($G = 77$ GPa) is to transmit 12 kW at a frequency of 25 Hz. Determine the required diameter of the shaft, knowing that the allowable shearing stress is 50 MPa and that the angle of twist must not exceed 10°.

3.70 A 2.5-m-long solid steel shaft ($G = 77$ GPa) of 30-mm diameter rotates at a frequency of 30 Hz. Determine the maximum power that the shaft may transmit, knowing that the allowable shearing stress is 50 MPa and that the angle of twist must not exceed 7.5°.

3.71 A 1.5-m-long solid steel shaft ($G = 77$ GPa) of 22-mm diameter is to transmit 12 kW. Determine the minimum speed at which the shaft may rotate, knowing that the allowable shearing stress is 30 MPa and that the angle of twist must not exceed 3.5°.

3.72 Knowing that the stepped shaft shown transmits a torque of magnitude $T = 300$ N · m, determine the maximum shearing stress in the shaft when the radius of the fillet is (*a*) $r = 0.8$ mm, (*b*) $r = 5$ mm.

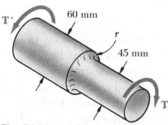

60 mm
T′
r
45 mm
T

Fig. P3.72 and P3.73

3.73 Knowing that the allowable shearing stress is 55 MPa for the stepped shaft shown, determine the magnitude T of the largest torque which may be transmitted by the shaft when the radius of the fillet is (*a*) $r = 1.5$ mm, (*b*) $r = 3$ mm.

3.74 The stepped shaft shown must transmit 120 kW at a speed of 480 rpm. Knowing that the allowable shearing stress in the shaft is 55 MPa, determine the smallest permissible radius r of the fillet.

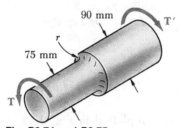

90 mm
T′
r
75 mm
T

Fig. P3.74 and P3.75

3.75 The stepped shaft shown must transmit 150 kW. Knowing that the allowable shearing stress in the shaft is 55 MPa and that the radius of the fillet is $r = 6$ mm, determine the smallest permissible frequency.

Fig. 3.32

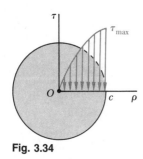

Fig. 3.33

Fig. 3.34

*3.9. PLASTIC DEFORMATIONS IN CIRCULAR SHAFTS

When we derived Eqs. (3.9) and (3.16), which define respectively the stress distribution and the angle of twist for a circular shaft subjected to a torque **T,** we assumed that Hooke's law applied throughout the shaft. If the yield strength is exceeded in some portion of the shaft, or if the material involved is a brittle material with a nonlinear shearing-stress-strain diagram, these relations cease to be valid. The purpose of this section is to develop a more general method—which may be used when Hooke's law does not apply—for determining the distribution of stresses in a solid circular shaft, and for computing the torque required to produce a given angle of twist.

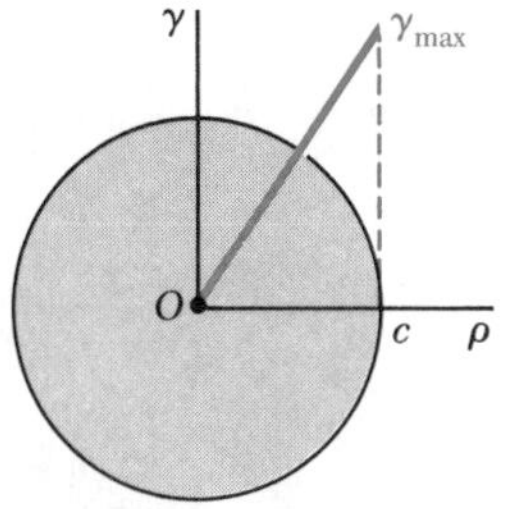

We first recall that no specific stress-strain relationship was assumed in Sec. 3.3, when we proved that the shearing strain γ varies linearly with the distance ρ from the axis of the shaft (Fig. 3.32). Thus, we may still use this property in our present analysis and write

$$\gamma = \frac{\rho}{c}\gamma_{\text{max}} \qquad (3.4)$$

where c is the radius of the shaft.

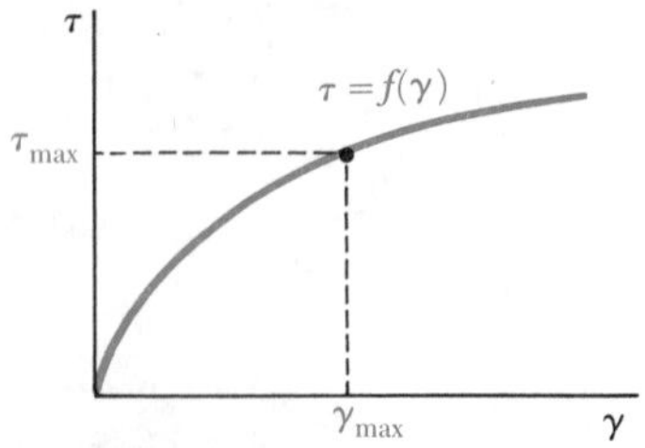

Assuming that the maximum value τ_{max} of the shearing stress τ has been specified, the plot of τ versus ρ may be obtained as follows. We first determine from the shearing-stress-strain diagram the value of γ_{max} corresponding to τ_{max} (Fig. 3.33), and carry this value into Eq. (3.4). Then, for each value of ρ, we determine the corresponding value of γ from Eq. (3.4) or Fig. 3.32 and obtain from the stress-strain diagram of Fig. 3.33 the shearing stress τ corresponding to this value of γ. Plotting τ against ρ yields the desired distribution of stresses (Fig. 3.34).

We now recall that, when we derived Eq. (3.1) in Sec. 3.2, we assumed no particular relation between shearing stress and strain. We may therefore use Eq. (3.1) to determine the torque **T** corresponding to the shearing-stress distribution obtained in Fig. 3.34. Considering an annular element of radius ρ and thickness $d\rho$, we express the element of area in Eq. (3.1) as $dA = 2\pi\rho \, d\rho$ and write

$$T = \int_0^c \rho\tau(2\pi\rho \, d\rho)$$

or

$$T = 2\pi \int_0^c \rho^2\tau \, d\rho \qquad (3.26)$$

where τ is the function of ρ plotted in Fig. 3.34.

If τ is a known analytical function of γ, Eq. (3.4) may be used to express τ as a function of ρ, and the integral in (3.26) may be determined analytically. Otherwise, the torque **T** may be obtained through a numerical integration. This computation becomes more meaningful if we note that the integral in Eq. (3.26) represents the second moment, or moment of inertia, with respect to the vertical axis of the area in Fig. 3.34 located above the horizontal axis and bounded by the stress-distribution curve.

An important value of the torque is the ultimate torque T_U which causes failure of the shaft. This value may be determined from the ultimate shearing stress τ_U of the material by choosing $\tau_{\max} = \tau_U$ and carrying out the computations indicated earlier. However, it is found more convenient in practice to determine T_U experimentally by twisting a specimen of a given material until it breaks. Assuming a fictitious linear distribution of stresses, Eq. (3.8) is then used to determine the corresponding maximum shearing stress R_T:

$$R_T = \frac{T_U c}{J} \qquad (3.27)$$

The fictitious stress R_T is called the *modulus of rupture in torsion* of the given material. It may be used to determine the ultimate torque T_U of a shaft made of the same material, but of different dimensions, by solving Eq. (3.27) for T_U. Since the actual and the fictitious linear stress distributions shown in Fig. 3.35 must yield the same value T_U for the ultimate torque, the areas they define must have the same moment of inertia with respect to the vertical axis. It is thus clear that the modulus of rupture R_T will always be larger than the actual ultimate shearing stress τ_U.

In some cases, we may wish to determine the stress distribution and the torque **T** corresponding to a given angle of twist ϕ. This may be done by recalling the expression obtained in Sec. 3.3 for the shearing strain γ in terms of ϕ, ρ, and the length L of the shaft:

$$\gamma = \frac{\rho \phi}{L} \qquad (3.2)$$

With ϕ and L given, we may determine from Eq. (3.2) the value of γ corresponding to any given value of ρ. Using the stress-strain diagram of the material, we may then obtain the corresponding value of the shearing stress τ and plot τ against ρ. Once the shearing-stress distribution has been obtained, the torque **T** may be determined analytically or numerically as explained earlier.

*3.10. CIRCULAR SHAFTS MADE OF AN ELASTOPLASTIC MATERIAL

Further insight into the plastic behavior of a shaft in torsion may be obtained by considering the idealized case of a *solid circular shaft made of an elastoplastic material*. The shearing-stress-strain diagram of such a material is shown in Fig. 3.36. Using this diagram, we may proceed as indicated earlier and find the stress distribution across a section of the shaft for any value of the torque **T**.

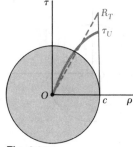

Fig. 3.35

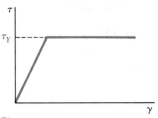

Fig. 3.36

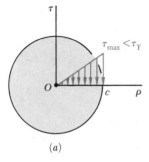

(a)

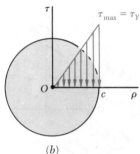

(b)

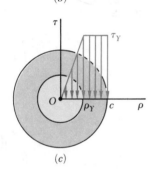

(c)

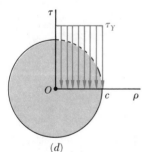

(d)

Fig. 3.37

As long as the shearing stress τ does not exceed the yield strength τ_Y, Hooke's law applies, and the stress distribution across the section is linear (Fig. 3.37a), with τ_{max} given by Eq. (3.8):

$$\tau_{max} = \frac{Tc}{J} \tag{3.8}$$

As the torque increases, τ_{max} eventually reaches the value τ_Y (Fig. 3.37b). Substituting this value into Eq. (3.8), and solving for the corresponding value of T, we obtain the value T_Y of the torque at the onset of yield:

$$T_Y = \frac{J}{c}\tau_Y \tag{3.28}$$

The value obtained is referred to as the *maximum elastic torque*, since it is the largest torque for which the deformation remains fully elastic. Recalling that for a solid circular shaft $J/c = \frac{1}{2}\pi c^3$, we have

$$T_Y = \tfrac{1}{2}\pi c^3 \tau_Y \tag{3.29}$$

As the torque further increases, a plastic region develops in the shaft, around an elastic core of radius ρ_Y (Fig. 3.37c). In the plastic region the stress is uniformly equal to τ_Y, while in the elastic core the stress varies linearly with ρ and may be expressed as

$$\tau = \frac{\tau_Y}{\rho_Y}\rho \tag{3.30}$$

As T increases, the plastic region expands until, at the limit, the deformation is fully plastic (Fig. 3.37d).

We shall use Eq. (3.26) to determine the value of the torque T corresponding to a given radius ρ_Y of the elastic core. Recalling that τ is given by Eq. (3.30) for $0 \leq \rho \leq \rho_Y$, and is equal to τ_Y for $\rho_Y \leq \rho \leq c$, we write

$$\begin{aligned}
T &= 2\pi \int_0^{\rho_Y} \rho^2 \left(\frac{\tau_Y}{\rho_Y}\rho\right)d\rho + 2\pi \int_{\rho_Y}^{c} \rho^2\,\tau_Y\,d\rho \\
&= \frac{1}{2}\pi\rho_Y^3\tau_Y + \frac{2}{3}\pi c^3\tau_Y - \frac{2}{3}\pi\rho_Y^3\tau_Y \\
T &= \frac{2}{3}\pi c^3\tau_Y\left(1 - \frac{1}{4}\frac{\rho_Y^3}{c^3}\right) \tag{3.31}
\end{aligned}$$

or, in view of Eq. (3.29),

$$T = \frac{4}{3}T_Y\left(1 - \frac{1}{4}\frac{\rho_Y^3}{c^3}\right) \tag{3.32}$$

where T_Y is the maximum elastic torque. We note that, as ρ_Y approaches zero, the torque approaches the limiting value

$$T_p = \frac{4}{3}T_Y \tag{3.33}$$

This value of the torque, which corresponds to a fully plastic deformation (Fig. 3.37d), is called the *plastic torque* of the shaft considered. We note that Eq. (3.33) is valid only for a *solid circular shaft made of an elastoplastic material*.

Since the distribution of *strain* across the section remains linear after the onset of yield, Eq. (3.2) remains valid and may be used to express the radius ρ_Y of the elastic core in terms of the angle of twist ϕ. If ϕ is large enough to cause a plastic deformation, the radius ρ_Y of the elastic core is obtained by making γ equal to the yield strain γ_Y in Eq. (3.2) and solving for the corresponding value ρ_Y of the distance ρ. We have

$$\rho_Y = \frac{L\gamma_Y}{\phi} \tag{3.34}$$

Let us denote by ϕ_Y the angle of twist at the onset of yield, i.e., when $\rho_Y = c$. Making $\phi = \phi_Y$ and $\rho_Y = c$ in Eq. (3.34), we have

$$c = \frac{L\gamma_Y}{\phi_Y} \tag{3.35}$$

Dividing (3.34) by (3.35), member by member, we obtain the following relation:†

$$\frac{\rho_Y}{c} = \frac{\phi_Y}{\phi} \tag{3.36}$$

If we carry into Eq. (3.32) the expression obtained for ρ_Y/c, we may express the torque T as a function of the angle of twist ϕ,

$$T = \frac{4}{3}T_Y\left(1 - \frac{1}{4}\frac{\phi_Y^3}{\phi^3}\right) \tag{3.37}$$

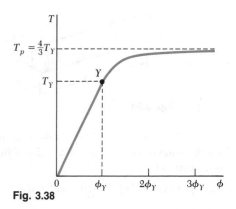

Fig. 3.38

where T_Y and ϕ_Y represent, respectively, the torque and the angle of twist at the onset of yield. Note that Eq. (3.37) may be used only for values of ϕ larger than ϕ_Y. For $\phi < \phi_Y$, the relation between T and ϕ is linear and given by Eq. (3.16). Combining both equations, we obtain the plot of T against ϕ represented in Fig. 3.38. We check that, as ϕ increases indefinitely, T approaches the limiting value $T_p = \frac{4}{3}T_Y$ corresponding to the case of a fully developed plastic zone (Fig. 3.37d). While the value T_p cannot actually be reached, we note from Eq. (3.37) that it is rapidly approached as ϕ increases. For $\phi = 2\phi_Y$, T is within about 3% of T_p, and for $\phi = 3\phi_Y$ within about 1 percent.

Since the plot of T against ϕ that we have obtained for an idealized elastoplastic material (Fig. 3.38) differs greatly from the shearing-stress-strain diagram of that material (Fig. 3.36), it is clear that the shearing-stress-strain diagram of an actual material cannot be obtained directly from a torsion test carried out on a solid circular rod made of that mate-

---

† Equation (3.36) applies to any ductile material with a well-defined yield point, since its derivation is independent of the shape of the stress-strain diagram beyond the yield point.

rial. However, a fairly accurate diagram may be obtained from a torsion test if the specimen used incorporates a portion consisting of a thin circular tube.† Indeed, we may assume that the shearing stress will have a constant value τ in that portion. Equation (3.1) thus reduces to

$$T = \rho A \tau$$

where ρ denotes the average radius of the tube and A its cross-sectional area. The shearing stress is thus proportional to the torque, and successive values of τ may be easily computed from the corresponding values of T. On the other hand, the values of the shearing strain γ may be obtained from Eq. (3.2) and from the values of ϕ and L measured on the tubular portion of the specimen.

Example 3.08

A solid circular shaft, 1.2 m long and 50 mm in diameter, is subjected to a 4.60-kN · m torque at each end (Fig. 3.39). Assuming the shaft to be made of an elastoplastic material with a yield strength in shear of 150 MPa and a modulus of rigidity of 80 GPa, determine (a) the radius of the elastic core, (b) the angle of twist of the shaft.

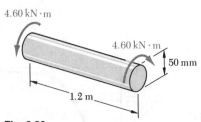

4.60 kN · m

4.60 kN · m

50 mm

1.2 m

Fig. 3.39

(a) *Radius of Elastic Core.* We first determine the torque T_Y at the onset of yield. Using Eq. (3.28) with $\tau_Y = 150$ MPa, $c = 25$ mm, and

$$J = \tfrac{1}{2}\pi c^4 = \tfrac{1}{2}\pi(25 \times 10^{-3} \text{ m})^4 = 614 \times 10^{-9} \text{ m}^4$$

we write

$$T_Y = \frac{J\tau_Y}{c} = \frac{(614 \times 10^{-9} \text{ m}^4)(150 \times 10^6 \text{ Pa})}{25 \times 10^{-3} \text{ m}} = 3.68 \text{ kN} \cdot \text{m}$$

Solving Eq. (3.32) for $(\rho_Y/c)^3$ and substituting the values of T and T_Y, we have

$$\left(\frac{\rho_Y}{c}\right)^3 = 4 - \frac{3T}{T_Y} = 4 - \frac{3(4.60 \text{ kN} \cdot \text{m})}{3.68 \text{ kN} \cdot \text{m}} = 0.250$$

$$\frac{\rho_Y}{c} = 0.630 \qquad \rho_Y = 0.630(25 \text{ mm}) = 15.8 \text{ mm}$$

(b) *Angle of Twist.* We first determine the angle of twist ϕ_Y at the onset of yield from Eq. (3.16):

$$\phi_Y = \frac{T_Y L}{JG} = \frac{(3.68 \times 10^3 \text{ N} \cdot \text{m})(1.2 \text{ m})}{(614 \times 10^{-9} \text{ m}^4)(80 \times 10^9 \text{ Pa})} = 89.9 \times 10^{-3} \text{ rad}$$

Solving Eq. (3.36) for ϕ and substituting the values obtained for ϕ_Y and ρ_Y/c, we write

$$\phi = \frac{\phi_Y}{\rho_Y/c} = \frac{89.9 \times 10^{-3} \text{ rad}}{0.630} = 142.7 \times 10^{-3} \text{ rad}$$

or

$$\phi = (142.7 \times 10^{-3} \text{ rad})\left(\frac{360°}{2\pi \text{ rad}}\right) = 8.18°$$

*3.11. RESIDUAL STRESSES IN CIRCULAR SHAFTS

We saw in the two preceding sections that a plastic region will develop in a shaft subjected to a large enough torque, and that the shearing stress τ at any given point in the plastic region may be obtained from the shearing-stress-strain diagram of Fig. 3.33. If the torque is removed, the resulting reduction of stress and strain at the point considered will take

† In order to minimize the possibility of failure by buckling, the specimen should be made so that the length of the tubular portion is no longer than its diameter.

place along a straight line (Fig. 3.40). As we shall see further in this section, the final value of the stress will not, in general, be zero. There will be a residual stress at most points, and that stress may be either positive or negative. We note that, as was the case for the normal stress, the shearing stress will keep decreasing until it has reached a value equal to its maximum value at C minus twice the yield strength of the material.

We shall consider again the idealized case of the elastoplastic material characterized by the shearing-stress-strain diagram of Fig. 3.36. Assuming that the relation between τ and γ at any point of the shaft remains linear as long as the stress does not decrease by more than $2\tau_Y$, we may use Eq. (3.16) to obtain the angle through which the shaft untwists as the torque decreases back to zero. As a result, the unloading of the shaft will be represented by a straight line on the T-ϕ diagram (Fig. 3.41). We note that the angle of twist does not return to zero after the torque has been removed. Indeed, the loading and unloading of the shaft result in a permanent deformation characterized by the angle

$$\phi_p = \phi - \phi' \tag{3.38}$$

where ϕ corresponds to the loading phase and may be obtained from T by solving Eq. (3.37), and where ϕ' corresponds to the unloading phase and may be obtained from Eq. (3.16).

The residual stresses in an elastoplastic material are obtained by applying the principle of superposition in a manner similar to that described in Sec. 2.19 for an axial loading. We consider, on one hand, the stresses due to the application of the given torque **T** and, on the other, the stresses due to the equal and opposite torque which is applied to unload the shaft. The first group of stresses reflects the elastoplastic behavior of the material during the loading phase (Fig. 3.42a), and the second group the linear behavior of the same material during the unloading phase (Fig. 3.42b). Adding the two groups of stresses, we obtain the distribution of the residual stresses in the shaft (Fig. 3.42c).

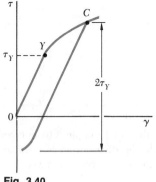

Fig. 3.40

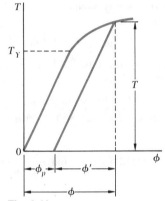

Fig. 3.41

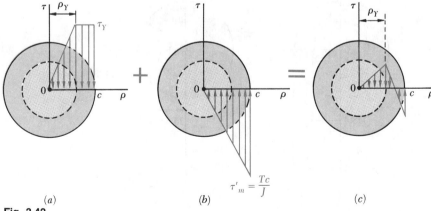

(a)   (b)   (c)

Fig. 3.42

We note from Fig. 3.42c that some residual stresses have the same sense as the original stresses, while others have the opposite sense. This was to be expected since, according to Eq. (3.1), the relation

$$\int \rho(\tau \, dA) = 0 \qquad (3.39)$$

must be verified after the torque has been removed.

Example 3.09

For the shaft of Example 3.08 determine (a) the permanent twist, (b) the distribution of residual stresses, after the 4.60-kN · m torque has been removed.

(a) *Permanent Twist.* We recall from Example 3.08 that the angle of twist corresponding to the given torque is $\phi = 8.18°$. The angle ϕ' through which the shaft untwists as the torque is removed is obtained from Eq. (3.16). Substituting the given data,

$$T = 4.60 \times 10^3 \text{ N} \cdot \text{m} \qquad L = 1.2 \text{ m} \qquad G = 80 \times 10^9 \text{ Pa}$$

and the value $J = 614 \times 10^{-9} \text{ m}^4$ obtained in the solution of Example 3.08, we have

$$\phi' = \frac{TL}{JG} = \frac{(4.60 \times 10^3 \text{ N} \cdot \text{m})(1.2 \text{ m})}{(614 \times 10^{-9} \text{ m}^4)(80 \times 10^9 \text{ Pa})} = 112.4 \times 10^{-3} \text{ rad}$$

or

$$\phi' = (112.4 \times 10^{-3} \text{ rad}) \frac{360°}{2\pi \text{ rad}} = 6.44°$$

The permanent twist is therefore

$$\phi_p = \phi - \phi' = 8.18° - 6.44° = 1.74°$$

(b) *Residual Stresses.* We recall from Example 3.08 that the yield strength is $\tau_Y = 150$ MPa and that the radius of the elastic core corresponding to the given torque is $\rho_Y = 15.8$ mm. The distribution of the stresses in the loaded shaft is thus as shown in Fig. 3.43a.

The distribution of stresses due to the opposite 4.60-kN · m torque required to unload the shaft is linear and as shown in Fig. 3.43b. The maximum stress in the distribution of the reverse stresses is obtained from Eq. (3.8):

$$\tau'_{max} = \frac{Tc}{J} = \frac{(4.60 \times 10^3 \text{ N} \cdot \text{m})(25 \times 10^{-3} \text{ m})}{614 \times 10^{-9} \text{ m}^4} = 187.3 \text{ MPa}$$

Superposing the two distributions of stresses, we obtain the residual stresses shown in Fig. 3.43c. We check that, even though the reverse stresses exceed the yield strength τ_Y, the assumption of a linear distribution of these stresses is valid, since they do not exceed $2\tau_Y$.

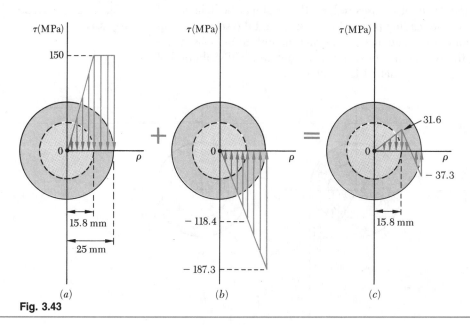

(a)　　　　　(b)　　　　　(c)

Fig. 3.43

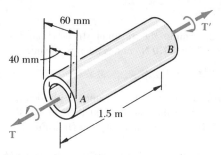

SAMPLE PROBLEM 3.7

Shaft AB is made of a mild steel which is assumed to be elastoplastic with $G = 80$ GPa and $\tau_Y = 150$ MPa. A torque $\mathbf{T}$ is applied and gradually increased in magnitude. Determine the magnitude of $\mathbf{T}$ and the corresponding angle of twist (a) when yield first occurs, (b) when the deformation has become fully plastic.

Solution. The geometric properties of the cross section are

$$c_1 = \tfrac{1}{2}(40 \text{ mm}) = 20 \text{ mm} \qquad c_2 = \tfrac{1}{2}(60 \text{ mm}) = 30 \text{ mm}$$
$$J = \tfrac{1}{2}\pi(c_2^4 - c_1^4) = \tfrac{1}{2}\pi[(30 \text{ mm})^4 - (20 \text{ mm})^4] = 1.02 \times 10^6 \text{ mm}^4$$

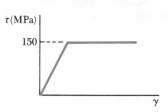

a. **Onset of Yield.** For $\tau_{max} = \tau_Y = 150$ MPa, we find

$$T_Y = \frac{\tau_Y J}{c_2} = \frac{(150 \text{ MPa})(1.02 \times 10^6 \text{ mm}^4)}{30 \text{ mm}} \qquad T_Y = 5105 \text{ N} \cdot \text{m} \quad \blacktriangleleft$$

Making $\rho = c_2$ and $\gamma = \gamma_Y$ in Eq. (3.2) and solving for ϕ, we obtain the value of ϕ_Y:

$$\phi_Y = \frac{\gamma_Y L}{c_2} = \frac{\tau_Y L}{c_2 G} = \frac{(150 \text{ MPa})(1.5 \text{ m})}{(30 \text{ mm})(80 \text{ GPa})} = 0.0937 \text{ rad} \qquad \phi_Y = 5.37° \quad \blacktriangleleft$$

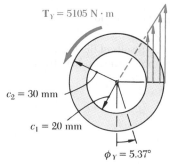

b. **Fully Plastic Deformation.** When the plastic zone reaches the inner surface, the stresses are uniformly distributed as shown. Using Eq. (3.26), we write

$$T_p = 2\pi\tau_Y \int_{c_1}^{c_2} \rho^2 \, d\rho = \tfrac{2}{3}\pi\tau_Y(c_2^3 - c_1^3)$$

$$= \tfrac{2}{3}\pi(150 \text{ MPa})[(30 \text{ mm})^3 - (20 \text{ mm})^3] \qquad T_p = 5970 \text{ N} \cdot \text{m} \quad \blacktriangleleft$$

When yield first occurs on the inner surface, the deformation is fully plastic; we have from Eq. (3.2):

$$\phi_f = \frac{\gamma_Y L}{c_1} = \frac{\tau_Y L}{c_1 G} = \frac{(150 \text{ MPa})(1.5 \text{ m})}{(20 \text{ mm})(80 \text{ GPa})} = 0.1406 \text{ rad} \qquad \phi_f = 8.06° \quad \blacktriangleleft$$

For larger angles of twist the torque remains constant; the T-ϕ diagram of the shaft is as shown.

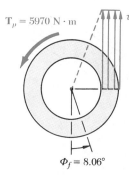

For the shaft of Sample Prob. 3.7, determine the residual stresses and the permanent angle of twist after the torque $T_p = 5970$ N · m has been removed.

Solution. Referring to Sample Prob. 3.7, we recall that when the plastic zone first reached the inner surface, the applied torque was $T_p = 5970$ N · m and the corresponding angle of twist was $\phi_f = 8.06°$. These values are shown in Fig. 1.

Elastic Unloading. We unload the shaft by applying a 5970-N · m torque in the sense shown in Fig. 2. During this unloading, the behavior of the material is linear. Recalling from Sample Prob. 3.7 the values found for c_1, c_2, and J, we obtain the following stresses and angle of twist:

$$\tau_{\max} = \frac{Tc_2}{J} = \frac{(5970 \text{ N} \cdot \text{m})(30 \text{ mm})}{1.02 \times 10^6 \text{ mm}^4} = 175.4 \text{ MPa}$$

$$\tau_{\min} = \tau_{\max} \frac{c_1}{c_2} = (175.4 \text{ MPa}) \frac{20 \text{ mm}}{30 \text{ mm}} = 116.9 \text{ MPa}$$

$$\phi' = \frac{TL}{JG} = \frac{(5970 \text{ N} \cdot \text{m})(1.5 \text{ m})}{(1.02 \times 10^6 \text{ mm}^4)(80 \text{ GPa})} = 0.1096 \text{ rad} = 6.28°$$

Residual Stresses and Permanent Twist. The results of the loading (Fig. 1) and the unloading (Fig. 2) are superposed (Fig. 3) to obtain the residual stresses and the permanent angle of twist ϕ_p.

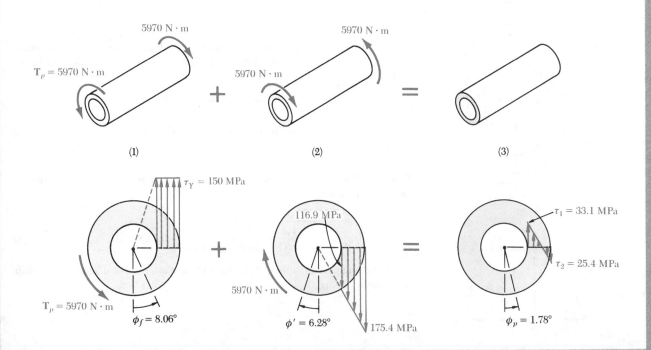

PROBLEMS

3.76 The solid shaft shown is made of a mild steel which is assumed to be elastoplastic with $G = 77$ GPa and $\tau_Y = 145$ MPa. Determine the maximum shearing stress and the radius of the elastic core caused by the application of a torque of magnitude (*a*) $T = 10$ kN · m, (*b*) $T = 15$ kN · m.

3.77 The solid shaft shown is made of a mild steel which is assumed to be elastoplastic with $G = 77$ GPa and $\tau_Y = 145$ MPa. Determine the maximum shearing stress and the radius of the elastic core caused by the application of a torque of magnitude (*a*) $T = 600$ N · m, (*b*) $T = 1000$ N · m.

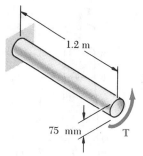

Fig. P3.76

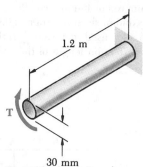

Fig. P3.77 and P3.78

3.78 The solid shaft shown is made of a mild steel which is assumed to be elastoplastic with $G = 77$ GPa and $\tau_Y = 145$ MPa. Determine the magnitude of the torque **T** for which the radius of the elastic core is (*a*) 7.5 mm, (*b*) 5 mm.

3.79 For the solid shaft of Prob. 3.76, determine the magnitude of the torque **T** for which the plastic zone is (*a*) 6 mm deep, (*b*) 20 mm deep.

3.80 For the solid shaft of Prob. 3.76, determine the angle of twist caused by the application of a torque of magnitude (*a*) $T = 9$ kN · m, (*b*) $T = 15$ kN · m.

3.81 For the solid shaft of Prob. 3.77, determine the angle of twist caused by the application of a torque of magnitude (*a*) $T = 600$ N · m, (*b*) $T = 1000$ N · m.

3.82 For the solid shaft of Prob. 3.77, determine (*a*) the magnitude of the torque **T** required to twist the shaft through an angle of 15°, (*b*) the radius of the corresponding elastic core.

3.83 A 0.9-mm-long solid shaft has a diameter of 60 mm and is made of a mild steel which is assumed to be elastoplastic with $G = 77$ GPa and $\tau_Y = 145$ MPa. Determine the torque required to twist the shaft through an angle of (*a*) 2.5°, (*b*) 5°.

3.84 The hollow shaft shown is made of a steel which is assumed to be elastoplastic with $G = 77$ MPa and $\tau_Y = 145$ MPa. Determine the magnitude T of the torque and the corresponding angle of twist (*a*) at the onset of yield, (*b*) when the plastic zone is 10 mm deep.

3.85 For the shaft of Prob. 3.84, determine (*a*) the angle of twist at which the shaft first becomes fully plastic, (*b*) the corresponding magnitude T of the torque. Sketch the T-ϕ curve for the shaft.

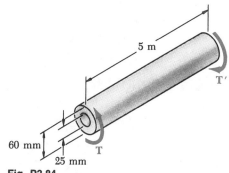

Fig. P3.84

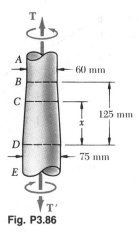

Fig. P3.86

3.86 A steel rod is machined to the shape shown to form a tapered solid shaft to which torques of magnitude $T = 8$ kN · m are applied. Assuming the steel to be elastoplastic with $G = 77$ GPa and $\tau_Y = 145$ MPa, determine (a) the radius of the elastic core in portion AB of the shaft, (b) the length of the portion CD which remains fully elastic.

3.87 If the torques applied to the tapered shaft of Prob. 3.86 are slowly increased, determine (a) the magnitude T of the largest torques which may be applied to the shaft, (b) the length of the portion CD which remains fully elastic.

3.88 Considering the partially plastic shaft of Fig. 3.37c, derive Eq. (3.32) by recalling that the integral in Eq. (3.26) represents the second moment about the τ axis of the area under the τ-ρ curve.

3.89 A solid aluminum rod of 40-mm diameter is subjected to a torque which produces in the rod a maximum shearing strain of 0.008. Using the τ-γ diagram shown for the aluminum alloy used, determine (a) the magnitude of the torque, (b) the angle of twist in a 750-mm length of the rod.

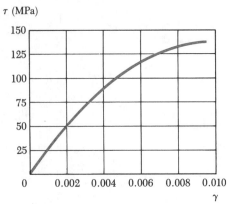

Fig. P3.89 and P3.92

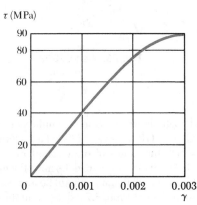

Fig. P3.90 and P3.91

3.90 A solid brass rod of 30-mm diameter is subjected to a torque which causes a maximum shearing stress of 90 MPa in the rod. Using the τ-γ diagram shown for the brass used, determine (a) the magnitude of the torque, (b) the angle of twist in a 600-mm length of the rod.

3.91 A solid brass rod of 20-mm diameter and 750-mm length is twisted through an angle of 10°. Using the τ-γ diagram shown for the brass used, determine (a) the magnitude of the torque applied to the rod, (b) the maximum shearing stress in the rod.

3.92 A hollow shaft of inner and outer diameters equal to 20 and 50 mm, respectively, is fabricated from an aluminum alloy for which the τ-γ diagram is shown. Determine (a) the torque required to twist a 600-mm length of this shaft through an angle of 10°, (b) the corresponding value of the maximum shearing stress in the shaft.

3.93 and 3.94 The curve shown in Fig. P3.89 and P3.92 may be approximated by the relation

$$\tau = 27.8 \times 10^9 \gamma - 1.390 \times 10^{12} \gamma^2$$

Using this relation and Eqs. (3.2) and (3.26), solve the problem indicated.

3.93 Prob. 3.89

3.94 Prob. 3.92

3.95 The solid shaft shown is made of a mild steel which is assumed to be elastoplastic with $G = 77$ GPa and $\tau_Y = 145$ MPa. Knowing that a 980-N $\cdot$ m torque **T** is applied to the shaft and then removed, determine (a) the maximum residual shearing stress, (b) the residual shearing stress at the surface of the shaft.

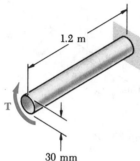

Fig. P3.95 and P3.96

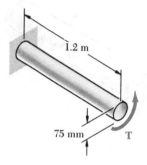

Fig. P3.97 and P3.98

3.96 The solid shaft shown is made of a mild steel which is assumed to be elastoplastic with $G = 77$ GPa and $\tau_Y = 145$ MPa. The torque **T** is increased in magnitude until the radius of the elastic core is 6.25 mm and is then removed. Determine (a) the maximum residual shearing stress, (b) the residual shearing stress at the surface of the shaft.

3.97 The solid shaft shown is made of a mild steel which is assumed to be elastoplastic with $G = 77$ GPa and $\tau_Y = 145$ MPa. Knowing that a 14-kN $\cdot$ m torque **T** is applied to the shaft and then removed, determine the magnitude and location of the maximum residual shearing stress in the shaft.

3.98 The solid shaft shown is made of a mild steel which is assumed to be elastoplastic with $G = 77$ GPa and $\tau_Y = 145$ MPa. The torque **T** is increased in magnitude until the shaft has been twisted through 5.5° and is then removed. Determine the magnitude and location of the maximum residual shearing stress in the shaft.

3.99 In Prob. 3.95, determine the permanent angle of twist of the shaft.

3.100 In Prob. 3.96, determine the permanent angle of twist of the shaft.

3.101 In Prob. 3.97, determine the permanent angle of twist of the shaft.

3.102 In Prob. 3.98, determine the permanent angle of twist of the shaft.

3.103   The hollow shaft shown is made of a steel which is assumed to be elastoplastic with $G = 77$ GPa and $\tau_Y = 145$ MPa.   The magnitude T of the torques is slowly increased until the plastic zone first reaches the inner surface of the shaft; the torques are then removed.   Determine (*a*) the magnitude and location of the maximum residual shearing stress, (*b*) the permanent angle of twist.

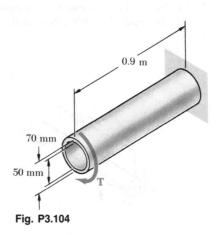

Fig. P3.104

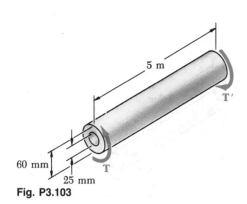

Fig. P3.103

3.104   The hollow shaft shown is made of a steel which is assumed to be elastoplastic with $G = 77$ GPa and $\tau_Y = 145$ MPa.   The magnitude T of the torque is slowly increased until the plastic zone first reaches the inner surface of the shaft; the torque is then removed.   Determine (*a*) the magnitude and location of the maximum residual shearing stress, (*b*) the permanent angle of twist.

3.105   After the solid shaft of Prob. 3.98 has been loaded and unloaded as described in that problem, a torque $\mathbf{T}_1$ of sense opposite to the original torque $\mathbf{T}$ is applied to the shaft.   Assuming no change in the value of τ_Y, determine the angle of twist ϕ_1 for which yield is initiated in this second loading and compare it with the angle ϕ_Y for which the shaft started to yield in the original loading.

3.106   After the solid shaft of Prob. 3.96 has been loaded and unloaded as described in that problem, a torque $\mathbf{T}_1$ of sense opposite to the original torque $\mathbf{T}$ is applied to the shaft.   Assuming no change in the value of τ_Y, determine the magnitude of $\mathbf{T}_1$ required to initiate yield in this second loading and compare it with the magnitude T_Y of the torque $\mathbf{T}$ which first caused the shaft to yield in the original loading.

3.107   A torque $\mathbf{T}$ applied to a solid rod made of an elastoplastic material is increased until the rod becomes fully plastic and then is removed.   (*a*) Show that the distribution of residual stresses is as represented in the figure.   (*b*) Determine the magnitude of the torque due to the stresses acting on the portion of rod located within a circle of radius c_0.

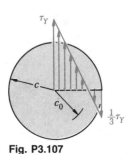

Fig. P3.107

*3.12. TORSION OF NONCIRCULAR MEMBERS

The formulas obtained in Secs. 3.3 and 3.4 for the distributions of strain and stress under a torsional loading apply only to members with a circular cross section. Indeed, their derivation was based on the assumption that the cross section of the member remained plane and undistorted, and we saw in Sec. 3.3 that the validity of this assumption depends upon the *axisymmetry* of the member, i.e., upon the fact that its appearance remains the same when it is viewed from a fixed position and rotated about its axis through an arbitrary angle.

A square bar, on the other hand, retains the same appearance only when it is rotated through 90° or 180°. Following a line of reasoning similar to that used in Sec. 3.3, one could show that the diagonals of the square cross section of the bar and the lines joining the midpoints of the sides of that section remain straight (Fig. 3.44). However, because of the lack of axisymmetry of the bar, any other line drawn in its cross section will deform when the bar is twisted, and the cross section itself will be warped out of its original plane.

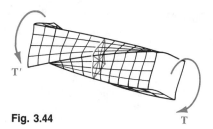

Fig. 3.44

It follows that Eqs. (3.4) and (3.6), which define respectively the distributions of strain and stress in an elastic circular shaft, cannot be used for noncircular members. For example, it would be wrong to assume that the shearing stress in the cross section of a square bar varies linearly with the distance from the axis of the bar and is therefore largest at the corners of the cross section. As we shall see presently, the shearing stress is actually zero at these points.

Consider a small cubic element located at a corner of the cross section of a square bar in torsion and select coordinate axes parallel to the edges of the element (Fig. 3.45a). Since the face of the element perpendicular to the y axis is part of the free surface of the bar, all stresses on this face must be zero. Referring to Fig. 3.45b, we write

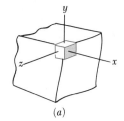

$$\tau_{yx} = 0 \qquad \tau_{yz} = 0 \qquad (3.40)$$

For the same reason all stresses on the face of the element perpendicular to the z axis must be zero, and we write

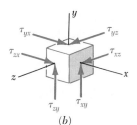

$$\tau_{zx} = 0 \qquad \tau_{zy} = 0 \qquad (3.41)$$

It follows from the first of Eqs. (3.40) and the first of Eqs. (3.41) that

$$\tau_{xy} = 0 \qquad \tau_{xz} = 0 \qquad (3.42)$$

Fig. 3.45

Thus, both components of the shearing stress on the face of the element perpendicular to the axis of the bar are zero. We conclude that there is no shearing stress at the corners of the cross section of the bar.

By twisting a rubber model of a square bar, one easily verifies that no deformations—and, thus, no stresses—occur along the edges of the bar, while the largest deformations—and, thus, the largest stresses—occur along the center line of each of the faces of the bar (Fig. 3.46).

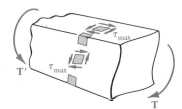

Fig. 3.46

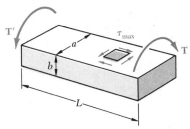

Fig. 3.47

The determination of the stresses in noncircular members subjected to a torsional loading is beyond the scope of this text. However, results obtained from the mathematical theory of elasticity for straight bars with a *uniform rectangular cross section* will be indicated here for convenience.† Denoting by L the length of the bar, by a and b, respectively, the wider and narrower side of its cross section, and by T the magnitude of the torques applied to the bar (Fig. 3.47), we find that the maximum shearing stress occurs along the center line of the *wider* face of the bar and is equal to

$$\tau_{max} = \frac{T}{c_1 a b^2} \tag{3.43}$$

The angle of twist, on the other hand, may be expressed as

$$\phi = \frac{TL}{c_2 a b^3 G} \tag{3.44}$$

The coefficients c_1 and c_2 depend only upon the ratio a/b and are given in Table 3.1 for a number of values of that ratio. Note that Eqs. (3.43) and (3.44) are valid only within the elastic range.

Table 3.1 Coefficients for Rectangular Bars in Torsion

a/b	c₁	c₂
1.0	0.208	0.1406
1.2	0.219	0.1661
1.5	0.231	0.1958
2.0	0.246	0.229
2.5	0.258	0.249
3.0	0.267	0.263
4.0	0.282	0.281
5.0	0.291	0.291
10.0	0.312	0.312
∞	0.333	0.333

We note from Table 3.1 that for $a/b \geq 5$, the coefficients c_1 and c_2 are equal. It may be shown that for such values of a/b, we have

$$c_1 = c_2 = \tfrac{1}{3}(1 - 0.630 b/a) \qquad (\textit{for } a/b \geq 5 \textit{ only}) \tag{3.45}$$

The distribution of shearing stresses in a noncircular member may be visualized more easily by using the *membrane analogy*. A homogeneous elastic membrane attached to a fixed frame and subjected to a uniform pressure on one of its sides happens to constitute an *analog* of the bar in torsion, i.e., the determination of the deformation of the membrane depends upon the solution of the same partial differential equation as the

---

† See S. P. Timoshenko and J. N. Goodier, *Theory of Elasticity*, 3d ed., McGraw-Hill, New York, 1970, sec. 109.

determination of the shearing stresses in the bar.† More specifically, if Q is a point of the cross section of the bar and Q' the corresponding point of the membrane (Fig. 3.48), the shearing stress τ at Q will have the same direction as the horizontal tangent to the membrane at Q', and its magnitude will be proportional to the maximum slope of the membrane at Q'.‡ Furthermore, the applied torque will be proportional to the volume between the membrane and the plane of the fixed frame. In the case of the membrane of Fig. 3.48, which is attached to a rectangular frame, the steepest slope occurs at the midpoint N' of the larger side of the frame. Thus, we verify that the maximum shearing stress in a bar of rectangular cross section will occur at the midpoint N of the larger side of that section.

The membrane analogy may be used just as effectively to visualize the shearing stresses in any straight bar of uniform, noncircular cross section. In particular, let us consider several thin-walled members with the cross sections shown in Fig. 3.49, which are subjected to the same torque. Using the membrane analogy to help us visualize the shearing stresses, we note that, since the same torque is applied to each member, the same volume will be located under each membrane, and the maximum slope will be about the same in each case. Thus, for a thin-walled member of uniform thickness and arbitrary shape, the maximum shearing stress is the same as for a rectangular bar with a very large value of a/b and may be determined from Eq. (3.43) with $c_1 = 0.333$.§

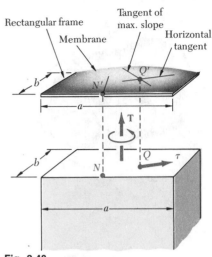

Fig. 3.48

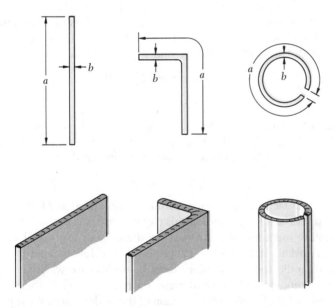

Fig. 3.49

---

† See ibid., sec. 107.

‡ This is the slope measured in a direction perpendicular to the horizontal tangent at Q'.

§ It could also be shown that the angle of twist may be determined from Eq. (3.44) with $c_2 = 0.333$.

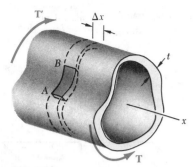

Fig. 3.50

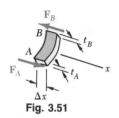

Fig. 3.51

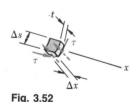

Fig. 3.52

Fig. 3.53

*3.13. THIN-WALLED HOLLOW SHAFTS

In the preceding section we saw that the determination of stresses in noncircular members generally requires the use of advanced mathematical methods. In the case of thin-walled hollow noncircular shafts, however, a good approximation of the distribution of stresses in the shaft may be obtained by a simple computation.

Consider a hollow cylindrical member of *noncircular* section subjected to a torsional loading (Fig. 3.50).† While the thickness t of the wall may vary within a transverse section, we shall assume that it remains small compared to the other dimensions of the member. We shall detach from the member the colored portion of wall AB bounded by two transverse planes at a distance Δx from each other, and by two longitudinal planes perpendicular to the wall. Since the portion AB is in equilibrium, the sum of the forces exerted on it in the longitudinal x direction must be zero (Fig. 3.51). But the only forces involved are the shearing forces $\mathbf{F}_A$ and $\mathbf{F}_B$ exerted on the ends of portion AB. We have therefore

$$\Sigma F_x = 0 \qquad\qquad F_A - F_B = 0 \qquad\qquad (3.46)$$

We shall now express F_A as the product of the longitudinal shearing stress τ_A on the small face at A and of the area $t_A\,\Delta x$ of that face:

$$F_A = \tau_A(t_A\,\Delta x)$$

We note that, while the shearing stress is independent of the x coordinate of the point considered, it may vary across the wall; thus τ_A represents the average value of the stress computed across the wall. Expressing F_B in a similar way and substituting for F_A and F_B into (3.46), we write

$$\tau_A(t_A\,\Delta x) - \tau_B(t_B\,\Delta x) = 0$$

or $\qquad\qquad\qquad\qquad \tau_A t_A = \tau_B t_B \qquad\qquad (3.47)$

Since A and B were chosen arbitrarily, Eq. (3.47) expresses that the product τt of the longitudinal shearing stress τ and of the wall thickness t is constant throughout the member. Denoting this product by q, we have

$$q = \tau t = \text{constant} \qquad\qquad (3.48)$$

We now detach a small element from the wall portion AB (Fig. 3.52). Since the upper and lower faces of this element are part of the free surface of the hollow member, the stresses on these faces are equal to zero. Recalling relations (1.18) and (1.19) of Sec. 1.8, it follows that the stress components indicated on the other faces by dashed arrows are also zero, while those represented by solid arrows are equal. Thus the shearing stress at any point of a transverse section of the hollow member is parallel to the wall surface (Fig. 3.53) and its average value computed across the wall satisfies Eq. (3.48).

---

† The wall of the member must enclose a single cavity and must not be slit open. In other words, the member should be topologically equivalent to a hollow circular shaft.

At this point we may note an analogy between the distribution of the shearing stresses τ in the transverse section of a thin-walled hollow shaft and the distribution of the velocities v in water flowing through a closed channel of unit depth and variable width. While the velocity v of the water varies from point to point on account of the variation in the width t of the channel, the rate of flow, $q = vt$, remains constant throughout the channel, just as τt in Eq. (3.48). Because of this analogy, the product $q = \tau t$ is referred to as the *shear flow* in the wall of the hollow shaft.

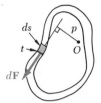

Fig. 3.54

We shall now derive a relation between the torque T applied to a hollow member and the shear flow q in its wall. We consider a small element of the wall section, of length ds (Fig. 3.54). The area of the element is $dA = t\,ds$, and the magnitude of the shearing force dF exerted on the element is

$$dF = \tau\,dA = \tau(t\,ds) = (\tau t)\,ds = q\,ds \qquad (3.49)$$

The moment dM_O of this force about an arbitrary point O within the cavity of the member may be obtained by multiplying dF by the perpendicular distance p from O to the line of action of $d\mathbf{F}$. We have

$$dM_O = p\,dF = p(q\,ds) = q(p\,ds) \qquad (3.50)$$

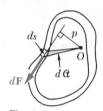

Fig. 3.55

But the product $p\,ds$ is equal to twice the area $d\mathcal{Q}$ of the colored triangle in Fig. 3.55. We thus have

$$dM_O = q(2d\mathcal{Q}) \qquad (3.51)$$

Since the integral around the wall section of the left-hand member of Eq. (3.51) represents the sum of the moments of all the elementary shearing forces exerted on the wall section, and since this sum is equal to the torque T applied to the hollow member, we have

$$T = \oint dM_O = \oint q(2d\mathcal{Q})$$

The shear flow q being a constant, we write

$$T = 2q\mathcal{Q} \qquad (3.52)$$

where $\mathcal{Q}$ is the area bounded by the center line of the wall cross section (Fig. 3.56).

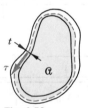

Fig. 3.56

The shearing stress τ at any given point of the wall may be expressed in terms of the torque T if we substitute for q from (3.48) into (3.52) and solve for τ the equation obtained. We have

$$\tau = \frac{T}{2t\mathcal{Q}} \qquad (3.53)$$

where t is the wall thickness at the point considered and $\mathcal{Q}$ the area bounded by the center line. We recall that τ represents the average value of the shearing stress across the wall. However, for elastic deformations the distribution of stresses across the wall may be assumed uniform, and Eq. (3.53) will yield the actual value of the shearing stress at a given point of the wall.

The angle of twist of a thin-walled hollow shaft may be obtained by using the method of energy (Chap. 10). Assuming an elastic deformation, it may be shown† that the angle of twist of a thin-walled shaft of length L and modulus of rigidity G is

$$\phi = \frac{TL}{4\alpha^2 G} \oint \frac{ds}{t} \tag{3.54}$$

where the integral is computed along the center line of the wall section.

Example 3.10

Structural aluminum tubing of 60 × 100-mm rectangular cross section was fabricated by extrusion. Determine the shearing stress in each of the four walls of a portion of such tubing when it is subjected to a torque of 3 kN · m, assuming (*a*) a uniform 4-mm wall thickness (Fig. 3.57*a*), (*b*) that, as a result of defective fabrication, walls *AB* and *AC* are 3-mm thick, and walls *BD* and *CD* are 5-mm thick (Fig. 3.57*b*).

(*a*) *Tubing of Uniform Wall Thickness.* The area bounded by the center line (Fig. 3.58) is

$$\alpha = (96 \text{ mm})(56 \text{ mm}) = 5.376 \times 10^{-3} \text{ m}^2$$

Since the thickness of each of the four walls is $t = 4$ mm, we find from Eq. (3.53) that the shearing stress in each wall is

$$\tau = \frac{T}{2t\alpha} = \frac{3 \times 10^3 \text{ N} \cdot \text{m}}{2(4 \times 10^{-3} \text{ m})(5.376 \times 10^{-3} \text{ m}^2)} = 69.8 \text{ MPa}$$

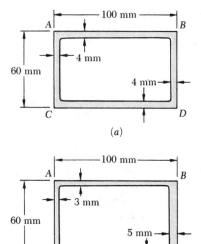

(*a*)

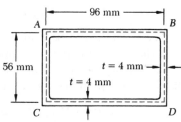

Fig. 3.58

(*b*) *Tubing with Variable Wall Thickness.* Observing that the area α bounded by the center line is the same as in part *a*, and substituting successively $t = 3$ mm and $t = 5$ mm into Eq. (3.53), we have

$$\tau_{AB} = \tau_{AC} = \frac{3 \times 10^3 \text{ N} \cdot \text{m}}{2(3 \times 10^{-3} \text{ m})(5.376 \times 10^{-3} \text{ m}^2)} = 93.0 \text{ MPa}$$

and

$$\tau_{BD} = \tau_{CD} = \frac{3 \times 10^3 \text{ N} \cdot \text{m}}{2(5 \times 10^{-3} \text{ m})(5.376 \times 10^{-3} \text{ m}^2)} = 55.8 \text{ MPa}$$

We note that the stress in a given wall depends only upon its thickness.

(*b*)

Fig. 3.57

---

† See Prob. 10.92.

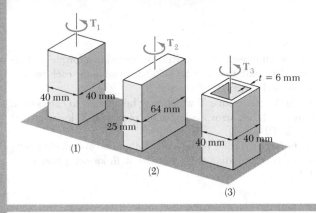

(1)

(2)

(3)

Using $\tau_{all} = 40$ MPa, determine the largest torque which may be applied to each of the brass bars and to the brass tube shown. Note that the two solid bars have the same cross-sectional area, and that the square bar and square tube have the same outside dimensions.

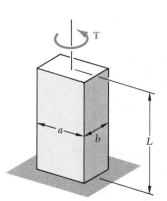

1. Bar with Square Cross Section. For a solid bar of rectangular cross section the maximum shearing stress is given by Eq. (3.43)

$$\tau_{max} = \frac{T}{c_1 a b^2}$$

where the coefficient c_1 is obtained from Table 3.1   We have

$$a = b = 0.040 \text{ m} \qquad \frac{a}{b} = 1.00 \qquad c_1 = 0.208$$

For $\tau_{max} = \tau_{all} = 40$ MPa, we have

$$\tau_{max} = \frac{T_1}{c_1 a b^2} \qquad 40 \text{ MPa} = \frac{T_1}{0.208(0.040 \text{ m})^3} \qquad T_1 = 532 \text{ N} \cdot \text{m} \quad \blacktriangleleft$$

2. Bar with Rectangular Cross Section. We now have

$$a = 0.064 \text{ m} \qquad b = 0.025 \text{ m} \qquad \frac{a}{b} = 2.56$$

Interpolating in Table 3.1: $c_1 = 0.259$

$$\tau_{max} = \frac{T_2}{c_1 a b^2} \qquad 40 \text{ MPa} = \frac{T_2}{0.259(0.064 \text{ m})(0.025 \text{ m})^2} \qquad T_2 = 414 \text{ N} \cdot \text{m} \quad \blacktriangleleft$$

3. Square Tube. For a tube of thickness t, the shearing stress is given by Eq. (3.53)

$$\tau = \frac{T}{2t\mathfrak{a}}$$

where $\mathfrak{a}$ is the area bounded by the center line of the cross section. We have

$$\mathfrak{a} = (0.034 \text{ m})(0.034 \text{ m}) = 1.156 \times 10^{-3} \text{ m}^2$$

We substitute $\tau = \tau_{all} = 40$ MPa and $t = 0.006$ m and solve for the allowable torque:

$$\tau = \frac{T}{2t\mathfrak{a}} \qquad 40 \text{ MPa} = \frac{T_3}{2(0.006 \text{ m})(1.156 \times 10^{-3} \text{ m}^2)} \qquad T_3 = 555 \text{ N} \cdot \text{m} \quad \blacktriangleleft$$

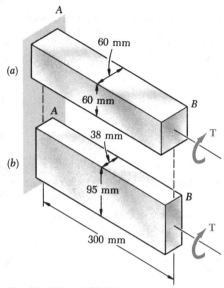

Fig. P3.108 and P3.109

PROBLEMS

3.108 Each of the two aluminum bars shown is subjected to a torque of magnitude $T = 1800$ N · m. Knowing that $G = 26$ GPa, determine for each bar the maximum shearing stress and the angle of twist at B.

3.109 Determine the largest torque **T** which may be applied to each of the two aluminum bars shown and the corresponding angle of twist at B, knowing that $\tau_{all} = 50$ MPa and $G = 26$ GPa.

3.110 Determine the largest torque **T** which may be applied to each of the two brass bars shown and the corresponding angle of twist at B, knowing that $\tau_{all} = 80$ MPa and $G = 38$ GPa.

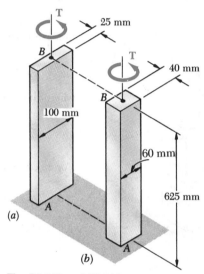

Fig. P3.110 and P3.111

3.111 Each of the two brass bars shown is subjected to a torque of magnitude $T = 1.4$ kN · m. Knowing that $G = 38$ GPa, determine for each bar the maximum shearing stress and the angle of twist at B.

3.112 Each of the three steel bars shown is subjected to a torque of magnitude $T = 275$ N · m. Knowing that the allowable shearing stress is 50 MPa, determine the required dimension b for each bar.

3.113 Each of the three steel bars shown is subjected to a torque of magnitude $T = 550$ N · m. Knowing that the allowable shearing stress is 50 MPa, determine the required dimension b for each bar.

3.114 Each of the three aluminum bars shown is to be twisted through an angle of 1.25°. Knowing that $b = 40$ mm, $\tau_{all} = 50$ MPa, and $G = 26$ GPa, determine the shortest allowable length of each bar.

3.115 Each of the three aluminum bars shown is to be twisted through an angle of 2°. Knowing that $b = 30$ mm, $\tau_{all} = 50$ MPa, and $G = 26$ GPa, determine the shortest allowable length of each bar.

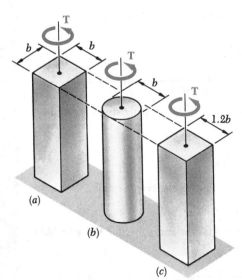

Fig. P3.112, P3.113, P3.114 and P3.115

3.116 Shafts *A* and *B* are made of the same material and have the same cross-sectional area, but *A* has a circular cross section and *B* has a square cross section. Determine the ratio of the maximum shearing stresses occurring in *A* and *B*, respectively, when the two shafts are subjected to the same torque $(T_A = T_B)$. Assume both deformations to be elastic.

3.117 Shafts *A* and *B* are made of the same material and have the same cross-sectional area, but *A* has a circular cross section and *B* has a square cross section. Determine the ratio of the maximum torques T_A and T_B which may be safely applied to *A* and *B*, respectively.

3.118 Shafts *A* and *B* are made of the same material and have the same length and same cross-sectional area, but *A* has a circular cross section and *B* has a square cross section. Determine the ratio of the maximum angles ϕ_A and ϕ_B through which shafts *A* and *B*, respectively, may be safely twisted

3.119 Shafts *A* and *B* are made of the same material and have the same length and same cross-sectional area, but *A* has a circular cross section and *B* has a square cross section. Determine the ratio of the angles ϕ_A and ϕ_B through which *A* and *B* are respectively twisted when the two shafts are subjected to the same torque $(T_A = T_B)$. Assume both deformations to be elastic.

3.120 A 3-m-long steel angle has an L 203 × 152 × 12.7 cross section. From Appendix C we find that the thickness of the section is 12.7 mm and that its area is 4350 mm². Knowing that $\tau_{all} = 50$ MPa, $G = 77$ GPa, and ignoring the effect of stress concentrations, determine (*a*) the largest torque **T** which may be applied, (*b*) the corresponding angle of twist.

3.121 A 4 kN · m torque is applied to a 3-m-long steel angle with an L 203 × 203 × 25.4 cross section. From Appendix C we find that the thickness of the section is 25.4 mm and that its area is 9680 mm². Knowing that $G = 77$ GPa, determine (*a*) the maximum shearing stress along line *a-a*, (*b*) the angle of twist.

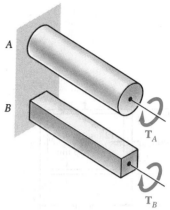

Fig. P3.116, P3.117, P3.118 and P3.119

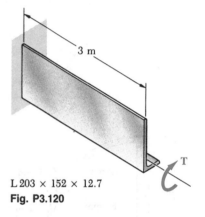

L 203 × 152 × 12.7
Fig. P3.120

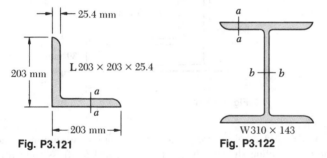

Fig. P3.121   **Fig. P3.122**

W310 × 143

3.122 A 3.5-m-long steel member with a W310 × 143 cross section is subjected to a 4.5 kN · m torque. Knowing that $G = 77$ GPa and referring to Appendix C for the dimensions of the cross section, determine (*a*) the maximum shearing stress along line *a-a*, (*b*) the maximum shearing stress along line *b-b*, (*c*) the angle of twist. (*Hint*: Consider the web and the flanges separately and obtain a relation between the torques exerted on the web and a flange, respectively, by expressing that the resulting angles of twist are equal.)

3.123 A 4-m-long steel member has a W310 × 60 cross section. Knowing that $G = 77$ GPa and that the maximum shearing stress is not to exceed 40 MPa, determine (*a*) the largest torque **T** which may be applied, (*b*) the corresponding angle of twist. Refer to Appendix C for the dimensions of the cross section and neglect the effect of stress concentrations. (See hint for Prob. 3.122.)

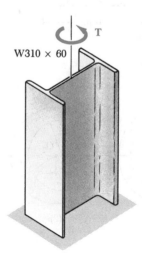

Fig. P3.123

3.124 A 7 kN · m torque is applied to a hollow aluminum shaft having the cross section shown. Neglecting the effect of stress concentrations, determine the shearing stress at points *a* and *b*.

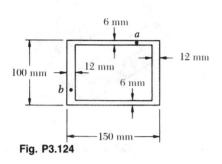

Fig. P3.124

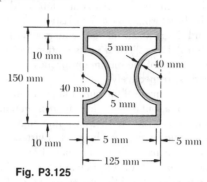

Fig. P3.125

3.125 A hollow brass shaft has the cross section shown. Knowing that the shearing stress must not exceed 80 MPa and neglecting the effect of stress concentrations, determine the largest torque which may be applied to the shaft.

3.126 A hollow brass shaft has the cross section shown. Knowing that the shearing stress must not exceed 80 MPa and neglecting the effect of stress concentrations, determine the largest torque which may be applied to the shaft.

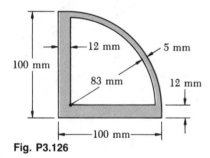

Fig. P3.126

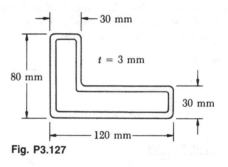

Fig. P3.127

3.127 A 1.2-kN · m torque is applied to a hollow aluminum member having the cross section shown. Neglecting the effect of stress concentrations, determine the shearing stress in the member.

3.128 A hollow cylindrical shaft was designed with the cross section shown in Fig. (*1*) to withstand a maximum torque T_0. Defective fabrication, however, resulted in a slight eccentricity *e* between the inner and outer cylindrical surfaces of the shaft, as shown in Fig. (*2*). (*a*) Express the maximum torque *T* which may be safely applied to the defective shaft in terms of T_0, *e*, and *t*. (*b*) Calculate the percent decrease in the allowable torque for values of the ratio *e*/*t* equal to 0.1, 0.5, and 0.9.

3.129 Knowing that the member of Prob. 3.127 is 900 mm long and that $G = 26$ GPa, determine the angle of twist caused by the 1.2-kN · m torque.

3.130 Knowing that the shaft of Prob. 3.124 is 1.2 m long and that $G = 26$ GPa, determine the angle of twist caused by the 7-kN · m torque.

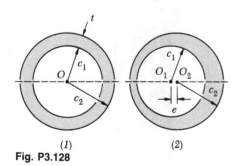

(*1*)   (*2*)

Fig. P3.128

3.131 A hollow cylindrical shaft of length L, mean radius c_m, and uniform thickness t is subjected to torques of magnitude T.   Consider, on the one hand, the values of the average shearing stress τ_{ave} and of the angle of twist ϕ obtained from the elastic torsion formulas developed in Secs. 3.4 and 3.5 and, on the other hand, the corresponding values obtained from the formulas developed in Sec. 3.13 for thin-walled hollow shafts.   (a) Show that the relative error introduced by using the thin-walled-shaft formulas rather than the elastic torsion formulas is the same for τ_{ave} and ϕ and that this relative error is positive and proportional to the square of the ratio t/c_m.   (b) Compute the percent error corresponding to values of t/c_m equal to 0.1, 0.2, and 0.4

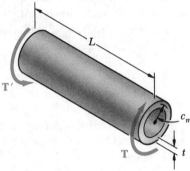

Fig. P3.131

3.132 A thin-walled tube has been fabricated by bending a metal plate of thickness t into a cylinder of radius c and bonding together the edges of the plate. A torque $\mathbf{T}$ is then applied to the tube, producing a shearing stress τ_1 and an angle of twist ϕ_1.   Denoting by τ_2 and ϕ_2, respectively, the shearing stress and the angle of twist which will develop if the bond suddenly fails, express the ratios τ_2/τ_1 and ϕ_2/ϕ_1 in terms of the ratio c/t.

Fig. P3.132

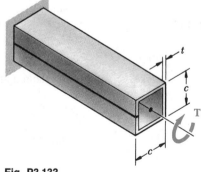

Fig. P3.133

3.133 Solve Prob. 3.132, assuming that the metal plate has been bent into the shape of a tube of square cross section of side c.

REVIEW AND SUMMARY

This chapter was devoted to the analysis and design of *shafts* subjected to twisting couples, or *torques*. Except for the last two sections of the chapter our discussion was limited to *circular shafts*.

Deformations in circular shafts

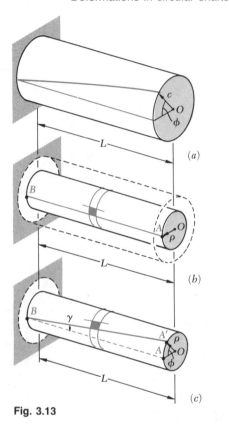

Fig. 3.13

In a preliminary discussion [Sec. 3.2], it was pointed out that the distribution of stresses in the cross section of a circular shaft is *statically indeterminate*. The determination of these stresses, therefore, requires a prior analysis of the *deformations* occurring in the shaft [Sec. 3.3]. Having demonstrated that in a circular shaft subjected to torsion, *every cross section remains plane and undistorted*, we derived the following expression for the *shearing strain* in a small element with sides parallel and perpendicular to the axis of the shaft and at a distance ρ from that axis:

$$\gamma = \frac{\rho\phi}{L} \tag{3.2}$$

where ϕ is the angle of twist for a length L of the shaft (Fig. 3.13). Equation (3.2) shows that the *shearing strain in a circular shaft varies linearly with the distance from the axis of the shaft*. It follows that the strain is maximum at the surface of the shaft, where ρ is equal to the radius c of the shaft. We wrote

$$\gamma_{\max} = \frac{c\phi}{L} \qquad \gamma = \frac{\rho}{c}\gamma_{\max} \tag{3.3,4}$$

Considering *shearing stresses* in a circular shaft *within the elastic range* [Sec. 3.4] and recalling Hooke's law for shearing stress and strain, $\tau = G\gamma$, we derived the relation

$$\tau = \frac{\rho}{c}\tau_{\max} \tag{3.6}$$

Shearing stresses in elastic range

which shows that within the elastic range, the *shearing stress τ in a circular shaft also varies linearly with the distance from the axis of the shaft*. Equating the sum of the moments of the elementary forces exerted on any section of the shaft to the magnitude T of the torque applied to the shaft, we derived the *elastic torsion formulas*

$$\tau_{\max} = \frac{Tc}{J} \qquad \tau = \frac{T\rho}{J} \tag{3.8,9}$$

where c is the radius of the cross section and J its centroidal polar moment of inertia. We noted that $J = \frac{1}{2}\pi c^4$ for a solid shaft and $J = \frac{1}{2}\pi(c_2^4 - c_1^4)$ for a hollow shaft of inner radius c_1 and outer radius c_2.

We noted that while the element a in Fig. 3.19 is in pure shear, the element c in the same figure is subjected to normal stresses of the same magnitude, Tc/J, two of the normal stresses being tensile and two compressive. This explains why in a torsion test ductile materials, which generally fail in shear, will break along a plane perpendicular to the axis of the specimen, while brittle materials, which are weaker in tension than in shear, will break along surfaces forming a 45° angle with that axis.

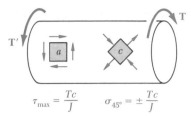

$$\tau_{max} = \frac{Tc}{J} \qquad \sigma_{45°} = \pm \frac{Tc}{J}$$

Fig. 3.19

In Sec. 3.5, we found that within the elastic range, the angle of twist ϕ of a circular shaft is proportional to the torque T applied to it (Fig. 3.21). Expressing ϕ in *radians*, we wrote

$$\phi = \frac{TL}{JG} \tag{3.16}$$

where L = length of shaft
J = polar moment of inertia of cross section
G = modulus of rigidity of material

Angle of twist

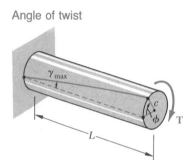

Fig. 3.21

If the shaft is subjected to torques at locations other than its ends or consists of several parts of various cross sections and possibly of different materials, the angle of twist of the shaft must be expressed as the *algebraic sum* of the angles of twist of its component parts [Sample Prob. 3.3]:

$$\phi = \sum_i \frac{T_i L_i}{J_i G_i} \tag{3.17}$$

We observed that when both ends of a shaft BE rotate (Fig. 3.25b), the angle of twist of the shaft is equal to the *difference* between the angles of rotation ϕ_B and ϕ_E of its ends. We also noted that when two shafts AD and BE are connected by gears A and B, the torques applied, respectively, by gear A on shaft AD and by gear B on shaft BE are *directly proportional* to the radii r_A and r_B of the two gears—since the forces applied on each other by the gear teeth at C are equal and opposite. On the other hand, the angles ϕ_A and ϕ_B through which the two gears rotate are *inversely proportional* to r_A and r_B—since the arcs CC' and CC'' described by the gear teeth are equal [Example 3.04 and Sample Prob. 3.4].

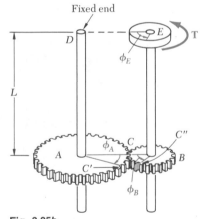

Fig. 3.25b

Statically indeterminate shafts

If the reactions at the supports of a shaft or the internal torques cannot be determined from statics alone, the shaft is said to be *statically indeterminate* [Sec. 3.16]. The equilibrium equations obtained from free-body diagrams must then be complemented by relations involving the deformations of the shaft and obtained from the geometry of the problem [Example 3.05, Sample Prob. 3.5].

Transmission shafts

In Sec. 3.7, we discussed the *design of transmission shafts*. We first observed that the power P transmitted by a shaft is

$$P = 2\pi f T \qquad (3.20)$$

where T is the torque exerted at each end of the shaft and f the *frequency* or speed of rotation of the shaft. The unit of frequency is the revolution per second (s^{-1}) or *hertz* (Hz). T is expressed in newton-meters $(N \cdot m)$ and P in *watts* (W). To design a shaft to transmit a given power P at a frequency f, one first solves Eq. (3.20) for T. Carrying this value and the maximum allowable value of τ for the material used into the elastic formula (3.8), one obtains the corresponding value of the parameter J/c, from which the required diameter of the shaft may be calculated [Examples 3.06 and 3.07].

Stress concentrations

In Sec. 3.8, we discussed *stress concentrations* in circular shafts. We saw that the stress concentrations resulting from an abrupt change in the diameter of a shaft may be reduced through the use of a *fillet* (Fig. 3.30). The maximum value of the shearing stress at the fillet is

$$\tau_{\max} = K \frac{Tc}{J} \qquad (3.25)$$

where the stress Tc/J is computed for the smaller-diameter shaft, and where K is a stress-concentration factor. Values of K were plotted in Fig. 3.31 on page 146 against the ratio r/d, where r is the radius of the fillet, for various values of D/d.

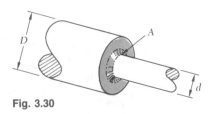

Fig. 3.30

Plastic deformations

Sections 3.9 through 3.11 were devoted to the discussion of *plastic deformations* and *residual stresses* in circular shafts. We first recalled that even when Hooke's law does not apply, the distribution of *strains* in a circular shaft is always linear [Sec. 3.9]. If the shearing-stress-strain diagram for the material is known, it is then possible to plot the shearing stress τ against the distance ρ from the axis of the shaft for any given value of $\tau_{\max}$ (Fig. 3.34). Summing the contributions to the torque of annular elements of radius ρ and thickness $d\rho$, we expressed the torque T as

$$T = \int_0^c \rho\tau(2\pi\rho \, d\rho) = 2\pi \int_0^c \rho^2\tau \, d\rho \qquad (3.26)$$

where τ is the function of ρ plotted in Fig. 3.34.

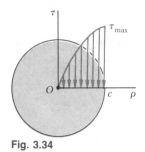

Fig. 3.34

An important value of the torque is the ultimate torque T_U which causes failure of the shaft. This value may be determined, either experimentally, or by carrying out the computations indicated above with τ_{max} chosen equal to the ultimate shearing stress τ_U of the material. From T_U, and assuming a linear stress distribution (Fig. 3.35), we determined the corresponding fictitious stress $R_T = T_U c/J$, known as the *modulus of rupture in torsion* of the given material.

Considering the idealized case of a *solid circular shaft* made of an *elastoplastic material* [Sec. 3.10], we first noted that, as long as τ_{max} does not exceed the yield strength τ_Y of the material, the stress distribution across a section of the shaft is linear (Fig. 3.37a). The torque T_Y corresponding to $\tau_{max} = \tau_Y$ (Fig. 3.37b) is known as the *maximum elastic torque*; for a solid circular shaft of radius c, we have

$$T_Y = \tfrac{1}{2}\pi c^3 \tau_Y \tag{3.29}$$

Modulus of rupture

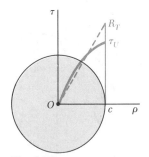

Fig. 3.35

Solid shaft of elastoplastic material

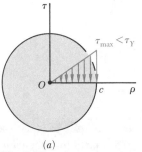 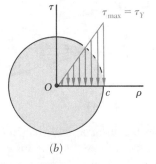 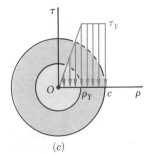 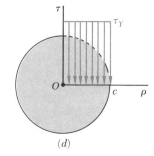

(a)　　　　(b)　　　　(c)　　　　(d)

Fig. 3.37

As the torque increases, a plastic region develops in the shaft around an elastic core of radius ρ_Y. The torque T corresponding to a given value of ρ_Y was found to be

$$T = \frac{4}{3}\,T_Y\!\left(1 - \frac{1}{4}\frac{\rho_Y^3}{c^3}\right) \tag{3.32}$$

We noted that as ρ_Y approaches zero, the torque approaches a limiting value T_p, called the *plastic torque* of the shaft considered:

$$T_p = \frac{4}{3}\,T_Y \tag{3.33}$$

Plotting the torque T against the angle of twist ϕ of a solid circular shaft (Fig. 3.38), we obtained the segment of straight line 0Y defined by Eq. (3.16), followed by a curve approaching the straight line $T = T_p$ and defined by the equation

$$T = \frac{4}{3}\,T_Y\!\left(1 - \frac{1}{4}\frac{\phi_Y^3}{\phi^3}\right) \tag{3.37}$$

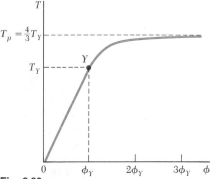

Fig. 3.38

Permanent deformation. Residual stresses

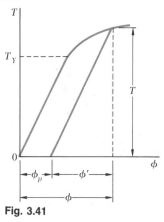

Fig. 3.41

Loading a circular shaft beyond the onset of yield and unloading it [Sec. 3.11] results in a *permanent deformation* characterized by the angle of twist $\phi_p = \phi - \phi'$, where ϕ corresponds to the loading phase described in the previous paragraph, and ϕ' to the unloading phase represented by a straight line in Fig. 3.41. There will also be *residual stresses* in the shaft, which may be determined by adding the maximum stresses reached during the loading phase and the reverse stresses corresponding to the unloading phase [Example 3.09].

The last two sections of the chapter dealt with the torsion of *noncircular members*. We first recalled that the derivation of the formulas for the distribution of strain and stress in circular shafts was based on the fact that due to the axisymmetry of these members, cross sections remain plane and undistorted. Since this property does not hold for noncircular members, such as the square bar of Fig. 3.44, none of the formulas derived earlier may be used in their analysis [Sec. 3.12].

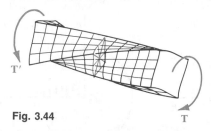

Fig. 3.44

Torsion of noncircular members

Bars of rectangular cross section

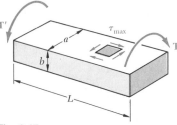

Fig. 3.47

Thin-walled hollow shafts

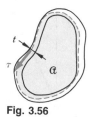

Fig. 3.56

It was indicated in Sec. 3.12 that in the case of straight bars with a *uniform rectangular cross section* (Fig. 3.47), the maximum shearing stress occurs along the center line of the *wider* face of the bar. Formulas for the maximum shearing stress and the angle of twist were given without proof. The *membrane analogy* for visualizing the distribution of stresses in a noncircular member was also discussed.

We next analyzed the distribution of stresses in *noncircular thin-walled hollow shafts* [Sec. 3.13]. We saw that the shearing stress is parallel to the wall surface and varies both across the wall and along the wall cross section. Denoting by τ the average value of the shearing stress computed across the wall at a given point of the cross section, and by t the thickness of the wall at that point (Fig. 3.56), we showed that the product $q = \tau t$, called the *shear flow*, is constant along the cross section.

Furthermore, denoting by T the torque applied to the hollow shaft and by $\mathcal{Q}$ the area bounded by the center line of the wall cross section, we expressed as follows the average shearing stress τ at any given point of the cross section:

$$\tau = \frac{T}{2t\mathcal{Q}} \tag{3.53}$$

REVIEW PROBLEMS

3.134 The torques shown are exerted on pulleys A, B, C, and D. Knowing that each shaft is solid, determine the maximum shearing stress (*a*) in shaft AB, (*b*) in shaft BC, (*c*) in shaft CD.

3.135 The torques shown are exerted on pulleys A, B, C, and D. Knowing that each shaft is solid and made of steel ($G = 77$ GPa), determine the angle through which pulley C rotates with respect to (*a*) pulley A, (*b*) pulley D.

3.136 The aluminum rod AB ($G = 26$ GPa) is bonded to the brass rod BD ($G = 39$ GPa). Knowing that portion CD of the brass rod is hollow and has an inner diameter of 40 mm, determine the angle of twist at A.

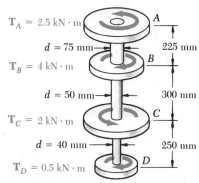

Fig. P3.134 and P3.135

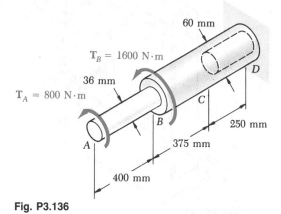

Fig. P3.136

3.137 For the composite rod of Prob. 3.136, determine the maximum shearing stress in portion (*a*) AB, (*b*) BC, (*c*) CD.

3.138 The shaft-disk-belt arrangement shown is used to transmit 2 kW from point A to point D. (*a*) Using an allowable shearing stress of 70 MPa, determine the required speed of shaft AB. (*b*) Solve part *a*, assuming that the diameters of shafts AB and CD are 18 mm and 15 mm, respectively.

3.139 The stepped shaft shown must transmit 60 kW at a speed of 720 rpm. Determine the minimum radius r of the fillet if an allowable stress of 55 MPa is not to be exceeded.

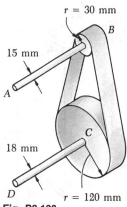

Fig. P3.138

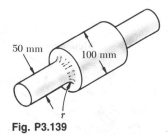

Fig. P3.139

3.140 Two solid steel shafts are connected to a flange coupling at B and to rigid supports at A and C. For the torque shown, determine the maximum shearing stress (*a*) in shaft AB, (*b*) in shaft BC.

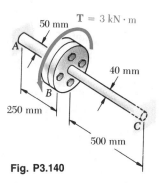

Fig. P3.140

3.141 Three solid steel shafts ($G = 77$ GPa), each of 18-mm diameter, are connected by the gears shown. For the given loading, determine (a) the angle through which end A of shaft AB rotates, (b) the angle through which end E of shaft EF rotates.

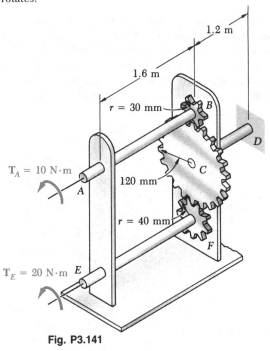

Fig. P3.141

3.142 Two solid aluminum shafts ($G = 26$ GPa) are connected by the gears shown. Knowing that ends B and D are fixed, determine for the given loading (a) the maximum shearing stress in shaft CD, (b) the angle through which end C rotates.

3.143 The composite shaft shown is twisted by applying a torque $\mathbf{T}$ at end A. Knowing that the maximum shearing stress in the aluminum shell is 60 MPa and that $G = 77$ GPa for steel and $G = 26$ GPa for aluminum, determine (a) the maximum shearing stress in the steel core, (b) the magnitude T of the torque applied at A.

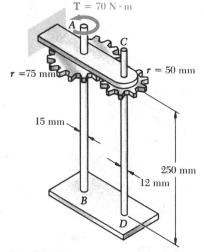

Fig. P3.142

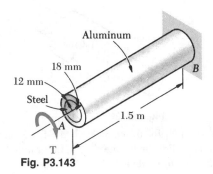

Fig. P3.143

3.144 The solid shaft shown is made of a mild steel which is assumed to be elastoplastic with $G = 77$ GPa and $\tau_Y = 150$ MPa. Determine (*a*) the magnitude of the torque **T** for which the radius of the elastic core is 12.5 mm, (*b*) the corresponding maximum shearing stress, (*c*) the corresponding angle of twist.

3.145 If the torque **T** applied to the shaft of Prob. 3.144 is removed, determine (*a*) the magnitude and location of the maximum residual stress in the shaft, (*b*) the permanent angle of twist of the shaft.

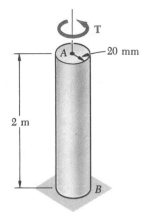

Fig. P3.144

The following problems are designed to be solved with a computer. Write each program so that it can be used with any consistent set of units.

3.C1 Shaft *AB* consists of *n* homogeneous cylindrical elements which may be solid or hollow. Its end *A* is fixed, while its end *B* is free, and it is subjected to the loading shown. The length of element *i* is denoted by L_i, its outer diameter by OD_i, its inner diameter by ID_i, its modulus of rigidity by G_i, and the torque applied to its right end by $\mathbf{T}_i$, the magnitude T_i of this torque being assumed to be positive if $\mathbf{T}_i$ is observed as counterclockwise from end *B* and negative otherwise. (Note that $ID_i = 0$ if the element is solid.) (*a*) Write a computer program which can be used to determine the maximum shearing stress in each element, the angle of twist of each element, and the angle of twist of the entire shaft. (*b*) Use this program to solve Probs. 3.6, 3.8, 3.28, 3.30, 3.134, and 3.136.

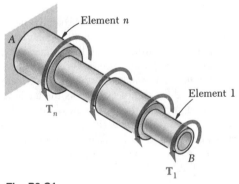

Fig. P3.C1

3.C2 The assembly shown consists of *n* cylindrical shafts, which may be solid or hollow, connected by gears and supported by brackets (not shown). End A_1 of the first shaft is free and is subjected to a torque $\mathbf{T}_0$, while end B_n of the last shaft is fixed. The length of shaft A_iB_i is denoted by L_i, its outer diameter by OD_i, its inner diameter by ID_i, and its modulus of rigidity by G_i. (Note that $ID_i = 0$ if the element is solid.) The radius of gear A_i is denoted by a_i, and the radius of gear B_i by b_i. (*a*) Write a computer program which can be used to determine the maximum shearing stress in each shaft, the angle of twist of each shaft, and the angle through which end A_1 rotates. (*b*) Use this program to solve Probs. 3.14, 3.18, and 3.34.

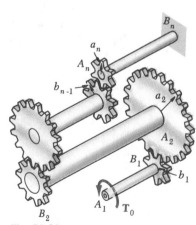

Fig. P3.C2

3.C3 Shaft AB consists of n homogeneous, cylindrical elements which may be solid or hollow. Both of its ends are fixed, and it is subjected to the loading shown. The length of element i is denoted by L_i, its outer diameter by OD_i, its inner diameter by ID_i, its modulus of rigidity by G_i, and the torque applied to its right end by $\mathbf{T}_i$, the magnitude T_i of this torque being assumed to be positive if $\mathbf{T}_i$ is observed as counterclockwise from end B and negative otherwise. (Note that $ID_i = 0$ if the element is solid and also that $T_1 = 0$.) (a) Write a computer program which can be used to determine the reactions at A and B, the maximum shearing stress in each element, and the angle of twist of each element. (b) Use this program to solve Probs. 3.46, 3.48, 3.50, and 3.140.

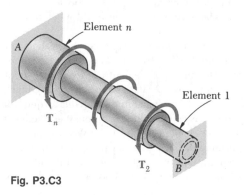

Fig. P3.C3

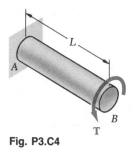

Fig. P3.C4

3.C4 The homogeneous, solid cylindrical shaft AB has a length L, a diameter d, a modulus of rigidity G, and a yield strength τ_Y. It is subjected to a torque $\mathbf{T}$ which is gradually increased from zero until the angle of twist of the shaft has reached a maximum value ϕ_m and then decreased back to zero. (a) Write a computer program which, for each of 16 values of ϕ_m equally spaced over a range extending from 0 to a value 3 times as large as the angle of twist at the onset of yield, can be used to determine the maximum value T_m of the torque, the radius of the elastic core, the maximum shearing stress, the permanent twist, and the residual shearing stress both at the surface of the shaft and at the interface of the elastic core and the plastic region. (b) Use this program to solve Probs. 3.78, 3.96, 3.98, 3.100, 3.102, and 3.144.

CHAPTER FOUR

PURE BENDING

4.1. INTRODUCTION

In the preceding chapters we analyzed the stresses and strains in prismatic members subjected to axial loads and to twisting couples. We shall now consider the analysis of prismatic members subjected to equal and opposite couples $\mathbf{M}$ and $\mathbf{M}'$ acting in the same longitudinal plane (Fig. 4.1); the member is then said to be in *pure bending*.

In the first part of the chapter, we shall analyze the stresses and deformations which exist in homogeneous members possessing a plane of symmetry. After proving that *transverse sections remain plane* during bending deformations (Sec. 4.4), we shall develop formulas which may be used to determine *normal stresses* and the *radius of curvature* for members in pure bending within the elastic range (Sec. 4.5).

In Sec. 4.7, we shall study the stresses and deformations in *composite members* made of more than one material and we shall learn to draw a *transformed section* which represents the section of a member made of a homogeneous material. The transformed section will be used to find the stresses and deformations in the original member. In Sec. 4.8, we shall discuss *stress concentrations* at locations where the cross section of a member undergoes a sudden change.

In the next part of the chapter, we shall consider the bending of members made of a material which does not follow Hooke's law (Sec. 4.9). In particular, we shall investigate the stresses and deformations in members made of an *elastoplastic material*. Starting with the *maximum elastic moment* M_Y, which corresponds to the onset of yield (Sec. 4.10),

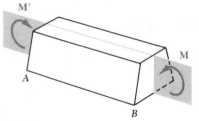

Fig. 4.1

we shall consider the effects of increasingly larger moments until we reach the *plastic moment M_p*, which occurs when the member has yielded fully. We shall also determine the *permanent deformations* and *residual stresses* which result from such loadings (Sec. 4.12).

In Sec. 4.13, we shall superpose the stresses due to pure bending and the stresses due to a centric loading (Chap. 1) to analyze cases of *eccentric loading* in a plane of symmetry. Our study of the bending of prismatic members will conclude with the analysis of *unsymmetric bending* (Sec. 4.14) and the study of the general case of *eccentric axial loading* (Sec. 4.15). The last section of the chapter is devoted to the analysis of the stresses in *curved members* (Sec. 4.16).

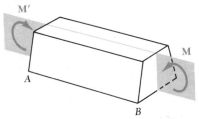

Fig. 4.1 (repeated)

4.2. PRISMATIC MEMBERS IN PURE BENDING

A member subjected to equal and opposite couples acting in the same longitudinal plane is said to be in *pure bending*. We observe that, if a section is passed through the member AB of Fig. 4.1, the conditions of equilibrium of the portion AC of the member require that the elementary forces exerted on AC by the other portion be equivalent to the couple $\mathbf{M}$ (Fig. 4.2). Thus, the internal forces in any cross section of a member in

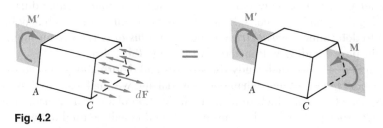

Fig. 4.2

pure bending are equivalent to a couple. The moment M of that couple is known as the *bending moment* in the section. We shall follow the usual convention and assign a positive sign to M when the member is bent as shown in Fig. 4.1, and a negative sign when the senses of the couples $\mathbf{M}$ and $\mathbf{M}'$ are reversed.

An example of a member in pure bending is furnished by *the portion BC* of the beam *AD* shown in Fig. 4.3*a*. Passing a section through an arbitrary point *E* located between *B* and *C*, and drawing the free-body diagrams of *AD* and *AE* (Fig. 4.3*b* and *c*), we verify that the internal forces acting in any cross section located between *B* and *C* must be equivalent to a 36-kN · m couple.

The relatively small number of engineering applications where pure bending is encountered does not in itself justify our devoting an entire chapter to this type of loading. The results that we shall obtain, however, may be applied to the analysis of other types of loading as well, such as *eccentric axial loadings* and *transverse loadings*. As we saw in Sec. 1.3, the internal forces in a given section of a member subjected to an eccentric axial load are equivalent to a force **P** applied at the centroid of the section and a couple **M** (Fig. 4.4). Using the principle of superposition, we shall be able, therefore, to combine our knowledge of the stresses under a *centric* axial load and the results of our forthcoming analysis of stresses in pure bending to obtain the distribution of stresses under an *eccentric* load. The study of pure bending will also play an important role in the study of beams, i.e., in the study of prismatic members subjected to *transverse loads*. Consider, for instance, a cantilever beam *AB* supporting a concentrated load **P** at its free end (Fig.4.5*a*). If we pass a section through *C* at a distance *x* from *A*, we note from the free-body diagram of *AC* (Fig. 4.5*b*) that the internal forces in the section consist of a force **P′** equal and opposite to **P** and a couple **M** of magnitude $M = Px$. As we shall see in Chap. 5, the distribution of shearing stresses in the section depends upon **P′**, while the distribution of normal stresses may be obtained from **M** and is the same as if the beam were in pure bending.

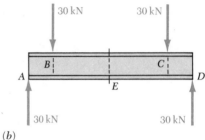

(a)

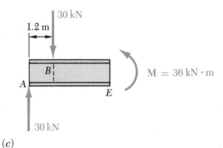

(b)

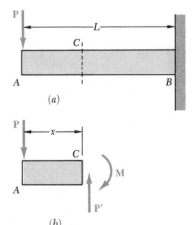

(c)

Fig. 4.3

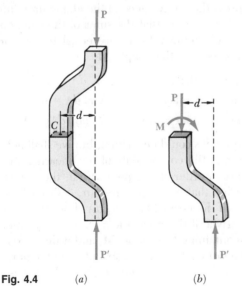

Fig. 4.4   (a)   (b)

Fig. 4.5

4.3. PRELIMINARY DISCUSSION OF THE STRESSES IN PURE BENDING

We shall use the methods of statics to derive the relations which must be satisfied by the stresses exerted on any cross section of a prismatic member in pure bending. Denoting by σ_x the normal stress at a given point of the cross section, and by τ_{xy} and τ_{xz} the components of the shearing stress,† we shall express that the system of the elementary internal forces exerted on the section is equivalent to the couple **M** (Fig. 4.6).

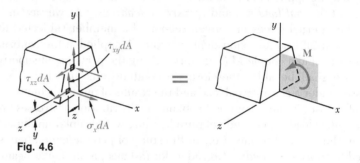

Fig. 4.6

We recall from statics that a couple **M** actually consists of two equal and opposite forces. The sum of the components of these forces in any direction is therefore equal to zero. Moreover, the moment of the couple is the same about *any* axis perpendicular to its plane, and is zero about any axis contained in that plane. Selecting arbitrarily the z axis as shown in Fig. 4.6, we express the equivalence of the elementary internal forces and of the couple **M** by writing that the sums of the components and of the moments of the elementary forces are equal to the corresponding components and moments of the couple **M**:

x components:	$\int \sigma_x \, dA = 0$	(4.1)
moments about y axis:	$\int z \sigma_x \, dA = 0$	(4.2)
moments about z axis:	$\int (-y \sigma_x \, dA) = M$	(4.3)

Three additional equations could be written, but we shall not write them, since they involve only the components of the shearing stress.†

Two remarks are in order at this point: First, the minus sign in Eq. (4.3) is due to the fact that a tensile stress ($\sigma_x > 0$) leads to a negative moment (clockwise) of the normal force $\sigma_x \, dA$ about the z axis. Second, Eq. (4.2) becomes trivial if the prismatic member is symmetric with respect to the plane containing the couple **M**, and if the y axis is chosen in that plane, as shown in Fig. 4.6. The distribution of the normal forces on the section will then clearly be symmetric about the y axis.

---

† As we shall see in the next section, the components of the shearing stress are both equal to zero.

Once more, we note that the actual distribution of stresses in a given cross section cannot be determined from statics alone. It is *statically indeterminate* and may be obtained only by analyzing the *deformations* produced in the member.

4.4. DEFORMATIONS IN A SYMMETRIC MEMBER IN PURE BENDING

We shall now analyze the deformations of a prismatic member possessing a plane of symmetry and subjected at its ends to equal and opposite couples **M** and **M'** acting in the plane of symmetry. The member will bend under the action of the couples, but will remain symmetric with respect to that plane (Fig. 4.7). Moreover, since the bending moment M is the same in any cross section, the member will bend uniformly. Thus, the line AB along which the upper face of the member intersects the plane of the couples will have a constant curvature. In other words, the line AB, which was originally a straight line, will be transformed into a circle of center C, and so will the line $A'B'$ (not shown in the figure) along which the lower face of the member intersects the plane of symmetry. We also note that the line AB will decrease in length when the member is bent as shown in the figure, i.e., when $M > 0$, while $A'B'$ will become longer.

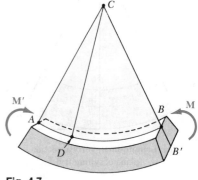

Fig. 4.7

Next we shall prove that any cross section perpendicular to the axis of the member remains plane, and that the plane of the section passes through C. If this were not the case, we could find a point E of the original section through D (Fig. 4.8a) which, after the member has been bent, would *not* lie in the plane perpendicular to the plane of symmetry which contains line CD (Fig. 4.8b). But, because of the symmetry of the member, there would be another point E' which would be transformed exactly in the same way. Let us assume that, after the beam has been bent, both points would be located to the left of the plane defined by CD, as shown in Fig. 4.8b. Since the bending moment M is the same throughout the member, a similar situation would prevail in any other cross section, and the points corresponding to E and E' would also move to the left. Thus, an observer at A would conclude that the loading causes the points E and E' in the various cross sections to move toward him. But an observer at B, to whom the loading looks the same, and who observes the points E and E' in the same positions (except that they are now inverted) would reach the opposite conclusion. This inconsistency leads us to conclude that E and E' will lie in the plane defined by CD and, therefore, that the section remains plane and passes through C. We should note, however, that this discussion does not rule out the possibility of deformations *within* the plane of the section (see Sec. 4.6).

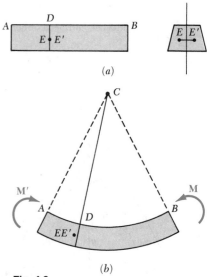

Fig. 4.8

Suppose that the member is divided into a large number of small cubic elements with faces respectively parallel to the three coordinate planes. The property we have established requires that these elements

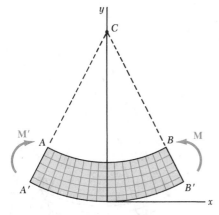

(a) Longitudinal, vertical section
(plane of symmetry)

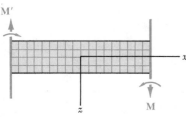

(b) Longitudinal, horizontal section

Fig. 4.9

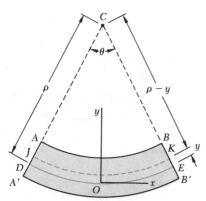

(a) Longitudinal, vertical section
(plane of symmetry)

Fig. 4.10

be transformed as shown in Fig. 4.9 when the member is subjected to the couples **M** and **M′**. Since all the faces represented in the two projections of Fig. 4.9 are at 90° to each other, we conclude that $\gamma_{xy} = \gamma_{zx} = 0$ and, thus, that $\tau_{xy} = \tau_{xz} = 0$. Regarding the three stress components that we have not yet discussed, namely, σ_y, σ_z, and τ_{yz}, we note that they must be zero on the surface of the member. Since, on the other hand, the deformations involved do not require any interaction between the elements of a given transverse cross section, we shall assume that these three stress components are equal to zero throughout the member. This assumption is verified, both from experimental evidence and from the theory of elasticity, for slender members undergoing small deformations.† We conclude that the only nonzero stress component exerted on any of the small cubic elements considered here is the normal component σ_x. Thus, at any point of a slender member in pure bending, we have a state of *uniaxial stress*. Recalling that, for $M > 0$, lines AB and $A'B'$ are observed, respectively, to decrease and increase in length, we note that the strain ϵ_x and the stress σ_x are negative in the upper portion of the member (*compression*) and positive in the lower portion (*tension*).

It follows from the above that there must exist a surface parallel to the upper and lower faces of the member, where ϵ_x and σ_x are zero. This surface is called the *neutral surface*. The neutral surface intersects the plane of symmetry along an arc of circle DE (Fig. 4.10a), and it intersects a transverse section along a straight line called the *neutral axis* of the section (Fig. 4.10b). We shall now select the origin of coordinates on the neutral surface, rather than on the lower face of the member as we did earlier, so that the distance from any point to the neutral surface will be measured by its coordinate y.

Denoting by ρ the radius of arc DE (Fig. 4.10a), by θ the central angle corresponding to DE, and observing that the length of DE is equal to the length L of the undeformed member, we write

$$L = \rho\theta \tag{4.4}$$

Considering now the arc JK located at a distance y above the neutral surface, we note that its length L' is

$$L' = (\rho - y)\theta \tag{4.5}$$

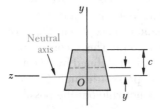

(b) Transverse section

† Also see Prob. 4.34.

Since the original length of arc JK was equal to L, the deformation of JK is

$$\delta = L' - L \tag{4.6}$$

or, substituting from (4.4) and (4.5) into (4.6),

$$\delta = (\rho - y)\theta - \rho\theta = -y\theta \tag{4.7}$$

The longitudinal strain ϵ_x in the elements of JK is obtained by dividing δ by the original length L of JK. We write

$$\epsilon_x = \frac{\delta}{L} = \frac{-y\theta}{\rho\theta}$$

or

$$\epsilon_x = -\frac{y}{\rho} \tag{4.8}$$

The minus sign is due to the fact that we have assumed the bending moment to be positive and, thus, the beam to be concave upward.

Because of the requirement that transverse sections remain plane, identical deformations will occur in all planes parallel to the plane of symmetry. Thus the value of the strain given by Eq. (4.8) is valid anywhere, and we conclude that the *longitudinal normal strain* ϵ_x *varies linearly,* throughout the member, *with the distance y from the neutral surface.*

The strain ϵ_x reaches its maximum absolute value when y itself is largest. Denoting by c the largest distance from the neutral surface (which corresponds to either the upper or the lower surface of the member), and by ϵ_m the *maximum absolute value* of the strain, we have

$$\epsilon_m = \frac{c}{\rho} \tag{4.9}$$

Solving (4.9) for ρ and substituting the value obtained into (4.8), we may also write

$$\epsilon_x = -\frac{y}{c}\epsilon_m \tag{4.10}$$

We shall conclude our analysis of the deformations of a member in pure bending by observing that we are still unable to compute the strain or stress at a given point of the member, since we have not yet located the neutral surface in the member. In order to locate this surface, we shall have to specify the stress-strain relation of the material used.†

---

† We may note, however, that if the member possesses both a vertical and a horizontal plane of symmetry (e.g., a member with a rectangular cross section), and if the stress-strain curve is the same in tension and compression, the neutral surface will coincide with the plane of symmetry (cf. Sec. 4.9).

4.5. STRESSES AND DEFORMATIONS IN THE ELASTIC RANGE

We shall now consider the case when the bending moment M is such that the normal stresses in the member remain below the yield strength σ_Y. This means that, for all practical purposes, the stresses in the member will remain below the proportional limit and the elastic limit as well. There will be no permanent deformation, and Hooke's law for uniaxial stress applies. Assuming the material to be homogeneous, and denoting by E its modulus of elasticity, we have in the longitudinal x direction

$$\sigma_x = E\epsilon_x \tag{4.11}$$

Recalling Eq. (4.10), and multiplying both members of that equation by E, we write

$$E\epsilon_x = -\frac{y}{c}(E\epsilon_m)$$

or, using (4.11),

$$\sigma_x = -\frac{y}{c}\sigma_m \tag{4.12}$$

where σ_m denotes the *maximum absolute value* of the stress. This result shows that, *in the elastic range, the normal stress varies linearly with the distance from the neutral surface* (Fig. 4.11).

It should be noted that, at this point, we do not know the location of the neutral surface, nor the maximum value σ_m of the stress. Both may be found if we recall the relations (4.1) and (4.3) which were obtained earlier from statics. Substituting first for σ_x from (4.12) into (4.1), we write

$$\int \sigma_x \, dA = \int \left(-\frac{y}{c}\sigma_m\right) dA = -\frac{\sigma_m}{c}\int y \, dA = 0$$

from which it follows that

$$\int y \, dA = 0 \tag{4.13}$$

This equation shows that the first moment of the cross section about its neutral axis must be zero.† In other words, for a member subjected to pure bending, and *as long as the stresses remain in the elastic range, the neutral axis passes through the centroid of the section.*

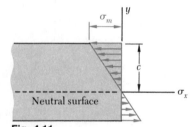

Fig. 4.11

---

† See Appendix A for a discussion of the moments of areas.

We now recall Eq. (4.3), which was derived in Sec. 4.3 with respect to an *arbitrary* horizontal z axis,

$$\int (-y\sigma_x \, dA) = M \qquad (4.3)$$

Specifying that the z axis should coincide with the neutral axis of the cross section, we substitute for σ_x from (4.12) into (4.3) and write

$$\int (-y)\left(-\frac{y}{c}\sigma_m\right) dA = M$$

or

$$\frac{\sigma_m}{c} \int y^2 \, dA = M \qquad (4.14)$$

Recalling that in the case of pure bending the neutral axis passes through the centroid of the cross section, we note that I is the moment of inertia, or second moment, of the cross section with respect to a centroidal axis perpendicular to the plane of the couple **M**.   Solving (4.14) for σ_m, we write therefore †

$$\sigma_m = \frac{Mc}{I} \qquad (4.15)$$

Substituting for σ_m from (4.15) into (4.12), we obtain the normal stress σ_x at any distance y from the neutral axis:

$$\sigma_x = -\frac{My}{I} \qquad (4.16)$$

Equations (4.15) and (4.16) are called the *elastic flexure formulas*, and the normal stress σ_x caused by the bending or "flexing" of the member is often referred to as the *flexural stress*.   We verify that the stress is compressive ($\sigma_x < 0$) above the neutral axis ($y > 0$) when the bending moment M is positive, and tensile when M is negative.

Returning to Eq. (4.15), we note that the ratio I/c depends only upon the geometry of the cross section.   This ratio is called the *elastic section modulus* and is denoted by S.   We have

$$\text{Elastic section modulus} = S = \frac{I}{c} \qquad (4.17)$$

---

† We recall that the bending moment was assumed to be positive.   If the bending moment is negative, M should be replaced in Eq. (4.15) by its absolute value $|M|$.

Substituting S for I/c into Eq. (4.15), we write this equation in the alternative form

$$\sigma_m = \frac{M}{S} \qquad (4.18)$$

Since the maximum stress σ_m is inversely proportional to the elastic section modulus S, it is clear that beams should be designed with as large a value of S as practicable. For example, in the case of a wooden beam with a rectangular cross section of width b and depth h, we have

$$S = \frac{I}{c} = \frac{\frac{1}{12}bh^3}{h/2} = \tfrac{1}{6}bh^2 = \tfrac{1}{6}Ah \qquad (4.19)$$

where A is the cross-sectional area of the beam. This shows that, of two beams with the same cross-sectional area A (Fig. 4.12), the beam with the larger depth h will have the larger section modulus and, thus, will be the more effective in resisting bending.†

In the case of structural steel, American standard beams (S-beams) and wide-flange beams (W-beams) are preferred to other shapes because a large portion of their cross section is located far from the neutral axis (Fig. 4.13). Thus, for a given cross-sectional area and a given depth, their design provides large values of I and, consequently, of S. Values of the elastic section modulus of commonly manufactured beams may be obtained from tables listing the various geometric properties of such beams. To determine the maximum stress σ_m in a given section of a standard beam, the engineer needs only to read the value of the elastic section modulus S in a table, and divide the bending moment M in the section by S.

The deformation of the member caused by the bending moment M is measured by the *curvature* of the neutral surface. The curvature is defined as the reciprocal of the radius of curvature ρ, and may be obtained by solving Eq. (4.9) for $1/\rho$:

$$\frac{1}{\rho} = \frac{\epsilon_m}{c} \qquad (4.20)$$

But, in the elastic range, we have $\epsilon_m = \sigma_m/E$. Substituting for ϵ_m into (4.20), and recalling (4.15), we write

$$\frac{1}{\rho} = \frac{\sigma_m}{Ec} = \frac{1}{Ec}\frac{Mc}{I}$$

or

$$\frac{1}{\rho} = \frac{M}{EI} \qquad (4.21)$$

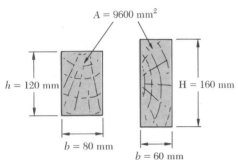

$A = 9600 \text{ mm}^2$

$h = 120 \text{ mm}$ $H = 160 \text{ mm}$

$b = 80 \text{ mm}$

$b = 60 \text{ mm}$

Fig. 4.12

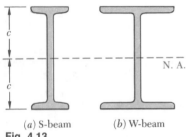

c

c

N. A.

(*a*) S-beam   (*b*) W-beam

Fig. 4.13

† However, large values of the ratio h/b may result in lateral instability of the beam.

Example 4.01

A steel bar of 20×60-mm rectangular cross section is subjected to two equal and opposite couples acting in the vertical plane of symmetry of the bar (Fig. 4.14). Determine the value of the bending moment M which causes the bar to yield. Assume $\sigma_Y = 250$ MPa.

Fig. 4.14

Since the neutral axis must pass through the centroid C of the cross section, we have $c = 30$ mm $= 30 \times 10^{-3}$ m (Fig. 4.15). On the other hand, the centroidal moment of inertia of the rectangular cross section is

$$I = \tfrac{1}{12}bh^3 = \tfrac{1}{12}(20 \times 10^{-3}\text{ m})(60 \times 10^{-3}\text{ m})^3 = 360 \times 10^{-9}\text{ m}^4$$

Solving Eq. (4.15) for M, and substituting the above data, we have

$$M = \frac{I}{c}\sigma_m = \frac{360 \times 10^{-9}\text{ m}^4}{30 \times 10^{-3}\text{ m}}(250 \times 10^6\text{ N/m}^2)$$

$$M = 3000\text{ N} \cdot \text{m} = 3\text{ kN} \cdot \text{m}$$

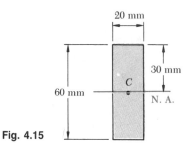

Fig. 4.15

Example 4.02

An aluminum rod with a semicircular cross section of radius $r = 12$ mm (Fig. 4.16) is bent into the shape of a circular arc of mean radius $\rho = 2.5$ m. Knowing that the flat face of the rod is turned toward the center of curvature of the arc, determine the maximum tensile and compressive stress in the rod. Use $E = 70$ GPa.

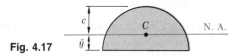

Fig. 4.16

We could use Eq. (4.21) to determine the bending moment M corresponding to the given radius of curvature ρ, and then Eq. (4.15) to determine σ_m. However, it is simpler to use Eq. (4.9) to determine ϵ_m, and Hooke's law to obtain σ_m.

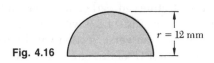

Fig. 4.17

The ordinate $\bar{y}$ of the centroid C of the semicircular cross section is

$$\bar{y} = \frac{4r}{3\pi} = \frac{4(12\text{ mm})}{3\pi} = 5.093\text{ mm}$$

The neutral axis passes through C (Fig. 4.17) and the distance c to the point of the cross section farthest away from the neutral axis is

$$c = r - \bar{y} = 12\text{ mm} - 5.093\text{ mm} = 6.907\text{ mm}$$

Using Eq. (4.9), we write

$$\epsilon_m = \frac{c}{\rho} = \frac{6.907 \times 10^{-3}\text{ m}}{2.5\text{ m}} = 2.763 \times 10^{-3}$$

and, applying Hooke's law,

$$\sigma_m = E\epsilon_m = (70 \times 10^9\text{ Pa})(2.763 \times 10^{-3}) = 193.4\text{ MPa}$$

Since this side of the rod faces away from the center of curvature, the stress obtained is a tensile stress. The maximum compressive stress occurs on the flat side of the rod. Using the fact that the stress is proportional to the distance from the neutral axis, we write

$$\sigma_{\text{comp}} = -\frac{\bar{y}}{c}\sigma_m = -\frac{5.093\text{ mm}}{6.907\text{ mm}}(193.4\text{ MPa}) = -142.6\text{ MPa}$$

4.6. DEFORMATIONS IN A TRANSVERSE CROSS SECTION

When we proved in Sec. 4.4 that the transverse cross section of a member in pure bending remains plane, we did not rule out the possibility of deformations within the plane of the section. That such deformations will exist is evident, if we recall from Sec. 2.11 that elements in a state of uniaxial stress, $\sigma_x \neq 0$, $\sigma_y = \sigma_z = 0$, are deformed in the transverse y and z directions, as well as in the axial x direction. The normal strains ϵ_y and ϵ_z depend upon Poisson's ratio ν for the material used and are expressed as

$$\epsilon_y = -\nu\epsilon_x \qquad \epsilon_z = -\nu\epsilon_x$$

or, recalling Eq. (4.8),

$$\epsilon_y = \frac{\nu y}{\rho} \qquad \epsilon_z = \frac{\nu y}{\rho} \tag{4.22}$$

The relations we have obtained show that the elements located above the neutral surface ($y > 0$) will expand in both the y and z directions, while the elements located below the neutral surface ($y < 0$) will contract. In the case of a member of rectangular cross section, the expansion and contraction of the various elements in the vertical direction will compensate, and no change in the vertical dimension of the cross section will be observed. As far as the deformations in the horizontal transverse z direction are concerned, however, the expansion of the elements located above the neutral surface and the corresponding contraction of the elements located below that surface will result in the various horizontal lines in the section being bent into arcs of circle (Fig. 4.18). The situation observed here is similar to that observed earlier in a longitudinal cross section. Comparing the second of Eqs. (4.22) with Eq. (4.8), we conclude that the neutral axis of the transverse section will be bent into a circle of radius $\rho' = \rho/\nu$. The center C' of this circle is located below the neutral surface (assuming $M > 0$), i.e., on the side opposite to the center of curvature C of the member. The reciprocal of the radius of curvature ρ' represents the curvature of the transverse cross section and is called the *anticlastic curvature*. We have

$$\text{Anticlastic curvature} = \frac{1}{\rho'} = \frac{\nu}{\rho} \tag{4.23}$$

In our discussion of the deformations of a symmetric member in pure bending, in this section and in the preceding ones, we have ignored the manner in which the couples $\mathbf{M}$ and $\mathbf{M}'$ were actually applied to the member. If *all* transverse sections of the member, from one end to the other, are to remain plane and free of shearing stresses, we must make

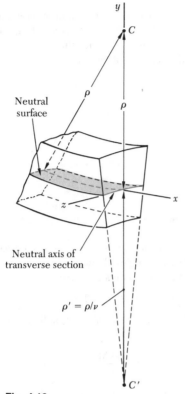

Fig. 4.18

sure that the couples are applied in such a way that the ends of the member themselves remain plane and free of shearing stresses. This may be accomplished by applying the couples **M** and **M′** to the member through the use of rigid and smooth plates (Fig. 4.19). The elementary forces exerted by the plates on the member will be normal to the end sections, and these sections, while remaining plane, will be free to deform as described earlier in this section.

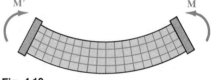

Fig. 4.19

We should note that these loading conditions cannot be actually realized, since they require each plate to exert tensile forces on the corresponding end section below its neutral axis, while allowing the section to freely deform in its own plane. The fact that the rigid-end-plates model of Fig. 4.19 cannot be physically realized, however, does not detract from its importance, which is to allow us to *visualize* the loading conditions corresponding to the relations derived in the preceding sections. Actual loading conditions may differ appreciably from this idealized model. By virtue of Saint-Venant's principle, however, the relations obtained may be used to compute stresses in engineering situations, as long as the section considered is not too close to the points where the couples are applied.

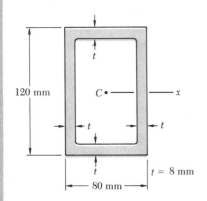

SAMPLE PROBLEM 4.1

The rectangular tube shown is extruded from an aluminum alloy for which $\sigma_Y = 150$ MPa, $\sigma_U = 300$ MPa, and $E = 70$ GPa. Neglecting the effect of fillets, determine (a) the bending moment M for which the factor of safety will be 3.00, (b) the corresponding radius of curvature of the tube.

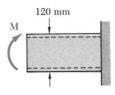

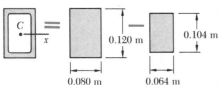

Moment of Inertia. Considering the cross-sectional area of the tube as the difference between the two rectangles shown and recalling the formula for the centroidal moment of inertia of a rectangle, we write

$$I = \tfrac{1}{12}(0.080)(0.120)^3 - \tfrac{1}{12}(0.064)(0.104)^3 \qquad I = 5.52 \times 10^{-6}\ \text{m}^4$$

Allowable Stress. For a factor of safety of 3.00 and an ultimate stress of 300 MPa, we have

$$\sigma_{\text{all}} = \frac{\sigma_U}{F.S.} = \frac{300\ \text{MPa}}{3.00} = 100\ \text{MPa}$$

Since $\sigma_{\text{all}} < \sigma_Y$, the tube remains in the elastic range and we may apply the results of Sec. 4.5.

a. Bending Moment. With $c = \tfrac{1}{2}(0.120\ \text{m}) = 0.060\ \text{m}$, we write

$$\sigma_{\text{all}} = \frac{Mc}{I} \qquad M = \frac{I}{c}\sigma_{\text{all}} = \frac{5.52 \times 10^{-6}\ \text{m}^4}{0.060\ \text{m}}(100\ \text{MPa})$$

$$M = 9.20\ \text{kN} \cdot \text{m} \quad \blacktriangleleft$$

b. Radius of Curvature. Recalling that $E = 70$ GPa, we substitute this value and the values obtained for I and M into Eq. (4.21) and find

$$\frac{1}{\rho} = \frac{M}{EI} = \frac{9.20\ \text{kN} \cdot \text{m}}{(70\ \text{GPa})(5.52 \times 10^{-6}\ \text{m}^4)} = 23.8 \times 10^{-3}\ \text{m}^{-1}$$

$$\rho = 42.0\ \text{m} \quad \blacktriangleleft$$

Alternative Solution. Since we know that the maximum stress is $\sigma_{\text{all}} = 100$ MPa, we may determine the maximum strain ϵ_m and then use Eq. (4.9),

$$\epsilon_m = \frac{\sigma_{\text{all}}}{E} = \frac{100\ \text{MPa}}{70\ \text{GPa}} = 1429\ \mu$$

$$\epsilon_m = \frac{c}{\rho}; \qquad \rho = \frac{c}{\epsilon_m} = \frac{0.060\ \text{m}}{1429\ \mu}$$

$$\rho = 42.0\ \text{m} \quad \blacktriangleleft$$

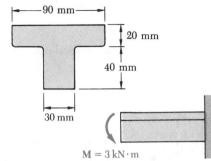

SAMPLE PROBLEM 4.2

A cast-iron machine part is acted upon by the 3-kN · m couple shown. Knowing that $E = 165$ GPa and neglecting the effect of fillets, determine (a) the maximum tensile and compressive stresses in the casting, (b) the radius of curvature of the casting.

$M = 3$ kN·m

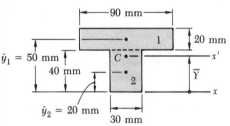

Centroid. We divide the T-shaped cross section into the two rectangles shown and write

Area, mm²		$\bar{y}$, mm	$\bar{y}A$, mm³	
1	$(20)(90) = 1800$	50	90×10^3	
2	$(40)(30) = 1200$	20	24×10^3	
	$\Sigma A = 3000$		$\Sigma \bar{y}A = 114 \times 10^3$	

$\bar{Y}\Sigma A = \Sigma \bar{y}A$
$\bar{Y}(3000) = 114 \times 10^6$
$\bar{Y} = 38$ mm

Centroidal Moment of Inertia. The parallel-axis theorem is used to determine the moment of inertia of each rectangle with respect to the axis x' which passes through the centroid of the composite section. Adding the moments of inertia of the rectangles, we write

$$I_{x'} = \Sigma(\bar{I} + Ad^2) = \Sigma(\tfrac{1}{12}bh^3 + Ad^2)$$
$$= \tfrac{1}{12}(90)(20)^3 + (90 \times 20)(12)^2 + \tfrac{1}{12}(30)(40)^3 + (30 \times 40)(18)^2$$
$$= 868 \times 10^3 \text{ mm}^4$$
$$I = 868 \times 10^{-9} \text{ m}^4$$

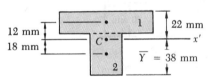

a. **Maximum Tensile Stress.** Since the applied couple bends the casting downward, the center of curvature is located below the cross section. The maximum tensile stress occurs at point A, which is farthest from the center of curvature.

$$\sigma_A = \frac{Mc_A}{I} = \frac{(3 \text{ kN} \cdot \text{m})(0.022 \text{ m})}{868 \times 10^{-9} \text{ m}^4} \qquad \sigma_A = +76.0 \text{ MPa} \blacktriangleleft$$

Maximum Compressive Stress. This occurs at point B; we have

$$\sigma_B = -\frac{Mc_B}{I} = -\frac{(3 \text{ kN} \cdot \text{m})(0.038 \text{ m})}{868 \times 10^{-9} \text{ m}^4} \qquad \sigma_B = -131.3 \text{ MPa} \blacktriangleleft$$

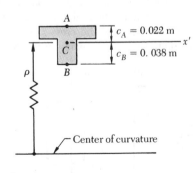

b. **Radius of Curvature.** From Eq. (4.21), we have

$$\frac{1}{\rho} = \frac{M}{EI} = \frac{3 \text{ kN} \cdot \text{m}}{(165 \text{ GPa})(868 \times 10^{-9} \text{ m}^4)}$$
$$= 20.95 \times 10^{-3} \text{ m}^{-1} \qquad \rho = 47.7 \text{ m} \blacktriangleleft$$

PROBLEMS

4.1 and 4.2 Knowing that the couple shown acts in a vertical plane, determine the stress at (*a*) point *A*, (*b*) point *B*.

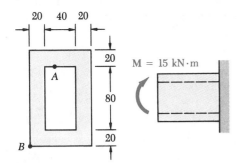

Dimensions in mm
Fig. P4.1

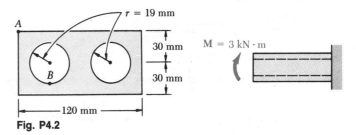

Fig. P4.2

4.3 Solve Prob. 4.2, assuming that the radius of each circular hole is 25 mm.

4.4 The steel beam shown is made of a grade of steel for which $\sigma_Y = 250$ MPa and $\sigma_U = 400$ MPa. Using a factor of safety of 2.50, determine the largest couple which may be applied to the beam when it is bent about the *z* axis.

4.5 Solve Prob. 4.4, assuming that the steel beam is bent about the *y* axis.

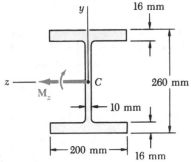

Fig. P4.4

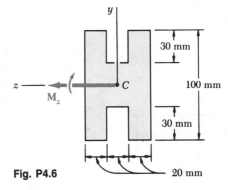

Fig. P4.6

4.6 A beam of the cross section shown is extruded from an aluminum alloy for which $\sigma_Y = 310$ MPa and $\sigma_U = 480$ MPa. Using a factor of safety 3.00, determine the largest couple which may be applied to the beam when it is bent about the *z* axis.

4.7 Solve Prob. 4.6, assuming that the beam is bent about the *y* axis.

4.8 Two vertical forces are applied to a beam of the cross section shown. Determine the maximum tensile and compressive stresses in portion *BC* of the beam.

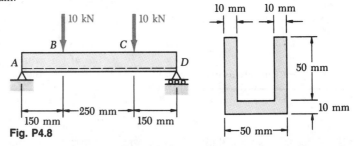

Fig. P4.8

4.9 and 4.10 Two vertical forces are applied to a beam of the cross section shown. Determine the maximum tensile and compressive stresses in portion BC of the beam.

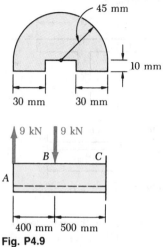

Fig. P4.9

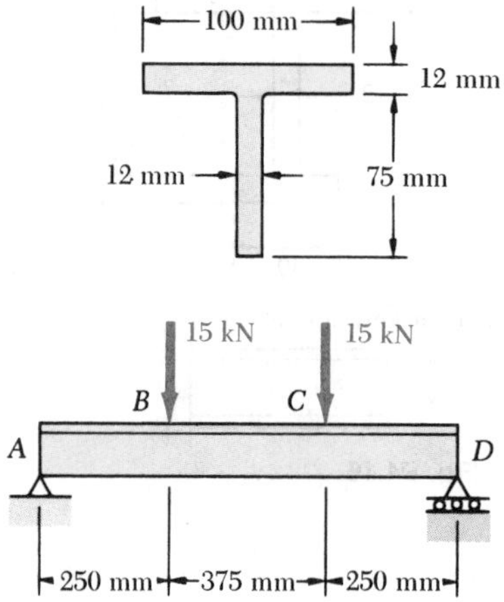

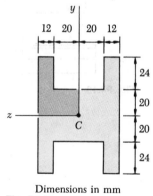

Fig. P4.10

4.11 Solve Prob. 4.10, assuming that the width of the flange is increased from 100 mm to 125 mm.

4.12 and 4.13 Knowing that a beam of the cross section shown is bent about a horizontal axis and that the bending moment is 5.5 kN · m, determine the magnitude of the total force acting (*a*) on the top flange, (*b*) on the shaded portion of the web.

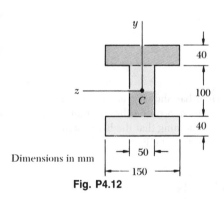

Dimensions in mm

Fig. P4.12

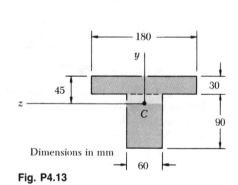

Dimensions in mm

Fig. P4.13

Fig. P4.14

4.14 Knowing that a beam of the cross section shown is bent about a horizontal axis and that the bending moment is 4 kN · m, determine the magnitude of the total force acting on the shaded portion of the beam.

4.15 Solve Prob. 4.14, assuming that the beam is bent about a vertical axis by a couple of moment 4 kN · m.

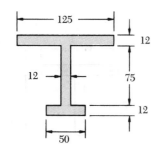

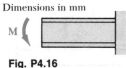

Dimensions in mm

M

Fig. P4.16

4.16 and 4.17 For the casting shown, determine the largest couple **M** which may be applied without exceeding either of the following allowable stresses: $\sigma_{all} = +40$ MPa, $\sigma_{all} = -105$ MPa.

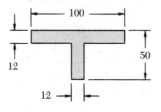

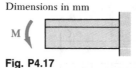

Dimensions in mm

M

Fig. P4.17

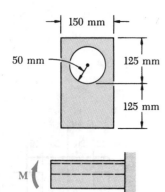

M

Fig. P4.18

4.18 and 4.19 Knowing that for the extruded beam shown the allowable stress is 120 MPa in tension and 150 MPa in compression, determine the largest couple **M** which may be applied.

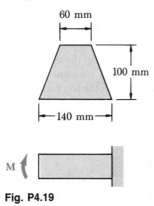

M

Fig. P4.19

4.20 A 3.5-kN · m couple is applied to the steel bar shown. (*a*) Assuming that the couple bends the bar about a horizontal axis, determine the maximum stress and the radius of curvature. (*b*) Solve part *a*, assuming that the bar is bent about a vertical axis by a 3.5-kN · m couple. Use $E = 200$ GPa.

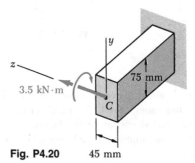

Fig. P4.20

4.21 A 8.5-kN · m couple is applied to the S 250 × 37.8 rolled-steel beam shown. (*a*) Assuming that the couple bends the beam about a horizontal axis, determine the maximum stress and the radius of curvature. (*b*) Solve part *a*, assuming that the beam is bent about a vertical axis by a 8.5-kN · m couple. Use $E = 200$ GPa.

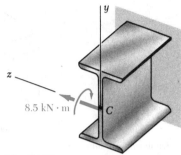

8.5 kN · m

Fig. P4.21

4.22 A steel band blade, which was originally straight, passes over 200 mm-diameter pulleys when mounted on a band saw. Determine the maximum stress in the blade, knowing that it is 0.45 mm thick and 16 mm wide. Use $E = 200$ GPa.

0.45 mm

Fig. P4.22

4.23 Straight rods of 6-mm diameter and 30-m length are stored by coiling the rods inside a drum of 1.25-m inside diameter. Assuming that the yield strength is not exceeded, determine (*a*) the maximum stress in a coiled rod, (*b*) the corresponding bending moment in a rod. Use $E = 200$ GPa.

Fig. P4.23

4.24 A 900-mm strip of steel is bent into a full circle by two couples applied as shown. Determine (*a*) the maximum thickness t of the strip if the allowable stress of the steel is 420 MPa, (*b*) the corresponding moment M of the couples. Use $E = 200$ GPa.

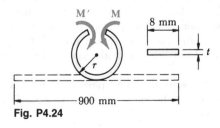

Fig. P4.24

4.25 A 40 × 300-mm plank may be reinforced by securely nailing to it the two 40 × 40 mm boards shown. Knowing that $\sigma_{all} = 9.3$ MPa and $E = 12$ GPa, determine the largest couple which may be applied and the corresponding radius of curvature, (a) for the original unreinforced plank, (b) for the reinforced plank.

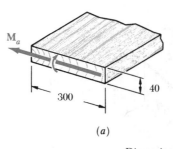

(a)

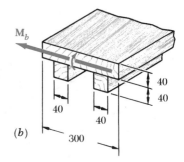

(b)

Fig. P4.25   Dimensions in mm

4.26 A couple **M** will be applied to a beam of rectangular cross section which is to be sawed from a log of circular cross section. Determine the ratio d/b, for which (a) the maximum stress σ_m will be as small as possible, (b) the radius of curvature of the beam will be maximum.

4.27 A thick-walled pipe is bent about a horizontal axis by a couple **M**. The pipe may be designed with or without four fins. (a) Using an allowable stress of 15 MPa, determine the largest couple which may be applied if the pipe is designed with four fins as shown. (b) Solve part a, assuming that the pipe is designed with no fins.

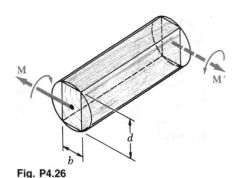

Fig. P4.26

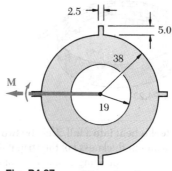

Fig. P4.27   Dimensions in mm

4.28 A portion of a square bar is removed by milling, so that its cross section is as shown. The bar is then bent about its horizontal diagonal by a couple **M.** Considering the case where $h = 0.9\,h_0$, express the maximum stress in the bar in the form $\sigma_m = k\sigma_0$, where σ_0 is the maximum stress that would have occurred if the original square bar had bent by the same couple **M,** and determine the value of k.

4.29 In Prob. 4.28, determine (a) the range of values of h for which $\sigma_m < \sigma_0$, (b) the value of h for which the maximum stress is as small as possible and the corresponding value of k.

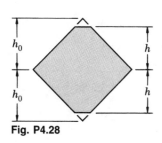

Fig. P4.28

4.30 A steel beam of the cross section shown is subjected to a couple **M** of moment 6 kN · m. Knowing that $E = 200$ GPa and $\nu = 0.30$, determine (*a*) the radius of curvature ρ, (*b*) the radius of curvature ρ' in a transverse plane, (*c*) the angle between the vertical sides of the flange of the beam.

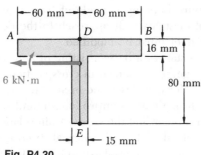

Fig. P4.30

4.31 For the bar and loading of Example 4.01, determine (*a*) the radius of curvature ρ, (*b*) the radius of curvature ρ' in a transverse plane, (*c*) the angle between the originally vertical sides of the bar. Use $E = 200$ GPa and $\nu = 0.30$.

4.32 A W 410 × 114 rolled-steel beam is subjected to a couple **M** of moment 360 kN · m. Knowing that $E = 200$ GPa and $\nu = 0.30$, determine (*a*) the change in width of the top flange, (*b*) the change in length of the top half *AC* of the web.

4.33 For the beam and loading of Prob. 4.30, determine (*a*) the change in width of the top edge *AB* of the flange, (*b*) the change in length of the center line *DE*.

4.34 It was assumed in Sec. 4.4 that the normal stresses σ_y in a member in pure bending are negligible. For an initially straight elastic member of rectangular cross section, (*a*) derive an approximate expression for σ_y as a function of y, (*b*) show that $(\sigma_y)_{max} \approx -(c/2\rho)(\sigma_x)_{max}$ and, thus, that σ_y may be neglected in all practical situations. (*Hint:* Consider the free-body diagram of the portion of beam located below the surface of ordinate y and assume that the distribution of the stresses σ_x is still linear.)

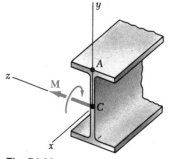

Fig. P4.32

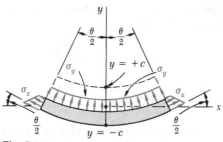

Fig. P4.34

4.7. BENDING OF MEMBERS MADE OF SEVERAL MATERIALS

The derivations given in Sec. 4.5 were based on the assumption of a homogeneous material with a given modulus of elasticity E. If the member subjected to pure bending is made of two or more materials with different moduli of elasticity, our approach to the determination of the stresses in the member must be modified.

Consider, for instance, a bar consisting of two portions of different materials bonded together as shown in cross section in Fig. 4.20. This composite bar will deform as described in Sec. 4.4, since its cross section remains the same throughout its entire length, and since no assumption was made in Sec. 4.4 regarding the stress-strain relationship of the material or materials involved. Thus, the normal strain ϵ_x still varies linearly with the distance y from the neutral axis of the section (Fig. 4.21a and b), and formula (4.8) holds:

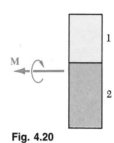

Fig. 4.20

$$\epsilon_x = -\frac{y}{\rho} \tag{4.8}$$

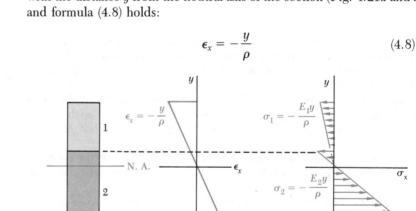

(a)   (b)   (c)

Fig. 4.21 Strain and stress distribution in bar made of two materials.

However, we cannot assume that the neutral axis passes through the centroid of the composite section, and one of the goals of the present analysis will be to determine the location of this axis.

Since the moduli of elasticity E_1 and E_2 of the two materials are different, the expressions obtained for the normal stress in each material will also be different. We write

$$\sigma_1 = E_1\epsilon_x = -\frac{E_1 y}{\rho}$$
$$\sigma_2 = E_2\epsilon_x = -\frac{E_2 y}{\rho} \tag{4.24}$$

and obtain a stress-distribution curve consisting of two segments of straight line (Fig. 4.21c). It follows from Eqs. (4.24) that the force dF_1

exerted on an element of area dA of the upper portion of the cross section is

$$dF_1 = \sigma_1 \, dA = -\frac{E_1 y}{\rho} \, dA \qquad (4.25)$$

while the force dF_2 exerted on an element of the same area dA of the lower portion is

$$dF_2 = \sigma_2 \, dA = -\frac{E_2 y}{\rho} \, dA \qquad (4.26)$$

But, denoting by n the ratio E_2/E_1 of the two moduli of elasticity, we may express dF_2 as

$$dF_2 = -\frac{(nE_1)y}{\rho} \, dA = -\frac{E_1 y}{\rho}(n \, dA) \qquad (4.27)$$

Comparing Eqs. (4.25) and (4.27), we note that the same force dF_2 would be exerted on an element of area $n \, dA$ of the first material. In other words, the resistance to bending of the bar would remain the same if both portions were made of the first material, providing that the width of each element of the lower portion were multiplied by the factor n. Note that this widening (if $n > 1$), or narrowing (if $n < 1$), must be effected *in a direction parallel to the neutral axis of the section*, since it is essential that the distance y of each element from the neutral axis remain the same. The new cross section obtained in this way is called the *transformed section* of the member (Fig. 4.22).

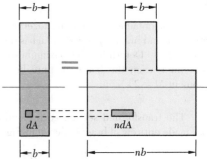

Fig. 4.22 Transformed section for composite bar.

Since the transformed section represents the cross section of a member made of a *homogeneous material* with a modulus of elasticity E_1, the method described in Sec. 4.5 may be used to determine the neutral axis of the section and the normal stress at various points of the section. The neutral axis will be drawn *through the centroid of the transformed section* (Fig. 4.23), and the stress σ_x at any point of the corresponding fictitious homogeneous member will be obtained from Eq. (4.16)

$$\sigma_x = -\frac{My}{I} \qquad (4.16)$$

where y is the distance from the neutral surface, and I *the moment of inertia of the transformed section* with respect to its centroidal axis.

To obtain the stress σ_1 at a point located in the upper portion of the cross section of the original composite bar, we shall simply compute the stress σ_x at the corresponding point of the transformed section. However, to obtain the stress σ_2 at a point in the lower portion of the cross section, we shall *multiply by n* the stress σ_x computed at the corresponding point of the transformed section. Indeed, as we saw earlier, the same elementary force dF_2 is applied to an element of area $n \, dA$ of the transformed section and to an element of area dA of the original section. Thus, the stress σ_2 at a point of the original section must be n times larger than the stress at the corresponding point of the transformed section.

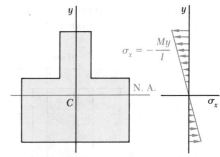

Fig. 4.23 Distribution of stresses in transformed section.

The deformations of a composite member may also be determined by using the transformed section. We recall that the transformed section represents the cross section of a member, made of a homogeneous material of modulus E_1, which deforms in the same manner as the composite member. Therefore, using Eq. (4.21), we write that the curvature of the composite member is

$$\frac{1}{\rho} = \frac{M}{E_1 I}$$

where I is the moment of inertia of the transformed section with respect to its neutral axis.

Example 4.03

A bar obtained by bonding together pieces of steel ($E_s = 200$ GPa) and brass ($E_b = 100$ GPa) has the cross section shown (Fig. 4.24). Determine the maximum stress in the steel and in the brass when the bar is in pure bending with a bending moment $M = 2$ kN · m.

The transformed section corresponding to an equivalent bar made entirely of brass is shown in Fig. 4.25. Since

$$n = \frac{E_s}{E_b} = \frac{200 \text{ GPa}}{100 \text{ GPa}} = 2$$

the width of the central portion of brass, which replaces the original steel portion, is obtained by multiplying the original width by 2.

Note that this change in dimension occurs in a direction parallel to the neutral axis. The moment of inertia of the transformed section about its centroidal axis is

$$I = \tfrac{1}{12}bh^3 = \tfrac{1}{12}(30 \times 10^{-3} \text{ m})(40 \times 10^{-3} \text{ m})^3 = 160 \times 10^{-9} \text{ m}^4$$

and the maximum distance from the neutral axis is $c = 20$ mm. Using Eq. (4.15), we find the maximum stress in the transformed section:

$$\sigma_m = \frac{Mc}{I} = \frac{(2 \times 10^3 \text{ N} \cdot \text{m})(20 \times 10^{-3} \text{ m})}{160 \times 10^{-9} \text{ m}^4} = 250 \text{ MPa}$$

The value obtained also represents the maximum stress in the brass portion of the original composite bar. The maximum stress in the steel portion, however, will be larger than the value obtained for the transformed section, since the area of the central portion must be reduced by the factor $n = 2$ when we return from the transformed section to the original one. We thus conclude that

$$(\sigma_{\text{brass}})_{\text{max}} = 250 \text{ MPa}$$
$$(\sigma_{\text{steel}})_{\text{max}} = 500 \text{ MPa}$$

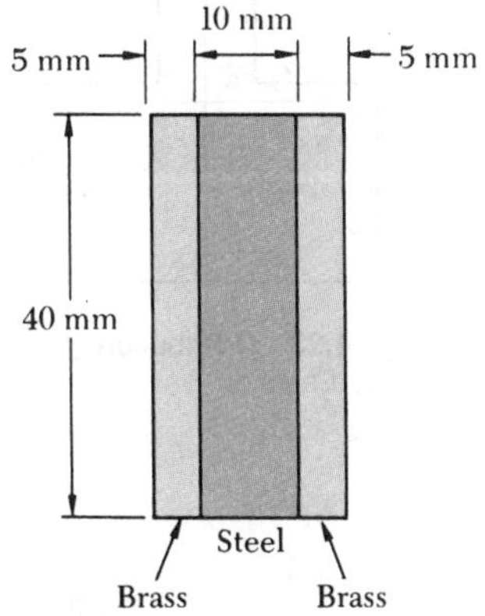

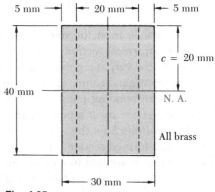

Fig. 4.24

Fig. 4.25

An important example of structural members made of two different materials is furnished by *reinforced concrete beams*. These beams, when subjected to positive bending moments, are reinforced by steel rods placed a short distance above their lower face (Fig. 4.26a). Since concrete is very weak in tension, it will crack below the neutral surface and the steel rods will carry the entire tensile load, while the upper part of the concrete beam will carry the compressive load.

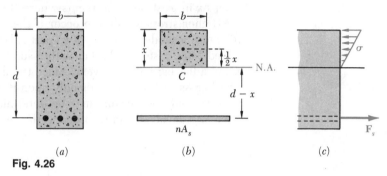

(a)   (b)   (c)

Fig. 4.26

To obtain the transformed section of a reinforced concrete beam, we replace the total cross-sectional area A_s of the steel bars by an equivalent area nA_s, where n is the ratio E_s/E_c of the moduli of elasticity of steel and concrete (Fig. 4.26b). On the other hand, since the concrete in the beam acts effectively only in compression, only the portion of the cross section located above the neutral axis should be used in the transformed section.

The position of the neutral axis is obtained by determining the distance x from the upper face of the beam to the centroid C of the transformed section. Denoting by b the width of the beam, and by d the distance from the upper face to the center line of the steel rods, we write that the first moment of the transformed section with respect to the neutral axis must be zero. Since the first moment of each of the two portions of the transformed section is obtained by multiplying its area by the distance of its own centroid from the neutral axis, we have

$$(bx)\frac{x}{2} - nA_s\,(d-x) = 0$$

or

$$\frac{1}{2}\,bx^2 + nA_sx - nA_sd = 0 \tag{4.28}$$

Solving this quadratic equation for x, we obtain both the position of the neutral axis in the beam, and the portion of the cross section of the concrete beam which is effectively used.

The determination of the stresses in the transformed section is carried out as explained earlier in this section (see Sample Prob. 4.4). The distribution of the compressive stresses in the concrete and the resultant $\mathbf{F}_s$ of the tensile forces in the steel rods are shown in Fig. 4.26c.

4.8. STRESS CONCENTRATIONS

The formula $\sigma_m = Mc/I$ was derived in Sec. 4.5 for a member with a plane of symmetry and a uniform cross section, and we saw in Sec. 4.6 that it was accurate throughout the entire length of the member only if the couples **M** and **M'** were applied through the use of rigid and smooth plates. Under other conditions of application of the loads, stress concentrations will exist near the points where the loads are applied.

Higher stresses will also occur if the cross section of the member undergoes a sudden change. Two particular cases of interest have been studied,† the case of a flat bar with a sudden change in width, and the case of a flat bar with grooves. Since the distribution of stresses in the critical cross sections depends only upon the geometry of the members, stress-concentration factors may be determined for various ratios of the parameters involved and recorded as shown in Figs. 4.27 and 4.28. The

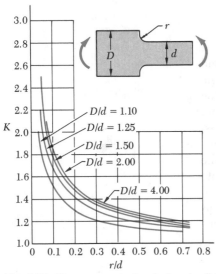

Fig. 4.27 Stress-concentration factors for *flat bars* with fillets under pure bending.†

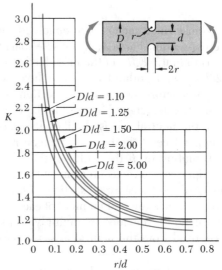

Fig. 4.28 Stress-concentration factors for *flat bars* with grooves under pure bending.†

† M. M. Frocht, "Photoelastic Studies in Stress Concentration," *Mechanical Engineering*, August 1936, pp. 485–489.

value of the maximum stress in the critical cross section may then be expressed as

$$\sigma_m = K\frac{Mc}{I} \tag{4.29}$$

where K is the stress-concentration factor, and where c and I refer to the critical section, i.e., to the section of width d in both of the cases considered here. An examination of Figs. 4.27 and 4.28 clearly shows the importance of using fillets and grooves of radius r as large as practical.

Finally, we should point out that, as was the case for axial loading and torsion, the values of the factors K have been computed under the assumption of a linear relation between stress and strain. In many applications, plastic deformations will occur and result in values of the maximum stress lower than those indicated by Eq. (4.29).

Example 4.04

Grooves 10 mm deep are to be cut in a steel bar which is 60 mm wide and 9 mm thick (Fig. 4.29). Determine the smallest allowable width of the grooves if the stress in the bar is not to exceed 150 MPa when the bending moment is equal to 180 N · m.

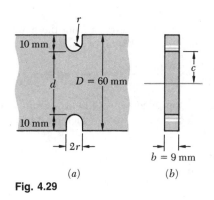

Fig. 4.29

We note from Fig. 4.29a that

$$d = 60 \text{ mm} - 2(10 \text{ mm}) = 40 \text{ mm}$$
$$c = \tfrac{1}{2}d = 20 \text{ mm} \qquad b = 9 \text{ mm}$$

The moment of inertia of the critical cross section about its neutral axis is

$$I = \tfrac{1}{12}bd^3 = \tfrac{1}{12}(9 \times 10^{-3} \text{ m})(40 \times 10^{-3} \text{ m})^3 = 48 \times 10^{-9} \text{ m}^4$$

The value of the stress Mc/I is thus

$$\frac{Mc}{I} = \frac{(180 \text{ N} \cdot \text{m})(20 \times 10^{-3} \text{ m})}{48 \times 10^{-9} \text{ m}^4} = 75 \text{ MPa}$$

Substituting this value for Mc/I into Eq. (4.29) and making $\sigma_m = 150$ MPa, we write

$$150 \text{ MPa} = K(75 \text{ MPa})$$
$$K = 2$$

We have, on the other hand,

$$\frac{D}{d} = \frac{60 \text{ mm}}{40 \text{ mm}} = 1.5$$

Using the curve of Fig. 4.28 corresponding to $D/d = 1.5$, we find that the value $K = 2$ corresponds to a value of r/d equal to 0.1. We have, therefore,

$$\frac{r}{d} = 0.1 \qquad r = 0.1d = 0.1(40 \text{ mm}) = 4 \text{ mm}$$

The smallest allowable width of the grooves is thus

$$2r = 2(4 \text{ mm}) = 8 \text{ mm}$$

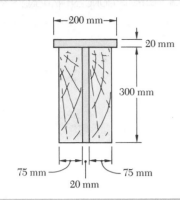

SAMPLE PROBLEM 4.3

A steel beam in the shape of a T has been strengthened by securely bolting to it the two oak timbers shown. The modulus of elasticity is 12.5 GPa for the wood and 200 GPa for the steel. Knowing that a bending moment $M = 50$ kN · m is applied to the composite beam, determine (a) the maximum stress in the wood, (b) the stress in the steel along the top edge.

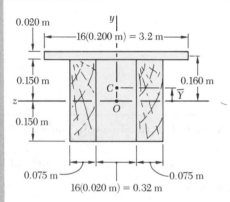

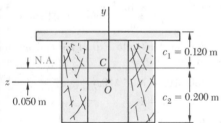

Transformed Section. We first compute the ratio

$$n = \frac{E_s}{E_w} = \frac{200 \text{ GPa}}{12.5 \text{ GPa}} = 16$$

Multiplying the horizontal dimensions of the steel portion of the section by $n = 16$, we obtain a transformed section made entirely of wood.

Neutral Axis. The neutral axis passes through the centroid of the transformed section. Since the section consists of two rectangles, we have

$$\overline{Y} = \frac{\Sigma \overline{y}A}{\Sigma A} = \frac{(0.160 \text{ m})(3.2 \text{ m} \times 0.020 \text{ m}) + 0}{3.2 \text{ m} \times 0.020 \text{ m} + 0.470 \text{ m} \times 0.300 \text{ m}} = 0.050 \text{ m}$$

Centroidal Moment of Inertia. Using the parallel-axis theorem:

$$I = \tfrac{1}{12}(0.470)(0.300)^3 + (0.470 \times 0.300)(0.050)^2$$
$$+ \tfrac{1}{12}(3.2)(0.020)^3 + (3.2 \times 0.020)(0.160 - 0.050)^2$$
$$I = 2.19 \times 10^{-3} \text{ m}^4$$

a. **Maximum Stress in Wood.** The wood farthest from the neutral axis is located along the bottom edge, where $c_2 = 0.200$ m.

$$\sigma_w = \frac{Mc_2}{I} = \frac{(50 \times 10^3 \text{ N} \cdot \text{m})(0.200 \text{ m})}{2.19 \times 10^{-3} \text{ m}^4} \qquad \sigma_w = 4.57 \text{ MPa} \blacktriangleleft$$

b. **Stress in Steel.** Along the top edge $c_1 = 0.120$ m. From the transformed section we obtain an equivalent stress in wood, which must be multiplied by n to obtain the stress in steel.

$$\sigma_s = n\frac{Mc_1}{I} = (16)\frac{(50 \times 10^3 \text{ N} \cdot \text{m})(0.120 \text{ m})}{2.19 \times 10^{-3} \text{ m}^4} \qquad \sigma_s = 43.8 \text{ MPa} \blacktriangleleft$$

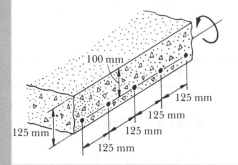

SAMPLE PROBLEM 4.4

A concrete floor slab is reinforced by 16-mm-diameter steel rods placed 25 mm above the lower face of the slab and spaced 125 mm on centers. The modulus of elasticity is 20 GPa for the concrete used and 200 GPa for the steel. Knowing that a bending moment of 12 kN · m is applied to each 1 metre width of the slab, determine (a) the maximum stress in the concrete, (b) the stress in the steel.

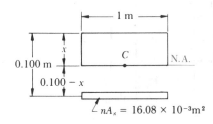

Transformed Section. We consider a portion of the slab 1 m wide, in which there are eight 16-mm-diameter rods having a total cross-sectional area

$$A_s = 8[\tfrac{1}{4}\pi(0.016 \text{ m})^2] = 1.608 \times 10^{-3} \text{ m}^2$$

Since concrete acts only in compression, all tensile forces are carried by the steel rods, and the transformed section consists of the two areas shown. One is the portion of concrete in compression (located above the neutral axis), and the other is the transformed steel area nA_s. We have

$$n = \frac{E_s}{E_c} = \frac{200 \text{ GPa}}{20 \text{ GPa}} = 10$$

$$nA_s = 10(1.608 \times 10^{-3} \text{ m}^2) = 16.08 \times 10^{-3} \text{ m}^2$$

Neutral Axis. The neutral axis of the slab passes through the centroid of the transformed section. Summing moments of the transformed area about the neutral axis, we write

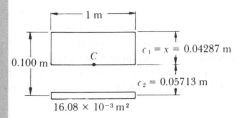

$$(1 \text{ m})x\left(\frac{x}{2}\right) - (16.08 \times 10^{-3} \text{ m}^2)(0.100 \text{ m} - x) = 0 \qquad x = 0.04287 \text{ m}$$

Moment of Inertia. The centroidal moment of inertia of the transformed area is

$$I = \tfrac{1}{3}(1)(0.04287)^3 + (16.08 \times 10^{-3})(0.100 - 0.04287)^2 = 78.75 \times 10^{-6} \text{ m}^4$$

a. **Maximum Stress in Concrete.** At the top of the slab, we have $c_1 = 0.04287$ m and

$$\sigma_c = \frac{Mc_1}{I} = \frac{(12 \text{ kN} \cdot \text{m})(0.04287 \text{ m})}{78.75 \times 10^{-6} \text{ m}^4} \qquad \sigma_c = 6.53 \text{ MPa} \quad \blacktriangleleft$$

b. **Stress in Steel.** For the steel, we have $c_2 = 0.100$ m $- x = 0.05713$ m, $n = 10$ and

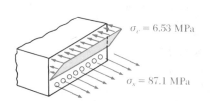

$$\sigma_s = n\frac{Mc_2}{I} = 10\frac{(12 \text{ kN} \cdot \text{m})(0.05713 \text{ m})}{78.75 \times 10^{-6} \text{ m}^4} \qquad \sigma_s = 87.1 \text{ MPa} \quad \blacktriangleleft$$

PROBLEMS

4.35 and 4.36 Two brass bars and two aluminum bars are securely bonded together to form the composite member shown. Using the data given below, determine the largest permissible bending moment when the member is bent about a horizontal axis.

	Aluminum	Brass
Modulus of elasticity	70 GPa	105 GPa
Allowable stress	100 MPa	160 MPa

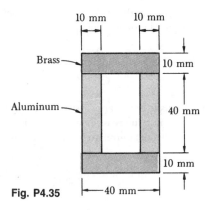

Fig. P4.35

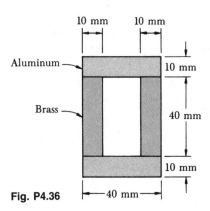

Fig. P4.36

4.37 For the composite member of Prob. 4.36, determine the largest permissible bending moment when the member is bent about a vertical axis.

4.38 The 150 × 250 mm timber beam has been strengthened by bolting to it two steel strips to form the composite member shown. Using the data given below, determine the largest permissible bending moment when the member is bent about a horizontal axis.

	Wood	Steel
Modulus of elasticity	13 GPa	200 GPa
Allowable stress	12 MPa	165 MPa

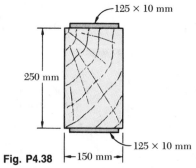

Fig. P4.38

4.39 For the composite member of Prob. 4.38, determine the largest permissible bending moment when the member is bent about a vertical axis.

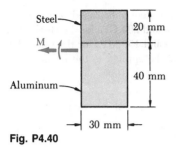

Fig. P4.40

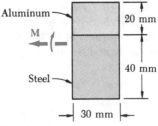

Fig. P4.41

Fig. P4.42

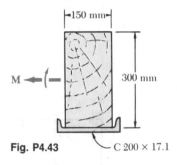

Fig. P4.43

4.40 and 4.41 A steel bar and an aluminum bar are bonded together to form the composite beam shown. The modulus of elasticity for aluminum is 70 GPa and for steel is 200 GPa. Knowing that the beam is bent about a horizontal axis by a couple of moment $M = 1500$ N $\cdot$ m, determine the maximum stress in (a) the aluminum, (b) the steel.

4.42 and 4.43 The 150×300 mm timber beam has been strengthened by bolting to it the steel reinforcement shown. The modulus of elasticity for wood is 12 GPa and for steel is 200 GPa. Knowing that the beam is bent about a horizontal axis by a couple of moment $M = 50$ kN $\cdot$ m, determine the maximum stress in (a) the wood, (b) the steel.

4.44 A concrete slab is reinforced by 16-mm-diameter steel rods placed on 140-mm centers as shown. The modulus of elasticity is 20 GPa for concrete and 200 GPa for steel. Using an allowable stress of 10 MPa for the concrete and 140 MPa for the steel, determine the largest bending moment per meter of width which may be applied to the slab.

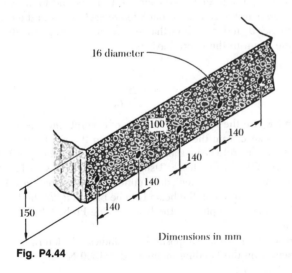

Dimensions in mm

Fig. P4.44

4.45 Solve Prob. 4.44, assuming that the spacing of the 16-mm-diameter rods is increased to 190 mm on centers.

4.46 A concrete beam is reinforced by three steel rods placed as shown. The modulus of elasticity is 20 GPa for the concrete and 200 GPa for the steel. Using an allowable stress of 10 MPa for the concrete and 150 MPa for the steel, determine the largest bending moment which may be applied to the beam.

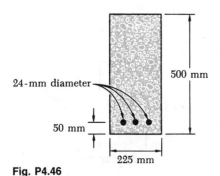

Fig. P4.46

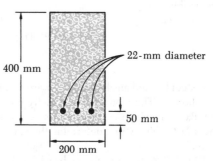

Fig. P4.47

4.47 The reinforced concrete beam shown is subjected to a bending moment of 55 kN · m. Knowing that the modulus of elasticity is 25 GPa for the concrete and 200 GPa for the steel, determine (a) the stress in the steel, (b) the maximum stress in the concrete.

4.48 Solve Prob. 4.47, assuming that the width of the concrete beam is increased to 250 mm.

4.49 The design of a reinforced concrete beam is said to be *balanced* if the maximum stresses in the steel and concrete are each equal to the corresponding allowable stresses σ_s and σ_c. Show that for a balanced design the distance x from the top of the beam to the neutral axis is

$$x = \frac{d}{1 + \dfrac{\sigma_s E_c}{\sigma_c E_s}}$$

where E_c and E_s are the moduli of elasticity of concrete and steel, respectively, and d is the distance from the top of the beam to the reinforcing steel.

4.50 For the concrete beam shown, the modulus of elasticity is 25 GPa for the concrete and 200 GPa for the steel. Using an allowable stress of 12.5 MPa for the concrete and 140 MPa for the steel, determine (a) the required area A_s of the steel reinforcement if the design of the beam is to be balanced, (b) the largest bending moment which may be applied to the beam. (See Prob. 4.49 for definition of a balanced beam.)

4.51 and 4.52 For the composite beam indicated, determine the radius of curvature caused by the bending moment $M = 1500$ N · m.

4.51 Beam of Prob. 4.41

4.52 Beam of Prob. 4.40

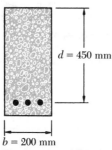

Fig. P4.49 and P4.50

4.53 and 4.54 For the composite beam indicated, determine the radius of curvature caused by the bending moment $M = 50$ kN · m.

   4.53 Beam of Prob. 4.43
   4.54 Beam of Prob. 4.42

4.55 and 4.56 Five metal strips, each 40 mm wide, are bonded together to form the composite beam shown. The modulus of elasticity is 210 GPa for the steel, 105 GPa for the brass, and 70 GPa for the aluminum. Knowing that the beam is bent about a horizontal axis by couples of moment $M = 1800$ N · m, determine the maximum stress (a) in the steel, (b) in the brass, (c) in the aluminum. (d) Also, determine the radius of curvature of the beam.

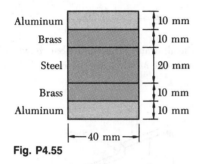

Fig. P4.55

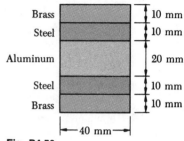

Fig. P4.56

4.57 A steel pipe and an aluminum pipe are securely bonded to form the composite beam shown. The modulus of elasticity is 200 GPa for the steel and 72 GPa for the aluminum. For a bending moment of 1 kN · m, determine the maximum stress (a) in the aluminum, (b) in the steel. (c) Also, determine the radius of curvature of the beam.

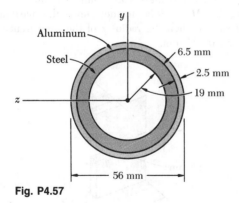

Fig. P4.57

4.58 Solve Prob. 4.57, assuming that the 6.5-mm-thick inner pipe is made of aluminum and that the 2.5-mm-thick shell is made of steel.

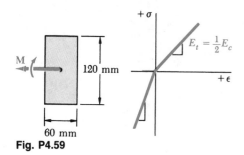

Fig. P4.59

4.59 The rectangular beam shown is made of a plastic for which the value of the modulus of elasticity in tension is one-half of its value in compression. For a bending moment of 1.5 kN · m, determine the maximum (a) tensile stress, (b) compressive stress.

*4.60 A rectangular beam is made of a material for which the modulus of elasticity is E_t in tension and E_c in compression. Show that the curvature in pure bending is

$$\frac{1}{\rho} = \frac{M}{E_r I}$$

where

$$E_r = \frac{4E_t E_c}{(\sqrt{E_t} + \sqrt{E_c})^2}$$

4.61 Knowing that the allowable stress in the bar shown is 70 MPa, determine the allowable bending moment M when the radius r of the fillets is (a) 20 mm, (b) 10 mm.

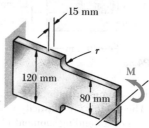

Fig. P4.61 and P4.62

4.62 Knowing that $M = 1.0$ kN · m, determine the maximum stress in the bar shown when the radius r of the fillets is (a) 15 mm, (b) 5 mm.

4.63 Knowing that $M = 400$ N · m, determine the maximum stress in the machine element shown when the radius r of the semicircular portions of the grooves is (a) 5 mm, (b) 15 mm.

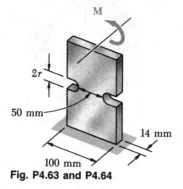

Fig. P4.63 and P4.64

4.64 Knowing that the allowable stress in the machine element shown is 90 MPa, determine the allowable bending moment M when the radius r of the semicircular portions of the grooves is (a) 10 mm, (b) 20 mm.

4.65   Semicircular grooves of radius r must be milled in the top and bottom portions of a steel bar. Using an allowable stress of 100 MPa, determine the largest bending moment which may be applied to the bar shown when the radius of the semicircular grooves is (*a*) 5 mm, (*b*) 20 mm.

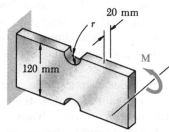

20 mm

r

M

120 mm

Fig. P4.65 and P4.66

4.66   Semicircular grooves of radius r must be milled in the top and bottom portions of a steel bar. Knowing that a couple of moment $M = 900$ N · m is applied as shown, determine the maximum stress in the bar for (*a*) $r = 12$ mm, (*b*) $r = 30$ mm.

4.67   A couple of moment $M = 2$ kN · m is to be applied to the end of a steel bar. Determine the maximum stress in the bar (*a*) if the bar is designed with grooves having semicircular portions of radius $r = 10$ mm, as shown in Fig. *a*, (*b*) if the bar is redesigned by removing the material above and below the dashed lines, as shown in Fig. *b*.

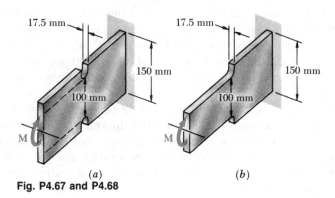

17.5 mm

150 mm

100 mm

M

(*a*)

17.5 mm

150 mm

100 mm

M

(*b*)

Fig. P4.67 and P4.68

4.68   The allowable stress used in the design of a steel bar is 80 MPa. Determine the largest couple **M** which may be applied to the bar (*a*) if the bar is designed with grooves having semicircular portions of radius $r = 15$ mm, as shown in Fig. *a*, (*b*) if the bar is redesigned by removing the material above and below the dashed lines, as shown in Fig. *b*.

*4.9. PLASTIC DEFORMATIONS

When we derived the fundamental relation $\sigma_x = -My/I$ in Sec. 4.5, we assumed that Hooke's law applied throughout the member. If the yield strength is exceeded in some portion of the member, or if the material involved is a brittle material with a nonlinear stress-strain diagram, this relation ceases to be valid. The purpose of this section is to develop a more general method for the determination of the distribution of stresses in a member in pure bending, which may be used when Hooke's law does not apply.

We first recall that no specific stress-strain relationship was assumed in Sec. 4.4, when we proved that the normal strain ϵ_x varies linearly with the distance y from the neutral surface. Thus, we may still use this property in our present analysis and write

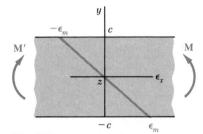

Fig. 4.30

$$\epsilon_x = -\frac{y}{c}\epsilon_m \tag{4.10}$$

where y represents the distance of the point considered from the neutral surface, and c the maximum value of y.

However, we cannot assume any more that, in a given section, the neutral axis passes through the centroid of that section, since this property was derived in Sec. 4.5 under the assumption of elastic deformations. In general, the neutral axis must be located by trial and error, until a distribution of stresses has been found, which satisfies Eqs. (4.1) and (4.3) of Sec. 4.3. However, in the particular case of a member possessing both a vertical and a horizontal plane of symmetry, and made of a material characterized by the same stress-strain relation in tension and in compression, the neutral axis will coincide with the horizontal axis of symmetry of the section. Indeed, the properties of the material require that the stresses be symmetric with respect to the neutral axis, i.e., with respect to *some* horizontal axis, and it is clear that this condition will be met, and Eq. (4.1) satisfied at the same time, only if that axis is the horizontal axis of symmetry itself.

Our analysis will first be limited to the special case we have just described. The distance y in Eq. (4.10) is thus measured from the horizontal axis of symmetry z at the cross section, and the distribution of strain ϵ_x is linear and symmetric with respect to that axis (Fig. 4.30). On the other hand, the stress-strain curve is symmetric with respect to the origin of coordinates (Fig. 4.31).

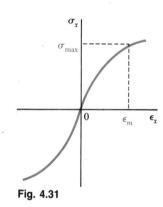

Fig. 4.31

The distribution of stresses in the cross section of the member, i.e., the plot of σ_x versus y, is obtained as follows. Assuming that σ_{max} has been specified, we first determine the corresponding value of ϵ_m from the stress-strain diagram and carry this value into Eq. (4.10). Then, for each value of y, we determine the corresponding value of ϵ_x from Eq. (4.10) or Fig. 4.30, and obtain from the stress-strain diagram of Fig. 4.31 the stress σ_x corresponding to this value of ϵ_x. Plotting σ_x against y yields the desired distribution of stresses (Fig. 4.32).

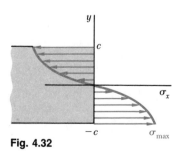

Fig. 4.32

We now recall that, when we derived Eq. (4.3) in Sec. 4.3, we assumed no particular relation between stress and strain. We may therefore use Eq. (4.3) to determine the bending moment M corresponding to the stress distribution obtained in Fig. 4.32. Considering the particular case of a member with a rectangular cross section of width b, we express the element of area in Eq. (4.3) as $dA = b \, dy$ and write

$$M = -b \int_{-c}^{c} y\sigma_x \, dy \qquad (4.30)$$

where σ_x is the function of y plotted in Fig. 4.32. Since σ_x is an odd function of y, we may write Eq. (4.30) in the alternative form

$$M = -2b \int_{0}^{c} y\sigma_x \, dy \qquad (4.31)$$

If σ_x is a known analytical function of ϵ_x, Eq. (4.10) may be used to express σ_x as a function of y, and the integral in (4.31) may be determined analytically. Otherwise, the bending moment M may be obtained through a numerical integration. This computation becomes more meaningful if we note that the integral in Eq. (4.31) represents the first moment with respect to the horizontal axis of the area in Fig. 4.32 which is located above the horizontal axis and is bounded by the stress-distribution curve and the vertical axis.

An important value of the bending moment is the ultimate bending moment M_U which causes failure of the member. This value may be determined from the ultimate strength σ_U of the material by choosing $\sigma_{\max} = \sigma_U$ and carrying out the computations indicated earlier. However, it is found more convenient in practice to determine M_U experimentally for a specimen of a given material. Assuming a fictitious linear distribution of stresses, Eq. (4.15) is then used to determine the corresponding maximum stress R_B:

$$R_B = \frac{M_U c}{I} \qquad (4.32)$$

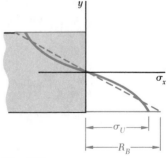

Fig. 4.33

The fictitious stress R_B is called the *modulus of rupture in bending* of the given material. It may be used to determine the ultimate bending moment M_U of a member made of the same material and having a cross section of the same shape, but of different dimensions, by solving Eq. (4.32) for M_U. Since, in the case of a member with a rectangular cross section, the actual and the fictitious linear stress distributions shown in Fig. 4.33 must yield the same value M_U for the ultimate bending moment, the areas they define must have the same first moment with respect to the horizontal axis. It is thus clear that the modulus of rupture R_B will always be larger than the actual ultimate strength σ_U.

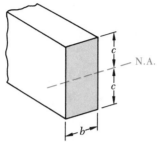

Fig. 4.34

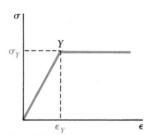

Fig. 4.35

*4.10. MEMBERS MADE OF AN ELASTOPLASTIC MATERIAL

In order to gain a better insight into the plastic behavior of a member in bending, we shall consider the case of a member made of an *elastoplastic material* and first assume the member to have a *rectangular cross section* of width b and depth $2c$ (Fig. 4.34). We recall from Sec. 2.18 that the stress-strain diagram for an idealized elastoplastic material is as shown in Fig. 4.35.

As long as the normal stress σ_x does not exceed the yield strength σ_Y, Hooke's law applies, and the stress distribution across the section is linear (Fig. 4.36a). The maximum value of the stress is

$$\sigma_m = \frac{Mc}{I} \tag{4.15}$$

As the bending moment increases, σ_m eventually reaches the value σ_Y (Fig. 4.36b). Substituting this value into Eq. (4.15), and solving for the corresponding value of M, we obtain the value M_Y of the bending moment at the onset of yield:

$$M_Y = \frac{I}{c}\sigma_Y \tag{4.33}$$

The moment M_Y is referred to as the *maximum elastic moment*, since it is the largest moment for which the deformation remains fully elastic. Recalling that, for the rectangular cross section considered here,

$$\frac{I}{c} = \frac{b(2c)^3}{12c} = \frac{2}{3}bc^2 \tag{4.34}$$

we write

$$M_Y = \frac{2}{3}bc^2\sigma_Y \tag{4.35}$$

As the bending moment further increases, plastic zones develop in the member, with the stress uniformly equal to $-\sigma_Y$ in the upper zone, and to $+\sigma_Y$ in the lower zone (Fig. 4.36c). Between the plastic zones, an elastic core subsists, in which the stress σ_x varies linearly with y,

$$\sigma_x = -\frac{\sigma_Y}{y_Y}y \tag{4.36}$$

where y_Y represents half the thickness of the elastic core. As M increases, the plastic zones expand until, at the limit, the deformation is fully plastic (Fig. 4.36d).

We shall use Eq. (4.31) to determine the value of the bending moment M corresponding to a given thickness $2y_Y$ of the elastic core. Recalling that σ_x is given by Eq. (4.36) for $0 \le y \le y_Y$, and is equal to $-\sigma_Y$ for $y_Y \le y \le c$, we write

$$M = -2b \int_0^{y_Y} y \left(-\frac{\sigma_Y}{y_Y} y \right) dy - 2b \int_{y_Y}^{c} y(-\sigma_Y)\, dy$$

$$= \frac{2}{3} b y_Y^2 \sigma_Y + b c^2 \sigma_Y - b y_Y^2 \sigma_Y$$

$$M = b c^2 \sigma_Y \left(1 - \frac{1}{3} \frac{y_Y^2}{c^2} \right) \tag{4.37}$$

or, in view of Eq. (4.35),

$$M = \frac{3}{2} M_Y \left(1 - \frac{1}{3} \frac{y_Y^2}{c^2} \right) \tag{4.38}$$

where M_Y is the maximum elastic moment. We note that as y_Y approaches zero, the bending moment approaches the limiting value

$$M_p = \tfrac{3}{2} M_Y \tag{4.39}$$

This value of the bending moment, which corresponds to a fully plastic deformation (Fig. 4.36d), is called the *plastic moment* of the member considered. We note that Eq. (4.39) is valid only for a *rectangular member made of an elastoplastic material.*

It should be kept in mind that the distribution of *strain* across the section remains linear after the onset of yield. Therefore, Eq. (4.8) of Sec. 4.4 remains valid and may be used to determine the half-thickness y_Y of the elastic core. We have

$$y_Y = \epsilon_Y \rho \tag{4.40}$$

where ϵ_Y is the yield strain and ρ the radius of curvature corresponding to a bending moment $M \geq M_Y$. When the bending moment is equal to M_Y, we have $y_Y = c$ and Eq. (4.40) yields

$$c = \epsilon_Y \rho_Y \tag{4.41}$$

where ρ_Y is the radius of curvature corrresponding to the maximum elastic moment M_Y. Dividing (4.40) by (4.41) member by member, we obtain the relation†

$$\frac{y_Y}{c} = \frac{\rho}{\rho_Y} \tag{4.42}$$

Substituting for y_Y/c from (4.42) into Eq. (4.38), we express the bending moment M as a function of the radius of curvature ρ of the neutral surface:

$$M = \frac{3}{2} M_Y \left(1 - \frac{1}{3} \frac{\rho^2}{\rho_Y^2} \right) \tag{4.43}$$

We should note that Eq. (4.43) is valid only after the onset of yield, i.e.,

---

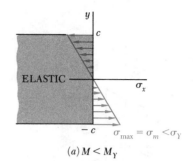

(a) $M < M_Y$

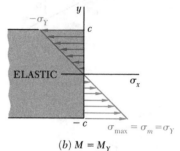

(b) $M = M_Y$

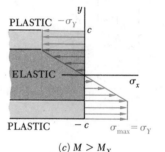

(c) $M > M_Y$

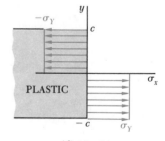

(d) $M = M_p$

Fig. 4.36

---

†Equation (4.42) applies to any member made of any ductile material with a well-defined yield point, since its derivation is independent of the shape of the cross section and of the shape of the stress-strain diagram beyond the yield point.

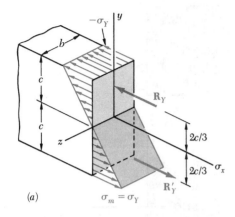

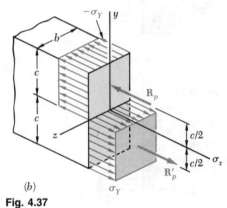

Fig. 4.37

for values of M larger than M_Y. For $M < M_Y$, Eq. (4.21) of Sec. 4.5 should be used.

We observe from Eq. (4.43) that the bending moment reaches the value $M_p = \frac{3}{2}M_Y$ only when $\rho = 0$. Since we clearly cannot have a zero radius of curvature at every point of the neutral surface, we conclude that a fully plastic deformation cannot develop in pure bending. As we shall see in Chap. 5, however, such a situation may occur at one point in the case of a beam under a transverse loading.

The stress distributions in a rectangular member corresponding respectively to the maximum elastic moment M_Y and to the limiting case of the plastic moment M_p have been represented in three dimensions in Fig. 4.37. Since, in both cases, the resultants of the elementary tensile and compressive forces must pass through the centroids of the volumes representing the stress distributions, and be equal in magnitude to these volumes, we check that

$$R_Y = \tfrac{1}{2}bc\sigma_Y \qquad R_p = bc\sigma_Y$$

and that the moments of the corresponding couples are, respectively,

$$M_Y = (\tfrac{4}{3}c)R_Y = \tfrac{2}{3}bc^2\sigma_Y \tag{4.44}$$

and

$$M_p = cR_p = bc^2\sigma_Y \tag{4.45}$$

We thus verify that, for a rectangular member, $M_p = \frac{3}{2}M_Y$ as required by Eq. (4.39).

For beams of *nonrectangular cross section*, the computation of the maximum elastic moment M_Y and of the plastic moment M_p will usually be simplified if a graphical method of analysis is used, as shown in Sample Prob. 4.5. It will be found in this more general case that the ratio $k = M_p/M_Y$ is generally not equal to $\frac{3}{2}$. For structural shapes such as wide-flange beams, for example, this ratio varies approximately from 1.08 to 1.14. Because it depends only upon the shape of the cross section, the ratio $k = M_p/M_Y$ is referred to as the *shape factor* of the cross section. We note that, if the shape factor k and the maximum elastic moment M_Y of a beam are known, the plastic moment M_p of the beam may be obtained by multiplying M_Y by k:

$$M_p = kM_Y \tag{4.46}$$

The ratio M_p/σ_Y obtained by dividing the plastic moment M_p of a member by the yield strength σ_Y of its material is called the *plastic section modulus* of the member and is denoted by Z. When the plastic section modulus Z and the yield strength σ_Y of a beam are known, the plastic moment M_p of the beam may be obtained by multiplying σ_Y by Z:

$$M_p = Z\sigma_Y \tag{4.47}$$

Recalling from Eq. (4.18) that $M_Y = S\sigma_Y$, and comparing this relation with Eq. (4.47), we note that the shape factor $k = M_p/M_Y$ of a given cross section may be expressed as the ratio of the plastic and elastic section moduli:

$$k = \frac{M_p}{M_Y} = \frac{Z\sigma_Y}{S\sigma_Y} = \frac{Z}{S} \tag{4.48}$$

Considering the particular case of a rectangular beam of width b and depth h, we note from Eqs. (4.45) and (4.47) that the *plastic section modulus* of a rectangular beam is

$$Z = \frac{M_p}{\sigma_Y} = \frac{bc^2\sigma_Y}{\sigma_Y} = bc^2 = \tfrac{1}{4}bh^2$$

On the other hand, we recall from Eq. (4.19) of Sec. 4.5 that the *elastic section modulus* of the same beam is

$$S = \tfrac{1}{6}bh^2$$

Substituting into Eq. (4.48) the values obtained for Z and S, we verify that the shape factor of a rectangular beam is

$$k = \frac{Z}{S} = \frac{\tfrac{1}{4}bh^2}{\tfrac{1}{6}bh^2} = \frac{3}{2}$$

Example 4.05

A member of uniform rectangular cross section 50 by 120 mm (Fig. 4.38) is subjected to a bending moment $M = 36.8$ kN · m. Assuming that the member is made of an elastoplastic material with a yield strength of 240 MPa and a modulus of elasticity of 200 GPa, determine (*a*) the thickness of the elastic core, (*b*) the radius of curvature of the neutral surface.

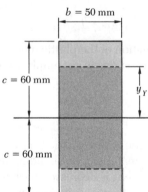

Fig. 4.38

(*a*) *Thickness of Elastic Core.* We first determine the maximum elastic moment M_Y. Substituting the given data into Eq. (4.34), we have

$$\frac{I}{c} = \frac{2}{3}bc^2 = \frac{2}{3}(50 \times 10^{-3} \text{ m})(60 \times 10^{-3} \text{ m})^2 = 120 \times 10^{-6} \text{ m}^3$$

and carrying this value, as well as $\sigma_Y = 240$ MPa, into Eq. (4.33),

$$M_Y = \frac{I}{c}\sigma_Y = (120 \times 10^{-6} \text{ m}^3)(240 \text{ MPa}) = 28.8 \text{ kN} \cdot \text{m}$$

Substituting the values of M and M_Y into Eq. (4.38), we have

$$36.8 \text{ kN} \cdot \text{m} = \frac{3}{2}(28.8 \text{ kN} \cdot \text{m})\left(1 - \frac{1}{3}\frac{y_Y^2}{c^2}\right)$$

$$\left(\frac{y_Y}{c}\right)^2 = 0.444 \qquad \frac{y_Y}{c} = 0.666$$

and, since $c = 60$ mm,

$$y_Y = 0.666(60 \text{ mm}) = 40 \text{ mm}$$

The thickness $2y_Y$ of the elastic core is thus 80 mm.

(*b*) *Radius of Curvature.* We note that the yield strain is

$$\epsilon_Y = \frac{\sigma_Y}{E} = \frac{240 \times 10^6 \text{ Pa}}{200 \times 10^9 \text{ Pa}} = 1.2 \times 10^{-3}$$

Solving Eq. (4.40) for ρ and substituting the values obtained for y_Y and ϵ_Y, we write

$$\rho = \frac{y_Y}{\epsilon_Y} = \frac{40 \times 10^{-3} \text{ m}}{1.2 \times 10^{-3}} = 33.3 \text{ m}$$

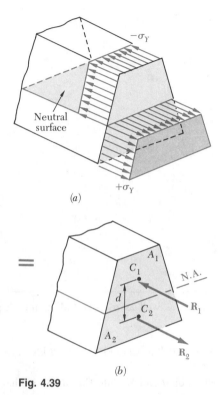

(a)

(b)

Fig. 4.39

*4.11. PLASTIC DEFORMATIONS OF MEMBERS WITH A SINGLE PLANE OF SYMMETRY

In our discussion of plastic deformations, we have assumed so far that the member in bending had two planes of symmetry, one containing the couples **M** and **M′**, and the other perpendicular to that plane. We shall now consider the more general case when the member possesses only one plane of symmetry containing the couples **M** and **M′**. However, we shall limit our analysis to the situation where the deformation is fully plastic, with the normal stress uniformly equal to $-\sigma_Y$ above the neutral surface, and to $+\sigma_Y$ below that surface (Fig. 4.39a).

As indicated in Sec. 4.9, the neutral axis cannot be assumed to coincide with the centroidal axis of the cross section when the cross section is not symmetric with respect to that axis. To locate the neutral axis, we shall consider the resultant $\mathbf{R}_1$ of the elementary compressive forces exerted on the portion A_1 of the cross section located above the neutral axis, and the resultant $\mathbf{R}_2$ of the tensile forces exerted on the portion A_2 located below the neutral axis (Fig. 4.39b). Since the forces $\mathbf{R}_1$ and $\mathbf{R}_2$ form a couple equivalent to the couple applied to the member, they must have the same magnitude. We have therefore $R_1 = R_2$, or $A_1\sigma_Y = A_2\sigma_Y$, from which we conclude that $A_1 = A_2$. In other words, *the neutral axis divides the cross section into portions of equal areas.* Note that the axis obtained in this fashion will *not*, in general, be a centroidal axis of the section.

We also observe that the lines of action of the resultants $\mathbf{R}_1$ and $\mathbf{R}_2$ pass through the centroids C_1 and C_2 of the two portions we have just defined. Denoting by d the distance between C_1 and C_2, and by A the total area of the cross section, we express the plastic moment of the member as

$$M_p = (\tfrac{1}{2}A\sigma_Y)\, d$$

An example of the actual computation of the plastic moment of a member with only one plane of symmetry is given in Sample Prob. 4.6.

*4.12. RESIDUAL STRESSES

We saw in the preceding sections that plastic zones will develop in a member made of an elastoplastic material if the bending moment is large enough. When the bending moment is decreased back to zero, the corresponding reduction in stress and strain at any given point may be represented by a straight line on the stress-strain diagram, as shown in Fig. 4.40. As we shall see presently, the final value of the stress at a point will not, in general, be zero. There will be a residual stress at most points, and that stress may or may not have the same sign as the maximum stress reached at the end of the loading phase.

Since the linear relation between σ_x and ϵ_x applies at all points of the member during the unloading phase, Eq. (4.16) may be used to obtain

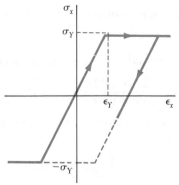

Fig. 4.40

the change in stress at any given point. In other words, the unloading phase may be handled by assuming the member to be fully elastic.

The residual stresses are obtained by applying the principle of superposition in a manner similar to that described in Sec. 2.19 for an axial centric loading and used again in Sec. 3.11 for torsion. We consider, on one hand, the stresses due to the application of the given bending moment M and, on the other, the reverse stresses due to the equal and opposite bending moment $-M$ which is applied to unload the member. The first group of stresses reflect the *elastoplastic* behavior of the material during the loading phase, and the second group the *linear* behavior of the same material during the unloading phase. Adding the two groups of stresses, we obtain the distribution of residual stresses in the member.

Example 4.06

For the member of Example 4.05, determine (a) the distribution of the residual stresses, (b) the radius of curvature, after the bending moment has been decreased from its maximum value of 36.8 kN · m back to zero.

(a) *Distribution of Residual Stresses.* We recall from Example 4.05 that the yield strength is $\sigma_Y = 240$ MPa and that the thickness of the elastic core is $2y_Y = 80$ mm. The distribution of the stresses in the loaded member is thus as shown in Fig. 4.41a.

The distribution of the reverse stresses due to the opposite 36.8-kN · m bending moment required to unload the member is linear and as shown in Fig. 4.41b. The maximum stress σ_m' in that distribution is obtained from Eq. (4.15). Recalling from Example 4.05 that $I/c = 120 \times 10^{-6}$ m³, we write

$$\sigma_m' = \frac{Mc}{I} = \frac{36.8 \text{ kN} \cdot \text{m}}{120 \times 10^{-6} \text{ m}^3} = 306.7 \text{ MPa}$$

Superposing the two distributions of stresses, we obtain the residual stresses shown in Fig. 4.41c. We check that, even though the reverse stresses exceed the yield strength σ_Y, the assumption of a linear distribution of the reverse stresses is valid, since they do not exceed $2\sigma_Y$.

(b) *Radius of Curvature After Unloading.* We may apply Hooke's law at any point of the core $|y| < 40$ mm, since no plastic deformation has occurred in that portion of the member. Thus, the residual strain at the distance $y = 40$ mm is

$$\epsilon_x = \frac{\sigma_x}{E} = \frac{-35.5 \times 10^6 \text{ Pa}}{200 \times 10^9 \text{ Pa}} = -177.5 \times 10^{-6}$$

Solving Eq. (4.8) for ρ and substituting the appropriate values of y and ϵ_x, we write

$$\rho = -\frac{y}{\epsilon_x} = \frac{40 \times 10^{-3} \text{ m}}{177.5 \times 10^{-6}} = 225 \text{ m}$$

The value obtained for ρ after the load has been removed represents a permanent deformation of the member.

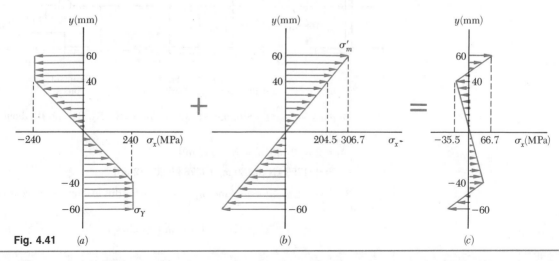

Fig. 4.41   (a)   (b)   (c)

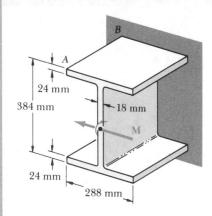

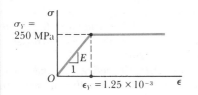

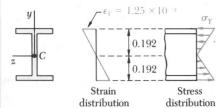

Strain
distribution

Stress
distribution

Dimensions in m

SAMPLE PROBLEM 4.5

Beam AB is fabricated from a mild steel which is assumed to be elastoplastic with $E = 200$ GPa and $\sigma_Y = 250$ MPa. Determine the bending moment M and the corresponding radius of curvature (a) when yield first occurs, (b) when the flanges have just become fully plastic.

a. Onset of Yield. The centroidal moment of inertia of the section is

$$I = \tfrac{1}{12}(0.288 \text{ m})(0.384 \text{ m})^3 - \tfrac{1}{12}(0.288 \text{ m} - 0.018 \text{ m})(0.336 \text{ m})^3$$
$$I = 505.5 \times 10^{-6} \text{ m}^4$$

Bending Moment. For $\sigma_{max} = \sigma_Y = 250$ MPa and $c = 0.192$ m, we have

$$M_Y = \frac{\sigma_Y I}{c} = \frac{(250 \text{ MPa})(505.5 \times 10^{-6} \text{ m}^4)}{0.192 \text{ m}} \qquad M_Y = 658 \text{ kN} \cdot \text{m} \quad \blacktriangleleft$$

Radius of Curvature. Noting that, at $c = 0.192$ m, the strain is $\epsilon_Y = \sigma_Y/E = (250 \text{ MPa})/(200 \text{ GPa}) = 1.25 \times 10^{-3}$, we have from Eq. (4.41)

$$c = \epsilon_Y \rho_Y \qquad 0.192 \text{ m} = (1.25 \times 10^{-3})\rho_Y \qquad \rho_Y = 153.6 \text{ m} \quad \blacktriangleleft$$

b. Flanges Fully Plastic. When the flanges have just become fully plastic, the strains and stresses in the section are as shown in the figure below.

We replace the elementary compressive forces exerted on the top flange and on the top half of the web by their resultants $\mathbf{R}_1$ and $\mathbf{R}_2$, and similarly replace the tensile forces by $\mathbf{R}_3$ and $\mathbf{R}_4$.

$$R_1 = R_4 = (250 \text{ MPa})(0.288 \text{ m})(0.024 \text{ m}) = 1728 \text{ kN}$$
$$R_2 = R_3 = \tfrac{1}{2}(250 \text{ MPa})(0.168 \text{ m})(0.018 \text{ m}) = 378 \text{ kN}$$

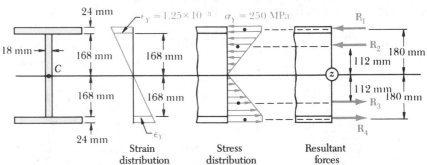

Strain
distribution

Stress
distribution

Resultant
forces

Bending Moment. Summing the moments of $\mathbf{R}_1$, $\mathbf{R}_2$, $\mathbf{R}_3$, and $\mathbf{R}_4$ about the z axis, we write

$$M = 2[R_1(0.180 \text{ m}) + R_2(0.112 \text{ m})]$$
$$= 2[(1728 \text{ kN})(0.180 \text{ m}) + (378 \text{ kN})(0.112 \text{ m})] \qquad M = 707 \text{ kN} \cdot \text{m} \quad \blacktriangleleft$$

Radius of Curvature. Since $y_Y = 0.168$ m for this loading, we have from Eq. (4.40)

$$y_Y = \epsilon_Y \rho \qquad 0.168 \text{ m} = (1.25 \times 10^{-3})\rho \qquad \rho = 134.4 \text{ m} \quad \blacktriangleleft$$

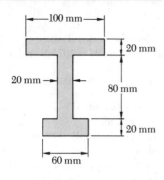

Determine the plastic moment of a member with the cross section shown, assuming the material to be elastoplastic with a yield strength of 240 MPa.

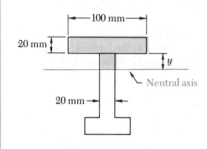

Neutral Axis. When the deformation is fully plastic, the neutral axis divides the cross section into two portions of equal areas. Since the total area is

$$A = (100)(20) + (80)(20) + (60)(20) = 4800 \text{ mm}^2$$

the area located above the neutral axis must be 2400 mm². We write

$$(20)(100) + 20y = 2400 \qquad y = 20 \text{ mm}$$

Note that the neutral axis does *not* pass through the centroid of the cross section.

Plastic Moment. The resultant $\mathbf{R}_i$ of the elementary forces exerted on the partial area A_i is equal to

$$R_i = A_i\sigma_Y$$

and passes through the centroid of that area. We have

$$R_1 = A_1\sigma_Y = [(0.100 \text{ m})(0.020 \text{ m})]240 \text{ MPa} = 480 \text{ kN}$$
$$R_2 = A_2\sigma_Y = [(0.020 \text{ m})(0.020 \text{ m})]240 \text{ MPa} = 96 \text{ kN}$$
$$R_3 = A_3\sigma_Y = [(0.020 \text{ m})(0.060 \text{ m})]240 \text{ MPa} = 288 \text{ kN}$$
$$R_4 = A_4\sigma_Y = [(0.060 \text{ m})(0.020 \text{ m})]240 \text{ MPa} = 288 \text{ kN}$$

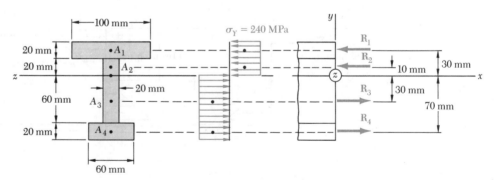

The plastic moment M_p may be obtained by summing the moments of the forces about the z axis.

$$M_p = (0.030 \text{ m})R_1 + (0.010 \text{ m})R_2 + (0.030 \text{ m})R_3 + (0.070 \text{ m})R_4$$
$$= (0.030 \text{ m})(480 \text{ kN}) + (0.010 \text{ m})(96 \text{ kN})$$
$$\quad + (0.030 \text{ m})(288 \text{ kN}) + (0.070 \text{ m})(288 \text{ kN})$$
$$= 44.16 \text{ kN} \cdot \text{m} \qquad\qquad M_p = 44.2 \text{ kN} \cdot \text{m} \blacktriangleleft$$

Note: Since the cross section is *not* symmetric about the z axis, the sum of the moments of $\mathbf{R}_1$ and $\mathbf{R}_2$ is *not* equal to the sum of the moments of $\mathbf{R}_3$ and $\mathbf{R}_4$.

SAMPLE PROBLEM 4.7

For the beam of Sample Prob. 4.5, determine the residual stresses and the permanent radius of curvature after the bending moment $M = 707$ kN · m has been removed.

Loading. In Sample Prob. 4.5 a bending moment $M = 707$ kN · m was applied and the stresses shown in Fig. (1) were obtained.

Elastic Unloading. The beam is unloaded by the application of a bending moment $M = -707$ kN · m (which is equal and opposite to the moment originally applied). During this unloading, the action of the beam is fully elastic; recalling from Sample Prob. 4.5 that $I = 505.5 \times 10^{-6}$ m⁴ we compute the maximum stress

$$\sigma'_m = \frac{Mc}{I} = \frac{(707 \text{ kN} \cdot \text{m})(0.192 \text{ m})}{505.5 \times 10^{-6} \text{ m}^4} = 268.53 \text{ MPa}$$

The stresses caused by the unloading are shown in Fig. (2).

Residual Stresses. We superpose the stresses due to the loading (Fig. 1) and to the unloading (Fig. 2) and obtain the residual stresses in the beam (Fig. 3).

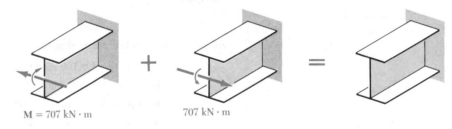

$M = 707$ kN · m   707 kN · m

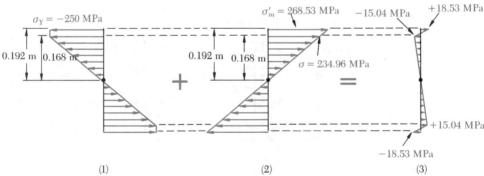

$\sigma_Y = -250$ MPa $\sigma'_m = 268.53$ MPa -15.04 MPa $+18.53$ MPa

0.192 m | 0.168 m   0.192 m   0.168 m $\sigma = 234.96$ MPa

$+15.04$ MPa

-18.53 MPa

(1)   (2)   (3)

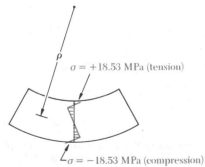

$\sigma = +18.53$ MPa (tension)

$\sigma = -18.53$ MPa (compression)

Permanent Radius of Curvature. At $y = 0.168$ m the residual stress is 15.04 MPa. Since no plastic deformation occurred at this point, Hooke's law may be used and we have $\epsilon_x = \sigma/E$. Recalling Eq. (4.8), we write

$$\rho = -\frac{y}{\epsilon_x} = -\frac{yE}{\sigma} = -\frac{(0.168 \text{ m})(200 \text{ GPa})}{-15.04 \text{ MPa}} = 2234 \text{ m}$$

$$\rho = 2234 \text{ m} \quad \blacktriangleleft$$

We note that the residual stress is tensile on the upper face of the beam and compressive on the lower face, even though the beam is concave upward.

A bar of the square cross section shown is made of a steel which is assumed to be elastoplastic with $E = 200$ GPa and $\sigma_Y = 290$ MPa. Determine the bending moment M for which (a) yield first occurs, (b) the plastic zones at the top and bottom of the bar are 4 mm thick.

A bar of the rectangular cross section shown is made of a steel which is assumed to be elastoplastic with $E = 200$ GPa and $\sigma_Y = 300$ MPa. Determine the bending moment M for which (a) yield first occurs, (b) the plastic zones at the top and bottom of the bar are 12 mm thick.

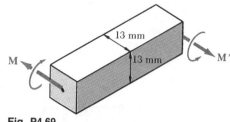

Fig. P4.69

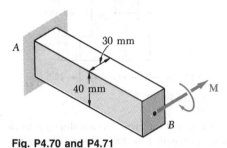

Fig. P4.70 and P4.71

Bar AB is made of a steel which is assumed to be elastoplastic with $E = 200$ GPa and $\sigma_Y = 240$ MPa. Determine the bending moment M for which the radius of curvature of the bar will be (a) 18 m, (b) 9 m.

The prismatic bar AB is made of a steel which is assumed to be elastoplastic with $E = 200$ GPa and $\sigma_Y = 250$ MPa. For a couple of moment 150 N · m parallel to the z axis and applied at end B as shown, determine (a) the thickness of the elastic core, (b) the radius of curvature of the bar.

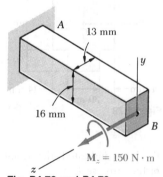

Fig. P4.72 and P4.73

Solve Prob. 4.72, assuming that the 150 N · m couple is parallel to the y axis.

The prismatic bar AB is made of a steel which is assumed to be elastoplastic with $E = 200$ GPa and $\sigma_Y = 250$ MPa. Knowing that $M = 450$ N · m, determine (a) the thickness of the elastic core, (b) the radius of curvature of the bar.

Solve Prob. 4.74, assuming that $M = 550$ N · m.

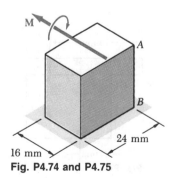

Fig. P4.74 and P4.75

4.76 and 4.77 A beam of the cross section shown is made of a steel which is assumed to be elastoplastic with $E = 200$ GPa and $\sigma_Y = 290$ MPa. For bending about the z axis, determine the bending moment and the radius of curvature at which (a) yield first occurs, (b) the plastic zones at the top and bottom of the beam are 25 mm thick.

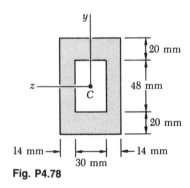

25 mm →‖ ‖← 25 mm
25 mm

Fig. P4.76

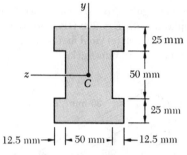

12.5 mm →‖ ‖← 50 mm →‖ ‖← 12.5 mm

Fig. P4.77

4.78 and 4.79 A beam of the cross section shown is made of a steel which is assumed to be elastoplastic with $E = 200$ GPa and $\sigma_Y = 300$ MPa. For bending about the z axis, determine the bending moment and the radius of curvature at which (a) yield first occurs, (b) the plastic zones at the top and bottom of the beam are 20 mm thick.

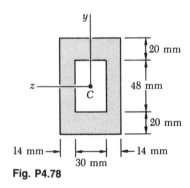

14 mm →‖ ‖← 14 mm
30 mm

Fig. P4.78

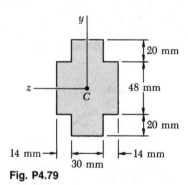

14 mm →‖ ‖← 14 mm
30 mm

Fig. P4.79

4.80 and 4.83 For the beam indicated, determine (a) the plastic moment M_p, (b) the shape factor of the cross section.

    4.80 Beam of Prob. 4.76
    4.81 Beam of Prob. 4.77
    4.82 Beam of Prob. 4.78
    4.83 Beam of Prob. 4.79

4.84 Determine the plastic moment M_p of a steel beam of the cross section shown, assuming the steel to be elastoplastic with a yield strength of 330 MPa.

4.85 Determine the plastic moment M_p of the beam of Prob. 4.16, assuming that the beam is made of a steel which is elastoplastic with a yield strength of 275 MPa.

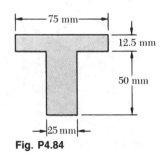

← 75 mm →

12.5 mm

50 mm

→‖ 25 mm ‖←

Fig. P4.84

4.86 Determine the plastic moment M_p of the beam of Prob. 4.8, assuming the steel to be elastoplastic with a yield strength of 250 MPa.

4.87 Determine the plastic moment M_p of a steel beam of the cross section shown, assuming the steel to be elastoplastic with a yield strength of 250 MPa.

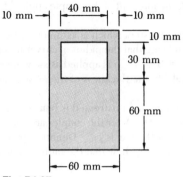

Fig. P4.87

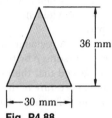

Fig. P4.88

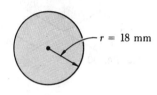

Fig. P4.89

4.88 and 4.89 Determine the plastic moment M_p of a steel beam of the cross section shown when the beam is bent about a horizontal axis. Assume the steel to be elastoplastic with a yield strength of 200 MPa.

4.90 and 4.91 For the beam indicated, a couple of moment equal to the plastic moment M_p of the beam is applied and then removed. Using a yield strength of 290 MPa, determine the residual stress at (a) $y = 25$ mm, (b) $y = 50$ mm.

    4.90 Beam of Prob. 4.76
    4.91 Beam of Prob. 4.77

4.92 and 4.93 For the beam indicated, a couple of moment equal to the plastic moment M_p of the beam is applied and then removed. Using a yield strength of 300 MPa, determine the residual stress at (a) $y = 24$ mm, (b) $y = 44$ mm.

    4.92 Beam of Prob. 4.78
    4.93 Beam of Prob. 4.79

4.94 and 4.95 A couple **M** is applied to the beam indicated about the z axis, causing plastic zones 25 mm thick to develop at the top and bottom of the beam. After the couple has been removed, determine (a) the residual stress at $y = 50$ mm, (b) the points where the residual stress is zero, (c) the radius of curvature corresponding to the permanent deformation of the beam.

    4.94 Beam of Prob. 4.76
    4.95 Beam of Prob. 4.77

4.96 and 4.97 A couple **M** is applied to the beam indicated about the z axis, causing plastic zones 20 mm thick to develop at the top and bottom of the beam. After the couple has been removed, determine (a) the residual stress at $y = 44$ mm, (b) the points where the residual stress is zero, (c) the radius of curvature corresponding to the permanent deformation of the beam

    4.96 Beam of Prob. 4.78
    4.97 Beam of Prob. 4.79

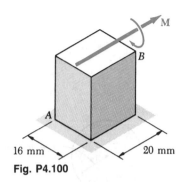

16 mm   20 mm

Fig. P4.100

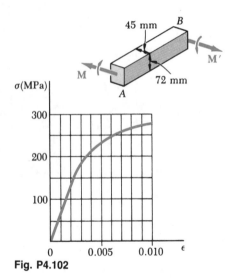

Fig. P4.102

4.98 A solid rectangular rod is made of a material which is assumed to be elastoplastic. Denoting by M_Y and ρ_Y, respectively, the bending moment and radius of curvature at the onset of yield, determine (a) the radius of curvature when a couple of moment $M = 1.25M_Y$ is applied to the rod, (b) the radius of curvature after the couple is removed. Check the results obtained by using the relation derived in Prob. 4.101.

4.99 Solve Prob. 4.98, assuming that the moment of the couple applied to the rod is $M = 1.45M_Y$.

***4.100** The prismatic bar AB is made of a steel which is assumed to be elastoplastic and for which $E = 200$ GPa. Knowing that the radius of curvature of the bar is 2.4 m when a couple of moment $M = 350$ N · m is applied as shown, determine (a) the yield strength σ_Y of the steel, (b) the thickness of the elastic core of the bar.

***4.101** A rectangular bar which is straight and unstressed is bent into an arc of circle of radius ρ by two couples of moment M. After the couples are removed, it is observed that the radius of curvature of the bar is ρ_R. Denoting by ρ_Y the radius of curvature of the bar at the onset of yield, show that the radii of curvature satisfy the following relation:

$$\frac{1}{\rho_R} = \frac{1}{\rho}\left\{1 - \frac{3}{2}\frac{\rho}{\rho_Y}\left[1 - \frac{1}{3}\left(\frac{\rho}{\rho_Y}\right)^2\right]\right\}$$

4.102 The prismatic bar AB is made of an aluminum alloy for which the tensile stress-strain diagram is as shown. Assuming that the diagram is the same in compression as in tension, determine (a) the radius of curvature of the bar when the maximum stress is 250 MPa, (b) the corresponding value of the bending moment. (*Hint:* For part b, plot σ versus y and use an approximate method of integration.)

4.103 For the bar of Prob. 4.102, determine (a) the maximum stress when the radius of curvature of the bar is 4 m, (b) the corresponding value of the bending moment. (See hint given in Prob. 4.102.)

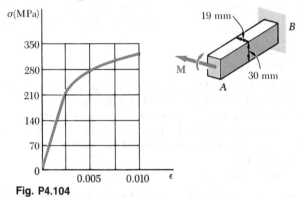

Fig. P4.104

4.104 The prismatic bar AB is made of an aluminum alloy for which the tensile stress-strain diagram is as shown. Assuming that the diagram is the same in compression as in tension, determine (a) the maximum stress in the bar when the radius of curvature is 2 m, (b) the corresponding value of the bending moment. (See hint given in Prob. 4.102.)

4.105 For the bar of Prob. 4.104, determine (*a*) the radius of curvature of the bar when the maximum stress is 310 MPa, (*b*) the corresponding value of the bending moment.   (See hint given in Prob. 4.102.)

4.106 A prismatic bar of rectangular cross section is made of an alloy for which the stress-strain diagram may be represented by the relation, $\epsilon = k\sigma^n$ for $\sigma > 0$, and $\epsilon = -|k\sigma^n|$ for $\sigma < 0$.   If a couple **M** is applied to the bar, show that the maximum stress is

$$\sigma_m = \frac{1 + 2n}{3n} \frac{Mc}{I}$$

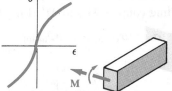

Fig. P4.106 and P4.107

4.107 A prismatic bar of rectangular cross section is made of an alloy for which the stress-strain diagram, in both tension and compression, may be represented by the relation $\epsilon = k\sigma^3$.   If a couple **M** is applied to the bar, show that the maximum stress is

$$\sigma_m = \frac{7}{9} \frac{Mc}{I}$$

4.13. ECCENTRIC AXIAL LOADING IN A PLANE OF SYMMETRY

We saw in Sec. 1.3 that the distribution of stresses in the cross section of a member under axial loading may be assumed uniform only if the line of action of the loads **P** and **P′** passes through the centroid of the cross section.   Such a loading is said to be *centric*.   We shall now analyze the distribution of stresses when the line of action of the loads does *not* pass through the centroid of the cross section, i.e., when the loading is *eccentric*.

In this section, our analysis will be limited to members which possess a plane of symmetry, and we shall assume that the loads are applied in the plane of symmetry of the member (Fig. 4.42*a*).   The internal forces acting on a given cross section may then be represented by a force **F** applied at the centroid *C* of the section and a couple **M** acting in the plane of symmetry of the member (Fig. 4.42*b*).   The conditions of equilibrium of the free body *AC* require that the force **F** be equal and opposite to **P′** and that the moment of the couple **M** be equal and opposite to the moment of **P′** about *C*.   Denoting by *d* the distance from the centroid *C* to the line of

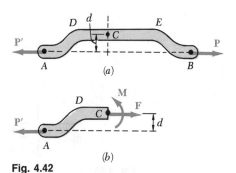

Fig. 4.42

action AB of the forces $\mathbf{P}$ and $\mathbf{P}'$, we have

$$F = P \quad \text{and} \quad M = Pd \tag{4.49}$$

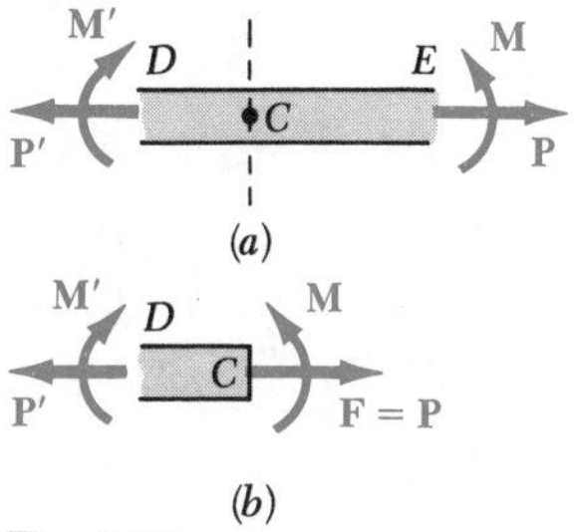

We now observe that the internal forces in the section would have been represented by the same force and couple if the straight portion DE of member AB had been detached from AB and subjected simultaneously to the centric loads $\mathbf{P}$ and $\mathbf{P}'$ and to the bending couples $\mathbf{M}$ and $\mathbf{M}'$ (Fig. 4.43). Thus, the stress distribution due to the original eccentric loading may be obtained by superposing the uniform stress distribution corresponding to the centric loads $\mathbf{P}$ and $\mathbf{P}'$ and the linear distribution corresponding to the bending couples $\mathbf{M}$ and $\mathbf{M}'$ (Fig. 4.44). We write

$$\sigma_x = (\sigma_x)_{\text{centric}} + (\sigma_x)_{\text{bending}}$$

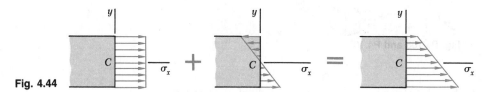

Fig. 4.43

Fig. 4.44

or, recalling Eqs. (1.1) and (4.16):

$$\sigma_x = \frac{P}{A} - \frac{My}{I} \tag{4.50}$$

where A is the area of the cross section and I its centroidal moment of inertia, and where y is measured from the centroidal axis of the cross section. The relation obtained shows that the distribution of stresses across the section is *linear but not uniform*. Depending upon the geometry of the cross section and the eccentricity of the load, the combined stresses may all have the same sign, as shown in Fig. 4.44, or some may be positive and others negative, as shown in Fig. 4.45. In the latter case, there will be a line in the section, along which $\sigma_x = 0$. This line represents the *neutral axis* of the section. We note that the neutral axis does *not* coincide with the centroidal axis of the section, since $\sigma_x \neq 0$ for $y = 0$.

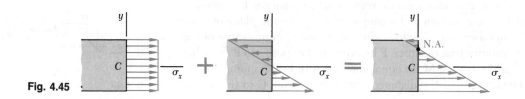

Fig. 4.45

The results obtained will be valid only to the extent that the conditions of applicability of the superposition principle (Sec. 2.12) and of Saint-Venant's principle (Sec. 2.16) are met. This means that the stresses involved must not exceed the proportional limit of the material, that the deformations due to bending must not appreciably affect the distance d in Fig. 4.42a, and that the cross section where the stresses are computed must not be too close to points D or E in the same figure. The first of these requirements clearly shows that the superposition method cannot be applied to plastic deformations.

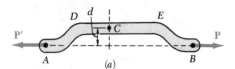

Fig. 4.42 (repeated)

Example 4.07

An open-link chain is obtained by bending low-carbon steel rods of 12 mm diameter into the shape shown (Fig. 4.46). Knowing that the chain carries a load of 800 N, determine (a) the largest tensile and compressive stresses in the straight portion of a link, (b) the distance between the centroidal and the neutral axis of a cross section.

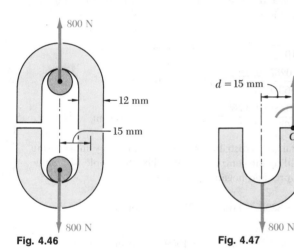

Fig. 4.46   **Fig. 4.47**

(a) *Largest Tensile and Compressive Stresses.* The internal forces in the cross section are equivalent to a centric force **P** and a bending couple **M** (Fig. 4.47) of magnitudes

$$P = 800 \text{ N}$$
$$M = Pd = (800 \text{ N})(0.015 \text{ m}) = 12 \text{ N} \cdot \text{m}$$

The corresponding stress distributions are shown in parts a and b of Fig. 4.48. The distribution due to the centric force **P** is uniform and equal to $\sigma_0 = P/A$. We have

$$A = \pi c^2 = \pi(0.006 \text{ m})^2 = 113.1 \times 10^{-6} \text{ m}^2$$

$$\sigma_0 = \frac{P}{A} = \frac{800 \text{ N}}{113.1 \times 10^{-6} \text{ m}^2} = 7.07 \text{ MPa}$$

The distribution due to the bending couple **M** is linear with a maximum stress $\sigma_m = Mc/I$. We write

$$I = \tfrac{1}{4}\pi c^4 = \tfrac{1}{4}\pi(0.006 \text{ m})^4 = 1.018 \times 10^{-9} \text{ m}^4$$

$$\sigma_m = \frac{Mc}{I} = \frac{(12 \text{ N} \cdot \text{m})(0.006 \text{ m})}{1.018 \times 10^{-9} \text{ m}^4} = 70.7 \text{ MPa}$$

Superposing the two distributions, we obtain the stress distribution corresponding to the given eccentric loading (Fig. 4.48c). The largest tensile and compressive stresses in the section are found to be respectively.

$$\sigma_t = \sigma_0 + \sigma_m = 7.1 + 70.7 = 77.8 \text{ MPa}$$
$$\sigma_c = \sigma_0 - \sigma_m = 7.1 - 70.7 = -63.6 \text{ MPa}$$

(b) *Distance Between Centroidal and Neutral Axes.* The distance y_0 from the centroidal to the neutral axis of the section is obtained by setting $\sigma_x = 0$ in Eq. (4.50) and solving for y_0:

$$0 = \frac{P}{A} - \frac{My_0}{I}$$

$$y_0 = \left(\frac{P}{A}\right)\left(\frac{I}{M}\right) = (7.07 \text{ MPa}) \frac{1.018 \times 10^{-9} \text{ m}^4}{12 \text{ N} \cdot \text{m}}$$

$$y_0 = 0.600 \text{ mm}$$

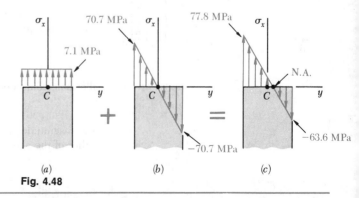

Fig. 4.48

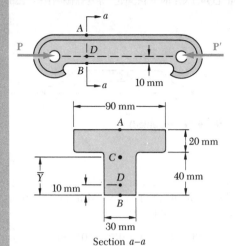

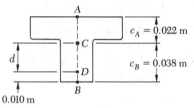

Section a–a

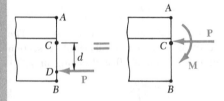

$c_A = 0.022$ m

$c_B = 0.038$ m

0.010 m

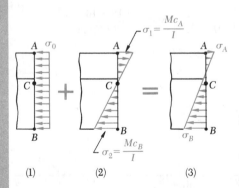

(1)   (2)   (3)

SAMPLE PROBLEM 4.8

Knowing that for the cast iron link shown the allowable stresses are 30 MPa in tension and 120 MPa in compression, determine the largest force **P** which may be applied to the link. (*Note:* The T-shaped cross section of the link has previously been considered in Sample Prob. 4.2.)

Properties of Cross Section. From Sample Prob. 4.2, we have

$$A = 3000 \text{ mm}^2 = 3 \times 10^{-3} \text{ m}^2 \qquad \overline{Y} = 38 \text{ mm} = 0.038 \text{ m}$$
$$I = 868 \times 10^{-9} \text{ m}^4$$

We now write: $d = (0.038 \text{ m}) - (0.010 \text{ m}) = 0.028 \text{ m}$

Force and Couple at C. We replace **P** by an equivalent force-couple system at the centroid C.

$$P = P \qquad M = P(d) = P(0.028 \text{ m}) = 0.028\, P$$

The force **P** acting at the centroid causes a uniform stress distribution (Fig. 1). The bending couple **M** causes a linear stress distribution (Fig. 2).

$$\sigma_0 = \frac{P}{A} = \frac{P}{3 \times 10^{-3}} = 333P \qquad \text{(Compression)}$$

$$\sigma_1 = \frac{Mc_A}{I} = \frac{(0.028P)(0.022)}{868 \times 10^{-9}} = 710P \qquad \text{(Tension)}$$

$$\sigma_2 = \frac{Mc_B}{I} = \frac{(0.028P)(0.038)}{868 \times 10^{-9}} = 1226P \qquad \text{(Compression)}$$

Superposition. The total stress distribution (Fig. 3) is found by superposing the stress distributions caused by the centric force **P** and by the couple **M**. Since tension is positive, and compression negative, we have

$$\sigma_A = -\frac{P}{A} + \frac{Mc_A}{I} = -333P + 710P = +377P \qquad \text{(Tension)}$$

$$\sigma_B = -\frac{P}{A} - \frac{Mc_B}{I} = -333P - 1226P = -1559P \qquad \text{(Compression)}$$

Largest Allowable Force. The magnitude of **P** for which the tensile stress at point A is equal to the allowable tensile stress of 30 MPa is found by writing

$$\sigma_A = 377P = 30 \text{ MPa} \qquad\qquad P = 79.6 \text{ kN} \quad \triangleleft$$

We also determine the magnitude of **P** for which the stress at B is equal to the allowable compressive stress of 120 MPa.

$$\sigma_B = -1559P = -120 \text{ MPa} \qquad\qquad P = 77.0 \text{ kN} \quad \triangleleft$$

The magnitude of the largest force **P** which may be applied without exceeding either of the allowable stresses is the smaller of the two values we have found.

$$P = 77.0 \text{ kN} \quad \blacktriangleleft$$

PROBLEMS

4.108 A short wooden post supports a 25 kN axial load as shown. Determine the stress at point A when (a) b = 0, (b) b = 38 mm, (c) b = 75 mm.

4.109 Two 10-kN forces are applied to a 20 × 60-mm rectangular bar as shown. Determine the stress at point A when (a) b = 0, (b) b = 15 mm, (c) b = 25 mm.

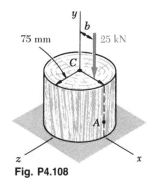

Fig. P4.108

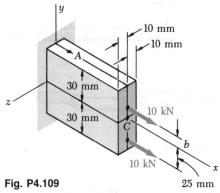
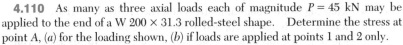

Fig. P4.109

4.110 As many as three axial loads each of magnitude P = 45 kN may be applied to the end of a W 200 × 31.3 rolled-steel shape. Determine the stress at point A, (a) for the loading shown, (b) if loads are applied at points 1 and 2 only.

4.111 As many as three axial loads each of magnitude P = 45 kN may be applied to the end of a W 200 × 31.3 rolled-steel shape. Determine the stress at point A, (a) for the loading shown, (b) if loads are applied at points 2 and 3 only.

4.112 A short 120 × 180-mm column supports the three axial loads shown. Knowing that section ABD is sufficiently far from the loads to remain plane, determine the stress at (a) corner A, (b) corner B.

4.113 Solve Prob. 4.112, assuming that the 20-kN load is removed.

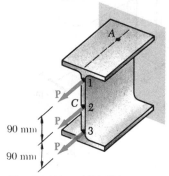

Fig. P4.110 and P4.111

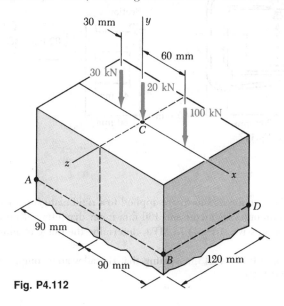

Fig. P4.112

4.114 An offset h must be introduced into a solid circular rod. Knowing that the maximum stress after the offset is introduced must not exceed three times the stress in the rod when it is straight, determine (a) the largest offset which may be used, (b) the corresponding maximum stress in the rod.

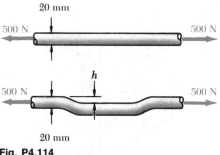

Fig. P4.114

4.115 Solve Prob. 4.114, assuming that the rod is replaced by a tube of 20-mm outer diameter and 16-mm inner diameter.

4.116 Knowing that the allowable stress in section ABD is 70 MPa, determine the largest force **P** which may be applied to the bracket shown.

4.117 Knowing that the allowable stress in section a-a is 75 MPa, determine the largest force which may be exerted by the press shown.

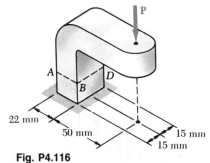

Fig. P4.116

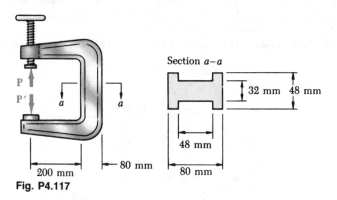

Fig. P4.117

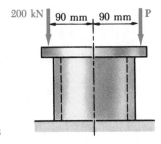

Fig. P4.118

4.118 The two forces shown are applied to a rigid plate supported by a steel pipe of 160-mm outer diameter and 130-mm inner diameter. Knowing that the allowable compressive stress is 75 MPa, determine the range of allowable values of P.

4.119 Solve Prob. 4.118, assuming that the allowable compressive stress is reduced to 60 MPa.

4.120 Knowing that the clamp shown has been tightened until $P = 400$ N, determine (a) the stress at point A, (b) the stress at point B, (c) the location of the neutral axis of section a–a.

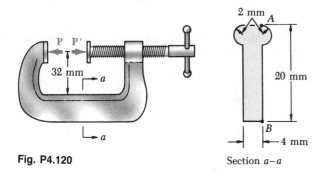

Fig. P4.120

Section a–a

4.121 Solve Prob. 4.120, assuming that the radius of each of the two semicircular portions of the cross section of the clamp is increased from 2 mm to 3 mm.

4.122 A short column is made by nailing four 25×100-mm planks to a 100×100-mm timber. (a) Determine the compressive stress in the column caused by the 70-kN centric axial load. Knowing that in each of the following cases the 70-kN axial load is applied at the center of the 100×100-mm timber, determine the largest compressive stress in the column if the indicated planks are removed; (b) plank 1, (c) planks 1 and 2, (d) planks 1, 2, and 3, (e) all planks.

4.123 In order to provide access to the interior of a hollow square tube of 6 mm wall thickness, the portion BD of one side of the tube has been removed. Knowing that the loading of the tube is equivalent to two equal and opposite 60 kN axial forces acting at the geometric centers A and E of the ends of the tube, determine (a) the maximum stress in section a-a, (b) the stress at point F. *Given*: Centroid of cross section is at C and $I_z = 1.812 \times 10^{-6}$ m^4.

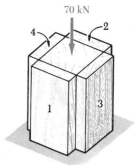

Fig. P4.122

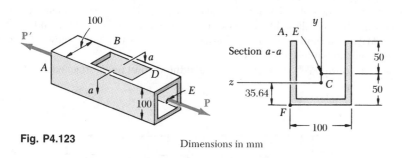

Fig. P4.123

Section a-a

Dimensions in mm

4.124 A milling operation was used to remove half of a 20-mm-diameter rod along part of its length. Knowing that the 4-kN forces act at the centers of the ends of the rod, determine (a) the stress at point A, (b) the stress at point B, (c) the location of the neutral axis of section a–a.

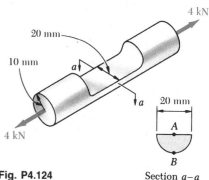

Fig. P4.124

Section a–a

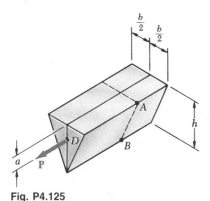

$\frac{b}{2}$ $\frac{b}{2}$

h

a

P

Fig. P4.125

4.125 The loading of the plastic prism shown is equivalent to an axial force acting at point D of the end of the prism. Show that the stresses at points A and B are

$$\sigma_A = 3\sigma_{\text{ave}}(1 - 2\alpha) \qquad \sigma_B = 3\sigma_{\text{ave}}(-1 + 4\alpha)$$

where $\alpha = a/h$ and $\sigma_{\text{ave}} = P/(\tfrac{1}{2}bh)$.

4.126 Knowing that the allowable stress is 90 MPa, determine the largest force P which may be applied to the machine element shown.

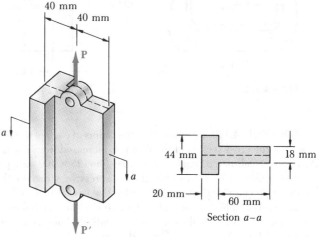

40 mm

40 mm

P

a

a

44 mm

18 mm

20 mm

60 mm

Section a–a

P'

Fig. P4.126

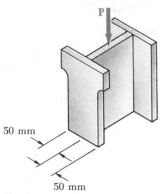

P

50 mm

50 mm

Fig. P4.127

4.127 Three steel plates, each of 20×140-mm cross section, are welded together to form a short H-shaped column. Later, for architectural reasons, a 20-mm strip is removed from each side of one of the flanges. Knowing that the load remains centric with respect to the original cross section and that the allowable stress is 150 MPa, determine the largest force P, (a) which could be applied to the original column, (b) which may be applied to the modified column.

4.128 A vertical rod is attached at point A to the cast iron hanger shown. Knowing that the allowable stresses in the hanger are $\sigma_{\text{all}} = 35$ MPa and $\sigma_{\text{all}} = -85$ MPa, determine the largest downward force and the largest upward force which may be exerted by the rod.

4.129 Solve Prob. 4.128, assuming that the vertical rod is attached at point B instead of point A.

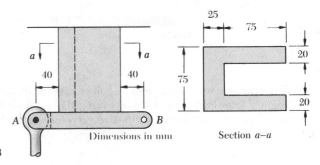

25

75

a a

40 40

75

20

20

A B

Dimensions in mm

Section a–a

Fig. P4.128

4.130 An eccentric axial force **P** is applied as shown to a steel bar of 75 × 120-mm cross section. The strains at A and B are measured and found to be

$$\epsilon_A = +600 \ \mu \qquad \epsilon_B = +450 \ \mu$$

Knowing that $E = 200$ GPa, determine (a) the magnitude of **P**, (b) the distance a from the top edge of the bar to the line of action of **P**.

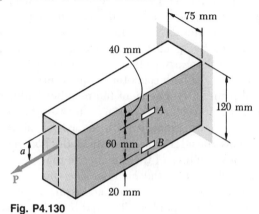

Fig. P4.130

4.131 Solve Prob. 4.130, assuming that the measured strains are

$$\epsilon_A = +175 \ \mu \qquad \epsilon_B = -50 \ \mu$$

4.132 A short length of a W 200 × 46.1 rolled-steel shape supports a rigid plate on which two loads **P** and **Q** are applied as shown.   The strains at two points A and B located on outside faces of the flanges have been measured and are

$$\epsilon_A = -550 \ \mu \qquad \epsilon_B = -680 \ \mu$$

Knowing that $E = 200$ GPa, determine the magnitude of each load.

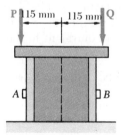

Fig. P4.132

4.133 Solve Prob. 4.132, assuming that the measured strains are

$$\epsilon_A = +35 \ \mu \qquad \epsilon_B = -450 \ \mu$$

4.134 The eccentric axial force **P** acts at point D, which must be located 30 mm below the top surface of the steel bar shown.   For $P = 90$ kN, determine (a) the depth d of the bar for which the tensile stress at point A is maximum, (b) the corresponding value of the stress at point A.

4.135 For the bar and loading of Prob. 4.134, determine the range of values of the depth d for which the stress at point A is tensile and is equal to or less than 30 MPa.

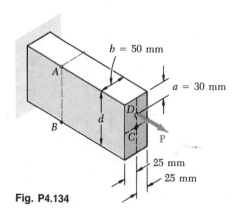

Fig. P4.134

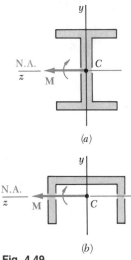

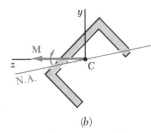

Fig. 4.49

4.14. UNSYMMETRIC BENDING

Our analysis of pure bending has been limited so far to members possessing at least one plane of symmetry and subjected to couples acting in that plane. Because of the symmetry of such members and of their loadings, we concluded that the members would remain symmetric with respect to the plane of the couples and thus bend in that plane (Sec. 4.4). This has been illustrated in Fig. 4.49; part *a* shows the cross section of a member possessing two planes of symmetry, one vertical and one horizontal, and part *b* the cross section of a member with a single, vertical plane of symmetry. In both cases the couple exerted on the section acts in the vertical plane of symmetry of the member and is represented by the horizontal couple vector **M**, and in both cases the neutral axis of the cross section is found to coincide with the axis of the couple.

We shall now consider situations where the bending couples do *not* act in a plane of symmetry of the member, either because they act in a different plane, or because the member does not possess any plane of symmetry. In such situations, we cannot assume that the member will bend in the plane of the couples. This has been illustrated in Fig. 4.50. In each part of the figure, the couple exerted on the section has again been assumed to act in a vertical plane and has been represented by a horizontal couple vector **M**. However, since the vertical plane is not a plane of symmetry, *we cannot expect the member to bend in that plane, or the neutral axis of the section to coincide with the axis of the couple.*

We propose to determine the precise conditions under which the neutral axis of a cross section of arbitrary shape coincides with the axis of the couple **M** representing the forces acting on that section. Such a section has been shown in Fig. 4.51, and both the couple vector **M** and the neutral axis have been assumed to be directed along the *z* axis. We recall from Sec. 4.3 that, if we then express that the elementary internal forces $\sigma_x \, dA$ form a system equivalent to the couple **M**, we obtain

x components: $\qquad \int \sigma_x \, dA = 0 \qquad (4.1)$

moments about y axis: $\qquad \int z\sigma_x \, dA = 0 \qquad (4.2)$

moments about z axis: $\qquad \int (-y\sigma_x \, dA) = M \qquad (4.3)$

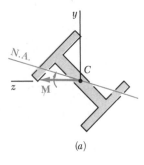

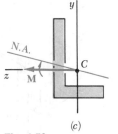

Fig. 4.50

As we saw earlier, when all the stresses are within the proportional limit, the first of these equations leads to the requirement that the neutral axis be a centroidal axis, and the last to the fundamental relation $\sigma_x = -My/I$. Since we had assumed in Sec. 4.3 that the cross section was symmetric with respect to the y axis, Eq. (4.2) was dismissed as trivial at that time. Now that we are considering a cross section of arbitrary shape, Eq. (4.2) becomes highly significant. Assuming the stresses to remain within the

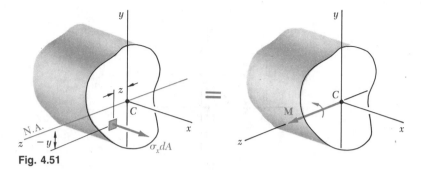

Fig. 4.51

proportional limit of the material, we may substitute $\sigma_x = -\sigma_m y/c$ into Eq. (4.2) and write

$$\int z\left(-\frac{\sigma_m y}{c}\right) dA = 0 \qquad \text{or} \qquad \int yz\, dA = 0 \qquad (4.51)$$

The integral $\int yz\, dA$ represents the product of inertia I_{yz} of the cross section with respect to the y and z axes, and will be zero if these axes are the *principal centroidal axes of the cross section.*† We thus conclude that the neutral axis of the cross section will coincide with the axis of the couple **M** representing the forces acting on that section *if, and only if, the couple vector **M** is directed along one of the principal centroidal axes of the cross section.*

We note that the cross sections shown in Fig. 4.49 are symmetric with respect to at least one of the coordinate axes. It follows that, in each case, the y and z axes are the principal centroidal axes of the section. Since the couple vector **M** is directed along one of the principal centroidal axes, we verify that the neutral axis will coincide with the axis of the couple. We also note that, if the cross sections are rotated through 90° (Fig. 4.52), the couple vector **M** will still be directed along a principal centroidal axis, and the neutral axis will again coincide with the axis of the couple, even though in case *b* the couple does *not* act in a plane of symmetry of the member.

In Fig. 4.50, on the other hand, neither of the coordinate axes is an axis of symmetry for the sections shown, and the coordinate axes are not principal axes. Thus, the couple vector **M** is not directed along a principal centroidal axis, and the neutral axis does not coincide with the axis of the couple. However, any given section possesses principal centroidal axes, even if it is unsymmetric, as the section shown in Fig. 4.50*c*, and these axes may be determined analytically or by using Mohr's circle.† If the couple vector **M** is directed along one of the principal centroidal axes of the section, the neutral axis will coincide with the axis of the couple (Fig. 4.53) and the equations derived in Secs. 4.4 and 4.5 for symmetric members may be used to determine the stresses in this case as well.

†See Ferdinand P. Beer and E. Russell Johnston, Jr., *Mechanics for Engineers* 4th ed., McGraw-Hill, New York, 1987, or *Vector Mechanics for Engineers*, 5th ed., McGraw-Hill, New York, 1988, secs. 9.8–9.10.

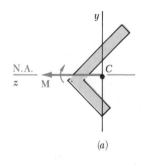

(a)

(b)

Fig. 4.52

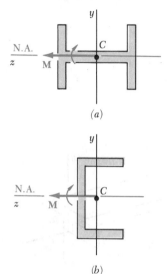

(a)

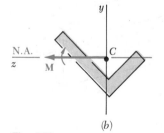

(b)

Fig. 4.53

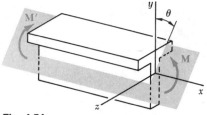

Fig. 4.54

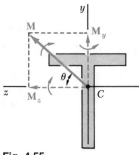

Fig. 4.55

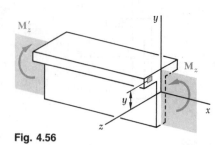

Fig. 4.56

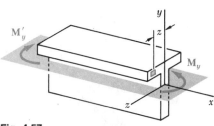

Fig. 4.57

As we shall see presently, the principle of superposition may be used to determine stresses in the most general case of unsymmetric bending. Consider first a member with a vertical plane of symmetry, which is subjected to bending couples **M** and **M'** acting in a plane forming an angle θ with the vertical plane (Fig. 4.54). The couple vector **M** representing the forces acting on a given cross section will form the same angle θ with the horizontal z axis (Fig. 4.55). Resolving the vector **M** into component vectors $\mathbf{M}_z$ and $\mathbf{M}_y$ along the z and y axes, respectively, we write

$$M_z = M \cos \theta \qquad M_y = M \sin \theta \qquad (4.52)$$

Since the y and z axes are the principal centroidal axes of the cross section, we may use Eq. (4.16) to determine the stresses resulting from the application of either of the couples represented by $\mathbf{M}_z$ and $\mathbf{M}_y$. The couple $\mathbf{M}_z$ acts in a vertical plane and bends the member in that plane (Fig. 4.56). The resulting stresses are

$$\sigma_x = -\frac{M_z y}{I_z} \qquad (4.53)$$

where I_z is the moment of inertia of the section about the principal centroidal z axis. The negative sign is due to the fact that we have compression above the xz plane ($y > 0$) and tension below ($y < 0$). On the other hand, the couple $\mathbf{M}_y$ acts in a horizontal plane and bends the member in that plane (Fig. 4.57). The resulting stresses are found to be

$$\sigma_x = +\frac{M_y z}{I_y} \qquad (4.54)$$

where I_y is the moment of inertia of the section about the principal centroidal y axis, and where the positive sign is due to the fact that we have tension to the left of the vertical xy plane ($z > 0$) and compression to its right ($z < 0$). The distribution of the stresses caused by the original couple **M** is obtained by superposing the stress distributions defined by Eqs. (4.53) and (4.54), respectively. We have

$$\sigma_x = -\frac{M_z y}{I_z} + \frac{M_y z}{I_y} \tag{4.55}$$

We note that the expression obtained may also be used to compute the stresses in an unsymmetric section, such as the one shown in Fig. 4.58, once the principal centroidal y and z axes have been determined. On the other hand, Eq. (4.55) is valid only if the conditions of applicability of the principle of superposition are met. In other words, it should not be used if the combined stresses exceed the proportional limit of the material, or if the deformations caused by one of the component couples appreciably affect the distribution of the stresses due to the other.

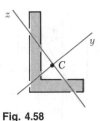

Fig. 4.58

Equation (4.55) shows that the distribution of stresses caused by unsymmetric bending is linear. However, as we have indicated earlier in this section, the neutral axis of the cross section will not, in general, coincide with the axis of the bending couple. Since the normal stress is zero at any point of the neutral axis, the equation defining that axis may be obtained by setting $\sigma_x = 0$ in Eq. (4.55). We write

$$-\frac{M_z y}{I_z} + \frac{M_y z}{I_y} = 0$$

or, solving for y and substituting for M_z and M_y from Eqs. (4.52),

$$y = \left(\frac{I_z}{I_y} \tan \theta\right) z \tag{4.56}$$

The equation obtained is that of a straight line of slope $m = (I_z/I_y) \tan \theta$. Thus, the angle ϕ that the neutral axis forms with the z axis (Fig. 4.59) is defined by the relation

$$\tan \phi = \frac{I_z}{I_y} \tan \theta \tag{4.57}$$

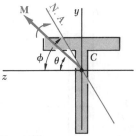

Fig. 4.59

where θ is the angle that the couple vector **M** forms with the same axis. Since I_z and I_y are both positive, ϕ and θ have the same sign. Furthermore, we note that $\phi > \theta$ when $I_z > I_y$, and $\phi < \theta$ when $I_z < I_y$. Thus, the neutral axis is always located between the couple vector **M** and the principal axis corresponding to the minimum moment of inertia.

Example 4.08

A 200-N · m couple is applied to a wooden beam, of rectangular cross section 40 by 90 mm, in a plane forming an angle of 30° with the vertical (Fig. 4.60). Determine (*a*) the maximum stress in the beam, (*b*) the angle that the neutral surface forms with the horizontal plane.

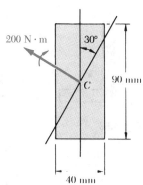

Fig. 4.60

(*a*) **Maximum Stress.** The components $\mathbf{M}_z$ and $\mathbf{M}_y$ of the couple vector are first determined (Fig. 4.61):

$$M_z = (200 \text{ N} \cdot \text{m}) \cos 30° = 173.2 \text{ N} \cdot \text{m}$$
$$M_y = (200 \text{ N} \cdot \text{m}) \sin 30° = 100 \text{ N} \cdot \text{m}$$

We also compute the moments of inertia of the cross section with respect to the z and y axes:

$$I_z = \tfrac{1}{12}(0.040 \text{ m})(0.090 \text{ m})^3 = 2.43 \times 10^{-6} \text{ m}^4$$
$$I_y = \tfrac{1}{12}(0.090 \text{ m})(0.040 \text{ m})^3 = 0.480 \times 10^{-6} \text{ m}^4$$

The largest tensile stress due to $\mathbf{M}_z$ occurs *along AB* and is

$$\sigma_1 = \frac{M_z y}{I_z} = \frac{(173.2 \text{ N} \cdot \text{m})(0.045 \text{ m})}{2.43 \times 10^{-6} \text{ m}^4} = 3.21 \text{ MPa}$$

The largest tensile stress due to $\mathbf{M}_y$ occurs along *AD* and is

$$\sigma_2 = \frac{M_y z}{I_y} = \frac{(100 \text{ N} \cdot \text{m})(0.020 \text{ m})}{0.480 \times 10^{-6} \text{ m}^4} = 4.17 \text{ MPa}$$

The largest tensile stress due to the combined loading, therefore, occurs at *A* and is

$$\sigma_{\text{max}} = \sigma_1 + \sigma_2 = 3.21 + 4.17 = 7.38 \text{ MPa}$$

The largest compressive stress has the same magnitude and occurs at *E*.

(*b*) **Angle of Neutral Surface with Horizontal Plane.** The angle ϕ that the neutral surface forms with the horizontal plane (Fig. 4.62) is obtained from Eq. (4.57):

$$\tan \phi = \frac{I_z}{I_y} \tan \theta = \frac{2.43 \times 10^{-6} \text{ m}^4}{0.480 \times 10^{-6} \text{ m}^4} \tan 30° = 2.92$$

$$\phi = 71.1°$$

The distribution of the stresses across the section is shown in Fig. 4.63.

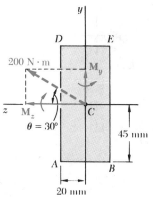

Fig. 4.61

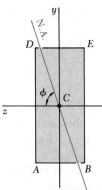

Fig. 4.62

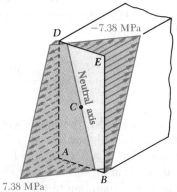

Fig. 4.63

4.15. GENERAL CASE OF ECCENTRIC AXIAL LOADING

In Sec. 4.13 we analyzed the stresses produced in a member by an eccentric axial load applied in a plane of symmetry of the member. We shall now study the more general case when the axial load is not applied in a plane of symmetry.

Consider a straight member AB subjected to equal and opposite eccentric axial forces $\mathbf{P}$ and $\mathbf{P}'$ (Fig. 4.64a), and let a and b denote the distances from the line of action of the forces to the principal centroidal axes of the cross section of the member. The eccentric force $\mathbf{P}$ is statically equivalent to the system consisting of a centric force $\mathbf{P}$ and of the two couples $\mathbf{M}_y$ and $\mathbf{M}_z$ of moments $M_y = Pa$ and $M_z = Pb$ represented in Fig. 4.64b. Similarly, the eccentric force $\mathbf{P}'$ is equivalent to the centric force $\mathbf{P}'$ and the couples $\mathbf{M}_y'$ and $\mathbf{M}_z'$.

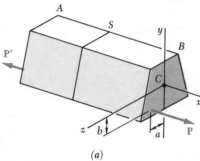
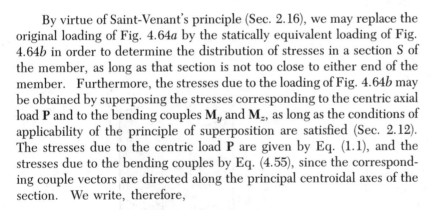
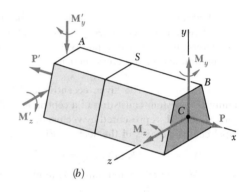

(a)

(b)

Fig. 4.64

By virtue of Saint-Venant's principle (Sec. 2.16), we may replace the original loading of Fig. 4.64a by the statically equivalent loading of Fig. 4.64b in order to determine the distribution of stresses in a section S of the member, as long as that section is not too close to either end of the member. Furthermore, the stresses due to the loading of Fig. 4.64b may be obtained by superposing the stresses corresponding to the centric axial load $\mathbf{P}$ and to the bending couples $\mathbf{M}_y$ and $\mathbf{M}_z$, as long as the conditions of applicability of the principle of superposition are satisfied (Sec. 2.12). The stresses due to the centric load $\mathbf{P}$ are given by Eq. (1.1), and the stresses due to the bending couples by Eq. (4.55), since the corresponding couple vectors are directed along the principal centroidal axes of the section. We write, therefore,

$$\sigma_x = \frac{P}{A} - \frac{M_z y}{I_z} + \frac{M_y z}{I_y} \qquad (4.58)$$

where y and z are measured from the principal centroidal axes of the section. The relation obtained shows that the distribution of stresses across the section is *linear*.

In computing the combined stress σ_x from Eq. (4.58), care should be taken to correctly determine the sign of each of the three terms in the right-hand member, since each of these terms may be positive or negative, depending upon the sense of the loads **P** and **P'** and the location of their line of action with respect to the principal centroidal axes of the cross section. Depending upon the geometry of the cross section and the location of the line of action of **P** and **P'**, the combined stresses σ_x obtained from Eq. (4.58) at various points of the section may all have the same sign, or some may be positive and others negative. In the latter case, there will be a line in the section, along which the stresses are zero. Setting $\sigma_x = 0$ in Eq. (4.58), we obtain the equation of a straight line, which represents the *neutral axis* of the section.

Example 4.09

A vertical 4.80-kN load is applied as shown on a wooden post of rectangular cross section, 80 by 120 mm (Fig. 4.65). (*a*) Determine the stress at points *A*, *B*, *C*, and *D*. (*b*) Locate the neutral axis of the cross section.

(*a*) *Stresses.* The given eccentric load is replaced by an equivalent system consisting of a centric load **P** and two couples **M**$_x$ and **M**$_z$ represented by vectors directed along the principal centroidal axes of the section (Fig. 4.66). We have

$$M_x = (4.80 \text{ kN})(40 \text{ mm}) = 192 \text{ N} \cdot \text{m}$$
$$M_z = (4.80 \text{ kN})(60 \text{ mm} - 35 \text{ mm}) = 120 \text{ N} \cdot \text{m}$$

We also compute the area and the centroidal moments of inertia of the cross section:

$$A = (0.080 \text{ m})(0.120 \text{ m}) = 9.60 \times 10^{-3} \text{ m}^2$$
$$I_x = \tfrac{1}{12}(0.120 \text{ m})(0.080 \text{ m})^3 = 5.12 \times 10^{-6} \text{ m}^4$$
$$I_z = \tfrac{1}{12}(0.080 \text{ m})(0.120 \text{ m})^3 = 11.52 \times 10^{-6} \text{ m}^4$$

The stress σ_0 due to the centric load **P** is negative and uniform across the section. We have

$$\sigma_0 = \frac{P}{A} = \frac{-4.80 \text{ kN}}{9.60 \times 10^{-3} \text{ m}^2} = -0.5 \text{ MPa}$$

The stresses due to the bending couples **M**$_x$ and **M**$_z$ are linearly distributed across the section, with maximum values equal, respectively, to

$$\sigma_1 = \frac{M_x z_{\text{max}}}{I_x} = \frac{(192 \text{ N} \cdot \text{m})(40 \text{ mm})}{5.12 \times 10^{-6} \text{ m}^4} = 1.5 \text{ MPa}$$

$$\sigma_2 = \frac{M_z x_{\text{max}}}{I_z} = \frac{(120 \text{ N} \cdot \text{m})(60 \text{ mm})}{11.52 \times 10^{-6} \text{ m}^4} = 0.625 \text{ MPa}$$

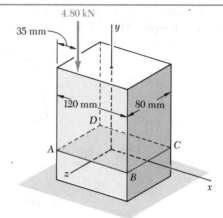

4.80 kN

Fig. 4.65

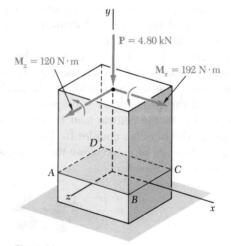

$M_z = 120 \text{ N} \cdot \text{m}$

$P = 4.80 \text{ kN}$

$M_x = 192 \text{ N} \cdot \text{m}$

Fig. 4.66

The stresses at the corners of the section are

$$\sigma_y = \sigma_0 \pm \sigma_1 \pm \sigma_2$$

where the signs must be determined from Fig. 4.66. Noting that the stresses due to $\mathbf{M}_x$ are positive at C and D, and negative at A and B, and that the stresses due to $\mathbf{M}_z$ are positive at B and C, and negative at A and D, we obtain

$$\sigma_A = -0.5 - 1.5 - 0.625 = -2.625 \text{ MPa}$$
$$\sigma_B = -0.5 - 1.5 + 0.625 = -1.375 \text{ MPa}$$
$$\sigma_C = -0.5 + 1.5 + 0.625 = +1.625 \text{ MPa}$$
$$\sigma_D = -0.5 + 1.5 - 0.625 = +0.375 \text{ MPa}$$

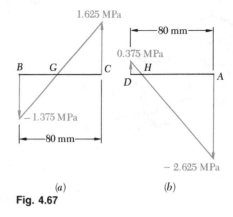

Fig. 4.67

(b) Neutral Axis. We note that the stress will be zero at a point G between B and C, and at a point H between D and A (Fig. 4.67). Since the stress distribution is linear, we write

$$\frac{BG}{80 \text{ mm}} = \frac{1.375}{1.625 + 1.375} \qquad BG = 36.7 \text{ mm}$$

$$\frac{HA}{80 \text{ mm}} = \frac{2.625}{2.625 + 0.375} \qquad HA = 70 \text{ mm}$$

The neutral axis may be drawn through points G and H (Fig. 4.68). The distribution of the stresses across the section is shown in Fig. 4.69.

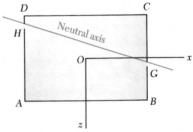

Fig. 4.68

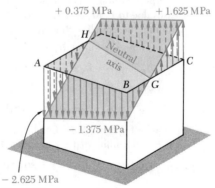

Fig. 4.69

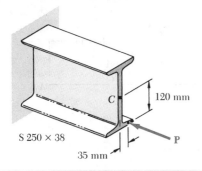

SAMPLE PROBLEM 4.9

A horizontal load **P** is applied as shown to a short section of an S 250 × 38 rolled-steel member. Knowing that the compressive stress in the member is not to exceed 80 MPa determine the largest permissible load **P**.

S 250 × 38

120 mm

35 mm

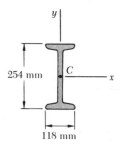

254 mm

118 mm

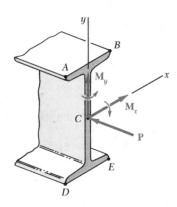

Properties of Cross Section. The following data are taken from Appendix C.

Area: $A = 4806 \text{ mm}^2 = 4.806 \times 10^{-3} \text{ m}^2$

Section moduli: $S_x = 406 \times 10^3 \text{ mm}^3 = 406 \times 10^{-6} \text{ m}^3$
$S_y = 48.0 \times 10^3 \text{ mm}^3 = 48.0 \times 10^{-6} \text{ m}^3$

Force and Couple at C. We replace **P** by an equivalent force-couple system at the centroid C of the cross section.

$$M_x = (0.120 \text{ m})P \qquad M_y = (0.035 \text{ m})P$$

We note that the couple vectors $\mathbf{M}_x$ and $\mathbf{M}_y$ are directed along the principal axes of the cross section.

Normal Stresses. The absolute values of the stresses at points A, B, D, and E due, respectively, to the centric load **P** and to the couples $\mathbf{M}_x$ and $\mathbf{M}_y$ are:

$$\sigma_1 = \frac{P}{A} = \frac{P}{4.806 \times 10^{-3} \text{ m}^2} = 208.1P$$

$$\sigma_2 = \frac{M_x}{S_x} = \frac{(0.120 \text{ m})P}{406 \times 10^{-6} \text{ m}^3} = 295.6P$$

$$\sigma_3 = \frac{M_y}{S_y} = \frac{(0.035 \text{ m})P}{48.0 \times 10^{-6} \text{ m}^3} = 729.2P$$

Superposition. The total stress at each point is found by superposing the stresses due to **P**, $\mathbf{M}_x$ and $\mathbf{M}_y$. We determine the sign of each stress by carefully examining the sketch of the force-couple system.

$$\sigma_A = -\sigma_1 + \sigma_2 + \sigma_3 = -208.1P + 295.6P + 729.2P = +816.7P$$

$$\sigma_B = -\sigma_1 + \sigma_2 - \sigma_3 = -208.1P + 295.6P - 729.2P = -641.7P$$

$$\sigma_D = -\sigma_1 - \sigma_2 + \sigma_3 = -208.1P - 295.6P + 729.2P = +225.5P$$

$$\sigma_E = -\sigma_1 - \sigma_2 - \sigma_3 = -208.1P - 295.6P - 729.2P = -1232.9P$$

Largest Permissible Load. The maximum compressive stress ocurrs at point E. Recalling that $\sigma_{\text{all}} = -80$ MPa, we write

$$\sigma_{\text{all}} = \sigma_E \qquad\qquad -80 \text{ MPa} = -1232.9P \qquad\qquad P = 64.9 \times 10^3 \text{ N}$$

$$P = 64.9 \text{ kN} \quad \blacktriangleleft$$

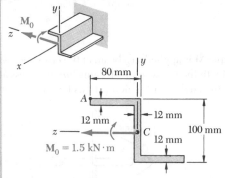

*SAMPLE PROBLEM 4.10

A couple of magnitude $M_0 = 1.5$ kN $\cdot$ m acting in a vertical plane is applied to a beam having the Z-shaped cross section shown. Determine (a) the stress at point A, (b) the angle that the neutral axis forms with the horizontal plane. The moments and product of inertia of the section with respect to the y and z axes have been computed and are as follows:

$$I_y = 3.25 \times 10^{-6} \text{ m}^4$$
$$I_z = 4.18 \times 10^{-6} \text{ m}^4$$
$$I_{yz} = 2.87 \times 10^{-6} \text{ m}^4.$$

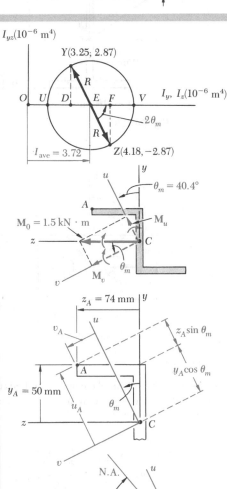

Principal Axes. We draw Mohr's circle and determine the orientation of the principal axes and the corresponding principal moments of inertia.

$$\tan 2\theta_m = \frac{FZ}{EF} = \frac{2.87}{0.465} \qquad 2\theta_m = 80.8° \qquad \theta_m = 40.4°$$

$$R^2 = (EF)^2 + (FZ)^2 = (0.465)^2 + (2.87)^2 \qquad R = 2.91 \times 10^{-6} \text{ m}^4$$

$$I_u = I_{\min} = OU = I_{\text{ave}} - R = 3.72 - 2.91 = 0.810 \times 10^{-6} \text{ m}^4$$

$$I_v = I_{\max} = OV = I_{\text{ave}} + R = 3.72 + 2.91 = 6.63 \times 10^{-6} \text{ m}^4$$

Loading. The applied couple $\mathbf{M}_0$ is resolved into components parallel to the principal axes.

$$M_u = M_0 \sin \theta_m = 1500 \sin 40.4° = 972 \text{ N} \cdot \text{m}$$
$$M_v = M_0 \cos \theta_m = 1500 \cos 40.4° = 1142 \text{ N} \cdot \text{m}$$

a. **Stress at A.** The perpendicular distances from each principal axis to point A are

$$u_A = y_A \cos \theta_m + z_A \sin \theta_m = 50 \cos 40.4° + 74 \sin 40.4° = 86.0 \text{ mm}$$
$$v_A = -y_A \sin \theta_m + z_A \cos \theta_m = -50 \sin 40.4° + 74 \cos 40.4° = 23.9 \text{ mm}$$

Considering separately the bending about each principal axis, we note that $\mathbf{M}_u$ produces a tensile stress at point A while $\mathbf{M}_v$ produces a compressive stress at the same point.

$$\sigma_A = +\frac{M_u v_A}{I_u} - \frac{M_v u_A}{I_v} = +\frac{(972 \text{ N} \cdot \text{m})(0.0239 \text{ m})}{0.810 \times 10^{-6} \text{ m}^4} - \frac{(1142 \text{ N} \cdot \text{m})(0.0860 \text{ m})}{6.63 \times 10^{-6} \text{ m}^4}$$

$$= +(28.68 \text{ MPa}) - (14.81 \text{ MPa}) \qquad \sigma_A = +13.87 \text{ MPa} \blacktriangleleft$$

b. **Neutral Axis.** Using Eq. (4.57), we find the angle ϕ that the neutral axis forms with the v axis.

$$\tan \phi = \frac{I_v}{I_u} \tan \theta_m = \frac{6.63}{0.810} \tan 40.4° \qquad \phi = 81.8°$$

The angle β formed by the neutral axis and the horizontal is

$$\beta = \phi - \theta_m = 81.8° - 40.4° = 41.4° \qquad \beta = 41.4° \blacktriangleleft$$

PROBLEMS

4.136 through 4.141   The couple **M** is applied to a beam of the cross section shown in a plane forming an angle β with the vertical.   Determine (*a*) the stress at point *A*, (*b*) the stress at point *B*, (*c*) the angle that the neutral axis forms with the horizontal plane.

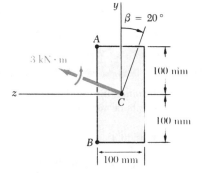

Fig. P4.136

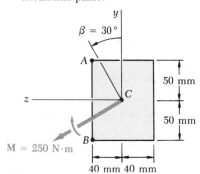

Fig. P4.137

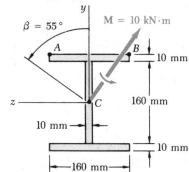

Fig. P4.138

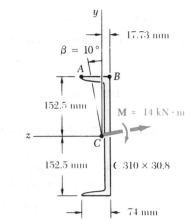

Fig. P4.139

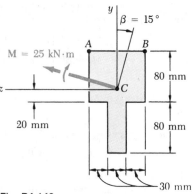

Fig. P4.140

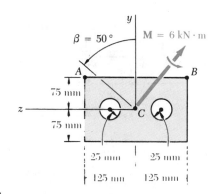

Fig. P4.141

4.142 through 4.144   The couple **M** acts in a vertical plane and is applied to a beam oriented as shown.   Determine (*a*) the angle that the neutral axis forms with the horizontal plane, (*b*) the maximum tensile stress in the beam.

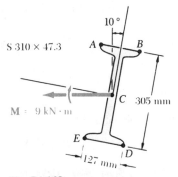

Fig. P4.142

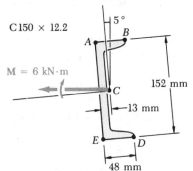

Fig. P4.143

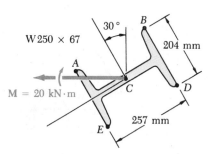

Fig. P4.144

4.145 through 4.147 The couple **M** acts in a vertical plane and is applied to a beam oriented as shown. Determine (*a*) the angle that the neutral axis forms with the horizontal plane, (*b*) the maximum tensile stress in the beam.

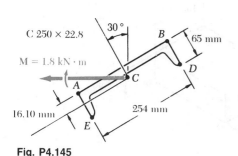

Fig. P4.145

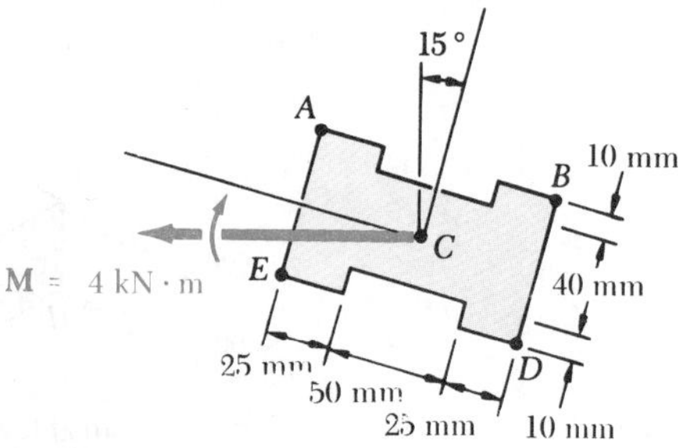

Fig. P4.146

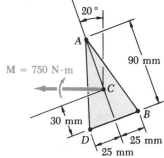

Fig. P4.147

*4.148 through *4.150 The couple **M** acts in a vertical plane and is applied to a beam of the cross section shown. Determine the stress at point *A*.

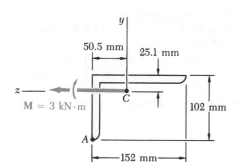

$I_y = 7.24 \times 10^6 \text{ mm}^4$
$I_z = 2.61 \times 10^6 \text{ mm}^4$
$I_{yz} = -2.54 \times 10^6 \text{ mm}^4$

Fig. P4.148

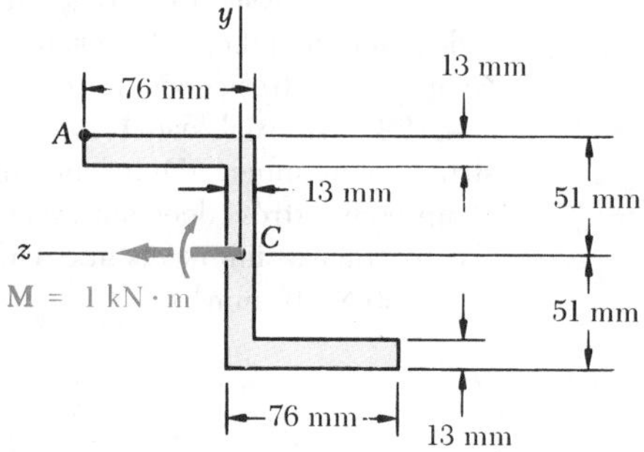

$I_y = 2.90 \times 10^6 \text{ mm}^4$
$I_z = 4.32 \times 10^6 \text{ mm}^4$
$I_{yz} = +2.73 \times 10^6 \text{ mm}^4$

Fig. P4.149

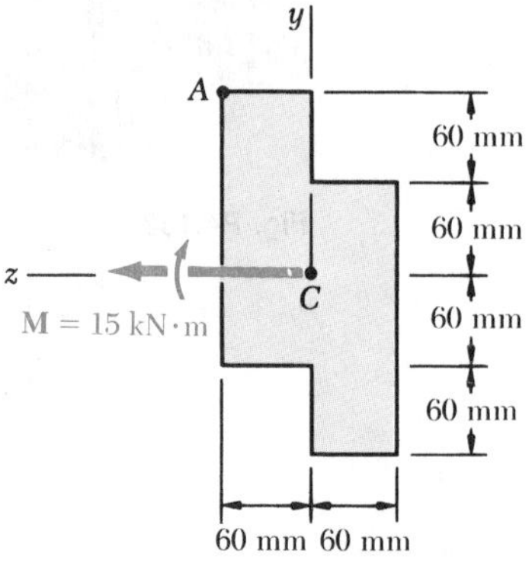

Fig. P4.150

*4.151 through *4.153 For the beam indicated, determine the stress at point *A*, assuming that the plane in which the couple **M** acts has been rotated 90° clockwise.

  4.151 Beam of Prob. 4.148
  4.152 Beam of Prob. 4.149
  4.153 Beam of Prob. 4.150

4.154 The 120-kN load may be applied at any point on the circumference of the 40-mm-radius circle shown. If $\theta = 20°$, determine (*a*) the stress at *A*, (*b*) the stress at *B*, (*c*) the point where the neutral axis intersects line *ABD*.

4.155 Knowing that the 120-kN load may be applied at any point on the circumference of the 40-mm-radius circle shown, determine (*a*) the value of θ for which the stress at *D* reaches its largest absolute value, (*b*) the corresponding stress at *A*, *B*, and *D*.

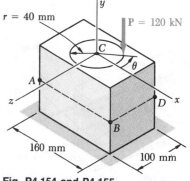

Fig. P4.154 and P4.155

4.156 For the loading shown, determine (*a*) the stress at points *A* and *B*, (*b*) the point where the neutral axis intersects line *ABD*.

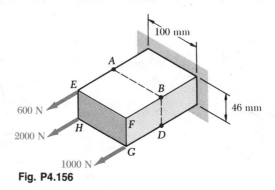

Fig. P4.156

4.157 Solve Prob. 4.156, assuming that the magnitude of the force applied at *G* is increased from 1000 N to 1800 N.

4.158 An axial load **P** is applied as shown to a short section of a W 200 × 35.9 rolled-steel member. Determine the largest distance *a* for which the maximum compressive stress does not exceed 125 MPa.

4.159 An axial load **P** is applied as shown to a short section of a T-shaped structural member. Determine (*a*) the largest distance *a* for which the maximum compressive stress does not exceed 120 MPa, (*b*) the corresponding point where the neutral axis intersects line *AB*. (*Given:* $A = 4450$ mm², $I_y = 9.16 \times 10^6$ mm⁴, $I_z = 6.00 \times 10^6$ mm⁴.)

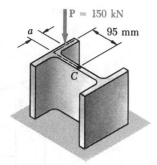

Fig. P4.158

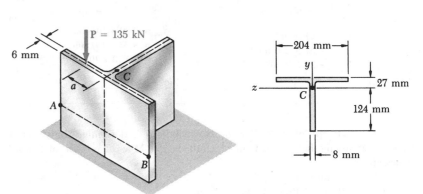

Fig. P4.159

4.160 Two horizontal loads are applied as shown to a short section of a C 250 × 22.8 rolled-steel channel. Knowing that the tensile stress in the member is not to exceed 70 MPa, determine the largest permissible load **P**.

4.161 Solve Prob. 4.160, assuming that the tensile stress in the member is not to exceed 80 MPa.

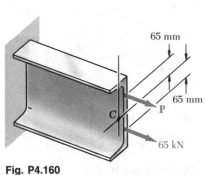

Fig. P4.160

4.162 A beam having the cross section shown is subjected to a couple M_0 which acts in a vertical plane.   Determine the largest permissible value of the bending moment M_0 if the maximum stress in the beam is not to exceed 80 MPa.   *Given:* $I_y = I_z = 4.70 \times 10^{-6} \text{ m}^4$, $A = 3.07 \times 10^{-3} \text{ m}^2$, $k_{min} = 25.0$ mm.   (*Hint:* By reason of symmetry, the principal axes form an angle of $45°$ with the coordinate axes.   Use the relations $I_{min} = A k^2_{min}$ and $I_{min} + I_{max} = I_y + I_z$.)

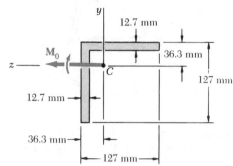

Fig. P4.162

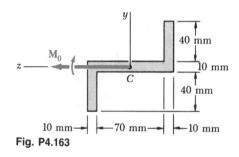

Fig. P4.163

4.163 The Z section shown is subjected to a couple M_0 acting in a vertical plane.   Determine the largest permissible value of the bending moment M_0 if the maximum stress in the beam is not to exceed 100 MPa.   *Given:* $I_{max} = 2.28 \times 10^6 \text{ mm}^4$, $I_{min} = 0.23 \times 10^6 \text{ mm}^4$, principal axes $\angle 25.7°$ and $\angle 64.3°$.

*4.164 A beam having the cross section shown is subjected to a couple M_0 which acts in a vertical plane.   Determine the largest permissible value of the bending moment M_0 if the maximum stress in the beam is not to exceed 100 MPa. *Given:* $I_y = I_z = b^4/36$ and $I_{yz} = b^4/72$.

*4.165 A beam having the cross section shown is subjected to a couple M_0 which acts in a vertical plane.   Determine the largest permissible value of the bending moment M_0 if the maximum stress in the beam is not to exceed 3.5 MPa.

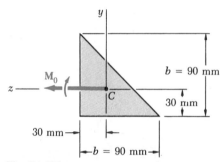

Fig. P4.164

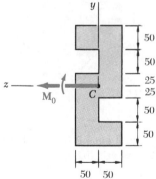

Fig. P4.165

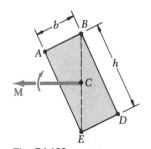

4.166 Show that if a solid rectangular beam is bent by a couple applied in the plane containing one diagonal of the rectangular cross section, the neutral axis will lie along the other diagonal.

Fig. P4.166

4.167 In a given section of an S 200 × 27.4 rolled-steel beam, the effect of gravity has been represented by a horizontal couple vector $\mathbf{M}_0$. Denoting by σ_0 the maximum stress in the beam when $\theta = 0$, determine the angle of inclination θ of the beam for which the maximum stress is $3\sigma_0$.

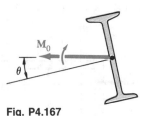

Fig. P4.167

4.168 (a) Show that the stress at corner A of a rectangular member will be zero if the vertical force $\mathbf{P}$ is applied at a point located on the line

$$\frac{x}{b/6} + \frac{z}{h/6} = 1$$

(b) Further show that if no tensile stress is to occur in the member, the force $\mathbf{P}$ must be applied at a point located within the area bounded by the line found in part a and three similar lines corresponding to the condition of zero stress at B, C, and D. (This area is known as the *kern* of the cross section.)

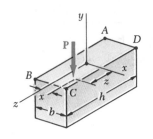

Fig. P4.168

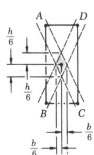

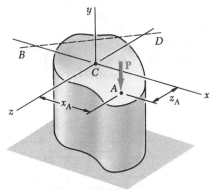

Fig. P4.169

4.169 (a) Show that if a vertical force $\mathbf{P}$ is applied at point A of the section shown, the equation of the neutral axis BD is

$$\left(\frac{x_A}{k_z^2}\right)x + \left(\frac{z_A}{k_x^2}\right)z = -1$$

where k_z and k_x denote the radius of gyration of the cross section with respect to the z axis and the x axis, respectively. (b) Further show that if a vertical force $\mathbf{Q}$ is applied at any point located on line BD, the stress at point A will be zero.

4.170 A beam of unsymmetric cross section is subjected to a couple $\mathbf{M}_z$ acting in a vertical plane. Show that the stress at point A, of coordinates y and z, is

$$\sigma_A = = -\frac{yI_y - zI_{yz}}{I_yI_z - I_{yz}^2}M_z$$

where I_y, I_z, and I_{yz} denote the moments and product of inertia of the cross section with respect to the centroidal y and z axes.

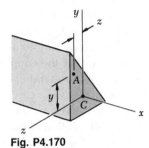

Fig. P4.170

4.171 The beam of Prob. 4.170 is subjected to a couple $\mathbf{M}_y$ acting in a horizontal plane. Show that the stress at point A is

$$\sigma_A = \frac{zI_z - yI_{yz}}{I_yI_z - I_{yz}^2}M_y$$

*4.16. BENDING OF CURVED MEMBERS

Our analysis of stresses due to bending has been restricted so far to straight members. In this section we shall consider the stresses caused by the application of equal and opposite couples to members which are initially curved. Our discussion will be limited to curved members of uniform cross section possessing a plane of symmetry in which the bending couples are applied, and we shall assume that all stresses remain below the proportional limit.

If the initial curvature of the member is small, i.e., if its radius of curvature is large compared to the depth of its cross section, a good approximation may be obtained for the distribution of stresses by assuming the member to be straight and using the formulas derived in Secs. 4.4 and 4.5.† However, when the radius of curvature and the dimensions of the cross section of the member are of the same order of magnitude, we must use a different method of analysis which was first introduced by the German engineer E. Winkler (1835–1888).

† See Prob. 4.177.

Consider the curved member of uniform cross section shown in Fig. 4.70. Its transverse section is symmetric with respect to the y axis (Fig. 4.70b) and, in its unstressed state, its upper and lower surfaces intersect the vertical xy plane along arcs of circle AB and FG centered at C (Fig. 4.70a). We now apply two equal and opposite couples $\mathbf{M}$ and $\mathbf{M'}$ in the plane of symmetry of the member (Fig. 4.70c). A reasoning similar to that of Sec. 4.4 would show that any transverse plane section containing C will remain plane, and that the various arcs of circle indicated in Fig. 4.70a will be transformed into circular and concentric arcs with a center C' different from C. More specifically, if the couples $\mathbf{M}$ and $\mathbf{M'}$ are directed as shown, the curvature of the various arcs of circle will increase, that is $A'C' < AC$. We also note that the couples $\mathbf{M}$ and $\mathbf{M'}$ will cause the length of the upper surface of the member to decrease ($A'B' < AB$) and the length of the lower surface to increase ($F'G' > FG$). We conclude that a *neutral surface* must exist in the member, the length of which remains constant. The intersection of the neutral surface with the xy plane has been represented in Fig. 4.70a by the arc DE of radius R, and in Fig. 4.70c by the arc $D'E'$ of radius R'. Denoting by θ and θ' the central angles corresponding respectively to DE and $D'E'$, we express the fact that the length of the neutral surface remains constant by writing

$$R\theta = R'\theta' \tag{4.59}$$

Considering now the arc of circle JK located at a distance y above the neutral surface, and denoting respectively by r and r' the radius of this

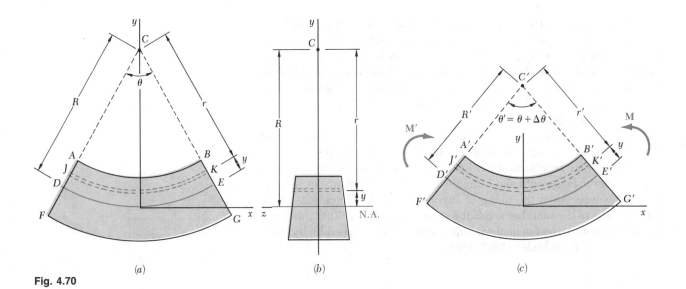

(a)   (b)   (c)

Fig. 4.70

arc before and after the bending couples have been applied, we express the deformation of JK as

$$\delta = r'\theta' - r\theta \tag{4.60}$$

Observing from Fig. 4.70 that

$$r = R - y \qquad r' = R' - y \tag{4.61}$$

and substituting these expressions into Eq. (4.60), we write

$$\delta = (R' - y)\theta' - (R - y)\theta$$

or, recalling Eq. (4.59) and setting $\theta' - \theta = \Delta\theta$,

$$\delta = -y\,\Delta\theta \tag{4.62}$$

The normal strain ϵ_x in the elements of JK is obtained by dividing the deformation δ by the original length $r\theta$ of arc JK. We write

$$\epsilon_x = \frac{\delta}{r\theta} = -\frac{y\,\Delta\theta}{r\theta}$$

or, recalling the first of the relations (4.61),

$$\epsilon_x = -\frac{\Delta\theta}{\theta}\frac{y}{R - y} \tag{4.63}$$

The relation obtained shows, that, while each transverse section remains plane, the normal strain ϵ_x *does not vary linearly* with the distance y from the neutral surface.

The normal stress σ_x may now be obtained from Hooke's law, $\sigma_x = E\epsilon_x$, by substituting for ϵ_x from Eq. (4.63). We have

$$\sigma_x = -\frac{E\,\Delta\theta}{\theta}\frac{y}{R - y} \tag{4.64}$$

or, alternatively, recalling the first of Eqs. (4.61),

$$\sigma_x = -\frac{E\,\Delta\theta}{\theta}\frac{R - r}{r} \tag{4.65}$$

Equation (4.64) shows that, like ϵ_x, the normal stress σ_x *does not vary linearly* with the distance y from the neutral surface. Plotting σ_x versus y, we obtain an arc of hyperbola (Fig. 4.71).

In order to determine the location of the neutral surface in the member and the value of the coefficient $E\,\Delta\theta/\theta$ used in Eqs. (4.64) and (4.65), we now recall that the elementary forces acting on any transverse section must be statically equivalent to the bending couple **M.** Expressing, as we did in Sec. 4.3 for a straight member, that the sum of the elementary forces acting on the section must be zero, and that the sum of their mo-

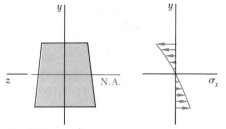

Fig. 4.71

ments about the transverse z axis must be equal to the bending moment M, we write the equations

$$\int \sigma_x \, dA = 0 \qquad (4.1)$$

and

$$\int (-y\sigma_x \, dA) = M \qquad (4.3)$$

Substituting for σ_x from (4.65) into Eq. (4.1), we write

$$-\int \frac{E \, \Delta\theta}{\theta} \cdot \frac{R - r}{r} \, dA = 0$$

$$\int \frac{R - r}{r} \, dA = 0$$

$$R \int \frac{dA}{r} - \int dA = 0$$

from which it follows that the distance R from the center of curvature C to the neutral surface is defined by the relation

$$R = \frac{A}{\displaystyle\int \frac{dA}{r}} \qquad (4.66)$$

We note that the value obtained for R is not equal to the distance $\bar{r}$ from C to the centroid of the cross section, since $\bar{r}$ is defined by a different relation, namely,

$$\bar{r} = \frac{1}{A} \int r \, dA \qquad (4.67)$$

We thus conclude that, in a curved member, *the neutral axis of a transverse section does not pass through the centroid of that section* (Fig. 4.72).† Expressions for the radius R of the neutral surface will be derived for some specific cross-sectional shapes in Example 4.10 and in Probs. 4.195 through 4.197. For convenience, these expressions are shown in Fig. 4.73.

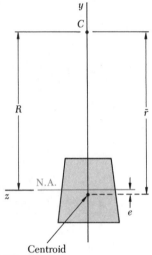

Centroid

Fig. 4.72

---

† However, an interesting property of the neutral surface may be noted if we write Eq. (4.66) in the alternative form

$$\frac{1}{R} = \frac{1}{A} \int \frac{1}{r} \, dA \qquad (4.66')$$

Equation (4.66′) shows that, if the member is divided into a large number of fibers of cross-sectional area dA, the curvature $1/R$ of the neutral surface will be equal to the average value of the curvature $1/r$ of the various fibers.

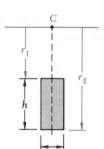

Rectangle

$$R = \frac{h}{\ln \dfrac{r_2}{r_1}}$$

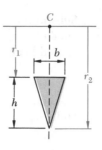

Circle

$$R = \tfrac{1}{2}(\bar{r} + \sqrt{\bar{r}^2 - c^2})$$

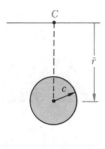

Triangle

$$R = \frac{\frac{1}{2}h}{\dfrac{r_2}{h}\ln \dfrac{r_2}{r_1} - 1}$$

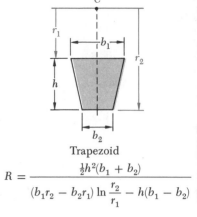

Trapezoid

$$R = \frac{\frac{1}{2}h^2(b_1 + b_2)}{(b_1 r_2 - b_2 r_1)\ln \dfrac{r_2}{r_1} - h(b_1 - b_2)}$$

Fig. 4.73 Radius of neutral surface for various cross-sectional shapes.

Substituting now for σ_x from (4.65) into Eq. (4.3), we write

$$\int \frac{E\,\Delta\theta}{\theta} \frac{R - r}{r} y\, dA = M$$

or, since $y = R - r$,

$$\frac{E\,\Delta\theta}{\theta} \int \frac{(R - r)^2}{r}\, dA = M$$

Expanding the square in the integrand, we obtain after reductions

$$\frac{E\,\Delta\theta}{\theta}\left[R^2 \int \frac{dA}{r} - 2RA + \int r\, dA \right] = M$$

Recalling Eqs. (4.66) and (4.67), we note that the first term in the brackets is equal to RA, while the last term is equal to $\bar{r}A$. We have, therefore,

$$\frac{E\,\Delta\theta}{\theta}(RA - 2RA + \bar{r}A) = M$$

and, solving for $E\,\Delta\theta/\theta$,

$$\frac{E\,\Delta\theta}{\theta} = \frac{M}{A(\bar{r} - R)} \tag{4.68}$$

Referring to Fig. 4.70, we note that $\Delta\theta > 0$ for $M > 0$. It follows that $\bar{r} - R > 0$, or $R < \bar{r}$, regardless of the shape of the section. Thus, the neutral axis of a transverse section is always located between the centroid of the section and the center of curvature of the member (Fig. 4.72). Setting $\bar{r} - R = e$, we write Eq. (4.68) in the form

$$\frac{E\,\Delta\theta}{\theta} = \frac{M}{Ae} \tag{4.69}$$

Substituting now for $E\,\Delta\theta/\theta$ from (4.69) into Eqs. (4.64) and (4.65), we obtain the following alternative expressions for the normal stress σ_x in a curved beam:

$$\sigma_x = -\frac{My}{Ae(R-y)} \qquad (4.70)$$

and

$$\sigma_x = \frac{M(r-R)}{Aer} \qquad (4.71)$$

We should note that the parameter e in the above equations is a small quantity obtained by subtracting two lengths of comparable size, R and $\bar{r}$. In order to determine σ_x with a reasonable degree of accuracy, it is therefore necessary to compute R and $\bar{r}$ very accurately, particularly when both of these quantities are large, i.e., when the curvature of the member is small. However, as we indicated earlier, it is possible in such a case to obtain a good approximation for σ_x by using the formula $\sigma_x = -My/I$ developed for straight members.

We shall now determine the change in curvature of the neutral surface caused by the bending moment M. Solving Eq. (4.59) for the curvature $1/R'$ of the neutral surface in the deformed member, we write

$$\frac{1}{R'} = \frac{1}{R}\frac{\theta'}{\theta}$$

or, setting $\theta' = \theta + \Delta\theta$ and recalling Eq. (4.69),

$$\frac{1}{R'} = \frac{1}{R}\left(1 + \frac{\Delta\theta}{\theta}\right) = \frac{1}{R}\left(1 + \frac{M}{EAe}\right)$$

from which it follows that the change in curvature of the neutral surface is

$$\frac{1}{R'} - \frac{1}{R} = \frac{M}{EAeR} \qquad (4.72)$$

Example 4.10

A curved rectangular bar has a mean radius $\bar{r} = 100$ mm and a cross section of width $b = 50$ mm and depth $h = 25$ mm (Fig. 4.74). Determine the distance e between the centroid and the neutral axis of the cross section.

We shall first derive the expression for the radius R of the neutral surface. Denoting by r_1 and r_2, respectively, the inner

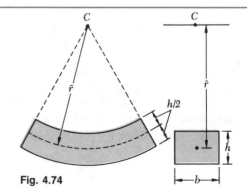

Fig. 4.74

and outer radius of the bar (Fig. 4.75), we use Eq. (4.66) and write

$$R = \frac{A}{\int_{r_1}^{r_2} \dfrac{dA}{r}} = \frac{bh}{\int_{r_1}^{r_2} \dfrac{b\,dr}{r}} = \frac{h}{\int_{r_1}^{r_2} \dfrac{dr}{r}}$$

$$R = \frac{h}{\ln \dfrac{r_2}{r_1}} \qquad (4.73)$$

For the given data, we have

$$r_1 = \bar{r} - \tfrac{1}{2}h = 100 - 12.5 = 87.5 \text{ mm}$$
$$r_2 = \bar{r} + \tfrac{1}{2}h = 100 + 12.5 = 112.5 \text{ mm}$$

Substituting for h, r_1, and r_2 into Eq. (4.73), we have

$$R = \frac{h}{\ln \dfrac{r_2}{r_1}} = \frac{25 \text{ mm}}{\ln \dfrac{112.5}{87.5}} = 99.477 \text{ mm}$$

The distance between the centroid and the neutral axis of the cross section (Fig. 4.76), is thus

$$e = \bar{r} - R = 100 - 99.477 = 0.523 \text{ mm}$$

We note that it was necessary to calculate R with five significant figures in order to obtain e with the usual degree of accuracy.

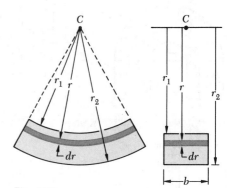

Fig. 4.75

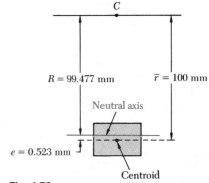

Fig. 4.76

Example 4.11

For the bar of Example 4.10, determine the largest tensile and compressive stresses, knowing that the bending moment in the bar is $M = 500 \text{ N} \cdot \text{m}$.

We shall use Eq. (4.71) with the given data,

$$M = 500 \text{ N} \cdot \text{m} \qquad A = bh = (50 \text{ mm})(25 \text{ mm}) = 1250 \text{ mm}^2$$

and the values obtained in Example 4.10 for R and e,

$$R = 99.48 \text{ mm} \qquad e = 0.523 \text{ mm}$$

Making first $r = r_2 = 112.5 \text{ mm}$ in Eq. (4.71), we write

$$\sigma_{\max} = \frac{M(r_2 - R)}{Aer_2}$$

$$= \frac{(500 \text{ N} \cdot \text{m})(112.5 - 99.48) \times 10^{-3} \text{ m}}{(1250 \times 10^{-6} \text{ m}^2)(0.523 \times 10^{-3} \text{ m})(112.5 \times 10^{-3} \text{ m})}$$

$$\sigma_{\max} = 88.5 \text{ MPa}$$

Making now $r = r_1 = 87.5 \text{ mm}$ in Eq. (4.71), we have

$$\sigma_{\min} = \frac{M(r_1 - R)}{Aer_1}$$

$$= \frac{(500 \text{ N} \cdot \text{m})(87.5 - 99.48) \times 10^{-3} \text{ m}}{(1250 \times 10^{-6} \text{ m}^2)(0.523 \times 10^{-3} \text{ m})(87.5 \times 10^{-3} \text{ m})}$$

$$\sigma_{\min} = -104.7 \text{ MPa}$$

Remark. We may wish to compare the values obtained for $\sigma_{\max}$ and $\sigma_{\min}$ with the result we would get for a straight bar. Using Eq. (4.15) of Sec. 4.4, we write

$$\sigma_{\max,\min} = \pm\frac{Mc}{I}$$

$$= \pm\frac{(500 \text{ N} \cdot \text{m})(12.5 \times 10^{-3} \text{ m})}{\frac{1}{12}(50 \times 10^{-3} \text{ m})(25 \times 10^{-3} \text{ m})^3} = \pm 96.0 \text{ MPa}$$

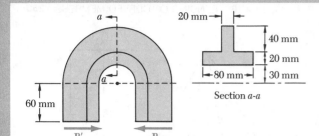

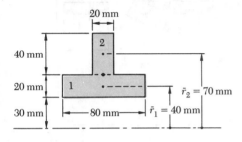

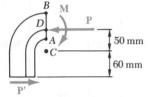

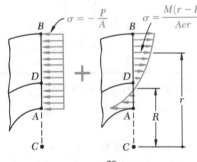

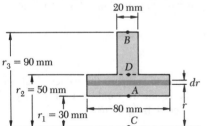

A machine component has a T-shaped cross section and is loaded as shown. Knowing that the allowable compressive stress is 50 MPa, determine the largest force **P** which may be applied to the component.

Centroid of the Cross Section. We locate the centroid D of the cross section

	A_i, mm²	$\bar{r}_i$, mm	$\bar{r}_iA_i$, mm³
1	$(20)(80) = 1600$	40	64×10^3
2	$(40)(20) = \ \ 800$	70	56×10^3
	$\Sigma A_i = 2400$		$\Sigma \bar{r}_i A_i = 120 \times 10^3$

$\bar{r}\Sigma A_i = \Sigma \bar{r}_i A_i$

$\bar{r}(2400) = 120 \times 10^3$

$\bar{r} = 50$ mm $= 0.050$ m

Force and Couple at D. The internal forces in section $a\text{-}a$ are equivalent to a force **P** acting at D and a couple **M** of moment

$$M = P(50 \text{ mm} + 60 \text{ mm}) = (0.110 \text{ m})P$$

Superposition. The centric force **P** causes a uniform compressive stress on section $a\text{-}a$. The bending couple **M** causes a varying stress distribution [Eq. (4.71)]. We note that the couple **M** tends to increase the curvature of the member and is therefore positive (cf. Fig. 4.70). The total stress at a point of section $a\text{-}a$ located at distance r from the center of curvature C is

$$\sigma = -\frac{P}{A} + \frac{M(r-R)}{Aer} \qquad (1)$$

Radius of Neutral Surface. We now determine the radius R of the neutral surface by using Eq. (4.66).

$$R = \frac{A}{\displaystyle\int \frac{dA}{r}} = \frac{2400 \text{ mm}^2}{\displaystyle\int_{r_1}^{r_2} \frac{(80 \text{ mm}) \, dr}{r} + \int_{r_2}^{r_3} \frac{(20 \text{ mm}) \, dr}{r}}$$

$$= \frac{2400}{80 \ln \dfrac{50}{30} + 20 \ln \dfrac{90}{50}} = \frac{2400}{40.866 + 11.756} = 45.61 \text{ mm}$$

$$= 0.04561 \text{ m}$$

We also compute: $e = \bar{r} - R = 0.050 \text{ m} - 0.04561 \text{ m} = 0.00439 \text{ m}$

Allowable Load. We observe that the largest compressive stress will occur at point A where $r = 0.030$ m. Recalling that $\sigma_{all} = 50$ MPa and using Eq. (1), we write

$$-50 \times 10^6 \text{ Pa} = -\frac{P}{2.4 \times 10^{-3} \text{ m}^2} + \frac{(0.110 \, P)(0.030 \text{ m} - 0.04561 \text{ m})}{(2.4 \times 10^{-3} \text{ m}^2)(0.00439 \text{ m})(0.030 \text{ m})}$$

$$-50 \times 10^6 = -417P - 5432P \qquad\qquad P = 8.55 \text{ kN} \ \blacktriangleleft$$

PROBLEMS

4.172 For the machine component and loading shown, determine the stress at point A when (a) $h = 50$ mm, (b) $h = 66$ mm.

4.173 For the curved bar and loading shown, determine the maximum compressive stress when (a) $r_1 = 45$ mm, (b) $r_1 = 15$ mm.

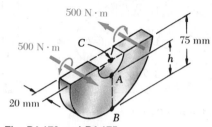

Fig. P4.172 and P4.175

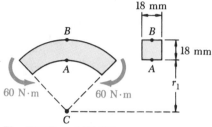

Fig. P4.173 and P4.174

4.174 For the curved bar and loading shown, determine the stress at points A and B when $r_1 = 8$ mm.

4.175 For the machine component and loading shown, determine the stress at points A and B when $h = 64$ mm.

4.176 The curved prismatic bar has a cross section of 5×7.5 mm and an inner radius $r_1 = 15$ mm. For the loading shown, determine the largest tensile and compressive stresses.

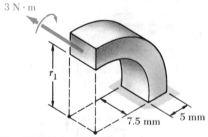

Fig. P4.176 and P4.177

4.177 For the curved bar and loading shown, determine the percent error introduced in the computation of the maximum stress by assuming that the bar is straight. Consider the case when (a) $r_1 = 5$ mm, (b) $r_1 = 25$ mm, (c) $r_1 = 150$ mm.

4.178 The curved portion of the bar shown has an inner radius of 7.5 mm. The line of action of the 250-N force is located at a distance a from the plane containing the center of curvature of the bar. Determine the largest compressive stress when (a) $a = 0$, (b) $a = 30$ mm.

4.179 Knowing that the allowable compressive stress is 85 MPa, determine the largest permissible distance a from the line of action of the 250-N force to the plane containing the center of curvature of the bar.

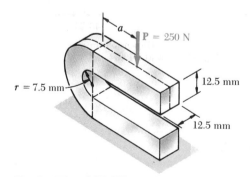

Fig. P4.178 and P4.179

4.180   Steel links having the cross section shown are available with different central angles β. Knowing that the allowable stress is 100 MPa, determine the largest force **P** which may be applied to a link for which $\beta = 90°$.

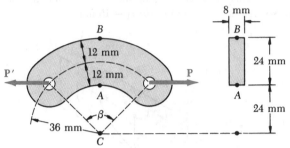

Fig. P4.180 and P4.181

4.181   Steel links having the cross section shown are available with different central angles β. Determine (*a*) the largest allowable value of β if no compressive stress is to occur in the section containing points *A* and *B*, (*b*) the corresponding magnitude and location of the maximum tensile stress when *P* = 3 kN.

4.182 and 4.183   A machine component has a T-shaped cross section which is oriented as shown. Knowing that $M = 2500$ N · m, determine the stress at (*a*) point *A*, (*b*) point *B*.

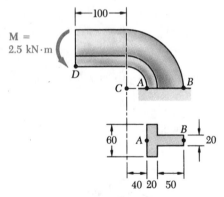

Dimensions in mm
Fig. P4.182 and P4.184

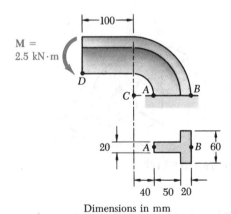

Dimensions in mm
Fig. P4.183 and P4.185

4.184 and 4.185   Assuming that the couple shown is replaced by a vertical 10-kN force attached at point *D* and acting downward, determine the stress at (*a*) point *A*, (*b*) point *B*.

4.186 Three plates are welded together to form the curved beam shown. For $M = 1$ kN · m, determine the stress at (a) point A, (b) point B, (c) the centroid of the cross section.

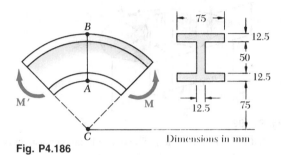

75

12.5

50

12.5

12.5   75

Dimensions in mm

Fig. P4.186

4.187 Solve Prob. 4.186, assuming that the curved beam has been redesigned using the same cross section with the inner radius increased from 75 mm to 125 mm.

4.188 The split ring shown has an inner radius $r_1 = 20$ mm and a circular cross section of diameter $d = 15$ mm. Knowing that each of the 500-N forces is applied at the centroid of the cross section, determine the stress at (a) point A, (b) point B.

4.189 Solve Prob. 4.188, assuming that the ring has an inner radius $r_1 = 15$ mm and a cross-sectional diameter $d = 20$ mm.

4.190 For the crane hook shown, determine the largest stress in section a-a when $b_1 = 35$ mm and $b_2 = 25$ mm.

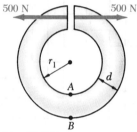

500 N   500 N

r_1

A

d

B

Fig. P4.188

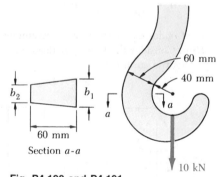

b_2 b_1

60 mm

Section a-a

60 mm

40 mm

a

a

10 kN

Fig. P4.190 and P4.191

4.191 For the crane hook shown, determine the largest stress in section a-a when $b_1 = 50$ mm and $b_2 = 10$ mm.

4.192 Knowing that in the half ring shown $h = 30$ mm, determine the stress at (a) point A, (b) point B.

4.193 Solve Prob. 4.192, assuming that $h = 40$ mm.

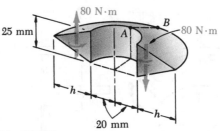

80 N·m

25 mm

B

−80 N·m

A

h

h

20 mm

Fig. P4.192

4.194 Show that if the cross section of a curved beam consists of two or more rectangles, the radius R of the neutral surface may be expressed as

$$R = \frac{A}{\ln\left[\left(\dfrac{r_2}{r_1}\right)^{b_1}\left(\dfrac{r_3}{r_2}\right)^{b_2}\left(\dfrac{r_4}{r_3}\right)^{b_3}\right]}$$

where A is the total area of the cross section.

4.195 through 4.197 Using Eq. (4.66), derive the expression for R given in Fig. 4.73 for

    *4.195 A circular cross section

    4.196 A trapezoidal cross section

    4.197 A triangular cross section

*4.198 For a curved bar of rectangular cross section subjected to a bending couple **M**, show that the radial stress at the neutral surface is

$$\sigma_r = \frac{M}{Ae}\left[1 - \frac{r_1}{R} - \ln\frac{R}{r_1}\right]$$

and compute the value of σ_r for the curved bar of Examples 4.10 and 4.11 on pages 262 and 263. (*Hint:* Consider the free-body diagram of the portion of the beam located above the neutral surface.)

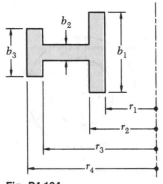

b_2

b_3

b_1

r_1

r_2

r_3

r_4

Fig. P4.194

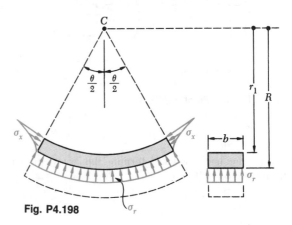

C

$\dfrac{\theta}{2}$ $\dfrac{\theta}{2}$

r_1

R

σ_x σ_x

b

σ_r

Fig. P4.198

σ_r

REVIEW AND SUMMARY

This chapter was devoted to the analysis of members in *pure bending*. That is, we considered the stresses and deformation in members subjected to equal and opposite couples **M** and **M′** acting in the same longitudinal plane (Fig. 4.1).

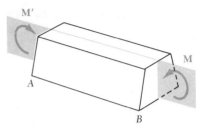

Fig. 4.1

Normal strain in bending

We first studied members possessing a plane of symmetry and subjected to couples acting in that plane. Considering possible deformations of the member, we proved that *transverse sections remain plane* as a member is deformed [Sec. 4.4]. We then noted that a member in pure bending has a *neutral surface* along which normal strains and stresses are zero and that the longitudinal *normal strain* ϵ_x *varies linearly* with the distance y from the neutral surface:

$$\epsilon_x = -\frac{y}{\rho} \tag{4.8}$$

where ρ is the *radius of curvature* of the neutral surface (Fig. 4.10a). The intersection of the neutral surface with a transverse section is known as the *neutral axis* of the section.

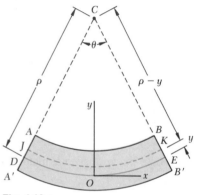

Fig. 4.10 a

For members made of a material which follows Hooke's law [Sec. 4.5], we found that the *normal stress* σ_x *varies linearly* with the distance from the neutral axis (Fig. 4.11). Denoting by σ_m the maximum stress we wrote

$$\sigma_x = -\frac{y}{c}\sigma_m \tag{4.12}$$

where c is the largest distance from the neutral axis to a point in the section.

Normal stress in elastic range

By setting the sum of the elementary forces, $\sigma_x \, dA$, equal to zero, we proved that the *neutral axis passes through the centroid* of the cross section of a member in pure bending. Then by setting the sum of the moments of the elementary forces equal to the bending moment, we derived the *elastic flexure formula* for the maximum normal stress

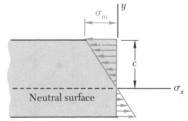

Fig. 4.11

Elastic flexure formula

$$\sigma_m = \frac{Mc}{I} \tag{4.15}$$

where I is the moment of inertia of the cross section with respect to the neutral axis. We also obtained the normal stress at any distance y from the neutral axis:

$$\sigma_x = -\frac{My}{I} \tag{4.16}$$

Elastic section modulus

Noting that I and c depend only on the geometry of the cross section, we introduced the *elastic section modulus*

$$S = \frac{I}{c} \qquad (4.17)$$

and then used the section modulus to write an alternative expression for the maximum normal stress:

$$\sigma_m = \frac{M}{S} \qquad (4.18)$$

Curvature of member

Recalling that the curvature of a member is the reciprocal of its radius of curvature, we expressed the *curvature* of the member as

$$\frac{1}{\rho} = \frac{M}{EI} \qquad (4.21)$$

Anticlastic curvature

In Sec. 4.6, we completed our study of the bending of homogeneous members possessing a plane of symmetry by noting that deformations occur in the plane of a transverse cross section and result in *anticlastic curvature* of the members.

Members made of several materials

Next we considered the bending of members made of several materials with *different moduli of elasticity* [Sec. 4.7]. While transverse sections remain plane, we found that, in general, the *neutral axis does not pass through the centroid* of the composite cross section (Fig. 4.21). Using the ratio of the moduli of elasticity of the materials, we obtained a

Transformed section

transformed section corresponding to an equivalent member made entirely of one material. We then used the methods previously developed to determine the stresses in this equivalent homogeneous member (Fig. 4.23) and then again used the ratio of the moduli of elasticity to determine the stresses in the composite beam [Sample Probs. 4.3 and 4.4].

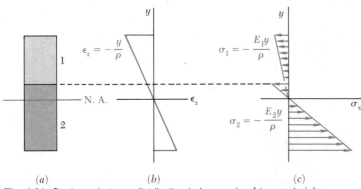

(a)   *(b)*   *(c)*

Fig. 4.21 Strain and stress distribution in bar made of two materials.

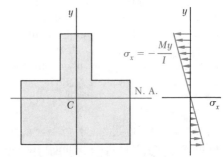

Fig. 4.23 Distribution of stresses in transformed section.

In Sec. 4.8, *stress concentrations* which occur in members in pure bending were discussed and charts giving stress concentration factors for flat bars with fillets and grooves were presented in Figs. 4.27 and 4.28 on page 208.

We next investigated members made of materials which do not follow Hooke's law [Sec. 4.9]. A rectangular beam made of an *elastoplastic material* (Fig. 4.35) was analyzed as the magnitude of the bending moment was increased. The *maximum elastic moment* M_Y occurred when yielding was initiated in the beam (Fig. 4.36). As the bending moment was further increased, plastic zones developed and the size of the elastic core of the member decreased [Sec. 4.10]. Finally the beam became fully plastic and we obtained the maximum or *plastic moment* M_p. In Sec. 4.12, we found that *permanent deformations* and *residual stresses* remain in a member after the loads which caused yielding have been removed.

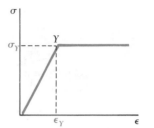

Fig. 4.35 Elastoplastic material

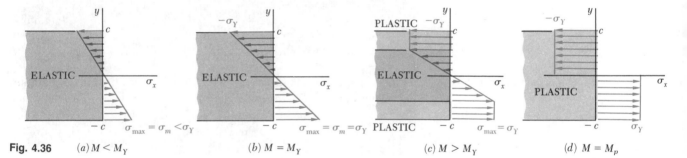

Fig. 4.36 (a) $M < M_Y$ (b) $M = M_Y$ (c) $M > M_Y$ (d) $M = M_p$

In Sec. 4.13, we studied the stresses in members loaded *eccentrically in a plane of symmetry*. Our analysis made use of methods developed earlier. We replaced the *eccentric load* by a force-couple system located at the centroid of the cross section (Fig. 4.42b) and then superposed stresses due to the centric load and the bending couple (Fig. 4.45):

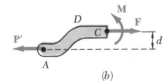

Fig. 4.42

$$\sigma_x = \frac{P}{A} - \frac{My}{I} \qquad (4.50)$$

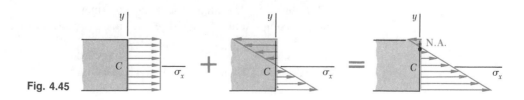

Fig. 4.45

Unsymmetric bending

The bending of members of *unsymmetric cross section* was considered next [Sec. 4.14]. We found that the flexure formula may be used, provided that the couple vector **M** is directed along one of the principal centroidal axes of the cross section. When necessary we resolved **M** into

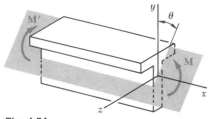

Fig. 4.54

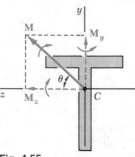

Fig. 4.55

components along the principal axes and superposed the stresses due to the component couples (Figs. 4.54 and 4.55).

$$\sigma_x = -\frac{M_z y}{I_z} + \frac{M_y z}{I_y} \qquad (4.55)$$

Orientation of neutral axis

For the couple **M** shown in Fig. 4.59, we determined the orientation of the neutral axis by writing

$$\tan \phi = \frac{I_z}{I_y} \tan \theta \qquad (4.57)$$

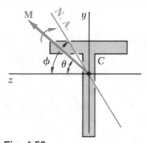

Fig. 4.59

General eccentric axial loading

The general case of *eccentric axial loading* was considered in Sec. 4.15, where we again replaced the load by a force-couple system located at the centroid. We then superposed the stresses due to the centric load and two component couples directed along the principal axes:

$$\sigma_x = \frac{P}{A} - \frac{M_z y}{I_z} + \frac{M_y z}{I_y} \qquad (4.58)$$

The chapter concluded with the analysis of stresses in *curved members* (Fig. 4.70a). While transverse sections remain plane when the member is subjected to bending, we found that the *stresses do not vary linearly* and the neutral surface does not pass through the centroid of the section. The distance R from the center of curvature of the member to the neutral surface was found to be

$$R = \frac{A}{\int \dfrac{dA}{r}} \quad (4.66)$$

where A is the area of the cross section. The normal stress at a distance y from the neutral surface was expressed as

$$\sigma_x = -\frac{My}{Ae(R - y)} \quad (4.70)$$

where M is the bending moment and e the distance from the centroid of the section to the neutral surface.

Curved members

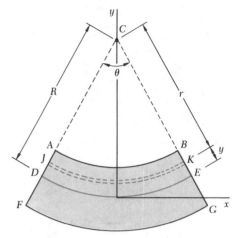

Fig. 4.70   (a)

REVIEW PROBLEMS

4.199 Knowing that the allowable stress for the beam shown is 80 MPa, determine the allowable bending moment M when (a) $D = 150$ mm, (b) $D = 180$ mm, (c) $D = 240$ mm.

4.200 Two steel rods are welded to a 2.5-mm-thick steel plate to form the machine element shown. Knowing that $Q = 20$ kN, determine the largest tensile stress in the element. (*Given:* $I_x = 158 \times 10^{-9}$ m^4.)

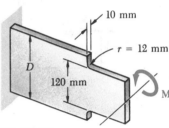

Fig. P4.199

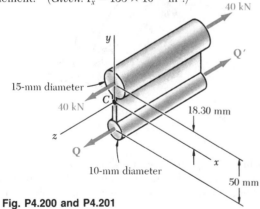

Fig. P4.200 and P4.201

4.201 Two steel rods are welded to a 2.5-mm-thick steel plate to form the machine element shown. Determine the range of values of the load Q for which the tensile stress in the element does not exceed 140 MPa. (*Given:* $I_x = 158 \times 10^{-9}$ m^4.)

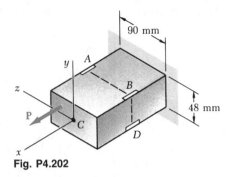

Fig. P4.202

4.202 A single force **P** is applied to the steel bar shown. Gages located at A, B, and D indicate the following strains:

$$\epsilon_A = 800\ \mu \qquad \epsilon_B = 400\ \mu \qquad \epsilon_D = 200\ \mu$$

Knowing that $E = 200$ GPa, determine (a) the magnitude of **P**, (b) the line of action of **P**.

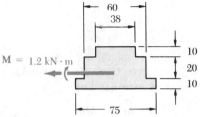

Fig. P4.203   Dimensions in mm

4.203 A couple **M** acts in a vertical plane and is applied to a beam having the cross section shown. Knowing that $E = 200$ GPa, determine (a) the maximum tensile and compressive stresses, (b) the radius of curvature of the beam.

4.204 The couple **M** is applied to a beam of the cross section shown in a plane forming an angle $\beta = 30°$ with the vertical. Determine (a) the angle that the neutral axis forms with the z axis, (b) the maximum tensile and compressive stresses in the beam. (*Given*: $I_y = 5.078 \times 10^{-6}$ m^4, $I_z = 6.660 \times 10^{-6}$ m^4.)

4.205 For the split ring and loading shown, determine the stress at (a) point A, (b) point B.

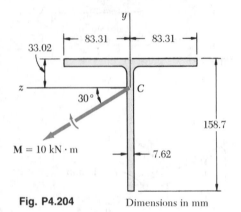

Fig. P4.204   Dimensions in mm

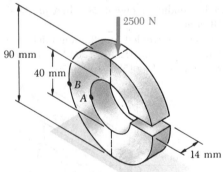

Fig. P4.205

4.206 Two brass strips are securely bonded to an aluminum extrusion as shown. Using the data given below, determine the largest permissible bending moment when the composite member is bent about a horizontal axis.

	Aluminum	Brass
Modulus of elasticity	70 GPa	105 GPa
Allowable stress	100 MPa	160 MPa

4.207 For the composite member of Prob. 4.206, determine the largest permissible bending moment when the member is bent about a vertical axis.

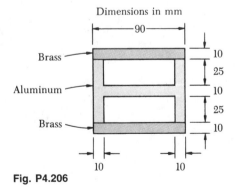

Fig. P4.206

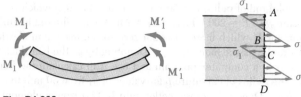

Fig. P4.208

4.208 Two thin strips of the same material and the same cross section are bent by couples of the same magnitude and glued together. After the two surfaces in contact have been securely bonded, the couples are removed. Denoting by σ_1 the maximum stress and by ρ_1 the radius of curvature of each strip while the couples were applied, determine (a) the final stresses at points A, B, C, and D, (b) the final radius of curvature.

4.209 The four bars shown have the same cross-sectional area. For the given loadings, show that (a) the maximum compressive stresses are in the ratio 4:5:7:9, (b) the maximum tensile stresses are in the ratio 2:3:5:3. (*Note:* The cross section of the triangular bar is an equilateral triangle.)

4.210 A strain gage located at point A indicates a strain of $-400\ \mu$ when a couple **M** is applied to the steel beam shown. Knowing that **M** acts in a vertical plane and using $E = 200$ GPa, determine (a) the moment M of the applied couple, (b) the corresponding radius of curvature of the beam.

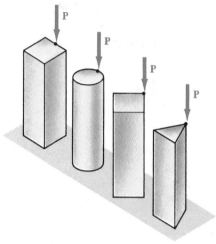

Fig. P4.209

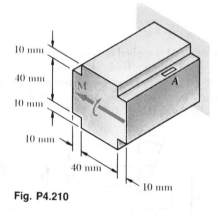

Fig. P4.210

The following problems are designed to be solved with a computer

4.C1 The 150×350 mm timber beam is strengthened by bolting to it a steel reinforcing plate. Using the data given below, write a computer program and use it to calculate the largest permissible bending moment which may be applied to the beam for values of w from 90 to 150 at 10-mm intervals and with values of t equal to 10, 12.5, and 15 mm.

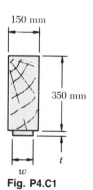

Fig. P4.C1

	Wood	**Steel**
Modulus of elasticity	14 GPa	200 GPa
Allowable stress	14 MPa	125 MPa

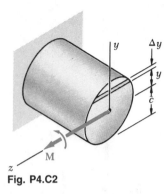

Fig. P4.C2

4.C2 A solid rod of radius $c = 50$ mm is made of a steel which is assumed to be elastoplastic with $E = 200$ GPa and $\sigma_Y = 240$ MPa. The rod is subjected to a bending moment M which increases from zero to the maximum elastic moment M_Y and then to the plastic moment M_p. Denoting by y_Y the half-thickness of the elastic core, write a computer program and use it to calculate the bending moment M and the radius of curvature ρ for values of y_Y from 50 mm to 0 at 10-mm intervals. (*Hint:* Divide the cross section into 100 horizontal elements of 1-mm height.)

4.C3 The couple **M** is applied to a beam of the cross section shown. Write a computer program which, for loads expressed in any consistent set of units, can be used to calculate the maximum tensile and compressive stresses in the beam. Use this program to solve Probs. 4.8, 4.10, and 4.203a.

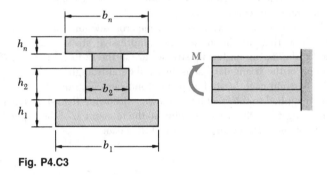

Fig. P4.C3

4.C4 A 1.5-kN · m couple **M** is applied to a beam of the cross section shown in a plane forming an angle β with the vertical. Knowing that the y and z axes are principal centroidal axes df the cross section, write a computer program and use it to calculate the stress at A, B, D, and E for values of β from 0 to 180° at 10° intervals. (*Given:* $\bar{z} = 33.02$ mm, $I_y = 1.145 \times 10^{-6}$ m^4, $I_z = 0.7 \times 10^{-6}$ m^4.)

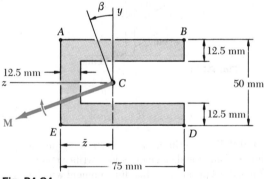

Fig. P4.C4

C H A P T E R   F I V E

TRANSVERSE LOADING

5.1. INTRODUCTION

In this chapter, we shall analyze both the normal and shearing stresses in prismatic members subjected to *transverse loads*. Assuming that the distribution of normal stresses due to bending is not affected by the presence of shear, we shall determine the shearing forces acting on horizontal sections of a beam (Secs. 5.3 and 5.4).

Next we shall determine the *shear flow* and the *horizontal shearing stresses* in beams. Recalling from Chap. 1 that at a given point horizontal and vertical shearing stresses are equal, we shall also determine the *shearing stresses on transverse sections* (Sec. 5.5).

In Secs. 5.6 and 5.7, we shall analyze the magnitude and distribution of shearing stresses in beams of rectangular cross section and in beams formed of rolled-steel shapes.

Considering the shear on an arbitrary longitudinal cut (Sec. 5.8), we shall determine the shear flow and shearing stresses along the cut. This will permit us in Sec. 5.9 to determine the shearing stresses at any point in symmetric thin-walled members. The effect of plastic deformations on the magnitude and distribution of shearing stresses will be discussed in Sec. 5.10.

In the next section of the chapter (Sec. 5.11), we shall combine our ability to analyze the stresses due to axial, torsional, bending, and transverse loadings to determine the normal and shearing stresses at a given point of a slender member under general loading conditions.

In the final section of the chapter (Sec. 5.12), we shall define and locate the *shear center* of thin-walled members. We shall then determine the shearing stresses in unsymmetrically loaded thin-walled members.

5.2. TRANSVERSE LOADING OF PRISMATIC MEMBERS

One of the most common examples of a transverse loading occurs when a horizontal member, referred to as a *beam*, is subjected to vertical loads. The loads may be *concentrated* (Fig. 5.1a) or *distributed* (Fig. 5.1b), or may consist of a combination of both.

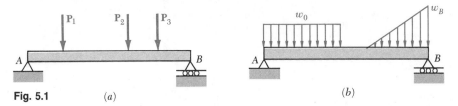

Fig. 5.1   (a)   (b)

Let us first consider a cantilever beam AB, that is, a beam with a fixed end B, subjected at its free end A to a single upward force $\mathbf{P}$ (Fig. 5.2a). We shall assume that the beam possesses a vertical, longitudinal plane of symmetry, and that the force $\mathbf{P}$ is applied in that plane. Passing a section through the beam at C, and considering the free-body diagram of portion AC (Fig. 5.2b), we observe that the internal forces exerted on AC must be equivalent to a shearing force $\mathbf{V}$ of magnitude $V = P$, and to a bending couple $\mathbf{M}$ of moment $M = Px$, where x is the distance from C to the free end A. Recalling the sign convention specified in Sec. 4.2, we note that the bending moment M is positive. Introducing now an equally arbitrary sign convention for the shear V, we assign to V a positive sign when the shearing force $\mathbf{V}$ is directed as shown in Fig. 5.2b, i.e., when the shearing force exerted on the portion of beam located to the left of the section is directed downward.

Six equations may be written to express that the elementary normal and shearing forces exerted on the section are equivalent to the shearing force $\mathbf{V}$ and the bending couple $\mathbf{M}$ that we have just determined (Fig. 5.3). Three of these equations involve only the normal forces $\sigma_x \, dA$ and have already been obtained in Sec. 4.3; they are Eqs. (4.1), (4.2), and (4.3), which express that the sum of the normal forces is zero and that the

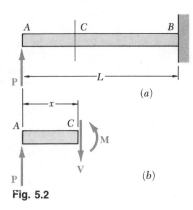

Fig. 5.2

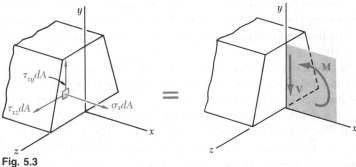

Fig. 5.3

sum of their moments about the y and z axes are equal to zero and M, respectively. We should note, however, that in the case of a transverse load, the bending moment M varies with the location of the section considered. Three more equations, relating to the shearing forces $\tau_{xy}\,dA$ and $\tau_{xz}\,dA$, may now be written. One of these involves the moments of the shearing forces about the x axis, and may be dismissed as trivial, in view of the symmetry of the beam with respect to the xy plane. The other two involve the y and z components of the elementary forces and are

$$y \text{ components:} \qquad \int \tau_{xy}\,dA = -V \qquad (5.1)$$

$$z \text{ components:} \qquad \int \tau_{xz}\,dA = 0 \qquad (5.2)$$

The first of these equations shows that vertical shearing stresses *must exist* in any transverse section of the beam considered here, and that they will be negative, i.e., directed downward. The second equation indicates that the *average* horizontal shearing stress in any section is zero. However, as we shall see later, this does not mean that the shearing stress τ_{xz} is necessarily zero everywhere.

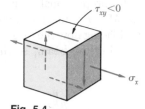

Fig. 5.4

We shall now consider a small cubic element located in the vertical plane of symmetry of the beam (where we know that τ_{xz} must be zero) and examine the stresses exerted on its faces (Fig. 5.4). As we have just seen, a normal stress σ_x and a shearing stress τ_{xy} are exerted on each of the two faces perpendicular to the x axis. But we know from Chaps. 1 and 3 that, when shearing stresses τ_{xy} are exerted on the vertical faces of an element, equal stresses must be exerted on the horizontal faces of the same element. We thus conclude that *longitudinal shearing stresses* must exist in any member subjected to a transverse loading. This may be verified by considering a cantilever beam made of separate planks clamped together at one end (Fig. 5.5a). When a transverse load **P** is applied to the free end of this composite beam, the planks are observed to slide with respect to each other (Fig. 5.5b). While sliding does not actually take place in a beam made of a homogeneous and cohesive material, the tendency to slide does exist, showing that stresses occur on horizontal longitudinal planes as well as on vertical transverse planes. On the other hand, if the same composite beam is subjected to a bending couple **M** at its free end (Fig. 5.5c), the various planks will bend into concentric arcs of circle and will not slide with respect to each other, thus verifying the fact that shear does not occur in a beam subjected to pure bending (cf. Sec. 4.4).

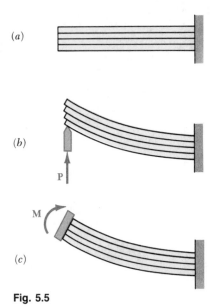

(a)

(b)

(c)

Fig. 5.5

5.3. BASIC ASSUMPTION REGARDING THE DISTRIBUTION OF THE NORMAL STRESSES

(a)

(b)

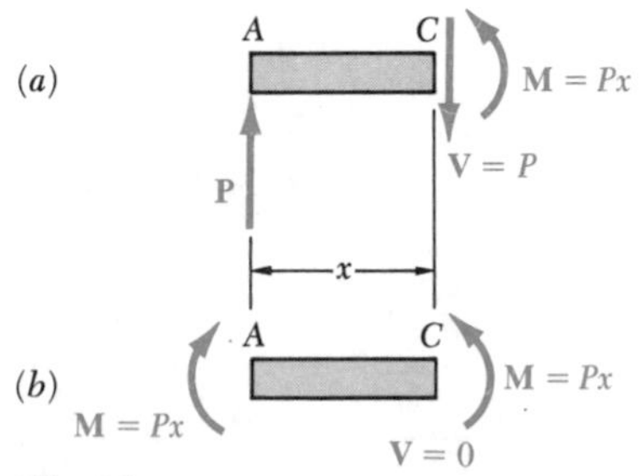

Fig. 5.6

In order to determine the distribution of stresses due to an axial loading, to torsion, or to pure bending, we found it convenient in the preceding chapters to first analyze the deformations corresponding to each of these various loading conditions. Since the deformations caused by a transverse loading are much more complex, we shall use here a different approach. *We shall assume that the distribution of the normal stresses in a given cross section is not affected by the deformations caused by the shearing stresses.* Under this assumption, which will be verified in Sec. 5.7, the distribution of the *normal stresses* in a given cross section must be the same when the beam is subjected to a transverse load **P** (Fig. 5.6*a*) or when it is subjected to a bending couple **M** of moment $M = Px$ (Fig. 5.6*b*). Indeed, both loadings create the same bending moment in the given section and, while they result in different values of the shear in that section ($V = P$ in one case, $V = 0$ in the other), this should not affect the normal stresses, in view of the assumption we have made. As long as the stresses do not exceed the proportional limit, we may therefore use Eq. (4.16) of Chap. 4, with $M = Px$, to determine the normal stresses under a transverse loading. Placing now the origin of the coordinate axes at the centroid of the free end of the beam, so that x will measure the

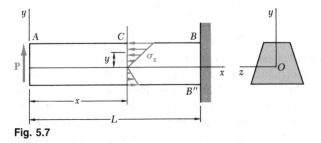

Fig. 5.7

distance of a given point from the load **P**, and y its elevation above the neutral surface (Fig. 5.7), we write

$$\sigma_x = -\frac{My}{I} = -\frac{Pxy}{I} \tag{5.3}$$

We note that the distribution of normal stresses in a given section ($x = $ constant) is linear, as was the case in pure bending. The stresses, however, are also proportional to the distance x from the load to the section considered. Thus, the maximum compressive stress in the beam will occur at B, where $x = L$, and the maximum tensile stress at B''.

Having determined the distribution of the normal stresses in the beam, we shall now proceed to determine the distribution of the shearing stresses τ_{xy} through the use of a simple equilibrium condition.

5.4. DETERMINATION OF THE SHEAR ON A HORIZONTAL PLANE

We recall from Fig. 5.4 that τ_{xy} represents both the vertical component of the shearing stress on a section perpendicular to the axis of the beam, and the longitudinal component of the shearing stress on a horizontal section. Our approach for determining τ_{xy} will be to analyze the shearing forces exerted on a *horizontal* section of the beam.

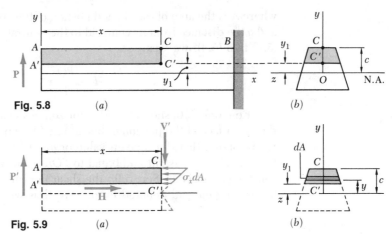

Fig. 5.8   (a)   (b)

Fig. 5.9   (a)   (b)

Considering again a cantilever beam AB subjected to a force **P** at its free end (Fig. 5.8), we detach from the beam the portion $ACC'A'$ obtained by passing a horizontal section $A'C'$ at a distance y_1 above the neutral surface, and a vertical section CC' at a distance x from the free end. The forces acting on the free body $ACC'A'$ are shown in Fig. 5.9. They include the portion **P'** of the load which is applied to the free body, either directly or through interaction between elements of the two portions of beam, the resultant **V'** of the shearing forces exerted on section CC', the normal forces $\sigma_x\, dA$ acting on the same section, and the resultant **H** of the horizontal shearing forces exerted on the lower face of the free body. Recalling Eq. (5.3), we have

$$\sigma_x\, dA = -\frac{Pxy}{I}\, dA$$

We now write the equation $\Sigma F_x = 0$ for the free body $ACC'A'$:

$$\Sigma F_x = 0: \qquad H - \int \frac{Pxy}{I}\, dA = 0$$

Solving this equation for H, and observing that x is constant over the cross section, we write

$$H = \frac{Px}{I} \int_{y=y_1}^{y=c} y\, dA \qquad (5.4)$$

We note that the integral obtained represents the *first moment* with respect to the neutral axis of that part of the area of the cross section which is located above the line $y = y_1$. Denoting this moment by Q, we write

$$Q = \int_{y=y_1}^{y=c} y \, dA \tag{5.5}$$

and observe† that

$$Q = A\bar{y} \tag{5.5'}$$

where A is the area of the shaded portion of the cross section in Fig. 5.9*b* and $\bar{y}$ the distance from its centroid to the neutral axis. Substituting from (5.5) into (5.4), we write

$$H = \frac{PQ}{I}x \tag{5.6}$$

Equation (5.6) shows that the horizontal shearing force **H** exerted on the lower face of the portion of beam $ACC'A'$ is proportional to its length x. It follows that, for a given value of y_1, *the horizontal shear per unit length, H/x, is constant* and equal to PQ/I. For a reason which will become apparent later (Sec. 5.9), the shear per unit length is referred to as the *shear flow* and denoted by q. We thus write

$$q = \frac{PQ}{I} \tag{5.7}$$

The same result would have been obtained if we had used as a free body the lower portion $A'C'C''A''$ of the beam, rather than the upper portion $ACC'A'$ (Fig. 5.10), since the shearing forces exerted by the two portions of beam on each other are equal and opposite. This leads us to observe that the first moment Q of the portion of the cross section located below the line $y = y_1$ (Fig. 5.10*b*) is equal in magnitude, and opposite in sign, to the first moment of the area located above that line (Fig. 5.8*b*). Indeed, the sum of these two moments is equal to the moment of the area of the entire cross section with respect to its centroidal axis and, thus, must be zero. This property may sometimes be used to simplify the computation of Q. We also note that Q is maximum for $y_1 = 0$, since the elements of the cross section located above the neutral axis contribute positively to the integral (5.5) which defines Q, while the elements located below that axis contribute negatively.

The application of the method we have described for the determination of the shear flow q in a beam is not limited to cantilever beams or to single concentrated loads. Consider, for instance, a simply supported beam subjected to several concentrated and distributed loads (Fig. 5.11*a*). We may use formula (5.7) to determine the shear flow produced at a given point C' by each of the concentrated loads applied to the portion of beam to the left of C', including the reaction at A. On the other hand, we may, by Saint-Venant's principle, replace the distributed load

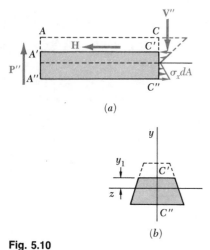

(a)

(b)

Fig. 5.10

---

† See Appendix A.

by an equivalent concentrated load of magnitude wa and compute the corresponding shear flow at C'. Adding the various values of q obtained in this way and applying the principle of superposition, we would obtain the shear flow at C' due to the given loading. However, since the relationship between P and q is linear, the same result may be obtained more directly by substituting for P in Eq. (5.7) the sum of the forces exerted on the portion of beam to the left of the section through C'. Recalling that this sum is equal to the shear V in the section (Fig. 5.11b), we write

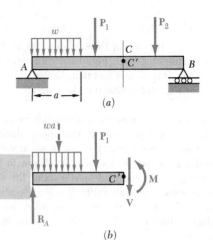

$$q = \frac{VQ}{I} \qquad (5.8)$$

We recall that Q is the first moment about the neutral axis of the portion of the cross section which is located either above or below point C', where the shear flow is to be computed, and I the moment of inertia of the *entire* cross-sectional area about the neutral axis.

Fig. 5.11

We may observe from Eq. (5.8) that q remains constant between two successive loads, since V itself remains constant. We also check that, in the case of a beam in pure bending, i.e., when the beam is subjected to equal and opposite couples, both the vertical shear, V, and the horizontal shear per unit length, q, are zero.†

---

† Equation (5.8) may be obtained directly for the loading condition shown in Fig. 5.11a by considering the equilibrium of the portion of beam located above the horizontal plane $C'D'$ and between the vertical planes CC' and DD' (Fig. 5.12a). Denoting by σ_C and σ_D the normal stress in sections CC' and DD', respectively, and by ΔH the shearing force exerted on the lower face of the free body (Fig. 5.12b), we write

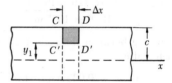

(a)

$$\Sigma F_x = 0: \qquad \Delta H - \int (\sigma_D - \sigma_C)dA = 0$$

Solving this equation for ΔH, and using Eq. (4.16) to express the normal stresses in terms of the bending moments at C and D, we have

$$\Delta H = \frac{M_D - M_C}{I} \int_{y=y_1}^{y=c} y\, dA$$

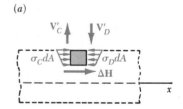

(b)

Fig. 5.12

Recalling Eq. (5.5), and denoting by ΔM the difference $M_D - M_C$ between the values of the bending moment at C and D, we write

$$\Delta H = \frac{(\Delta M)Q}{I}$$

or, dividing both members by the length Δx of the beam element, and letting Δx approach zero,

$$\frac{dH}{dx} = \frac{dM}{dx}\frac{Q}{I}$$

But dH/dx represents the horizontal shear per unit length, or shear flow, q, and, as we shall see in Chap. 7, the derivative dM/dx of the bending moment is equal to the shear V. We have, therefore,

$$q = \frac{VQ}{I} \qquad (5.8)$$

Example 5.01

A beam is made of three planks, 20 by 100 mm in cross section, nailed together (Fig. 5.13). Knowing that the spacing between nails is 25 mm and that the vertical shear in the beam is $V = 500$ N, determine the shearing force in each nail.

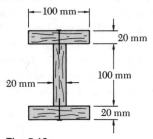

Fig. 5.13

We first determine the horizontal force per unit length, q, exerted on the lower face of the upper plank. We use Eq. (5.8), where Q represents the first moment with respect to the neutral axis of the shaded area A' shown in Fig. 5.14a, and where I is the moment of inertia about the same axis of the entire cross-sectional area (Fig. 5.14b). Recalling that the first moment of an area with respect to a given axis is equal to the product of the area and of the distance from its centroid to the axis,† we have

$$Q = A'\overline{y} = (0.020 \text{ m} \times 0.100 \text{ m})(0.060 \text{ m}) = 120 \times 10^{-6} \text{ m}^3$$
$$I = \tfrac{1}{12}(0.020 \text{ m})(0.100 \text{ m})^3$$
$$+2[\tfrac{1}{12}(0.100 \text{ m})(0.020 \text{ m})^3$$
$$+(0.020 \text{ m} \times 0.100 \text{ m})(0.060 \text{ m})^2]$$
$$= 1.667 \times 10^{-6} + 2(0.0667 + 7.2)10^{-6} = 16.20 \times 10^{-6} \text{ m}^4$$

Substituting into Eq. (5.8), we write

$$q = \frac{VQ}{I} = \frac{(500 \text{ N})(120 \times 10^{-6} \text{ m}^3)}{16.20 \times 10^{-6} \text{ m}^4} = 3704 \text{ N/m}$$

Since the spacing between the nails is 25 mm, the shearing force in each nail is

$$F = (0.025 \text{ m})q = (0.025 \text{ m})(3704 \text{ N/m}) = 92.6 \text{ N}$$

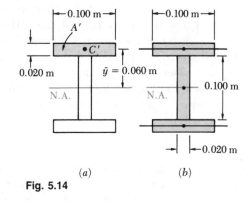

(a)   (b)

Fig. 5.14

5.5. DETERMINATION OF THE SHEARING STRESSES τ_{xy} IN A BEAM

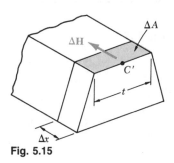

Fig. 5.15

Let us consider again a beam with a vertical plane of symmetry, subjected to various concentrated or distributed loads applied in that plane. We saw in the preceding section that, if V denotes the vertical shear in a given cross section, the horizontal shear flow q at a point C' of that cross section is

$$q = \frac{VQ}{I} \qquad (5.8)$$

where Q is the first moment defined by Eq. (5.5), and I the moment of inertia of the cross-sectional area about the neutral axis. The shearing

† See Appendix A.

force ΔH exerted on a portion of length Δx of the horizontal cut through C' (Fig. 5.15) is thus

$$\Delta H = q \, \Delta x = \frac{VQ}{I} \Delta x \qquad (5.9)$$

Dividing (5.9) by the area $\Delta A = t \, \Delta x$, where t is the width of the cut, we obtain the average value of the shearing stress τ_{yx}:

$$\tau_{ave} = \frac{\Delta H}{\Delta A} = \frac{VQ}{I} \frac{\Delta x}{t \, \Delta x}$$

or

$$\tau_{ave} = \frac{VQ}{It} \qquad (5.10)$$

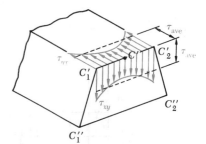

Fig. 5.16

We note that, since the shearing stresses τ_{xy} and τ_{yx} exerted respectively on a transverse and a horizontal plane through C' are equal, the expression obtained also represents the average value of τ_{xy} along the line $C_1'C_2'$ (Fig. 5.16).

We observe that $\tau_{yx} = 0$ on the upper and lower faces of the beam, since no forces are exerted on these faces. It follows that $\tau_{xy} = 0$ along the upper and lower edges of the transverse section (Fig. 5.17). We also note that, while Q is maximum for $y = 0$ (see Sec. 5.4), we *cannot* conclude that τ_{ave} will be maximum along the neutral axis, since τ_{ave} depends upon the width t of the section as well as upon Q.

As long as the width of the beam cross section remains small, compared to its depth, the shearing stress varies only slightly along the line $C_1'C_2'$, and Eq. (5.10) may be used to compute τ_{xy} at any point along $C_1'C_2'$. Actually, τ_{xy} is larger at points C_1' and C_2' than at C', but the theory of elasticity shows † that, for a beam of rectangular section of width b and depth h, and as long as $b \le \frac{1}{4}h$, the value of the shearing stress at points C_1 and C_2 (Fig. 5.18) does not exceed by more than 0.8% the average value of the stress computed along the neutral axis.‡

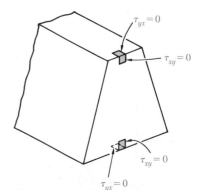

Fig. 5.17

† See S. P. Timoshenko and J. N. Goodier, *Theory of Elasticity*, McGraw Hill, New York, 3d ed., 1970, sec. 124.

‡ On the other hand, for large values of b/h, the value τ_{max} of the stress at C_1 and C_2 may be many times larger than the average value τ_{ave} computed along the neutral axis, as we may see from the following table:

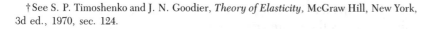

b/h	0.25	0.5	1	2	4	6	10	20	50
τ_{max}/τ_{ave}	1.008	1.033	1.126	1.396	1.988	2.582	3.770	6.740	15.65
τ_{min}/τ_{ave}	0.996	0.983	0.940	0.856	0.805	0.800	0.800	0.800	0.800

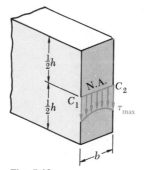

Fig. 5.18

5.6. SHEARING STRESSES τ_{xy} IN COMMON TYPES OF BEAMS

We saw in the preceding section that, for a *narrow rectangular beam*, i.e., for a beam of rectangular section of width b and depth h with $b \leq \frac{1}{4}h$, the variation of the shearing stress τ_{xy} across the width of the beam is less than 0.8% of τ_{ave}. We may, therefore, use Eq. (5.10) in practical applications to determine the shearing stress at any point of the cross section of a narrow rectangular beam and write

$$\tau_{xy} = \frac{VQ}{It} \tag{5.11}$$

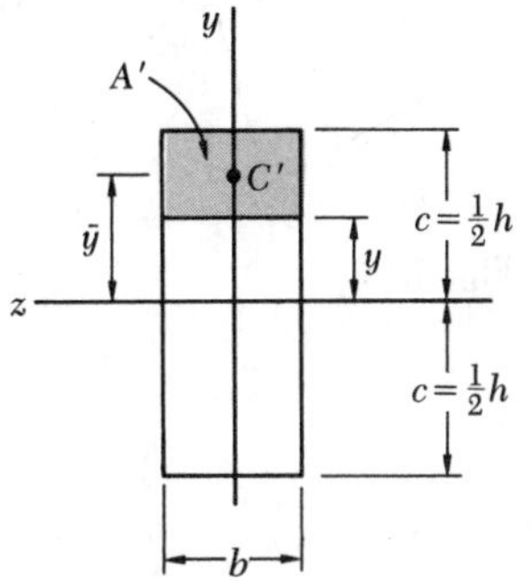

where t is equal to the width b of the beam, and where Q is the first moment with respect to the neutral axis of the shaded area A' (Fig. 5.19).

Observing that the distance from the neutral axis to the centroid C' of A' is $\bar{y} = \frac{1}{2}(c + y)$, and using Eq. (5.5′), we write

$$Q = A'\bar{y} = b(c - y)\tfrac{1}{2}(c + y) = \tfrac{1}{2}b(c^2 - y^2) \tag{5.12}$$

Recalling, on the other hand, that $I = bh^3/12 = \frac{2}{3}bc^3$, we have

$$\tau_{xy} = \frac{VQ}{Ib} = \frac{3}{4}\frac{c^2 - y^2}{bc^3}V$$

or, noting that the cross-sectional area of the beam is $A = 2bc$,

$$\tau_{xy} = \frac{3}{2}\frac{V}{A}\left(1 - \frac{y^2}{c^2}\right) \tag{5.13}$$

Equation (5.13) shows that the distribution of shearing stresses in a transverse section of a rectangular beam is *parabolic* (Fig. 5.20). As we have already observed in the preceding section, the shearing stresses are zero at the top and bottom of the cross section ($y = \pm c$). Making $y = 0$ in Eq. (5.13), we obtain the value of the maximum shearing stress in a given section of a *narrow rectangular beam*:

$$\tau_{max} = \frac{3}{2}\frac{V}{A} \tag{5.14}$$

The relation obtained shows that the maximum value of the shearing stress in a beam of rectangular cross section is 50% larger than the value V/A which would be obtained by wrongly assuming a uniform stress distribution across the entire cross section.

In the case of an *American standard beam* (S-beam) or a *wide-flange beam* (W-beam), Eq. (5.10) may be used to determine the average value of the shearing stress τ_{xy} over a section aa' or bb' of the transverse cross section of the beam (Figs. 5.21a and b). We write

$$\tau_{ave} = \frac{VQ}{It} \tag{5.10}$$

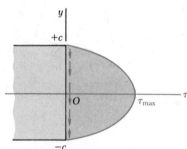

Fig. 5.19

Fig. 5.20

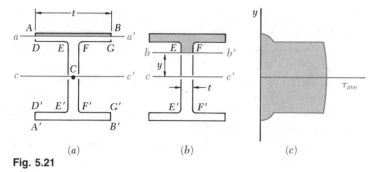

Fig. 5.21

where V is the vertical shear, t the width of the section at the elevation considered, Q the first moment of the shaded area with respect to the neutral axis cc', and I the moment of inertia of the entire cross-sectional area about cc'. Plotting τ_{ave} against the vertical distance y, we obtain the curve shown in Fig. 5.21c. We note the discontinuities existing in this curve, which reflect the difference between the values of t corresponding respectively to the flanges $ABGD$ and $A'B'G'D'$ and to the web $EFF'E'$.

In the case of the web, the shearing stress τ_{xy} varies only very slightly across the section bb', and may be assumed equal to its average value τ_{ave}. This is not true, however, for the flanges. For example, considering the horizontal line $DEFG$, we note that τ_{xy} is zero between D and E and between F and G, since these two segments are part of the free surface of the beam. On the other hand the value of τ_{xy} between E and F may be obtained by making $t = EF$ in Eq. (5.10). In practice, one usually assumes that the entire shear load is carried by the web, and that a good approximation of the maximum value of the shearing stress in the cross section may be obtained by dividing V by the cross-sectional area of the web.

We should note, however, that while the vertical component τ_{xy} of the shearing stress in the flanges may be neglected, its horizontal component τ_{xz} has a significant value, which will be determined in Sec. 5.9.

*5.7. FURTHER DISCUSSION OF THE DISTRIBUTION OF STRESSES IN A NARROW RECTANGULAR BEAM

Consider a narrow cantilever beam of rectangular cross section of width b and depth h subjected to a load $\mathbf{P}$ at its free end (Fig. 5.22). Since the shear V in the beam is constant and equal in magnitude to the load $\mathbf{P}$, Eq. (5.13) yields

$$\tau_{xy} = \frac{3}{2}\frac{P}{A}\left(1 - \frac{y^2}{c^2}\right) \qquad (5.15)$$

We note from Eq. (5.15) that the shearing stresses depend only upon the distance y from the neutral surface. They are independent, therefore, of

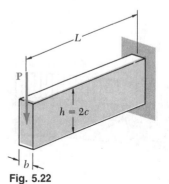

Fig. 5.22

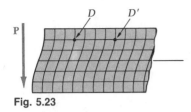

Fig. 5.23

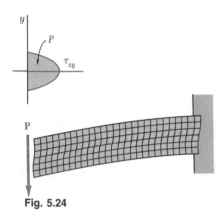

Fig. 5.24

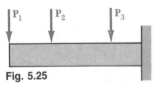

Fig. 5.25

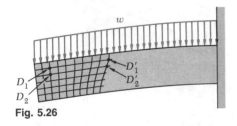

Fig. 5.26

the distance from the point of application of the load; it follows that all elements located at the same distance from the neutral surface undergo the same shear deformation (Fig. 5.23). While plane sections do *not* remain plane, the distance between two corresponding points D and D' located in different sections remains the same. This indicates that the normal strains ϵ_x, and thus the normal stresses σ_x, are unaffected by the shearing stresses, and that the assumption made in Sec. 5.3 is justified for the loading condition of Fig. 5.22.

We conclude that our analysis of the stresses in a cantilever beam of rectangular cross section, subjected to a concentrated load **P** at its free end, is valid. The correct values of the shearing stresses in the beam are given by Eq. (5.15), and the normal stresses are obtained from Eq. (5.3). With the load **P** now directed downward, the latter equation yields

$$\sigma_x = +\frac{Pxy}{I} \tag{5.16}$$

The validity of this statement, however, depends upon the end conditions. If Eq. (5.15) is to apply everywhere, then the load P must be distributed parabolically over the free-end section. Moreover, the fixed-end support must be of such a nature that it will allow the type of shear deformation indicated in Fig. 5.23. The resulting model (Fig. 5.24) is highly unlikely to be encountered in practice. However, it follows from Saint-Venant's principle that, for other modes of application of the load and for other types of fixed-end supports, Eqs. (5.15) and (5.16) still provide us with the correct distribution of stresses, except close to either end of the beam.

When a beam of rectangular cross section is subjected to several concentrated loads (Fig. 5.25), the principle of superposition may be used to determine the normal and shearing stresses in sections located between the points of application of the loads. However, since the loads **P**$_2$, **P**$_3$, etc., are applied on the surface of the beam and cannot be assumed to be distributed parabolically throughout the cross section, the results obtained cease to be valid in the immediate vicinity of the points of application of the loads.

When the beam is subjected to a distributed load (Fig. 5.26), the shear varies with the distance from the end of the beam, and so does the shearing stress at a given elevation y. The resulting shear deformations are such that the distance between two corresponding points of different cross sections, such as D_1 and D_1', or D_2 and D_2', will depend upon their elevation. This indicates that the assumption that plane sections remain plane, under which Eqs. (5.15) and (5.16) were derived, must be rejected for the loading condition of Fig. 5.26. The error involved, however, is small for the values of the span-depth ratio encountered in practice.

We should also note that, in portions of the beam located under a concentrated or distributed load, normal stresses σ_y will be exerted on the horizontal faces of a cubic element of material, in addition to the stresses τ_{xy} shown in Fig. 5.4. This will be discussed in Chap. 7.

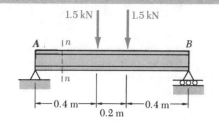

Beam AB is made of three planks glued together and is subjected, in its plane of symmetry, to the loading shown. Knowing that the width of each glued joint is 20 mm, determine the average shearing stress in each joint at section n-n of the beam. The location of the centroid of the section is given in the sketch and the centroidal moment of inertia is known to be $I = 8.63 \times 10^{-6}$ m^4.

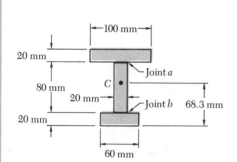

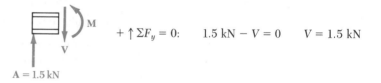

Since the beam and loading are both symmetric with respect to the center of the beam, we have $\mathbf{A} = \mathbf{B} = 1.5$ kN $\uparrow$. Considering the portion of the beam to the left of section n-n as a free body, we write

$$+\uparrow \Sigma F_y = 0: \qquad 1.5 \text{ kN} - V = 0 \qquad V = 1.5 \text{ kN}$$

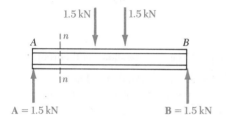

We pass the section a-a through the glued joint and separate the cross-sectional area into two parts. We choose to determine Q by computing the first moment with respect to the neutral axis of the area above section a-a.

$$Q = A\bar{y}_1 = [(0.100 \text{ m})(0.020 \text{ m})](0.0417 \text{ m}) = 83.4 \times 10^{-6} \text{ m}^3$$

Recalling that the width of the glued joint is $t = 0.020$ m, we use Eq. (5.10) to determine the average shearing stress in the joint.

$$\tau_{\text{ave}} = \frac{VQ}{It} = \frac{(1500 \text{ N})(83.4 \times 10^{-6} \text{ m}^3)}{(8.63 \times 10^{-6} \text{ m}^4)(0.020 \text{ m})}$$

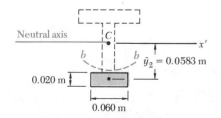

We now pass section b-b and compute Q by using the area below the section.

$$Q = A\bar{y}_2 = [(0.060 \text{ m})(0.020 \text{ m})](0.0583 \text{ m}) = 70.0 \times 10^{-6} \text{ m}^3$$

$$\tau_{\text{ave}} = \frac{VQ}{It} = \frac{(1500 \text{ N})(70.0 \times 10^{-6} \text{ m}^3)}{(8.63 \times 10^{-6} \text{ m}^4)(0.020 \text{ m})}$$

SAMPLE PROBLEM 5.2

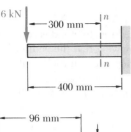

A machine part has a T-shaped cross section and is acted upon in its plane of symmetry by the single force shown. Determine (a) the maximum compressive stress at section n-n, (b) the maximum shearing stress.

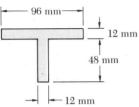

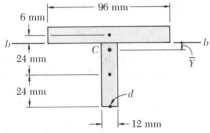

Neutral Axis. The neutral axis passes through the centroid C of the cross section. Using the axis b-b as a reference axis and choosing the positive sense downward, we write

$$\bar{Y} = \frac{\Sigma A \bar{y}}{\Sigma A} = \frac{(96 \text{ min})(12 \text{ mm})(-6 \text{ mm}) + (48 \text{ mm})(12 \text{ mm})(24 \text{ mm})}{(96 \text{ mm})(12 \text{ mm}) + (48 \text{ mm})(12 \text{ mm})}$$

$$= \frac{6912 \text{ mm}^3}{1728 \text{ mm}^2} = 4.00 \text{ mm}$$

Centroidal Moment of Inertia. Using the parallel-axis theorem:

$$I = \tfrac{1}{12}(96 \text{ mm})(12 \text{ mm})^3 + (96 \text{ mm})(12 \text{ mm})(6 \text{ mm} + 4 \text{ mm})^2$$
$$+ \tfrac{1}{12}(12 \text{ mm})(48 \text{ mm})^3 + (12 \text{ mm})(48 \text{ mm})(24 \text{ mm} - 4 \text{ mm})^2$$
$$I = 470 \times 10^3 \text{ mm}^4 = 470 \times 10^{-9} \text{ m}^4$$

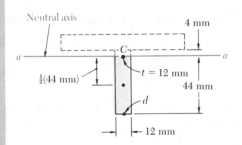

(a) Maximum Compressive Stress. At section n-n the bending moment is $M = (6 \text{ kN})(0.3 \text{ m}) = 1800 \text{ N} \cdot \text{m}$. The maximum compressive stress occurs at point d where c = 48 mm − 4 mm = 44 mm. We write

$$\sigma_m = \frac{Mc}{I} = \frac{(1800 \text{ N} \cdot \text{m})(0.044 \text{ m})}{470 \times 10^{-9} \text{ m}^4} \qquad \sigma_m = 168.5 \text{ MPa} \qquad \blacktriangleleft$$

(b) Maximum Shearing Stress. The maximum value of Q occurs at the neutral axis. Since in this cross section the width t is minimum at the neutral axis, the maximum shearing stress will occur there. Passing section a-a at the neutral axis, we separate the cross-sectional area into two portions. Choosing the area below section a-a, we have

$$Q = [(44 \text{ mm})(12 \text{ mm})] \frac{44 \text{ mm}}{2} = 11.62 \times 10^3 \text{ mm}^3 = 11.62 \times 10^{-6} \text{ m}^3$$

Noting that V = 6 kN and t = 12 mm, we write

Distribution of
Shearing Stresses

$$\tau_m = \frac{VQ}{It} = \frac{(6000 \text{ N})(11.62 \times 10^{-6} \text{ m}^3)}{(470 \times 10^{-9} \text{ m}^4)(0.012 \text{ m})} \qquad \tau_m = 12.36 \text{ MPa} \qquad \blacktriangleleft$$

PROBLEMS

5.1 Three boards, each of 40 mm × 90 mm rectangular cross section, are nailed together to form a beam which is subjected to a vertical shear of 1 kN. Knowing that the spacing between each pair of nails is 60 mm, determine the shearing force in each nail.

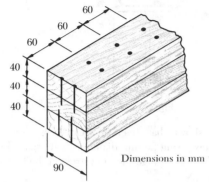

Fig. P5.1

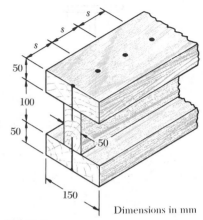

Fig. P5.2

5.2 Three boards, each 50 mm thick, are nailed together to form a beam which is subjected to a vertical shear. Knowing that the allowable shearing force in each nail is 600 N, determine the allowable shear if the spacing between the nails is $s = 75$ mm.

5.3 Solve Prob. 5.2, assuming that the spacing between the nails is increased to $s = 100$ mm.

5.4 Three boards are nailed together to form the beam shown, which is subjected to a vertical shear. Knowing that the spacing between the nails is $s = 75$ mm and that the allowable shearing force in each nail is 400 N, determine the allowable shear when $w = 120$ mm.

5.5 Solve Prob. 5.4, assuming that the width of the top and bottom boards is changed to $w = 100$ mm.

5.6 The American Standard rolled-steel beam shown has been reinforced by attaching to it two 16 × 200-mm plates, using 18-mm-diameter bolts spaced longitudinally every 120 mm. Knowing that the average allowable shearing stress in the bolts is 90 MPa, determine the largest permissible vertical shearing force.

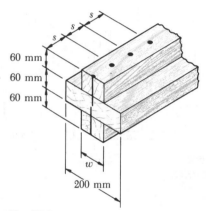

Fig. P5.4

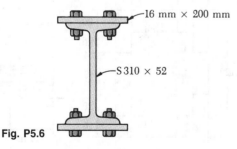

Fig. P5.6

5.7 Solve Prob. 5.6, assuming that two 12 × 200-mm plates are used to reinforce the rolled-steel beam.

5.8 The composite beam shown has been formed by bolting together two W 150 × 29.8 rolled-steel sections, using 16 mm-diameter bolts spaced longitudinally every 150 mm. Knowing that the average allowable shearing stress in the bolts is 75 MPa, determine the largest permissible vertical shear.

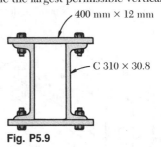

400 mm × 12 mm

C 310 × 30.8

Fig. P5.9

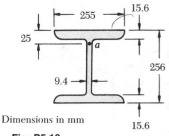

Fig. P5.8

5.9 The beam shown has been fabricated by connecting two channel shapes and two plates, using 20 mm-diameter bolts spaced longitudinally every 200 mm. Determine the average shearing stress in the bolts caused by a vertical shearing force of 100 kN.

5.10 through 5.13 For the beam and loading shown, consider section *n-n* and determine (*a*) the largest normal stress, (*b*) the shearing stress at point *a*, (*c*) the largest shearing stress.

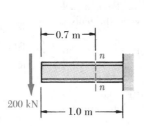

Dimensions in mm

Fig. P5.10

0.7 m

n

n

200 kN

1.0 m

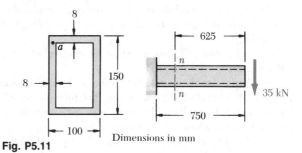

8

a

8

150

100

Dimensions in mm

Fig. P5.11

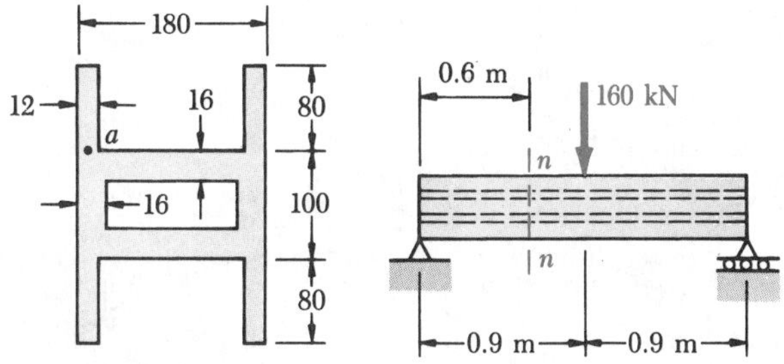

625

n

n

750

35 kN

180

12

16

a

16

80

100

80

Dimensions in mm

Fig. P5.12

0.6 m

160 kN

n

n

0.9 m   0.9 m

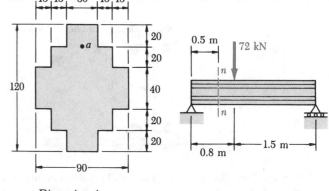

15 15   30   15 15

20

20

40

20

20

120

a

90

Dimensions in mm

Fig. P5.13

0.5 m   72 kN

n

n

0.8 m   1.5 m

5.14 and 5.15   For the wide-flange beam and loading shown, determine in a section located at mid-span, (*a*) the largest normal stress, (*b*) the largest shearing stress, using the approximation $\tau_m = V/A_{web}$, suggested in Sec. 5.6.

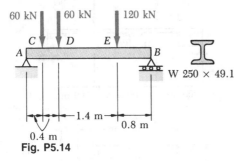

Fig. P5.14

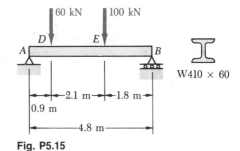

Fig. P5.15

5.16 and 5.17   For the wide-flange beam and loading shown, determine in a section located halfway between points *D* and *E*, (*a*) the largest normal stress, (*b*) the largest shearing stress, using the approximation $\tau_m = V/A_{web}$, suggested in Sec. 5.6.

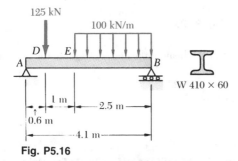

Fig. P5.16

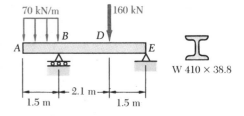

Fig. P5.17

5.18   Two rectangular plates are welded to the 310-mm-wide-flange beam as shown.   Determine the largest allowable vertical shear if the shearing stress in the beam is not to exceed 90 MPa.

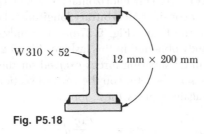

Fig. P5.18

5.19   Solve Prob. 5.18, assuming that the two plates are (*a*) replaced by 8 × 200-mm plates, (*b*) removed.

5.20   For a timber beam having the cross section shown, determine the largest allowable vertical shear if the shearing stress is not to exceed 1 MPa.

5.21   Solve Prob. 5.20, assuming that $w = 50$ mm.

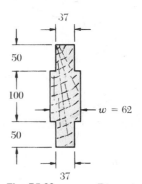

Fig. P5.20   Dimensions in mm

5.22 and 5.23 For the beam and loading shown, consider section *n-n* and determine the shearing stress at (*a*) point *a*, (*b*) point *b*.

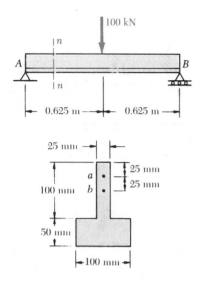

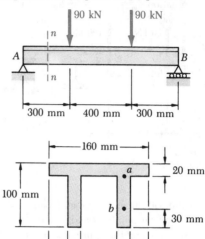

Fig. P5.22

Fig. P5.23

5.24 and 5.25 For the beam indicated, determine the largest shearing stress in section *n-n*.

    5.24 Beam of Prob. 5.23

    5.25 Beam of Prob. 5.22

5.8. SHEAR ON AN ARBITRARY LONGITUDINAL CUT

    In Sec. 5.4 we considered a cantilever beam *AB* subjected at its free end to a force **P** applied in its vertical plane of symmetry, and determined the shearing force **H** exerted on a horizontal section of a portion *AC* of the beam. We shall now consider an arbitrary longitudinal cut *A'C'C"* of the same portion *AC* of the beam (Fig. 5.27*a*). The only horizontal forces acting on the free body obtained in this fashion (Fig. 5.27*b*) are the resultant **H** of the horizontal shearing forces exerted on the longitudinal cut and the normal forces $\sigma_x \, dA$ acting on the vertical section at *C*. Recalling Eq. (5.3), we may again express these forces as

$$\sigma_x \, dA = -\frac{Pxy}{I} dA$$

Writing $\Sigma F_x = 0$, we obtain the same equation as in Sec. 5.4, namely

$$H - \int \frac{Pxy}{I} dA = 0 \qquad (5.4)$$

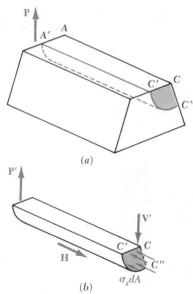

Fig. 5.27

Solving for H, we have again

$$H = \frac{PQ}{I}x \qquad (5.6)$$

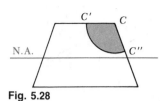

Fig. 5.28

where Q represents the first moment of the shaded area in Fig. 5.28 with respect to the neutral axis of the section, and I the moment of inertia of the entire section.

It follows that the horizontal shearing force per unit length or shear flow, q, exerted on the longitudinal cut may still be obtained from Eq. (5.7). In the more general case of a beam subjected in its vertical plane of symmetry to several concentrated or distributed loads, we have

$$q = \frac{VQ}{I} \qquad (5.8)$$

where V denotes the vertical shear in the section considered.

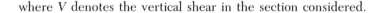

Example 5.02

A square box beam is made of two 20 × 80-mm planks and two 20 × 120-mm planks, nailed together as shown (Fig. 5.29). Knowing that the spacing between nails is 30 mm and that the beam is subjected to a vertical shear of magnitude $V = 1200$ N, determine the shearing force in each nail.

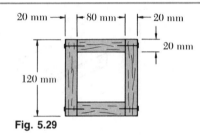

Fig. 5.29

We isolate the upper plank and consider the total force per unit length, q, exerted on its two edges. We use Eq. (5.8), where Q represents the first moment with respect to the neutral axis of the shaded area A' shown in Fig. 5.30a, and where I is the moment of inertia about the same axis of the entire cross-sectional area of the box beam (Fig. 5.30b). We have

$$Q = A'\bar{y} = (0.020 \text{ m})(0.080 \text{ m})(0.050 \text{ m}) = 80 \times 10^{-6} \text{ m}^3$$

Recalling that the moment of inertia of a square of side a about a centroidal axis is $I = \frac{1}{12}a^4$, we write

$$I = \tfrac{1}{12}(0.120 \text{ m})^4 - \tfrac{1}{12}(0.080 \text{ m})^4 = 13.87 \times 10^6 \text{ m}^4$$

Substituting into Eq. (5.8), we write

$$q = \frac{VQ}{I} = \frac{(1200 \text{ N})(80 \times 10^{-6} \text{ m}^3)}{13.87 \times 10^{-6} \text{ m}^4)} = 6920 \text{ N/m}$$

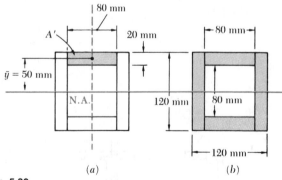

Fig. 5.30

Because both the beam and the upper plans are symmetric with respect to the vertical plane of loading, equal forces are exerted on both edges of the plank. The force per unit length on each of of these edges is thus $\frac{1}{2}q = \frac{1}{2}(6920 \text{ N/m}) = 3460 \text{ N/m}$. Since the spacing between nails is 30 mm, the shearing force in each nail is

$$F = (0.030 \text{ m})(3460 \text{ N/m}) = 103.8 \text{ N}$$

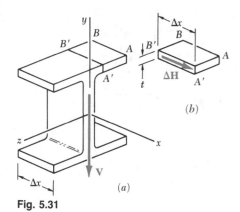

Fig. 5.31

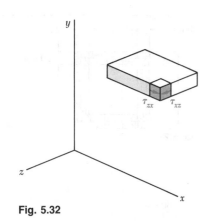

Fig. 5.32

5.9. SHEARING STRESSES IN THIN-WALLED MEMBERS

We saw in the preceding section that Eq. (5.8) may be used to determine the shear flow q on any longitudinal cut of a member subjected to a transverse loading in its vertical plane of symmetry. This means that the application of Eq. (5.10) may be similarly extended. Consider, for example, a segment of length Δx of a wide-flange beam (Fig. 5.31a), and let V be the vertical shear in the transverse section shown. If we detach a portion $ABB'A'$ of the upper flange, the horizontal shearing force ΔH exerted on the cut (Fig. 5.31b) may be expressed as

$$\Delta H = q \, \Delta x \qquad (5.17)$$

where q is the shear flow on the cut. Substituting for q from Eq. (5.8) and dividing both members of (5.17) by the area $\Delta A = t \, \Delta x$, we obtain again

$$\tau_{\text{ave}} = \frac{VQ}{It} \qquad (5.10)$$

We should note, however, that τ_{ave} now represents the average value of the shearing stress τ_{zx} exerted on the vertical cut. Since the thickness t of the flange is small, there is very little variation of τ_{zx} across the cut. Thus, recalling that $\tau_{xz} = \tau_{zx}$ (Fig. 5.32), we conclude that the horizontal component τ_{xz} of the shearing stress at any point of a transverse section of the flange may be obtained from Eq. (5.10), where Q is the first moment of the shaded area about the neutral axis (Fig. 5.33a). We recall that a similar result was obtained in Sec. 5.6 for the vertical component τ_{xy} of the shearing stress in the web (Fig. 5.33b).

Equation (5.10) may be used to determine the shearing stresses in other thin-walled members, such as box beams (Fig. 5.34) and half pipes (Fig. 5.35), as long as the loads are applied in a plane of symmetry of the member. In each case, the cut must be perpendicular to the surface of

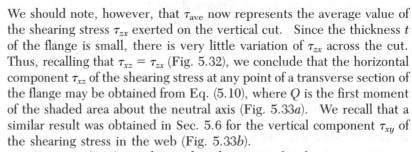

Fig. 5.33

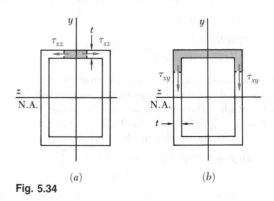

Fig. 5.34

the member, and Eq. (5.10) will yield the component of the shearing stress in the direction of the tangent to that surface. (The other component may be assumed equal to zero, in view of the proximity of the two free surfaces.)

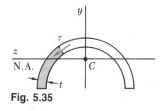

Fig. 5.35

Comparing Eqs. (5.8) and (5.10), we note that the product of the shearing stress τ at a given point of the section and of the thickness t of the section at that point is equal to q. Since V and I are constant in any given section, q depends only upon the first moment Q and, thus, may easily be sketched on the section. In the case of a box beam, for example (Fig. 5.36), we note that q grows smoothly from zero at A to a maximum value at C and C' on the neutral axis, and then decreases back to zero as E is reached. We also note that there is no sudden variation in the magnitude of q as we pass a corner at B, D, B', or D', and that the sense of q in the horizontal portions of the section may be easily obtained from its sense in the vertical portions (which is the same as the sense of the shear $\mathbf{V}$). In the case of a wide-flange section (Fig. 5.37), the values of q in portions AB and $A'B$ of the upper flange are distributed symmetrically. As we turn at B into the web, the values of q corresponding to the two halves of the flange must be combined to obtain the value of q at the top of the web. After reaching a maximum value at C on the neutral axis, q decreases, and at D splits into two equal parts corresponding to the two halves of the lower flange. The name of *shear flow* commonly used to refer to the shear per unit length, q, reflects the similarity between the properties of q that we have just described and some of the characteristics of a fluid flow through an open channel or pipe.†

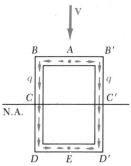

Fig. 5.36. Variation of q in box-beam section.

So far we have assumed that all the loads were applied in a plane of symmetry of the member. In the case of members possessing two planes of symmetry, such as the wide-flange beam of Fig. 5.33 or the box beam of Fig. 5.34, any load applied through the centroid of a given cross section may be resolved into components along the two axes of symmetry of the section. Each component will cause the member to bend in a plane of symmetry, and the corresponding shearing stresses may be obtained from Eq. (5.10). As we shall see in Sec. 5.11, the principle of superposition may then be used to determine the resulting stresses.

However, if the member considered possesses no plane of symmetry, or if it possesses a single plane of symmetry and is subjected to a load which is not contained in that plane, the member is observed to *bend and twist* at the same time, except when the load is applied at a specific point, called the *shear center*. Note that the shear center generally does *not* coincide with the centroid of the cross section. The determination of the shear center of various thin-walled shapes is discussed in Sec. 5.12.

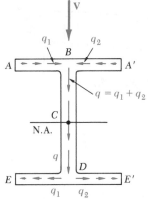

Fig. 5.37. Variation of q in wide-flange-beam section.

---

† We recall that the concept of shear flow was used to analyze the distribution of shearing stresses in thin-walled hollow shafts (Sec. 3.13). However, while the shear flow in a hollow shaft is constant, the shear flow in a member under a transverse loading is not.

*5.10. PLASTIC DEFORMATIONS

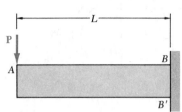

Fig. 5.38 $(PL \le M_Y)$.

Consider a cantilever beam AB of length L and rectangular cross section, subjected at its free end A to a concentrated load $\mathbf{P}$ (Fig. 5.38). The largest value of the bending moment occurs at the fixed end B and is equal to $M = PL$. As long as this value does not exceed the maximum elastic moment M_Y, that is, as long as $PL \le M_Y$, the normal stress σ_x will not exceed the yield strength σ_Y anywhere in the beam. However, as P is increased beyond the value M_Y/L, yield is initiated at points B and B' and spreads toward the free end of the beam. Assuming the material to be elastoplastic, and considering a cross section CC' located at a distance x from the free end A of the beam (Fig. 5.39), we obtain the half-thickness y_Y of the elastic core in that section by making $M = Px$ in Eq. (4.38) of Sec. 4.10. We have

$$Px = \frac{3}{2} M_Y \left(1 - \frac{1}{3} \frac{y_Y^2}{c^2} \right) \tag{5.18}$$

where c is the half-depth of the beam. Plotting y_Y against x, we obtain the boundary between the elastic and plastic zones.

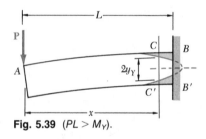

Fig. 5.39 $(PL > M_Y)$.

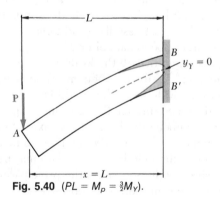

Fig. 5.40 $(PL = M_p = \frac{3}{2}M_Y)$.

As long as $PL < \frac{3}{2}M_Y$, the parabola defined by Eq. (5.18) intersects the line BB', as shown in Fig. 5.39. However, when PL reaches the value $\frac{3}{2}M_Y$, that is, when $PL = M_p$, where M_p is the plastic moment defined in Sec. 4.10, Eq. (5.18) yields $y_Y = 0$ for $x = L$, which shows that the vertex of the parabola is now located in section BB', and that this section has become fully plastic (Fig. 5.40). Recalling Eq. (4.40), we also note that the radius of curvature ρ of the neutral surface at that point is equal to zero, indicating the presence of a sharp bend in the beam at its fixed end. We say that a *plastic hinge* has developed at that point. The load $P = M_p/L$ is the largest load which may be supported by the beam.

The above discussion was based only on the analysis of the normal stresses in the beam. We shall now examine the distribution of the shearing stresses in a section which has become partly plastic. Consider the portion of beam $CC''D''D$ located between the transverse sections CC' and DD', and above the horizontal plane $D''C''$ (Fig. 5.41*a*). If this portion is located entirely in the plastic zone, the normal stresses exerted on the faces CC'' and DD'' will be uniformly distributed and equal to the yield strength σ_Y (Fig. 5.41*b*). The equilibrium of the free body $CC''D''D$

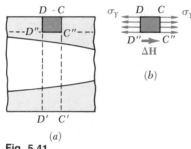

(*a*)

Fig. 5.41

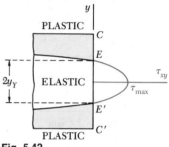

Fig. 5.42

thus requires that the horizontal shearing force $\Delta\mathbf{H}$ exerted on its lower face be equal to zero. It follows that the average value of the horizontal shearing stress τ_{yx} across the beam at C'' is zero, as well as the average value of the vertical shearing stress τ_{xy}. We thus conclude that the vertical shear $V = P$ in section CC' must be distributed entirely over the portion EE' of that section which is located within the elastic zone (Fig. 5.42). It may be shown† that the distribution of the shearing stresses over EE' is the same as in an elastic rectangular beam of the same width b as beam AB, and of depth equal to the thickness $2y_Y$ of the elastic zone. Denoting by A' the area $2by_Y$ of the elastic portion of the cross section, we have

$$\tau_{xy} = \frac{3}{2}\frac{P}{A'}\left(1 - \frac{y^2}{y_Y^2}\right) \qquad (5.19)$$

The maximum value of the shearing stress occurs for $y = 0$ and is

$$\tau_{max} = \frac{3}{2}\frac{P}{A'} \qquad (5.20)$$

As the area A' of the elastic portion of the section decreases, τ_{max} increases and eventually reaches the yield strength in shear τ_Y. Thus, shear contributes to the ultimate failure of the beam. A more exact analysis of this mode of failure should take into account the combined effect of the normal and shearing stresses.

† See Prob. 5.57.

SAMPLE PROBLEM 5.3

Knowing that the vertical shear is 250 kN in a W 250 × 101 rolled-steel beam, determine the horizontal shearing stress in the top flange at a point *a* located 110 mm from the edge of the beam.   The dimensions and other geometric data of the rolled-steel section are given in Appendix C.

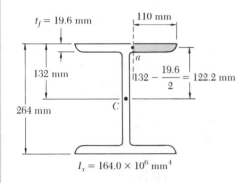

$$I_x = 164.0 \times 10^6 \text{ mm}^4$$

Solution.   We isolate the shaded portion of the flange by cutting along the dotted line which passes through point *a*.

$$Q = (110 \text{ mm})(19.6 \text{ mm})(122.2 \text{ mm}) = 263.5 \times 10^3 \text{ mm}^3 = 263.5 \times 10^{-6} \text{ m}^3$$

$$\tau = \frac{VQ}{It} = \frac{(250 \text{ kN})(263.5 \times 10^{-6} \text{ m}^3)}{(164.0 \times 10^{-6} \text{ m}^4)(0.0196 \text{ m})} \qquad \tau = 20.5 \text{ MPa} \quad \blacktriangleleft$$

SAMPLE PROBLEM 5.4

Solve Sample Prob. 5.3, assuming that 18 × 300-mm plates have been attached to the flanges of the W 250 × 101 beam by continuous fillet welds as shown:

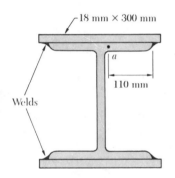

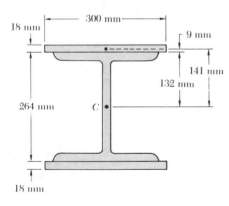

Solution.   For the composite beam the centroidal moment of inertia is

$$I = 164.0 \times 10^6 \text{ mm}^4 + 2[\tfrac{1}{12}(300 \text{ mm})(18 \text{ mm})^3 + (300 \text{ mm})(18 \text{ mm})(141 \text{ mm})^2]$$
$$I = 379 \times 10^6 \text{ mm}^4 = 379 \times 10^{-6} \text{ m}^4$$

Since the top plate and the flange are connected only at the welds, we may find the shearing stress at *a* by passing a section through the flange at *a*, *between* the plate and the flange, and again through the flange at the symmetric point *a'*.

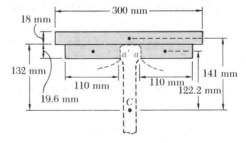

For the shaded area that we have isolated, we have

$$t = 2t_f = 2(19.6 \text{ mm}) = 39.2 \text{ mm} = 0.0392 \text{ m}$$
$$Q = 2[(100 \text{ mm})(19.6 \text{ mm})(122.2 \text{ mm})] + (300 \text{ mm})(18 \text{ mm})(141 \text{ mm})$$
$$Q = 1.288 \times 10^6 \text{ mm}^3 = 1.288 \times 10^{-3} \text{ m}^3$$
$$\tau = \frac{VQ}{It} = \frac{(250 \text{ kN})(1.288 \times 10^{-3} \text{ m}^3)}{(379 \times 10^{-6} \text{ m}^4)(0.0392 \text{ m})} \qquad \tau = 21.7 \text{ MPa} \quad \blacktriangleleft$$

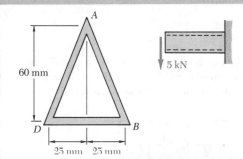

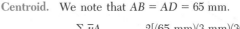

$5\ \mathrm{kN}$

SAMPLE PROBLEM 5.5

The thin-walled extruded beam shown is made of aluminum and has a uniform 3-mm wall thickness. Knowing that the shear in the beam is 5 kN, determine (a) the shearing stress at point A, (b) the maximum shearing stress in the beam. *Note:* The dimensions given are to lines midway between the outer and inner surfaces of the beam.

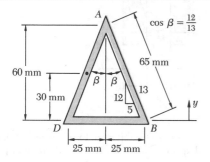

$\cos \beta = \frac{12}{13}$

Centroid. We note that $AB = AD = 65$ mm.

$$\overline{Y} = \frac{\Sigma \, \overline{y}A}{\Sigma \, A} = \frac{2[(65 \text{ mm})(3 \text{ mm})(30 \text{ mm})]}{2[(65 \text{ mm})(3 \text{ mm})] + (50 \text{ mm})(3 \text{ mm})}$$

$$\overline{Y} = 21.67 \text{ mm}$$

Centroidal Moment of Inertia. Each side of the thin-walled beam may be considered as a parallelogram and we recall that for the case shown $I_{nn} = bh^3/12$ where b is measured parallel to the axis nn.

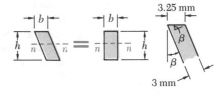

$$b = (3 \text{ mm})/\cos \beta = (3 \text{ mm})/(12/13) = 3.25 \text{ mm}$$
$$I = \Sigma(\overline{I} + Ad^2) = 2[\tfrac{1}{12}(3.25 \text{ mm})(60 \text{ mm})^3 + (3.25 \text{ mm})(60 \text{ mm})(8.33 \text{ mm})^2]$$
$$\qquad + [\tfrac{1}{12}(50 \text{ mm})(3 \text{ mm})^3 + (50 \text{ mm})(3 \text{ mm})(21.67 \text{ mm})^2]$$
$$I = 214.6 \times 10^3 \text{ mm}^4 \qquad I = 0.2146 \times 10^{-6} \text{ m}^4$$

(a) Shearing Stress at A. If a shearing stress τ_A occurs at A, the shear flow will be $q_A = \tau_A t$ and must be directed in one of the two ways shown. But the cross section and the loading are symmetric about a vertical line through A, and thus the shear flow must also be symmetric. Since neither of the possible shear flows is symmetric, we conclude that

$$\tau_A = 0 \quad \blacktriangleleft$$

(b) Maximum Shearing Stress. Since the wall thickness is constant, the maximum shearing stress occurs at the neutral axis, where Q is maximum. Since we know that the shearing stress at A is zero, we cut the section along the dashed line shown and isolate the shaded portion of the beam. In order to obtain the largest shearing stress, the cut at the neutral axis is made perpendicular to the sides, and is of length $t = 3$ mm.

$$Q = [(3.25 \text{ mm})(38.33 \text{ mm})] \left(\frac{38.33 \text{ mm}}{2} \right) = 2387 \text{ mm}^3$$

$$Q = 2.387 \times 10^{-6} \text{ m}^3$$

$$\tau_E = \frac{VQ}{It} = \frac{(5 \text{ kN})(2.387 \times 10^{-6} \text{ m}^3)}{(0.2146 \times 10^{-6} \text{ m}^4)(0.003 \text{ m})} \qquad \tau_{max} = \tau_E = 18.54 \text{ MPa} \quad \blacktriangleleft$$

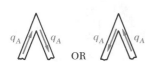

OR

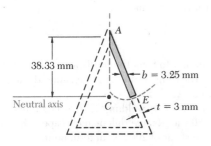

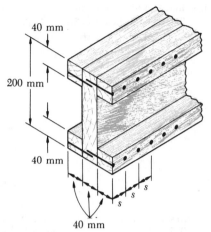

Fig. P5.26

5.26 The built-up wooden beam shown is subjected to a vertical shear of 5 kN. Knowing that the longitudinal spacing of the nails is $s = 35$ mm and that each nail is 65 mm long, determine the shearing force in each nail.

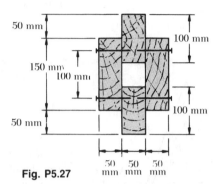

Fig. P5.27

5.27 The built-up wooden beam shown is subjected to a vertical shear of 6 kN. Knowing that the longitudinal spacing of the nails is $s = 60$ mm and that each nail is 90 mm long, determine the shearing force in each nail.

5.28 The built-up beam shown is made by gluing together several planks of wood. Knowing that the beam is subjected to a vertical shear of 5 kN, determine the average shearing stress in the glued joint (a) at A, (b) at B.

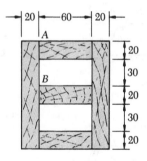

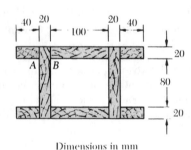

Dimensions in mm
Fig. P5.28

Dimensions in mm
Fig. P5.29

5.29 Several planks are glued together to form the box beam shown. Knowing that the beam is subjected to a vertical shear of 3 kN, determine the average shearing stress in the glued joint (a) at A, (b) at B.

5.30 The composite beam shown is made by welding C 200 × 17.1 rolled-steel channels to the flanges of a W 250 × 80 wide-flange rolled-steel shape. Knowing that the beam is subjected to a vertical shear of 200 kN, determine the horizontal shearing force per metre at each weld.

5.31 For the composite beam of Prob. 5.30, determine the shearing stress at point a of the flange of the wide-flange shape.

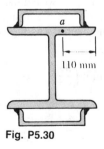

Fig. P5.30

5.32 An extruded beam has the cross section shown and is subjected to a vertical shear of 50 kN. For $t = 6$ mm, determine the shearing stress at (a) point a, (b) point b.

5.33 Solve Prob. 5.32, assuming that the dimension t is increased to 8 mm.

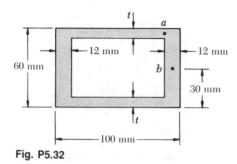

Fig. P5.32

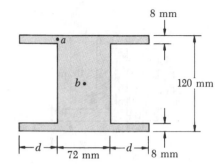

Fig. P5.34

5.34 The vertical shear is 25 kN in a beam having the cross section shown. For $d = 50$ mm, determine the shearing stress at (a) point a, (b) point b.

5.35 For the beam and loading of Prob. 5.34, determine (a) the distance d for which $\tau_a = \tau_b$, (b) the corresponding stress at points a and b.

5.36 An extruded beam has the cross section shown and a uniform wall thickness of 5 mm. Knowing that a given vertical shear **V** causes a maximum shearing stress $\tau = 62$ MPa, determine the shearing stress at the four points indicated.

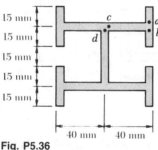

Fig. P5.36

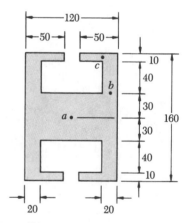

Dimensions in mm
Fig. P5.37

5.37 Knowing that a vertical shear **V** causes a maximum shearing stress $\tau = 75$ MPa in an extruded beam having the cross section shown, determine the corresponding shearing stress at the three points indicated.

5.38 Solve Prob. 5.37, assuming that the beam is subjected to a horizontal shear **V**.

5.39 Solve Prob. 5.36, assuming that the beam is subjected to a horizontal shear **V**.

5.40 Four $102 \times 102 \times 9.5$ steel angle shapes and a 12×400-mm steel plate are bolted together to form a beam with the cross section shown. The bolts are of 22-mm diameter and are spaced longitudinally every 120 mm. Knowing that the beam is subjected to a vertical shear of 240 kN, determine the average shearing stress in each bolt.

5.41 Solve Prob. 5.40, assuming that the height of the plate is increased to 600 mm.

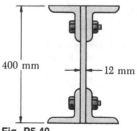

Fig. P5.40

5.42 A beam consists of five planks of 38 mm × 150 mm cross section connected by steel bolts with a longitudinal spacing of 200 mm. Knowing that the shear in the beam is vertical and equal to 10 kN and that the allowable average shearing stress in each bolt is 50 MPa, determine the smallest permissible bolt diameter which may be used.

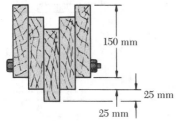

Fig. P5.42

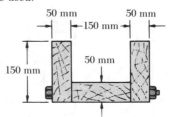

Fig. P5.43

5.43 Three planks are connected as shown by 16 mm-diameter bolts spaced every 300 mm along the longitudinal axis of the beam. For a vertical shear of 12 kN, determine the average shearing stress in the bolts.

5.44 The cross section of two extruded beams, each with a uniform wall thickness of 6 mm, are shown. For a vertical shear of 10 kN, determine the maximum shearing stress in each beam.

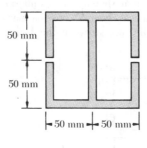

(a)

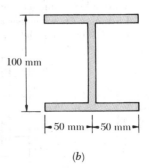

(b)

Fig. P5.44

5.45 An extruded beam has the cross section shown and a uniform wall thickness of 3 mm. For a vertical shear of 10 kN, determine (a) the shearing stress at point A, (b) the maximum shearing stress in the beam. Also sketch the shear flow in the cross section.

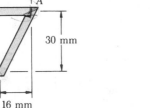

Fig. P5.45

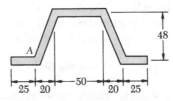

Dimensions in mm

Fig. P5.46

5.46 A plate of 4-mm thickness is bent as shown and then used as a beam. For a vertical shear of 12 kN, determine (a) the shearing stress at point A, (b) the maximum shearing stress in the beam. Also sketch the shear flow in the cross section.

5.47 and 5.48 An extruded beam has the cross section shown and a uniform wall thickness of 4 mm.   For a vertical shear of 12 kN, determine the shearing stress at the four points indicated and sketch the shear flow in the cross section.

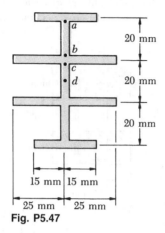

Fig. P5.47

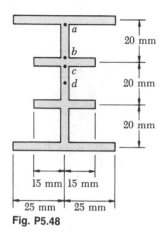

Fig. P5.48

5.49

The cross section of an extruded beam is a hollow square of side $a = 75$ mm and thickness $t = 6$ mm.   For a vertical shear of 60 kN, determine the maximum shearing stress in the beam and sketch the shear flow in the cross section.

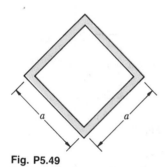

Fig. P5.49

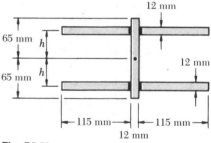

Fig. P5.50

5.50 The design of a beam requires welding four horizontal plates to a vertical 12×130 mm plate as shown.   For a vertical shear **V**, determine the dimension h for which the shear flow through the welded surface is maximum.

5.51 Determine the shearing stress at point A of a thin-walled pipe of the cross section shown caused by a vertical shear **V**.

5.52 For a beam made of two or more materials with different moduli of elasticity, show that Eq. (5.10)

$$\tau = \frac{VQ}{It}$$

remains valid provided that both Q and I are computed using the transformed section of the beam (see Sec. 4.7) and provided that t is the actual width of the beam at the point where τ is computed.

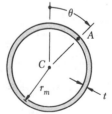

Fig. P5.51

5.53 and 5.54   A composite beam is made by attaching the timber and steel portions shown by 15 mm-diameter bolts spaced longitudinally every 200 mm.   The modulus of elasticity is 13 GPa for wood and 200 GPa for steel.   For a vertical shear of 18 kN, determine (*a*) the average shearing stress in the bolts, (*b*) the shearing stress at the center of the cross section.   (*Hint:* Use the method indicated in Prob. 5.52.)

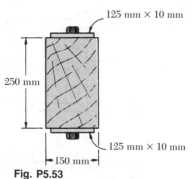

250 mm

125 mm × 10 mm

125 mm × 10 mm

150 mm

Fig. P5.53

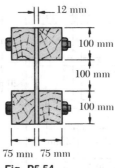

12 mm

100 mm

100 mm

100 mm

75 mm   75 mm

Fig. P5.54

5.55 and 5.56   A steel bar (E_s = 200 GPa) and an aluminum bar (E_a = 70 GPa) are bonded together to form the composite beam shown.   For a vertical shear of 20 kN, determine (*a*) the average stress at the bonded surface, (*b*) the maximum shearing stress in the beam.   (*Hint:* Use the method indicated in Prob. 5.52.)

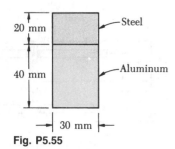

20 mm — Steel

40 mm — Aluminum

30 mm

Fig. P5.55

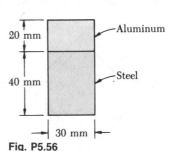

20 mm — Aluminum

40 mm — Steel

30 mm

Fig. P5.56

5.57   Consider the cantilever beam *AB* discussed in Sec. 5.10 and the portion *ACKJ* of the beam which is located to the left of the transverse section *CC'* and above the horizontal plane *JK*, where *K* is a point at a distance $y < y_Y$ above the neutral axis (see figure).   (*a*) Recalling that $\sigma_x = \sigma_Y$ between *C* and *E* and $\sigma_x = (\sigma_Y/y_Y) y$ between *E* and *K*, show that the magnitude of the horizontal shearing force **H** exerted on the lower face of the portion of beam *ACKJ* is

$$H = b\sigma_Y\left(c - \frac{1}{2}y_Y - \frac{1}{2}\frac{y^2}{y_Y}\right)$$

(*b*) Observing that the shearing stress at *K* is

$$\tau_{xy} = \lim_{\Delta A \to 0}\frac{\Delta H}{\Delta A} = \lim_{\Delta x \to 0}\frac{1}{b}\frac{\Delta H}{\Delta x} = \frac{1}{b}\frac{\partial H}{\partial x}$$

and recalling that y_Y is a function of *x* defined by Eq. (5.18), derive Eq. (5.19).

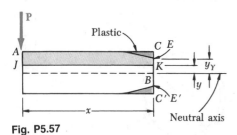

P

Plastic

A

J

C   E

K

y_Y

B

y

C   E'

x

Neutral axis

Fig. P5.57

5.11. STRESSES UNDER COMBINED LOADINGS

In Chaps. 1 and 2 we learned to determine the stresses caused by a centric axial load. In Chap. 3, we analyzed the distribution of stresses in a cylindrical member subjected to a twisting couple. In Chap. 4, we determined the stresses caused by bending couples and, in this chapter, the stresses produced by transverse loads. As we shall see presently, we may combine the knowledge we have acquired to determine the stresses in slender structural members or machine components under fairly general loading conditions.

Consider, for example, the bent member *ABDE* of circular cross section, which is subjected to several forces (Fig. 5.43). In order to determine the stresses produced at a point *K* by the given loads, we shall first pass a section through *K* and determine the force-couple system at the centroid *C* of the section, which is required to maintain the equilibrium of portion *ABK*.† This system represents the internal forces in the section and, in general, consists of three force components and three couple vectors, which we shall assume directed as shown (Fig. 5.44).

We note that the force **P** is a centric axial force which produces normal stresses in the section. The couple vectors M_y and M_z cause the member to bend and also produce normal stresses in the section. The normal stress σ_x at point *K* is the sum of the stresses produced by the force and couples shown in Fig. 5.45*a*, and may be determined as indicated in Sec. 4.15. On the other hand, the twisting couple **T** and the shearing forces V_y and V_z produce shearing stresses in the section. The components τ_{xy} and τ_{xz} of the shearing stress at *K* may thus be obtained by adding the corresponding components of the stresses produced at *K* by each of the forces and couple shown in Fig. 5.45*b*.‡

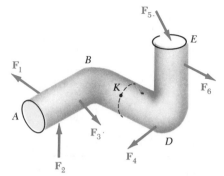

Fig. 5.43

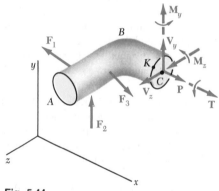

Fig. 5.44

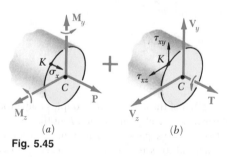

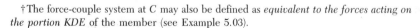

(a)   (b)

Fig. 5.45

---

†The force-couple system at *C* may also be defined as *equivalent to the forces acting on the portion KDE* of the member (see Example 5.03).

‡We should note that our present knowledge enables us to determine the effect of the shearing forces V_y and V_z only if these forces are applied along axes of symmetry of the cross section. Also, the effect of the twisting couple **T** may be evaluated only in the cases of circular shafts, of members with a rectangular cross section (Sec. 3.12), or of thin-walled hollow members (Sec. 3.13).

The results obtained will be valid only to the extent that the conditions of applicability of the superposition principle (Sec. 2.12) and of Saint-Venant's principle (Sec. 2.16) are met. This means that the stresses involved must not exceed the proportional limit of the material, that the deformations due to one of the loadings must not affect the determination of the stresses due to the others, and that the section used in our analysis must not be too close to the points of application of the given forces. It is clear from the first of these requirements that the method presented here cannot be applied to plastic deformations.

Example 5.03

Two forces $\mathbf{P}_1$ and $\mathbf{P}_2$, of magnitude $P_1 = 15$ kN and $P_2 = 18$ kN, are applied as shown to the end A of bar AB, which is welded to a cylindrical member BD of radius $c = 20$ mm (Fig. 5.46). Knowing that the distance from A to the axis of member BD is $a = 50$ mm, determine the normal and shearing stresses at points H and K of the transverse section of member BD located at a distance $b = 60$ mm from end B. Assume all stresses to remain below the proportional limit of the material.

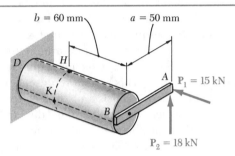

Fig. 5.46

Internal Forces in Section HK. We first replace the forces $\mathbf{P}_1$ and $\mathbf{P}_2$ by an equivalent system of forces and couples applied at the center C of the section containing points H and K (Fig. 5.47). This system, which represents the internal forces in the section, consists of the following forces and couples:

1. A centric axial force $\mathbf{F}$ equal to the force $\mathbf{P}_1$, of magnitude
$$F = P_1 = 15 \text{ kN}$$

2. A shearing force $\mathbf{V}$ equal to the force $\mathbf{P}_2$, of magnitude
$$V = P_2 = 18 \text{ kN}$$

3. A twisting couple $\mathbf{T}$ of torque T equal to the moment of $\mathbf{P}_2$ about the axis of member BD:
$$T = P_2 a = (18 \text{ kN})(50 \text{ mm}) = 900 \text{ N} \cdot \text{m}$$

4. A bending couple $\mathbf{M}_y$ of magnitude M_y equal to the moment of $\mathbf{P}_1$ about a vertical axis through C:
$$M_y = P_1 a = (15 \text{ kN})(50 \text{ mm}) = 750 \text{ N} \cdot \text{m}$$

5. A bending couple $\mathbf{M}_z$ of magnitude M_z equal to the moment of $\mathbf{P}_2$ about a transverse, horizontal axis through C:
$$M_z = P_2 b = (18 \text{ kN})(60 \text{ mm}) = 1080 \text{ N} \cdot \text{m}$$

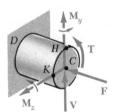

Fig. 5.47

Each of these forces and couples may produce a normal or shearing stress at points H and K of the section. Our purpose is to compute separately each of these stresses, and then to add the normal stresses and add the shearing stresses at each of the two points. First, however, we shall determine the geometric properties of the section. We have

$$A = \pi c^2 = \pi(0.020 \text{ m})^2 = 1.257 \times 10^{-3} \text{ m}^2$$

$$I_y = I_z = \tfrac{1}{4}\pi c^4 = \tfrac{1}{4}\pi(0.020 \text{ m})^4 = 125.7 \times 10^{-9} \text{ m}^4$$

$$J_C = \tfrac{1}{2}\pi c^4 = \tfrac{1}{2}\pi(0.020 \text{ m})^4 = 251.3 \times 10^{-9} \text{ m}^4$$

Stresses at H. We observe that normal stresses σ_x are produced at H by the centric force **F** and the bending couple $\mathbf{M}_z$, and that a horizontal shearing stress τ_{xz} is caused by the twisting couple **T** (Fig. 5.48). On the other hand, the bending couple $\mathbf{M}_y$ does not produce any normal stress at H, since H is located on the corresponding neutral axis, and the vertical shearing force **V** does not produce any shearing stress at H, since H is located at the top of the section. Determining the sign of each stress from the figure, we write

$$\sigma_x = (\sigma_x)_{\text{centric}} + (\sigma_x)_{\text{bending}} = -\frac{F}{A} - \frac{M_z c}{I_z}$$

$$= -\frac{15\ \text{kN}}{1.257 \times 10^{-3}\ \text{m}^2} - \frac{(1080\ \text{N} \cdot \text{m})(0.020\ \text{m})}{125.7 \times 10^{-9}\ \text{m}^4}$$

$$= -11.9\ \text{MPa} - 171.9\ \text{MPa}$$

$$\sigma_x = -183.8\ \text{MPa}$$

and $\quad \tau_{xz} = (\tau_{xz})_{\text{twist}} = \dfrac{Tc}{J_C} = \dfrac{(900\ \text{N} \cdot \text{m})(0.020\ \text{m})}{251.3 \times 10^{-9}\ \text{m}^4}$

$$\tau_{xz} = 71.6\ \text{MPa}$$

Stresses at K. We note that normal stresses σ_x are produced at K by the centric force **F** and the bending couple $\mathbf{M}_y$, and that vertical shearing stresses τ_{xy} are caused by the twisting couple **T** and the shearing force **V** (Fig. 5.49). We write

$$\sigma_x = -\frac{F}{A} + \frac{M_y c}{I_y} = -11.9\ \text{MPa} + \frac{(750\ \text{N} \cdot \text{m})(0.020\ \text{m})}{125.7 \times 10^{-9}\ \text{m}^4}$$

$$= -11.9\ \text{MPa} + 119.3\ \text{MPa}$$

$$\sigma_x = +107.4\ \text{MPa}$$

In order to compute the shearing stress due to **V**, we must determine the first moment Q and width t of the shaded area shown in Fig. 5.49. Recalling that $\bar{y} = 4c/3\pi$ for a semicircle of radius c, we have

$$Q = A'\bar{y} = \left(\frac{1}{2}\pi c^2\right)\left(\frac{4c}{3\pi}\right) = \frac{2}{3}c^3 = \frac{2}{3}(0.020\ \text{m})^3$$

$$= 5.33 \times 10^{-6}\ \text{m}^3$$

and $\quad t = 2c = 2(0.020\ \text{m}) = 0.040\ \text{m}$

We thus write

$$(\tau_{xy})_V = +\frac{VQ}{I_z t} = +\frac{(18 \times 10^3\ \text{N})(5.33 \times 10^{-6}\ \text{m}^3)}{(125.7 \times 10^{-9}\ \text{m}^4)(0.040\ \text{m})}$$

$$= +19.1\ \text{MPa}$$

Noting that $(\tau_{xy})_{\text{twist}} = (\tau_{xz})_{\text{twist}}$, we have

$$\tau_{xy} = (\tau_{xy})_V - (\tau_{xy})_{\text{twist}} = +19.1\ \text{MPa} - 71.6\ \text{MPa}$$

$$\tau_{xy} = -52.5\ \text{MPa}$$

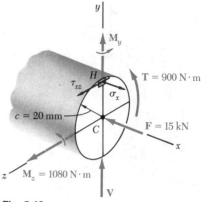

Fig. 5.48

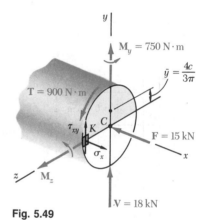

Fig. 5.49

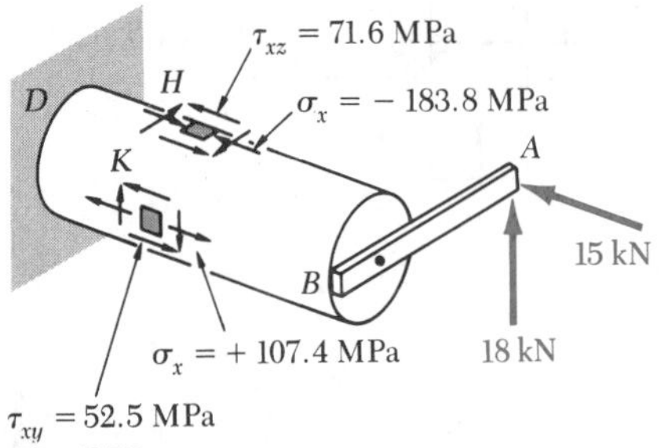

$\tau_{xz} = 71.6\ \text{MPa}$
$\sigma_x = -183.8\ \text{MPa}$
$\sigma_x = +107.4\ \text{MPa}$
$\tau_{xy} = 52.5\ \text{MPa}$

Fig. 5.50

Square elements located at H and K on the surface of the cylindrical member are used in Fig. 5.50 to summarize the results obtained. Note that shearing stresses acting on the longitudinal sides of the elements have been included.

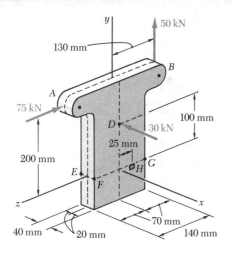

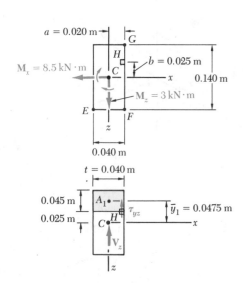

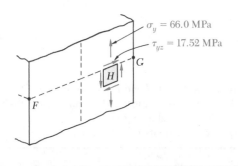

SAMPLE PROBLEM 5.6

Three forces are applied as shown at points A, B, and D of a short steel post. Knowing that the horizontal cross section of the post is a 40×140-mm rectangle, determine the normal and shearing stress at point H.

Internal Forces in Section _EFG_. We replace the three applied forces by an equivalent force-couple system at the center C of the rectangular section EFG. We have

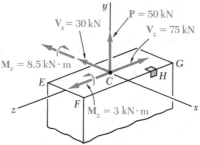

$$V_x = -30 \text{ kN} \qquad P = 50 \text{ kN} \qquad V_z = -75 \text{ kN}$$
$$M_x = (50 \text{ kN})(0.130 \text{ m}) - (75 \text{ kN})(0.200 \text{ m}) = -8.5 \text{ kN} \cdot \text{m}$$
$$M_z = (30 \text{ kN})(0.100 \text{ m}) = 3 \text{ kN} \cdot \text{m}$$

We note that there is no twisting couple about the y axis. The geometric properties of the rectangular section are

$$A = (0.040 \text{ m})(0.140 \text{ m}) = 5.6 \times 10^{-3} \text{ m}^2$$
$$I_x = \tfrac{1}{12}(0.040 \text{ m})(0.140 \text{ m})^3 = 9.15 \times 10^{-6} \text{ m}^4$$
$$I_z = \tfrac{1}{12}(0.140 \text{ m})(0.040 \text{ m})^3 = 0.747 \times 10^{-6} \text{ m}^4$$

Normal Stress at _H_. We note that normal stresses σ_y are produced by the centric force $\mathbf{P}$ and by the bending couples $\mathbf{M}_x$ and $\mathbf{M}_z$. We determine the sign of each stress by carefully examining the sketch of the force-couple system at C.

$$\sigma_y = +\frac{P}{A} + \frac{|M_z|a}{I_z} - \frac{|M_x|b}{I_x}$$

$$= \frac{50 \text{ kN}}{5.6 \times 10^{-3} \text{ m}^2} + \frac{(3 \text{ kN} \cdot \text{m})(0.020 \text{ m})}{0.747 \times 10^{-6} \text{ m}^4} - \frac{(8.5 \text{ kN} \cdot \text{m})(0.025 \text{ m})}{9.15 \times 10^{-6} \text{ m}^4}$$

$$\sigma_y = 8.93 \text{ MPa} + 80.3 \text{ MPa} - 23.2 \text{ MPa} \qquad \sigma_y = 66.0 \text{ MPa} \quad \blacktriangleleft$$

Shearing Stress at _H_. Considering first the shearing force $\mathbf{V}_x$, we note that $Q = 0$ with respect to the z axis, since H is on the edge of the cross section. Thus $\mathbf{V}_x$ produces no shearing stress at H. The shearing force $\mathbf{V}_z$ does produce a shearing stress at H and we write

$$Q = A_1\bar{y}_1 = [(0.040 \text{ m})(0.045 \text{ m})](0.0475 \text{ m}) = 85.5 \times 10^{-6} \text{ m}^3$$

$$\tau_{yz} = \frac{V_zQ}{I_xt} = \frac{(75 \text{ kN})(85.5 \times 10^{-6} \text{ m}^3)}{(9.15 \times 10^{-6} \text{ m}^4)(0.040 \text{ m})} \qquad \tau_{yz} = 17.52 \text{ MPa} \quad \blacktriangleleft$$

Summary. We now show both σ_y and τ_{yz} on a square element at H. Note that the direction of the shearing stress on the upper face of the element is the same as that of the shearing force $\mathbf{V}_z$ exerted on the cross section.

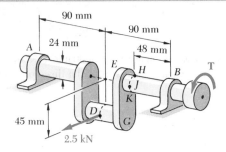

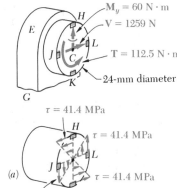

$\mathbf{M}_y = 60 \text{ N} \cdot \text{m}$
$V = 1259 \text{ N}$
$T = 112.5 \text{ N} \cdot \text{m}$
24-mm diameter

(a)

$\tau = 41.4 \text{ MPa}$
$\tau = 41.4 \text{ MPa}$
$\tau = 41.4 \text{ MPa}$
$\tau = 41.4 \text{ MPa}$

(b)

$\tau = 3.68 \text{ MPa}$
$\tau = 0$
$\tau = 3.68 \text{ MPa}$

(c)

$\sigma = 0$
$\sigma = 44.2 \text{ MPa}$
$\sigma = 44.2 \text{ MPa}$
$\sigma = 0$

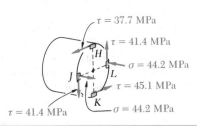

$\tau = 37.7 \text{ MPa}$
$\tau = 41.4 \text{ MPa}$
$\sigma = 44.2 \text{ MPa}$
$\tau = 45.1 \text{ MPa}$
$\tau = 41.4 \text{ MPa}$
$\sigma = 44.2 \text{ MPa}$

SAMPLE PROBLEM 5.7

A horizontal 2.5-kN force acts at point D of crankshaft AB which is held in static equilibrium by a twisting couple $\mathbf{T}$ and by reactions at A and B. Knowing that the bearings are self-aligning and exert no couples on the shaft, determine the normal and shearing stresses at points H, J, K, and L located at the ends of the vertical and horizontal diameters of a transverse section located 48 mm to the left of bearing B.

Free Body. Entire Crankshaft. $A = B = 1250 \text{ N}$

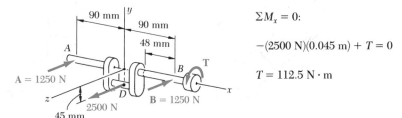

$\Sigma M_x = 0$:

$-(2500 \text{ N})(0.045 \text{ m}) + T = 0$

$T = 112.5 \text{ N} \cdot \text{m}$

Internal Forces in Transverse Section. We replace the reaction $\mathbf{B}$ and the twisting couple $\mathbf{T}$ by an equivalent force-couple system at the center C of the transverse section containing H, J, K, and L.

$$V = B = 1250 \text{ N} \qquad T = 112.5 \text{ N} \cdot \text{m}$$

$$M_y = (1250 \text{ N})(0.048 \text{ m}) = 60 \text{ N} \cdot \text{m}$$

The geometric properties of the 24-mm-diameter section are

$$A = \pi(0.012 \text{ m})^2 = 452 \times 10^{-6} \text{ m}^4 \qquad I = \tfrac{1}{4}\pi(0.012 \text{ m})^4 = 16.29 \times 10^{-9} \text{ m}^4$$

$$c = 12 \text{ mm} = 0.012 \text{ m} \qquad J = \tfrac{1}{2}\pi(0.012 \text{ m})^4 = 32.6 \times 10^{-9} \text{ m}^4$$

Stresses Produced by Twisting Couple T. Using Eq. (3.8), we determine the shearing stresses at points H, J, K, and L and show them in Fig. (a).

$$\tau = \frac{Tc}{J} = \frac{(112.5 \text{ N} \cdot \text{m})(0.012 \text{ m})}{32.6 \times 10^{-9} \text{ m}^4} = 41.4 \text{ MPa}$$

Stresses Produced by Shearing Force V. The shearing force $\mathbf{V}$ produces no shearing stresses at points J and L. At points H and K we first compute Q for a semicircle about a vertical diameter and then determine the shearing stress produced by the shear force $V = 1250 \text{ N}$. These stresses are shown in Fig. (b).

$$Q = \left(\frac{1}{2}\pi c^2\right)\left(\frac{4c}{3\pi}\right) = \frac{2}{3}c^3 = \frac{2}{3}(0.012 \text{ m})^3 = 1.152 \times 10^{-6} \text{ m}^3$$

$$\tau = \frac{VQ}{It} = \frac{(1250 \text{ N})(1.152 \times 10^{-6} \text{ m}^3)}{(16.29 \times 10^{-9} \text{ m}^4)(0.024 \text{ m})} = 3.68 \text{ MPa}$$

Stresses Produced by the Bending Couple $\mathbf{M}_y$. Since the bending couple $\mathbf{M}_y$ acts in a horizontal plane, it produces no stresses at H and K. Using Eq. (4.15), we determine the normal stresses at points J and L and show them in Fig. (c).

$$\sigma = \frac{|M_y|c}{I} = \frac{(60 \text{ N} \cdot \text{m})(0.012 \text{ m})}{16.29 \times 10^{-9} \text{ m}^4} = 44.2 \text{ MPa}$$

Summary. We add the stresses shown and obtain the total normal and shearing stresses at points H, J, K, and L.

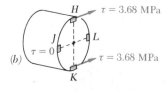

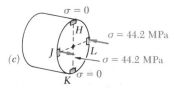

PROBLEMS

5.58 The lever AB has a rectangular cross section of 10 mm × 30 mm. Knowing that $\theta = 40°$, determine the normal and shearing stresses at the three points indicated.

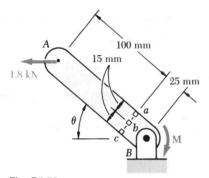

Fig. P5.58

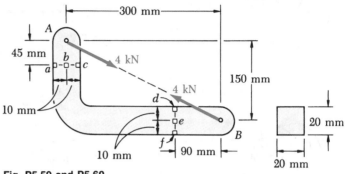

Fig. P5.59 and P5.60

5.59 Two 4-kN forces are applied to an L-shaped machine element AB as shown. Determine the normal and shearing stresses at (a) point a, (b) point b, (c) point c.

5.60 Two 4-kN forces are applied to an L-shaped machine element AB as shown. Determine the normal and shearing stresses at (a) point d, (b) point e, (c) point f.

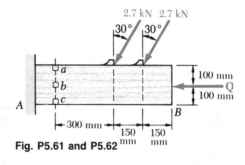

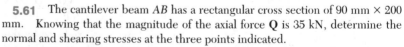

Fig. P5.61 and P5.62

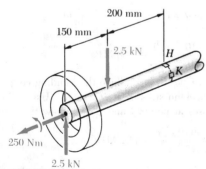

Fig. P5.63

5.61 The cantilever beam AB has a rectangular cross section of 90 mm × 200 mm. Knowing that the magnitude of the axial force Q is 35 kN, determine the normal and shearing stresses at the three points indicated.

5.62 The cantilever beam AB has a rectangular cross section of 90 mm × 200 mm. Determine (a) the value of the axial force Q for which the normal stress at point c is −55 MPa, (b) the corresponding normal and shearing stresses at point b.

5.63 The axle of an automobile is acted upon by the forces and couple shown. Knowing that the diameter of the axle is 32 mm, determine the normal and shearing stresses at (a) point H, (b) point K.

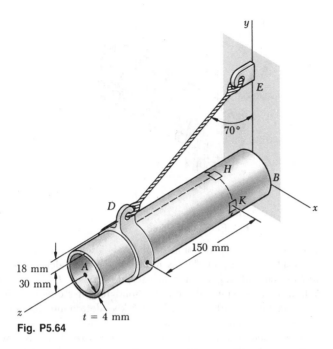

Fig. P5.64

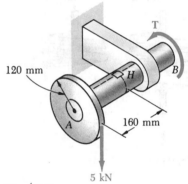

Fig. P5.65

5.64 The standard-weight pipe *AB* has a 60-mm outer diameter and a 4-mm wall thickness. Knowing that the tension in cable *DE* is 18 kN, determine the normal and shearing stresses at (*a*) point *H*, (*b*) point *K*.

5.65 A vertical 5-kN force is applied to the rim of a 120-mm-radius disk as shown. Knowing that shaft *AB* has a diameter of 80 mm, determine the normal and shearing stresses at point *H*.

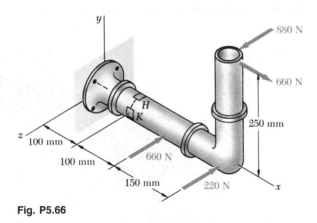

Fig. P5.66

5.66 Several forces are applied to the pipe assembly shown. Knowing that the pipe has inner and outer diameters equal to 41.0 mm and 48.5 mm, respectively, determine the normal and shearing stresses at (*a*) point *H*, (*b*) point *K*.

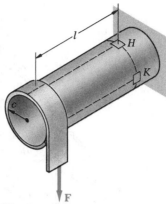

Fig. P5.67 and P5.68

5.67 A thin strap is wrapped around a solid rod of radius c as shown. Knowing that a vertical force of magnitude F is applied to the strap, determine in terms of F, c, and l the normal and shearing stresses at (a) point H, (b) point K.

5.68 A thin strap is wrapped around a solid rod of radius $c = 20$ mm as shown. Knowing that $l = 100$ mm and $F = 5$ kN, determine the normal and shearing stresses at (a) point H, (b) point K.

5.69 A close-coiled spring is made of a circular wire of radius c which is formed into a helix of radius R. Determine the maximum shearing stress produced by the two equal and opposite forces $\mathbf{P}$ and $\mathbf{P}'$. (*Hint:* First determine the shear $\mathbf{V}$ and the torque $\mathbf{T}$ in a transverse cross section.)

5.70 Rod AB is connected by ball-and-socket joints to collar A and to the 90-mm-radius disk. The disk is welded to pipe DE, which has an 80-mm outer diameter and a 6-mm wall thickness. Knowing that in the position shown rod AB is in compression and exerts an 8.6-kN force on point B, determine the normal and shearing stresses at point H.

5.71 For the system of Prob. 5.70, determine the normal and shearing stresses at point K.

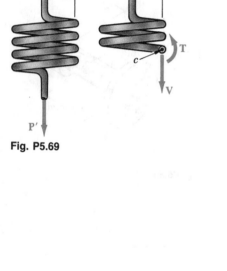

Fig. P5.69

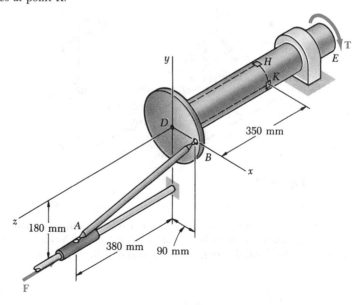

Fig. P5.70

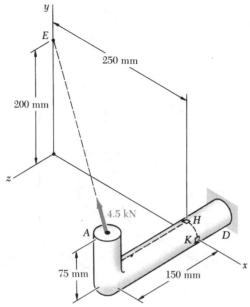

Fig. P5.72

5.72   A 4.5-kN force is applied at point A of the cast-iron member shown. Knowing that the member has a diameter of 45 mm, determine the normal and shearing stresses at point H located on the top surface of the member.

5.73   For the member and loading of Prob. 5.72, determine the normal and shearing stresses at point K.

5.74   Three forces are applied to the bar shown.   Determine the normal and shearing stresses at the three points indicated.

5.75   Solve Prob. 5.74, assuming that h = 300 mm.

5.76   Three forces are applied to the bar shown.   Determine the normal and shearing stresses at the three points indicated.

5.77   Solve Prob. 5.76, assuming that the 750-N force is directed vertically upward.

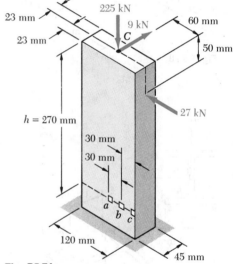

Fig. P5.74

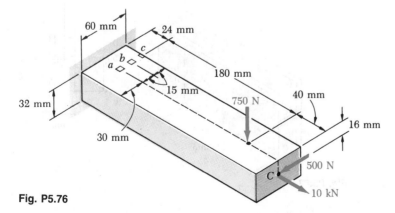

Fig. P5.76

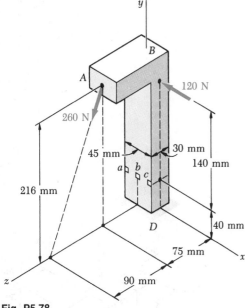

Fig. P5.78

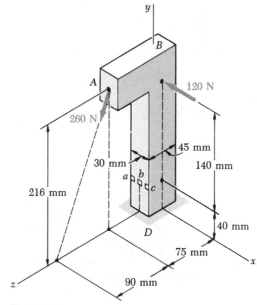

Fig. P5.79

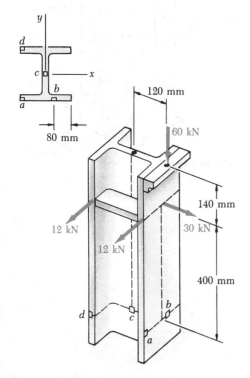

Fig. P5.80 and P5.81

5.78 and 5.79 Two forces are applied to the machine component *ABD* as shown. Knowing that the cross section of the component is a 30 × 45-mm rectangle, determine the normal and shearing stresses at the three points indicated.

5.80 Four forces are applied to a W 200 × 46.1 rolled-steel shape as shown. Determine the normal and shearing stresses at points *a* and *b*.

5.81 Four forces are applied to a W 200 × 46.1 rolled-steel shape as shown. Determine the normal and shearing stresses at points *c* and *d*.

5.82 Three steel plates, each 13 mm thick, are welded together to form a cantilever beam. For the loading shown, determine the normal and shearing stresses at points *a* and *b*. (*Given*: $I_x = 7.321 \times 10^{-6}\ \text{m}^4$ and $I_y = 3.930 \times 10^{-6}\ \text{m}^4$.)

5.83 For the beam and loading of Prob. 5.82, determine the normal and shearing stresses at points *d* and *e*.

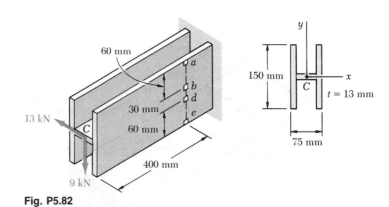

Fig. P5.82

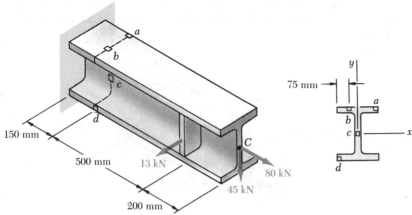

Fig. P5.84 and P5.85

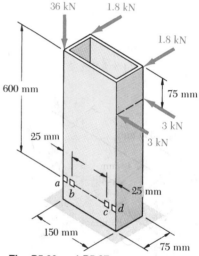

5.84   Three forces are applied to a W 250 × 49.1 rolled-steel shape as shown. Determine the normal and shearing stresses at points *a* and *b*.

5.85   Three forces are applied to a W 250 × 49.1 rolled-steel shape as shown. Determine the normal and shearing stresses at points *c* and *d*.

5.86   Knowing that the structural tubing shown has a uniform wall thickness of 6 mm, determine (*a*) the normal stress at point *a*, (*b*) the shearing stress at point *b*.

5.87   Knowing that the structural tubing shown has a uniform wall thickness of 6 mm, determine (*a*) the shearing stress at point *c*, (*b*) the normal stress at point *d*.

5.88   Knowing that the structural tubing shown has a uniform wall thickness of 6 mm, determine (*a*) the normal stress at point *a*, (*b*) the shearing stress at point *b*.

5.89   Knowing that the structural tubing shown has a uniform wall thickness of 6 mm, determine (*a*) the shearing stress at point *c*, (*b*) the normal stress at point *d*.

Fig. P5.86 and P5.87

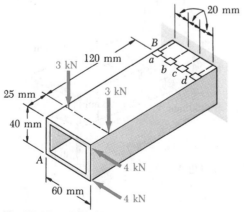

Fig. P5.88 and P5.89

5.90   A 10-kN force is applied to the free end of the cantilever beam *AB* in a direction perpendicular to the longitudinal axis of the beam.   Determine the normal stress at point (*a*) $\beta = 0$, (*b*) $\beta = 5°$.

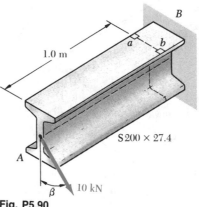

Fig. P5.90

5.91   For the beam and loading of Prob. 5.90, determine (*a*) the value of β for which the normal stress at point *b* is zero, (*b*) the corresponding normal stress at point *a*.

5.92   The cantilever beam *AB* will be installed so that the 60-mm side forms an angle β between 0 and 90° with the vertical.   Knowing that the 600-N vertical force is applied at the center of the free end of the beam, determine the normal stress at point *a* when (*a*) $\beta = 0$, (*b*) $\beta = 90°$.   (*c*) Also, determine the value of β for which the normal stress at point *a* is maximum and the corresponding value of that stress.

5.93   For the beam and loading of Prob. 5.92, determine (*a*) the value of β for which the normal stress at point *b* is zero, (*b*) the corresponding normal stress at point *a*.

*5.94   Knowing that the structural tube shown has a uniform wall thickness of 6 mm, determine the shearing stress at each of the three points indicated.

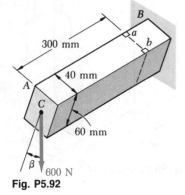

Fig. P5.92

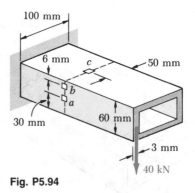

Fig. P5.94

*5.95 Knowing that the structural tube shown has a uniform wall thickness of 8 mm, determine the shearing stress at each of the two points indicated.

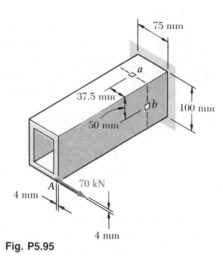

Fig. P5.95

*5.96 Solve Prob. 5.95, assuming that the 70-kN force applied at point *A* acts vertically downward.

*5.97 A 5-kN force **P** is applied to a wire which is wrapped around the bar *AB* as shown. Knowing that the cross section of the bar is a square of side $d = 40$ mm, determine the shearing stress at point *a*.

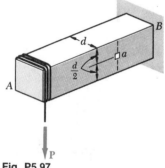

Fig. P5.97

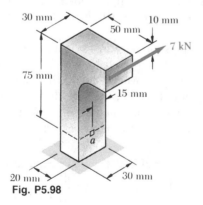

Fig. P5.98

*5.98 Determine the shearing stress at point *a* of the post shown.

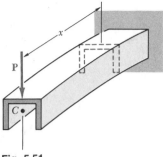

Fig. 5.51

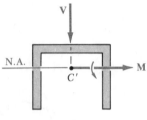

$(V = P, M = Px)$

Fig. 5.52

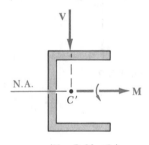

$(V = P, M = Px)$

Fig. 5.53

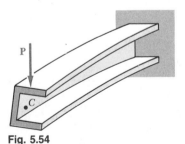

Fig. 5.54

*5.12. UNSYMMETRIC LOADING OF THIN-WALLED MEMBERS; SHEAR CENTER

Our analysis of the effects of transverse loadings in Secs. 5.2 through 5.10 was limited to members possessing a vertical plane of symmetry and to loads applied in that plane. The members were observed to bend in the plane of loading (Fig. 5.51) and, in any given cross section, the bending couple **M** and the shear **V** (Fig. 5.52) were found to result in normal and shearing stresses defined, respectively, by the formulas†

$$\sigma_x = -\frac{My}{I} \tag{4.16}$$

and

$$\tau_{\text{ave}} = \frac{VQ}{It} \tag{5.10}$$

In this section, we shall discuss the effects of transverse loadings on *thin-walled members which do not possess a vertical plane of symmetry.* Let us assume, for example, that the channel member of Fig. 5.51 has been rotated through 90° and that the line of action of **P** still passes through the centroid of the end section. The couple vector **M** representing the bending moment in a given cross section is still directed along a principal axis of the section (Fig. 5.53), and the neutral axis will coincide with that axis (cf. Sec. 4.14). Equation (4.16), therefore, is applicable and may be used to compute the normal stresses in the section. However, Eq. (5.10) cannot be used to determine the shearing stresses in the section, since this equation was derived for a member possessing a vertical plane of symmetry (cf. Sec. 5.9). Actually, the member will be observed to *bend and twist* under the applied load (Fig. 5.54), and the resulting distribution of shearing stresses will be quite different from that defined by Eq. (5.10).

The following question now arises: Is it possible to apply the vertical load **P** in such a way that the channel member of Fig. 5.54 will *bend without twisting* and, if so, where should the load **P** be applied? If the member bends without twisting, then the shearing stress at any point of a given cross section may be obtained from Eq. (5.10), where Q is the first moment of the shaded area with respect to the neutral axis (Fig. 5.55a),

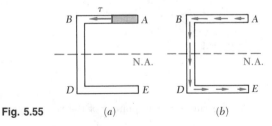

Fig. 5.55   *(a)*   *(b)*

† More general loading conditions were considered in Sec. 5.11 but, since all members involved were assumed to possess two planes of symmetry, both the bending couple and the shearing force exerted on a given cross section could be resolved into components along the axes of symmetry of the section.

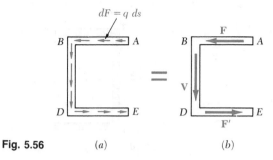

Fig. 5.56   (a)   (b)

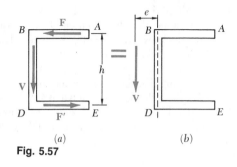

(a)   (b)
Fig. 5.57

and the distribution of stresses will look as shown in Fig. 5.55b, with $\tau = 0$ at both A and E. We note that the shearing force exerted on a small element of cross-sectional area $dA = t\ ds$ is $dF = \tau\ dA = \tau t\ ds$, or $dF = q\ ds$ (Fig. 5.56a), where q is the shear flow $q = \tau t = VQ/I$ at the point considered. The resultant of the shearing forces exerted on the elements of the upper flange AB of the channel is found to be a horizontal force $\mathbf{F}$ (Fig. 5.56b) of magnitude

$$F = \int_A^B q\ ds \qquad (5.21)$$

Because of the symmetry of the channel section about its neutral axis, the resultant of the shearing forces exerted on the lower flange DE is a force $\mathbf{F}'$ of the same magnitude as $\mathbf{F}$ but of opposite sense. We conclude that the resultant of the shearing forces exerted on the web BD must be equal to the vertical shear $\mathbf{V}$ in the section:

$$V = \int_B^D q\ ds \qquad (5.22)$$

We now observe that the forces $\mathbf{F}$ and $\mathbf{F}'$ form a couple of moment Fh, where h is the distance between the center lines of the flanges AB and DE (Fig. 5.57a). This couple may be eliminated if the vertical shear $\mathbf{V}$ is moved to the left through a distance e such that the moment of $\mathbf{V}$ about B is equal to Fh (Fig. 5.57b). We write $Ve = Fh$ or

$$e = \frac{Fh}{V} \qquad (5.23)$$

and conclude that, when the force $\mathbf{P}$ is applied at a distance e to the left of the center line of the web BD, the member bends in a vertical plane without twisting (Fig. 5.58).

The point O where the line of action of $\mathbf{P}$ intersects the axis of symmetry of the end section is called the *shear center* of that section. We note that, in the case of an oblique load $\mathbf{P}$ (Fig. 5.59a), the member will also be free of any twist if the load $\mathbf{P}$ is applied at the shear center of the section. Indeed, the load $\mathbf{P}$ may then be resolved into two components $\mathbf{P}_z$ and $\mathbf{P}_y$ (Fig. 5.59b) corresponding respectively to the loading conditions of Figs. 5.51 and 5.58, neither of which causes the member to twist.

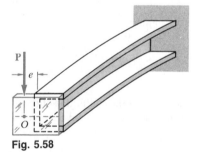

Fig. 5.58

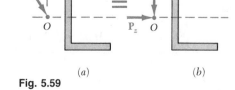

(a)   (b)
Fig. 5.59

Example 5.04

Determine the shear center O of a channel section of uniform thickness (Fig. 5.60), knowing that $b = 100$ mm, $h = 150$ mm, and $t = 3$ mm.

Assuming that the member does not twist, we first determine the shear flow q in flange AB at a distance s from A (Fig. 5.61). Recalling Eq. (5.8) and observing that the first moment Q of the shaded area with respect to the neutral axis is $Q = (st)(h/2)$, we write

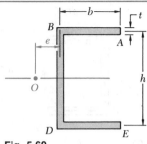

Fig. 5.60

$$q = \frac{VQ}{I} = \frac{Vsth}{2I} \qquad (5.24)$$

where V is the vertical shear and I the moment of inertia of the section with respect to the neutral axis.

Recalling Eq. (5.21), we determine the magnitude of the shearing force $\mathbf{F}$ exerted on flange AB by integrating the shear flow q from A to B:

$$F = \int_0^b q\, ds = \int_0^b \frac{Vsth}{2I}\, ds = \frac{Vth}{2I} \int_0^b s\, ds$$

$$F = \frac{Vthb^2}{4I} \qquad (5.25)$$

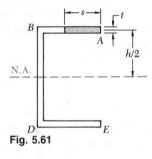

Fig. 5.61

The distance e from the center line of the web BD to the shear center O may now be obtained from Eq. (5.23):

$$e = \frac{Fh}{V} = \frac{Vthb^2}{4I} \frac{h}{V} = \frac{th^2b^2}{4I} \qquad (5.26)$$

The moment of inertia I of the channel section may be expressed as follows:

$$I = I_{\text{web}} + 2I_{\text{flange}}$$

$$= \frac{1}{12}th^3 + 2\left[\frac{1}{12}bt^3 + bt\left(\frac{h}{2}\right)^2\right]$$

Neglecting the term containing t^3, which is very small, we have

$$I = \tfrac{1}{12}th^3 + \tfrac{1}{2}tbh^2 = \tfrac{1}{12}th^2(6b + h) \qquad (5.27)$$

Substituting this expression into (5.26), we write

$$e = \frac{3b^2}{6b + h} = \frac{b}{2 + \dfrac{h}{3b}} \qquad (5.28)$$

We note that the distance e does not depend upon t and may vary from 0 to $b/2$, depending upon the value of the ratio $h/3b$. For the given channel section, we have

$$\frac{h}{3b} = \frac{150 \text{ mm}}{3(100 \text{ mm})} = 0.5$$

and

$$e = \frac{100 \text{ mm}}{2 + 0.5} = 40 \text{ mm}$$

Example 5.05

For the channel section of Example 5.04, determine the distribution of the shearing stresses caused by an 800-N vertical shear **V** exerted at the shear center O (Fig. 5.62).

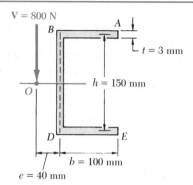

Fig. 5.62   e = 40 mm

Since **V** is applied at the shear center, there is no torsion, and the stresses in flange AB are obtained from Eq. (5.24) of Example 5.04.   We have

$$\tau = \frac{q}{t} = \frac{VQ}{It} = \frac{Vh}{2I} s \qquad (5.29)$$

which shows that the stress distribution in flange AB is linear. Letting $s = b$ and substituting for I from Eq. (5.27), we obtain the value of the shearing stress at B:

$$\tau_B = \frac{Vhb}{2(\frac{1}{12}th^2)(6b + h)} = \frac{6Vb}{th(6b + h)} \qquad (5.30)$$

Letting $V = 800$ N, and using the given dimensions, we have

$$\tau_B = \frac{6(800 \text{ N})(0.100 \text{ m})}{(0.003 \text{ m})(0.150 \text{ m})(6 \times 0.100 \text{ m} + 0.150 \text{ m})}$$

$$= 1.422 \text{ MPa}$$

The distribution of the shearing stresses in the web BD is parabolic, as in the case of a W-beam, and the maximum stress occurs at the neutral axis.   Computing the first moment of the upper half of the cross section with respect to the neutral axis (Fig. 5.63), we write

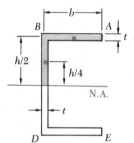

Fig. 5.63

$$Q = bt(\tfrac{1}{2}h) + \tfrac{1}{2}ht(\tfrac{1}{4}h) = \tfrac{1}{8}ht(4b + h) \qquad (5.31)$$

Substituting for I and Q from (5.27) and (5.31), respectively, into the expression for the shearing stress, we have

$$\tau_{max} = \frac{VQ}{It} = \frac{V(\tfrac{1}{8}ht)(4b + h)}{\tfrac{1}{12}th^2(6b + h)t} = \frac{3V(4b + h)}{2th(6b + h)}$$

or, with the given data,

$$\tau_{max} = \frac{3(800 \text{ N})(4 \times 0.100 \text{ m} + 0.150 \text{ m})}{2(0.003 \text{ m})(0.150 \text{ m})(6 \times 0.100 \text{ m} + 0.150 \text{ m})}$$

$$= 1.956 \text{ MPa}$$

The distribution of the shearing stresses over the entire channel section has been plotted in Fig. 5.64.

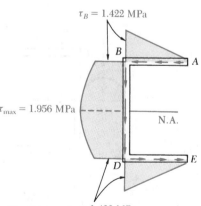

Fig. 5.64 $\tau_D = 1.422$ MPa

Example 5.06

For the channel section of Example 5.04, and neglecting stress concentrations, determine the maximum shearing stress caused by a 800-N vertical shear **V** applied at the centroid C of the section, which is located 29 mm to the right of the center line of the web BD (Fig. 5.65).

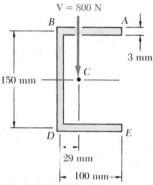

V = 800 N

B   A

150 mm

3 mm

C

D   E

29 mm

100 mm

Fig. 5.65

The shear center O of the cross section was determined in Example 5.04 and found to be at a distance $e = 40$ mm to the left of the center line of the web BD. We shall replace the shear **V** (Fig. 5.66a) by an equivalent force-couple system at the shear center O (Fig. 5.66b). This system consists of an 800-N force **V** and of a torque **T** of magnitude

$$T = V(OC) = (800 \text{ N})(40 \text{ mm} + 29 \text{ mm})$$
$$T = 55.2 \text{ N} \cdot \text{m}$$

The 800-N force **V** causes the member to bend, and the corresponding distribution of shearing stresses in the section (Fig. 5.66c) was determined in Example 5.05. We recall that the maximum value of the stress due to this force was found to be

$$(\tau_{max})_{bending} = 1.956 \text{ MPa}$$

The torque **T** causes the member to twist, and the corresponding distribution of stresses is shown in Fig. 5.66d. We recall from Sec. 3.12 that the membrane analogy shows that, in a thin-walled member of uniform thickness, the stress caused by a torque **T** is maximum along the edge of the section. Using Eqs. (3.45) and (3.43) with

$$a = 100 \text{ mm} + 150 \text{ mm} + 100 \text{ mm} = 350 \text{ mm}$$
$$b = t = 3 \text{ mm} \qquad b/a = 0.00857$$

we have

$$c_1 = \tfrac{1}{3}(1 - 0.630 b/a) = \tfrac{1}{3}(1 - 0.630 \times 0.00857) = 0.332$$

$$(\tau_{max})_{torsion} = \frac{T}{c_1 ab^2} = \frac{55.2 \text{ N} \cdot \text{m}}{(0.331)(350 \text{ mm})(3 \text{ mm})^2} = 52.8 \text{ MPa}$$

The maximum stress due to the combined bending and torsion occurs at the neutral axis, on the inside surface of the web, and is

$$\tau_{max} = 1.956 \text{ MPa} + 52.8 \text{ MPa} = 54.8 \text{ MPa}$$

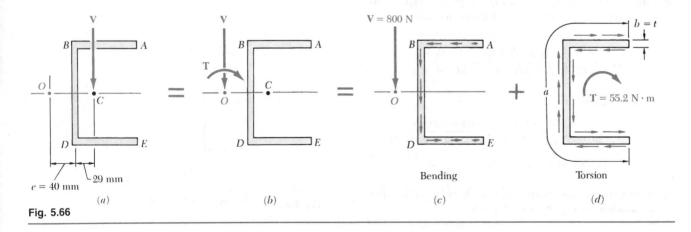

Bending   Torsion

e = 40 mm   29 mm

(a)   (b)   (c)   (d)

Fig. 5.66

Turning our attention to thin-walled members possessing no plane of symmetry, we shall consider the case of an angle shape subjected to a vertical load **P**. If the member is oriented in such a way that the load **P** is perpendicular to one of the principal centroidal axes Cz of the cross section, the couple vector **M** representing the bending moment in a given section will be directed along Cz (Fig. 5.67), and the neutral axis will coincide with that axis (cf. Sec. 4.14). Equation (4.16), therefore, is applicable and may be used to compute the normal stresses in the section. We now propose to determine where the load **P** should be applied if Eq. (5.10) is to define the shearing stresses in the section, i.e., if the member is to *bend without twisting*.

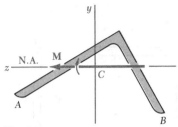

Fig. 5.67

Let us *assume* that the shearing stresses in the section are defined by Eq. (5.10). As in the case of the channel member considered earlier, the elementary shearing forces exerted on the section may be expressed as $dF = q\,ds$, with $q = VQ/I$, where Q represents a first moment with respect to the neutral axis (Fig. 5.68a). We note that the resultant of the

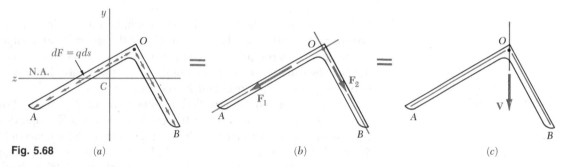

Fig. 5.68 (a) (b) (c)

shearing forces exerted on portion OA of the cross section is a force $\mathbf{F}_1$ directed along OA, and that the resultant of the shearing forces exerted on portion OB is a force $\mathbf{F}_2$ along OB (Fig. 5.68b). Since both $\mathbf{F}_1$ and $\mathbf{F}_2$ pass through point O at the corner of the angle, it follows that their own resultant, which is the shear **V** in the section, must also pass through O (Fig. 5.68c). We conclude that the member will not be twisted if the line of action of the load **P** passes through the corner O of the section in which it is applied.

The same reasoning may be applied when the load **P** is perpendicular to the other principal centroidal axis Cy of the angle section. And, since any load **P** applied at the corner O of a cross section may be resolved into components perpendicular to the principal axes, it follows that the member will not be twisted if each load is applied at the corner O of a cross section. We thus conclude that O is the shear center of the section.

Angle shapes with one vertical and one horizontal leg are encountered in many structures. It follows from the preceding discussion that such members will not be twisted if vertical loads are applied along the center line of their vertical leg. We note from Fig. 5.69 that the resultant of the elementary shearing forces exerted on the vertical portion OA of a

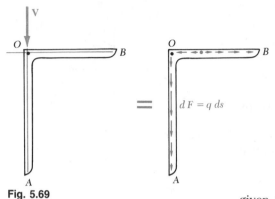

Fig. 5.69

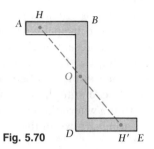

Fig. 5.70

given section will be equal to the shear **V**, while the resultant of the shearing forces on the horizontal portion OB will be zero:

$$\int_O^A q\, ds = V \qquad \int_O^B q\, ds = 0$$

This does *not* mean, however, that there will be no shearing stress in the horizontal leg of the member. By resolving the shear **V** into components perpendicular to the principal centroidal axes of the section and computing the shearing stress at every point, we would verify that τ is zero at only one point between O and B (see Sample Prob. 5.8).

Another type of thin-walled member frequently encountered in practice is the Z shape. While the cross section of a Z shape does not possess any axis of symmetry, it does possess a *center of symmetry* O (Fig. 5.70). This means that, to any point H of the cross section corresponds another point H' such that the segment of straight line HH' is bisected by O. Clearly, the center of symmetry O coincides with the centroid of the cross section. As we shall see presently, point O is also the shear center of the cross section.

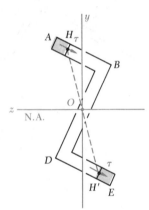

Fig. 5.71

As we did earlier in the case of an angle shape, we shall assume that the loads are applied in a plane perpendicular to one of the principal axes of the section, so that this axis is also the neutral axis of the section (Fig. 5.71). We further assume that the shearing stresses in the section are defined by Eq. (5.10), i.e., that the member is bent without being twisted. Denoting by Q the first moment about the neutral axis of portion AH of the cross section, and by Q' the first moment of portion EH', we note that $Q' = -Q$. Thus the shearing stresses at H and H' have the same magnitude and the same direction, and the shearing forces exerted on small elements of area dA located respectively at H and H' are equal forces, which have equal and opposite moments about O (Fig. 5.72). Since this is true for any pair of symmetric elements, it follows that the resultant of the shearing forces exerted on the section has a zero moment about O. This means that the shear **V** in the section is directed along a line which passes through O. Since this analysis may be repeated when the loads are applied in a plane perpendicular to the other principal axis, we conclude that point O is the shear center of the section.

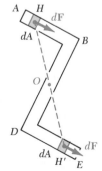

Fig. 5.72

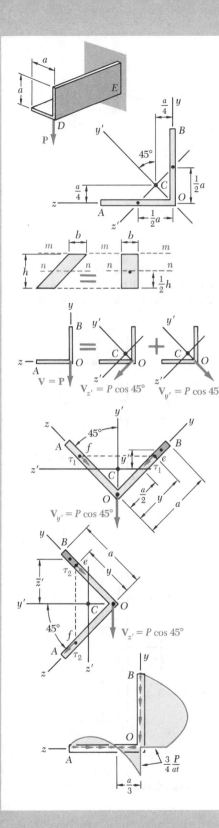

Determine the distribution of shearing stresses in the thin-walled angle shape DE of uniform thickness t for the loading shown.

Shear Center. We recall from Sec. 5.12 that the shear center of the cross section of a thin-walled angle shape is located at its corner. Since the load **P** is applied at D, it causes bending but no twisting of the shape.

Principal Axes. We locate the centroid C of a given cross section AOB. Since the y' axis is an axis of symmetry, the y' and z' axes are the principal centroidal axes of the section. We recall that for the parallelogram shown $I_{nn} = \frac{1}{12}bh^3$ and $I_{mm} = \frac{1}{3}bh^3$. Considering each leg of the section as a parallelogram, we now determine the centroidal moments of inertia $I_{y'}$ and $I_{z'}$:

$$I_{y'} = 2\left[\frac{1}{3}\left(\frac{t}{\cos 45°}\right)(a\cos 45°)^3\right] = \frac{1}{3}ta^3$$

$$I_{z'} = 2\left[\frac{1}{12}\left(\frac{t}{\cos 45°}\right)(a\cos 45°)^3\right] = \frac{1}{12}ta^3$$

Superposition. The shear **V** in the section is equal to the load **P**. We resolve it into components parallel to the principal axes.

Shearing Stresses Due to $V_{y'}$. We shall determine the shearing stress at point e of coordinate y:

$$\bar{y}' = \tfrac{1}{2}(a+y)\cos 45° - \tfrac{1}{2}a\cos 45° = \tfrac{1}{2}y\cos 45°$$

$$Q = t(a-y)\bar{y}' = \tfrac{1}{2}t(a-y)y\cos 45°$$

$$\tau_1 = \frac{V_{y'}Q}{I_{z'}t} = \frac{(P\cos 45°)[\frac{1}{2}t(a-y)y\cos 45°]}{(\frac{1}{12}ta^3)t} = \frac{3P(a-y)y}{ta^3}$$

The shearing stress at point f is represented by a similar function of z.

Shearing Stresses Due to $V_{z'}$. We again consider point e:

$$\bar{z}' = \tfrac{1}{2}(a+y)\cos 45°$$

$$Q = (a-y)t\bar{z}' = \tfrac{1}{2}(a^2-y^2)t\cos 45°$$

$$\tau_2 = \frac{V_{z'}Q}{I_{y'}t} = \frac{(P\cos 45°)[\frac{1}{2}(a^2-y^2)t\cos 45°]}{(\frac{1}{3}ta^3)t} = \frac{3P(a^2-y^2)}{4ta^3}$$

The shearing stress at point f is represented by a similar function of z.

Combined Stresses. *Along the Vertical Leg.* The shearing stress at point e is

$$\tau_e = \tau_2 + \tau_1 = \frac{3P(a^2-y^2)}{4ta^3} + \frac{3P(a-y)y}{ta^3} = \frac{3P(a-y)}{4ta^3}[(a+y)+4y]$$

$$\boxed{\tau_e = \frac{3P(a-y)(a+5y)}{4ta^3}} \quad \blacktriangleleft$$

Along the Horizontal Leg. The shearing stress at point f is

$$\tau_f = \tau_2 - \tau_1 = \frac{3P(a^2-z^2)}{4ta^3} - \frac{3P(a-z)z}{ta^3} = \frac{3P(a-z)}{4ta^3}[(a+z)-4z]$$

$$\boxed{\tau_f = \frac{3P(a-z)(a-3z)}{4ta^3}} \quad \blacktriangleleft$$

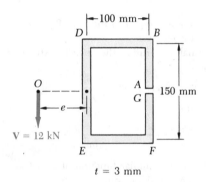

Fig. P5.99

PROBLEMS

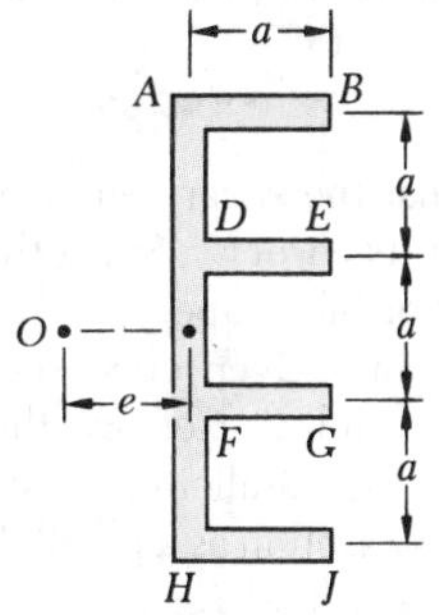

5.99 and 5.100 Determine the location of the shear center O of a thin-walled beam of uniform thickness having the cross section shown.

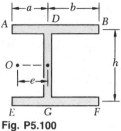

Fig. P5.100

5.101 through 5.106 For an extruded beam having the cross section shown, determine (a) the location of the shear center O, (b) the distribution of the shearing stresses caused by the vertical shearing force **V** shown applied at O.

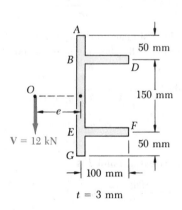

Fig. P5.101

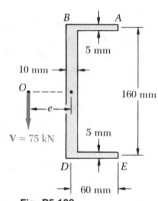

Fig. P5.102

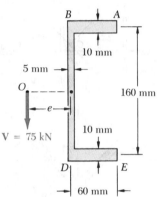

Fig. P5.103

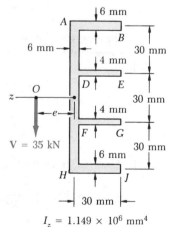

$I_z = 1.149 \times 10^6 \text{ mm}^4$

Fig. P5.105

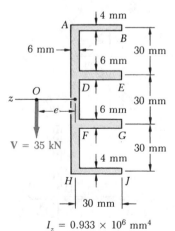

$I_z = 0.933 \times 10^6 \text{ mm}^4$

Fig. P5.106

Fig. P5.104

5.107 through 5.112   Determine the location of the shear center O of a thin-walled beam of uniform thickness having the cross section shown.

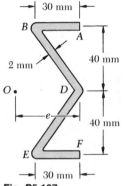

Fig. P5.107

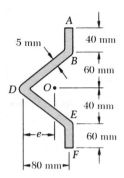

Fig. P5.108

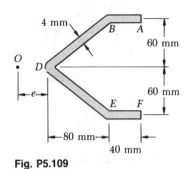

Fig. P5.109

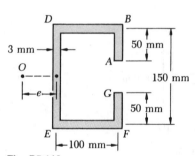

Fig. P5.110

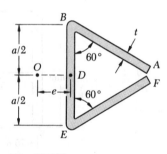

Fig. P5.111

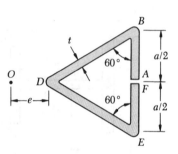

Fig. P5.112

5.113 and 5.114   Determine the location of the shear center O of a thin-walled beam of uniform thickness having the cross section shown.

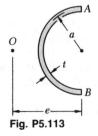

Fig. P5.113

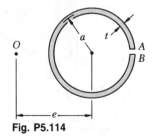

Fig. P5.114

5.115 and 5.116 A thin-walled beam of uniform thickness has the cross section shown. Determine the dimension b for which the shear center O of the beam is located at the point indicated.

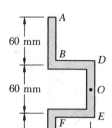

Fig. P5.115

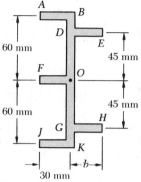

Fig. P5.116

5.117 Determine the location of the shear center O of a thin-walled beam having the cross section shown.

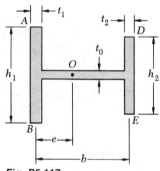

Fig. P5.117

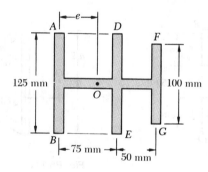

Fig. P5.118

5.118 Determine the location of the shear center O of a thin-walled beam of uniform thickness having the cross section shown.

5.119 Solve Prob. 5.117 when $t_0 = t_1 = t_2 = 8$ mm, $h_1 = 75$ mm, $h_2 = 60$ mm, and $b = 50$ mm.

*5.120 A steel plate, 160 mm wide and 8 mm thick, is bent to form the channel shown. Knowing that the vertical load **P** acts at a point in the midplane of the web of the channel, determine (*a*) the torque **T** which would cause the channel to twist in the same way that it does under the load **P**, (*b*) the maximum shearing stress in the channel caused by the load **P**.

*5.121 Solve Prob. 5.120, assuming that a 6-mm-thick plate is bent to form the channel shown.

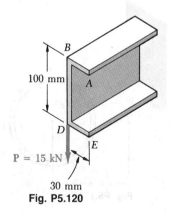

P = 15 kN

30 mm

Fig. P5.120

*5.122 The cantilever beam AB consists of half of a thin-walled pipe of 30-mm mean radius and 5-mm wall thickness. Knowing that the vertical load **P** passes through the centroid C of the cross section of the beam, determine (*a*) the torque which would cause the beam to twist in the same way that it does under the load **P**, (*b*) the maximum shearing stress in the beam caused by the load **P**.

*5.123 Solve Prob. 5.122, assuming that the wall thickness of the cantilever beam is increased to 6 mm.

5.124 For the angle shape and loading of Sample Prob. 5.8, check that $\int q\, dz = 0$ along the horizontal leg of the angle.

5.125 For the angle shape and loading of Sample Prob. 5.8, check that $\int q\, dy = P$ along the vertical leg of the angle.

5.126 For the angle shape and loading of Sample Prob. 5.8, (*a*) determine the points where the shearing stress is maximum and the corresponding values of the stress, (*b*) verify that the points obtained are located on the neutral axis corresponding to the given loading.

5.127 The cantilever beam shown consists of a channel of uniform thickness. Knowing that the load **P** acts at the shear center O and in the plane of the end section of the beam, determine the shearing stress at (*a*) point B', (*b*) point D'.

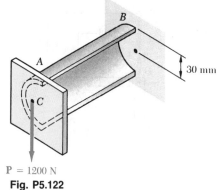

$P = 1200$ N

Fig. P5.122

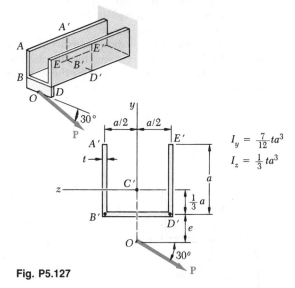

$$I_y = \tfrac{7}{12} ta^3$$
$$I_z = \tfrac{1}{3} ta^3$$

Fig. P5.127

*5.128 For the cantilever beam and loading of Prob. 5.127, determine the point where the shearing stress is maximum and the corresponding value of the stress (*a*) along line $B'D'$ in the web, (*b*) along line $D'E'$ in a leg.

*5.129 For the cantilever beam and loading of Prob. 5.127, using the results of Prob. 5.128, verify that the points where the shearing stress is maximum are located on the neutral axis corresponding to the given loading.

*5.130 Determine the distribution of the shearing stresses along line $D'B'$ in the horizontal leg of the angle shape for the loading shown. The x' and y' axes are the principal centroidal axes of the cross section.

*5.131 For the angle shape and loading of Prob. 5.130, determine the distribution of the shearing stresses along line $D'A'$ in the vertical leg.

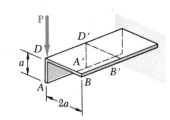

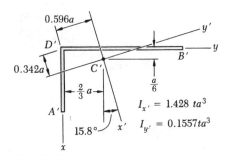

$I_{x'} = 1.428\, ta^3$
$I_{y'} = 0.1557 ta^3$

Fig. P5.130

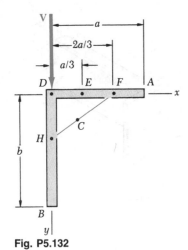

Fig. P5.132

*5.132 Show that for the given angle shape and loading, (*a*) the distribution of the shearing stresses in each leg is parabolic, with $\tau = 0$ at *A* and *B*, (*b*) the shearing stress in the horizontal leg is also zero at point *E* of abscissa $x = a/3$, and is maximum at point *F* of abscissa $x = 2a/3$. (*Hint:* The resultant of the shearing forces in the horizontal leg must be zero.)

*5.133 For the angle shape and loading of Prob. 5.132, show that the shearing stress in the vertical leg is maximum at point *H* of ordinate $y = 2b/(4 + \beta)$, where β is the ratio A_h/A_v of the cross-sectional areas of the horizontal and vertical legs. (*Hint:* Points *F* and *H*, where the shearing stress is maximum are located on the neutral axis, and this axis passes through the centroid *C* of the cross section.)

*5.134 For the angle shape and loading of Prob. 5.132, show that the maximum shearing stress in the vertical leg is

$$\tau_m = \frac{3}{2}\frac{V}{bt}\frac{\left(1 + \frac{1}{2}\beta\right)^2}{(1 + \beta)\left(1 + \frac{1}{4}\beta\right)}$$

where *t* is the thickness of the vertical leg and β is the ratio A_h/A_v of the cross-sectional areas of the horizontal and vertical legs. (*Hint:* Use the results obtained in part *a* of Prob. 5.132 and in Prob. 5.133, and the fact that the resultant of the shearing forces in the vertical leg is equal to **V**.)

*5.135 and 5.136 The cantilever beam shown consists of a Z shape of uniform thickness. For the loading shown, determine the distribution of shearing stresses in the vertical leg *DE*. The *x'* and *y'* axes are the principal centroidal axes of the cross section.

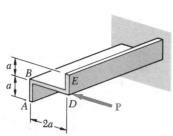

Fig. P5.135 and P5.137

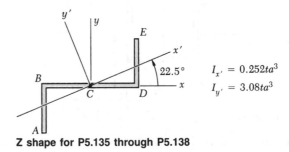

Z shape for P5.135 through P5.138

$I_{x'} = 0.252ta^3$
$I_{y'} = 3.08ta^3$

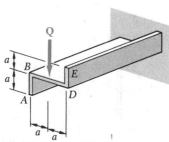

Fig. P5.136 and P5.138

*5.137 and 5.138 The cantilever beam shown consists of a Z shape of uniform thickness. For the loading shown, determine the distribution of shearing stresses in the horizontal web *BD*. The *x'* and *y'* axes are the principal centroidal axes of the cross section.

REVIEW AND SUMMARY

This chapter was devoted to the study of the internal forces and stresses in prismatic members caused by *transverse loadings*.

In Sec. 5.2, we considered a cantilever beam with a constant shear V caused by a single load at its free end and found that longitudinal shearing forces and stresses exist in the beam. Having determined in Sec. 5.3 the distribution of the normal stresses due to bending under the assumption that these stresses are not affected by the presence of shear, we considered in Sec. 5.4 the portion of beam shown in Fig. 5.8 and determined the shear H exerted on its lower horizontal face (Fig. 5.9):

$$H = \frac{PQ}{I}x \qquad (5.6)$$

Distribution of normal stresses

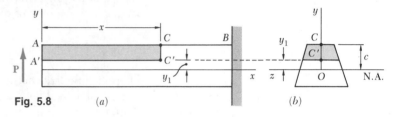

Fig. 5.8   (*a*)   (*b*)

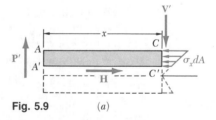 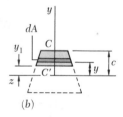

Fig. 5.9   (*a*)   (*b*)

We then defined the *shear flow q* at a given point C' of a transverse section as the horizontal shear per unit length at that point. We wrote

Shear flow

$$q = \frac{VQ}{I} \qquad (5.8)$$

where V is the shear at the section considered, I the centroidal moment of inertia of the entire cross section of the beam, and Q the first moment with respect to the neutral axis of the area of the cross section above point C'. We were then able to calculate the internal horizontal forces in beams subjected to shear [Example 5.01].

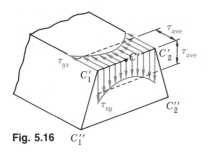

Fig. 5.16

Vertical shearing stresses

Maximum shearing stress in a rectangular beam

Maximum shearing stress in wide-flange beams

Shear flow and shearing stresses on longitudinal cuts

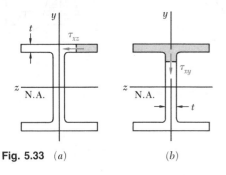

Fig. 5.33 *(a)*   *(b)*

Plastic deformations

In Sec. 5.5, we determined the average horizontal shearing stress at a given point C' (Fig. 5.16):

$$\tau_{\text{ave}} = \frac{VQ}{It} \qquad (5.10)$$

where t is the width of the cross section at point C'. Recalling from Chap. 1 that at a given point the shearing stresses on a transverse and a horizontal plane are of equal magnitude, we noted that the above expression also represents the average vertical shearing stress along a given horizontal line $C_1'C_2'$.

In Secs. 5.6 and 5.7, we analyzed the shearing stresses along horizontal cuts in a beam of rectangular cross section. We found that the distribution of shearing stresses is parabolic and that the maximum stress occurs at the center of the section:

$$\tau_{\text{max}} = \frac{3}{2}\frac{V}{A} \qquad (5.14)$$

where A is the area of the rectangular section. For wide-flange beams, we found that a good approximation of the maximum shearing stress may be obtained by dividing the shear V by the area of the web.

Considering in Sec. 5.8 an arbitrary longitudinal cut along a prismatic member, we found that the shear flow across the cut could still be obtained from Eq. (5.8). We then discussed [Sec. 5.9] the distribution of stresses in symmetric thin-walled members and found that Eq. (5.10) was still valid:

$$\tau_{\text{ave}} = \frac{VQ}{It} \qquad (5.10)$$

In the case of a wide-flange beam, for instance, Eq. (5.10) may be used to determine the shearing stress τ_{xz} in the flange and the shearing stress τ_{xy} in the web (Fig. 5.33). In both instances, t represents the thickness of the wall at the cut and Q the first moment with respect to the neutral axis of the area of the section isolated by the cut.

In Sec. 5.10, we considered the effect of plastic deformations on the magnitude and distribution of shearing stresses. From Chap. 4, we recalled that once plastic deformation has been initiated, additional loading causes plastic zones to penetrate into the elastic core of a beam. After demonstrating that shearing stresses can occur only in the elastic core of a beam, we noted that both an increase in loading and the resulting decrease in the size of the elastic core contribute to an increase in shearing stresses.

In preceding chapters we learned to determine the stresses in prismatic members caused by axial loadings [Chap. 1], torsion [Chap. 3], and bending [Chap. 4], and in this chapter, we determined the stresses produced by transverse loadings. In Sec. 5.11, we combined this knowledge to determine the stresses in members under more general loading conditions.

Stresses under combined loadings

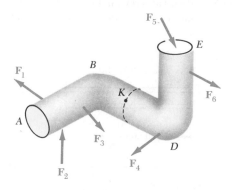

Fig. 5.43

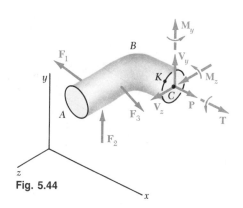

Fig. 5.44

For instance, in order to determine the stresses at point K of the bent member $ABDE$ shown in Fig. 5.43, the applied loads were replaced by an equivalent force-couple system at the centroid C of the cross section containing point K (Fig. 5.44). The corresponding normal or shearing stresses at K were then determined for each of the forces and couples applied at C. At point K, the total normal stress was obtained by combining the stresses caused by the axial load and by the bending couples. Similarly, shearing stresses caused by the torque and by the shearing forces were combined to obtain the shearing stress at K.

For a prismatic member which is not loaded in a plane of symmetry we found that, in general, both bending and twisting occur. In Sec. 5.12, we learned to locate a point O of the cross section, known as the *shear center*, where loads must be applied if the member is to bend without twisting (Fig. 5.58). We found that for loads applied at the shear center of a section, the following formulas remain valid:

Shear center

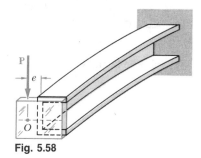

Fig. 5.58

$$\sigma_x = -\frac{My}{I} \quad \text{and} \quad \tau_{ave} = \frac{VQ}{It}$$

Using the principle of superposition, we were able to determine the stresses in unsymmetric thin-walled members such as channels, angles, and extruded beams [Example 5.06 and Sample Prob. 5.8].

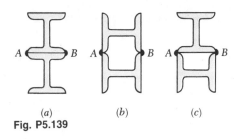

(a)   (b)   (c)

Fig. P5.139

REVIEW PROBLEMS

5.139 Two W 130 × 23.8 rolled sections may be welded at A and B in any of the three ways shown to form a composite beam. Knowing that for each weld the allowable horizontal shearing force is 480 kN·m, determine for each arrangement the maximum allowable vertical shear in the composite beam.

5.140 Two full-size 2×150-mm planks are connected by lag screws to form a beam which is subjected to a vertical shear of 1.8 kN. Knowing that the allowable shearing force is 1.6 kN in each lag screw, determine (a) the largest permissible longitudinal spacing of the lag screws, (b) the corresponding maximum shearing stress in the beam.

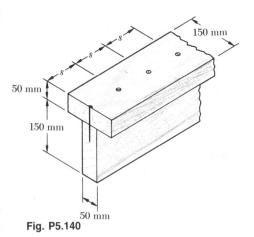

Fig. P5.140

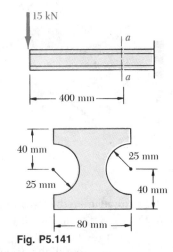

Fig. P5.141

5.141 For the beam and loading shown, consider section a-a and determine (a) the largest normal stress, (b) the largest shearing stress.

5.142 A column is fabricated by bolting together two C 250 × 22.8 channels and two plates as shown. Determine the shearing stress at point a caused by a shearing force of 120 kN applied along the y axis.

5.143 A column is fabricated by bolting together two C 250 × 22.8 channels and two plates as shown. Determine the shearing stress at point b caused by a shearing force of 120 kN applied along the x axis.

5.144 Three forces are applied to a W 150 × 29.8 rolled beam as shown. Determine the normal and shearing stresses at the three points indicated.

5.145 Solve Prob. 5.144, assuming that the 12.5 kN vertical load is removed.

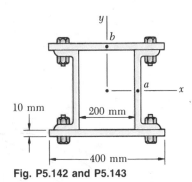

Fig. P5.142 and P5.143

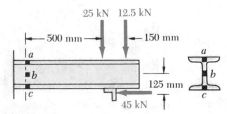

Fig. P5.144

5.146 The box beam shown is made by gluing together four planks. Knowing that in the glued joints the average allowable shearing stress is 825 kPa, determine (a) the largest permissible vertical shear in the beam, (b) the corresponding maximum shearing stress in the beam.

5.147 For the box beam of Prob. 5.146, determine (a) the largest permissible horizontal shear in the beam, (b) the corresponding maximum shearing stress in the beam.

5.148 Two forces are applied to the pipe AB as shown. Knowing that the pipe has inner and outer diameters equal to 35 and 42 mm, respectively, determine the normal and shearing stresses at (a) point a, (b) point b.

5.149 Solve Prob. 5.148, assuming that the 1500-N force is directed upward.

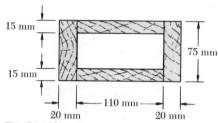

Fig. P5.146

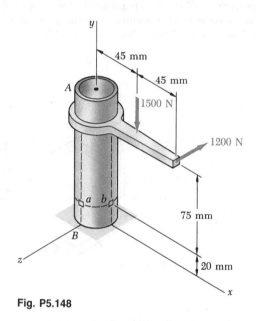

Fig. P5.148

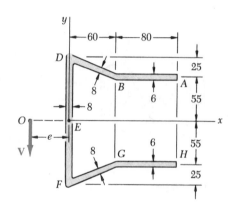

Dimensions in mm

Fig. P5.150

5.150 For an extruded beam having the cross section shown, determine (a) the location of the shear center O, (b) the distribution of the shearing stresses caused by a vertical 2.5-kN force **V** applied at O.

The following problems are designed to be solved with a computer.

5.C1 A beam having the cross section shown is subjected to a vertical shear **V**. Write a computer program which, for loads and dimensions expressed in any consistent set of units, can be used to calculate the shearing stress along the line between any two adjacent rectangular areas which form the cross section. Use this program to solve (a) Prob. 5.22, (b) Prob 5.23, (c) Prob. 5.11b and c, (d) Prob. 5.12b and c, (e) Prob. 5.13b and c.

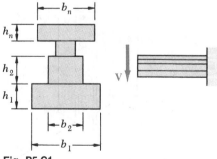

Fig. P5.C1

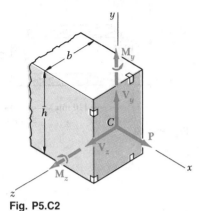

Fig. P5.C2

5.C2 Write a computer program which can be used to calculate the shearing and normal stresses at points with given coordinates y and z located on the surface of a machine part having a rectangular cross section. The internal forces are known to be equivalent to the force-couple system shown, which is located at the centroid of the cross section. Write the program so that the loads and dimensions may be expressed in any consistent set of units. Use this program to calculate the stresses in (*a*) Prob. 5.74, (*a*) Prob. 5.76.

5.C3 Write a computer program which can be used to calculate the shearing and normal stresses at points *A*, *B*, *C*, and *D* located on the outer surface of a machine part having the annular cross section shown. The internal forces are known to be equivalent to the force-couple system shown which is located at the centroid of the cross section. Write the program so that the loads and dimensions may be expressed in any consistent set of units. Use this program to calculate the stresses in (*a*) Prob. 5.64, (*b*) Prob. 5.66, (*c*) Probs. 5.70 and 5.71, (*d*) Probs. 5.72 and 5.73.

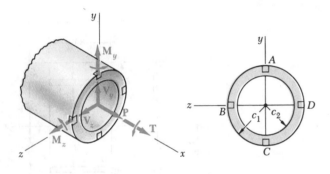

Fig. P5.C3

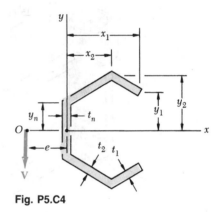

Fig. P5.C4

5.C4 The cross section of an extruded beam is symmetric with respect to the *x* axis and consists of several straight segments as shown. Write a computer program which can be used to determine (*a*) the location of the shear center *O*, (*b*) the distribution of shearing stresses caused by a vertical force applied at *O*. Use this program to solve Probs. 5.102, 5.108, and 5.150.

5.C5 The short post *AB* is acted upon by the force **F**, which lies in the *xy* plane and forms an angle β with the horizontal. Write a computer program and use it to calculate, for the data given below, (*a*) the normal stress at point *a* for values of β from 0 to 60° at 5° intervals, (*b*) the largest compressive stress occurring at point *a*, (*c*) the corresponding value of β.

Shape	F	L	d
W 150 × 18	15 kN	750 mm	100 mm
W 100 × 19.3	10 kN	500 mm	125 mm

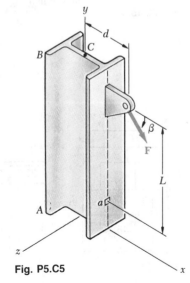

Fig. P5.C5

C H A P T E R S I X

TRANSFORMATIONS OF STRESS AND STRAIN

6.1. INTRODUCTION

We saw in Sec. 1.8 that the most general state of stress at a given point Q may be represented by six components. Three of these components, σ_x, σ_y, and σ_z, define the normal stresses exerted on the faces of a small cubic element centered at Q and of the same orientation as the coordinate axes (Fig. 6.1a), and the other three, τ_{xy}, τ_{yz}, and τ_{zx},† the components of the shearing stresses on the same element. As we remarked at the time, the same state of stress will be represented by a different set of components if the coordinate axes are rotated (Fig. 6.1b). We propose in the first part of this chapter to determine how the components of stress are transformed under a rotation of the coordinate axes. The second part of the chapter will be devoted to a similar analysis of the transformation of the components of strain.

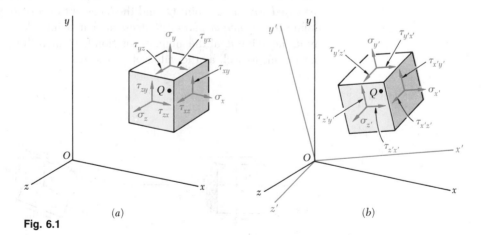

(a)   (b)

Fig. 6.1

† We recall that $\tau_{yx} = \tau_{xy}$, $\tau_{zy} = \tau_{yz}$, and $\tau_{xz} = \tau_{zx}$.

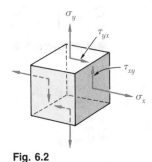

Fig. 6.2

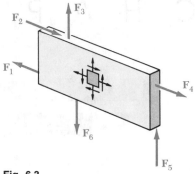

Fig. 6.3

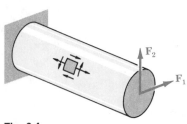

Fig. 6.4

Our discussion of the transformation of stress will deal mainly with *plane stress*, i.e., with a situation in which two of the faces of the cubic element are free of any stress. If the z axis is chosen perpendicular to these faces, we have $\sigma_z = \tau_{zx} = \tau_{zy} = 0$, and the only remaining stress components are σ_x, σ_y, and τ_{xy} (Fig. 6.2). Such a situation occurs in a thin plate subjected to forces acting in the midplane of the plate (Fig. 6.3). It also occurs on the free surface of a structural element or machine component, i.e., at any point of the surface of that element or component which is not subjected to an external force (Fig. 6.4).

Considering in Sec. 6.2 a state of plane stress at a given point Q characterized by the stress components σ_x, σ_y, and τ_{xy} associated with the element shown in Fig. 6.5a, we shall learn to determine the components $\sigma_{x'}$, $\sigma_{y'}$, and $\tau_{x'y'}$ associated with that element after it has been rotated through an angle θ about the z axis (Fig. 6.5b). In Sec. 6.3, we shall determine the value θ_p of θ for which the stresses $\sigma_{x'}$ and $\sigma_{y'}$ are, respectively, maximum and minimum; these values of the normal stress are the *principal stresses* at point Q, and the faces of the corresponding element define the *principal planes of stress* at that point. We shall also determine the value θ_s of the angle of rotation for which the shearing stress is maximum, as well as the value of that stress.

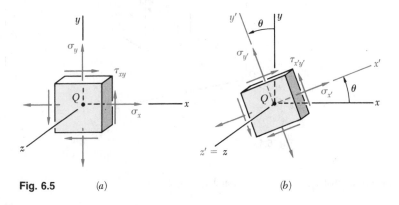

Fig. 6.5   (a)   (b)

An alternative method for the solution of problems involving the transformation of plane stress, based on the use of *Mohr's circle*, will be presented in Sec. 6.4. We should note that, whatever method we select, we shall be in a position to determine both the maximum normal stress and the maximum shearing stress produced by combined loadings at any point on the free surface of a structural element or machine component. Indeed, we learned in Sec. 5.12 to determine the stress components produced by combined loadings on an element such as the one shown in Fig. 6.4; from these components we shall now be able to determine the maximum normal and shearing stresses at that point.

In Sec. 6.5, we shall consider a *three-dimensional state of stress* at a given point and develop a formula for the determination of the normal stress on a plane of arbitrary orientation at that point. In Sec. 6.6, we shall consider the rotations of a cubic element about each of the principal axes of stress and note that the corresponding transformations of stress may be described by three different Mohr's circles. We shall also observe that, in the case of a state of *plane stress* at a given point, the maximum value of the shearing stress obtained earlier by considering rotations in the plane of stress may not necessarily represent the maximum shearing stress at that point. This will bring us to distinguish between *in-plane* and *out-of-plane* maximum shearing stresses.

Yield criteria for ductile materials under plane stress will be developed in Sec. 6.7. To predict whether a material will yield at some critical point under given loading conditions, we shall determine the principal stresses σ_a and σ_b at that point and check whether σ_a, σ_b, and the yield strength σ_Y of the material satisfy some criterion. Two criteria in common use are: the *maximum-shearing-strength criterion* and the *maximum-distortion-energy criterion*. In Sec. 6.8, *fracture criteria* for brittle materials under plane stress will be developed in a similar fashion; they will involve the principal stresses σ_a and σ_b at some critical point and the ultimate strength σ_U of the material. The two criteria we shall discuss are the *maximum-normal-stress* criterion and *Mohr's criterion*.

Thin-walled pressure vessels provide an important application of the analysis of plane stress. In Sec. 6.9, we shall discuss stresses in both cylindrical and spherical pressure vessels.

Sections 6.10 and 6.11 will be devoted to a discussion of the *transformation of plane strain* and to *Mohr's circle for plane strain*. In Sec. 6.12, we shall consider the three-dimensional analysis of strain and see how Mohr's circles may be used to determine the maximum shearing strain at a given point. Two particular cases are of special interest and should not be confused: the case of *plane strain* and the case of *plane stress*.

Finally, in Sec. 6.13, we shall discuss the use of *strain gages* to measure the normal strain on the surface of a structural element or machine component. We shall see how the components ϵ_x, ϵ_y, and γ_{xy} characterizing the state of strain at a given point may be computed from the measurements made with three strain gages forming a *strain rosette*.

6.2. TRANSFORMATION OF PLANE STRESS

Let us assume that a state of plane stress exists at point Q (with $\sigma_z = \tau_{zx} = \tau_{zy} = 0$), and that it is defined by the stress components σ_x, σ_y, and τ_{xy} associated with the element shown in Fig. 6.5a. We propose to determine the stress components $\sigma_{x'}$, $\sigma_{y'}$, and $\tau_{x'y'}$ associated with the

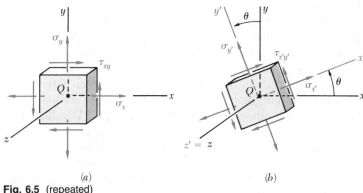

(a) (b)

Fig. 6.5 (repeated)

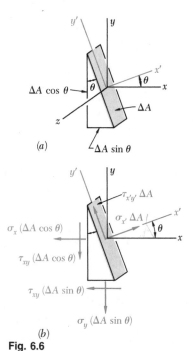

(a)

(b)

Fig. 6.6

element after it has been rotated through an angle θ about the z axis (Fig. 6.5b), and to express these components in terms of σ_x, σ_y, τ_{xy}, and θ.

In order to determine the normal stress $\sigma_{x'}$ and the shearing stress $\tau_{x'y'}$ exerted on the face perpendicular to the x' axis, we shall consider a prismatic element with faces respectively perpendicular to the x, y, and x' axes (Fig. 6.6a). We observe that, if the area of the oblique face is denoted by ΔA, the areas of the vertical and horizontal faces are respectively equal to $\Delta A \cos\theta$ and $\Delta A \sin\theta$. It follows that the *forces* exerted on the three faces are as shown in Fig. 6.6b. (No forces are exerted on the triangular faces of the element, since the corresponding normal and shearing stresses have all been assumed equal to zero.) Using components along the x' and y' axes, we write the following equilibrium equations:

$$\Sigma F_{x'} = 0: \quad \sigma_{x'} \Delta A - \sigma_x(\Delta A \cos\theta)\cos\theta - \tau_{xy}(\Delta A \cos\theta)\sin\theta$$
$$-\sigma_y(\Delta A \sin\theta)\sin\theta - \tau_{xy}(\Delta A \sin\theta)\cos\theta = 0$$

$$\Sigma F_{y'} = 0: \quad \tau_{x'y'} \Delta A + \sigma_x(\Delta A \cos\theta)\sin\theta - \tau_{xy}(\Delta A \cos\theta)\cos\theta$$
$$-\sigma_y(\Delta A \sin\theta)\cos\theta + \tau_{xy}(\Delta A \sin\theta)\sin\theta = 0$$

Solving the first equation for $\sigma_{x'}$ and the second for $\tau_{x'y'}$, we have

$$\sigma_{x'} = \sigma_x \cos^2 \theta + \sigma_y \sin^2 \theta + 2\tau_{xy} \sin \theta \cos \theta \qquad (6.1)$$

$$\tau_{x'y'} = -(\sigma_x - \sigma_y) \sin \theta \cos \theta + \tau_{xy}(\cos^2 \theta - \sin^2 \theta) \qquad (6.2)$$

Recalling the trigonometric relations

$$\sin 2\theta = 2 \sin \theta \cos \theta \qquad \cos 2\theta = \cos^2 \theta - \sin^2 \theta \qquad (6.3)$$

and

$$\cos^2 \theta = \frac{1 + \cos 2\theta}{2} \qquad \sin^2 \theta = \frac{1 - \cos 2\theta}{2} \qquad (6.4)$$

we write Eq. (6.1) as follows:

$$\sigma_{x'} = \sigma_x \frac{1 + \cos 2\theta}{2} + \sigma_y \frac{1 - \cos 2\theta}{2} + \tau_{xy} \sin 2\theta$$

or

$$\sigma_{x'} = \frac{\sigma_x + \sigma_y}{2} + \frac{\sigma_x - \sigma_y}{2} \cos 2\theta + \tau_{xy} \sin 2\theta \qquad (6.5)$$

Using the relations (6.3), we write Eq. (6.2) as

$$\tau_{x'y'} = -\frac{\sigma_x - \sigma_y}{2} \sin 2\theta + \tau_{xy} \cos 2\theta \qquad (6.6)$$

The expression for the normal stress $\sigma_{y'}$ may be obtained by replacing θ in Eq. (6.5) by the angle $\theta + 90°$ that the y' axis forms with the x axis. Since $\cos (2\theta + 180°) = -\cos 2\theta$ and $\sin (2\theta + 180°) = -\sin 2\theta$, we have

$$\sigma_{y'} = \frac{\sigma_x + \sigma_y}{2} - \frac{\sigma_x - \sigma_y}{2} \cos 2\theta - \tau_{xy} \sin 2\theta \qquad (6.7)$$

Adding Eqs. (6.5) and (6.7) member to member, we obtain

$$\sigma_{x'} + \sigma_{y'} = \sigma_x + \sigma_y \qquad (6.8)$$

Since $\sigma_z = \sigma_{z'} = 0$, we thus verify in the case of plane stress that the sum of the normal stresses exerted on a cubic element of material is independent of the orientation of that element.[†]

† Cf. footnote on page 78.

6.3. PRINCIPAL STRESSES; MAXIMUM SHEARING STRESS

The equations (6.5) and (6.6) obtained in the preceding section are the parametric equations of a circle. This means that, if we choose a set of rectangular axes and plot a point M of abscissa $\sigma_{x'}$ and ordinate $\tau_{x'y'}$ for any given value of the parameter θ, all the points thus obtained will lie on a circle. To establish this property we shall eliminate θ from Eqs. (6.5) and (6.6); this is done by first transposing $(\sigma_x + \sigma_y)/2$ in Eq. (6.5) and squaring both members of the equation, then squaring both members of Eq. (6.6), and finally adding member to member the two equations obtained in this fashion. We have

$$\left(\sigma_{x'} - \frac{\sigma_x + \sigma_y}{2}\right)^2 + \tau_{x'y'}^2 = \left(\frac{\sigma_x - \sigma_y}{2}\right)^2 + \tau_{xy}^2 \qquad (6.9)$$

Setting

$$\sigma_{\text{ave}} = \frac{\sigma_x + \sigma_y}{2} \quad \text{and} \quad R = \sqrt{\left(\frac{\sigma_x - \sigma_y}{2}\right)^2 + \tau_{xy}^2} \qquad (6.10)$$

we write the identity (6.9) in the form

$$(\sigma_{x'} - \sigma_{\text{ave}})^2 + \tau_{x'y'}^2 = R^2 \qquad (6.11)$$

which is the equation of a circle of radius R centered at the point C of abscissa σ_{ave} and ordinate 0 (Fig. 6.7). It may be observed that, due to the symmetry of the circle about the horizontal axis, the same result would have been obtained if, instead of plotting M, we had plotted a point N of abscissa $\sigma_{x'}$ and ordinate $-\tau_{x'y'}$ (Fig. 6.8). This property will be used in Sec. 6.4.

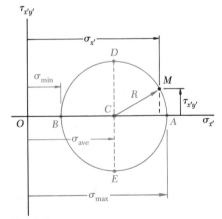

Fig. 6.7

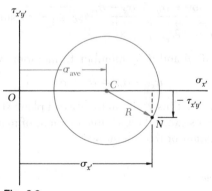

Fig. 6.8

The two points A and B where the circle of Fig. 6.7 intersects the horizontal axis are of special interest: Point A corresponds to the maximum value of the normal stress $\sigma_{x'}$, while point B corresponds to its minimum value. Besides, both points correspond to a zero value of the shearing stress $\tau_{x'y'}$. Thus, the values θ_p of the parameter θ which correspond to points A and B may be obtained by setting $\tau_{x'y'} = 0$ in Eq. (6.6). We write†

$$\tan 2\theta_p = \frac{2\tau_{xy}}{\sigma_x - \sigma_y} \tag{6.12}$$

This equation defines two values $2\theta_p$ which are 180° apart, and thus two values θ_p which are 90° apart. Either of these values may be used to determine the orientation of the corresponding element (Fig. 6.9). The planes containing the faces of the element obtained in this way are called the *principal planes of stress* at point Q, and the corresponding values σ_{max} and σ_{min} of the normal stress exerted on these planes are called the *principal stresses* at Q. Since the two values θ_p defined by Eq. (6.12) were obtained by setting $\tau_{x'y'} = 0$ in Eq. (6.6), it is clear that no shearing stress is exerted on the principal planes.

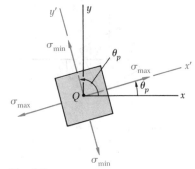

Fig. 6.9

We observe from Fig. 6.7 that

$$\sigma_{max} = \sigma_{ave} + R \qquad \text{and} \qquad \sigma_{min} = \sigma_{ave} - R \tag{6.13}$$

Substituting for σ_{ave} and R from (6.10), we write

$$\sigma_{max, min} = \frac{\sigma_x + \sigma_y}{2} \pm \sqrt{\left(\frac{\sigma_x - \sigma_y}{2}\right)^2 + \tau_{xy}^2} \tag{6.14}$$

Unless it is possible to tell by inspection which of the two principal planes is subjected to σ_{max} and which is subjected to σ_{min}, it is necessary to substitute one of the values θ_p into Eq. (6.5) in order to determine which of the two corresponds to the maximum value of the normal stress.

Referring again to the circle of Fig. 6.7, we note that the points D and E located on the vertical diameter of the circle correspond to the largest numerical value of the shearing stress $\tau_{x'y'}$. Since the abscissa of points D and E is $\sigma_{ave} = (\sigma_x + \sigma_y)/2$, the values θ_s of the parameter θ corresponding to these points are obtained by setting $\sigma_{x'} = (\sigma_x + \sigma_y)/2$ in Eq. (6.5). It follows that the sum of the last two terms in that equation

---

†This relation may also be obtained by differentiating $\sigma_{x'}$ in Eq. (6.5) and setting the derivative equal to zero: $d\sigma_{x'}/d\theta = 0$.

must be zero. Thus, for $\theta = \theta_s$, we write†

$$\frac{\sigma_x - \sigma_y}{2} \cos 2\theta_s + \tau_{xy} \sin 2\theta_s = 0$$

or

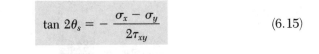

$$\tan 2\theta_s = -\frac{\sigma_x - \sigma_y}{2\tau_{xy}} \tag{6.15}$$

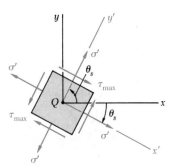

Fig. 6.10

This equation defines two values $2\theta_s$ which are 180° apart, and thus two values θ_s which are 90° apart. Either of these values may be used to determine the orientation of the element corresponding to the maximum shearing stress (Fig. 6.10). Observing from Fig. 6.7 that the maximum value of the shearing stress is equal to the radius R of the circle, and recalling the second of Eqs. (6.10), we write

$$\tau_{max} = \sqrt{\left(\frac{\sigma_x - \sigma_y}{2}\right)^2 + \tau_{xy}^2} \tag{6.16}$$

As observed earlier, the normal stress corresponding to the condition of maximum shearing stress is

$$\sigma' = \sigma_{ave} = \frac{\sigma_x + \sigma_y}{2} \tag{6.17}$$

Comparing Eqs. (6.12) and (6.15), we note that $\tan 2\theta_s$ is the negative reciprocal of $\tan 2\theta_p$. This means that the angles $2\theta_s$ and $2\theta_p$ are 90° apart and, therefore, that the angles θ_s and θ_p are 45° apart. We thus conclude that *the planes of maximum shearing stress are at 45° to the principal planes*. This confirms the results obtained earlier in Sec. 1.8 in the case of a centric axial loading (Fig. 1.36) and in Sec. 3.4 in the case of a torsional loading (Fig. 3.19).

We should be aware that our analysis of the transformation of plane stress has been limited to rotations *in the plane of stress*. If the cubic element of Fig. 6.5 is rotated about an axis other than the z axis, its faces may be subjected to shearing stresses larger than the stress defined by Eq. (6.16). As we shall see in Sec. 6.5, this will occur when the principal stresses defined by Eq. (6.14) have the same sign, i.e., when they are either both tensile or both compressive. In such cases, the value given by Eq. (6.16) is referred to as the maximum *in-plane* shearing stress.

---

†This relation may also be obtained by differentiating $\tau_{x'y'}$ in Eq. (6.6) and setting the derivative equal to zero: $d\tau_{x'y'}/d\theta = 0$.

Example 6.01

For the state of plane stress shown in Fig. 6.11, determine (a) the principal planes, (b) the principal stresses, (c) the maximum shearing stress and the corresponding normal stress.

(a) *Principal Planes.* Following the usual sign convention, we write the stress components as

$$\sigma_x = +50 \text{ MPa} \qquad \sigma_y = -10 \text{ MPa} \qquad \tau_{xy} = +40 \text{ MPa}$$

Substituting into Eq. (6.12), we have

$$\tan 2\theta_p = \frac{2\tau_{xy}}{\sigma_x - \sigma_y} = \frac{2(+40)}{50 - (-10)} = \frac{80}{60}$$

$$2\theta_p = 53.1° \quad \text{and} \quad 180° + 53.1° = 233.1°$$

$$\theta_p = 26.6° \quad \text{and} \quad 116.6°$$

(b) *Principal Stresses.* Formula (6.14) yields

$$\sigma_{\text{max, min}} = \frac{\sigma_x + \sigma_y}{2} \pm \sqrt{\left(\frac{\sigma_x - \sigma_y}{2}\right)^2 + \tau_{xy}^2} = 20 \pm \sqrt{(30)^2 + (40)^2}$$

$$\sigma_{\text{max}} = 20 + 50 = 70 \text{ MPa} \qquad \sigma_{\text{min}} = 20 - 50 = -30 \text{ MPa}$$

The principal planes and principal stresses are sketched in Fig. 6.12. Making $\theta = 26.6°$ in Eq. (6.5), we check that the normal stress exerted on face BC of the element is the maximum stress:

$$\sigma_{x'} = \frac{50 - 10}{2} + \frac{50 + 10}{2} \cos 53.1° + 40 \sin 53.1°$$

$$= 20 + 30 \cos 53.1° + 40 \sin 53.1° = 70 \text{ MPa} = \sigma_{\text{max}}$$

(c) *Maximum Shearing Stress.* Formula (6.16) yields

$$\tau_{\text{max}} = \sqrt{\left(\frac{\sigma_x - \sigma_y}{2}\right)^2 + \tau_{xy}^2} = \sqrt{(30)^2 + (40)^2} = 50 \text{ MPa}$$

Since σ_{max} and σ_{min} have opposite signs, the value obtained for τ_{max} actually represents the maximum value of the shearing stress at the point considered. The orientation of the planes of maximum shearing stress and the sense of the shearing stresses are best determined by passing a section along the diagonal plane AC of the element of Fig. 6.12. Since the faces AB and BC of the element are contained in the principal planes, the diagonal plane AC must be one of the planes of maximum shearing stress (Fig. 6.13). Furthermore, the equilibrium conditions for the prismatic element ABC require that the shearing stress exerted on AC be directed as shown. The cubic element corresponding to the maximum shearing stress is shown in Fig. 6.14. The normal stress on each of the four faces of the element is given by Eq. (6.17):

$$\sigma' = \sigma_{\text{ave}} = \frac{\sigma_x + \sigma_y}{2} = \frac{50 - 10}{2} = 20 \text{ MPa}$$

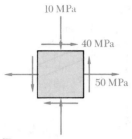

Fig. 6.11

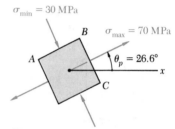

Fig. 6.12

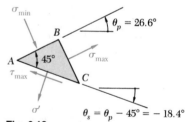

Fig. 6.13

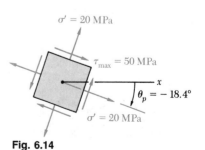

Fig. 6.14

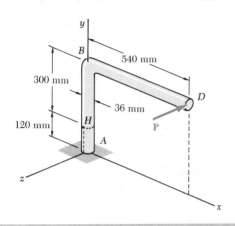

SAMPLE PROBLEM 6.1

A single horizontal force **P** of magnitude 900 N is applied to end D of lever ABD. Knowing that portion AB of the lever has a diameter of 36 mm, determine (a) the normal and shearing stresses on an element located at point H and having sides parallel to the x and y axes, (b) the principal planes and the principal stresses at point H.

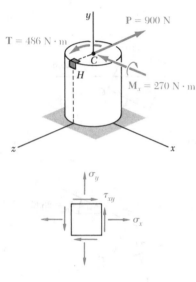

Solution. We replace the force **P** by an equivalent force-couple system at the center C of the transverse section containing point H:

$$P = 900 \text{ N} \qquad T = (900 \text{ N})(0.540 \text{ m}) = 486 \text{ N} \cdot \text{m}$$
$$M_x = (900 \text{ N})(0.300 \text{ m}) = 270 \text{ N} \cdot \text{m}$$

(a) **Stresses σ_x, σ_y, τ_{xy} at Point H.** Using the sign convention shown in Fig. 6.2, we determine the sense and the sign of each stress component by carefully examining the sketch of the force-couple system at point C:

$$\sigma_x = 0 \qquad \sigma_y = +\frac{Mc}{I} = +\frac{(270 \text{ N} \cdot \text{m})(0.018 \text{ m})}{\frac{1}{4}\pi(0.018 \text{ m})^4} \qquad \sigma_y = 58.9 \text{ MPa} \blacktriangleleft$$

$$\tau_{xy} = \frac{Tc}{J} = +\frac{(486 \text{ N} \cdot \text{m})(0.018 \text{ m})}{\frac{1}{2}\pi(0.018 \text{ m})^4} \qquad \tau_{xy} = 53.1 \text{ MPa} \blacktriangleleft$$

We note that the shearing force **P** does not cause any shearing stress at point H.

(b) **Principal Planes and Principal Stresses.** Substituting the values of the stress components into Eq. (6.12), we determine the orientation of the principal planes:

$$\tan 2\theta_p = \frac{2\tau_{xy}}{\sigma_x - \sigma_y} = \frac{2(53.1)}{0 - 58.9} = -1.80$$

$$2\theta_p = -61.0° \qquad \text{and} \qquad 180° - 61.0° = +119°$$

$$\theta_p = -30.5° \qquad \text{and} \qquad +59.5° \blacktriangleleft$$

Substituting into Eq. (6.14), we determine the magnitudes of the principal stresses:

$$\sigma_{\text{max, min}} = \frac{\sigma_x + \sigma_y}{2} \pm \sqrt{\left(\frac{\sigma_x - \sigma_y}{2}\right)^2 + \tau_{xy}^2}$$

$$= \frac{0 + 58.9}{2} \pm \sqrt{\left(\frac{0 - 58.9}{2}\right)^2 + (53.1)^2} = +29.45 \pm 60.72$$

$$\sigma_{\text{max}} = +90.2 \text{ MPa} \blacktriangleleft$$
$$\sigma_{\text{min}} = -32.3 \text{ MPa} \blacktriangleleft$$

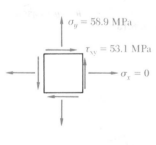

Considering face ab of the element shown, we make $\theta_p = -30.5°$ in Eq. (6.5) and find $\sigma_{x'} = -31.3$ MPa. We conclude that the principal stresses are as shown.

PROBLEMS

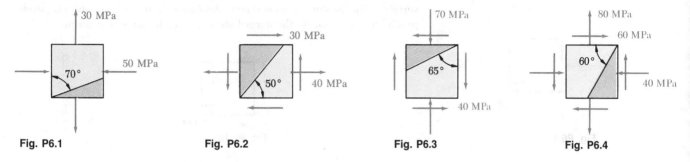

Fig. P6.1   **Fig. P6.2**   **Fig. P6.3**   **Fig. P6.4**

6.1 through 6.4   For the given state of stress, determine the normal and shearing stresses exerted on the oblique face of the shaded triangular element shown.   Use a method of analysis based on the equilibrium equations of that element, as was done in the derivations presented in Sec. 6.2.

6.5 through 6.8   For the given state of stress, determine (*a*) the principal planes, (*b*) the principal stresses.

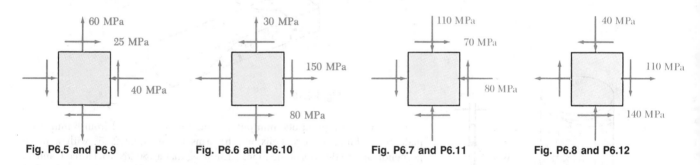

Fig. P6.5 and P6.9   **Fig. P6.6 and P6.10**   **Fig. P6.7 and P6.11**   **Fig. P6.8 and P6.12**

6.9 through 6.12   For the given state of stress, determine (*a*) the orientation of the planes of maximum in-plane shearing stress, (*b*) the maximum in-plane shearing stress, (*c*) the corresponding normal stress.

6.13 through 6.16   For the given state of stress, determine the normal and shearing stresses after the element shown has been rotated through (*a*) 40° counterclockwise, (*b*) 15° clockwise.

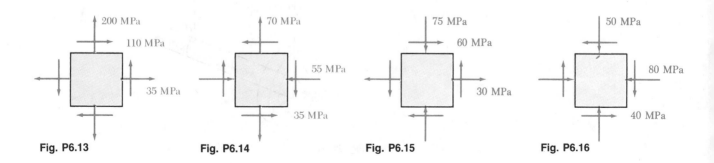

Fig. P6.13   **Fig. P6.14**   **Fig. P6.15**   **Fig. P6.16**

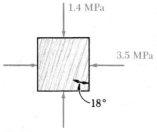

1.4 MPa

3.5 MPa

18°

Fig. P6.17

6.17 and 6.18 The grain of a wooden member forms an angle of 18° with the vertical. For the state of stress shown, determine (*a*) the in-plane shearing stress parallel to the grain, (*b*) the normal stress perpendicular to the grain.

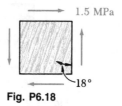

1.5 MPa

18°

Fig. P6.18

6.19 Two wooden members of 80 × 120-mm uniform rectangular cross section are joined by the simple glued scarf splice shown. Knowing that the maximum allowable stresses in the joint are, respectively, 400 kPa in tension (perpendicular to the splice) and 600 kPa in shear (parallel to the splice), determine the largest axial load **P** which may be applied.

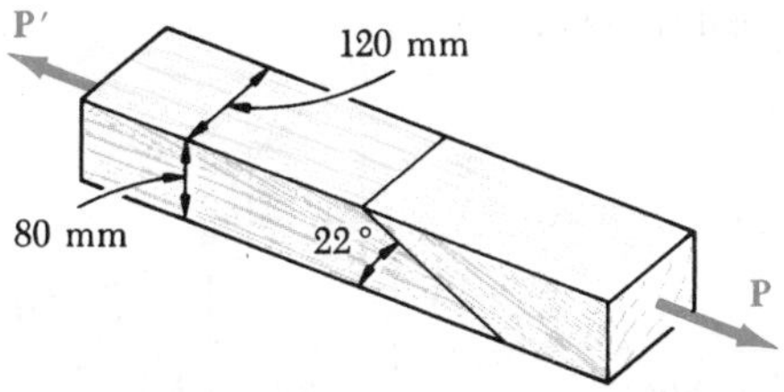

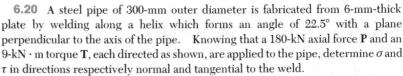

P′

120 mm

80 mm

22°

P

Fig. P6.19

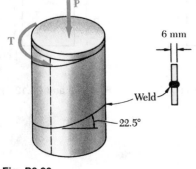

P

T

6 mm

Weld

22.5°

Fig. P6.20

6.20 A steel pipe of 300-mm outer diameter is fabricated from 6-mm-thick plate by welding along a helix which forms an angle of 22.5° with a plane perpendicular to the axis of the pipe. Knowing that a 180-kN axial force **P** and an 9-kN · m torque **T**, each directed as shown, are applied to the pipe, determine σ and τ in directions respectively normal and tangential to the weld.

6.21 The axle of an automobile is acted upon by the forces and couple shown. Knowing that the diameter of the axle is 32 mm, determine (*a*) the principal planes and principal stresses at point *H* located on top of the axle, (*b*) the maximum shearing stress at the same point.

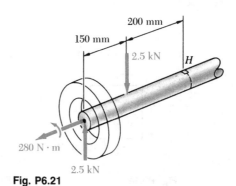

200 mm

150 mm

2.5 kN

H

280 N · m

2.5 kN

Fig. P6.21

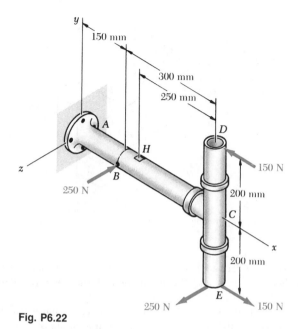

Fig. P6.22

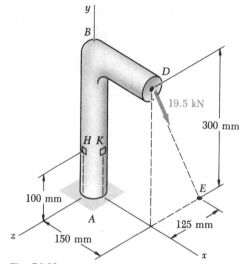

Fig. P6.23

6.22 Several forces are applied to the pipe assembly shown. Knowing that the inner and outer diameters of the pipe are equal to 38 and 45 mm, respectively, determine (*a*) the principal planes and principal stresses at point *H* located at the top of the outside surface of the pipe, (*b*) the maximum shearing stress at the same point.

6.23 A 19.5-kN force is applied at point *D* of the cast-iron post shown. Knowing that the post has a diameter of 60 mm, determine the principal stresses and the maximum shearing stress at (*a*) point *H*, (*b*) point *K*.

6.24 Two forces are applied as shown to rod *AB*, which is welded to the 50-mm-diameter cylinder *DE*. Assuming that all stresses remain below the proportional limit, determine the principal stresses and the maximum shearing stress at (*a*) point *H*, (*b*) point *K*.

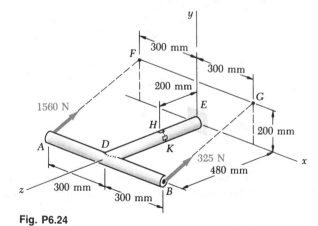

Fig. P6.24

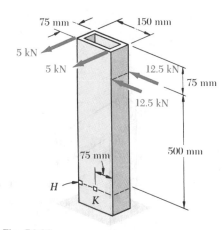

Fig. P6.26

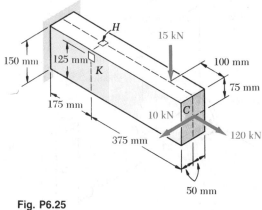

Fig. P6.25

6.25 Three forces are applied as shown to a cantilever beam. Determine the maximum shearing stress and the orientation of the corresponding planes at (a) point H, (b) point K.

6.26 Knowing that the structural tubing shown has a uniform wall thickness of 6 mm, determine for the given loading the maximum shearing stress and the orientation of the corresponding planes at (a) point H, (b) point K.

*6.27 A vertical 18-kN force is applied at end A of bar AB, which is welded to an extruded aluminum tube of uniform 6-mm wall thickness. Determine the principal stresses and the maximum shearing stress at (a) point H, (b) point K.

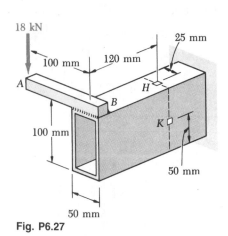

Fig. P6.27

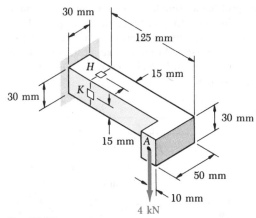

Fig. P6.28

*6.28 A vertical 4-kN force is applied at point A of the machine component shown. Determine the principal stresses and the maximum shearing stress at (a) point H, (b) point K.

6.4. MOHR'S CIRCLE FOR PLANE STRESS

The circle used in the preceding section to derive some of the basic formulas relating to the transformation of plane stress was first introduced by the German engineer Otto Mohr (1835–1918) and is known as *Mohr's circle* for plane stress. As we shall see presently, this circle may be used to obtain an alternative method for the solution of the various problems considered in Secs. 6.2 and 6.3. This method is based on simple geometric considerations and does not require the use of specialized formulas. While originally designed for graphical solutions, it lends itself well to the use of a calculator.

Consider a square element of a material subjected to plane stress (Fig. 6.15a), and let σ_x, σ_y, and τ_{xy} be the components of the stress exerted on the element. We shall plot a point X of coordinates σ_x and $-\tau_{xy}$, and a point Y of coordinates σ_y and $+\tau_{xy}$ (Fig. 6.15b). If τ_{xy} is positive, as assumed in Fig. 6.15a, point X is located below the σ axis and point Y above, as shown in Fig. 6.15b. If τ_{xy} is negative, X is located above the σ axis and Y below. Joining X and Y by a straight line, we define the point C of intersection of line XY with the σ axis and draw the circle of center C and diameter XY. Noting that the abscissa of C and the radius of the circle are respectively equal to the quantities σ_{ave} and R defined by Eqs. (6.10), we conclude that the circle obtained is Mohr's circle for plane stress. Thus the abscissas of points A and B where the circle intersects the σ axis represent respectively the principal stresses σ_{max} and σ_{min} at the point considered.

We also note that, since $\tan (XCA) = 2\tau_{xy}/(\sigma_x - \sigma_y)$, the angle XCA is equal in magnitude to one of the angles $2\theta_p$ which satisfy Eq. (6.12). Thus, the angle θ_p which defines in Fig. 6.15a the orientation of the principal plane corresponding to point A in Fig. 6.15b may be obtained by dividing in half the angle XCA measured on Mohr's circle. We further observe that if $\sigma_x > \sigma_y$ and $\tau_{xy} > 0$, as in the case considered here, the rotation which brings CX into CA is counterclockwise. But, in that case, the angle θ_p obtained from Eq. (6.12) and defining the direction of the normal Oa to the principal plane is positive; thus the rotation bringing Ox into Oa is also counterclockwise. We conclude that the senses of rotation in both parts of Fig. 6.15 are the same; if a counterclockwise rotation through $2\theta_p$ is required to bring CX into CA on Mohr's circle, a counterclockwise rotation through θ_p will bring Ox into Oa in Fig. 6.15a.†

Since Mohr's circle is uniquely defined, the same circle may be obtained by considering the stress components $\sigma_{x'}$, $\sigma_{y'}$, and $\tau_{x'y'}$, corresponding to the x' and y' axes shown in Fig. 6.16a. The point X' of

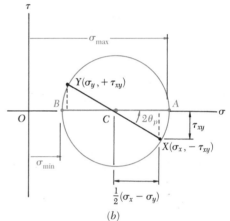

(a)

(b)

Fig. 6.15

---

† This is due to the fact that we are using the circle of Fig. 6.8 rather than the circle of Fig. 6.7 as Mohr's circle.

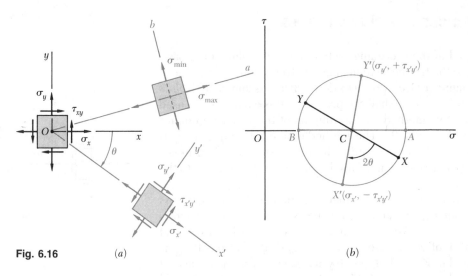

Fig. 6.16 (a)   (b)

coordinates $\sigma_{x'}$ and $-\tau_{x'y'}$, and the point Y' of coordinates $\sigma_{y'}$ and $+\tau_{x'y'}$, are therefore located on Mohr's circle, and the angle $X'CA$ in Fig. 6.16b must be equal to twice the angle $x'Oa$ in Fig. 6.16a. Since, as noted before, the angle XCA is twice the angle xOa, it follows that the angle XCX' in Fig. 6.16b is twice the angle xOx' in Fig. 6.16a. Thus the diameter $X'Y'$ defining the normal and shearing stresses $\sigma_{x'}$, $\sigma_{y'}$, and $\tau_{x'y'}$ may be obtained by rotating the diameter XY through an angle equal to twice the angle θ formed by the x' and x axes in Fig. 6.16a. We note that the rotation which brings the diameter XY into the diameter $X'Y'$ in Fig. 6.16b has the same sense as the rotation which brings the xy axes into the $x'y'$ axes in Fig. 6.16a.

The property we have just indicated may be used to verify the fact that the planes of maximum shearing stress are at 45° to the principal planes. Indeed, we recall that points D and E on Mohr's circle corre-

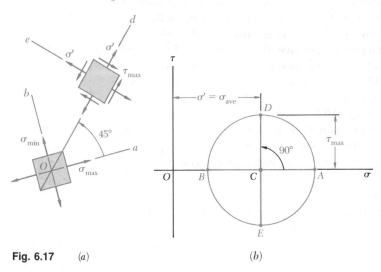

Fig. 6.17 (a)   (b)

spond to the planes of maximum shearing stress, while A and B correspond to the principal planes (Fig. 6.17b). Since the diameters AB and DE of Mohr's circle are at 90° to each other, it follows that the faces of the corresponding elements are at 45° to each other (Fig. 6.17a).

The construction of Mohr's circle for plane stress is greatly simplified if we consider separately each face of the element used to define the stress components. From Figs. 6.15 and 6.16 we observe that, when the shearing stress exerted *on a given face* tends to rotate the element *clockwise*, the point on Mohr's circle corresponding to that face is located *above* the σ axis. When the shearing stress on a given face tends to rotate the element *counterclockwise*, the point corresponding to that face is located *below* the σ axis (Fig. 6.18).† As far as the normal stresses are concerned, the usual convention holds, i.e., a tensile stress is considered as positive and is plotted to the right, while a compressive stress is considered as negative and is plotted to the left.

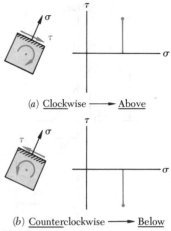

(a) Clockwise ⟶ Above

(b) Counterclockwise ⟶ Below

Fig. 6.18

Example 6.02

For the state of plane stress already considered in Example 6.01, (a) construct Mohr's circle, (b) determine the principal stresses, (c) determine the maximum shearing stress and the corresponding normal stress.

(a) *Construction of Mohr's Circle.* We note from Fig. 6.19a that the normal stress exerted on the face oriented toward the x axis is tensile (positive) and that the shearing stress exerted on that face tends to rotate the element counterclockwise. Point X of Mohr's circle, therefore, will be plotted to the right of the vertical axis and below the horizontal axis (Fig. 6.19b). A similar inspection of the normal stress and shearing

stress exerted on the upper face of the element shows that point Y should be plotted to the left of the vertical axis and above the horizontal axis. Drawing the line XY, we obtain the center C of Mohr's circle; its abscissa is

$$\sigma_{\text{ave}} = \frac{\sigma_x + \sigma_y}{2} = \frac{50 + (-10)}{2} = 20 \text{ MPa}$$

Since the sides of the shaded triangle are

$$CF = 50 - 20 = 30 \text{ MPa} \quad \text{and} \quad FX = 40 \text{ MPa}$$

the radius of the circle is

$$R = CX = \sqrt{(30)^2 + (40)^2} = 50 \text{ MPa}$$

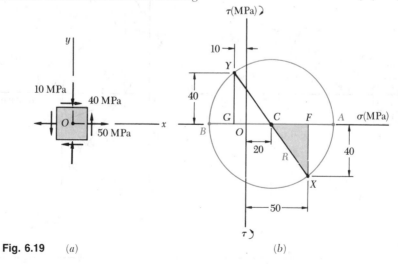

Fig. 6.19 (a)   (b)

†The following jingle may be helpful in remembering this convention: "In the kitchen, the *clock* is above, and the *counter* is below."

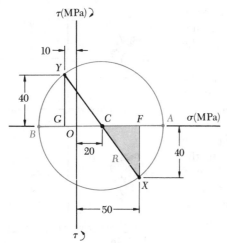

Fig. 6.19b (repeated)

(b) Principal Planes and Principal Stresses. The principal stresses are

$$\sigma_{\max} = OA = OC + CA = 20 + 50 = 70 \text{ MPa}$$
$$\sigma_{\min} = OB = OC - BC = 20 - 50 = -30 \text{ MPa}$$

Recalling that the angle ACX represents $2\theta_p$ (Fig. 6.19*b*), we write

$$\tan 2\theta_p = \frac{FX}{CF} = \frac{40}{30}$$
$$2\theta_p = 53.1° \qquad \theta_p = 26.6°$$

Since the rotation which brings CX into CA in Fig. 6.20*b* is counterclockwise, the rotation which brings Ox into the axis Oa corresponding to $\sigma_{\max}$ in Fig. 6.20*a* is also counterclockwise.

(c) Maximum Shearing Stress. Since a further rotation of 90° counterclockwise brings CA into CD in Fig. 6.20*b*, a further rotation of 45° counterclockwise will bring the axis Oa into the axis Od corresponding to the maximum shearing stress in Fig. 6.20*a*. We note from Fig. 6.20*b* that $\tau_{\max} = R = 50$ MPa and that the corresponding normal stress is $\sigma' = \sigma_{\text{ave}} = 20$ MPa. Since point D is located above the σ axis in Fig. 6.20*b*, the shearing stresses exerted on the faces perpendicular to Od in Fig. 6.20*a* must be directed so that they will tend to rotate the element clockwise.

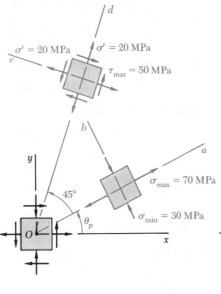

(*a*)

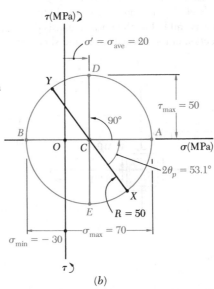

(*b*)

Fig. 6.20

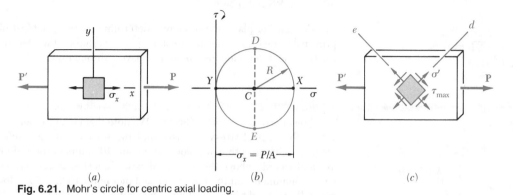

(a)   *(b)*   *(c)*

Fig. 6.21. Mohr's circle for centric axial loading.

Mohr's circle provides a convenient way of checking the results obtained earlier for stresses under a centric axial loading (Sec. 1.8) and under a torsional loading (Sec. 3.4). In the first case (Fig. 6.21*a*), we have $\sigma_x = P/A$, $\sigma_y = 0$, and $\tau_{xy} = 0$. The corresponding points X and Y define a circle of radius $R = P/2A$ which passes through the origin of coordinates (Fig. 6.21*b*). Points D and E yield the orientation of the planes of maximum shearing stress (Fig. 6.21*c*), as well as the values of τ_{max} and of the corresponding normal stresses σ':

$$\tau_{max} = \sigma' = R = \frac{P}{2A} \tag{6.18}$$

In the case of torsion (Fig. 6.22*a*), we have $\sigma_x = \sigma_y = 0$ and $\tau_{xy} = \tau_{max} = Tc/J$. Points X and Y, therefore, are located on the τ axis, and Mohr's circle is a circle of radius $R = Tc/J$ centered at the origin (Fig. 6.22*b*). Points A and B define the principal planes (Fig. 6.22*c*) and the principal stresses:

$$\sigma_{max, min} = \pm R = \pm \frac{Tc}{J} \tag{6.19}$$

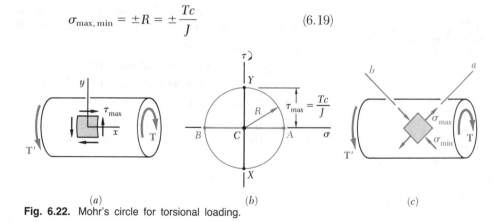

(a)   *(b)*   *(c)*

Fig. 6.22. Mohr's circle for torsional loading.

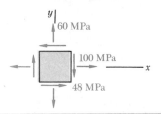

60 MPa

100 MPa

48 MPa

SAMPLE PROBLEM 6.2

For the state of plane stress shown, determine (a) the principal planes and the principal stresses, (b) the stress components exerted on the element obtained by rotating the given element counterclockwise through 30°.

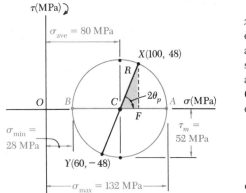

$\sigma_{ave} = 80$ MPa

$X(100, 48)$

R

$2\theta_p$

$\sigma_{min} = 28$ MPa

$\tau_m = 52$ MPa

$Y(60, -48)$

$\sigma_{max} = 132$ MPa

$\theta_p = 33.7°$

$\sigma_{min} = 28$ MPa

$\sigma_{max} = 132$ MPa

a

$\phi = 180° - 60° - 67.4°$
$\phi = 52.6°$

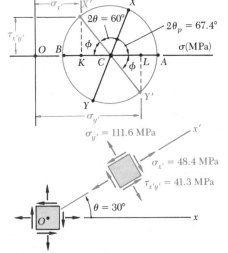

$2\theta = 60°$

$2\theta_p = 67.4°$

$\sigma_{y'} = 111.6$ MPa

$\sigma_{x'} = 48.4$ MPa

$\tau_{x'y'} = 41.3$ MPa

$\theta = 30°$

Construction of Mohr's Circle. We note that on a face perpendicular to the x axis, the normal stress is tensile and the shearing stress tends to rotate the element clockwise; thus we plot X at a point 100 units to the right of the vertical axis and 48 units above the horizontal axis. In a similar fashion, we examine the stress components on the upper face and plot point $Y(60, -48)$. Joining points X and Y by a straight line, we define the center C of Mohr's circle. The abscissa of C, which represents σ_{ave}, and the radius R of the circle may be measured directly or calculated as follows:

$$\sigma_{ave} = OC = \tfrac{1}{2}(\sigma_x + \sigma_y) = \tfrac{1}{2}(100 + 60) = 80 \text{ MPa}$$
$$R = \sqrt{(CF)^2 + (FX)^2} = \sqrt{(20)^2 + (48)^2} = 52 \text{ MPa}$$

(a) Principal Planes and Principal Stresses. We rotate the diameter XY clockwise through $2\theta_p$ until it coincides with the diameter AB. We have

$$\tan 2\theta_p = \frac{XF}{CF} = \frac{48}{20} = 2.4 \qquad 2\theta_p = 67.4° \downarrow \qquad \theta_p = 33.7° \downarrow \quad \blacktriangleleft$$

The principal stresses are represented by the abscissas of points A and B:

$$\sigma_{max} = OA = OC + CA = 80 + 52 \qquad \sigma_{max} = +132 \text{ MPa} \quad \blacktriangleleft$$
$$\sigma_{min} = OB = OC - BC = 80 - 52 \qquad \sigma_{min} = +\ 28 \text{ MPa} \quad \blacktriangleleft$$

Since the rotation which brings XY into AB is clockwise, the rotation which brings Ox into the axis Oa corresponding to σ_{max} is also clockwise; we obtain the orientation shown for the principal planes.

(b) Stress Components on Element Rotated 30° ↖. Points X' and Y' on Mohr's circle which correspond to the stress components on the rotated element are obtained by rotating XY counterclockwise through $2\theta = 60°$. We find

$$\phi = 180° - 60° - 67.4° \qquad \phi = 52.6° \quad \blacktriangleleft$$
$$\sigma_{x'} = OK = OC - KC = 80 - 52 \cos 52.6° \qquad \sigma_{x'} = +\ 48.4 \text{ MPa} \quad \blacktriangleleft$$
$$\sigma_{y'} = OL = OC + CL = 80 + 52 \cos 52.6° \qquad \sigma_{y'} = +111.6 \text{ MPa} \quad \blacktriangleleft$$
$$\tau_{x'y'} = KX' = 52 \sin 52.6° \qquad \tau_{x'y'} = \ 41.3 \text{ MPa} \quad \blacktriangleleft$$

Since X' is located above the horizontal axis, the shearing stress on the face perpendicular to Ox' tends to rotate the element clockwise.

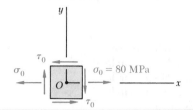

A state of plane stress consists of a tensile stress $\sigma_0 = 80$ MPa exerted on vertical surfaces and of unknown shearing stresses. Determine (a) the magnitude of the shearing stress τ_0 for which the largest normal stress is 100 MPa, (b) the corresponding maximum shearing stress.

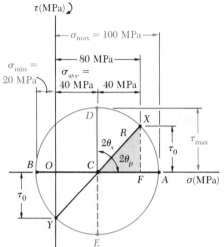

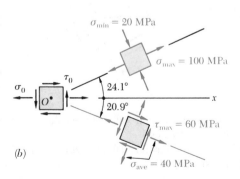

(a)

(b)

Construction of Mohr's Circle. We shall assume that the shearing stresses act in the senses shown. Thus, the shearing stress τ_0 on a face perpendicular to the x axis tends to rotate the element clockwise and we plot the point X of coordinates 80 MPa and τ_0 above the horizontal axis. Considering a horizontal face of the element, we observe that $\sigma_y = 0$ and that τ_0 tends to rotate the element counterclockwise; thus, we plot point Y at a distance τ_0 below O.

We note that the abscissa of the center C of Mohr's circle is

$$\sigma_{ave} = \tfrac{1}{2}(\sigma_x + \sigma_y) = \tfrac{1}{2}(80 + 0) = 40 \text{ MPa}$$

The radius R of the circle is determined by observing that the maximum normal stress, $\sigma_{max} = 100$ MPa, is represented by the abscissa at point A and writing

$$\sigma_{max} = \sigma_{ave} + R$$
$$100 \text{ MPa} = 40 \text{ MPa} + R \qquad R = 60 \text{ MPa}$$

(a) **Shearing Stress τ_0.** Considering the right triangle CFX, we find

$$\cos 2\theta_p = \frac{CF}{CX} = \frac{CF}{R} = \frac{40 \text{ MPa}}{60 \text{ MPa}} \qquad 2\theta_p = 48.2° \downarrow \qquad \theta_p = 24.1° \downarrow$$
$$\tau_0 = FX = R \sin 2\theta_p = (60 \text{ MPa}) \sin 48.2° \qquad \tau_0 = 44.7 \text{ MPa} \blacktriangleleft$$

(b) **Maximum Shearing Stress.** The coordinates of point D of Mohr's circle represent the maximum shearing stress and the corresponding normal stress.

$$\tau_{max} = R = 60 \text{ MPa} \qquad\qquad \tau_{max} = 60 \text{ MPa} \blacktriangleleft$$
$$2\theta_s = 90° - 2\theta_p = 90° - 48.2° = 41.8° \uparrow \qquad \theta_s = 20.9° \uparrow$$

The maximum shearing stress is exerted on an element which is oriented as shown in Fig. a. (The element upon which the principal stresses are exerted is also shown.)

Note. If our original assumption regarding the sense of τ_0 was reversed, we would obtain the same circle and the same answers, but the orientation of the elements would be as shown in Fig. b.

PROBLEMS

6.29 Solve Probs. 6.5 and 6.9, using Mohr's circle.
6.30 Solve Probs. 6.6 and 6.10, using Mohr's circle.
6.31 Solve Probs. 6.7 and 6.11, using Mohr's circle.
6.32 Solve Probs. 6.8 and 6.12, using Mohr's circle.
6.33 Solve Prob. 6.13, using Mohr's circle.
6.34 Solve Prob. 6.14, using Mohr's circle.
6.35 Solve Prob. 6.15, using Mohr's circle.
6.36 Solve Prob. 6.16, using Mohr's circle.
6.37 Solve Prob. 6.17, using Mohr's circle.
6.38 Solve Prob. 6.18, using Mohr's circle.
6.39 Solve Prob. 6.19, using Mohr's circle.
6.40 Solve Prob. 6.20, using Mohr's circle.
6.41 Solve Prob. 6.21, using Mohr's circle.
6.42 Solve Prob. 6.22, using Mohr's circle.
6.43 Solve Prob. 6.23, using Mohr's circle.
6.44 Solve Prob. 6.24, using Mohr's circle.
6.45 Solve Prob. 6.25, using Mohr's circle.
6.46 Solve Prob. 6.26, using Mohr's circle.

6.47 through 6.50 Determine the principal planes and the principal stresses for the state of plane stress resulting from the superposition of the two states of plane stress shown.

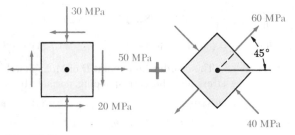

Fig. P6.47

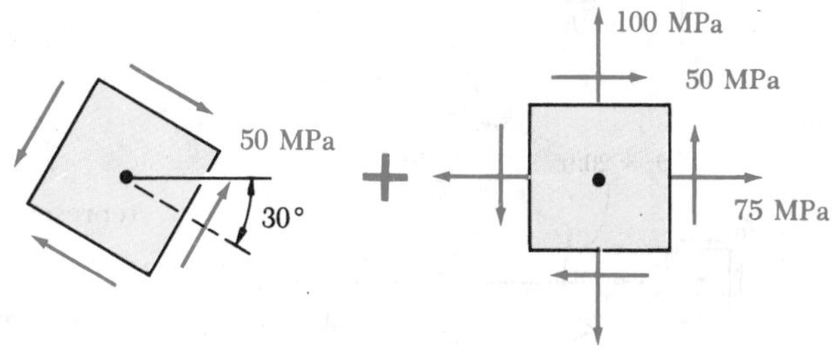

Fig. P6.48

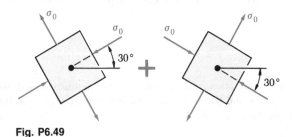

Fig. P6.49

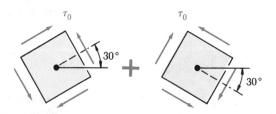

Fig. P6.50

6.51 Solve Prob. 6.50, assuming that in the second state of stress shown, all stresses are replaced by equal and opposite stresses.

6.52 Solve Prob. 6.49, assuming that in the second state of stress shown, all stresses are replaced by equal and opposite stresses.

6.53   For the element shown, determine the range of values of τ_{xy} for which the maximum tensile stress is equal to or less than 82.5 MPa.

6.54   For the element shown, determine the range of values of τ_{xy} for which the largest in-plane shearing stress is equal to or less than 150 MPa.

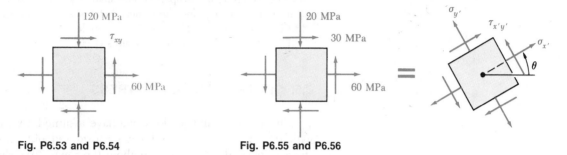

Fig. P6.53 and P6.54   Fig. P6.55 and P6.56

6.55   For the state of plane stress shown, determine the range of values of θ for which the magnitude of the shearing stress $\tau_{x'y'}$ is equal to or less than 45 MPa.

6.56   For the state of plane stress shown, determine the range of values of θ for which the normal stress $\sigma_{x'}$ is equal to or less than +65 MPa.

6.57   For a state of plane stress it is known that the normal and shearing stresses are directed as shown and that $\sigma_x = 84$ MPa, $\sigma_y = 42$ MPa, and $\sigma_{min} = 28$ MPa. Determine (*a*) the orientation of the principal planes, (*b*) the principal stress σ_{max}, (*c*) the maximum in-plane shearing stress.

6.58   For a state of plane stress it is known that the normal and shearing stresses are directed as shown and that $\sigma_x = 30$ MPa, $\sigma_y = 80$ MPa, and $\sigma_{max} = 120$ MPa.   Determine (*a*) the orientation of the principal planes, (*b*) the principal stress σ_{min}, (*c*) the maximum in-plane shearing stress.

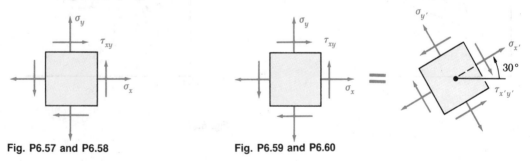

Fig. P6.57 and P6.58   Fig. P6.59 and P6.60

6.59   For a state of plane stress it is known that for each of two given orientations of the coordinate axes, the normal and shearing stresses are directed as shown and the magnitudes of the normal stresses σ_x, σ_y, and $\sigma_{x'}$ are, respectively, 80 MPa, 30 MPa, and 75 MPa.   Determine (*a*) the principal planes and principal stresses, (*b*) the maximum in-plane shearing stress.

6.60   For a state of plane stress it is known that for each of two given orientations of the coordinate axes, the normal and shearing stresses are directed as shown and the magnitudes of the normal stresses σ_x, σ_y, and $\sigma_{x'}$ are, respectively, 75 MPa, 15 MPa, and 90 MPa.   Determine (*a*) the principal planes and principal stresses, (*b*) the maximum in-plane shearing stress.

6.61 (a) Prove that the expression $\sigma_{x'}\sigma_{y'} - \tau_{x'y'}^2$, where $\sigma_{x'}$, $\sigma_{y'}$, and $\tau_{x'y'}$ are the components of stress along the rectangular axes x' and y', is independent of the orientation of these axes. Also, show that the given expression represents the square of the tangent drawn from the origin of the coordinates to Mohr's circle. (b) Using the invariance property established in part a, express the shearing stress τ_{xy} in terms of σ_x, σ_y, and the principal stresses $\sigma_{\max}$ and $\sigma_{\min}$.

*6.62 Solve Prob. 6.28, using Mohr's circle.

*6.63 Solve Prob. 6.27, using Mohr's circle.

6.5. GENERAL STATE OF STRESS

In the preceding sections, we have assumed a state of plane stress with $\sigma_z = \tau_{zx} = \tau_{zy} = 0$, and have considered only transformations of stress associated with a rotation about the z axis. We shall now consider the general state of stress represented in Fig. 6.1a and the transformation of stress associated with the rotation of axes shown in Fig. 6.1b. We shall, however, limit our analysis to the determination of the *normal stress* σ_n on a plane of arbitrary orientation.

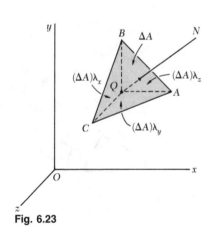

Fig. 6.23

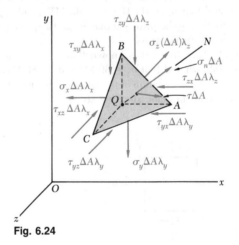

Fig. 6.24

Consider the tetrahedron shown in Fig. 6.23. Three of its faces are parallel to the coordinate planes, while its fourth face, ABC, is perpendicular to the line QN. Denoting by ΔA the area of face ABC, and by λ_x, λ_y, λ_z the direction cosines of line QN, we find that the areas of the faces perpendicular to the x, y, and z axes are, respectively, $(\Delta A)\lambda_x$, $(\Delta A)\lambda_y$, and $(\Delta A)\lambda_z$. If the state of stress at point Q is defined by the stress components σ_x, σ_y, σ_z, τ_{xy}, τ_{yz}, and τ_{zx}, then the *forces* exerted on the faces parallel to the coordinate planes may be obtained by multiplying the appropriate stress components by the area of each face (Fig. 6.24). On the other hand, the forces exerted on face ABC consist of a normal force of magnitude $\sigma_n \Delta A$ directed along QN, and of a shearing force of magnitude $\tau \Delta A$ perpendicular to QN but of otherwise unknown direction. Note that, since QBC, QCA, and QAB respectively face the negative x,y, and z axes, the forces exerted on them must be shown with negative senses.

We shall now express that the sum of the components along QN of all the forces acting on the tetrahedron is zero. Observing that the component along QN of a force parallel to the x axis is obtained by multiplying the magnitude of that force by the direction cosine λ_x, and that the components of forces parallel to the y and z axes are obtained in a similar way, we write

$$\Sigma F_n = 0: \quad \sigma_n \Delta A - (\sigma_x \Delta A\ \lambda_x)\lambda_x - (\tau_{xy} \Delta A\ \lambda_x)\lambda_y - (\tau_{xz} \Delta A\ \lambda_x)\lambda_z$$
$$-(\tau_{yx} \Delta A\ \lambda_y)\lambda_x - (\sigma_y \Delta A\ \lambda_y)\lambda_y - (\tau_{yz} \Delta A\ \lambda_y)\lambda_z$$
$$-(\tau_{zx} \Delta A\ \lambda_z)\lambda_x - (\tau_{zy} \Delta A\ \lambda_z)\lambda_y - (\sigma_z \Delta A\ \lambda_z)\lambda_z = 0$$

Dividing through by ΔA and solving for σ_n, we have

$$\sigma_n = \sigma_x\lambda_x^2 + \sigma_y\lambda_y^2 + \sigma_z\lambda_z^2 + 2\tau_{xy}\lambda_x\lambda_y + 2\tau_{yz}\lambda_y\lambda_z + 2\tau_{zx}\lambda_z\lambda_x \quad (6.20)$$

We note that the expression obtained for the normal stress σ_n is a *quadratic form* in λ_x, λ_y, and λ_z. It follows that we may select the coordinate axes in such a way that the right-hand member of Eq. (6.20) reduces to the three terms containing the squares of the direction cosines.[†] Denoting these axes by a, b, and c, the corresponding normal stresses by σ_a, σ_b, and σ_c, and the direction cosines of QN with respect to these axes by λ_a, λ_b, and λ_c, we write

$$\sigma_n = \sigma_a\lambda_a^2 + \sigma_b\lambda_b^2 + \sigma_c\lambda_c^2 \quad (6.21)$$

The coordinate axes a, b, c are referred to as the *principal axes of stress*. Since their orientation depends upon the state of stress at Q, and thus upon the position of Q, they have been represented in Fig. 6.25 as attached to Q. The corresponding coordinate planes are known as the *principal planes of stress*, and the corresponding normal stresses σ_a, σ_b, and σ_c as the *principal stresses* at Q.[‡]

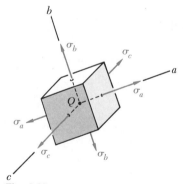

Fig. 6.25

[†] In Sec. 9.16 of F. P. Beer and E. R. Johnston, *Vector Mechanics for Engineers*, 5th ed., McGraw-Hill Book Company, 1988, a similar quadratic form is found to represent the moment of inertia of a rigid body with respect to an arbitrary axis. It is shown in Sec. 9.17 that this form is associated with a *quadric surface*, and that reducing the quadratic form to terms containing only the squares of the direction cosines is equivalent to determining the principal axes of that surface.

[‡] For a discussion of the determination of the principal planes of stress and of the principal stresses, see S. P. Timoshenko and J. N. Goodier, *Theory of Elasticity*, 3d ed., McGraw-Hill Book Company, 1970, sec. 77.

6.6. APPLICATION OF MOHR'S CIRCLE TO THE THREE-DIMENSIONAL ANALYSIS OF STRESS

If the element shown in Fig. 6.25 is rotated about one of the principal axes at Q, say the c axis (Fig. 6.26), the corresponding transformation of stress may be analyzed by means of Mohr's circle as if it were a transformation of plane stress. Indeed, the shearing stresses exerted on the faces perpendicular to the c axis remain equal to zero, and the normal stress σ_c is perpendicular to the plane ab in which the transformation takes place and, thus, does not affect this transformation. We may therefore use the circle of diameter AB to determine the normal and shearing stresses ex-

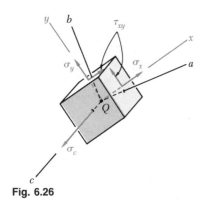

Fig. 6.26

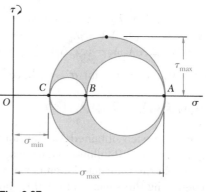

Fig. 6.27

erted on the faces of the element as it is rotated about the c axis (Fig. 6.27). Similarly, circles of diameter BC and CA may be used to determine the stresses on the element as it is rotated about the a and b axes, respectively. While our analysis will be limited to rotations about the principal axes, it could be shown that any other transformation of axes would lead to stresses represented in Fig. 6.27 by a point located within the shaded area. Thus, the radius of the largest of the three circles yields the maximum value of the shearing stress at point Q. Noting that the diameter of that circle is equal to the difference between σ_{max} and σ_{min}, we write

$$\tau_{max} = \tfrac{1}{2}|\sigma_{max} - \sigma_{min}| \qquad (6.22)$$

where σ_{max} and σ_{min} represent the *algebraic* values of the maximum and minimum stresses at point Q.

Let us now return to the particular case of *plane stress*, which was discussed in Secs. 6.2 through 6.4. We recall that, if the x and y axes are selected in the plane of stress, we have $\sigma_z = \tau_{zx} = \tau_{zy} = 0$. This means that the z axis, i.e., the axis perpendicular to the plane of stress, is one of the three principal axes of stress. In a Mohr-circle diagram, this axis corresponds to the origin O, where $\sigma = \tau = 0$. We also recall that the

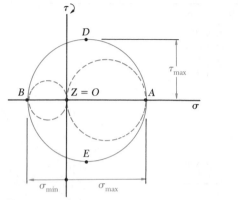

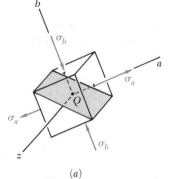

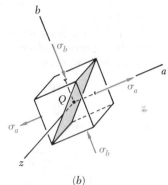

Fig. 6.28

Fig. 6.29

other two principal axes correspond to points A and B where Mohr's circle for the xy plane intersects the σ axis. If A and B are located on opposite sides of the origin O (Fig. 6.28), the corresponding principal stresses represent the maximum and minimum normal stresses at point Q, and the maximum shearing stress is equal to the maximum "in-plane" shearing stress. As noted in Sec. 6.3, the planes of maximum shearing stress correspond to points D and E of Mohr's circle and are at 45° to the principal planes corresponding to points A and B. They are, therefore, the shaded diagonal planes shown in Figs. 6.29a and b.

If, on the other hand, A and B are on the same side of O, that is, if σ_a and σ_b have the same sign, then the circle defining σ_{max}, σ_{min}, and τ_{max} is *not* the circle corresponding to a transformation of stress within the xy plane. If $\sigma_a > \sigma_b > 0$, as assumed in Fig. 6.30, we have $\sigma_{max} = \sigma_a$, $\sigma_{min} = 0$, and τ_{max} is equal to the radius of the circle defined by points O and A, that is, $\tau_{max} = \frac{1}{2}\sigma_{max}$. We also note that the normals Qd' and Qe' to the planes of maximum shearing stress are obtained by rotating the axis Qa through 45° within the za plane. Thus, the planes of maximum shearing stress are the shaded diagonal planes shown in Figs. 6.31a and b.

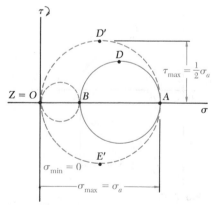

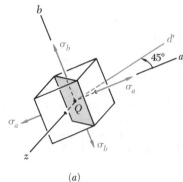

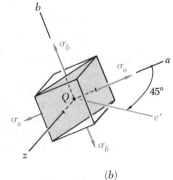

Fig. 6.30

Fig. 6.31

Example 6.03

For the state of plane stress shown in Fig. 6.32, determine (a) the three principal planes and principal stresses, (b) the maximum shearing stress.

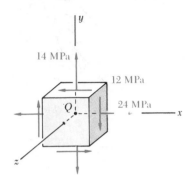

Fig. 6.32

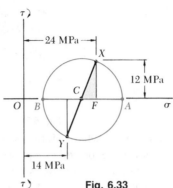

Fig. 6.33

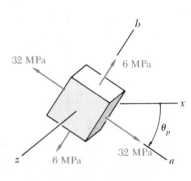

Fig. 6.34

(a) *Principal Planes and Principal Stresses.* We construct Mohr's circle for the transformation of stress in the xy plane (Fig. 6.33). Point X is plotted 24 units to the right of the τ axis and 12 units above the σ axis (since the corresponding shearing stress tends to rotate the element clockwise). Point Y is plotted 14 units to the right of the τ axis and 12 units below the σ axis. Drawing the line XY, we obtain the center C of Mohr's circle for the xy plane: its abscissa is

$$\sigma_{ave} = \frac{\sigma_x + \sigma_y}{2} = \frac{24 + 14}{2} = 19 \text{ MPa}$$

Since the sides of the right triangle CFX are $CF = 24 - 19 = 5$ MPa and $FX = 12$ MPa, the radius of the circle is

$$R = CX = \sqrt{(5)^2 + (12)^2} = 13 \text{ MPa}$$

The principal stresses in the plane of stress are

$$\sigma_a = OA = OC + CA = 19 + 13 = 32 \text{ MPa}$$
$$\sigma_b = OB = OC - BC = 19 - 13 = 6 \text{ MPa}$$

Since the faces of the element which are perpendicular to the z axis are free of stress, these faces define one of the principal planes, and the corresponding principal stress is $\sigma_z = 0$. The other two principal planes are defined by points A and B on Mohr's circle. The angle θ_p through which the element should be rotated about the z axis to bring its faces to coincide with these planes (Fig. 6.34) is half the angle ACX. We have

$$\tan 2\theta_p = \frac{FX}{CF} = \frac{12}{5}$$

$$2\theta_p = 67.4° \downarrow \qquad \theta_p = 33.7° \downarrow$$

(b) *Maximum Shearing Stress.* We now draw the circles of diameter OB and OA, which correspond respectively to rotations of the element about the a and b axes (Fig. 6.35). We note that the maximum shearing stress is equal to the radius of the circle of diameter OA. We thus have

$$\tau_{max} = \tfrac{1}{2}\sigma_a = \tfrac{1}{2}(32 \text{ MPa}) = 16 \text{ MPa}$$

Since points D' and E', which define the planes of maximum shearing stress, are located at the ends of the vertical diameter of the circle corresponding to a rotation about the b axis, the faces of the element of Fig. 6.34 may be brought to coincide with the planes of maximum shearing stress through a rotation of 45° about the b axis.

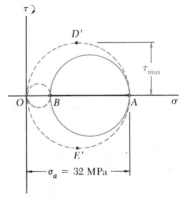

Fig. 6.35

*6.7. YIELD CRITERIA FOR DUCTILE MATERIALS UNDER PLANE STRESS

Structural elements and machine components made of a ductile material are usually designed so that the material will not yield under the expected loading conditions. When the element or component is under uniaxial stress (Fig. 6.36), the value of the normal stress σ_x which will cause the material to yield may be obtained readily from a tensile test conducted on a specimen of the same material, since the test specimen and the structural element or machine component are in the same state of stress. Thus, regardless of the actual mechanism which causes the material to yield, we may state that the element or component will be safe as long as $\sigma_x < \sigma_Y$, where σ_Y is the yield strength of the test specimen.

On the other hand, when a structural element or machine component is in a state of plane stress (Fig. 6.37a), it is found convenient to use one of the methods developed earlier to determine the principal stresses σ_a and σ_b at any given point (Fig. 6.37b). The material may then be regarded as being in a state of biaxial stress at that point. Since this state is different from the state of uniaxial stress found in a specimen subjected to a tensile test, it is clearly not possible to predict directly from such a test whether or not the structural element or machine component under investigation will fail. Some criterion regarding the actual mechanism of failure of the material must first be established, which will make it possible to compare the effects of both states of stress on the material. The purpose of this section is to present the two yield criteria most frequently used for ductile materials.

Maximum-Shearing-Stress Criterion. This criterion is based on the observation that yield in ductile materials is caused by slippage of the material along oblique surfaces and is due primarily to shearing stresses (cf. Sec. 2.3). According to this criterion, a given structural component is safe as long as the maximum value τ_{max} of the shearing stress in that component remains smaller than the corresponding value of the shearing stress in a tensile-test specimen of the same material as the specimen starts to yield.

Recalling from Sec. 1.7 that the maximum value of the shearing stress under a centric axial load is equal to half the value of the corresponding normal, axial stress, we conclude that the maximum shearing stress in a tensile-test specimen is $\frac{1}{2}\sigma_Y$ as the specimen starts to yield. On the other hand, we saw in Sec. 6.6 that, for plane stress, the maximum value τ_{max} of the shearing stress is equal to $\frac{1}{2}|\sigma_{max}|$ if the principal stresses are either both positive or both negative, and to $\frac{1}{2}|\sigma_{max} - \sigma_{min}|$ if the maximum stress is positive and the minimum stress negative. Thus, if the principal stresses σ_a and σ_b have the same sign, the maximum-shearing-stress criterion gives

$$|\sigma_a| < \sigma_Y \qquad |\sigma_b| < \sigma_Y \qquad (6.23)$$

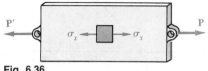

Fig. 6.36

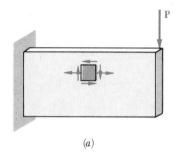

(a)

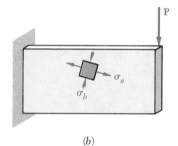

(b)

Fig. 6.37

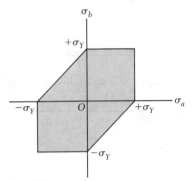

Fig. 6.38

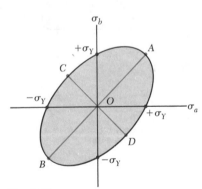

Fig. 6.39

If the principal stresses σ_a and σ_b have opposite signs, the maximum-shearing-stress criterion yields

$$|\sigma_a - \sigma_b| < \sigma_Y \tag{6.24}$$

The relations obtained have been represented graphically in Fig. 6.38. Any given state of stress will be represented in that figure by a point of coordinates σ_a and σ_b, where σ_a and σ_b are the two principal stresses. If this point falls within the area shown in the figure, the structural component is safe. If it falls outside this area, the component will fail as a result of yield in the material. The hexagon associated with the initiation of yield in the material is known as *Tresca's hexagon* after the French engineer Henri Edouard Tresca (1814–1885).

Maximum-Distortion-Energy Criterion. This criterion is based on the determination of the distortion energy in a given material, i.e., of the energy associated with changes in shape in that material (as opposed to the energy associated with changes in volume in the same material). According to this criterion, also known as the *von Mises criterion*, after the German-American applied mathematician Richard von Mises (1883–1953), a given structural component is safe as long as the maximum value of the distortion energy per unit volume in that material remains smaller than the distortion energy per unit volume required to cause yield in a tensile-test specimen of the same material. As we shall see in Sec. 10.6, the distortion energy per unit volume in an isotropic material under plane stress is

$$u_d = \frac{1}{6G}(\sigma_a^2 - \sigma_a\sigma_b + \sigma_b^2) \tag{6.25}$$

where σ_a and σ_b are the principal stresses and G the modulus of rigidity. In the particular case of a tensile-test specimen which is starting to yield, we have $\sigma_a = \sigma_Y$, $\sigma_b = 0$, and $(u_d)_Y = \sigma_Y^2/6G$. Thus, the maximum-distortion-energy criterion indicates that the structural component is safe as long as $u_d < (u_d)_Y$, or

$$\sigma_a^2 - \sigma_a\sigma_b + \sigma_b^2 < \sigma_Y^2 \tag{6.26}$$

i.e., as long as the point of coordinates σ_a and σ_b falls within the area shown in Fig. 6.39. This area is bounded by the ellipse of equation

$$\sigma_a^2 - \sigma_a\sigma_b + \sigma_b^2 = \sigma_Y^2 \tag{6.27}$$

which intersects the coordinate axes at $\sigma_a = \pm\sigma_Y$ and $\sigma_b = \pm\sigma_Y$. We may verify that the major axis of the ellipse bisects the first and third quadrants and extends from A ($\sigma_a = \sigma_b = \sigma_Y$) to B ($\sigma_a = \sigma_b = -\sigma_Y$), while its minor axis extends from C ($\sigma_a = -\sigma_b = -0.577\sigma_Y$) to D ($\sigma_a = -\sigma_b = 0.577\sigma_Y$).

The maximum-shearing-stress criterion and the maximum-distortion-energy criterion are compared in Fig. 6.40. We note that the ellipse passes through the vertices of the hexagon. Thus, for the states of stress represented by these six points, the two criteria give the same results. For any other state of stress, the maximum-shearing-stress criterion is more conservative than the maximum-distortion-energy criterion, since the hexagon is located within the ellipse.

A state of stress of particular interest is that associated with yield in a torsion test. We recall from Fig. 6.22 of Sec. 6.4 that, for torsion, $\sigma_{\min} = -\sigma_{\max}$; thus, the corresponding points in Fig. 6.40 are located on the bisector of the second and fourth quadrants. It follows that yield occurs in a torsion test when $\sigma_a = -\sigma_b = \pm 0.5\sigma_Y$ according to the maximum-shearing-stress criterion, and when $\sigma_a = -\sigma_b = \pm 0.577\sigma_Y$ according to the maximum-distortion-energy criterion. But, recalling again Fig. 6.22, we note that σ_a and σ_b must be equal in magnitude to $\tau_{\max}$, that is, to the value obtained from a torsion test for the yield strength τ_Y of the material. Since the values of the yield strength σ_Y in tension and of the yield strength τ_Y in shear are given for various ductile materials in Appendix B, we may compute the ratio τ_Y/σ_Y for these materials and verify that the values obtained range from 0.55 to 0.60. Thus, the maximum-distortion-energy criterion appears somewhat more accurate than the maximum-shearing-stress criterion as far as predicting yield in torsion is concerned.

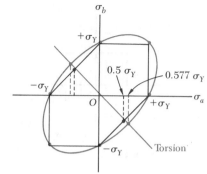

Fig. 6.40

*6.8. FRACTURE CRITERIA FOR BRITTLE MATERIALS UNDER PLANE STRESS

As we saw in Chap. 2, brittle materials are characterized by the fact that, when subjected to a tensile test, they fail suddenly through rupture—or fracture—without any prior yielding. When a structural element or machine component made of a brittle material is under uniaxial tensile stress, the value of the normal stress which causes it to fail is equal to the ultimate strength σ_U of the material as determined from a tensile test, since both the tensile-test specimen and the element or component under investigation are in the same state of stress. However, when a structural element or machine component is in a state of plane stress, it is found convenient to first determine the principal stresses σ_a and σ_b at any given point, and to use one of the criteria indicated in this section to predict whether or not the structural element or machine component will fail.

Maximum-Normal-Stress Criterion. According to this criterion, a given structural component fails when the maximum normal stress in that component reaches the ultimate strength σ_U obtained from the tensile test of a specimen of the same material. Thus, the structural component will be safe as long as the absolute values of the principal stresses σ_a and σ_b are both less than σ_U:

$$|\sigma_a| < \sigma_U \qquad |\sigma_b| < \sigma_U \qquad (6.28)$$

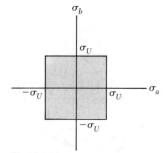

Fig. 6.41

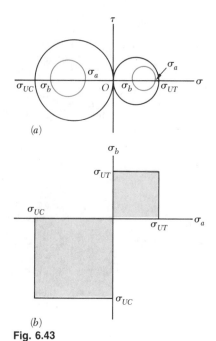

(a)

(b)

Fig. 6.43

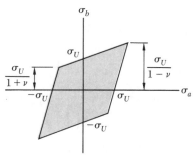

Fig. 6.42

The maximum-normal-stress criterion may be expressed graphically as shown in Fig. 6.41. If the point obtained by plotting the values σ_a and σ_b of the principal stresses falls within the square area shown in the figure, the structural component is safe. If it falls outside that area, the component will fail.

The maximum-normal-stress criterion, also known as *Coulomb's criterion*, after the French physicist Charles Augustin de Coulomb (1736–1806), suffers from an important shortcoming, since it is based on the assumption that the ultimate strength of the material is the same in tension and in compression. As we noted in Sec. 2.3, this is seldom the case, because of the presence of flaws in the material, such as microscopic cracks or cavities, which tend to weaken the material in tension, while not appreciably affecting its resistance to compressive failure. Besides, this criterion makes no allowance for effects other than those of the normal stresses on the failure mechanism of the material.†

Mohr's Criterion. This criterion, suggested by the German engineer Otto Mohr, may be used to predict the effect of a given state of plane stress on a brittle material, when results of various types of tests are available for that material.

Let us first assume that a tensile test and a compressive test have been conducted on a given material, and that the values σ_{UT} and σ_{UC} of the ultimate strength in tension and in compression have been determined for that material. The state of stress corresponding to the rupture of the tensile-test specimen may be represented on a Mohr-circle diagram by the circle intersecting the horizontal axis at O and σ_{UT} (Fig. 6.43a). Similarly, the state of stress corresponding to the failure of the compressive-test specimen may be represented by the circle intersecting the horizontal axis at O and σ_{UC}. Clearly, a state of stress represented by a circle entirely contained in either of these circles will be safe. Thus, if both principal stresses are positive, the state of stress is safe as long as $\sigma_a < \sigma_{UT}$ and $\sigma_b < \sigma_{UT}$; if both principal stresses are negative, the state of

---

†Another failure criterion known as the *maximum-normal-strain criterion*, or Saint-Venant's criterion, was widely used during the nineteenth century. According to this criterion, a given structural component is safe as long as the maximum value of the normal strain in that component remains smaller than the value ϵ_U of the strain at which a tensile-test specimen of the same material will fail. But, as will be shown in Sec. 6.12, the strain is maximum along one of the principal axes of stress, if the deformation is elastic and the material homogeneous and isotropic. Thus, denoting by ϵ_a and ϵ_b the values of the normal strain along the principal axes in the plane of stress, we write

$$|\epsilon_a| < \epsilon_U \qquad |\epsilon_b| < \epsilon_U \qquad (6.29)$$

Making use of the generalized Hooke's law (Sec. 2.12), we could express these relations in terms of the principal stresses σ_a and σ_b and the ultimate strength σ_U of the material. We would find that, according to the maximum-normal-strain criterion, the structural component is safe as long as the point obtained by plotting σ_a and σ_b falls within the area shown in Fig. 6.42, where ν is Poisson's ratio for the given material.

stress is safe as long as $|\sigma_a| < |\sigma_{UC}|$ and $|\sigma_b| < |\sigma_{UC}|$. Plotting the point of coordinates σ_a and σ_b (Fig. 6.43b), we verify that the state of stress is safe as long as that point falls within one of the square areas shown in that figure.

In order to analyze the cases when σ_a and σ_b have opposite signs, we shall now assume that a torsion test has been conducted on the material and that its ultimate strength in shear, τ_U, has been determined. Drawing the circle centered at O representing the state of stress corresponding to the failure of the torsion-test specimen (Fig. 6.44a), we observe that any state of stress represented by a circle entirely contained in that circle is also safe. Mohr's criterion is a logical extension of this observation: According to Mohr's criterion, a state of stress is safe if it is represented by a circle located entirely within the area bounded by the envelope of the circles corresponding to the available data. The remaining portions of the principal-stress diagram may now be obtained by drawing various circles tangent to this envelope, determining the corresponding values of σ_a and σ_b, and plotting the points of coordinates σ_a and σ_b (Fig. 6.44b).

More accurate diagrams may be drawn when additional test results, corresponding to various states of stress, are available. If, on the other hand, the only available data consists of the ultimate strengths σ_{UT} and σ_{UC}, the envelope in Fig. 6.44a is replaced by the tangents AB and $A'B'$ to the circles corresponding respectively to failure in tension and failure in compression (Fig. 6.45a). From the similar triangles drawn in that figure, we note that the abscissa of the center C of a circle tangent to AB and $A'B'$ is linearly related to its radius R. Since $\sigma_a = OC + R$ and $\sigma_b = OC - R$, it follows that σ_a and σ_b are also linearly related. Thus, the shaded area corresponding to this simplified Mohr's criterion is bounded by straight lines in the second and fourth quadrants (Fig. 6.45b).

Note that in order to determine whether a structural component will be safe under a given loading, the state of stress should be calculated at all critical points of the component, i.e., at all points where stress concentrations are likely to occur. This may be done in a number of cases by using the stress-concentration factors given in Figs. 2.59, 3.31, 4.27, and 4.28. There are many instances, however, when the theory of elasticity must be used to determine the state of stress at a critical point.

Special care should be taken when *macroscopic cracks* have been detected in a structural component. While it may be assumed that the test specimen used to determine the ultimate tensile strength of the material contained the same type of flaws (i.e., *microscopic* cracks or cavities) as the structural component under investigation, the specimen was certainly free of any detectable macroscopic cracks. When a crack is detected in a structural component, it is necessary to determine whether that crack will tend to propagate under the expected loading condition and cause the component to fail, or whether it will remain stable. This requires an analysis involving the energy associated with the growth of the crack. Such an analysis is beyond the scope of this text and should be carried out by the methods of fracture mechanics.

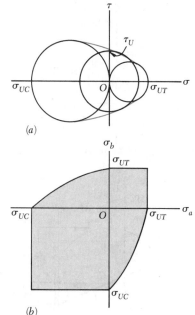

(a)

(b)

Fig. 6.44

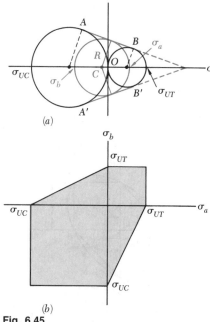

(a)

(b)

Fig. 6.45

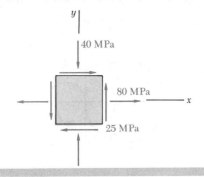

40 MPa

80 MPa

25 MPa

SAMPLE PROBLEM 6.4

The state of plane stress shown occurs at a critical point of a steel machine component. As a result of several tensile tests, it has been found that the tensile yield strength is $\sigma_Y = 250$ MPa for the grade of steel used. Determine the factor of safety with respect to yield, using (a) the maximum-shearing-stress criterion, and (b) the maximum-distortion-energy criterion.

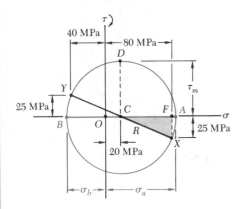

Mohr's Circle. We construct Mohr's circle for the given state of stress and find

$$\sigma_{ave} = OC = \tfrac{1}{2}(\sigma_x + \sigma_y) = \tfrac{1}{2}(80 - 40) = 20 \text{ MPa}$$
$$\tau_m = R = \sqrt{(CF)^2 + (FX)^2} = \sqrt{(60)^2 + (25)^2} = 65 \text{ MPa}$$

Principal Stresses

$$\sigma_a = OC + CA = 20 + 65 = +85 \text{ MPa}$$
$$\sigma_b = OC - BC = 20 - 65 = -45 \text{ MPa}$$

(a) **Maximum-Shearing-Stress Criterion.** Since for the grade of steel used the tensile strength is $\sigma_Y = 250$ MPa, the corresponding shearing stress at yield is

$$\tau_Y = \tfrac{1}{2}\sigma_Y = \tfrac{1}{2}(250 \text{ MPa}) = 125 \text{ MPa}$$

For $\tau_m = 65$ MPa: $\qquad$ F.S. $= \dfrac{\tau_Y}{\tau_m} = \dfrac{125 \text{ MPa}}{65 \text{ MPa}} \qquad$ F.S. $= 1.92$ ◀

(b) **Maximum-Distortion-Energy Criterion.** Introducing a factor of safety into Eq. (6.26), we write

$$\sigma_a^2 - \sigma_a\sigma_b + \sigma_b^2 = \left(\frac{\sigma_Y}{\text{F.S.}}\right)^2$$

For $\sigma_a = +85$ MPa, $\sigma_b = -45$ MPa, and $\sigma_Y = 250$ MPa, we have

$$(85)^2 - (85)(-45) + (45)^2 = \left(\frac{250}{\text{F.S.}}\right)^2$$

$$114.3 = \frac{250}{\text{F.S.}} \qquad \text{F.S.} = 2.19 \ ◀$$

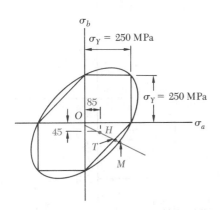

Comment. For a ductile material with $\sigma_Y = 250$ MPa, we have drawn the hexagon associated with the maximum-shearing-stress criterion and the ellipse associated with the maximum-distortion-energy criterion. The given state of plane stress is represented by point H of coordinates $\sigma_a = 85$ MPa and $\sigma_b = -45$ MPa. We note that the straight line drawn through points O and H intersects the hexagon at point T and the ellipse at point M. For each criterion, the value obtained for F.S. may be verified by measuring the line segments indicated and computing their ratios:

(a) F.S. $= \dfrac{OT}{OH} = 1.92 \qquad$ (b) F.S. $= \dfrac{OM}{OH} = 2.19$

PROBLEMS

6.64 For the state of plane stress shown, determine the maximum shearing stress when (a) $\tau_{xy} = 42$ MPa, (b) $\tau_{xy} = 96$ MPa. (*Hint:* Consider both in-plane and out-of-plane shearing stresses.)

6.65 Solve Prob. 6.64, when (a) $\tau_{xy} = 75$ MPa, (b) $\tau_{xy} = 30$ MPa.

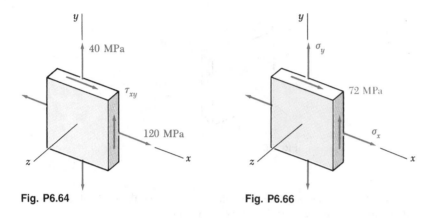

Fig. P6.64 Fig. P6.66

6.66 For the state of plane stress shown, determine the maximum shearing stress when (a) $\sigma_x = 84$ MPa and $\sigma_y = 24$ MPa, (b) $\sigma_x = 126$ MPa and $\sigma_y = 84$ MPa. (*Hint:* Consider both in-plane and out-of-plane shearing stresses.)

6.67 Solve Prob. 6.66, when (a) $\sigma_x = 120$ MPa and $\sigma_y = 60$ MPa, (b) $\sigma_x = 72$ MPa and $\sigma_y = 30$ MPa.

6.68 and 6.69 For the state of stress shown, determine the maximum shearing stress when (a) $\tau_{yz} = 105$ MPa, (b) $\tau_{yz} = 48$ MPa, (c) $\tau_{yz} = 0$.

6.70 and 6.71 For the state of stress shown, determine the maximum shearing stress when (a) $\sigma_z = 0$, (b) $\sigma_z = +60$ MPa, (c) $\sigma_z = -60$ MPa.

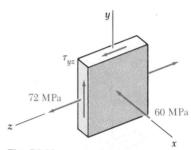

Fig. P6.68

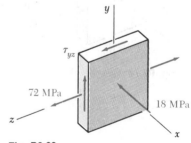

Fig. P6.69

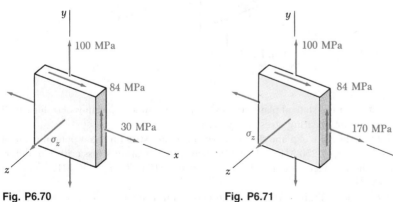

Fig. P6.70 Fig. P6.71

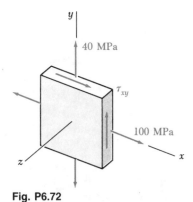

Fig. P6.72

6.72 For the state of plane stress shown, determine the value of τ_{xy} for which the maximum shearing stress is (*a*) 60 MPa, (*b*) 78 MPa.

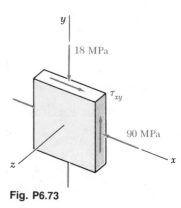

Fig. P6.73

6.73 For the state of plane stress shown, determine the value of τ_{xy} for which the maximum shearing stress is (*a*) 60 MPa, (*b*) 49.5 MPa.

6.74 For the state of plane stress shown, determine two values of σ_y for which the maximum shearing stress is 60 MPa.

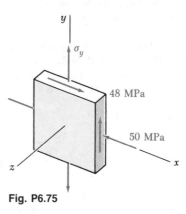

Fig. P6.75

6.75 For the state of plane stress shown, determine two values of σ_y for which the maximum shearing stress is 73 MPa.

6.76 The state of plane stress shown occurs in a structural component made of a steel with $\sigma_Y = 250$ MPa. Using the maximum-shearing-stress criterion, determine whether yield occurs when (*a*) $\sigma_y = 160$ MPa, (*b*) $\sigma_y = 40$ MPa, (*c*) $\sigma_y = -40$ MPa.

6.77 Solve Prob. 6.76, using the maximum-distortion-energy criterion.

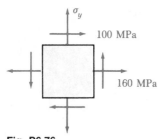

Fig. P6.76

6.78 The state of plane stress shown occurs in a machine component made of a steel with $\sigma_Y = 216$ MPa. Using the maximum-distortion-energy criterion, determine whether yield occurs when (a) $\tau_{xy} = 90$ MPa, (b) $\tau_{xy} = 108$ MPa, (c) $\tau_{xy} = 126$ MPa.

6.79 Solve Prob. 6.78, using the maximum-shearing-stress criterion.

6.80 and 6.81 The state of plane stress shown occurs in a steel member made of a grade of steel with a tensile yield strength of 300 MPa. Determine the factor of safety with respect to yield, using (a) the maximum-shearing-stress criterion, (b) the maximum-distortion-energy criterion.

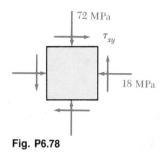

Fig. P6.78

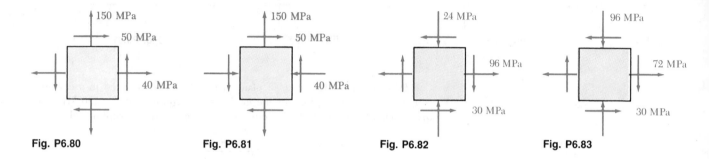

Fig. P6.80 **Fig. P6.81** **Fig. P6.82** **Fig. P6.83**

6.82 and 6.83 The state of plane stress shown occurs in an aluminum member made of an alloy with a tensile yield strength of 210 MPa. Determine the factor of safety with respect to yield, using (a) the maximum-shearing-stress criterion, (b) the maximum-distortion-energy criterion.

6.84 A 36-mm-diameter shaft is made of a grade of steel with a 250-MPa tensile yield strength. Using the maximum-shearing-stress criterion, determine the magnitude of the torque **T** at which yield first occurs when $P = 200$ kN.

6.85 Solve Prob. 6.84, using the maximum-distortion-energy criterion.

Fig. P6.84

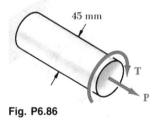

Fig. P6.86

6.86 A 45-mm-diameter shaft is made of a grade of steel with a 250-MPa tensile yield strength. Using the maximum-distortion-energy criterion, determine the magnitude of the force **P** at which yield first occurs when $T = 1.7$ kN · m.

6.87 Solve Prob. 6.86, using the maximum-shearing-stress criterion.

6.88 and 6.89 The state of plane stress shown is expected in an aluminum casting. Knowing that for the aluminum alloy used $\sigma_{UT} = 80$ MPa and $\sigma_{UC} = 200$ MPa, and using Mohr's criterion, determine whether rupture will occur.

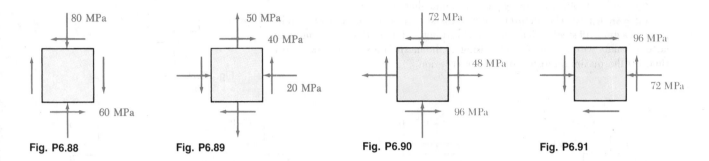

Fig. P6.88   Fig. P6.89   Fig. P6.90   Fig. P6.91

6.90 and 6.91 The state of plane stress shown is expected in a cast-iron component. Knowing that for the grade of cast iron used $\sigma_{UT} = 150$ MPa and $\sigma_{UC} = 300$ MPa, and using Mohr's criterion, determine whether rupture will occur.

Fig. P6.92   Fig. P6.93

6.92 and 6.93 The state of stress shown will occur in a cast-aluminum component. Knowing that for the aluminum alloy used $\sigma_{UT} = 60$ MPa and $\sigma_{UC} = 120$ MPa, and using Mohr's criterion, determine the shearing stress τ_0 at which rupture may be expected.

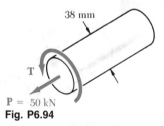

Fig. P6.94

6.94 A cast-aluminum rod is made of an alloy for which $\sigma_{UT} = 70$ MPa and $\sigma_{UC} = 140$ MPa and is subjected to a torque **T** and a 50-kN axial force **P**. Using Mohr's criterion, determine the magnitude of the torque **T** for which rupture of the rod will occur.

6.95 Solve Prob. 6.94, assuming that the sense of the force **P** is reversed.

6.9. STRESSES IN THIN-WALLED PRESSURE VESSELS

Thin-walled pressure vessels provide an important application of the analysis of plane stress. Since their walls offer little resistance to bending, it may be assumed that the internal forces exerted on a given portion of wall are tangent to the surface of the vessel (Fig. 6.46). The resulting stresses on an element of wall will thus be contained in a plane tangent to the surface of the vessel.

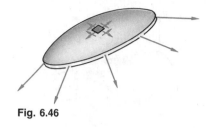

Fig. 6.46

We shall limit our analysis of stresses in thin-walled pressure vessels to the two types of vessels most frequently encountered: cylindrical pressure vessels and spherical pressure vessels.

Consider a cylindrical vessel of inside radius r and wall thickness t containing a fluid under pressure (Fig. 6.47). We propose to determine the stresses exerted on a small element of wall with sides respectively parallel and perpendicular to the axis of the cylinder. Because of the axisymmetry of the vessel and its contents, it is clear that no shearing stress is exerted on the element. The normal stresses σ_1 and σ_2 shown in Fig. 6.47 are therefore principal stresses. The stress σ_1 is known as the *hoop stress*, because it is the type of stress found in hoops used to hold together the various slats of a wooden barrel, and the stress σ_2 is called the *longitudinal stress*.

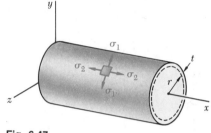

Fig. 6.47

In order to determine the hoop stress σ_1, we shall detach a portion of the vessel and its contents bounded by the xy plane and by two planes parallel to the yz plane at a distance Δx from each other (Fig. 6.48). The forces parallel to the z axis acting on the free body defined in this fashion consist of the elementary internal forces $\sigma_1\, dA$ on the wall sections, and of the elementary pressure forces $p\, dA$ exerted on the portion of fluid included in the free body. Note that p denotes the *gage pressure* of the fluid, i.e., the excess of the inside pressure over the outside atmospheric pressure. The resultant of the internal forces $\sigma_1\, dA$ is equal to the product of σ_1 and of the cross-sectional area $2t\, \Delta x$ of the wall, while the resultant of the pressure forces $p\, dA$ is equal to the product of p and of the area $2r\, \Delta x$. Writing the equilibrium equation $\Sigma F_z = 0$, we have

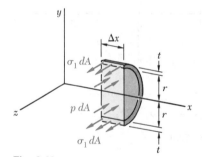

Fig. 6.48

$$\Sigma F_z = 0: \qquad \sigma_1(2t\, \Delta x) - p(2r\, \Delta x) = 0$$

and, solving for the hoop stress σ_1,

$$\sigma_1 = \frac{pr}{t} \tag{6.30}$$

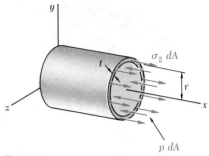

Fig. 6.49

To determine the longitudinal stress σ_2, we shall now pass a section perpendicular to the x axis and consider the free body consisting of the portion of the vessel and its contents located to the left of the section (Fig. 6.49). The forces acting on this free body are the elementary internal forces $\sigma_2 \, dA$ on the wall section and the elementary pressure forces $p \, dA$ exerted on the portion of fluid included in the free body. Noting that the area of the fluid section is πr^2 and that the area of the wall section may be obtained by multiplying the circumference $2\pi r$ of the cylinder by its wall thickness t, we write the equilibrium equation†

$$\Sigma F_x = 0: \qquad\qquad \sigma_2(2\pi r t) - p(\pi r^2) = 0$$

and, solving for the longitudinal stress σ_2,

$$\sigma_2 = \frac{pr}{2t} \tag{6.31}$$

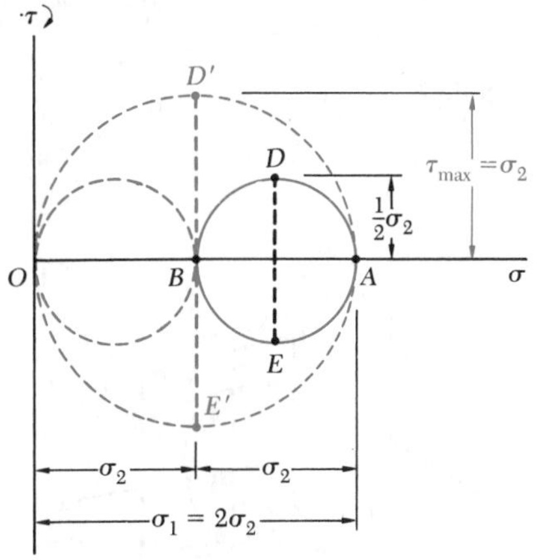

We note from Eqs. (6.30) and (6.31) that the hoop stress σ_1 is twice as large as the longitudinal stress σ_2:

$$\sigma_1 = 2\sigma_2 \tag{6.32}$$

Drawing Mohr's circle through the points A and B which correspond respectively to the principal stresses σ_1 and σ_2 (Fig. 6.50), and recalling that the maximum in-plane shearing stress is equal to the radius of this circle, we have

$$\tau_{\text{max(in plane)}} = \tfrac{1}{2}\sigma_2 = \frac{pr}{4t} \tag{6.33}$$

This stress corresponds to points D and E and is exerted on an element obtained by rotating the original element of Fig. 6.47 through 45° *within the plane* tangent to the surface of the vessel. The maximum shearing stress in the wall of the vessel, however, is larger. It is equal to the

Fig. 6.50

---

† Using the mean radius of the wall section, $r_m = r + \tfrac{1}{2}t$, in computing the resultant of the forces on that section, we would obtain a more accurate value of the longitudinal stress, namely,

$$\sigma_2 = \frac{pr}{2t} \frac{1}{1 + \dfrac{t}{2r}} \tag{6.31'}$$

However, for a thin-walled pressure vessel, the term $t/2r$ is sufficiently small to allow the use of Eq. (6.31) for engineering design and analysis. If a pressure vessel is not thin-walled (i.e., if $t/2r$ is not small), the stresses σ_1 and σ_2 vary across the wall and must be determined by the methods of the theory of elasticity.

radius of the circle of diameter OA and corresponds to a rotation of 45° about a longitudinal axis and *out of the plane* of stress.† We have

$$\tau_{max} = \sigma_2 = \frac{pr}{2t} \qquad (6.34)$$

We shall now consider a spherical vessel of inside radius r and wall thickness t, containing a fluid under a gage pressure p. We observe that, for reasons of symmetry, the stresses exerted on the four faces of a small element of wall must be equal (Fig. 6.51). We have

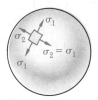

$$\sigma_1 = \sigma_2 \qquad (6.35)$$

Fig. 6.51

To determine the value of the stress, we pass a section through the center C of the vessel and consider the free body consisting of the portion of the vessel and its contents located to the left of the section (Fig. 6.52). The equation of equilibrium for this free body is the same as for the free body of Fig. 6.49. We thus conclude that, for a spherical vessel,

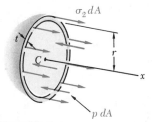

Fig. 6.52

$$\sigma_1 = \sigma_2 = \frac{pr}{2t} \qquad (6.36)$$

Since the principal stresses σ_1 and σ_2 are equal, Mohr's circle for transformations of stress within the plane tangent to the surface of the vessel reduces to a point (Fig. 6.53); we conclude that the in-plane normal stress is constant and that the in-plane maximum shearing stress is zero. The maximum shearing stress in the wall of the vessel, however, is not zero; it is equal to the radius of the circle of diameter OA and corresponds to a rotation of 45° out of the plane of stress. We have

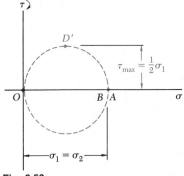

Fig. 6.53

$$\tau_{max} = \tfrac{1}{2}\sigma_1 = \frac{pr}{4t} \qquad (6.37)$$

---

† It should be observed that, while the third principal stress is zero on the outside surface of the vessel, it is equal to $-p$ on the inside surface, and is represented by a point $C(-p, 0)$ on a Mohr-circle diagram. Thus, close to the inside surface of the vessel, the maximum shearing stress is equal to the radius of a circle of diameter CA, and we have

$$\tau_{max} = \frac{1}{2}(\sigma_1 + p) = \frac{pr}{2t}\left(1 + \frac{t}{r}\right)$$

For a thin-walled vessel, however, the term t/r is small, and we may neglect the variation of τ_{max} across the wall section. This remark also applies to spherical pressure vessels.

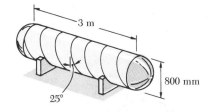

SAMPLE PROBLEM 6.5

A compressed-air tank is supported by two cradles as shown: one of the cradles is designed so that it does not exert any longitudinal force on the tank. The cylindrical body of the tank is fabricated from a 10-mm steel plate by butt welding along a helix which forms an angle of 25° with a transverse plane. The end caps are spherical and have a uniform wall thickness of 8 mm. For an internal gage pressure of 1.2 MPa, determine (a) the normal stress and the maximum shearng stress in the spherical caps, (b) the stressses in directions perpendicular and parallel to the helical weld.

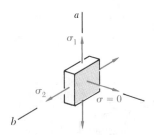

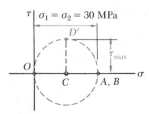

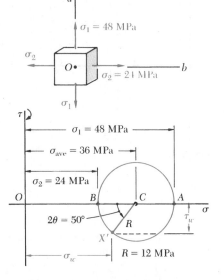

(a) **Spherical Cap.** Using Eq. (6.36), we write

$$p = 1.2 \text{ MPa} \qquad r = 400 \text{ mm} \qquad t = 8 \text{ mm}$$

$$\sigma_1 = \sigma_2 = \frac{pr}{2t} = \frac{(1.2 \text{ MPa})(0.400 \text{ m})}{2(0.008 \text{ m})} \qquad \sigma = 30 \text{ MPa} \quad \blacktriangleleft$$

We note that for stresses in a plane tangent to the cap, Mohr's circle reduces to a point (A, B) on the horizontal axis and that all in-plane shearing stresses are zero. On the surface of the cap the third principal stress is zero and corresponds to point O. On a Mohr's circle of diameter AO, point D' represents the maximum shearing stress; it occurs on planes forming an angle of 45° with a plane tangent to the cap.

$$\tau_{max} = \tfrac{1}{2}(30 \text{ MPa}) \qquad\qquad \tau_{max} = 15 \text{ MPa} \quad \blacktriangleleft$$

(b) **Cylindrical Body of the Tank.** We first determine the hoop stress σ_1 and the longitudinal stress σ_2. Using Eqs. (6.30) and (6.32), we write

$$\sigma_1 = \frac{pr}{t} = \frac{(1.20 \text{ MPa})(0.400 \text{ m})}{0.010 \text{ m}} = 48 \text{ MPa}$$

$$\sigma_2 = \tfrac{1}{2}\sigma_1 = 24 \text{ MPa}$$

Stresses at the Weld. Noting that both the hoop stress and the longitudinal stress are principal stresses, we draw Mohr's circle as shown.

An element having a face parallel to the weld is obtained by rotating the face perpendicular to the axis Ob counterclockwise through 25°. Therefore, on Mohr's circle we locate the point X' corresponding to the stress components on the weld by rotating radius CB counterclockwise through $2\theta = 50°$.

$$\sigma_w = \sigma_{ave} - R \cos 50° = 36 \text{ MPa} - (12 \text{ MPa}) \cos 50° \qquad \sigma_w = +28.3 \text{ MPa} \quad \blacktriangleleft$$
$$\tau_w = R \sin 50° = (12 \text{ MPa}) \sin 50° \qquad\qquad\qquad \tau_w = 9.19 \text{ MPa} \quad \blacktriangleleft$$

Since X' is below the horizontal axis, τ_w tends to rotate the element counterclockwise.

PROBLEMS

6.96 A spherical gas container made of steel has a 5-m outside diameter and a 10-mm uniform wall thickness. Knowing that the internal pressure is 400 kPa, determine the maximum normal stress and the maximum shearing stress in the container.

6.97 A basketball has a 300-mm diameter and a 3-mm wall thickness. Determine the normal stress in the wall when the basketball is inflated to a 120-kPa gage pressure.

6.98 A spherical pressure vessel of 500-mm inside diameter and 6-mm wall thickness is fabricated from a steel with a 400-MPa ultimate strength in tension. Determine the factor of safety with respect to tensile failure when the gage pressure is 5.5 MPa.

6.99 A spherical pressure vessel of 500 mm inside diameter is to be fabricated from a steel with a 400 MPa ultimate strength in tension. Knowing that the maximum gage pressure is to be 8 MPa and that a factor of safety of 3.00 is desired, determine the smallest wall thickness which may be used.

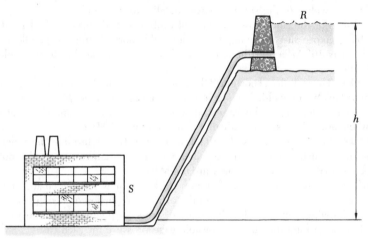

Fig. P6.100

6.100 A penstock of 750-mm outside diameter and 12-mm wall thickness connects a reservoir R with a generating station S. Knowing that $h = 300$ m, determine the maximum normal stress and the maximum shearing stress in the penstock under static conditions. (Density of water $= 1000$ kg/m^3.)

6.101 In Prob. 6.100, determine the maximum allowable value of h, knowing that the penstock is made of a steel with a 450-MPa ultimate strength in tension and that a factor of safety of 4.50 is desired.

6.102 The unpressurized cylindrical storage tank shown has a 5-mm wall thickness and is made of a steel with a 400-MPa ultimate strength in tension. Determine the maximum height h to which it may be filled with water if a factor of safety of 4.00 is desired. (Density of water $= 1$ Mg/m^3.)

6.103 For the storage tank of Prob. 6.102, determine the maximum normal stress and the maximum shearing stress in the cylindrical wall when the tank is filled to capacity ($h = 20$ m).

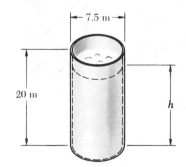

Fig. P6.102

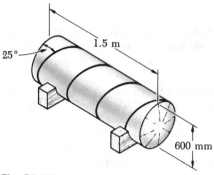

Fig. P6.104

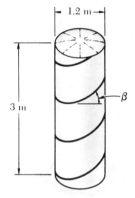

Fig. P6.106

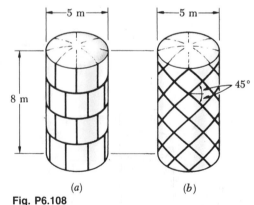

(a)   (b)

Fig. P6.108

6.104 The cylindrical portion of the compressed-air tank shown is fabricated of 5-mm-thick plate welded along a helix forming a 25° angle with a transverse plane.   For a gage pressure of 4 MPa, determine (a) the normal stress perpendicular to the weld, (b) the shearing stress parallel to the weld.

6.105 For the compressed-air tank of Prob. 6.104, determine the largest allowable gage pressure, knowing that the allowable normal stress perpendicular to the weld is 165 MPa and the allowable shearing stress parallel to the weld is 110 MPa.

6.106 The cylindrical portion of the compressed-air tank shown is fabricated of 10-mm-thick plate welded along a helix forming an angle $\beta = 28°$ with the horizontal.   Determine the largest allowable gage pressure, knowing that the allowable normal stress perpendicular to the weld is 165 MPa.

6.107 For the compressed-air tank of Prob. 6.106, determine the range of values of β which may be used if the shearing stress parallel to the weld is not to exceed 50 MPa when the gage pressure is 4 MPa.

6.108 Square plates, each of 16-mm thickness, may be welded together in either of the two ways shown to form the cylindrical portion of a compressed-air tank.   Knowing that the allowable normal stress perpendicular to the weld is 65 MPa, determine the largest allowable gage pressure for each case.

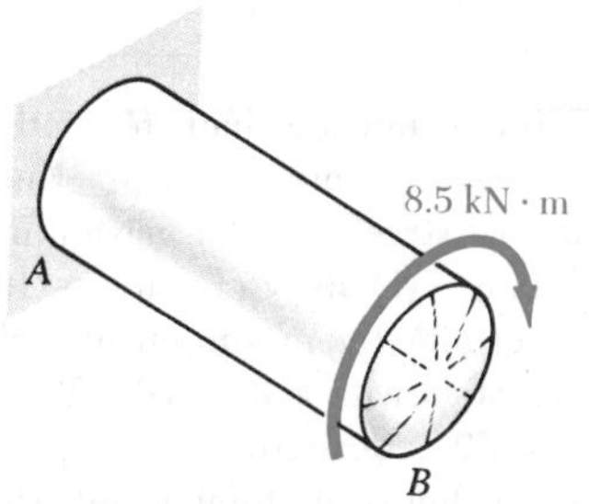

Fig. P6.109

6.109 An 8.5-kN · m torque is applied to the end of a tank containing oil under a pressure of 2.75 MPa.   Knowing that the tank has a 250-mm outside diameter and a 6-mm wall thickness, determine the maximum normal stress and the maximum shearing stress in the cylindrical wall of the tank.

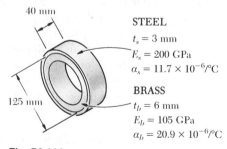

Fig. P6.110

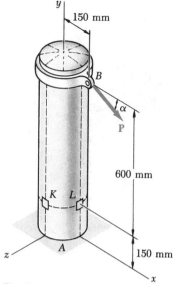

Fig. P6.111

6.110 A pressure vessel of 300-mm outside diameter is fabricated from 6-mm-thick steel plate by welding along the helix forming an angle of 22.5° with the horizontal.   Knowing that the gage pressure inside the vessel is 1.7 MPa and that a 250-kN axial force **P** is applied at the center of the end plate, determine (*a*) the normal stress perpendicular to the weld, (*b*) the shearing stress parallel to the weld.

6.111 The compressed-air tank *AB* has a 250-mm outside diameter and an 8-mm wall thickness.   It is fitted with a collar by which a 40-kN force **P** may be applied at *B* at an angle $\alpha = 30°$ with the horizontal.   Knowing that the gage pressure inside the tank is 5 MPa, determine the maximum normal stress and the maximum shearing stress at (*a*) point *K*, (*b*) point *L*.

6.112 Solve Prob. 6.111, assuming that the 40-kN force **P** is horizontal ($\alpha = 0$).

6.113 The compressed-air vessel *AB* has an outside diameter of 462 mm and a uniform wall thickness of 6 mm.   Knowing that the gage pressure inside the vessel is 120 kPa, determine the maximum normal stress and the maximum in-plane shearing stress at (*a*) point *K*, (*b*) point *L*.

6.114 A brass ring of 125-mm outer diameter and 6-mm thickness fits exactly inside a steel ring of 125-mm inner diameter and 3-mm thickness when the temperature of both rings is 10°C.   Knowing that the temperature of both rings is then raised to 50°C, determine (*a*) the tensile stress in the steel ring, (*b*) the corresponding pressure exerted by the brass ring on the steel ring.

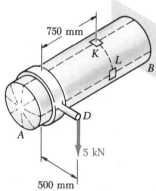

Fig. P6.113

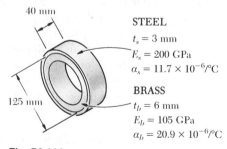

STEEL

$t_s = 3$ mm

$E_s = 200$ GPa

$\alpha_s = 11.7 \times 10^{-6}/°C$

BRASS

$t_b = 6$ mm

$E_b = 105$ GPa

$\alpha_b = 20.9 \times 10^{-6}/°C$

Fig. P6.114

6.115 Solve Prob. 6.114, assuming that the brass ring is 3 mm thick and the steel ring is 6 mm thick.

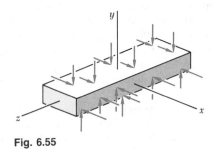

Fig. 6.54

Fixed support

Fixed support

*6.10. TRANSFORMATION OF PLANE STRAIN

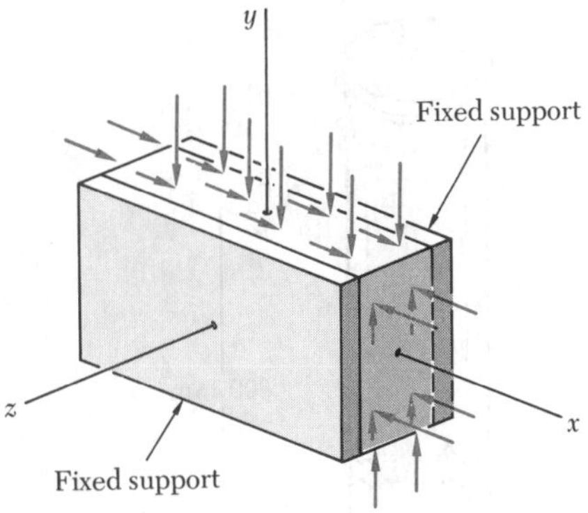

We shall now discuss transformations of *strain* under a rotation of the coordinate axes. Our analysis will first be limited to states of *plane strain*, i.e., to situations where the deformations of the material take place within parallel planes, and are the same in each of these planes. If the z axis is chosen perpendicular to the planes in which the deformations take place, we have $\epsilon_z = \gamma_{zx} = \gamma_{zy} = 0$, and the only remaining strain components are ϵ_x, ϵ_y, and γ_{xy}. Such a situation occurs in a plate subjected along its edges to uniformly distributed loads and restrained from expanding or contracting laterally by smooth, rigid, and fixed supports (Fig. 6.54). It would also be found in a bar of infinite length subjected on its sides to uniformly distributed loads since, by reason of symmetry, the elements located in a given transverse plane may not move out of that plane. This idealized model shows that, in the actual case of a long bar subjected to uniformly distributed transverse loads (Fig. 6.55), a state of plane strain exists in any given transverse section which is not located too close to either end of the bar.†

Fig. 6.55

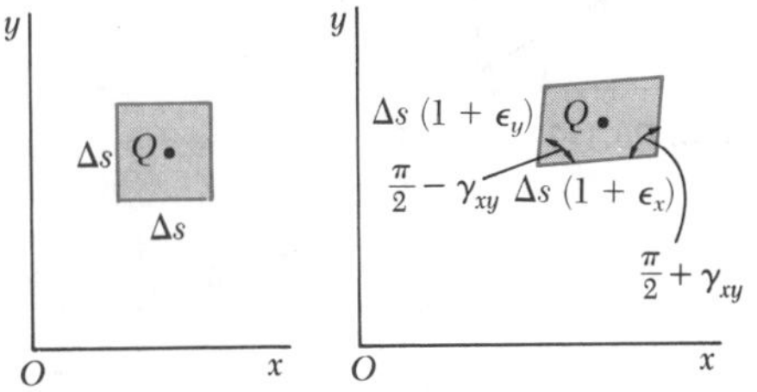

Let us assume that a state of plane strain exists at point Q (with $\epsilon_z = \gamma_{zx} = \gamma_{zy} = 0$), and that it is defined by the strain components ϵ_x, ϵ_y, and γ_{xy} associated with the x and y axes. As we know from Secs. 2.12 and 2.14, this means that a square element of center Q, with sides of length Δs respectively parallel to the x and y axes, is deformed into a parallelogram with sides of length respectively equal to $\Delta s\,(1 + \epsilon_x)$ and $\Delta s\,(1 + \epsilon_y)$, forming angles of $\frac{\pi}{2} - \gamma_{xy}$ and $\frac{\pi}{2} + \gamma_{xy}$ with each other (Fig. 6.56). We recall that, as a result of the deformations of the other elements located in the xy plane, the element considered may also undergo a rigid-body motion, but such a motion is irrelevant to the determination of the strains at point Q and will be ignored in this analysis. Our purpose is to

Fig. 6.56

Δs $Q\bullet$ Δs

$\Delta s\,(1 + \epsilon_y)$ $Q\bullet$

$\frac{\pi}{2} - \gamma_{xy}$ $\Delta s\,(1 + \epsilon_x)$

$\frac{\pi}{2} + \gamma_{xy}$

† It should be observed that a state of *plane strain* and a state of *plane stress* (cf. Sec. 6.1) do not occur simultaneously, except for ideal materials with a Poisson ratio equal to zero. The constraints placed on the elements of the plate of Fig. 6.54 and of the bar of Fig. 6.55 result in a stress σ_z different from zero. On the other hand, in the case of the plate of Fig. 6.3, the absence of any lateral restraint results in $\sigma_z = 0$ and $\epsilon_z \neq 0$.

determine in terms of ϵ_x, ϵ_y, γ_{xy}, and θ the stress components $\epsilon_{x'}$, $\epsilon_{y'}$, and $\gamma_{x'y'}$ associated with the frame of reference $x'y'$ obtained by rotating the x and y axes through the angle θ. As shown in Fig. 6.57, these new strain components define the parallelogram into which a square with sides respectively parallel to the x' and y' axes is deformed.

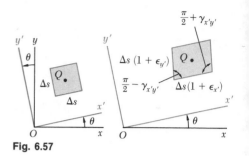

Fig. 6.57

We shall first derive an expression for the normal strain $\epsilon(\theta)$ along a line AB forming an arbitrary angle θ with the x axis. To do so, we shall consider the right triangle ABC which has AB for hypotenuse (Fig. 6.58a), and the oblique triangle $A'B'C'$ into which triangle ABC is deformed (Fig. 6.58b). Denoting by Δs the length of AB, we express the length of $A'B'$ as $\Delta s\,[1 + \epsilon(\theta)]$. Similarly, denoting by Δx and Δy the lengths of sides AC and CB, we express the lengths of $A'C'$ and $C'B'$ as $\Delta x\,(1 + \epsilon_x)$ and $\Delta y\,(1 + \epsilon_y)$, respectively. Recalling from Fig. 6.56 that the right angle at C in Fig. 6.58a deforms into an angle equal to $\frac{\pi}{2} + \gamma_{xy}$ in Fig. 6.58b, and applying the law of cosines to triangle $A'B'C'$, we write

$$(A'B')^2 = (A'C')^2 + (C'B')^2 - 2(A'C')(C'B')\cos\left(\frac{\pi}{2} + \gamma_{xy}\right)$$

$$
\begin{aligned}
(\Delta s)^2[1 + \epsilon(\theta)]^2 = (\Delta x)^2(1 + \epsilon_x)^2 + (\Delta y)^2(1 + \epsilon_y)^2 \\
- 2(\Delta x)(1 + \epsilon_x)(\Delta y)(1 + \epsilon_y)\cos\left(\frac{\pi}{2} + \gamma_{xy}\right) \quad (6.38)
\end{aligned}
$$

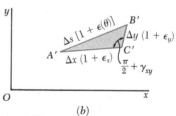

(a)

But from Fig. 6.58a we have

$$\Delta x = (\Delta s)\cos\theta \qquad \Delta y = (\Delta s)\sin\theta \qquad (6.39)$$

and we note that, since γ_{xy} is very small,

$$\cos\left(\frac{\pi}{2} + \gamma_{xy}\right) = -\sin\gamma_{xy} \approx -\gamma_{xy} \qquad (6.40)$$

(b)

Fig. 6.58

Substituting from Eqs. (6.39) and (6.40) into Eq. (6.38), recalling that $\cos^2\theta + \sin^2\theta = 1$, and neglecting second-order terms in $\epsilon(\theta)$, ϵ_x, ϵ_y, and γ_{xy}, we write

$$\epsilon(\theta) = \epsilon_x\cos^2\theta + \epsilon_y\sin^2\theta + \gamma_{xy}\sin\theta\cos\theta \qquad (6.41)$$

Equation (6.41) enables us to determine the normal strain $\epsilon(\theta)$ in any direction AB in terms of the strain components ϵ_x, ϵ_y, γ_{xy}, and the angle θ that AB forms with the x axis. We check that, for $\theta = 0$, Eq. (6.41) yields $\epsilon(0) = \epsilon_x$ and that, for $\theta = 90°$, it yields $\epsilon(90°) = \epsilon_y$. On the other hand, making $\theta = 45°$ in Eq. (6.41), we obtain the normal strain in the direction of the bisector OB of the angle formed by the x and y axes (Fig. 6.59). Denoting this strain by ϵ_{OB}, we write

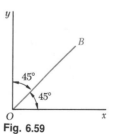

$$\epsilon_{OB} = \epsilon(45°) = \tfrac{1}{2}(\epsilon_x + \epsilon_y + \gamma_{xy}) \qquad (6.42)$$

Fig. 6.59

Solving Eq. (6.42) for γ_{xy}, we have

$$\gamma_{xy} = 2\epsilon_{OB} - (\epsilon_x + \epsilon_y) \qquad (6.43)$$

This relation makes it possible to express the *shearing strain* associated with a given pair of rectangular axes in terms of the *normal strains* measured along these axes and their bisector. It will play a fundamental role in our present derivation and will also be used in Sec. 6.13 in connection with the experimental determination of shearing strains.

Recalling that the main purpose of this section is to express the strain components associated with the frame of reference $x'y'$ of Fig. 6.57 in terms of the angle θ and the strain components ϵ_x, ϵ_y, and γ_{xy} associated with the x and y axes, we note that the normal strain $\epsilon_{x'}$ along the x' axis is given by Eq. (6.41). Using the trigonometric relations (6.3) and (6.4), we write this equation in the alternative form

$$\epsilon_{x'} = \frac{\epsilon_x + \epsilon_y}{2} + \frac{\epsilon_x - \epsilon_y}{2} \cos 2\theta + \frac{\gamma_{xy}}{2} \sin 2\theta \qquad (6.44)$$

Replacing θ by $\theta + 90°$, we obtain the normal strain along the y' axis. Since $\cos(2\theta + 180°) = -\cos 2\theta$ and $\sin(2\theta + 180°) = -\sin 2\theta$, we have

$$\epsilon_{y'} = \frac{\epsilon_x + \epsilon_y}{2} - \frac{\epsilon_x - \epsilon_y}{2} \cos 2\theta - \frac{\gamma_{xy}}{2} \sin 2\theta \qquad (6.45)$$

Adding Eqs. (6.44) and (6.45) member to member, we obtain

$$\epsilon_{x'} + \epsilon_{y'} = \epsilon_x + \epsilon_y \qquad (6.46)$$

Since $\epsilon_z = \epsilon_{z'} = 0$, we thus verify in the case of plane strain that the sum of the normal strains associated with a cubic element of material is independent of the orientation of that element.[†]

Replacing now θ by $\theta + 45°$ in Eq. (6.44), we obtain an expression for the normal strain along the bisector OB' of the angle formed by the x' and y' axes. Since $\cos(2\theta + 90°) = -\sin 2\theta$ and $\sin(2\theta + 90°) = \cos 2\theta$, we have

$$\epsilon_{OB'} = \frac{\epsilon_x + \epsilon_y}{2} - \frac{\epsilon_x - \epsilon_y}{2} \sin 2\theta + \frac{\gamma_{xy}}{2} \cos 2\theta \qquad (6.47)$$

Writing Eq. (6.43) with respect to the x' and y' axes, we express the shearing strain $\gamma_{x'y'}$ in terms of the normal strains measured along the x' and y' axes and the bisector OB':

$$\gamma_{x'y'} = 2\epsilon_{OB'} - (\epsilon_{x'} + \epsilon_{y'}) \qquad (6.48)$$

Substituting from Eqs. (6.46) and (6.47) into (6.48), we obtain

$$\gamma_{x'y'} = -(\epsilon_x - \epsilon_y) \sin 2\theta + \gamma_{xy} \cos 2\theta \qquad (6.49)$$

[†] Cf. footnote on page 78.

Equations (6.44), (6.45), and (6.49) are the desired equations defining the transformation of plane strain under a rotation of axes in the plane of strain. Dividing all terms in Eq. (6.49) by 2, we write this equation in the alternative form

$$\frac{\gamma_{x'y'}}{2} = -\frac{\epsilon_x - \epsilon_y}{2}\sin 2\theta + \frac{\gamma_{xy}}{2}\cos 2\theta \qquad (6.49')$$

and observe that Eqs. (6.44), (6.45), and (6.49') for the transformation of plane strain closely resemble the equations derived in Sec. 6.2 for the transformation of plane stress. While the former may be obtained from the latter by replacing the normal stresses by the corresponding normal strains, it should be noted, however, that the shearing stresses τ_{xy} and $\tau_{x'y'}$ should be replaced by *half* of the corresponding shearing strains, i.e., by $\frac{1}{2}\gamma_{xy}$ and $\frac{1}{2}\gamma_{x'y'}$, respectively.

*6.11. MOHR'S CIRCLE FOR PLANE STRAIN

Since the equations for the transformation of plane strain are of the same form as the equations for the transformation of plane stress, the use of Mohr's circle may be extended to the analysis of plane strain. Given the strain components ϵ_x, ϵ_y, and γ_{xy} defining the deformation represented in Fig. 6.56, we plot a point $X(\epsilon_x, -\frac{1}{2}\gamma_{xy})$ of abscissa equal to the normal strain ϵ_x and of ordinate equal to minus half the shearing strain γ_{xy}, and a point $Y(\epsilon_y, +\frac{1}{2}\gamma_{xy})$ (Fig. 6.60). Drawing the diameter XY, we define the center C of Mohr's circle for plane strain. The abscissa of C and the radius R of the circle are respectively equal to

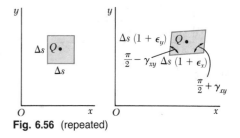

Fig. 6.56 (repeated)

$$\epsilon_{ave} = \frac{\epsilon_x + \epsilon_y}{2} \qquad \text{and} \qquad R = \sqrt{\left(\frac{\epsilon_x - \epsilon_y}{2}\right)^2 + \left(\frac{\gamma_{xy}}{2}\right)^2} \qquad (6.50)$$

We note that if γ_{xy} is positive, as assumed in Fig. 6.56, points X and Y are plotted, respectively, below and above the horizontal axis in Fig. 6.60. But, in the absence of any overall rigid-body rotation, the side of the element in Fig. 6.56 which is associated with ϵ_x is observed to rotate counterclockwise, while the side associated with ϵ_y is observed to rotate clockwise. Thus, if the shear deformation causes a given side to rotate *clockwise*, the corresponding point on Mohr's circle for plane strain is plotted *above* the horizontal axis, and if the deformation causes the side to rotate *counterclockwise*, the corresponding point is plotted *below* the horizontal axis. We note that this convention matches the convention used to draw Mohr's circle for plane stress.

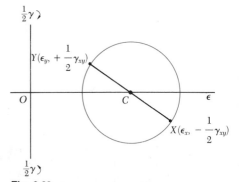

Fig. 6.60

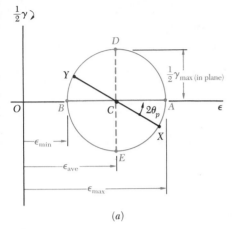

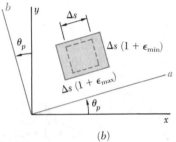

Fig. 6.61

Points A and B where Mohr's circle intersects the horizontal axis correspond to the *principal strains* ϵ_{max} and ϵ_{min} (Fig. 6.61a). We find

$$\epsilon_{max} = \epsilon_{ave} + R \qquad \text{and} \qquad \epsilon_{min} = \epsilon_{ave} - R \qquad (6.51)$$

where ϵ_{ave} and R are defined by Eqs. (6.50). The corresponding value θ_p of the angle θ is obtained by observing that the shearing strain is zero for A and B. Setting $\gamma_{x'y'} = 0$ in Eq. (6.49), we have

$$\tan 2\theta_p = \frac{\gamma_{xy}}{\epsilon_x - \epsilon_y} \qquad (6.52)$$

The corresponding axes a and b in Fig. 6.61b are the *principal axes of strain*. The angle θ_p which defines the direction of the principal axis Oa in Fig. 6.61b corresponding to point A in Fig. 6.61a is equal to half of the angle XCA measured on Mohr's circle, and the rotation which brings Ox into Oa has the same sense as the rotation which brings the diameter XY of Mohr's circle into the diameter AB.

We recall from Sec. 2.14 that, in the case of the elastic deformation of a homogeneous, isotropic material, Hooke's law for shearing stress and strain applies and yields $\tau_{xy} = G\gamma_{xy}$ for any pair of rectangular x and y axes. Thus, $\gamma_{xy} = 0$ when $\tau_{xy} = 0$, which indicates that the principal axes of strain coincide with the principal axes of stress.

The maximum in-plane shearing strain is defined by points D and E in Fig. 6.61a. It is equal to the diameter of Mohr's circle. Recalling the second of Eqs. (6.50), we write

$$\gamma_{max(\text{in plane})} = 2R = \sqrt{(\epsilon_x - \epsilon_y)^2 + \gamma_{xy}^2} \qquad (6.53)$$

Finally, we note that the points X' and Y' which define the components of strain corresponding to a rotation of the coordinate axes through an angle θ (Fig. 6.57) may be obtained by rotating the diameter XY of Mohr's circle in the same sense through an angle 2θ (Fig. 6.62).

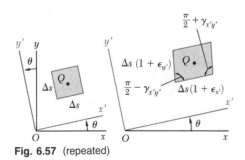

Fig. 6.57 (repeated)

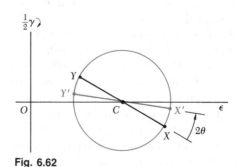

Fig. 6.62

Example 6.04

In a material in a state of plane strain, it is known that the horizontal side of a 10×10-mm square elongates by 4 μm, while its vertical side remains unchanged, and that the angle at the lower left corner increases by 0.4×10^{-3} rad (Fig. 6.63). Determine (a) the principal axes and principal strains, (b) the maximum shearing strain and the corresponding normal strain.

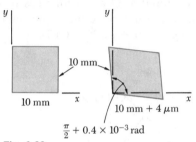

Fig. 6.63

(a) *Principal Axes and Principal Strains.* We first determine the coordinates of points X and Y on Mohr's circle for strain. We have

$$\epsilon_x = \frac{+4 \times 10^{-6} \text{ m}}{10 \times 10^{-3} \text{ m}} = +400 \ \mu \qquad \epsilon_y = 0 \qquad \left| \frac{\gamma_{xy}}{2} \right| = 200 \ \mu$$

Since the side of the square associated with ϵ_x rotates *clockwise*, point X of coordinates ϵ_x and $|\gamma_{xy}/2|$ is plotted *above* the horizontal axis. Since $\epsilon_y = 0$ and the corresponding side rotates *counterclockwise*, point Y is plotted directly *below* the origin (Fig. 6.64). Drawing the diameter XY, we determine the center C of Mohr's circle and its radius R. We have

$$OC = \frac{\epsilon_x + \epsilon_y}{2} = 200 \ \mu \qquad OY = 200 \ \mu$$

$$R = \sqrt{(OC)^2 + (OY)^2} = \sqrt{(200 \ \mu)^2 + (200 \ \mu)^2} = 283 \ \mu$$

The principal strains are defined by the abscissas of points A and B. We write

$$\epsilon_a = OA = OC + R = 200 \ \mu + 283 \ \mu = 483 \ \mu$$
$$\epsilon_b = OB = OC - R = 200 \ \mu - 283 \ \mu = -83 \ \mu$$

The principal axes Oa and Ob are shown in Fig. 6.65. Since $OC = OY$, the angle at C in triangle OCY is $45°$. Thus, the angle $2\theta_p$ which brings XY into AB is $45°\downarrow$ and the angle θ_p bringing Ox into Oa is $22.5°\downarrow$.

(b) *Maximum Shearing Strain.* Points D and E define the maximum in-plane shearing strain which, since the principal strains have opposite signs, is also the actual maximum shearing strain (see Sec. 6.12). We have

$$\frac{\gamma_{max}}{2} = R = 283 \ \mu \qquad \gamma_{max} = 566 \ \mu$$

The corresponding normal strains are both equal to

$$\epsilon' = OC = 200 \ \mu$$

The axes of maximum shearing strain are shown in Fig. 6.66.

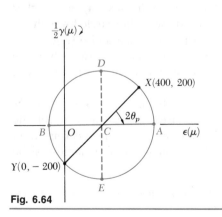

Fig. 6.64

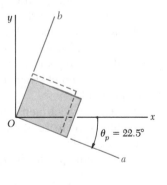

Fig. 6.65

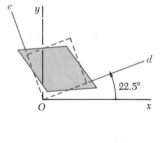

Fig. 6.66

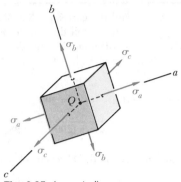

Fig. 6.25 (repeated)

*6.12. THREE-DIMENSIONAL ANALYSIS OF STRAIN

We saw in Sec. 6.5 that, in the most general case of stress, we can determine three coordinate axes a, b, and c, called the principal axes of stress. A small cubic element with faces respectively perpendicular to these axes is free of shearing stresses (Fig. 6.25), i.e., we have $\tau_{ab} = \tau_{bc} = \tau_{ca} = 0$. As recalled in the preceding section, Hooke's law for shearing stress and strain applies when the deformation is elastic and the material homogeneous and isotropic. It follows that, in such a case, $\gamma_{ab} = \gamma_{bc} = \gamma_{ca} = 0$, i.e., the axes a, b, and c are also *principal axes of strain*. A small cube of side equal to unity, centered at Q and with faces respectively perpendicular to the principal axes, is deformed into a rectangular parallelepiped of sides $1 + \epsilon_a$, $1 + \epsilon_b$, and $1 + \epsilon_c$ (Fig. 6.67).

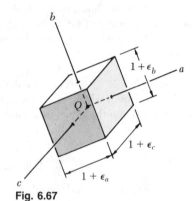

Fig. 6.67

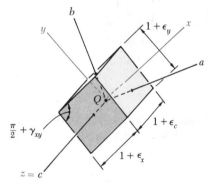

Fig. 6.68

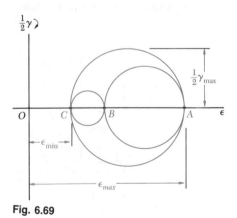

Fig. 6.69

If the element of Fig 6.67 is rotated about one of the principal axes at Q, say the c axis (Fig. 6.68), the method of analysis developed earlier for the transformation of plane strain may be used to determine the strain components ϵ_x, ϵ_y, and γ_{xy} associated with the faces perpendicular to the c axis, since the derivation of this method did not involve any of the other strain components.† We may, therefore, draw Mohr's circle through the points A and B corresponding to the principal axes a and b (Fig. 6.69). Similarly, circles of diameters BC and CA may be used to analyze the transformation of strain as the element is rotated about the a and b axes, respectively.

The three-dimensional analysis of strain by means of Mohr's circle is limited here to rotations about principal axes (as was the case for the analysis of stress) and is used to determine the maximum shearing strain $\gamma_{\max}$ at point Q. Since $\gamma_{\max}$ is equal to the diameter of the largest of the

---

†We may note that the other four faces of the element remain rectangular and that the edges parallel to the c axis remain unchanged.

three circles shown in Fig. 6.69, we have

$$\gamma_{\max} = |\epsilon_{\max} - \epsilon_{\min}|$$

(6.54)

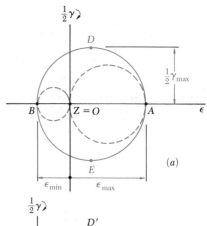

where $\epsilon_{\max}$ and $\epsilon_{\min}$ represent the *algebraic* values of the maximum and minimum strains at point Q.

Returning to the particular case of *plane strain*, and selecting the x and y axes in the plane of strain, we have $\epsilon_z = \gamma_{zx} = \gamma_{zy} = 0$. Thus, the z axis is one of the three principal axes at Q, and the corresponding point in the Mohr-circle diagram is the origin O, where $\epsilon = \gamma = 0$. If the points A and B which define the principal axes within the plane of strain fall on opposite sides of O (Fig. 6.70a), the corresponding principal strains represent the maximum and minimum normal strains at point Q, and the maximum shearing strain is equal to the maximum in-plane shearing strain corresponding to points D and E. If, on the other hand, A and B are on the same side of O (Fig. 6.70b), that is, if ϵ_a and ϵ_b have the same sign, then the maximum shearing strain is defined by points D' and E' on the circle of diameter OA, and we have $\gamma_{\max} = \epsilon_{\max}$.

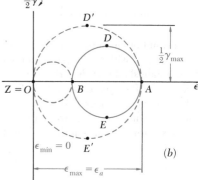

Fig. 6.70. Mohr's circles for plane strain.

We shall now consider the particular case of *plane stress* encountered in a thin plate or on the free surface of a structural element or machine component (Sec. 6.1). Selecting the x and y axes in the plane of stress, we have $\sigma_z = \tau_{zx} = \tau_{zy} = 0$ and verify that the z axis is a principal axis of stress. As we saw earlier, if the deformation is elastic and if the material is homogeneous and isotropic, it follows from Hooke's law that $\gamma_{zx} = \gamma_{zy} = 0$; thus, the z axis is also a principal axis of strain, and Mohr's circle may be used to analyze the transformation of strain in the xy plane. However, as we shall see presently, it does *not* follow from Hooke's law that $\epsilon_z = 0$; indeed, a state of plane stress does not, in general, result in a state of plane strain.†

Denoting by a and b the principal axes within the plane of stress, and by c the principal axis perpendicular to that plane, we let $\sigma_x = \sigma_a$, $\sigma_y = \sigma_b$, and $\sigma_z = 0$ in Eqs. (2.28) for the generalized Hooke's law (Sec. 2.12) and write

$$\epsilon_a = \frac{\sigma_a}{E} - \frac{\nu\sigma_b}{E}$$

(6.55)

$$\epsilon_b = -\frac{\nu\sigma_a}{E} + \frac{\sigma_b}{E}$$

(6.56)

$$\epsilon_c = -\frac{\nu}{E}(\sigma_a + \sigma_b)$$

(6.57)

Adding Eqs. (6.55) and (6.56) member to member, we have

$$\epsilon_a + \epsilon_b = \frac{1-\nu}{E}(\sigma_a + \sigma_b)$$

(6.58)

† See footnote on page 384.

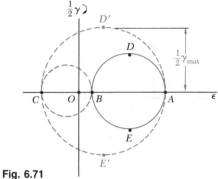

Fig. 6.71

Solving Eq. (6.58) for $\sigma_a + \sigma_b$ and substituting into Eq. (6.57), we write

$$\epsilon_c = -\frac{\nu}{1-\nu}(\epsilon_a + \epsilon_b) \qquad (6.59)$$

The relation obtained defines the third principal strain in terms of the "in-plane" principal strains. We note that, if B is located between A and C on the Mohr-circle diagram (Fig. 6.71), the maximum shearing strain is equal to the diameter CA of the circle corresponding to a rotation about the b axis, out of the plane of stress.

Example 6.05

As a result of measurements made on the surface of a machine component with strain gages oriented in various ways, it has been established that the principal strains on the free surface are $\epsilon_a = +400\,\mu$ and $\epsilon_b = -50\,\mu$. Knowing that Poisson's ratio for the given material is $\nu = 0.30$, determine (a) the maximum in-plane shearing strain, (b) the true value of the maximum shearing strain near the surface of the component.

(a) *Maximum In-Plane Shearing Strain.* We draw Mohr's circle through the points A and B corresponding to the given principal strains (Fig. 6.72). The maximum in-plane shearing strain is defined by points D and E and is equal to the diameter of Mohr's circle:

$$\gamma_{\text{max(in plane)}} = 400\,\mu + 50\,\mu = 450\,\mu$$

(b) *Maximum Shearing Strain.* We first determine the third principal strain ϵ_c. Since we have a state of plane *stress* on the surface of the machine component, we use Eq. (6.59) and write

$$\epsilon_c = -\frac{\nu}{1-\nu}(\epsilon_a + \epsilon_b)$$

$$= -\frac{0.30}{0.70}(400\,\mu - 50\,\mu) = -150\,\mu$$

Drawing Mohr's circles through A and C and through B and C (Fig. 6.73), we find that the maximum shearing strain is equal to the diameter of the circle of diameter CA:

$$\gamma_{\text{max}} = 400\,\mu + 150\,\mu = 550\,\mu$$

We note that, even though ϵ_a and ϵ_b have opposite signs, the maximum in-plane shearing strain does not represent the true maximum shearing strain.

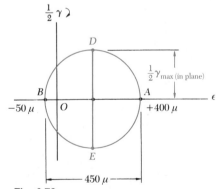

Fig. 6.72

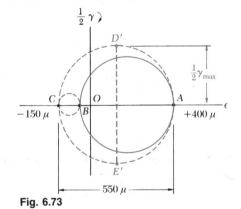

Fig. 6.73

*6.13. MEASUREMENTS OF STRAIN; STRAIN ROSETTE

The normal strain may be determined in any given direction on the surface of a structural element or machine component by scribing two gage marks A and B across a line drawn in the desired direction and measuring the length of the segment AB before and after the load has been applied.   If L is the undeformed length of AB and δ its deformation, the normal strain along AB is $\epsilon_{AB} = \delta/L$.

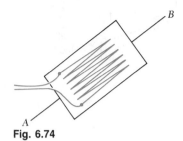

Fig. 6.74

A more convenient and more accurate method for the measurement of normal strains is provided by electrical strain gages.   A typical electrical strain gage consists of a length of thin wire arranged as shown in Fig. 6.74 and cemented to two pieces of paper.   In order to measure the strain ϵ_{AB} of a given material in the direction AB, the gage is cemented to the surface of the material, with the wire folds running parallel to AB.   As the material elongates, the wire increases in length and decreases in diameter, causing the electrical resistance of the gage to increase.   By measuring the current passing through a properly calibrated gage, the strain ϵ_{AB} may be determined accurately and continuously as the load is increased.

The strain components ϵ_x and ϵ_y may be determined at a given point of the free surface of a material by simply measuring the normal strain along x and y axes drawn through that point.   Recalling Eq. (6.43) of Sec. 6.10, we note that a third measurement of normal strain, made along the bisector OB of the angle formed by the x and y axes, enables us to determine the shearing strain γ_{xy} as well (Fig. 6.75):

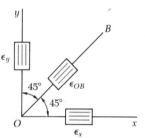

$$\gamma_{xy} = 2\epsilon_{OB} - (\epsilon_x + \epsilon_y) \qquad (6.43)$$

Fig. 6.75

It should be noted that the strain components ϵ_x, ϵ_y, and γ_{xy} at a given point could be obtained from normal strain measurements made along *any three lines* drawn through that point (Fig. 6.76).   Denoting respectively by θ_1, θ_2, and θ_3 the angle each of the three lines forms with the x axis, by ϵ_1, ϵ_2, and ϵ_3 the corresponding strain measurements, and substituting into Eq. (6.41), we write the three equations

$$\epsilon_1 = \epsilon_x \cos^2 \theta_1 + \epsilon_y \sin^2 \theta_1 + \gamma_{xy} \sin \theta_1 \cos \theta_1$$
$$\epsilon_2 = \epsilon_x \cos^2 \theta_2 + \epsilon_y \sin^2 \theta_2 + \gamma_{xy} \sin \theta_2 \cos \theta_2 \qquad (6.60)$$
$$\epsilon_3 = \epsilon_x \cos^2 \theta_3 + \epsilon_y \sin^2 \theta_3 + \gamma_{xy} \sin \theta_3 \cos \theta_3$$

which may be solved simultaneously for ϵ_x, ϵ_y, and γ_{xy}.†

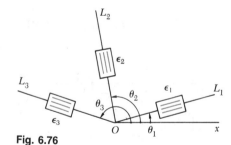

Fig. 6.76

The arrangement of strain gages used to measure the three normal strains ϵ_1, ϵ_2, and ϵ_3 is known as a *strain rosette*.   The rosette used to measure normal strains along the x and y axes and their bisector is referred to as a 45° rosette.   Another rosette frequently used is the 60° rosette (see Sample Prob. 6.7).

---

† It should be noted that the free surface on which the strain measurements are made is in a state of *plane stress*, while Eqs. (6.41) and (6.43) were derived for a state of *plane strain*.   However, as observed earlier, the normal to the free surface is a principal axis of strain and the derivations given in Sec. 6.10 remain valid.

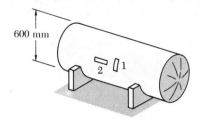

600 mm

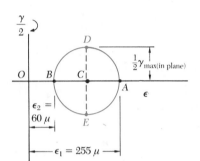

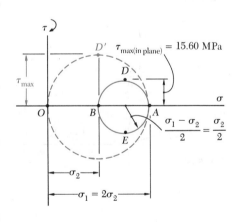

SAMPLE PROBLEM 6.6

A cylindrical storage tank used to transport gas under pressure has an inside diameter of 600 mm and a wall thickness of 20 mm. Strain gages attached to the surface of the tank in transverse and longitudinal directions indicate strains of 255 μ and 60 μ respectively. Knowing that a torsion test has shown that the modulus of rigidity of the material used in the tank is $G = 80$ GPa, determine (a) the gage pressure inside the tank, (b) the principal stresses and the maximum shearing stress in the wall of the tank.

(a) **Gage Pressure Inside Tank.** We note that the given strains are the principal strains at the surface of the tank. Plotting the corresponding points A and B, we draw Mohr's circle for strain. The maximum in-plane shearing strain is equal to the diameter of the circle.

$$\gamma_{\text{max(in plane)}} = \epsilon_1 - \epsilon_2 = 255\,\mu - 60\,\mu = 195\,\mu$$

From Hooke's law for shearing stress and strain, we have

$$\tau_{\text{max(in plane)}} = G\gamma_{\text{max(in plane)}}$$
$$= (80 \times 10^9 \text{ Pa})(195 \times 10^{-6}) = 15.60 \text{ MPa}$$

Substituting this value and the given data in Eq. (6.33), we write

$$\tau_{\text{max(in plane)}} = \frac{pr}{4t} \qquad 15.60 \text{ MPa} = \frac{p(0.3 \text{ m})}{4(0.02 \text{ m})}$$

Solving for the gage pressure p, we have

$$p = 4.16 \text{ MPa} \quad \blacktriangleleft$$

(b) **Principal Stresses and Maximum Shearing Stress.** Recalling that, for a thin-walled cylindrical pressure vessel, $\sigma_1 = 2\sigma_2$, we draw Mohr's circle for stress and obtain

$$\sigma_2 = 2\tau_{\text{max(in plane)}} = 2(15.60 \text{ MPa}) \qquad \sigma_2 = 31.2 \text{ MPa} \quad \blacktriangleleft$$
$$\sigma_1 = 2\sigma_2 = 2(31.2 \text{ MPa}) \qquad \sigma_1 = 62.4 \text{ MPa} \quad \blacktriangleleft$$

The maximum shearing stress is equal to the radius of the circle of diameter OA and corresponds to a rotation of 45° about a longitudinal axis.

$$\tau_{\text{max}} = \tfrac{1}{2}\sigma_1 = \sigma_2 = 31.2 \text{ MPa} \qquad \tau_{\text{max}} = 31.2 \text{ MPa} \quad \blacktriangleleft$$

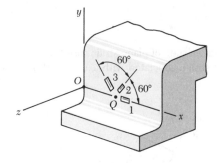

SAMPLE PROBLEM 6.7

Using a 60° rosette, the following strains have been determined at point Q on the surface of a steel machine base:

$$\epsilon_1 = 40\ \mu \qquad \epsilon_2 = 980\ \mu \qquad \epsilon_3 = 330\ \mu$$

Using the coordinate axes shown, determine at point Q, (a) the strain components ϵ_x, ϵ_y, and γ_{xy}, (b) the principal strains, (c) the maximum shearing strain. (Use $\nu = 0.29$.)

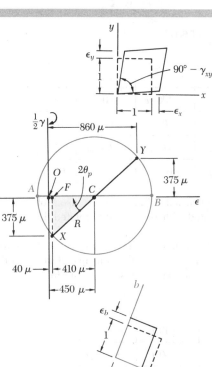

(a) **Strain Components ϵ_x, ϵ_y, γ_{xy}.** For the coordinate axes shown

$$\theta_1 = 0 \qquad \theta_2 = 60° \qquad \theta_3 = 120°$$

Substituting these values into Eqs. (6.60), we have

$$\epsilon_1 = \epsilon_x(1) \qquad\quad + \epsilon_y(0) \qquad\quad + \gamma_{xy}(0)(1)$$
$$\epsilon_2 = \epsilon_x(0.500)^2 \quad + \epsilon_y(0.866)^2 + \gamma_{xy}(0.866)(0.500)$$
$$\epsilon_3 = \epsilon_x(-0.500)^2 + \epsilon_y(0.866)^2 + \gamma_{xy}(0.866)(-0.500)$$

Solving these equations for ϵ_x, ϵ_y, and γ_{xy}, we obtain

$$\epsilon_x = \epsilon_1 \qquad \epsilon_y = \tfrac{1}{3}(2\epsilon_2 + 2\epsilon_3 - \epsilon_1) \qquad \gamma_{xy} = \frac{\epsilon_2 - \epsilon_3}{0.866}$$

Substituting the given values for ϵ_1, ϵ_2, and ϵ_3, we have

$$\epsilon_x = 40\ \mu \qquad \epsilon_y = \tfrac{1}{3}[2(980) + 2(330) - 40] = +860\ \mu \quad \blacktriangleleft$$
$$\gamma_{xy} = (980 - 330)/0.866 = 750\ \mu \quad \blacktriangleleft$$

These strains are indicated on the element shown.

(b) **Principal Strains.** We note that the side of the element associated with ϵ_x rotates counterclockwise, thus we plot point X below the horizontal axis, i.e., $X(40, -375)$. We then plot $Y(860, +375)$ and draw Mohr's circle.

$$\epsilon_{\text{ave}} = \tfrac{1}{2}(860\ \mu + 40\ \mu) = 450\ \mu$$
$$R = \sqrt{(375\ \mu)^2 + (410\ \mu)^2} = 556\ \mu$$

$$\tan 2\theta_p = \frac{375\ \mu}{410\ \mu} \qquad 2\theta_p = 42.4° \downarrow \qquad \theta_p = 21.2° \downarrow$$

Points A and B correspond to the principal strains. We have

$$\epsilon_a = \epsilon_{\text{ave}} - R = 450\ \mu - 556\ \mu \qquad\qquad\qquad \epsilon_a = -106\ \mu \quad \blacktriangleleft$$
$$\epsilon_b = \epsilon_{\text{ave}} + R = 450\ \mu + 556\ \mu \qquad\qquad\qquad \epsilon_b = +1006\ \mu \quad \blacktriangleleft$$

Since $\sigma_z = 0$ on the surface, we may use Eq. (6.59) to find the principal strain ϵ_c:

$$\epsilon_c = -\frac{\nu}{1 - \nu}(\epsilon_a + \epsilon_b) = -\frac{0.29}{1 - 0.29}(-106\ \mu + 1006\ \mu) \qquad \epsilon_c = -368\ \mu \quad \blacktriangleleft$$

(c) **Maximum Shearing Strain.** Plotting point C and drawing Mohr's circle through points B and C, we obtain point D' and write

$$\tfrac{1}{2}\gamma_{\text{max}} = \tfrac{1}{2}(1006\ \mu + 368\ \mu) \qquad\qquad \gamma_{\text{max}} = 1374\ \mu \quad \blacktriangleleft$$

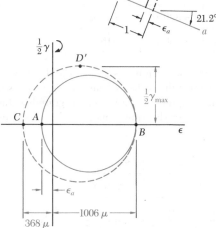

PROBLEMS

6.116 though 6.119   For the given state of plane strain, use the methods of Sec. 6.10 to determine the state of strain associated with axes x' and y' rotated through the given angle θ.

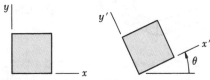

Fig. P6.116 through P6.123

	ϵ_x	ϵ_y	γ_{xy}	θ
6.116 and 6.120	$+500\ \mu$	$-300\ \mu$	0	$25°\ \nwarrow$
6.117 and 6.121	$+350\ \mu$	0	$+120\ \mu$	$15°\ \swarrow$
6.118 and 6.122	$-240\ \mu$	$+320\ \mu$	$-330\ \mu$	$65°\ \nwarrow$
6.119 and 6.123	$+240\ \mu$	$+160\ \mu$	$+150\ \mu$	$60°\ \swarrow$

6.120 through 6.123   For the given state of plane strain, use Mohr's circle to determine the state of strain associated with axes x' and y' rotated through the given angle θ.

6.124 through 6.127   For the given state of plane strain, use Mohr's circle to determine (a) the orientation and magnitude of the principal strains, (b) the maximum in-plane shearing strain, (c) the maximum shearing strain.

	ϵ_x	ϵ_y	γ_{xy}
6.124	$+320\ \mu$	$+160\ \mu$	$+300\ \mu$
6.125	$+300\ \mu$	$+60\ \mu$	$+100\ \mu$
6.126	$+80\ \mu$	$+320\ \mu$	$-70\ \mu$
6.127	$-180\ \mu$	$-260\ \mu$	$+315\ \mu$

6.128 and 6.129   The following state of strain has been measured at a point on the surface of a thin plate.   Knowing that the surface of the plate is unstressed and that Poisson's ratio is 0.30, determine (a) the orientation and magnitude of the principal strains, (b) the maximum in-plane shearing strain, (c) the maximum shearing strain.

	ϵ_x	ϵ_y	γ_{xy}
6.128	$+48\ \mu$	$+92\ \mu$	$+240\ \mu$
6.129	$-280\ \mu$	0	$-165\ \mu$

6.130 A square $ABCD$ of 50-mm side is scribed on the surface of a thin plate while the plate is unloaded. After the plate is loaded, the lengths of sides AB and AD are observed to have increased, respectively, by 7.5 μm and 12.5 μm, while angle DAB is observed to have decreased by 0.240×10^{-3} rad. Knowing that Poisson's ratio is $\frac{1}{3}$, determine (a) the orientation and magnitude of the principal strains, (b) the maximum in-plane shearing strain, (c) the maximum shearing strain.

6.131 A square $ABCD$ of 15-mm side is scribed on the surface of a thin plate while the plate is unloaded. After the plate is loaded, the length of side AB is observed to have decreased by 12 μm while the length of AD has remained unchanged and angle DAB has increased by 0.180×10^{-3} rad. Knowing that Poisson's ratio is $\frac{1}{3}$, determine (a) the orientation and magnitude of the principal strains, (b) the maximum in-plane shearing strain, (c) the maximum shearing strain.

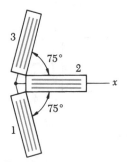

Fig. P6.130 and P6.131

6.132 The strains determined by the use of a rosette attached as shown to the surface of a structural component are

$$\epsilon_1 = +110\ \mu \qquad \epsilon_2 = +212.5\ \mu \qquad \epsilon_3 = +240\ \mu$$

Determine (a) the orientation and magnitude of the principal strains in the plane of the rosette, (b) the maximum in-plane shearing strain.

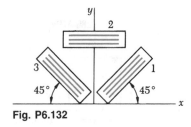

Fig. P6.132

6.133 The strains determined by the use of a rosette attached as shown to the surface of a machine element are

$$\epsilon_1 = -93.1\ \mu \qquad \epsilon_2 = +385\ \mu \qquad \epsilon_3 = +210\ \mu$$

Determine (a) the orientation and magnitude of the principal strains in the plane of the rosette, (b) the maximum in-plane shearing strain.

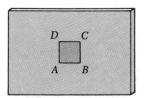

Fig. P6.133

6.134 The following readings have been obtained from a strain rosette attached to the surface of a machine component:

$$\epsilon_1 = -57.5\ \mu \qquad \epsilon_2 = -57.5\ \mu$$
$$\epsilon_4 = +220\ \mu$$

(a) What should be the reading of gage 3? (b) Determine the principal strains and the maximum shearing strain in the plane of the rosette.

6.135 The following readings have been obtained from a strain rosette attached to the surface of a machine component:

$$\epsilon_2 = -90\ \mu \qquad \epsilon_3 = -190\ \mu$$
$$\epsilon_4 = -240\ \mu$$

(a) What should be the reading of gage 1? (b) Determine the principal strains and the maximum shearing strain in the plane of the rosette.

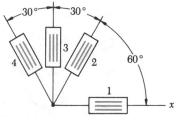

Fig. P6.134 and P6.135

6.136 through 6.139 For the problem indicated, and assuming $\nu = 0.30$, determine (a) the three principal strains at the point where the measurements are taken, (b) the maximum shearing strain at that point. (*Note:* It is recalled that the free surface of a machine or structural component is in a state of plane stress.)

      6.136 Prob. 6.132.
      6.137 Prob. 6.133.
      6.138 Prob. 6.134.
      6.139 Prob. 6.135.

6.140 and 6.141 The given state of strain has been determined on the surface of a cast-iron machine base. Knowing that $E = 70$ GPa and $G = 28$ GPa, determine the principal planes and principal stresses (a) by determining the corresponding state of plane stress [use Eq. (2.36), page 81; Eq. (2.43), page 84; and the first two equations of Prob. 2.76, page 87] and then using Mohr's circle for stress, (b) by using Mohr's circle for strain to determine the orientation and magnitude of the principal strains and then determining the corresponding stresses.

      6.140 $\epsilon_x = +210\,\mu$ $\epsilon_y = +50\,\mu$
                     $\gamma_{xy} = -300\,\mu$
      6.141 $\epsilon_x = -360\,\mu$ $\epsilon_y = -200\,\mu$
                     $\gamma_{xy} = +330\,\mu$

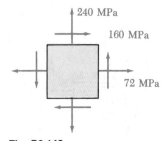

Fig. P6.142

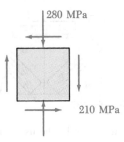

Fig. P6.143

Fig. P6.144

6.142 and 6.143 The given state of plane stress is known to exist on the surface of a high-strength-steel machine component. Knowing that $E = 200$ GPa and $G = 77$ GPa, determine the orientation and magnitude of the three principal strains (a) by determining the corresponding state of strain [use Eq. (2.43), page 84, and Eqs. (2.38), page 81] and then using Mohr's circle for strain, (b) by using Mohr's circle for stress to determine the principal planes and principal stresses and then determining the corresponding strains.

6.144 A single strain gage is cemented to a solid 100-mm-diameter steel shaft at an angle of 25° with a line parallel to the axis of the shaft. Knowing that $G = 80$ GPa, determine the torque **T** indicated by a gage reading of 300 μ.

6.145 A single strain gage forming an angle of 18° with a horizontal plane is used to determine the gage pressure in the cylindrical steel tank shown. The cylindrical wall of the tank is 6 mm thick, has a 600-mm inside diameter, and is made of a steel with $E = 200$ GPa and $\nu = 0.30$. Determine the pressure in the tank indicated by a strain-gage reading of 280 μ.

6.146 A cantilever steel bar AB with a 40 × 150-mm rectangular cross section supports a load **P** at end B. A single strain gage forming an angle of 45° with the horizontal is attached to the surface of the bar at a point C located 400 mm from end B and 50 mm below the bar's upper edge. Knowing that $E = 200$ GPa and $\nu = 0.30$, determine the magnitude of the load **P** indicated by a gage reading of 240 μ.

Fig. P6.145

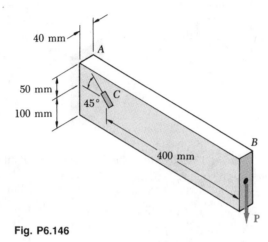

Fig. P6.146

6.147 For the bar of Prob. 6.146, determine the reading which would be obtained for the same loading if the gage were located at a point 25 mm directly above C and had the same orientation.

6.148 We denote by ϵ_1, ϵ_2, and ϵ_3 the three strain measurements made with a 60° rosette. (a) Show that the sum of these measurements and the sum of the squares of their differences are both independent of the orientation of the rosette and are equal, respectively, to

$$\epsilon_1 + \epsilon_2 + \epsilon_3 = 3\epsilon_{\text{ave}} \tag{1}$$

and

$$(\epsilon_1 - \epsilon_2)^2 + (\epsilon_2 - \epsilon_3)^2 + (\epsilon_3 - \epsilon_1)^2 = 4.5R^2 \tag{2}$$

where ϵ_{ave} is the abscissa of the center of the corresponding Mohr's circle for strain and R is the radius of that circle. (b) Knowing that the following measurements were made with the rosette at a given point of a machine component,

$$\epsilon_1 = +250\ \mu \qquad \epsilon_2 = +683\ \mu \qquad \epsilon_3 = -183\ \mu$$

determine the principal in-plane strains and the maximum in-plane shearing strain at that point.

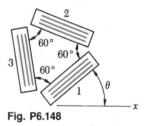

Fig. P6.148

REVIEW AND SUMMARY

The first part of this chapter was devoted to a study of the *transformation of stress* under a rotation of axes and to its application to the solution of engineering problems, and the second part to a similar study of the *transformation of strain*.

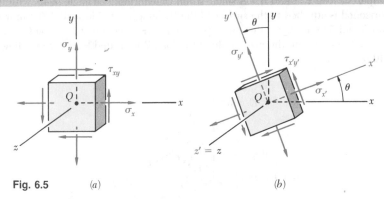

Fig. 6.5 (a) (b)

Transformation of plane stress

Considering first a state of *plane stress* at a given point Q [Sec. 6.2] and denoting by σ_x, σ_y, and τ_{xy} the stress components associated with the element shown in Fig. 6.5a, we derived the following formulas defining the components $\sigma_{x'}$, $\sigma_{y'}$, and $\tau_{x'y'}$ associated with that element after it had been rotated through an angle θ about the z axis (Fig. 6.5b):

$$\sigma_{x'} = \frac{\sigma_x + \sigma_y}{2} + \frac{\sigma_x - \sigma_y}{2}\cos 2\theta + \tau_{xy}\sin 2\theta \qquad (6.5)$$

$$\sigma_{y'} = \frac{\sigma_x + \sigma_y}{2} - \frac{\sigma_x - \sigma_y}{2}\cos 2\theta - \tau_{xy}\sin 2\theta \qquad (6.7)$$

$$\tau_{x'y'} = -\frac{\sigma_x - \sigma_y}{2}\sin 2\theta + \tau_{xy}\cos 2\theta \qquad (6.6)$$

Principal planes. Principal stresses

In Sec. 6.3, we determined the values θ_p of the angle of rotation which correspond to the maximum and minimum values of the normal stress at point Q. We wrote

$$\tan 2\theta_p = \frac{2\tau_{xy}}{\sigma_x - \sigma_y} \qquad (6.12)$$

The two values obtained for θ_p are 90° apart (Fig. 6.9) and define the *principal planes of stress* at point Q. The corresponding values of the normal stress are called the *principal stresses* at Q; we obtained

$$\sigma_{max,\ min} = \frac{\sigma_x + \sigma_y}{2} \pm \sqrt{\left(\frac{\sigma_x - \sigma_y}{2}\right)^2 + \tau_{xy}^2} \qquad (6.14)$$

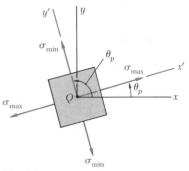

Fig. 6.9

We also noted that the corresponding value of the shearing stress is zero. Next, we determined the values θ_s of the angle θ for which the largest value of the shearing stress occurs. We wrote

$$\tan 2\theta_s = -\frac{\sigma_x - \sigma_y}{2\tau_{xy}} \qquad (6.15)$$

The two values obtained for θ_s are 90° apart (Fig. 6.10). We also noted that the planes of maximum shearing stress are at 45° to the principal planes. The maximum value of the shearing stress for a rotation *in the plane of stress* is

$$\tau_{max} = \sqrt{\left(\frac{\sigma_x - \sigma_y}{2}\right)^2 + \tau_{xy}^2} \qquad (6.16)$$

and the corresponding value of the normal stresses is

$$\sigma' = \sigma_{ave} = \frac{\sigma_x + \sigma_y}{2} \qquad (6.17)$$

Maximum in-plane shearing stress

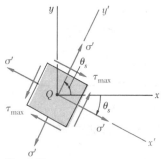

Fig. 6.10

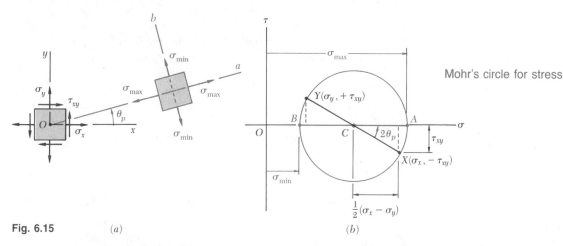

Fig. 6.15   *(a)*   *(b)*

Mohr's circle for stress

We saw in Sec. 6.4 that *Mohr's circle* provides an alternative method, based on simple geometric considerations, for the analysis of the transformation of plane stress. Given the state of stress shown in black in Fig. 6.15a, we plot point X of coordinates σ_x, $-\tau_{xy}$ and point Y of coordinates σ_y, $+\tau_{xy}$ (Fig. 6.15b). Drawing the circle of diameter XY, we obtain Mohr's circle. The abscissas of the points of intersection A and B of the circle with the horizontal axis represent the principal stresses, and the angle of rotation bringing the diameter XY into AB is twice the angle θ_p defining the principal planes in Fig. 6.15a, with both angles having the same sense. We also noted that diameter DE defines the maximum shearing stress and the orientation of the corresponding plane (Fig. 6.17b) [Example 6.02, Sample Probs. 6.2 and 6.3].

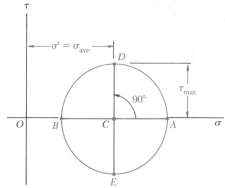

Fig. 6.17b

General state of stress

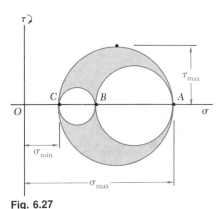

Fig. 6.27

Considering a *general state of stress* characterized by six stress components [Sec. 6.5], we showed that the normal stress on a plane of arbitrary orientation may be expressed as a quadratic form of the direction cosines of the normal to that plane. This proves the existence of three *principal axes of stress* and three *principal stresses* at any given point. Rotating a small cubic element about each of the three principal axes [Sec. 6.6], we drew the corresponding Mohr's circles which yield the values of σ_{max}, σ_{min}, and τ_{max} (Fig. 6.27). In the particular case of *plane stress*, and if the x and y axes are selected in the plane of stress, point C coincides with the origin O. If A and B are located on opposite sides of O, the maximum shearing stress is equal to the maximum "in-plane" shearing stress as determined in Sec. 6.3 or 6.4. If A and B are located on the same side of O, this will not be the case. If $\sigma_a > \sigma_b > 0$, for instance, the maximum shearing stress is equal to $\frac{1}{2}\sigma_a$ and corresponds to a rotation out of the plane of stress (see page 365).

Yield criteria
for ductile materials

Yield criteria for ductile materials under plane stress were developed in Sec. 6.7. To predict whether a structural or machine component will fail at some critical point due to yield in the material, we first determine the principal stresses σ_a and σ_b at that point for the given loading condition. We then plot the point of coordinates σ_a and σ_b. If this point falls within a certain area, the component is safe; if it falls outside, the component will fail. The area used with the *maximum-shearing-strength criterion* is shown in Fig. 6.38 and the area used with the *maximum-distortion-energy criterion* in Fig. 6.39. We note that both areas depend upon the value of the yield strength σ_Y of the material.

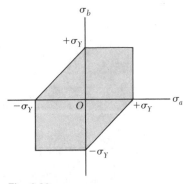

Fig. 6.38

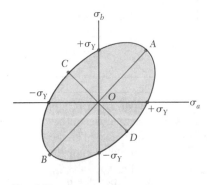

Fig. 6.39

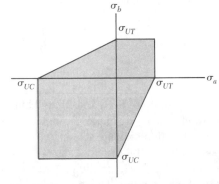

Fig. 6.45*b*

Fracture criteria
for brittle materials

Fracture criteria for brittle materials under plane stress were developed in Sec. 6.8 in a similar fashion. The most commonly used is *Mohr's criterion*, which utilizes the results of various types of test available for a given material. The shaded area shown in Fig. 6.45*b* is used when the ultimate strengths σ_{UT} and σ_{UC} have been determined, respectively, from a tension and a compression test. Again, the principal stresses σ_a and σ_b are determined at a given point of the structural or machine component being investigated. If the corresponding point falls within the shaded area, the component is safe; if it falls outside, the component will rupture.

In Sec. 6.9, we discussed the stresses in *thin-walled pressure vessels* and derived formulas relating the stresses in the walls of the vessels and the *gage pressure p* in the fluid they contain. In the case of a *cylindrical vessel* of inside radius r and thickness t (Fig. 6.47), we obtained the following expressions for the *hoop stress σ_1* and the *longitudinal stress σ_2*:

Cylindrical pressure vessels

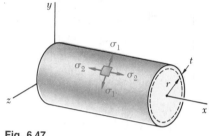

Fig. 6.47

$$\sigma_1 = \frac{pr}{t} \qquad \sigma_2 = \frac{pr}{2t} \qquad (6.30,\ 6.31)$$

We also found that the *maximum shearing* stress occurs out of the plane of stress and is

$$\tau_{max} = \sigma_2 = \frac{pr}{2t} \qquad (6.34)$$

In the case of a *spherical vessel* of inside radius r and thickness t (Fig. 6.51), we found that the two principal stresses are equal:

Spherical pressure vessels

Fig. 6.51

$$\sigma_1 = \sigma_2 = \frac{pr}{2t} \qquad (6.36)$$

Again, the *maximum shearing* stress occurs out of the plane of stress; it is

$$\tau_{max} = \tfrac{1}{2}\sigma_1 = \frac{pr}{4t} \qquad (6.37)$$

The last part of the chapter was devoted to the *transformation of strain*. In Secs. 6.10 and 6.11, we discussed the transformation of *plane strain* and introduced *Mohr's circle for plane strain*. The discussion was similar to the corresponding discussion of the transformation of stress, except that, where the shearing stress τ was used, we now used $\tfrac{1}{2}\gamma$, that is, *half the shearing strain*. The formulas obtained for the transformation of strain under a rotation of axes through an angle θ were

Transformation of plane strain

$$\epsilon_{x'} = \frac{\epsilon_x + \epsilon_y}{2} + \frac{\epsilon_x - \epsilon_y}{2}\cos 2\theta + \frac{\gamma_{xy}}{2}\sin 2\theta \qquad (6.44)$$

$$\epsilon_{y'} = \frac{\epsilon_x + \epsilon_y}{2} - \frac{\epsilon_x - \epsilon_y}{2}\cos 2\theta - \frac{\gamma_{xy}}{2}\sin 2\theta \qquad (6.45)$$

$$\gamma_{x'y'} = -(\epsilon_x - \epsilon_y)\sin 2\theta + \gamma_{xy}\cos 2\theta \qquad (6.49)$$

Mohr's circle for strain

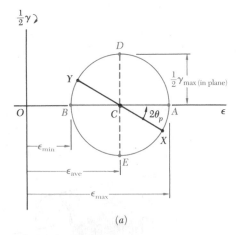

(a)

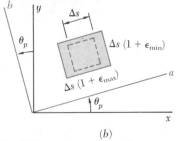

(b)

Fig. 6.61

Strain gages

Strain rosette

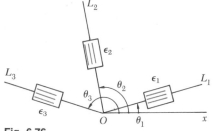

Fig. 6.76

Using Mohr's circle for strain (Fig. 6.61), we also obtained the following relations defining the angle of rotation θ_p corresponding to the *principal axes of strain* and the values of the *principal strains* ϵ_{max} and ϵ_{min}:

$$\tan 2\theta_p = \frac{\gamma_{xy}}{\epsilon_x - \epsilon_y} \tag{6.52}$$

$$\epsilon_{max} = \epsilon_{ave} + R \quad \text{and} \quad \epsilon_{min} = \epsilon_{ave} - R \tag{6.51}$$

where

$$\epsilon_{ave} = \frac{\epsilon_x + \epsilon_y}{2} \quad \text{and} \quad R = \sqrt{\left(\frac{\epsilon_x - \epsilon_y}{2}\right)^2 + \left(\frac{\gamma_{xy}}{2}\right)^2} \tag{6.50}$$

The *maximum shearing strain* for a rotation in the plane of strain was found to be

$$\gamma_{max(in\ plane)} = 2R = \sqrt{(\epsilon_x - \epsilon_y)^2 + \gamma_{xy}^2} \tag{6.53}$$

Section 6.12 was devoted to the three-dimensional analysis of strain, with application to the determination of the maximum shearing strain in the particular cases of plane strain and plane stress. In the case of *plane stress*, we also found that the principal strain ϵ_c in a direction perpendicular to the plane of stress could be expressed as follows in terms of the "in-plane" principal strains ϵ_a and ϵ_b:

$$\epsilon_c = -\frac{\nu}{1 - \nu}(\epsilon_a + \epsilon_b) \tag{6.59}$$

Finally, we discussed in Sec. 6.13 the use of *strain gages* to measure the normal strain on the surface of a structural element or machine component. Considering a *strain rosette* consisting of three gages aligned along lines forming, respectively, angles θ_1, θ_2, and θ_3 with the x axis (Fig. 6.76), we wrote the following relations among the measurements ϵ_1, ϵ_2, ϵ_3 of the gages and the components ϵ_x, ϵ_y, γ_{xy} characterizing the state of strain at that point:

$$\epsilon_1 = \epsilon_x \cos^2\theta_1 + \epsilon_y \sin^2\theta_1 + \gamma_{xy} \sin\theta_1 \cos\theta_1$$

$$\epsilon_2 = \epsilon_x \cos^2\theta_2 + \epsilon_y \sin^2\theta_2 + \gamma_{xy} \sin\theta_2 \cos\theta_2 \tag{6.60}$$

$$\epsilon_3 = \epsilon_x \cos^2\theta_3 + \epsilon_y \sin^2\theta_3 + \gamma_{xy} \sin\theta_3 \cos\theta_3$$

These equations may be solved for ϵ_x, ϵ_y, and γ_{xy}, once ϵ_1, ϵ_2, and ϵ_3 have been determined.

REVIEW PROBLEMS

6.149 The Mohr circle shown corresponds to the state of stress given in Figs. 6.5a and b, page 400. Noting that $\sigma_{x'} = OC + (CX') \cos (2\theta_p - 2\theta)$ and that $\tau_{x'y'} = (CX') \sin (2\theta_p - 2\theta)$, derive the expressions for $\sigma_{x'}$ and $\tau_{x'y'}$ given in Eqs. (6.5) and (6.6), respectively. [*Hint:* Use $\sin (A + B) = \sin A \cos B + \cos A \sin B$ and $\cos (A + B) = \cos A \cos B - \sin A \sin B$.]

6.150 Knowing that the diameter of the steel rod *ABD* is 30 mm, determine (*a*) the principal planes, (*b*) the principal stresses, (*c*) the maximum shearing stress, at point *H*.

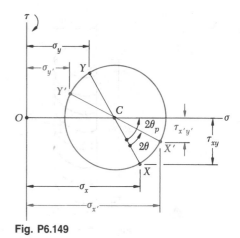

Fig. P6.149

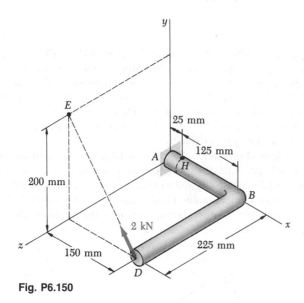

Fig. P6.150

6.151 For the rod of Prob. 6.150, and knowing that for the steel used $E = 200$ GPa and $\nu = 0.30$, determine the three principal strains and the maximum shearing strain at point *H*.

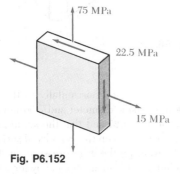

Fig. P6.152

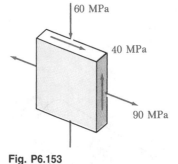

Fig. P6.153

6.152 and 6.153 For the state of plane stress shown, determine (*a*) the principal planes, (*b*) the principal stresses, (*c*) the maximum shearing stress.

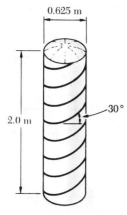

Fig. P6.155

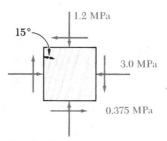

Fig. P6.154

6.154 The grain of a wooden member forms an angle of 15° with the vertical. For the state of plane stress shown, determine (*a*) the in-plane shearing stress parallel to the grain, (*b*) the normal stress perpendicular to the grain.

6.155 The cylindrical pressure vessel shown is fabricated of 6-mm-thick plate welded along a helix forming an angle of 30° with the horizontal. Knowing that the allowable normal stress perpendicular to the weld is 80 MPa, determine the largest gage pressure which may be used in the vessel.

6.156 A spherical pressure vessel is to be fabricated from 10-mm-thick steel plate with a 360-MPa ultimate strength. Knowing that a maximum allowable gage pressure of 4.50 MPa is desired and that a factor of safety of 4.00 is required, determine the maximum allowable inside diameter of the vessel.

6.157 For the state of stress shown, determine the maximum shearing stress when (*a*) $\sigma_y = +72$ MPa, (*b*) $\sigma_y = -72$ MPa.

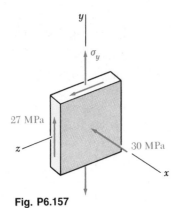

Fig. P6.157

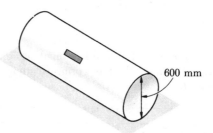

Fig. P6.158

6.158 A strain gage is attached horizontally to the cylindrical surface of a pressure vessel of 600-mm outside diameter and 7.50-mm wall thickness. Knowing that $E = 200$ GPa and $\nu = 0.25$ and that the strain gage reads 120 μ, determine (*a*) the three principal strains on the cylindrical surface of the vessel, (*b*) the principal stresses in the wall, (*c*) the gage pressure inside the vessel.

6.159 A cast-aluminum rod is made of an alloy for which $\sigma_{UT} = 80$ MPa and $\sigma_{UC} = 200$ MPa. Knowing that the magnitude T of the applied torques is slowly increased and using Mohr's criterion, determine the shearing stress τ_0 which can be expected at rupture.

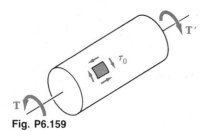

Fig. P6.159

6.160 The state of plane stress shown occurs in a steel member made of a grade of steel with a tensile yield strength of 270 MPa. Determine the factor of safety with respect to yield, using (*a*) the maximum-shearing-stress criterion, (*b*) the maximum-distortion-energy criterion.

The following problems are designed to be solved with a computer.

6.C1 A state of plane stress is defined by the stress components σ_x, σ_y, and τ_{xy} associated with the element shown in Fig. *a*. (*a*) Write a computer program which can be used to calculate the stress components $\sigma_{x'}$, $\sigma_{y'}$, and $\tau_{x'y'}$ associated with the element after it has been rotated through an angle θ about the z axis (Fig. *b*). (*b*) Use this program to solve Probs. 6.14, 6.16, and 6.18.

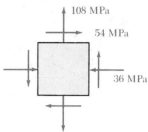

Fig. P6.160

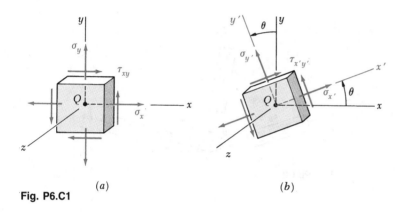

(*a*)

(*b*)

Fig. P6.C1

6.C2 A state of plane stress is defined by the stress components σ_x, σ_y, and τ_{xy} associated with the element shown in Fig. *a* of Prob. 6.C1. (*a*) Write a computer program which can be used to determine the principal axes, the principal stresses, the maximum in-plane shearing stress, and the maximum shearing stress. (*b*) Use this program to solve Probs. 6.6, 6.8, 6.64, and 6.66.

6.C3 (*a*) Write a computer program which, for a given state of plane stress and a given yield strength of a ductile material, can be used to determine whether the material will yield. The program should use both the maximum-shearing-strength criterion and the maximum-distortion-energy criterion. It should also print the values of the principal stresses and, if the material does not yield, calculate the factor of safety. (*b*) Use this program to solve Probs. 6.76, 6.78, 6.80, and 6.82.

6.C4 (*a*) Write a computer program based on Mohr's fracture criterion for brittle materials which, for a given state of plane stress and given values of the ultimate strength of a material in tension and in compression, can be used to determine whether rupture will occur. The program should also print the values of the principal stresses. (*b*) Use this program to solve Probs. 6.88 through 6.91.

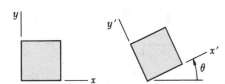

Fig. P6.C5

6.C5 A state of plane strain is defined by the strain components ϵ_x, ϵ_y, and γ_{xy} associated with the x and y axes. (*a*) Write a computer program which can be used to calculate the strain components $\epsilon_{x'}$, $\epsilon_{y'}$, and $\gamma_{x'y'}$ associated with the frame of reference $x'y'$ obtained by rotating the x and y axes through an angle θ. (*b*) Use this program to solve Probs. 6.116 through 6.119.

6.C6 A state of plane strain is defined by the strain components ϵ_x, ϵ_y, and γ_{xy} associated with the x and y axes. (*a*) Write a computer program which can be used to determine the orientation and magnitude of the principal strains, the maximum in-plane shearing strain, and the maximum shearing strain. (*b*) Use this program to solve Probs. 6.124 through 6.127.

6.C7 A state of strain is defined by the strain components ϵ_x, ϵ_y, and γ_{xy} measured at a point on the surface of a thin plate. (*a*) Write a computer program which can be used to determine the orientation and magnitude of the principal strains, the maximum in-plane shearing strain, and the maximum shearing strain. (*b*) Use this program to solve Probs. 6.128 through 6.131.

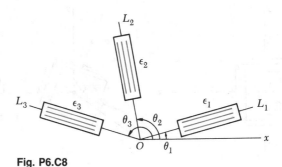

Fig. P6.C8

6.C8 A rosette consisting of three gages forming, respectively, angles θ_1, θ_2, and θ_3 with the x axis is attached to the free surface of a machine component made of a material with a given Poisson's ratio ν. (*a*) Write a computer program which, for given readings ϵ_1, ϵ_2, and ϵ_3 of the gages, can be used to calculate the strain components associated with the x and y axes and to determine the orientation and magnitude of the three principal strains, the maximum in-plane shearing strain, and the maximum shearing strain. (*b*) Use this program to solve Probs. 6.132, 6.134, 6.136, and 6.138.

C H A P T E R   S E V E N

DESIGN OF BEAMS AND SHAFTS FOR STRENGTH

7.1. INTRODUCTION

Our main concern in this chapter will be the design of beams for strength. We shall find out how the material and the cross section of a beam of a given span should be selected if the beam is not to fail under a given loading. Another aspect of the design of beams is the prediction of the deflection of a given beam under a given loading; this other aspect will be considered in Chaps. 8 and 9.†

We shall first consider *prismatic beams*, i.e., straight beams with a uniform cross section, and note that their design depends primarily upon the determination of the largest values of the bending moment and shear created in the beam by a given loading (Sec. 7.2).

The determination of these values and of the critical sections of the beam in which they occur is greatly facilitated if we draw a *shear diagram* and a *bending-moment diagram* representing, respectively, the variation of the shear V and of the bending moment M along the beam. The values of V and M at various points may be obtained either by drawing free-body diagrams of successive portions of the beam (Sec. 7.3) or from relations among load, shear, and bending moment that we shall derive in Sec. 7.4.

Another method for the determination of the maximum values of the shear and bending moment is based on expressing V and M in terms of *singularity functions* (Sec. 7.5). This approach will be expanded in Chap. 8 to facilitate the determination of the slope and deflection of beams.

---

† Still another aspect of the design of beams is the consideration of buckling, which may occur in beams which are not adequately braced to prevent lateral movement. In this chapter we shall assume that all beams are laterally braced and that lateral buckling is not possible.

409

In Sec. 7.6, we shall discuss the distribution of *principal stresses* across a beam section and question whether in some instances the maximum normal stress might be larger than the value *Mc/I* occurring on the surface of the beam.

Taking into account the various points brought up in the preceding sections, we shall learn in Sec. 7.7 a step-by-step procedure leading to the most economical design of a prismatic beam. Particular attention will be given to wooden beams with a rectangular cross section and rolled-steel S-beams and W-beams.

In Sec. 7.8, we shall discuss the design of *nonprismatic beams*, i.e., the design of beams with a variable cross section. By selecting the shape and size of the variable cross section so that its elastic section modulus $S = I/c$ varies along the length of the beam in the same way as the bending moment M, we shall be able to design beams for which the maximum normal stress at each section is equal to the allowable stress. Such beams are said to be of *constant strength*.

In Sec. 7.9, we shall consider the design of transmission shafts subjected to transverse loads as well as to torques. We shall take into account the effect of both the normal stresses due to bending and the shearing stresses due to torsion.

Finally, in Sec. 7.10, we shall analyze the state of stress in a beam at a point located under a concentrated or a distributed load.

7.2. BASIC CONSIDERATIONS FOR THE DESIGN OF PRISMATIC BEAMS

We saw in Chap. 5 that the transverse loading of a prismatic beam may cause both normal and shearing stresses on any given transverse section of the beam. We recall from Sec. 4.5 that, within the elastic range, the normal stress σ_x on a given section varies linearly with the distance y from the neutral axis and is largest at the point of the section farthest from that axis ($y = \pm c$). Since σ_x also depends upon the bending moment M in the section, its largest value will occur in the section where $|M|$ is maximum. Thus, the maximum absolute value σ_m of the normal (longitudinal) stress σ_x in a prismatic beam may be obtained by substituting $|M|_{max}$ for M in Eq. (4.15):

$$\sigma_m = \frac{|M|_{max}c}{I} \tag{7.1}$$

The corresponding stress may be tensile, as shown in Fig. 7.1, or compressive. As we shall see in Sec. 7.4, the section of maximum bending moment in the uniformly loaded, simply supported beam of Fig. 7.1 is located at the center of the beam.

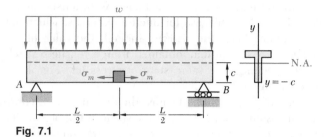

Fig. 7.1

We similarly recall from Chap. 5 that, for most of the common types of beams (e.g., rectangular beams, S-beams, W-beams), the shearing stress τ_{xy} on a given transverse section is maximum at the neutral axis. Since τ_{xy} also depends upon the shear V in the section, its largest value will occur on the neutral axis of the section where $|V|$ is maximum. Thus, the maximum value τ_m of the shearing stress τ_{xy} in most prismatic beams may be obtained by substituting $|V|_{\max}$ for V in Eq. (5.10). We have

$$\tau_m = \frac{|V|_{\max}Q}{It} \tag{7.2}$$

where t is the width of the cross section at the neutral axis, and Q the first moment about that axis of the portion of the cross-sectional area located either above or below the neutral axis. As we shall see in Sec. 7.4, the sections of maximum shear in the uniformly loaded, simply supported beam of Fig. 7.2 are located close to the beam supports.

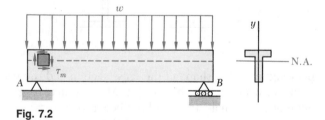

Fig. 7.2

The design of a beam is usually controlled by the maximum absolute value $|M|_{\max}$ of the bending moment in the beam. The material and the shape and dimensions of the cross section of the beam must be selected in such a way that the value of σ_m obtained from Eq. (7.1) will not exceed the allowable normal stress σ_{all}. We should note that this criterion is valid,

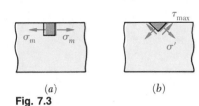

Fig. 7.3

Fig. 7.4

not only for brittle materials, which actually fail in tension along a transverse plane (Fig. 7.3a), but also for ductile materials, which fail along a plane of maximum shear (Fig. 7.3b). Indeed, the value of σ_{all} for the given material is based on data obtained from a tensile test which produces in the specimen the same state of stress (cf. Fig. 1.36) as that represented in Fig. 7.3.

In the case of short, stubby beams, the design may be controlled by the maximum absolute value $|V|_{max}$ of the shear in the beam. The material and the shape and dimensions of the cross section of the beam must be selected in such a way that the value obtained for τ_m from Eq. (7.2) will not exceed the allowable shearing stress τ_{all}. Again, this criterion is valid for both brittle and ductile materials, since the state of stress at a point located on the neutral axis of the cross section (Fig. 7.4) is the same as that found in a specimen subjected to a torsion test (cf. Fig. 3.19).

It is clear from the foregoing discussion that the design of a beam should begin with the determination of the critical sections where the absolute values of the bending moment and shear are respectively maximum, and with the determination of the corresponding values $|M|_{max}$ and $|V|_{max}$. These computations will be considerably simplified if we plot the values of the shear V and of the bending moment M in the various sections against the distance x measured from one end of the beam. The graphs obtained in this way are called, respectively, the *shear diagram* and the *bending-moment diagram* and will be discussed in Secs. 7.3 and 7.4. Another method for the determination of the maximum values of the shear and bending moment will be presented in Sec. 7.5; it is based on the use of *singularity functions*.

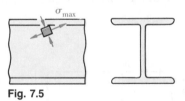

Fig. 7.5

Depending upon the shape of the cross section and the value of the shear V in the critical section where $|M| = |M|_{max}$, it may happen that the largest value of the normal stress will not occur for $y = \pm c$, but at some other point within the section. As we shall see in Secs. 7.6 and 7.7, a combination of large values of σ_x and τ_{xy} near the junction of the web with the flanges of an S-beam or W-beam may result in a value of σ_{max} on one of the principal planes (Fig. 7.5), which is larger than the value of σ_m obtained form Eq. (7.1).

7.3. SHEAR AND BENDING-MOMENT DIAGRAMS

As indicated in the preceding section, the determination of the maximum absolute values of the shear and of the bending moment in a beam are greatly facilitated if V and M are plotted against the distance x measured from one end of the beam. Besides, as we shall see in Chap. 8, the knowledge of M as a function of x is essential to the determination of the deflection of a beam.

In the examples and sample problems of this section, we shall obtain the shear and bending-moment diagrams by determining the values of V and M at selected points of the beam. These values will be found in the usual way, i.e., by passing a section through the point where they are to be determined and considering the equilibrium of the portion of beam located on either side of the section (Fig. 7.6).

We shall recall at this point the sign convention introduced earlier for V and M (Secs. 5.2 and 4.2): The shear V and the bending moment M at a given point of a beam are said to be positive when the internal forces and couples acting on each portion of the beam are directed as shown in Fig. 7.7a. This convention may be more easily remembered if we observe that

1. The shear at any given point of a beam is positive when the *external* forces (loads and reactions) acting on the beam tend to shear off the beam at that point as indicated in Fig. 7.7b.
2. The bending moment at any given point of a beam is positive when the *external* forces acting on the beam tend to bend the beam at that point as indicated in Fig. 7.7c.

It may also help to note, as we shall verify in the next example, that the situation described in Fig. 7.7, and corresponding to positive values of the shear and the bending moment, is precisely the situation which occurs in the left half of a simply supported beam carrying a single concentrated load at its midpoint.

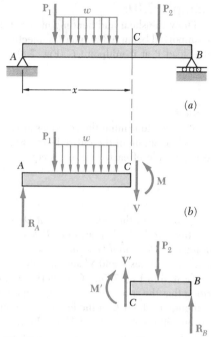

Fig. 7.6

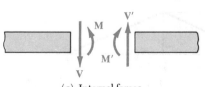

(a) Internal forces
(positive shear and positive bending moment)
Fig. 7.7

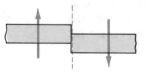

(b) Effect of external forces
(positive shear)

(c) Effect of external forces
(positive bending moment)

Example 7.01

Draw the shear and bending-moment diagrams for a simply supported beam AB of span L subjected to a single concentrated load $\mathbf{P}$ at it midpoint C (Fig. 7.8).

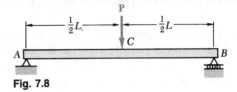

Fig. 7.8

We first determine the reactions at the supports from the free-body diagram of the entire beam (Fig. 7.9a); we find that the magnitude of each reaction is equal to $P/2$.

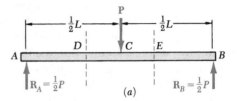

(a)

Next we cut the beam at a point D between A and C and draw the free-body diagrams of AD and DB (Fig. 7.9b). Assuming that shear and bending moment are positive, we direct the internal forces $\mathbf{V}$ and $\mathbf{V}'$ and the internal couples $\mathbf{M}$ and $\mathbf{M}'$ as indicated in Fig. 7.7a. Considering the free body AD and writing that the sum of the vertical components and the sum of the moments about D of the forces acting on the free body are zero, we find $V = +P/2$ and $M = +Px/2$. Both the shear and the bending moment are therefore positive; this may be checked by observing that the reaction at A tends to shear off and to bend the beam at D as indicated in Figs. 7.7b and c. We may plot V and M between A and C (Figs. 7.9d and e); the shear has a constant value $V = P/2$, while the bending moment increases linearly from $M = 0$ at $x = 0$ to $M = PL/4$ at $x = L/2$.

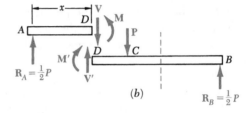

(b)

Cutting, now, the beam at a point E between C and B and considering the free body EB (Fig. 7.9c), we write that the sum of the vertical components and the sum of the moments about E of the forces acting on the free body are zero. We obtain $V = -P/2$ and $M = P(L - x)/2$. The shear is therefore negative and the bending moment positive; this may be checked by observing that the reaction at B bends the beam at E as indicated in Fig. 7.7c but tends to shear it off in a manner opposite to that shown in Fig. 7.7b. We can complete, now, the shear and bending-moment diagrams of Figs. 7.9d and e; the shear has a constant value $V = -P/2$ between C and B, while the bending moment decreases linearly from $M = PL/4$ at $x = L/2$ to $M = 0$ at $x = L$.

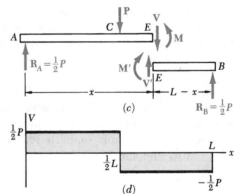

(c)

(d)

(e)

Fig. 7.9

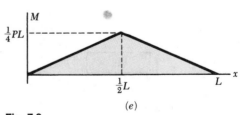

We note from the foregoing example that, when a beam is subjected only to concentrated loads, the shear is constant between loads and the bending moment varies linearly between loads. In such situations, therefore, the shear and bending-moment diagrams may easily be drawn, once the values of V and M have been obtained at sections selected just to the left and just to the right of the points where the loads and reactions are applied (see Sample Prob. 7.1).

Example 7.02

Draw the shear and bending-moment diagrams for a cantilever beam AB of span L supporting a uniformly distributed load w (Fig. 7.10).

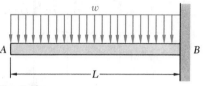

Fig. 7.10

We cut the beam at a point C between A and B and draw the free-body diagram of AC (Fig. 7.11a), directing $\mathbf{V}$ and $\mathbf{M}$ as indicated in Fig. 7.7a. Denoting by x the distance from A to C and replacing the distributed load over AC by its resultant wx applied at the midpoint of AC, we write

$$+\uparrow \Sigma F_y = 0: \qquad -wx - V = 0 \qquad V = -wx$$

$$+\curvearrowleft \Sigma M_C = 0: \qquad wx\left(\frac{x}{2}\right) + M = 0 \qquad M = -\frac{1}{2}wx^2$$

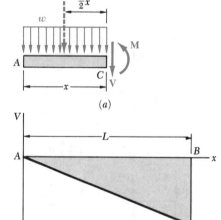

We note that the shear diagram is represented by an oblique straight line (Fig. 7.11b) and the bending-moment diagram by a parabola (Fig. 7.11c). The maximum values of V and M both occur at B, where we have

$$V_B = -wL \qquad M_B = -\tfrac{1}{2}wL^2$$

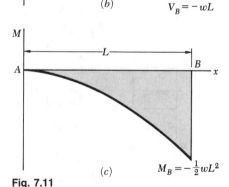

Fig. 7.11

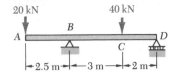

20 kN   40 kN
      B
A ▭▭▭▭▭▭▭▭▭▭ D
    △   C △
|←2.5 m→|←3 m→|←2 m→|

SAMPLE PROBLEM 7.1

Draw the shear and bending-moment diagrams for the beam and loading shown.

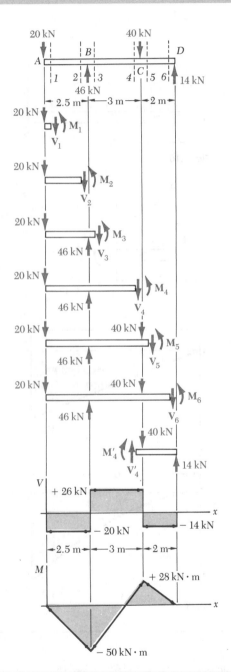

Solution. The reactions are determined by considering the entire beam as a free body; they are

$$\mathbf{R}_B = 46 \text{ kN} \uparrow \qquad \mathbf{R}_D = 14 \text{ kN} \uparrow$$

We first determine the internal forces just to the right of the 20-kN load at A. Considering the stub of beam to the left of section 1 as a free body and assuming V and M to be positive (according to the standard convention), we write

$+\uparrow \Sigma F_y = 0:$ $-20 \text{ kN} - V_1 = 0$ $V_1 = -20 \text{ kN}$

$+\backslash\Sigma M_1 = 0:$ $(20 \text{ kN})(0 \text{ m}) + M_1 = 0$ $M_1 = 0$

We next consider as a free body the portion of beam to the left of section 2 and write

$+\uparrow \Sigma F_y = 0:$ $-20 \text{ kN} - V_2 = 0$ $V_2 = -20 \text{ kN}$

$+\backslash\Sigma M_2 = 0:$ $(20 \text{ kN})(2.5 \text{ m}) + M_2 = 0$ $M_2 = -50 \text{ kN} \cdot \text{m}$

The shear and bending moment at sections 3, 4, 5, and 6 are determined in a similar way from the free-body diagrams shown. We obtain

$$V_3 = +26 \text{ kN} \qquad M_3 = -50 \text{ kN} \cdot \text{m}$$
$$V_4 = +26 \text{ kN} \qquad M_4 = +28 \text{ kN} \cdot \text{m}$$
$$V_5 = -14 \text{ kN} \qquad M_5 = +28 \text{ kN} \cdot \text{m}$$
$$V_6 = -14 \text{ kN} \qquad M_6 = 0$$

For several of the latter sections, the results may be more easily obtained by considering as a free body the portion of the beam to the right of the section. For example, considering the portion of the beam to the right of section 4, we write

$+\uparrow \Sigma F_y = 0:$ $V_4 - 40 \text{ kN} + 14 \text{ kN} = 0$ $V_4 = +26 \text{ kN}$

$+\backslash\Sigma M_4 = 0:$ $-M_4 + (14 \text{ kN})(2 \text{ m}) = 0$ $M_4 = +28 \text{ kN} \cdot \text{m}$

We may now plot the six points shown on the shear and bending-moment diagrams. As indicated in Sec. 7.3, the shear is of constant value between concentrated loads, and the bending moment varies linearly; we obtain therefore the shear and bending-moment diagrams shown.

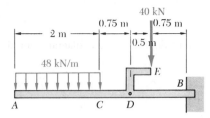

40 kN

0.75 m | 0.75 m

2 m

0.5 m

48 kN/m

E

B

A   C   D

SAMPLE PROBLEM 7.2

Draw the shear and bending-moment diagrams for the cantilever beam AB. The distributed load of 48 kN/m extends over 2 m of the beam and the 40-kN load is applied at E.

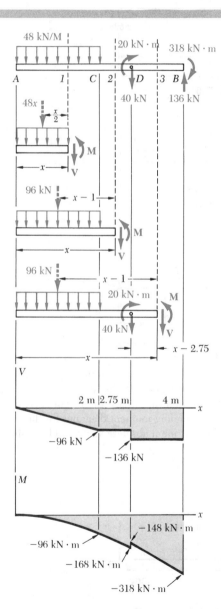

48 kN/M

20 kN · m

318 kN · m

A   1   C   2   D   3   B

48x

$\frac{x}{2}$

M

x

V

96 kN

x − 1

M

x

V

96 kN

x − 1

20 kN · m

M

40 kN

V

x

x − 2.75

V

2 m 2.75 m

4 m

x

−96 kN

−136 kN

M

x

−148 kN · m

−96 kN · m

−168 kN · m

−318 kN · m

40 kN

136 kN

Solution. The 40-kN load is replaced by an equivalent force-couple system acting on the beam at point D. The reaction at B is determined by considering the entire beam as a free body.

From A to C. We determine the internal forces at a distance x from point A by considering the portion of beam to the left of section *1*. That part of the distributed load acting on the free body is replaced by its resultant, and we write

$$+\uparrow \Sigma F_y = 0: \qquad -48x - V = 0 \qquad V = -48x \text{ kN}$$
$$+\curvearrowleft \Sigma M_1 = 0: \qquad 48x(\tfrac{1}{2}x) + M = 0 \qquad M = -24x^2 \text{ kN} \cdot \text{m}$$

Since the free-body diagram shown may be used for all values of x smaller than 2 m, the expressions obtained for V and M are valid in the region $0 < x < 2$ m.

From C to D. Considering the portion of beam to the left of section *2* and again replacing the distributed load by its resultant, we obtain

$$+\uparrow \Sigma F_y = 0: \qquad -96 - V = 0 \qquad V = -96 \text{ kN}$$
$$+\curvearrowleft \Sigma M_2 = 0: \qquad 96(x-1) + M = 0 \qquad M = 96 - 96x \qquad \text{kN} \cdot \text{m}$$

These expressions are valid in the region 2 m $< x <$ 2.75 m.

From D to B. Using the portion of beam to the left of section *3*, we obtain for the region 2.75 m $< x <$ 4 m

$$V = -136 \text{ kN} \qquad M = 226 - 136x \qquad \text{kN} \cdot \text{m}$$

The shear and bending-moment diagrams for the entire beam may now be plotted. We note that the couple of moment 20 kN · m applied at point D introduces a discontinuity into the bending-moment diagram.

PROBLEMS

7.1 through 7.6 Draw the shear and bending-moment diagrams for the beam and loading shown.

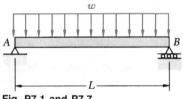

Fig. P7.1 and P7.7

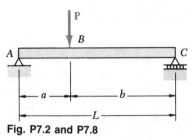

Fig. P7.2 and P7.8

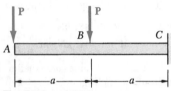

Fig. P7.3 and P7.9

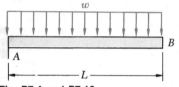

Fig. P7.4 and P7.10

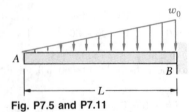

Fig. P7.5 and P7.11

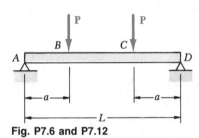

Fig. P7.6 and P7.12

7.7 through 7.12 Determine the equations of the shear and bending-moment curves for the beam and loading shown. (Place the origin at *A*.)

7.13 and 7.14 Draw the shear and bending-moment diagrams for the beam and loading shown, and determine the maximum absolute value (*a*) of the shear, (*b*) of the bending moment.

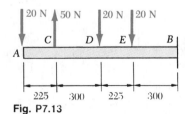

Fig. P7.13

Dimensions in mm

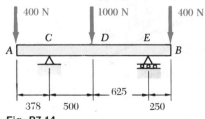

Fig. P7.14

7.15 through 7.18 Draw the shear and bending-moment diagrams for the beam and loading shown, and determine the maximum absolute value (*a*) of the shear, (*b*) of the bending moment.

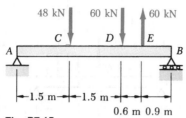

Fig. P7.15

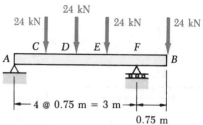

Fig. P7.16

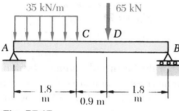

Fig. P7.17

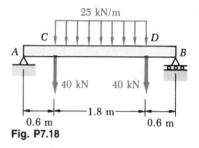

Fig. P7.18

7.19 and 7.20 Draw the shear and bending-moment diagrams for the beam *AB*.

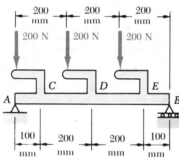

Fig. P7.19

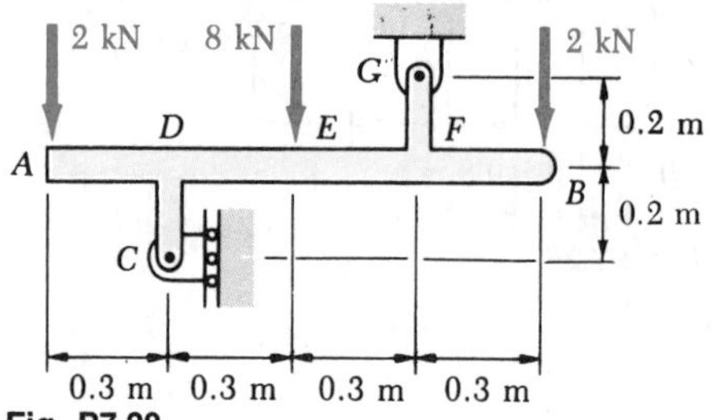

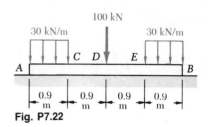

Fig. P7.20

7.21 and 7.22 Assuming that the upward reaction of the ground is uniformly distributed, draw the shear and bending-moment diagrams for the beam and loading shown and determine the maximum absolute value (*a*) of the shear, (*b*) of the bending moment.

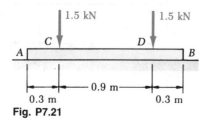

Fig. P7.21

Fig. P7.22

7.23 and 7.24 For the beam and loading shown, determine (*a*) the maximum normal stress on a transverse section at *C*, (*b*) the maximum shearing stress on a transverse section just to the left of *B*.

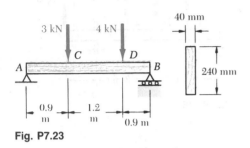

Fig. P7.23

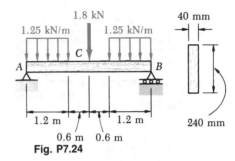

Fig. P7.24

7.25 and 7.26 For the beam and loading shown, determine (*a*) the maximum normal stress on a transverse section at *D*, (*b*) the maximum shearing stress on a transverse section just to the right of *A*. (Assume that $\tau_m = V/A_{web}$.)

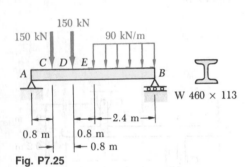

Fig. P7.25

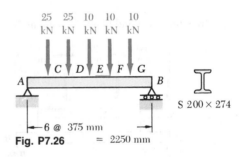

Fig. P7.26

7.27 and 7.28 Draw the shear and bending-moment diagrams for the beam and loading shown and determine the maximum normal stress due to bending.

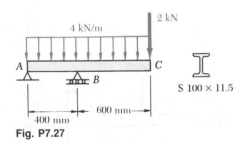

Fig. P7.27

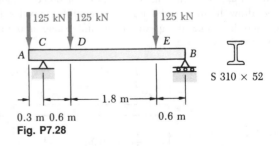

Fig. P7.28

7.29 and 7.30 Draw the shear and bending-moment diagrams for the beam and loading shown and determine the maximum normal stress due to bending.

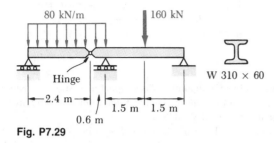

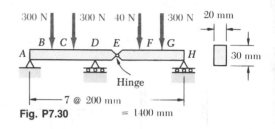

Fig. P7.29

Fig. P7.30

7.31 Determine (*a*) the magnitude *P* of the two upward forces for which the maximum absolute value of the bending moment in the beam is as small as possible, (*b*) the corresponding maximum normal stress due to bending. (*Hint:* Draw the bending-moment diagram and then equate the absolute values of the largest positive and negative bending moments obtained.)

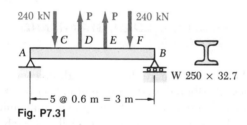

Fig. P7.31

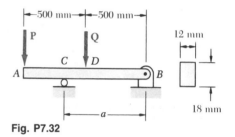

Fig. P7.32

7.32 Knowing that *P* = *Q* = 500 N, determine (*a*) the distance *a* for which the maximum absolute value of the bending moment in the beam is as small as possible, (*b*) the corresponding maximum normal stress due to bending. (See hint of Prob. 7.31.)

7.33 Solve Prob. 7.32, assuming that *P* = 400 N and *Q* = 600 N.

7.34 Determine (*a*) the distance *a* for which the maximum absolute value of the bending moment in the beam is as small as possible, (*b*) the corresponding maximum normal stress due to bending. (See hint of Prob. 7.31.)

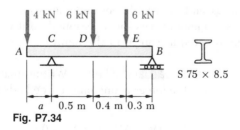

Fig. P7.34

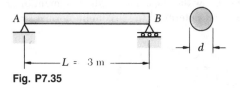

Fig. P7.35

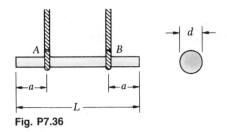

Fig. P7.36

7.35 A solid steel rod of diameter d is supported as shown. Knowing that for steel $\gamma = 7850$ kg/m^3, determine the smallest diameter d which may be used if the normal stress is not to exceed 35 MPa.

*7.36 A uniform steel rod of diameter d is to be picked up by crane cables attached at A and B. (a) Determine the distance a from the ends of the rod to the points where the cables should be attached if the maximum absolute value of the bending moment in the beam is to be as small as possible. (b) Determine the corresponding maximum normal stress due to bending, knowing that $d = 20$ mm and $L = 4$ m and that for steel $\rho = 7850$ kg/m^3. (See hint of Prob. 7.31.)

7.4. RELATIONS AMONG LOAD, SHEAR, AND BENDING MOMENT

When a beam carries more than two or three concentrated loads, or when it carries distributed loads, the method outlined in Sec. 7.3 for plotting shear and bending moment may prove quite cumbersome. The construction of the shear diagram and, especially, of the bending-moment diagram will be greatly facilitated if certain relations existing among load, shear, and bending moment are taken into consideration.

Let us consider a simply supported beam AB carrying a distributed load w per unit length (Fig. 7.12a), and let C and C' be two points of the beam at a distance Δx from each other. The shear and bending moment at C will be denoted by V and M, respectively, and will be assumed positive; the shear and bending moment at C' will be denoted by $V + \Delta V$ and $M + \Delta M$.

We now shall detach the portion of beam CC' and draw its free-body diagram (Fig. 7.12b). The forces exerted on the free body include a load of magnitude $w\,\Delta x$ and internal forces and couples at C and C'. Since shear and bending moment have been assumed positive, the forces and couples will be directed as shown in the figure.

Relations between Load and Shear. Writing that the sum of the vertical components of the forces acting on the free body CC' is zero, we have

$$+\uparrow \Sigma F_y = 0: \qquad V - (V + \Delta V) - w\,\Delta x = 0$$
$$\Delta V = -w\,\Delta x$$

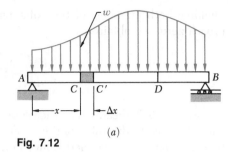

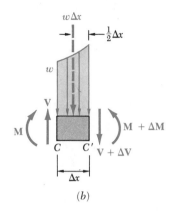

Fig. 7.12

(a)

(b)

Dividing both members of the equation by Δx and then letting Δx approach zero, we obtain

$$\frac{dV}{dx} = -w \qquad (7.3)$$

Equation (7.3) indicates that, for a beam loaded as shown in Fig. 7.12a, the slope dV/dx of the shear curve is negative; the numerical value of the slope at any point is equal to the load per unit length at that point.

Integrating (7.3) between points C and D, we write

$$V_D - V_C = -\int_{x_C}^{x_D} w \, dx \qquad (7.4)$$

$$V_D - V_C = -(\text{area under load curve between } C \text{ and } D) \qquad (7.4')$$

Note that this result could also have been obtained by considering the equilibrium of the portion of beam CD, since the area under the load curve represents the total load applied between C and D.

It should be observed that Eq. (7.3) is not valid at a point where a concentrated load is applied; the shear curve is discontinuous at such a point, as seen in Sec. 7.3. Similarly, Eqs. (7.4) and (7.4') cease to be valid when concentrated loads are applied between C and D, since they do not take into account the sudden change in shear caused by a concentrated load. Equations (7.4) and (7.4'), therefore, should be applied only between successive concentrated loads.

Relations between Shear and Bending Moment. Returning to the free-body diagram of Fig. 7.12b, and writing now that the sum of the moments about C' is zero, we have

$$+\!\uparrow\!\Sigma M_{C'} = 0: \quad (M + \Delta M) - M - V \, \Delta x + w \, \Delta x \frac{\Delta x}{2} = 0$$

$$\Delta M = V \, \Delta x - \frac{1}{2} w \, (\Delta x)^2$$

Dividing both members of the equation by Δx and then letting Δx approach zero, we obtain

$$\frac{dM}{dx} = V \tag{7.5}$$

Equation (7.5) indicates that the slope dM/dx of the bending-moment curve is equal to the value of the shear. This is true at any point where the shear has a well-defined value, i.e., at any point where no concentrated load is applied. Equation (7.5) also shows that $V = 0$ at points where M is maximum. This property facilitates the determination of the points where the beam is likely to fail under bending.

Integrating (7.5) between points C and D, we write

$$M_D - M_C = \int_{x_C}^{x_D} V \, dx \tag{7.6}$$

$$M_D - M_C = \text{area under shear curve between } C \text{ and } D \tag{7.6'}$$

Note that the area under the shear curve should be considered positive where the shear is positive and negative where the shear is negative. Equations (7.6) and (7.6′) are valid even when concentrated loads are applied between C and D, as long as the shear curve has been correctly drawn. The equations cease to be valid, however, if a couple is applied at a point between C and D, since they do not take into account the sudden change in bending moment caused by a couple (see Sample Prob. 7.6).

Example 7.03

Draw the shear and bending-moment diagrams for the simply supported beam shown in Fig. 7.13 and determine the maximum value of the bending moment.

From the free-body diagram of the entire beam, we determine the magnitude of the reactions at the supports (Fig. 7.13b):

$$R_A = R_B = \tfrac{1}{2}wL$$

Next, we draw the shear diagram. Close to the end A of the beam, the shear is equal to R_A, that is, to $\tfrac{1}{2}wL$, as we may check by considering as a free body a very small portion of the beam. Using Eq. (7.4), we may then determine the shear V at any distance x from A; we write

$$V - V_A = -\int_0^x w \, dx = -wx$$

$$V = V_A - wx = \tfrac{1}{2}wL - wx = w(\tfrac{1}{2}L - x)$$

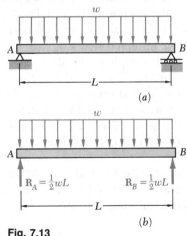

Fig. 7.13

The shear curve is thus an oblique straight line which crosses the x axis at $x = L/2$ (Fig. 7.14a). Considering, now, the bending moment, we first observe that $M_A = 0$. The value M of the bending moment at any distance x from A may then be obtained from Eq. (7.6); we have

$$M - M_A = \int_0^x V\, dx$$

$$M = \int_0^x w(\tfrac{1}{2}L - x)\, dx = \tfrac{1}{2}w(Lx - x^2)$$

The bending-moment curve is a parabola. The maximum value of the bending moment occurs when $x = L/2$, since V (and thus dM/dx) is zero for that value of x. Substituting $x = L/2$ in the last equation, we obtain $M_{max} = wL^2/8$ (Fig. 7.14b).

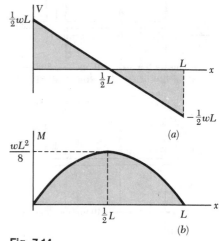

Fig. 7.14

In most engineering applications, one needs to know the value of the bending moment only at a few specific points. Once the shear diagram has been drawn, and after M has been determined at one of the ends of the beam, the value of the bending moment may then be obtained at any given point by computing the area under the shear curve and using Eq. (7.6'). For instance, since $M_A = 0$ for the beam of Example 7.03, the maximum value of the bending moment for that beam may be obtained simply by measuring the area of the shaded triangle in the shear diagram of Fig. 7.14a. We have

$$M_{max} = \frac{1}{2}\frac{L}{2}\frac{wL}{2} = \frac{wL^2}{8}$$

We note that, in this example, the load curve is a horizontal straight line, the shear curve an oblique straight line, and the bending-moment curve a parabola. If the load curve had been an oblique straight line (first degree), the shear curve would have been a parabola (second degree) and the bending-moment curve a cubic (third degree). The shear and bending-moment curves will always be, respectively, one and two degrees higher than the load curve. With this in mind, we should be able to sketch the shear and bending-moment diagrams without actually determining the functions $V(x)$ and $M(x)$, once a few values of the shear and bending moment have been computed. The sketches obtained will be more accurate if we make use of the fact that, at any point where the curves are continuous, the slope of the shear curve is equal to $-w$ and the slope of the bending-moment curve is equal to V.

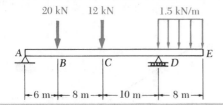

20 kN   12 kN   1.5 kN/m

A━━━━━━━━━━━━━━━━━━━━E
  |B |C   △D

├─6 m─┼─8 m──┼──10 m──┼─8 m─┤

Draw the shear and bending-moment diagrams for the beam and loading shown.

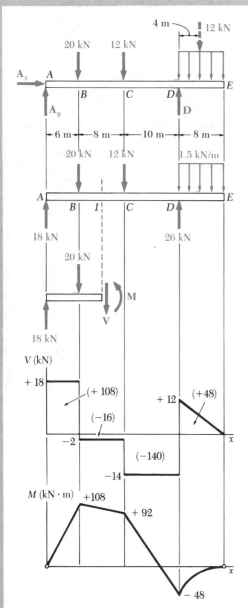

Solution.   Considering the entire beam as a free body, we determine the reactions:

$$+\!\uparrow \; \Sigma M_A = 0\text{:}$$
$$D(24\text{ m}) - (20\text{ kN})(6\text{ m}) - (12\text{ kN})(14\text{ m}) - (12\text{ kN})(28\text{ m}) = 0$$
$$D = +26\text{ kN} \qquad\qquad\qquad \mathbf{D} = 26\text{ kN} \uparrow$$

$$+\!\uparrow \Sigma F_y = 0\text{:} \qquad A_y - 20\text{ kN} - 12\text{ kN} + 26\text{ kN} - 12\text{ kN} = 0$$
$$A_y = +18\text{ kN} \qquad\qquad \mathbf{A}_y = 18\text{ kN} \uparrow$$

$$\xrightarrow{+}\Sigma F_x = 0\text{:} \qquad A_x = 0 \qquad\qquad\qquad \mathbf{A}_x = 0$$

We also note that at both A and E the bending moment is zero; thus two points (indicated by dots) are obtained on the bending-moment diagram.

Shear Diagram.   Since $dV/dx = -w$, we find that between concentrated loads and reactions the slope of the shear diagram is zero (i.e., the shear is constant).   The shear at any point is determined by dividing the beam into two parts and considering either part as a free body.   For example, using the portion of beam to the left of section *1*, we obtain the shear between B and C:

$$+\!\uparrow \Sigma F_y = 0\text{:} \qquad +18\text{ kN} - 20\text{ kN} - V = 0 \qquad\qquad V = -2\text{ kN}$$

We also find that the shear is $+12$ kN just to the right of D and zero at end E. Since the slope $dV/dx = -w$ is constant between D and E, the shear diagram between these two points is a straight line.

Bending-Moment Diagra.n.   We recall that the area under the shear curve between two points is equal to the change in bending moment between the same two points.   For convenience, the area of each portion of the shear diagram is computed and is indicated in parentheses on the diagram.   Since the bending moment M_A at the left end is known to be zero, we write

$$
\begin{array}{ll}
M_B - M_A = +108 & M_B = +108\text{ kN} \cdot \text{m} \\
M_C - M_B = -\;16 & M_C = +92\text{ kN} \cdot \text{m} \\
M_D - M_C = -140 & M_D = -48\text{ kN} \cdot \text{m} \\
M_E - M_D = +\;48 & M_E = 0
\end{array}
$$

Since M_E is known to be zero, a check of the computations is obtained.
      Between the concentrated loads and reactions the shear is constant; thus, the slope dM/dx is constant and the bending-moment diagram is drawn by connecting the known points with straight lines.   Between D and E where the shear diagram is an oblique straight line, the bending-moment diagram is a parabola.
      From the V and M diagrams we note that $V_{\max} = 18$ kN and $M_{\max} = 108$ kN $\cdot$ m.

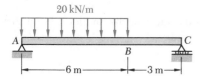

20 kN/m

A

B

C

├─────6 m─────┤├─3 m─┤

Draw the shear and bending-moment diagrams for the beam and loading shown and determine the location and magnitude of the maximum bending moment.

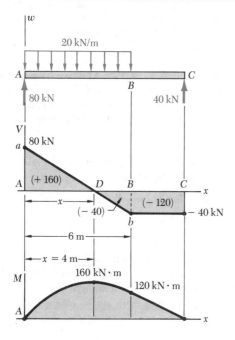

Solution. Considering the entire beam as a free body, we obtain the reactions

$$R_A = 80 \text{ kN} \uparrow \qquad R_C = 40 \text{ kN} \uparrow$$

Shear Diagram. The shear just to the right of A is $V_A = +80$ kN. Since the change in shear between two points is equal to *minus* the area under the load curve between the same two points, we obtain V_B by writing

$$V_B - V_A = -(20 \text{ kN/m})(6 \text{ m}) = -120 \text{ kN}$$
$$V_B = -120 + V_A = -120 + 80 = -40 \text{ kN}$$

The slope $dV/dx = -w$ being constant between A and B, the shear diagram between these two points is represented by a straight line. Between B and C, the area under the load curve is zero; therefore,

$$V_C - V_B = 0 \qquad V_C = V_B = -40 \text{ kN}$$

and the shear is constant between B and C.

Bending-Moment Diagram. We note that the bending moment at each end of the beam is zero. In order to determine the maximum bending moment, we locate the section D of the beam where $V = 0$. We write

$$V_D - V_A = -wx$$
$$0 - 80 \text{ kN} = -(20 \text{ kN/m})x$$

and, solving for x:
$$x = 4 \text{ m} \blacktriangleleft$$

The maximum bending moment occurs at point D, where we have $dM/dx = V = 0$. The areas of the various portions of the shear diagram are computed and are given (in parentheses) on the diagram. Since the area of the shear diagram between two points is equal to the change in bending moment between the same two points, we write

$$M_D - M_A = +160 \text{ kN} \cdot \text{m} \qquad M_D = +160 \text{ kN} \cdot \text{m}$$
$$M_B - M_D = - 40 \text{ kN} \cdot \text{m} \qquad M_B = +120 \text{ kN} \cdot \text{m}$$
$$M_C - M_B = -120 \text{ kN} \cdot \text{m} \qquad M_C = 0$$

The bending-moment diagram consists of an arc of parabola followed by a segment of straight line; the slope of the parabola at A is equal to the value of V at that point.

The maximum bending moment is

$$M_{\max} = M_D = +160 \text{ kN} \cdot \text{m} \blacktriangleleft$$

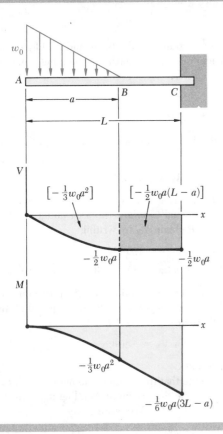

SAMPLE PROBLEM 7.5

Sketch the shear and bending-moment diagrams for the cantilever beam shown.

Solution. *Shear Diagram.* At the free end of the beam, we find $V_A = 0$. Between A and B, the area under the load curve is $\frac{1}{2}w_0 a$; we find V_B by writing

$$V_B - V_A = -\tfrac{1}{2}w_0 a \qquad V_B = -\tfrac{1}{2}w_0 a$$

Between B and C, the beam is not loaded; thus $V_C = V_B$. At A, we have $w = w_0$ and, according to Eq. (7.3), the slope of the shear curve is $dV/dx = -w_0$, while at B the slope is $dV/dx = 0$. Between A and B, the loading decreases linearly, and the shear diagram is parabolic. Between B and C, $w = 0$, and the shear diagram is a horizontal line.

Bending-Moment Diagram. The bending moment M_A at the free end of the beam is zero. We compute the area under the shear curve and write

$$M_B - M_A = -\tfrac{1}{3}w_0 a^2 \qquad M_B = -\tfrac{1}{3}w_0 a^2$$
$$M_C - M_B = -\tfrac{1}{2}w_0 a(L - a)$$
$$M_C = -\tfrac{1}{6}w_0 a(3L - a)$$

The sketch of the bending-moment diagram is completed by recalling that $dM/dx = V$. We find that between A and B the diagram is represented by a cubic curve with zero slope at A, and between B and C by a straight line.

SAMPLE 7.6

The simple beam AC is loaded by a couple of moment T applied at point B. Draw the shear and bending-moment diagrams of the beam.

Solution. The entire beam is taken as a free body, and we obtain

$$\mathbf{R}_A = \frac{T}{L} \uparrow \qquad \mathbf{R}_C = \frac{T}{L} \downarrow$$

The shear at any section is constant and equal to T/L. Since a couple is applied at B, the bending-moment diagram is discontinuous at B; the bending moment decreases suddenly by an amount equal to T.

PROBLEMS

7.37 Using the methods of Sec. 7.4, solve Prob. 7.1.
7.38 Using the methods of Sec. 7.4, solve Prob. 7.2.
7.39 Using the methods of Sec. 7.4, solve Prob. 7.3.
7.40 Using the methods of Sec. 7.4, solve Prob. 7.4.
7.41 Using the methods of Sec. 7.4, solve Prob. 7.5.
7.42 Using the methods of Sec. 7.4, solve Prob. 7.6.
7.43 Using the methods of Sec. 7.4, solve Prob. 7.15.
7.44 Using the methods of Sec. 7.4, solve Prob. 7.14.
7.45 Using the methods of Sec. 7.4, solve Prob. 7.17.
7.46 Using the methods of Sec. 7.4, solve Prob. 7.18.
7.47 through 7.50 Draw the shear and bending-moment diagrams for the beam and loading shown and determine the maximum absolute value (*a*) of the shear, (*b*) of the bending moment.

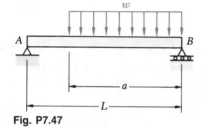

Fig. P7.47

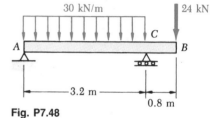

Fig. P7.48

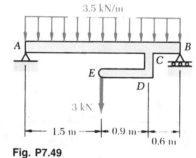

Fig. P7.49

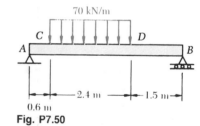

Fig. P7.50

7.51 and 7.52 Determine the equations of the shear and bending-moment curves for the given beam and loading. Also determine the maximum absolute value of the bending moment in the beam.

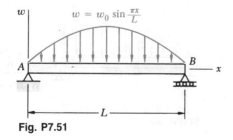

Fig. P7.51

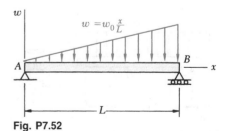

Fig. P7.52

7.53 and 7.54 Determine the equations of the shear and bending-moment curves for the given beam and loading. Also determine the maximum absolute value of the bending moment in the beam.

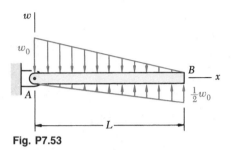

Fig. P7.53

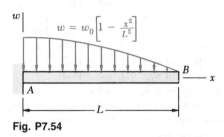

$$w = w_0\left[1 - \frac{x^2}{L^2}\right]$$

Fig. P7.54

7.55 Using the methods of Sec. 7.4, solve Prob. 7.23.
7.56 Using the methods of Sec. 7.4, solve Prob. 7.24.
7.57 Using the methods of Sec. 7.4, solve Prob. 7.25.
7.58 Using the methods of Sec. 7.4, solve Prob. 7.26.
7.59 through 7.62 For the beam and loading shown, determine (a) the maximum absolute value of the bending moment, (b) the maximum normal stress due to bending.

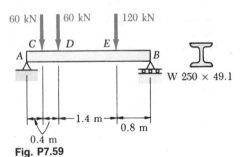

Fig. P7.59

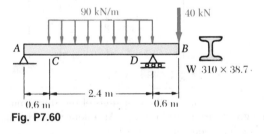

Fig. P7.60

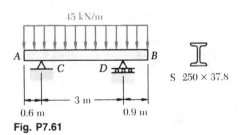

Fig. P7.61

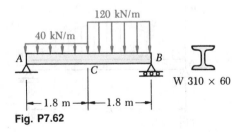

Fig. P7.62

7.63 through 7.66 For the beam and loading shown, determine (*a*) the largest normal stress due to bending, (*b*) the largest shearing stress on a transverse section.

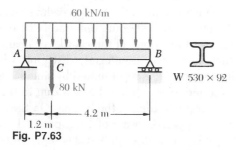

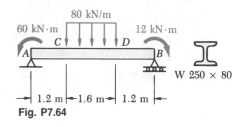

Fig. P7.63

Fig. P7.64

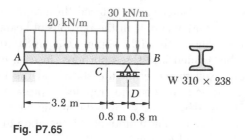

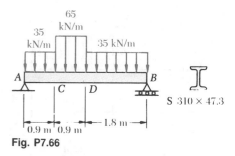

Fig. P7.65

Fig. P7.66

*7.67 The beam *AB* supports a uniformly distributed load of 7 kN/m and two concentrated loads **P** and **Q**. The normal stress due to bending on the bottom edge of the lower flange is +100 MPa at *D* and +70 MPa at *E*. (*a*) Draw the shear and bending-moment diagrams for the beam. (*b*) Determine the maximum normal stress due to bending which occurs in the beam.

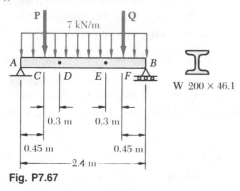

Fig. P7.67

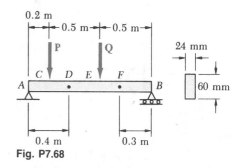

Fig. P7.68

*7.68 The beam *AB* supports two concentrated loads **P** and **Q**. The normal stress due to bending on the bottom edge of the beam is +55 MPa at *D* and +37.5 MPa at *F*. (*a*) Draw the shear and bending-moment diagrams for the beam. (*b*) Determine the maximum normal stress due to bending which occurs in the beam.

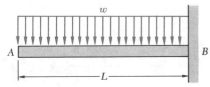

Fig. 7.10 (repeated)

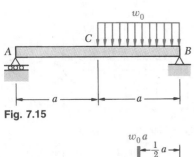

Fig. 7.15

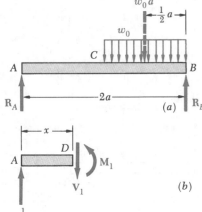

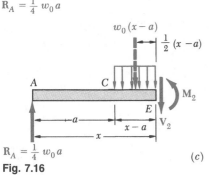

Fig. 7.16

*7.5. USING SINGULARITY FUNCTIONS TO DETERMINE SHEAR AND BENDING MOMENT IN A BEAM

Reviewing the work done in the two preceding sections, we note that the shear and bending moment could only rarely be described by single analytical functions. In the case of the cantilever beam of Example 7.02 (Fig. 7.10), which supported a uniformly distributed load w, the shear and bending moment *could* be represented by single analytical functions, namely, $V = -wx$ and $M = -\frac{1}{2}wx^2$; this was due to the fact that *no discontinuity* existed in the loading of the beam. On the other hand, in the case of the simply supported beam of Example 7.01, which was loaded only at its midpoint C, the load **P** applied at C represented a *singularity* in the beam loading. This singularity resulted in discontinuities in the shear and bending moment and required the use of different analytical functions to represent V and M in the portions of beam located, respectively, to the left and to the right of point C. In Sample Prob. 7.2, the beam had to be divided into three portions, in each of which different functions were used to represent the shear and the bending moment. This situation led us to rely on the graphical representation of the functions V and M provided by the shear and bending-moment diagrams and, later in Sec. 7.4, on a graphical method of integration to determine V and M from the distributed load w.

The purpose of this section is to show how the use of *singularity functions* makes it possible to represent the shear V and the bending moment M by single mathematical expressions.

Consider the simply supported beam AB, of length $2a$, which carries a uniformly distributed load w_0 extending from its midpoint C to its right-hand support B (Fig. 7.15). We first draw the free-body diagram of the entire beam (Fig. 7.16a); replacing the distributed load by an equivalent concentrated load and summing moments about B, we write

$$+\curvearrowleft\Sigma M_B = 0: \qquad (w_0 a)(\tfrac{1}{2}a) - R_A(2a) = 0 \qquad R_A = \tfrac{1}{4}w_0 a$$

Next we cut the beam at a point D between A and C. From the free-body diagram of AD (Fig. 7.16b) we conclude that, over the interval $0 < x < a$, the shear and bending moment are expressed, respectively, by the functions

$$V_1(x) = \tfrac{1}{4}w_0 a \qquad \text{and} \qquad M_1(x) = \tfrac{1}{4}w_0 ax$$

Cutting, now, the beam at a point E between C and B, we draw the free-body diagram of portion AE (Fig. 7.16c). Replacing the distributed load by an equivalent concentrated load, we write

$$+\uparrow\Sigma F_y = 0: \qquad\qquad \tfrac{1}{4}w_0 a - w_0(x-a) - V_2 = 0$$
$$+\curvearrowleft\Sigma M_E = 0: \qquad -\tfrac{1}{4}w_0 ax + w_0(x-a)[\tfrac{1}{2}(x-a)] + M_2 = 0$$

and conclude that, over the interval $a < x < 2a$, the shear and bending moment are expressed, respectively, by the functions

$$V_2(x) = \tfrac{1}{4}w_0a - w_0(x - a) \qquad \text{and} \qquad M_2(x) = \tfrac{1}{4}w_0ax - \tfrac{1}{2}w_0(x - a)^2$$

As we pointed out earlier in this section, the fact that the shear and bending moment are represented by different functions of x, depending upon whether x is smaller or larger than a, is due to the discontinuity in the loading of the beam. However, the functions $V_1(x)$ and $V_2(x)$ may be represented by the single expression

$$V(x) = \tfrac{1}{4}w_0a - w_0\langle x - a\rangle \tag{7.7}$$

if we specify that the second term should be included in our computations when $x \geq a$ and ignored when $x < a$. In other words, *the brackets $\langle\ \rangle$ should be replaced by ordinary parentheses $(\ )$ when $x \geq a$ and by zero when $x < a$.* With the same convention, the bending moment may be represented at any point of the beam by the single expression

$$M(x) = \tfrac{1}{4}w_0ax - \tfrac{1}{2}w_0\langle x - a\rangle^2 \tag{7.8}$$

From the convention we have adopted, it follows that brackets $\langle\ \rangle$ may be differentiated or integrated as ordinary parentheses. Instead of calculating the bending moment from free-body diagrams, we could have used the method indicated in Sec. 7.4 and integrated the expression obtained for $V(x)$:

$$M(x) - M(0) = \int_0^x V(x)\,dx = \int_0^x \tfrac{1}{4}w_0a\,dx - \int_0^x w_0\langle x - a\rangle\,dx$$

After integration, and observing that $M(0) = 0$, we obtain as before

$$M(x) = \tfrac{1}{4}w_0ax - \tfrac{1}{2}w_0\langle x - a\rangle^2$$

Furthermore, using the same convention again, we note that the distributed load at any point of the beam may be expressed as

$$w(x) = w_0\langle x - a\rangle^0 \tag{7.9}$$

Indeed, the brackets should be replaced by zero for $x < a$ and by parentheses for $x \geq a$; we thus check that $w(x) = 0$ for $x < a$ and, defining the zero power of any number as unity, that $\langle x - a\rangle^0 = (x - a)^0 = 1$ and $w(x) = w_0$ for $x \geq a$. From Sec. 7.4 we recall that the shear could have been obtained by integrating the function $-w(x)$. Observing that $V = \tfrac{1}{4}w_0a$ for $x = 0$, we write

$$V(x) - V(0) = -\int_0^x w(x)\,dx = -\int_0^x w_0\langle x - a\rangle^0\,dx$$
$$V(x) - \tfrac{1}{4}w_0a = w_0\langle x - a\rangle^1$$

Solving for $V(x)$ and dropping the exponent 1, we obtain again

$$V(x) = \tfrac{1}{4}w_0a - w_0\langle x - a\rangle$$

The expressions $\langle x - a \rangle^0$, $\langle x - a \rangle$, $\langle x - a \rangle^2$ are called *singularity functions*. By definition, we have, for $n \geq 0$,

$$\langle x - a \rangle^n = \begin{cases} (x - a)^n & \text{when } x \geq a \\ 0 & \text{when } x < a \end{cases} \tag{7.10}$$

We may also note that whenever the quantity between brackets is positive or zero, the brackets should be replaced by ordinary parentheses, and whenever that quantity is negative, the bracket itself is equal to zero.

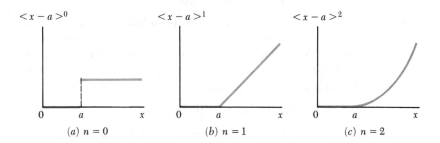

Fig. 7.17 Singularity functions $\langle x - y \rangle^n$.

(a) $n = 0$ (b) $n = 1$ (c) $n = 2$

The three singularity functions corresponding respectively to $n = 0$, $n = 1$, and $n = 2$ have been plotted in Fig. 7.17. We note that the function $\langle x - a \rangle^0$ is discontinuous at $x = a$ and is in the shape of a "step." For that reason it is referred to as the *step function*. According to (7.10), and with the zero power of any number defined as unity, we have†

$$\langle x - a \rangle^0 = \begin{cases} 1 & \text{when } x \geq a \\ 0 & \text{when } x < a \end{cases} \tag{7.11}$$

It follows from the definition of singularity functions that

$$\int \langle x - a \rangle^n \, dx = \frac{1}{n + 1} \langle x - a \rangle^{n+1} \qquad \text{for } n \geq 0 \tag{7.12}$$

and

$$\frac{d}{dx} \langle x - a \rangle^n = n \langle x - a \rangle^{n-1} \qquad \text{for } n \geq 1 \tag{7.13}$$

Most of the beam loadings encountered in engineering practice may be broken down into the basic loadings shown in Fig. 7.18. Whenever applicable, the corresponding functions $w(x)$, $V(x)$, and $M(x)$ have been expressed in terms of singularity functions and plotted against a color background. A heavier color background was used to indicate for each loading the expression that is most easily derived or remembered and from which the other functions may be obtained by integration.

---

† Since $\langle x - a \rangle^0$ is discontinuous at $x = a$, it may be argued that this function should be left undefined for $x = a$ or that it should be assigned both of the values 0 and 1 for $x = a$. However, defining $\langle x - a \rangle^0$ as equal to 1 when $x = a$, as stated in (7.11), has the advantage of being unambiguous and, thus, readily applicable to computer programming (cf. page 437).

Loading	Shear	Bending Moment

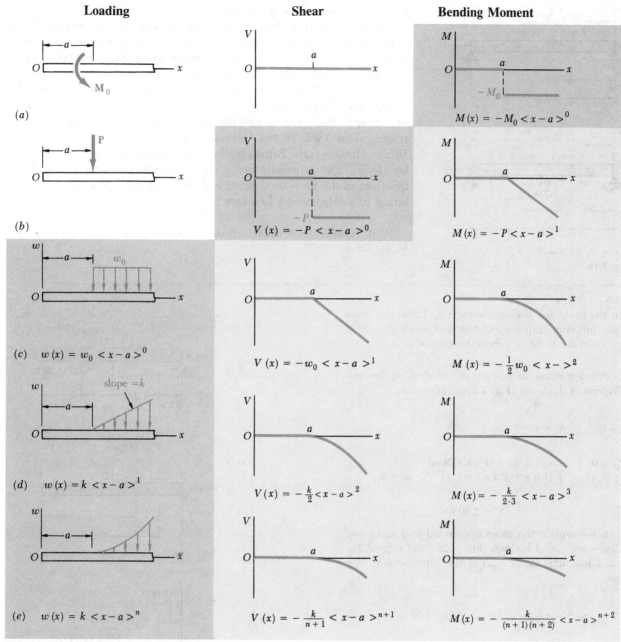

(a)

$$M(x) = -M_0 <x-a>^0$$

(b)

$$V(x) = -P<x-a>^0$$

$$M(x) = -P<x-a>^1$$

(c) $w(x) = w_0 <x-a>^0$

$$V(x) = -w_0 <x-a>^1$$

$$M(x) = -\tfrac{1}{2}w_0 <x->^2$$

(d) $w(x) = k<x-a>^1$

slope $= k$

$$V(x) = -\tfrac{k}{2}<x-a>^2$$

$$M(x) = -\tfrac{k}{2\cdot3}<x-a>^3$$

(e) $w(x) = k<x-a>^n$

$$V(x) = -\tfrac{k}{n+1}<x-a>^{n+1}$$

$$M(x) = -\tfrac{k}{(n+1)(n+2)}<x-a>^{n+2}$$

Fig. 7.18 Basic loadings and corresponding shears and bending moments expressed in terms of singularity functions.

After a given beam loading has been broken down into the basic loadings of Fig. 7.18, the functions $V(x)$ and $M(x)$ representing the shear and bending moment at any point of the beam may be obtained by adding the corresponding functions associated with each of the basic loadings and

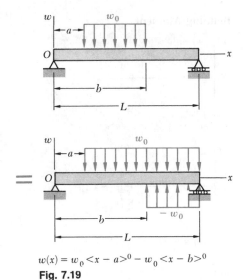

$$w(x) = w_0 <x - a>^0 - w_0 <x - b>^0$$

Fig. 7.19

reactions. Since all the distributed loadings shown in Fig. 7.18 are open-ended to the right, a distributed loading which does not extend to the right end of the beam or which is discontinuous should be replaced as shown in Fig. 7.19 by an equivalent combination of open-ended loadings. (See also Example 7.04 and Sample Prob. 7.7.)

As we shall see in Sec. 8.6, the use of singularity functions also greatly simplifies the determination of beam deflections. It was in connection with that problem that the approach used in this section was first suggested in 1862 by the German mathematician A. Clebsch (1833–1872). However, the British mathematician and engineer W. H. Macaulay (1853–1936) is usually given credit for introducing the singularity functions in the form used here, and the brackets $\langle\ \rangle$ are generally referred to as *Macaulay's brackets.*[†]

[†]W. H. Macaulay, "Note on the Deflection of Beams," *Messenger of Mathematics*, vol. 48, pp. 129–130, 1919.

Example 7.04

For the beam and loading shown (Fig. 7.20a) and using singularity functions, express the shear and bending moment as functions of the distance x from the support at A.

We first determine the reaction at A by drawing the free-body diagram of the beam (Fig. 7.20b) and writing

$$\xrightarrow{+}\Sigma F_x = 0: \qquad A_x = 0$$

$$+\curvearrowleft\Sigma M_B = 0: \qquad -A_y(3.6 \text{ m}) + (1.2 \text{ kN})(3 \text{ m})$$
$$+ (1.8 \text{ kN})(2.4 \text{ m}) + 1.44 \text{ kN} \cdot \text{m} = 0$$

$$A_y = 2.60 \text{ kN}$$

Next, we replace the given distributed loading by two equivalent open-ended loadings (Fig. 7.20c) and express the distributed load $w(x)$ as the sum of the corresponding step functions:

$$w(x) = +w_0\langle x - 0.6\rangle^0 - w_0\langle x - 1.8\rangle^0$$

The function $V(x)$ is obtained by integrating $w(x)$, reversing the $+$ and $-$ signs, and adding to the result the constants A_y and $-P\langle x - 0.6\rangle^0$ representing the respective contributions to the shear of the reaction at A and of the concentrated load. (No other constant of integration is required.) Since the concentrated couple does not directly affect the shear, it should be

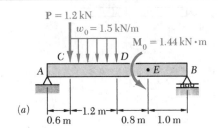

(a)

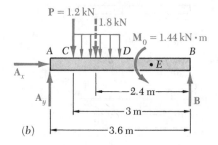

(b)

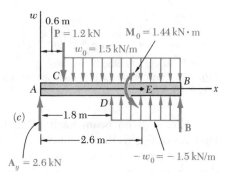

(c)

Fig. 7.20

ignored in this computation. We write

$$V(x) = -w_0\langle x - 0.6\rangle^1 + w_0\langle x - 1.8\rangle^1 + A_y - P\langle x - 0.6\rangle^0$$

In a similar way, the function $M(x)$ is obtained by integrating $V(x)$ and adding to the result the constant $-M_0\langle x - 2.6\rangle^0$ representing the contribution of the concentrated couple to the bending moment. We have

$$M(x) = -\tfrac{1}{2}w_0\langle x - 0.6\rangle^2 + \tfrac{1}{2}w_0\langle x - 1.8\rangle^2 \\ + A_y x - P\langle x - 0.6\rangle^1 - M_0\langle x - 2.6\rangle^0$$

Example 7.05

For the beam and loading of Example 7.04, determine the numerical values of the shear and bending moment at the midpoint D.

Making $x = 1.8$ m in the expressions found for $V(x)$ and $M(x)$ in Example 7.04, we obtain

$$V(1.8) = -1.5\langle 1.2\rangle^1 + 1.5\langle 0\rangle^1 + 2.6 - 1.2\langle 1.2\rangle^0$$

$$M(1.8) = -0.75\langle 1.2\rangle^2 + 0.75\langle 0\rangle^2 \\ + 2.6(1.8) - 1.2\langle 1.2\rangle^1 - 1.44\langle -0.8\rangle^0$$

Substituting the numerical values of the reaction and loads into the expressions obtained for $V(x)$ and $M(x)$ and being careful *not* to compute any product or expand any square involving a bracket, we obtain the following expressions for the shear and bending moment at any point of the beam:

$$V(x) = -1.5\langle x - 0.6\rangle^1 + 1.5\langle x - 1.8\rangle^1 + 2.6 - 1.2\langle x - 0.6\rangle^0$$

$$M(x) = -0.75\langle x - 0.6\rangle^2 + 0.75\langle x - 1.8\rangle^2 \\ + 2.6x - 1.2\langle x - 0.6\rangle^1 - 1.44\langle x - 2.6\rangle^0$$

Recalling that whenever a quantity between brackets is positive or zero, the brackets should be replaced by ordinary parentheses, and whenever the quantity is negative, the bracket itself is equal to zero, we write

$$V(1.8) = -1.5(1.2)^1 + 1.5(0)^1 + 2.6 - 1.2(1.2)^0 \\ = -1.5(1.2) + 1.5(0) + 2.6 - 1.2(1) \\ = -1.8 + 0 + 2.6 - 1.2 \\ V(1.8) = -0.4 \text{ kN}$$

and

$$M(1.8) = -0.75(1.2)^2 + 0.75(0)^2 \\ + 2.6(1.8) - 1.2(1.2)^1 - 1.44(0) \\ = -1.08 + 0 + 4.68 - 1.44 - 0 \\ M(1.8) = +2.16 \text{ kN} \cdot \text{m}$$

Application to Computer Programming. Singularity functions are particularly well suited to the use of computers. First we note that the step function $\langle x - a\rangle^0$, which we shall represent by the symbol STP, may be defined by an IF/THEN/ELSE statement as being equal to 1 for $X \geq A$ and to 0 otherwise. Any other singularity function $\langle x - a\rangle^n$, with $n \geq 1$, may then be expressed as the product of the ordinary algebraic function $(x - a)^n$ and the step function $\langle x - a\rangle^0$. Thus, the singularity function $\langle x - a\rangle^n$ will be programmed as

$$(X - A)^\wedge N * STP \qquad (BASIC)$$

or

$$(X - A) ** N * STP \qquad (FORTRAN)$$

When k different singularity functions are involved, such as $\langle x - a_i\rangle^n$, where $i = 1, 2, \ldots, k$, then the corresponding k step functions STP(I), where $I = 1, 2, \ldots, K$, may be defined by a loop containing a single IF/THEN/ELSE statement.

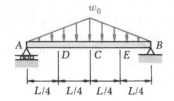

SAMPLE PROBLEM 7.7

For the beam and loading shown, determine (a) the equations defining the shear and bending moment at any point, (b) the shear and bending moment at points C, D, and E.

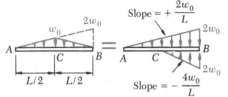

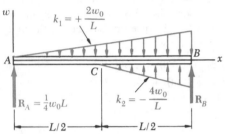

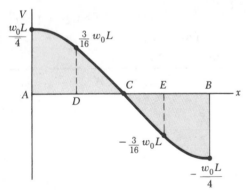

Reactions. The total load is $\frac{1}{2}w_0L$; because of symmetry, each reaction is equal to half that value, namely, $\frac{1}{4}w_0L$.

Distributed Load. The given distributed loading is replaced by two equivalent open-ended loadings as shown. Using a singularity function to express the second loading, we write

$$w(x) = k_1 x + k_2\langle x - \tfrac{1}{2}L\rangle = \frac{2w_0}{L}x - \frac{4w_0}{L}\langle x - \tfrac{1}{2}L\rangle \tag{1}$$

(a) Equations for Shear and Bending Moment. We obtain $V(x)$ by integrating (1), changing the signs, and adding a constant equal to R_A:

$$V(x) = -\frac{w_0}{L}x^2 + \frac{2w_0}{L}\langle x - \tfrac{1}{2}L\rangle^2 + \tfrac{1}{4}w_0L \tag{2} \quad \blacktriangleleft$$

We obtain $M(x)$ by integrating (2); since there is no concentrated couple, no constant of integration is needed:

$$M(x) = -\frac{w_0}{3L}x^3 + \frac{2w_0}{3L}\langle x - \tfrac{1}{2}L\rangle^3 + \tfrac{1}{4}w_0Lx \tag{3} \quad \blacktriangleleft$$

(b) Shear and Bending Moment at C, D, and E

At Point C: Making $x = \tfrac{1}{2}L$ in Eqs. (2) and (3) and recalling that whenever a quantity between brackets is positive or zero, the brackets may be replaced by parentheses, we have

$$V_C = -\frac{w_0}{L}(\tfrac{1}{2}L)^2 + \frac{2w_0}{L}\langle 0\rangle^2 + \tfrac{1}{4}w_0L \qquad V_C = 0 \quad \blacktriangleleft$$

$$M_C = -\frac{w_0}{3L}(\tfrac{1}{2}L)^3 + \frac{2w_0}{3L}\langle 0\rangle^3 + \tfrac{1}{4}w_0L(\tfrac{1}{2}L) \qquad M_C = \frac{1}{12}w_0L^2 \quad \blacktriangleleft$$

At Point D: Making $x = \tfrac{1}{4}L$ in Eqs. (2) and (3) and recalling that a bracket containing a negative quantity is equal to zero, we write

$$V_D = -\frac{w_0}{L}(\tfrac{1}{4}L)^2 + \frac{2w_0}{L}\langle -\tfrac{1}{4}L\rangle^2 + \tfrac{1}{4}w_0L \qquad V_D = \frac{3}{16}w_0L \quad \blacktriangleleft$$

$$M_D = -\frac{w_0}{3L}(\tfrac{1}{4}L)^3 + \frac{2w_0}{3L}\langle -\tfrac{1}{4}L\rangle^3 + \tfrac{1}{4}w_0L(\tfrac{1}{4}L) \qquad M_D = \frac{11}{192}w_0L^2 \quad \blacktriangleleft$$

At Point E: Making $x = \tfrac{3}{4}L$ in Eqs. (2) and (3), we have

$$V_E = -\frac{w_0}{L}(\tfrac{3}{4}L)^2 + \frac{2w_0}{L}\langle\tfrac{1}{4}L\rangle^2 + \tfrac{1}{4}w_0L \qquad V_E = -\frac{3}{16}w_0L \quad \blacktriangleleft$$

$$M_E = -\frac{w_0}{3L}(\tfrac{3}{4}L)^3 + \frac{2w_0}{3L}\langle\tfrac{1}{4}L\rangle^3 + \tfrac{1}{4}w_0L(\tfrac{3}{4}L) \qquad M_E = \frac{11}{192}w_0L^2 \quad \blacktriangleleft$$

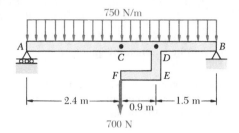

750 N/m

The rigid bar DEF is welded at point D to the steel beam AB. For the loading shown, determine (a) the equations defining the shear and bending moment at any point of the beam, (b) the location and magnitude of the largest bending moment.

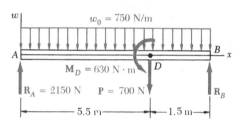

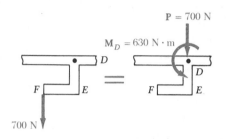

Reactions. We consider the beam and bar as a free body and observe that the total load is 430 N. Because of symmetry, each reaction is equal to 2150 N.

Modified Loading Diagram. We replace the 700-N load applied to F by an equivalent force-couple system at D. We thus obtain a loading diagram consisting of a concentrated couple, three concentrated loads (including the two reactions), and a uniformly distributed load

$$w(x) = 750 \text{ N/m} \qquad (1)$$

(a) **Equations for Shear and Bending Moment.** We obtain $V(x)$ by integrating (1), changing the sign, and adding constants representing the respective contributions of $\mathbf{R}_A$ abd $\mathbf{P}$ to the shear. Since $\mathbf{P}$ affecdts $V(x)$ only for values of x larger than 3.3 m, we use a step function to express its contribution.

$$V(x) = -750x + 2150 - 700\langle x - 3.3\rangle^0 \qquad (2) \blacktriangleleft$$

We obtain $M(x)$ by integrating (2) and using a step function to represent the contribution of the concentrated couple $\mathbf{M}_D$:

$$M(x) = -375x^2 + 2150 - 700\langle x - 3.3\rangle^1 - 630\langle x - 3.3\rangle^0 \qquad (3) \blacktriangleleft$$

(b) **Largest Bending Moment.** Since M is maximum or minimum when $V = 0$, we set $V = 0$ in (2) and solve that equation for x to find the location of the largest bending moment. Considering first values of x less than 3.3 m and noting that for such values the bracket is equal to zero, we write

$$-750x + 2150 = 0 \qquad x = 2.87 \text{ m}$$

Considering now values of x larger than 3.3 m, for which the bracket is equal to 1, we have

$$-750x + 2150 - 700 = 0 \qquad x = 1.93 \text{ m}$$

Since this value is *not* larger than 3.3 m, it must be rejected. Thus, the value of x corresponding to the largest bending moment is

$$x_m = 2.87 \text{ m} \blacktriangleleft$$

Substituting this value for x into Eq. (3), we obtain

$$M_{max} = -375(2.87)^2 + 2150(2.87) - 700\langle -0.43\rangle^1 - 630\langle -0.43\rangle^0$$

and, recalling that brackets containing a negative quantity are equal to zero,

$$M_{max} = -375(2.87)^2 + 2150 (2.87) M_{max} = 3082 \text{ N} \cdot \text{m} \blacktriangleleft$$

The bending-moment diagram has been plotted. Note the discontinuity at point D due to the concentrated couple applied at that point. The values of M just to the left and just to the right of D were obtained by making $x = 3.3$ in Eq. (3) and replacing the stop function $\langle x - 3.3\rangle^0$ by 0 and 1, respectively.

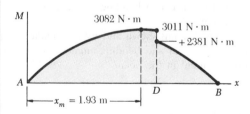

439

PROBLEMS

In the following problems, place the origin of coordinates at point A.

7.69 through 7.72 (*a*) For the beam and loading shown and using singularity functions, write the equations defining the shear and bending-moment curves. (*b*) Determine the maximum bending moment in the beam.

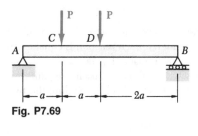

Fig. P7.69

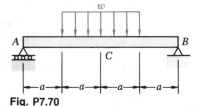

Fig. P7.70

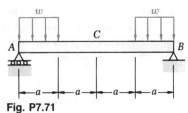

Fig. P7.71

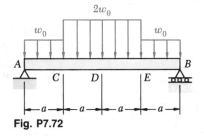

Fig. P7.72

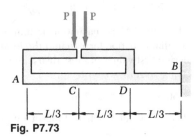

Fig. P7.73

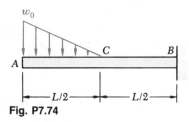

Fig. P7.74

7.73 and 7.74 (*a*) For the beam and loading shown and using singularity functions, write the equations defining the shear and bending-moment curves. (*b*) Determine the bending moment at point B.

7.75 and 7.76 (*a*) For the beam and loading shown and using singularity functions, write the equations defining the shear and bending-moment curves. (*b*) Determine the maximum bending moment in the beam.

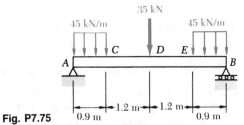

Fig. P7.75

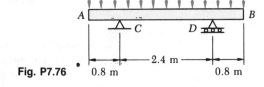

Fig. P7.76

7.77 and 7.78 (*a*) For the beam and loading shown and using singularity functions, write the equations defining the shear and bending-moment curves. (*b*) Determine the maximum bending moment in the beam.

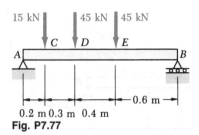

Fig. P7.77

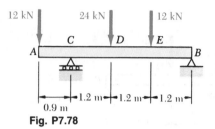

Fig. P7.78

7.79 through 7.82 (*a*) For the beam and loading shown and using singularity functions, write the equations defining the shear and bending-moment curves. (*b*) Determine the location and magnitude of the maximum bending moment in the beam and the maximum normal stress due to bending.

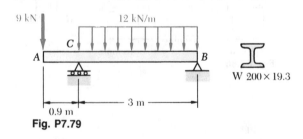

Fig. P7.79

W 200 × 19.3

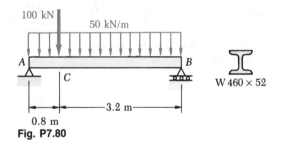

Fig. P7.80

W 460 × 52

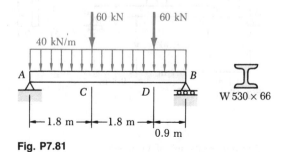

Fig. P7.81

W 530 × 66

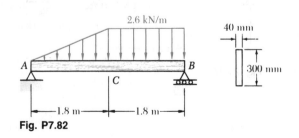

Fig. P7.82

7.83 and 7.84†   For the beam and loading shown and using singularity functions, write a computer program and use it to calculate the shearing force and bending moment at the specified intervals ΔL starting at point A and ending at the right-hand support.

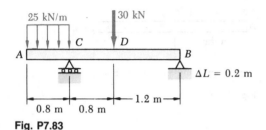

Fig. P7.83

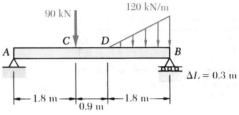

Fig. P7.84

7.85 through 7.88†   (a) For the beam and loading shown and using singularity functions, write a computer program which can be used to tabulate the shearing force and bending moment at the specified intervals ΔL starting at point A and ending at the right-hand support.   (b) Using the tabulation of part a, locate by interpolation the point where $V = 0$ and then calculate the maximum bending moment in the beam and the maximum normal stress due to bending.

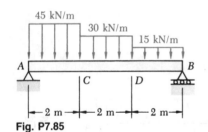

Fig. P7.85

W 530 × 66
$\Delta L = 0.5$ m

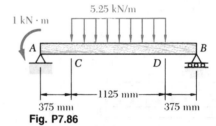

Fig. P7.86

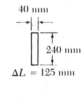

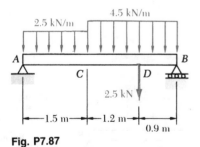

Fig. P7.87

W 250 × 22.3
$\Delta L = 0.3$ m

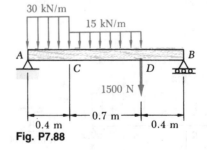

Fig. P7.88

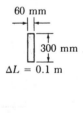

†Problems 7.83 through 7.88 require the use of a computer.

7.6. PRINCIPAL STRESSES IN A BEAM

Let us consider a prismatic beam AB subjected to some arbitrary transverse loading (Fig. 7.21); we shall denote by V and M, respectively, the shear and bending moment in a section through a given point C. We recall from Chaps. 4 and 5 that, within the elastic limit, the stresses exerted on a small element with faces perpendicular, respectively, to the x and y axes reduce to the normal stresses $\sigma_m = Mc/I$ if the element is at the free surface of the beam, and to the shearing stresses $\tau_m = VQ/It$ if the element is at the neutral surface (Fig. 7.22).

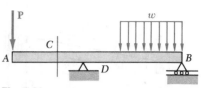

Fig. 7.21

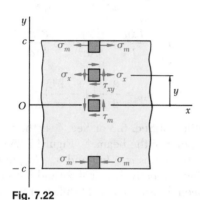

Fig. 7.22

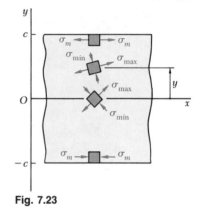

Fig. 7.23

At any other point of the cross section, an element of material is subjected simultaneously to the normal stresses

$$\sigma_x = -\frac{My}{I} \tag{7.14}$$

where y is the distance from the neutral surface, and to the shearing stresses

$$\tau_{xy} = -\frac{VQ}{It} \tag{7.15}$$

where Q is the first moment about the neutral axis of the portion of the cross-sectional area located above the point where the stresses are computed, and t the width of the cross section at that point. Using either of the methods of analysis presented in Chap. 6, we may obtain the principal stresses at any point of the cross section (Fig. 7.23).

The following question now arises: Can the maximum normal stress σ_{max} at some point within the cross section be larger than the value of $\sigma_m = Mc/I$ computed at the surface of the beam? If it can, then the determination of the largest normal stress in the beam will involve a great deal more than the computation of $|M|_{max}$ and the use of Eq. (7.1). We shall try to obtain an answer to this question by investigating the distribu-

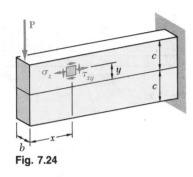

Fig. 7.24

tion of the principal stresses in a narrow rectangular cantilever beam subjected to a concentrated load **P** at its free end (Fig. 7.24).

We recall from Chap. 5 that the normal and shearing stresses at a distance x from the load **P** and a distance y above the neutral surface are given, respectively, by Eq. (5.16) and Eq. (5.15). Since the moment of inertia of the cross section is

$$I = \frac{bh^3}{12} = \frac{(bh)(2c)^2}{12} = \frac{Ac^2}{3}$$

where A is the cross-sectional area and c the half-depth of the beam, we write

$$\sigma_x = \frac{Pxy}{I} = \frac{Pxy}{\frac{1}{3}Ac^2} = 3\frac{P}{A}\frac{xy}{c^2} \tag{7.16}$$

and

$$\tau_{xy} = \frac{3}{2}\frac{P}{A}\left(1 - \frac{y^2}{c^2}\right) \tag{7.17}$$

Using the method of Sec. 6.3 or Sec. 6.4, the value of σ_{max} may be determined at any point of the beam.† Figure 7.25 shows the results of the computation of the ratios σ_{max}/σ_m and σ_{min}/σ_m in two sections of the beam, corresponding respectively to $x = 2c$ and $x = 8c$. In each section, these ratios have been determined at 11 different points, and the orientation of the principal axes has been indicated at each point.

It is clear that σ_{max} does not exceed σ_m in either of the two sections considered in Fig. 7.25 and that, if it does exceed σ_m elsewhere, it will be in sections close to the load **P**, where σ_m is small compared to τ_m.‡ But, for sections close to the load **P**, Saint-Venant's principle does not apply, Eqs. (7.16) and (7.17) cease to be valid, except in the very unlikely case of a load distributed parabolically over the end section (cf. Sec. 5.7), and more advanced methods of analysis taking into account the effect of stress concentrations should be used. We thus conclude that, for beams of rectangular cross section, and within the scope of the theory presented in this text, the maximum normal stress may be obtained from Eq. (7.1).

The above conclusion is also valid for many beams of nonrectangular cross section. However, when the width of the cross section varies in such a way that large shearing stresses τ_{xy} will occur at points close to the surface of the beam, where σ_x is also large, a value of the principal stress σ_{max} larger than σ_m may result at such points. One should be particularly aware of this possibility when selecting S-beams or W-beams, and compute the principal stress σ_{max} at the junction of the web with the flanges of the beam in the section where $|M|$ is maximum (see Sample Prob. 7.11).

† See Prob. 7.126.
‡ See Prob. 7.127.

	x = 2c			x = 8c		
y/c	σ_{min}/σ_m		σ_{max}/σ_m	σ_{min}/σ_m		σ_{max}/σ_m
1.0	0		1.000	0		1.000
0.8	−0.010		0.810	−0.001		0.801
0.6	−0.040		0.640	−0.003		0.603
0.4	−0.090		0.490	−0.007		0.407
0.2	−0.160		0.360	−0.017		0.217
0	−0.250		0.250	−0.062		0.062
−0.2	−0.360		0.160	−0.217		0.017
−0.4	−0.490		0.090	−0.407		0.007
−0.6	−0.640		0.040	−0.603		0.003
−0.8	−0.810		0.010	−0.801		0.001
−1.0	−1.000		0	−1.000		0

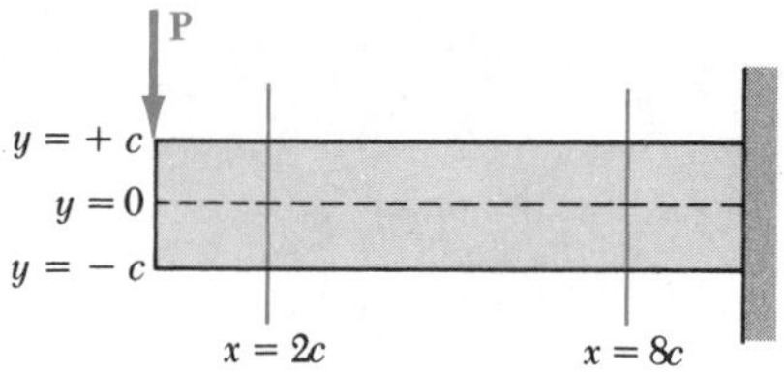

Fig. 7.25. Distribution of principal stresses in two transverse sections of a rectangular cantilever beam supporting a single concentrated load.

Returning to Fig. 7.25, we note that the directions of the principal axes were determined at 11 points in each of the two sections considered. If this analysis were extended to a larger number of sections and a larger number of points in each section, it would be possible to draw two orthogonal systems of curves on the side of the beam (Fig. 7.26). One system would consist of curves tangent to the principal axes corresponding to σ_{max}, and the other of curves tangent to the principal axes corresponding to σ_{min}. The curves obtained in this manner are known as the *stress trajectories*. A trajectory of the first group (solid lines) defines at each of its points the direction of the largest tensile stress, while a trajectory of the second group (dashed lines) defines the direction of the largest compressive stress.†

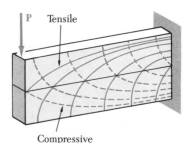

Fig. 7.26. Stress trajectories.

†A brittle material, such as concrete, will fail in tension along planes which are perpendicular to the tensile-stress trajectories. Thus, to be effective, steel reinforcing bars should be placed so that they intersect these planes. On the other hand, stiffeners attached to the web of a plate girder will be effective in preventing buckling only if they intersect planes perpendicular to the compressive-stress trajectories.

7.7. DESIGN OF PRISMATIC BEAMS

We saw in Secs. 7.2 and 7.6 that the design of a beam is usually controlled by the maximum absolute value $|M|_{\max}$ of the bending moment in the beam, and that, in the critical section where $|M|_{\max}$ occurs, the largest normal stress is usually found at the surface of the beam and may, therefore, be obtained from Eq. (7.1). We recall from the preceding section, however, that for certain shapes of beams $\sigma_{\max}$ may occur at points located elsewhere within the beam. Also, there may be situations where the design of the beam is controlled by the maximum absolute value $|V|_{\max}$ of the shear in the beam, rather than by $|M|_{\max}$.

A proper procedure for the design of a beam should take into account these various points. It also should lead to the most economical design. This means that, among beams of the same material, and other things being equal, the beam with the smallest weight per unit length (i.e., the smallest cross-sectional area) should be selected, since this beam will be the least expensive. The design procedure will include the following steps:†

1. From a table of properties of materials or from design specifications, we determine the values of σ_{all} and τ_{all} for the material selected, or compute these values by dividing σ_U and τ_U by the appropriate factor of safety (Sec. 1.9). Assuming for the time being that the value of σ_{all} is the same in tension and in compression, we proceed as follows.

2. We draw the shear and bending-moment diagrams corresponding to the specified loading conditions, and determine the maximum absolute values $|V|_{\max}$ and $|M|_{\max}$.

3. Assuming that the design of the beam is controlled by the value of the normal stress at $y = \pm c$ in the section of maximum bending moment, we determine the minimum allowable value of the section modulus $S = I/c$. Substituting σ_{all} for σ_m in Eq. (7.1) and solving for $S = I/c$, we have

$$S_{\min} = \frac{|M|_{\max}}{\sigma_{\text{all}}} \tag{7.18}$$

4. Of the available beam sections, we consider only those with a section modulus $S > S_{\min}$, and select from this group the section

---

†We assume that all beams considered in this chapter are adequately braced to prevent lateral buckling, and that bearing plates are provided under concentrated loads to prevent local buckling (crippling) of the web.

with the smallest weight per unit length. This is the most economical of the sections for which $\sigma_m \leq \sigma_{all}$. Note that this is not necessarily the section with the smallest value of S (see Example 7.06). In some cases, the selection of a section may be limited by other considerations, such as the allowable depth of the cross section, or the allowable deflection of the beam (Chaps. 8 and 9).

5. We now check the resistance to shear of the beam we have tentatively selected. Substituting the appropriate data for Q, I, and t into Eq. (7.2), we determine the maximum value τ_m of the shearing stress τ_{xy} in the beam:

$$\tau_m = \frac{|V|_{max}Q}{It} \tag{7.19}$$

From Sec. 5.6 we recall that, in the case of a rectangular beam, the maximum shearing stress is expressed as

$$\tau_m = \frac{3}{2}\frac{|V|_{max}}{A} \tag{7.20}$$

and that, for S-beams and W-beams, it is customary to assume that the entire shear load is uniformly distributed over the web, so that

$$\tau_m = \frac{|V|_{max}}{A_{web}} \tag{7.21}$$

If the value obtained for τ_m is smaller than τ_{all}, the beam is acceptable. If $\tau_m > \tau_{all}$, a stronger beam should be selected.†

6. In the case of S-beams and W-beams, we should check that the value of σ_{max} at the junction of the web with the flanges in the section of maximum bending moment does not exceed σ_{all}. A rough estimate of σ_{max} is usually sufficient, and the actual computation of σ_{max} from the stress components σ_x and τ_{xy} at those points is rarely necessary.

The foregoing discussion was limited to materials for which σ_{all} is the same in tension and in compression. If σ_{all} is different in tension and in compression, we should make sure to select the beam section in such a

---

†Note that $\tau_m > \tau_{all}$ simply indicates that $|V|_{max}$ is too large for the cross section tentatively selected. It does *not* mean that the material will necessarily fail in shear; actually, a brittle material would fail in tension along a principal plane of stress. Similarly, $\sigma_m > \sigma_{all}$ simply indicates that $|M|_{max}$ is too large for the cross section considered; while a brittle material would actually fail in tension, a ductile material would fail in shear along a plane of maximum shear (cf. Sec. 7.2).

way that $\sigma_m < \sigma_{all}$ for both tensile and compressive stresses. If the cross section is not symmetric about its neutral axis, the largest tensile and the largest compressive stresses will not necessarily occur in the section where $|M|$ is maximum. One may occur where M is maximum and the other where M is minimum. Thus, step 2 should include the determination of both M_{max} and M_{min}, and step 3 should be modified to take into account both tensile and compressive stresses.

Example 7.06

Select a wide-flange beam to support the 65-kN load as shown in Fig. 7.27. The allowable normal and shearing stresses for the steel used are respectively 170 MPa and 100 MPa.†

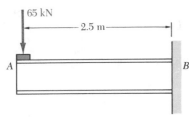

Fig. 7.27

1. The allowable stresses are given:

$$\sigma_{all} = 170 \text{ MPa} \qquad \tau_{all} = 100 \text{ MPa}$$

2. The shear is constant in the beam and the bending moment is maximum at B. We have

$$|V|_{max} = |V| = 65 \text{ kN}$$
$$|M|_{max} = (65 \text{ kN})(2.5 \text{ m}) = 162.5 \text{ kN} \cdot \text{m}$$

3. The minimum allowable section modulus is

$$S_{min} = \frac{|M|_{max}}{\sigma_{all}} = \frac{162.5 \text{ kN} \cdot \text{m}}{170 \text{ MPa}} = 955.9 \times 10^{-6} \text{ m}^3$$
$$= 956 \times 10^3 \text{ mm}^3$$

4. Referring to the table of *Properties of Rolled-Steel Shapes* in Appendix C, we note that the shapes are arranged in groups of the same depth, and that in each group they are listed in order of decreasing mass. We choose in each group the lightest beam having a section

modulus $S = I/c$ at least as large as S_{min}. These shapes are W 530 × 66, W 460 × 74, W 410 × 60, W 360 × 64, W 310 × 74, W 250 × 80. The most economical is the W 410 × 60, since it has a mass of only 60 kg/m, even though it has a larger section modulus than two of the other shapes. We also note from the table that the dead load is 0.584 kN/m; the total weight of the beam, therefore, is (0.584 kN/m)(2.5 m) = 1.46 kN. This weight is small compared to the 65-kN load and will be neglected in our analysis.

5. From the table, we find that the web of a W 410 × 60 section is 7.7 mm thick. Since the section is 407 mm deep, the area of the web may be assumed equal to 3134 mm². Substituting into Eq. (7.21), we find

$$\tau_m = \frac{|V|_{max}}{A_{web}} = \frac{65 \text{ kN}}{3.134 \times 10^{-3} \text{ m}^2} = 20.7 \text{ MPa}$$

and check that $\tau_m < \tau_{all}$. The beam is acceptable.

6. We shall now check whether the maximum normal stress at a point just below the upper flange in the section of maximum bending moment exceeds the allowable normal stress.‡ Since the flange is 12.8 mm thick, we have at that point

$$y = \tfrac{1}{2}(407 \text{ mm}) - 12.8 \text{ mm} = 190.7 \text{ mm}$$

From the table, we also note that the moment of inertia of the section about the neutral axis is $I = 216 \times 10^6 \text{ mm}^4$. Thus, the longitudinal stress σ_x at the point considered is

$$\sigma_x = \frac{My}{I} = \frac{(162.5 \text{ kN} \cdot \text{m})(190.7 \times 10^{-3} \text{ m})}{216 \times 10^{-6} \text{ m}^4} = 143.5 \text{ MPa}$$

Since τ_{xy} at that point is less than τ_m, that is, less than 20.7 MPa, we may easily check from a sketch of Mohr's circle that $\sigma_{max} < 170$ MPa.

---

†See footnote on page 446.

‡The use of this point is conservative, since we omit the effect of the fillet between the flange and the web.

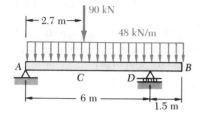

SAMPLE PROBLEM 7.9

The overhanging beam AB supports a uniformly distributed load of 48 kN/m and a concentrated load of 90 kN at C. Knowing that for the grade of steel to be used σ_{all} = 165 MPa and τ_{all} = 100 MPa, select the wide-flange shape which should be used.

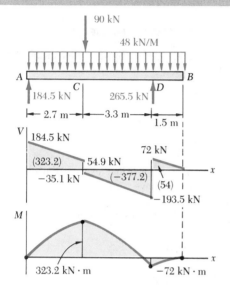

Shear and Bending-Moment Diagrams. Using the methods of Secs. 7.2 and 7.3, we draw the diagrams and observe that

$$|M|_{max} = 323.2 \text{ kN} \cdot \text{m} \qquad |V|_{max} = 193.5 \text{ kN}$$

Section Modulus. For $|M|_{max}$ = 323.2 kN/m and σ_{all} = 165 MPa, the minimum acceptable section modulus of the rolled-steel shape is

$$S_{min} = \frac{|M|_{max}}{\sigma_{all}} = \frac{323.2 \text{ kN} \cdot \text{m}}{165 \text{ MPa}} = 1959 \times 10^{-6} \text{ m}^3 = 1959 \times 10^3 \text{ mm}^3$$

Selection of Wide-Flange Shape. From the table of *Properties of Rolled-Steel Shapes* in Appendix C, we compile a list of shapes which have a section modulus larger than S_{min} and are also the lightest shape in a given depth group.

Shape	S (10^3 mm^3)
W 610 × 101	2530
W 530 × 92	2080
W 460 × 113	2390
W 410 × 114	2200
W 360 × 122	2020
W 310 × 143	2150

We now select the lightest shape available, namely   W 530 × 92 ◄

Shearing Stress. Assuming that the maximum shear is uniformly distributed over the web area of a W 530 × 92, we write

$$\tau_m = \frac{V_{max}}{A_{web}} = \frac{193.5 \text{ kN}}{5.44 \times 10^{-3} \text{ m}^2} = 35.6 \text{ MPa} < 100 \text{ MPa}$$

Maximum Normal Stress in the Web. We check whether the maximum stress at point b in the critical section where M is maximum will exceed σ_{all} = 165 MPa. We write

$$\sigma_a = \frac{M_{max}}{S} = \frac{323.2 \text{ kN} \cdot \text{m}}{2080 \times 10^{-6} \text{ m}^3} = 155.4 \text{ MPa}$$

$$\sigma_b = \sigma_a \frac{y_b}{c} = (155.4 \text{ MPa}) \frac{250.9 \text{ mm}}{266.5 \text{ mm}} = 146.3 \text{ MPa}$$

$$\tau_b = \frac{V}{A_{web}} = \frac{54.9 \text{ kN}}{5.44 \times 10^{-3} \text{ m}^2} = 10.1 \text{ MPa}$$

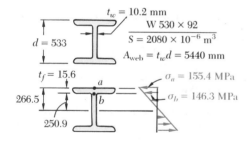

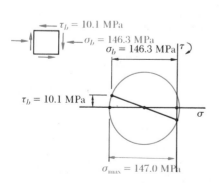

A sketch of Mohr's circle will quickly show that σ_{max} < 165 MPa.

Comment. We note from the table that the dead load, or weight per unit length of the shape chosen is 0.907 kN/m, that is, about 2% of the given distributed loading (48 kN/m). Since the actual section modulus of the W 530×92 (2080× 10^3 mm^3) is about 6% higher than the required section modulus (1959 × 10^3 mm^3), we conclude that σ_{max} < 165 MPa even when the weight of the beam is included.

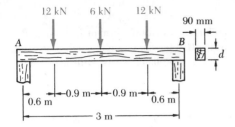

SAMPLE PROBLEM 7.8

A timber beam AB of span 3 m and width 90 mm is to support the three concentrated loads shown. Knowing that for the grade of timber used $\sigma_{all} = 12$ MPa and $\tau_{all} = 850$ kPa, determine the minimum required depth d of the beam.

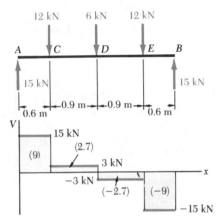

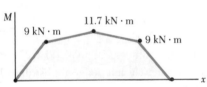

Maximum Shear and Bending Moment. After drawing the shear and bending-moment diagrams, we note that

$$M_{max} = 11.7 \text{ kN} \cdot \text{m}$$
$$V_{max} = 15 \text{ kN}$$

Design Based on Allowable Normal Stress. We first express the elastic section modulus S in terms of the depth d. We have

$$I = \frac{1}{12} bd^3 \qquad S = \frac{I}{c} = \frac{1}{6} bd^2 = \frac{1}{6}(0.090 \text{ m})d^2 = 0.015d^2$$

For $M_{max} = 11.7$ kN · m and $\sigma_{all} = 12$ MPa, we write

$$S = \frac{M_{max}}{\sigma_{all}} \qquad 0.015d^2 = \frac{11.7 \text{ kN} \cdot \text{m}}{12 \text{ MPa}}$$
$$d^2 = 65.0 \times 10^{-3}$$
$$d = 0.255 \text{ m} = 255 \text{ mm}$$

We have satisfied the requirement that $\sigma_{all} \leq 12$ MPa.
  Check Shearing Stress. For $V_{max} = 15$ kN and $d = 255$ mm, we find

$$\tau_m = \frac{3}{2} \frac{V_{max}}{A} = \frac{3}{2} \frac{15 \text{ kN}}{(0.090 \text{ m})(0.255 \text{ m})} \qquad \tau_m = 980 \text{ kPa}$$

Since $\tau_{all} = 850$ kPa, the depth $d = 255$ mm is *not* acceptable and we must redesign the beam on the basis of the requirement that $\tau_m \leq 850$ kPa.

Design Based on Allowable Shearing Stress. Since we now know that the allowable shearing stress controls the design, we write

$$\tau_m = \tau_{all} = \frac{3}{2} \frac{V_{max}}{A} \qquad 850 \text{ kPa} = \frac{3}{2} \frac{15 \text{ kN}}{(0.090 \text{ m})d}$$
$$d = 0.294 \text{ m} \qquad d = 294 \text{ mm} \blacktriangleleft$$

The normal stress is, of course, less than $\sigma_{all} = 12$ MPa and the depth of 294 mm is fully acceptable.

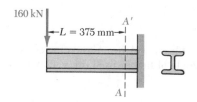

160 kN

$-L = 375$ mm$-$

A'

A

A 160-kN force is applied as shown at the end of a W 200 × 52 rolled-steel beam. Neglecting the effect of fillets and of stress concentrations, determine whether the normal stresses in the beam satisfy a design specification that they be equal to or less than 125 MPa at section A-A'.

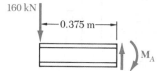

160 kN

-0.375 m$-$

M_A

V_A

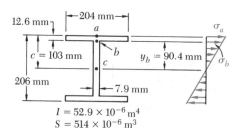

12.6 mm
-204 mm$-$
a
b
$c = 103$ mm
c
$y_b = 90.4$ mm
206 mm
-7.9 mm
σ_a
σ_b

$I = 52.9 \times 10^{-6}$ m^4
$S = 514 \times 10^{-6}$ m^3

12.6 mm
-204 mm$-$
a
b
103 mm
96.7 mm
c

Shear and Bending Moment. At section A-A', we have

$$M_A = (160 \text{ kN})(0.375 \text{ m}) = 60 \text{ kN} \cdot \text{m}$$
$$V_A = 160 \text{ kN}$$

Normal Stresses on Transverse Plane. Referring to the table of *Properties of Rolled-Steel Shapes* in Appendix C, we obtain the data shown and then determine the stresses σ_a and σ_b.

At point a:

$$\sigma_a = \frac{M_A}{S} = \frac{60 \text{ kN} \cdot \text{m}}{514 \times 10^{-6} \text{ m}^3} = 116.7 \text{ MPa}$$

At point b:

$$\sigma_b = \sigma_a \frac{y_b}{c} = (116.7 \text{ MPa}) \frac{90.4 \text{ mm}}{103 \text{ mm}} = 102.4 \text{ MPa}$$

We note that all normal stresses on the transverse plane are less than 125 MPa.

Shearing Stresses on Transverse Plane

At point a:

$$Q = 0 \qquad \tau_a = 0$$

At point b:

$$Q = (204 \times 12.6)(96.7) = 249 \times 10^3 \text{ mm}^3 = 249 \times 10^{-6} \text{ m}^3$$
$$\tau_b = \frac{V_A Q}{It} = \frac{(160 \text{ kN})(249 \times 10^{-6} \text{ m}^3)}{(52.9 \times 10^{-6} \text{ m}^4)(0.0079 \text{ m})} = 95.3 \text{ MPa}$$

Principal Stress at Point b. The state of stress at point b consists of the normal stress $\sigma_b = 102.4$ MPa and the shearing stress $\tau_b = 95.3$ MPa. We draw Mohr's circle and find

$$\sigma_{\max} = \frac{1}{2}\sigma_b + R = \frac{1}{2}\sigma_b + \sqrt{\left(\frac{1}{2}\sigma_b\right)^2 + \tau_b^2}$$
$$= \frac{102.4}{2} + \sqrt{\left(\frac{102.4}{2}\right)^2 + (95.3)^2}$$
$$\sigma_{\max} = 159.4 \text{ MPa}$$

The specification, $\sigma_{\max} \le 125$ MPa, is *not* satisfied ◄

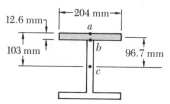

τ_b
σ_b
τ
$\sigma_{\max}$
Y
$\sigma_{\min}$
$\sigma_{\max}$
A
O
C
B
σ
R
τ_b
$\frac{\sigma_b}{2}$
X
σ_b

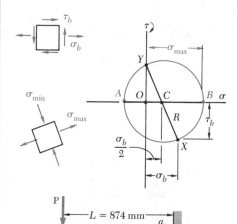

P
$-L = 874$ mm$-$
a
W 200 × 52
b
c

Comment. For this beam and loading, the principal stress at point b is 36% larger than the normal stress at point a. If the length L were longer than 874 mm, the maximum normal stress would occur at point a.

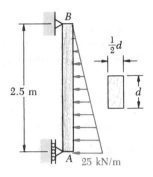

Fig. P7.89

PROBLEMS

7.89 through 7.92 For the beam and loading shown, design the cross section of the beam, knowing that for the grade of timber used $\sigma_{all} = 12$ MPa and $\tau_{all} = 825$ kPa.

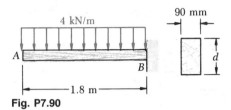

Fig. P7.90

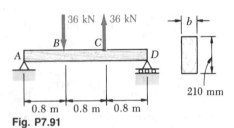

Fig. P7.91

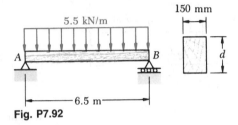

Fig. P7.92

7.93 through 7.96 For the beam and loading shown, design the cross section of the beam, knowing that for the grade of timber used $\sigma_{all} = 12$ MPa and $\tau_{all} = 0.9$ MPa

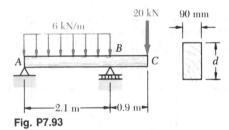

Fig. P7.93

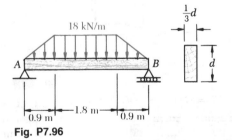

Fig. P7.94

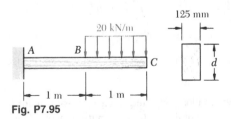

Fig. P7.95

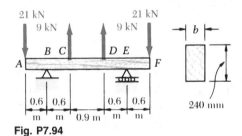

Fig. P7.96

7.97 through 7.100 Knowing that $\sigma_{all} = 160$ MPa and $\tau_{all} = 100$ MPa, select the metric wide-flange shape which should be used to support the loading shown.

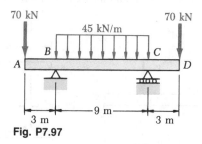

Fig. P7.97

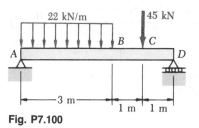

Fig. P7.98

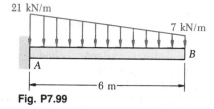

Fig. P7.99

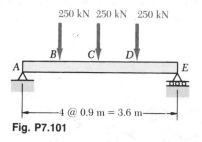

Fig. P7.100

7.101 through 7.104 Knowing that $\sigma_{all} = 165$ MPa and $\tau_{all} = 100$ MPa, select the wide-flange shape which should be used to support the loading shown.

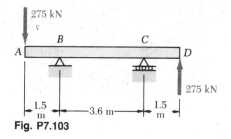

Fig. P7.101

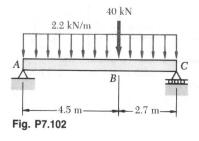

Fig. P7.102

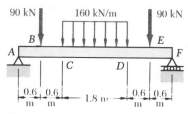

Fig. P7.103

Fig. P7.104

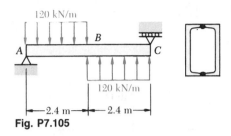

Fig. P7.105

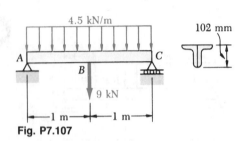

Fig. P7.107

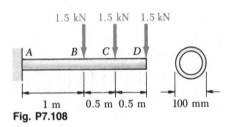

Fig. P7.108

7.105 Two rolled-steel channels are to be welded together and used to support the loading shown. Knowing that $\sigma_{all} = 140$ MPa and $\tau_{all} = 90$ MPa, determine the lightest channels which may be used.

7.106 Three plates are welded together to form the beam shown. Knowing that $\sigma_{all} = 150$ MPa, determine the minimum flange width b which may be used.

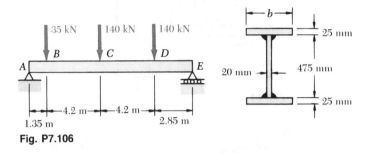

Fig. P7.106

7.107 Two L 102×76 rolled-steel angles are bolted together to support the loading shown. Knowing that $\sigma_{all} = 140$ MPa and $\tau_{all} = 85$ MPa, determine the minimum angle thickness which may be used.

7.108 A steel pipe must support the loading shown. Knowing that the stock of pipes available have wall thicknesses in increments of 3 mm from 6 mm to 24 mm and that $\sigma_{all} = 150$ MPa and $\tau_{all} = 90$ MPa, determine the minimum wall thickness t which may be used.

7.109 The four 4.5-m beams carry a load of 21.5 kN/m and are supported as shown by the 15-m girder AD and by the concrete walls. Knowing that $\sigma_{all} = 140$ MPa and $\tau_{all} = 90$ MPa, select the wide-flange shape which should be used (a) for the beams, (b) for the girder AD.

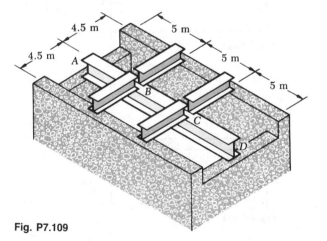

Fig. P7.109

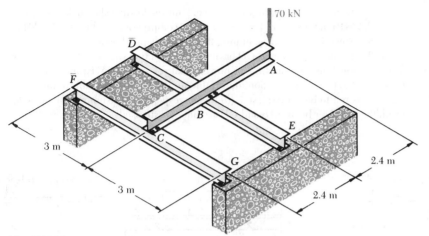

Fig. P7.110

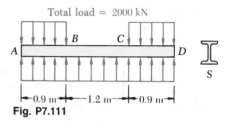

Total load = 2000 kN

Fig. P7.111

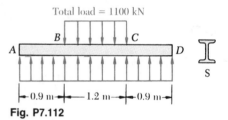

Total load = 1100 kN

Fig. P7.112

7.110 Beam *ABC* is bolted to beams *DBE* and *FCG*. Knowing that σ_{all} = 165 MPa and τ_{all} = 100 MPa, select the wide-flange shape which should be used (*a*) for beam *ABC*, (*b*) for beam *DBE*, (*c*) for beam *FCG*.

7.111 and 7.112 Knowing that σ_{all} = 150 MPa and τ_{all} = 100 MPa, select the lightest S-beam which may be used to support the given distributed loading while resting on the ground.

7.113 The 100-mm-wide bar supports an 11-kN load and rests on a foundation which exerts a linearly distributed load as shown. (*a*) Determine the values of w_A and w_B corresponding to equilibrium. (*b*) Knowing that σ_{all} = 170 MPa, determine the minimum thickness *t* which may be used.

7.114 Solve Prob. 7.113, assuming that the distance *a* is changed to 0.55 m.

7.115 and 7.116 Determine the allowable uniformly distributed load *w* for the beam shown, knowing that σ_{all} = +70 MPa in tension, σ_{all} = −130 MPa in compression, and τ_{all} = 60 MPa.

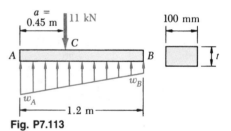

Fig. P7.113

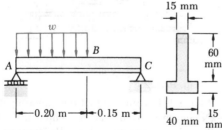

Fig. P7.115

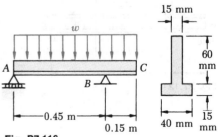

Fig. P7.116

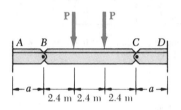

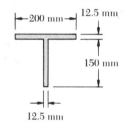

Fig. P7.117

7.117 Determine the allowable value of **P** for the loading shown, knowing that $\sigma_{all} = +55$ MPa in tension, $\sigma_{all} = -125$ MPa in compression, and $\tau_{all} = 27.5$ MPa.

7.118 Solve Prob. 7.117, assuming that the T-shaped beam is inverted.

7.119 Beams *AB*, *BC*, and *CD* have the cross section shown and are joined by pinned connections at *B* and *C*. Knowing that $\sigma_{all} = +150$ MPa in tension and $\sigma_{all} = -110$ MPa in compression, determine (*a*) the largest allowable value of **P** if beam *BC* is not to be overstressed, (*b*) the corresponding maximum distance *a* for which the cantilver beams *AB* and *CD* are not overstressed.

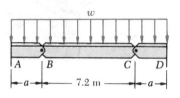

Fig. P7.119

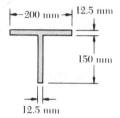

Fig. P7.120

7.120 Beams *AB*, *BC*, and *CD* have the cross section shown and are joined by pinned connections at *B* and *C*. Knowing that $\sigma_{all} = +150$ MPa in tension and $\sigma_{all} = -110$ MPa in compression, determine (*a*) the largest allowable value of *w* if beam *BC* is not to be overstressed, (*b*) the corresponding maximum distance *a* for which the cantilever beams *AB* and *CD* are not overstressed.

7.121 A uniformly distributed load of 66 kN/m must be supported on the beams *AB* and *CD* as shown. Knowing that $\sigma_{all} = 140$ MPa and $\tau_{all} = 85$ MPa, determine (*a*) the smallest allowable length *l* of beam *CD* if the W 460 × 74 rolled-steel shape is not to be overstressed, (*b*) the rolled-steel shape which should be used for beam *CD*. Neglect the weight of both beams.

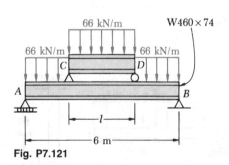

Fig. P7.121

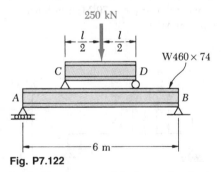

Fig. P7.122

7.122 Solve Prob. 7.121, assuming that the uniformly distributed load is replaced by a 250-kN concentrated load at the center of the 6-m span.

7.123 Solve Prob. 7.121, assuming that a W 530 × 66 rolled-steel shape is used for beam *AB*.

7.124 and 7.125 For the beam and loading shown, determine the width b of the beam and the distance x for which the maximum normal stress σ_m and the maximum shearing stress τ_m in the beam are simultaneously equal to

$$\sigma_{\text{all}} = 11 \text{ MPa} \qquad \tau_{\text{all}} = 0.75 \text{ MPa}$$

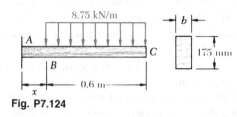

Fig. P7.124

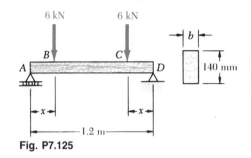

Fig. P7.125

7.126 Using Eqs. (7.16) and (7.17), show that at any given point of the beam of Fig. 7.24, (a) the directions of the principal axes of stress are defined by the relation

$$\tan 2\theta_p = \frac{c^2 - y^2}{xy}$$

(b) the maximum value of the normal stress is

$$\sigma_{\text{max}} = \frac{3}{2} \frac{P}{A} \frac{xy}{c^2} \left[1 + \sqrt{1 + \left(\frac{c^2 - y^2}{xy} \right)^2} \right]$$

*7.127 Using the equation derived in part b of Prob. 7.126, show that, in any given transverse section of the beam of Fig. 7.24 for which $x \geq 0.5444c$, the largest value of σ_{max} occurs for $y = c$. (*Hint:* Denoting by σ_m the value of σ_{max} coresponding to $y = c$, and setting $x/c = u$ and $y/c = v$, write an equation defining the values of v for which $\sigma_{\text{max}} = \sigma_m$, and show that, for $u \geq 0.5444$, the only real root of this equation is $v = 1$.)

7.128 For the beam and loading in Sample Prob. 7.11, determine the maximum shearing stress at points a, b, and c.

7.129 For the beam and loading shown, determine the maximum shearing stress at points a, b, and c.

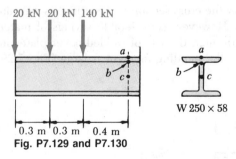

Fig. P7.129 and P7.130

7.130 For the beam and loading shown, determine the maximum normal stress at points a, b, and c.

*7.8. BEAMS OF CONSTANT STRENGTH

Our analysis has been limited so far to prismatic beams, i.e., to beams with a constant cross section. As we saw in the preceding section, prismatic beams are designed so that the stresses in the critical sections are at most equal to the allowable values of the normal and shearing stresses. It follows that, in all other sections, the stresses will be smaller, possibly much smaller, than the allowable stresses. A prismatic beam, therefore, is almost always overdesigned, and considerable savings of material may be realized by using nonprismatic beams, i.e., beams of variable cross section.

Since the maximum normal stresses σ_m usually control the design of a beam, the design of a nonprismatic beam will be optimum if the modulus $S = I/c$ of every cross section satisfies the equation $\sigma_m = M/S$, where σ_m is equal to the allowable stress. Solving for S, we write

$$S = \frac{M}{\sigma_{\text{all}}} \tag{7.22}$$

where M is the bending moment in the given section. A beam designed in this manner is referred to as a *beam of constant strength*.

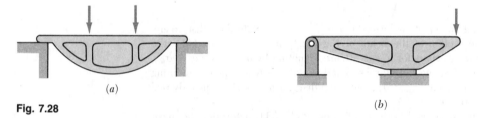

(a)

(b)

Fig. 7.28

For a forged or cast machine component, it is not difficult to vary the cross section of the component along its length and to eliminate most of the unnecessary material (Fig. 7.28). For a structural member, it is impractical to vary the cross section in a continuous way along the length of the member. However, considerable savings of material may still be accomplished through the use of welded cover plates in portions of the member where the bending moment is large (Fig. 7.29).

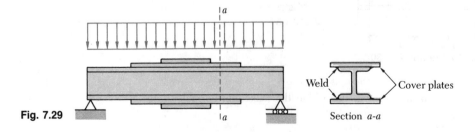

Fig. 7.29

Weld

Cover plates

Section *a-a*

Example 7.07

A simply supported beam *AB* of span *L* is to be constructed by gluing together plywood strips of the same constant width *b* and of variable length *l* (Fig. 7.30). If the beam is designed to be of constant strength when supporting a single concentrated load at its midpoint, express the length *l* of a strip in terms of its distance *h* from the flat surface, the span *L*, and the depth h_0 of a beam of constant cross section designed to support the same load.

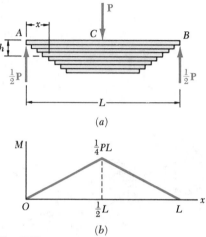

(a)

(b)

Fig. 7.31

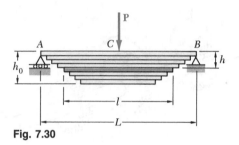

Fig. 7.30

For $x \leq \frac{1}{2}L$, the bending moment (Fig. 7.31) is

$$M = \tfrac{1}{2}Px \qquad (7.23)$$

On the other hand, the section modulus for a rectangular section (Fig. 7.32) is

$$S = \frac{I}{c} = \frac{bh^3/12}{h/2} = \frac{bh^2}{6} \qquad (7.24)$$

Substituting from (7.23) and (7.24) into (7.22) we write

$$\frac{bh^2}{6} = \frac{\tfrac{1}{2}Px}{\sigma_{\text{all}}} \qquad (7.25)$$

This relation must also be satisfied for $x = L/2$ by a beam of constant cross section $h = h_0$:

$$\frac{bh_0^2}{6} = \frac{(\tfrac{1}{2}P)(\tfrac{1}{2}L)}{\sigma_{\text{all}}} \qquad (7.26)$$

Dividing (7.25) by (7.26), member by member, we write

$$\frac{h^2}{h_0^2} = \frac{2x}{L} \qquad (7.27)$$

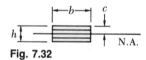

Fig. 7.32

The length *l* of a strip located at a distance *h* from the surface may be expressed as

$$l = L - 2x = L\left(1 - \frac{2x}{L}\right) \qquad (7.28)$$

Substituting for $2x/L$ from (7.27) into (7.28), we have

$$l = L\left(1 - \frac{h^2}{h_0^2}\right) \qquad (7.29)$$

We should note that the result obtained does not take into account the resistance of the beam to shear; the actual design should provide for several of the upper strips to extend over the entire span.

Gluing the plywood strips together is of critical importance, since horizontal shearing stresses are required if the beam is to act as a single unit rather then as many separate beams (cf. Chap. 5). Thus, there is only a superficial resemblance between the glued beam of this example and a typical leaf spring consisting of several independent leaves.

*7.9. DESIGN OF TRANSMISSION SHAFTS

When we discussed the design of transmission shafts in Sec. 3.7, we considered only the stresses due to the torques exerted on the shafts. However, if the power is transferred to and from the shaft by means of gears or sprocket wheels (Fig. 7.33a), the forces exerted on the gear teeth or sprockets are equivalent to force-couple systems applied at the centers of the corresponding cross sections (Fig. 7.33b). This means that the shaft is subjected to a transverse loading, as well as to a torsional loading.

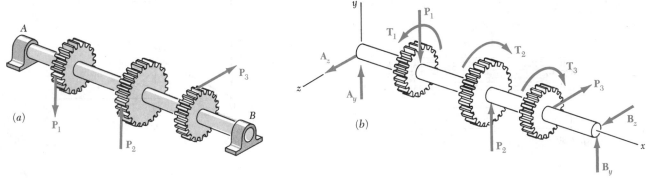

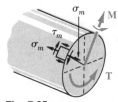

Fig. 7.33

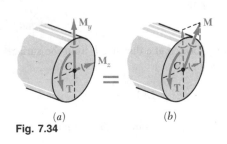

Fig. 7.34

The shearing stresses produced in the shaft by the transverse loads are usually much smaller than those produced by the torques and will be neglected in this analysis.† The normal stresses due to the transverse loads, however, may be quite large and, as we shall see presently, their contribution to the maximum shearing stress τ_{max} should be taken into account.

Consider the cross section of the shaft at some point C. We represent the torque $\mathbf{T}$ and the bending couples $\mathbf{M}_y$ and $\mathbf{M}_z$ acting respectively in a horizontal and a vertical plane by the couple vectors shown (Fig. 7.34a). Since any diameter of the section is a principal axis of inertia for the section, we may replace $\mathbf{M}_y$ and $\mathbf{M}_z$ by their resultant $\mathbf{M}$ (Fig. 7.34b) in order to compute the normal stresses σ_x exerted on the section. We thus find that σ_x is maximum at the end of the diameter perpendicular to the vector representing $\mathbf{M}$ (Fig. 7.35). Recalling that the values of the normal stresses at that point are, respectively, $\sigma_m = Mc/I$ and zero, while the shearing stress is $\tau_m = Tc/J$, we plot the corresponding points X and Y on a Mohr-circle diagram (Fig. 7.36) and determine the value of the maximum shearing stress:

Fig. 7.35

$$\tau_{max} = R = \sqrt{\left(\frac{\sigma_m}{2}\right)^2 + (\tau_m)^2} = \sqrt{\left(\frac{Mc}{2I}\right)^2 + \left(\frac{Tc}{J}\right)^2}$$

† For an application where the shearing stresses produced by the transverse loads must be considered, see Probs. 7.165 through 7.167.

Recalling that, for a circular or annular cross section, $2I = J$, we write

$$\tau_{max} = \frac{c}{J}\sqrt{M^2 + T^2} \qquad (7.30)$$

It follows that the minimum allowable value of the ratio J/c for the cross section of the shaft is

$$\frac{J}{c} = \frac{(\sqrt{M^2 + T^2}\,)_{max}}{\tau_{all}} \qquad (7.31)$$

where the numerator in the right-hand member of the expression obtained represents the maximum value of $\sqrt{M^2 + T^2}$ in the shaft, and τ_{all} the allowable shearing stress. Expressing the bending moment M in terms of its components in the two coordinate planes, we may also write

$$\frac{J}{c} = \frac{(\sqrt{M_y^2 + M_z^2 + T^2}\,)_{max}}{\tau_{all}} \qquad (7.32)$$

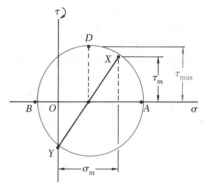

Fig. 7.36

Equations (7.31) and (7.32) may be used to design both solid and hollow circular shafts and should be compared with Eq. (3.22) of Sec. 3.7, which was obtained under the assumption of a torsional loading only.

The determination of the maximum value of $\sqrt{M_y^2 + M_z^2 + T^2}$ will be facilitated if the bending-moment diagrams corresponding to M_y and M_z are drawn, as well as a third diagram representing the values of T along the shaft (see Sample Prob. 7.12).

*7.10. STRESSES UNDER APPLIED LOADS

When we analyzed in Chap. 5 the stresses exerted on a small element of material in a beam subjected to a transverse loading, we assumed that the element was not located directly under an applied load. In fact, to apply Saint-Venant's principle, we had to assume that the element considered was not located close to a load or to a support (Sec. 5.7). Under these conditions, we found that the stresses consisted of normal stresses $\sigma_x = -My/I$ exerted on the faces of the element perpendicular to the axis of the beam, and of shearing stresses τ_{xy} of average value $\tau_{ave} = VQ/It$ (Fig. 7.37).

As we indicated in Sec. 5.7, these expressions may still be used to determine the values of the stresses σ_x and τ_{xy} exerted on an element located under a distributed load, since the error involved is small for the values of the span-depth ratio encountered in practice. However, it is clear that an element located under a distributed load will also be subjected to normal stresses σ_y on its horizontal faces (Fig. 7.38).

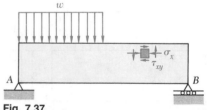

Fig. 7.37

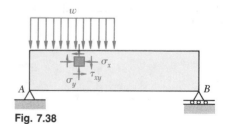

Fig. 7.38

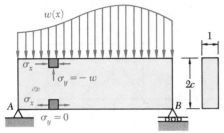

Fig. 7.39

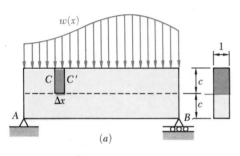

Fig. 7.40

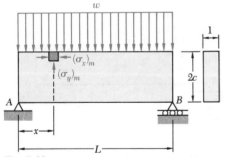

Fig. 7.41

In the case of a rectangular beam of unit width and depth $2c$ subjected to a distributed load $w(x)$, the normal stresses σ_y vary from $\sigma_y = -w$ at the upper surface of the beam to $\sigma_y = 0$ at the lower surface (Fig. 7.39). The value of σ_y at the neutral surface may be determined by considering the equilibrium of the portion of beam located above the neutral surface and bounded by transverse planes through points C and C' (Fig. 7.40a). Considering only the vertical forces, and noting that the shearing forces exerted on the vertical faces of the free body are respectively equal to half of the shear at C and C', we write (Fig. 7.40b).

$$+\downarrow \Sigma F_y = 0: \qquad w\,\Delta x + \sigma_y\,\Delta x - \frac{1}{2}V + \frac{1}{2}(V + \Delta V) = 0$$

$$\sigma_y = -w - \frac{1}{2}\frac{\Delta V}{\Delta x}$$

But the limit of $\Delta V/\Delta x$ as Δx approaches zero represents the derivative dV/dx of the shear, and we recall from Sec. 7.4 that $dV/dx = -w$. We thus conclude that, at the neutral surface, $\sigma_y = -\frac{1}{2}w$.[†]

In most engineering applications, the vertical stress σ_y is small, compared to the longitudinal stress σ_x, and contributes very little to the value of the maximum stress $\sigma_{\max}$ at any given point. For example, in the case of a rectangular beam AB of length L, unit width, and depth $2c$, supporting a uniformly distributed load w (Fig. 7.41), the largest value of σ_y is w, as observed earlier, while the largest value of σ_x at a distance x from end A is

$$(\sigma_x)_m = \frac{Mc}{I} = \frac{\frac{1}{2}wx(L-x)c}{\frac{2}{3}c^3} = \frac{3}{4}\frac{w}{c^2}x(L-x)$$

The ratio of the maximum values of the two normal stresses is thus

$$\frac{(\sigma_y)_m}{(\sigma_x)_m} = \frac{4}{3}\frac{c^2}{x(L-x)} \qquad (7.34)$$

For a span-depth ratio of 10, that is, for $L/c = 20$, one finds that $(\sigma_y)_m/(\sigma_x)_m$ is equal to 0.0370 at $x = 0.10L$, to 0.0178 at $x = 0.25L$, and to 0.0133 at midspan.

[†] The normal stress σ_y does *not* vary linearly with the distance y from the neutral surface. Considering the equilibrium of the portion of beam located between the transverse planes through C and C', and above the horizontal plane of ordinate y, we could show (see Prob. 7.177) that

$$\sigma_y = \frac{1}{4}w\left(\frac{y^3}{c^3} - 3\frac{y}{c} - 2\right) \qquad (7.33)$$

In the case of a concentrated load **P** (Fig. 7.42), the formulas obtained in Chap. 5 for σ_x and τ_{xy} cease to be valid in the neighborhood of the point of application of the load. On the other hand, it is clear that an element of material located directly under the load will be subjected to a very large normal stress σ_y. However, in practice, a load is always distributed over a small area, and bearing plates are often used to spread a large load over a portion of the beam (Fig. 7.43). Moreover, yield will take place in ductile materials under very large concentrated loads and will also result in spreading the load over a wider area, thus reducing the value of σ_y.

The stress concentrations caused by a load **P** are large close to the surface of the beam, but decrease rapidly as one gets deeper into the beam. For example, in the theoretical and extreme case of a load **P** concentrated over an infinitesimal portion of the surface of a rectangular beam of unit width and depth $2c$, the mathematical theory of elasticity shows that the normal stress at a point of the neutral surface located directly under the load is $\sigma_y = -0.456P/c$, which is almost equal to the value that would be obtained under a uniformly distributed load $w = P/c$ (Fig. 7.44). We also note from Fig. 7.44 that the longitudinal normal stress σ_x is not zero at that surface $(y = 0)$, and thus verify that the formula $\sigma_x = -My/I$ does not apply under a concentrated load. Finally, we observe that, while the shear V is discontinuous at the point of application of the concentrated load **P**, the variation of the shearing stress along the neutral surface is represented by a smooth curve, and that τ_{xy} is equal to zero directly under the load.†

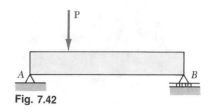

Fig. 7.42

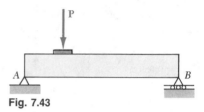

Fig. 7.43

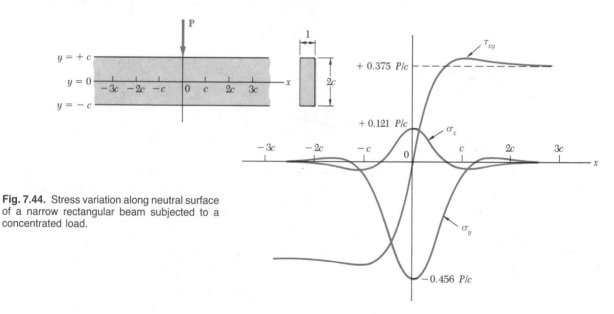

Fig. 7.44. Stress variation along neutral surface of a narrow rectangular beam subjected to a concentrated load.

†See S. P. Timoshenko and J. N. Goodier, *Theory of Elasticity*, 3d ed., McGraw-Hill Book Company, 1970, sec. 40.

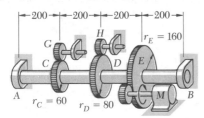

├─200─┼─200─┼─200─┼─200─┤

Dimensions in mm

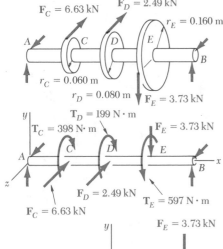

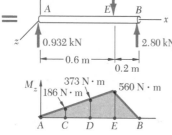

SAMPLE PROBLEM 7.12

The solid shaft AB rotates at 480 rpm and transmits 30 kW from the motor M to machine tools connected to gears G and H; 20 kW is taken off at gear G and 10 kW at gear H. Knowing that $\tau_{all} = 50$ MPa, determine the smallest permissible diameter for shaft AB.

Solution. Noting that $f = 480$ rpm $= 8$ Hz, we determine the torque exerted on gear E:

$$T_E = \frac{P}{2\pi f} = \frac{30 \text{ kW}}{2\pi(8 \text{ Hz})} = 597 \text{ N} \cdot \text{m}$$

The corresponding tangential force acting on the gear is

$$F_E = \frac{T_E}{r_E} = \frac{597 \text{ N} \cdot \text{m}}{0.16 \text{ m}} = 3.73 \text{ kN}$$

A similar analysis of gears C and D yields

$$T_C = \frac{20 \text{ kW}}{2\pi(8 \text{ Hz})} = 398 \text{ N} \cdot \text{m} \qquad F_C = 6.63 \text{ kN}$$

$$T_D = \frac{10 \text{ kW}}{2\pi(8 \text{ Hz})} = 199 \text{ N} \cdot \text{m} \qquad F_D = 2.49 \text{ kN}$$

We now replace the forces on the gears by equivalent force-couple systems.

Bending-Moment and Torque Diagrams

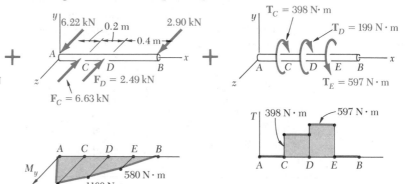

Critical Transverse Section. By computing $\sqrt{M_y^2 + M_z^2 + T^2}$ at all potentially critical sections, we find that its maximum value occurs just to the right of D:

$$\left(\sqrt{M_y^2 + M_z^2 + T^2}\right)_{\max} = \sqrt{(1160)^2 + (373)^2 + (597)^2} = 1357 \text{ N} \cdot \text{m}$$

Diameter of Shaft. For $\tau_{all} = 50$ MPa, Eq. (7.32) yields

$$\frac{J}{c} = \frac{\left(\sqrt{M_y^2 + M_z^2 + T^2}\right)_{\max}}{\tau_{all}} = \frac{1357 \text{ N} \cdot \text{m}}{50 \text{ MPa}} = 27.14 \times 10^{-6} \text{ m}^3$$

For a solid circular shaft of radius c, we have

$$\frac{J}{c} = \frac{\pi}{2} c^3 = 27.14 \times 10^{-6} \qquad c = 0.02585 \text{ m} = 25.85 \text{ mm}$$

Diameter $= 2c = 51.7$ mm ◄

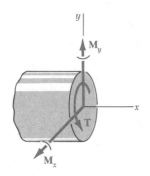

PROBLEMS

7.131 through 7.136 Beam AB has a rectangular cross section of constant width b and variable depth h. Knowing that the beam is to be of constant strength, express h in terms of x, L, and h_0.

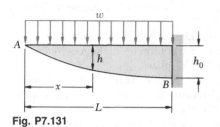

Fig. P7.131

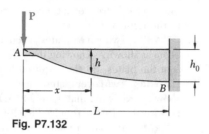

Fig. P7.132

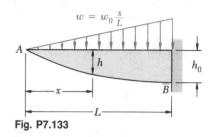

Fig. P7.133

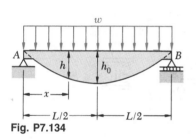

Fig. P7.134

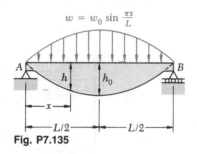

Fig. P7.135

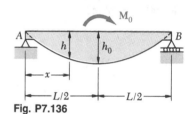

Fig. P7.136

7.137 Express the length l of a plywood strip in Example 7.07 in terms of h, L, and h_0, assuming that the beam is to be of constant strength when supporting a uniformly distributed load.

7.138 The cantilever beam AB is of constant depth h and of variable width b. Knowing that the beam is to be of constant strength, express b in terms of b_0, x, and L if the only load on the beam is (*a*) a vertical force at A, (*b*) a uniformly distributed load along line AB.

7.139 and 7.140 Beam AB is a solid of revolution of variable diameter d. For portion AC of the beam express d in terms of d_0, x, and L, knowing that the beam is to be of constant strength when supporting the loading shown.

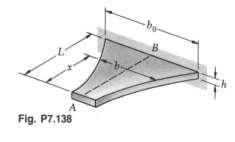

Fig. P7.138

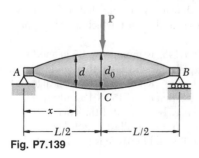

Fig. P7.139

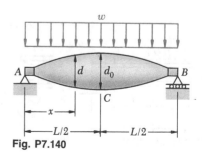

Fig. P7.140

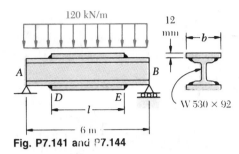

Fig. P7.141 and P7.144

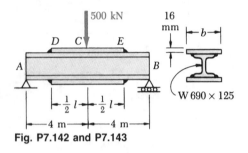

Fig. P7.142 and P7.143

7.141 Two cover plates, each 12 mm thick, are welded to a W 530 × 92 beam as shown. Knowing that $l = 4.2$ m and $b = 250$ mm, determine the maximum normal stress on a tranverse section (a) at the center of the beam, (b) just to the left of D.

7.142 Two cover plates, each 16 mm thick, are welded to a W 690 × 125 beam as shown. Knowing that $l = 4.4$ m and $b = 300$ mm, determine the maximum normal stress on a transverse section (a) at the center C of the beam, (b) just to the right of E.

7.143 Two cover plates, each 16 mm thick, are welded to a W 690 × 125 beam as shown. Knowing that $\sigma_{all} = 160$ MPa, determine (a) the required width b of the plates, (b) the required length l of the plates.

7.144 Two cover plates, each 12 mm thick, are welded to a W 530 × 92 beam as shown. Knowing that $\sigma_{all} = 140$ MPa, determine (a) the required width b of the cover plates, (b) the required length l of the plates.

7.145 Two cover plates, each 18 mm thick and 220 mm wide, are welded to a W 410 × 85 beam as shown. Knowing that $P = 175$ kN, determine the maximum normal stress on a transverse section (a) just to the left of B, (b) at C, (c) just to the right of D.

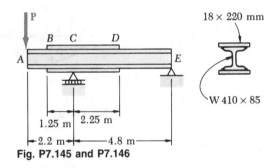

Fig. P7.145 and P7.146

7.146 Two cover plates, each 18 mm thick and 220 mm wide, are welded to a W 410 × 85 beam as shown. Knowing that $\sigma_{all} = 150$ MPa and $\tau_{all} = 90$ MPa, determine the largest vertical force **P** which may be applied at point A of the beam shown.

7.147 Solve Prob. 7.146, assuming that portion CD of the cover plates is 2.5 m long.

7.148 Two cover plates, each 12 mm thick and 250 mm wide, are welded to a W 360 × 79 beam as shown. Knowing that $\sigma_{all} = 150$ MPa and $\tau_{all} = 90$ MPa, determine the largest vertical force **P** which may be applied at point D when $a = 1.8$ m and $l = 3$ m.

7.149 Solve Prob. 7.148, assuming $a = 1.5$ m and $l = 3$ m.

7.150 Two cover plates, each 12 mm thick and 250 mm wide, are welded to a W 360 × 79 beam as shown. Knowing that a vertical force **P** may be applied at any point along the beam and that $\sigma_{all} = 150$ MPa and $\tau_{all} = 90$ MPa, determine (a) the largest force **P** which may be applied if the cover plates extend over the entire length AF of the beam, (b) the smallest length l of the cover plates for which the magnitude of the force **P** found in part a is not decreased.

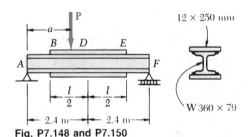

Fig. P7.148 and P7.150

7.151 and 7.152 For the beam and loading shown, determine (*a*) the transverse section at which the maximum normal stress occurs, (*b*) the corresponding maximum normal stress. (In Prob. 7.151, assume $h_1 > 2h_0$.)

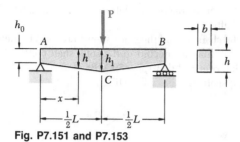

Fig. P7.151 and P7.153

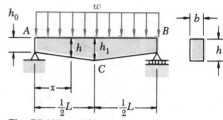

Fig. P7.152 and P7.154

7.153 For the beam and loading shown and using the given data, determine (*a*) the transverse section at which the maximum stress occurs, (*b*) the largest force **P** which may be applied. *Data:* $\sigma_{all} = 100$ MPa, $L = 1.5$ m, $h_0 = 40$ mm, $h_1 = 100$ mm, $b = 22$ mm.

7.154 For the beam and loading shown and using the given data, determine (*a*) the transverse section at which the maximum stress occurs, (*b*) the largest distributed load *w* which may be applied. *Data:* $\sigma_{all} = 140$ MPa, $L = 1.2$ m, $h_0 = 50$ mm, $h_1 = 80$ mm, $b = 40$ mm.

7.155 The 5-kN force is vertical, the force **Q** is horizontal, and both forces act in planes perpendicular to the solid shaft *AD*. Knowing that $\tau_{all} = 45$ MPa, determine the smallest permissible diameter for shaft *AD*.

7.156 The gear-and-shaft system shown is subjected to a 340 N · m torque $\mathbf{T}_A$ while pulley *D* is held fixed. Knowing that $\tau_{all} = 70$ MPa, determine the smallest permissible diameter of the solid shaft *AB*.

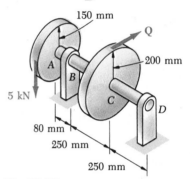

Fig. P7.155

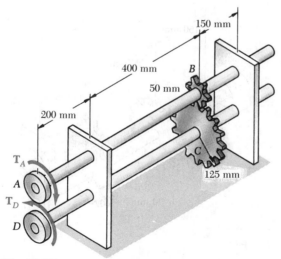

Fig. P7.156

7.157 For the gear-and-shaft system and loading of Prob. 7.156, determine the smallest permissible diameter of the solid shaft *CD*.

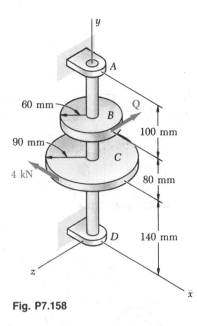

Fig. P7.158

7.158 The 4-kN force is parallel to the x axis, and the force **Q** is parallel to the z axis. Knowing that $\tau_{all} = 60$ MPa, determine the smallest permissible diameter of the solid shaft AD.

7.159 Assuming that the shaft of Prob. 7.158 is hollow and that the inner diameter is half the outer diameter, determine the smallest permissible outer diameter of the shaft.

7.160 The solid shaft AB and the gears shown are used to transmit 11 kW from the motor M to a machine connected to the solid shaft CDE. Knowing that the motor rotates at 360 rpm and that $\tau_{all} = 50$ MPa, determine the smallest permissible diameter of (a) shaft AB, (b) shaft CDE.

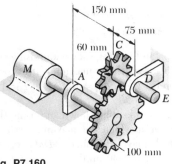

Fig. P7.160

7.161 The solid shaft AB rotates at 450 rpm and transmits 20 kW from the motor M to machine tools connected to gears F and G. Knowing that $\tau_{all} = 55$ MPa and assuming that 8 kW is taken off at gear F and 12 kW is taken off at gear G, determine the smallest permissible diameter of shaft AB.

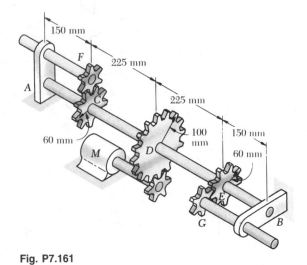

Fig. P7.161

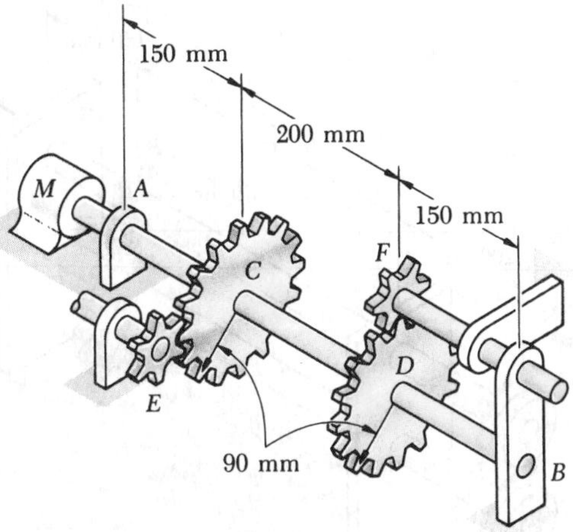

Fig. P7.162

7.162 The solid shaft AB rotates at 720 rpm and transmits 50 kW from the motor M to machine tools connected to gears E and F. Knowing that $\tau_{all} = 50$ MPa and assuming that 20 kW is taken off at gear E and 30 kW is taken off at gear F, determine the smallest permissible diameter of shaft AB.

7.163 Solve Prob. 7.162, assuming that 25 kW is taken off at each gear.

7.164 Solve Prob. 7.161, assuming that the entire 20 kW is taken off at gear G.

7.165 For the shaft AB and lever AC, show that the effect of the shearing stresses caused by the transverse load must be included in order to obtain the correct value of the maximum shearing stress when

$$\frac{L}{d} < \frac{1}{3}\sqrt{1 + \frac{6l}{d}}$$

7.166 Knowing that $\tau_{all} = 50$ MPa, determine the smallest permissible diameter of the solid shaft AB when $P = 665$ N, $l = 450$ mm, and $L = 75$ mm.   Include the effect of the shearing stresses caused by the transverse load.

7.167 Knowing that $\tau_{all} = 50$ MPa, determine the smallest permissible diameter of the solid shaft AB when $P = 2$ kN, $l = 300$ mm, and $L = 75$ mm.   Include the effect of the shearing stresses caused by the transverse load.

7.168 Neglecting the effect of fillets and of stress concentrations, determine the smallest permissible diameters of the solid rods BC and CD.   Use $\tau_{all} = 60$ MPa.

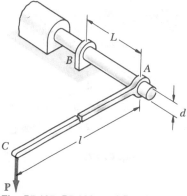

Fig. P7.165, P7.166, and P7.167

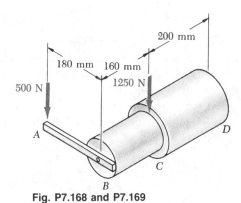

Fig. P7.168 and P7.169

7.169 Knowing that rods BC and CD are of diameter 24 mm and 36 mm, respectively, determine the maximum shearing stress in each rod.   Neglect the effect of fillets and of stress concentrations.

7.170 Knowing that the diameter of the conical taper shown is 40 mm at A and 60 mm at B, determine the maximum shearing stress (a) at A, (b) at B.

7.171 Using $\tau_{all} = 50$ MPa, design a conical taper for the loading shown by determining the required diameter (a) at A, (b) at B.

7.172 Using the notation of Sec. 7.9 and neglecting the effect of shearing stresses caused by transverse loads, show that the maximum normal stress in a circular shaft may be expressed as follows:

$$\sigma_{max} = \frac{c}{J}[(M_y^2 + M_z^2)^{1/2} + (M_y^2 + M_z^2 + T^2)^{1/2}]_{max}$$

7.173 Using the expression given in Prob. 7.172, determine (a) the transverse section of the shaft of Sample Prob. 7.12 at which the maximum normal stress occurs, (b) the value of σ_{max}, knowing that the diameter of the shaft is 51.7 mm.

7.174 For the rods and loading of Prob. 7.169 and using the expression given in Prob. 7.172, determine the maximum normal stress in (a) rod BC, (b) rod CD.

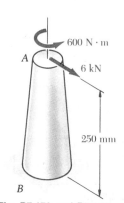

Fig. P7.170 and P7.171

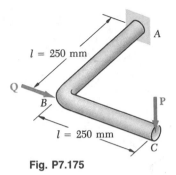

Fig. P7.175

7.175 An 18-mm-diameter rod has been bent into the shape shown. Knowing that $P = 200$ N and $Q = 0$, determine the maximum shearing stress and the maximum normal stress at end A. Neglect the effect of shearing stresses caused by the transverse load.

7.176 Solve Prob. 7.175, assuming that $P = Q = 200$ N.

*7.177 Derive Eq. (7.33), which defines the vertical stress σ_y in a rectangular beam of unit width and depth $2c$ subjected to a distributed load $w(x)$.

REVIEW AND SUMMARY

This chapter was devoted to the design of beams and shafts for strength. We learned to select the material and the cross section of a beam or shaft of a given span so that it would not fail under a given loading.

Considerations for the design of prismatic beams

We first discussed the basic criteria which must be satisfied in the design of *prismatic beams* [Sec. 7.2]. On one hand, the maximum normal stress σ_m in the beam must not exceed the allowable stress σ_{all} for the material and on the other, the maximum shearing stress τ_m must not exceed τ_{all}. We noted that σ_m is obtained by substituting for M in Eq. (4.15) the maximum value of the bending moment in the beam, and τ_m by substituting for V in Eq. (5.10) the maximum value of the shear:

$$\sigma_m = \frac{|M|_{\max}c}{I} \qquad \tau_m = \frac{|V|_{\max}Q}{It} \qquad (7.1, \ 7.2)$$

Thus, the maximum normal stress occurs in the section where $|M|$ is largest, at the point farthest from the neutral axis, while the maximum shearing stress occurs on the neutral axis in the section where $|V|$ is largest (Figs. 7.1 and 7.2).

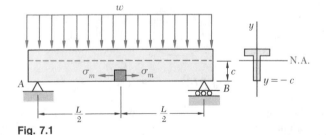

Fig. 7.1

Fig. 7.2

The determination of the maximum absolute values of the shear and bending moment and of the critical sections of the beam in which they occur is greatly simplified if we draw a *shear diagram* and a *bending-moment diagram*. These diagrams represent, respectively, the variation of the shear and of the bending moment along the beam and were obtained by determining the values of V and M at selected points of the beam [Sec. 7.3]. These values were found by passing a section through the point where they were to be determined and drawing the free-body diagram of either of the portions of beam obtained in this fashion. To avoid any confusion regarding the sense of the shearing force **V** and of the bending couple **M** (which act in opposite sense on the two portions of the beam), we followed the sign convention adopted earlier in the text and illustrated in Fig. 7.7a [Examples 7.01 and 7.02, Sample Probs. 7.1 and 7.2].

The construction of the shear and bending-moment diagrams is facilitated if the following relations are taken into account [Sec. 7.4]. Denoting by w the distributed load per unit length (assumed positive if directed downward), we wrote

$$\frac{dV}{dx} = -w \qquad \frac{dM}{dx} = V \qquad (7.3, 7.5)$$

or, in integrated form,

$$V_D - V_C = -(\text{area under load curve between } C \text{ and } D) \qquad (7.4')$$
$$M_D - M_C = \text{area under shear curve between } C \text{ and } D \qquad (7.6')$$

Equation (7.4') makes it possible to draw the shear diagram of a beam from the curve representing the distributed load on that beam and the value of V at one end of the beam. Similarly, Eq. (7.6') makes it possible to draw the bending-moment diagram from the shear diagram and the value of M at one end of the beam. However, concentrated loads introduce discontinuities in the shear diagram and concentrated couples in the bending-moment diagram, none of which is accounted for in these equations [Sample Probs. 7.3 and 7.6]. Finally, we noted from Eq. (7.5) that the points of the beam where the bending moment is maximum or minimum are also the points where the shear is zero [Sample Prob. 7.4].

In Sec. 7.5, we discussed another method for the determination of the maximum values of the shear and bending moment based on the use of the *singularity functions* $\langle x - a \rangle^n$. By definition, and for $n \geq 0$, we had

$$\langle x - a \rangle^n = \begin{cases} (x - a)^n & \text{when } x \geq a \\ 0 & \text{when } x < a \end{cases} \qquad (7.10)$$

We noted that whenever the quantity between brackets is positive or zero, the brackets should be replaced by ordinary parentheses, and whenever that quantity is negative, the bracket itself is equal to zero.

Shear and bending-moment diagrams

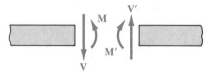

Fig. 7.7a. Positive shear and bending moment.

Relations among load, shear, and bending moment

Singularity functions

Step function

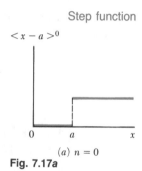

$\langle x - a \rangle^0$

(a) $n = 0$
Fig. 7.17a

Using singularity functions to express shear and bending moment

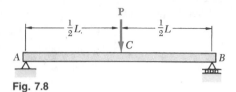

Fig. 7.8

Equivalent open-ended loadings

We also noted that singularity functions may be integrated and differentiated as ordinary binomials. Finally, we observed that the singularity function corresponding to $n = 0$ is discontinuous at $x = a$ (Fig. 7.17a). This function is called the *step function*. We wrote

$$\langle x - a \rangle^0 = \begin{cases} 1 & \text{when } x \geq a \\ 0 & \text{when } x < a \end{cases} \qquad (7.11)$$

The use of singularity functions makes it possible to represent the shear or the bending moment in a beam by a single expression, valid at any point of the beam. For example, the contribution to the shear of the concentrated load **P** applied at the midpoint C of a simply supported beam (Fig. 7.8) may be represented by $-P\langle x - \frac{1}{2}L \rangle^0$, since this expression is equal to zero to the left of C, and to $-P$ to the right of C. Adding the contribution of the reaction $R_A = \frac{1}{2}P$ at A, we express the shear at any point of the beam as

$$V(x) = \tfrac{1}{2}P - P\langle x - \tfrac{1}{2}L \rangle^0$$

The bending moment is obtained by integrating this expression:

$$M(x) = \tfrac{1}{2}Px - P\langle x - \tfrac{1}{2}L \rangle^1$$

The singularity functions representing, respectively, the load, shear, and bending moment corresponding to various basic loadings were given in Fig. 7.18 on page 435. We noted that a distributed loading which does not extend to the right end of the beam, or which is discontinuous, should be replaced by an equivalent combination of open-ended loadings. For instance, a uniformly distributed load extending from $x = a$ to $x = b$ (Fig. 7.19) should be expressed as

$$w(x) = w_0\langle x - a \rangle^0 - w_0\langle x - b \rangle^0$$

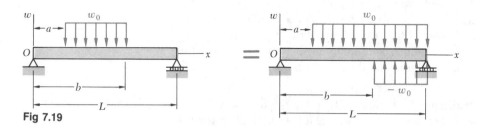

Fig 7.19

The contribution of this load to the shear and to the bending moment may be obtained through two successive integrations. Care should be taken, however, to also include in the expression for $V(x)$ the contribution of concentrated loads and reactions, and to include in the expression for $M(x)$ the contribution of concentrated couples [Examples 7.04 and 7.05, Sample Probs. 7.7 and 7.8]. We also observed that singularity functions are particularly well suited to the use of computers.

In Sec. 7.6, we analyzed the distribution of the *principal stresses* across the section of a prismatic beam and questioned whether the maximum normal stress $\sigma_{\max}$ at some point of the section could be larger than the value $\sigma_m = Mc/I$ computed at the surface of the beam. We concluded that in the case of a rectangular beam, and for any practical loading condition, the maximum normal stress was σ_m. When designing an S-beam or a W-beam, however, one should compute the principal stress $\sigma_{\max}$ at the junction of the web with the flanges of the beam (Fig. 7.5) in the section where $|M|$ is maximum and compare it with σ_m in the same section.

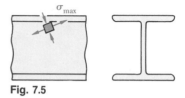

Fig. 7.5

A proper procedure for the design of a prismatic beam was described in Sec. 7.7 and is summarized below. This procedure takes into account the various points discussed in the preceding sections and leads to the most economical design, i.e., to the selection, all other things being equal, of the beam with the smallest weight per unit length, since this beam is the least expensive.

Having determined σ_{all} and τ_{all} for the material used, and assuming the design of the beam to be controlled by σ_m in the section where the bending moment is maximum, we compute the minimum allowable value of the section modulus $S = I/c$:

$$S_{\min} = \frac{|M|_{\max}}{\sigma_{\text{all}}} \tag{7.18}$$

From the group of sections with a section modulus $S > S_{\min}$, we select the section with the smallest weight per unit length. We then compute τ_m for that section and check that it does not exceed τ_{all}; if it does, a stronger beam should be selected. Finally, in the case of S-beams and W-beams, we should check that $\sigma_{\max}$ at the junction of the web with the flanges does not exceed σ_{all} [Example 7.06, Sample Prob. 7.9].

We were concerned so far only with prismatic beams, i.e., beams of uniform cross section. Considering in Sec. 7.8 the design of nonprismatic beams, i.e., beams of variable cross section, we saw that by selecting the shape and size of the cross section so that its elastic section modulus $S = I/c$ varied along the beam in the same way as the bending moment M, we were able to design beams for which σ_m at each section was equal to σ_{all}. Such beams, called *beams of constant strength*, clearly provide a more effective use of the material than prismatic beams. Their section modulus at any section along the beam was defined by the relation

$$S = \frac{M}{\sigma_{\text{all}}} \tag{7.22}$$

Design of transmission shafts

In Sec. 7.9, we considered the design of *transmission shafts* subjected to *transverse loads* as well as to torques. Taking into account the effect of both the normal stresses due to the bending moment M and the shearing stresses due to the torque T, we found that the minimum allowable value of the ratio J/c for the cross section of a circular shaft (either solid or hollow) was

$$\frac{J}{c} = \frac{(\sqrt{M^2 + T^2}\,)_{\max}}{\tau_{\text{all}}} \tag{7.31}$$

Stresses under applied loads

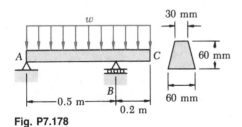

Fig 7.38

Finally, in Sec. 7.10, we analyzed the state of stress in a beam at a point located under a concentrated or a distributed load. In the case of a rectangular beam subjected to a *distributed load*, we noted the presence under the load of a normal stress σ_y, in addition to the longitudinal normal stress σ_x and the shearing stress τ_{xy} (Fig. 7.38). The normal stress σ_y varies in a nonlinear fashion from $-w$ at the upper surface of the beam to zero at its lower surface. In most engineering applications, this stress is small, compared with the longitudinal stress σ_x and contributes very little to the value of $\sigma_{\max}$ at any given point. In the case of a *concentrated load*, the formulas obtained in Chap. 5 for σ_x and τ_{xy} cease to be valid in the neighborhood of the point of application of the load, and an element of material located directly under the load is subjected to a very large normal stress σ_y. However, in practice, a load is always distributed over a small area, and while the stress concentrations are large close to the surface of the beam, they decrease rapidly as one goes deeper into the beam.

REVIEW PROBLEMS

7.178 A solid steel bar has a trapezoidal cross section and supports a uniformly distributed load w as shown. Knowing that $\sigma_{\text{all}} = +75$ MPa in tension and $\sigma_{\text{all}} = -100$ MPa in compression, determine the allowable load w which can be supported.

7.179 Solve Prob. 7.178, assuming that the bar is inverted.

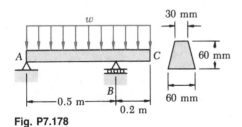

Fig. P7.178

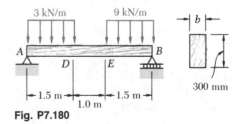

Fig. P7.180

7.180 For the beam and loading shown, design the cross section of the beam, knowing that for the grade of timber used $\sigma_{\text{all}} = 12$ MPa and $\tau_{\text{all}} = 1$ MPa.

7.181 (*a*) For the beam and loading shown, determine the distance *a* if the maximum absolute value of the bending moment in the beam is to be as small as possible. (*b*) Draw the corresponding bending-moment diagram.

7.182 Two brass rods, *AB* of 50-mm diameter and *BC* of 75-mm diameter, will be brazed together at *B* to form a nonuniform rod of total length 2.4 m which will be supported at *C*. Knowing that the specific weight of brass is 8700 kg/m³, determine (*a*) the length of rod *AB* for which the maximum normal stress in *ABC* is minimum, (*b*) the corresponding value of the maximum normal stress.

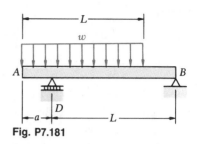

Fig. P7.181

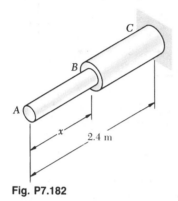

Fig. P7.182

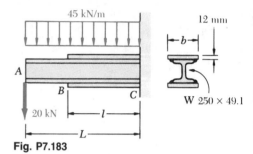

Fig. P7.183

7.183 Two cover plates of 12-mm thickness are to be welded as shown to a W 250 × 49.1 rolled-steel beam of length *L* = 2.4 m. Knowing that σ_{all} = 140 MPa and τ_{all} = 80 MPa, determine (*a*) the required width *b* of the cover plates, (*b*) the corresponding smallest permissible length *l* of the plates.

7.184 A transverse force **P** is applied as shown at end *A* of the conical taper *AB*. Denoting by d_0 the diameter of the taper at end *A*, show that the maximum normal stress occurs at point *H*, which is contained in a transverse section of diameter $d = \frac{3}{2}d_0$.

7.185 Knowing that σ_{all} = 160 MPa and τ_{all} = 100 MPa, select the metric wide-flange shape which should be used to support the loading shown.

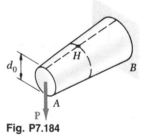

Fig. P7.184

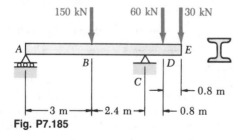

Fig. P7.185

7.186 Two solid steel rods are welded together to form the support shown. Knowing that τ_{all} = 50 MPa, determine the smallest permissible diameter of the rod *BD*.

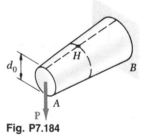

Fig. P7.186

7.187 For the beam and loading shown, draw the shear and bending-moment diagrams and design the cross section of the beam, knowing that $\sigma_{all} = 9$ MPa and $\tau_{all} = 0.6$ MPa.

7.188 Solve Prob. 7.187, if each of the 20 kN loads may be applied separately to the beam.

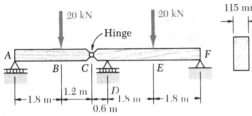

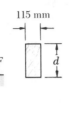

Fig. P7.187

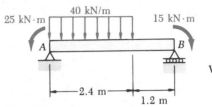

Fig. P7.189

7.189 For the beam and loading shown, determine (*a*) the largest normal stress due to bending, (*b*) the largest shearing stress on a transverse section.

The following problems are designed to be solved with a computer.

7.C1 Beam *AB* is 4.5 m long and has a rectangular cross section 90 mm wide and 300 mm deep. For the loading shown, write a computer program and use it to calculate the shear and bending moment from end *A* to end *B* at 0.15 m intervals, the maximum normal stress due to bending, the largest shearing stress on a transverse section, and the sections in which each of these maximum stresses occurs, when (*a*) $a_1 = 0.6$ m and $a_2 = 1.2$ m, (*b*) $a_1 = 1.5$ m and $a_2 = 2.7$ m.

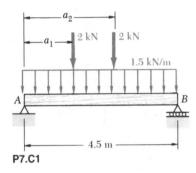

P7.C1

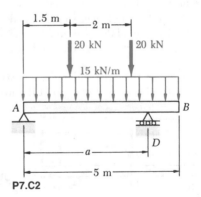

P7.C2

7.C2 A uniformly distributed load and two concentrated loads are applied to the overhanging beam *ADB*. Write a computer program and use it to calculate the shear and bending moment from end *A* to end *B* at 0.5-m intervals when (*a*) $a = 5$ m, (*b*) $a = 4$ m, (*c*) $a = 3.5$ m.

7.C3 Several concentrated loads may be applied to the simple beam *AB*. Write a computer program which can be used to calculate the shear between loads and the bending moment at each load. Use this program for the beam and loading of (*a*) Prob. 7.15, (*b*) Prob. 7.59.

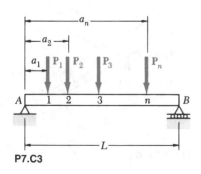

P7.C3

7.C4 Two 150-kN loads are maintained 2.4 m apart as they are moved slowly across beam AB. Write a computer program and use it to calculate the bending moment under each load and at the midpoint C for values of x from 0 to 9.6 m at intervals $\Delta x = 0.6$ m.

*7.C5 through 7.C8 Write a computer program which can be used to plot the shear and bending-moment diagrams for the beam and loading shown. Using this program and a plotting interval $\Delta L \leq L/100$, plot the V and M diagrams for the beam and loading indicated.

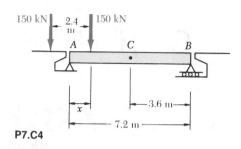

P7.C4

*7.C5 Beam and loading of (a) Prob. 7.17, (b) Prob. 7.63.

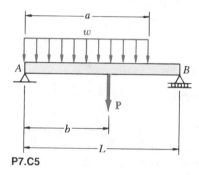

P7.C5

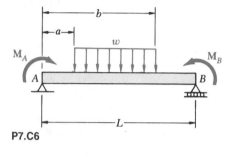

P7.C6

*7.C6 Beam and loading of (a) Prob. 7.64, (b) Prob. 7.189.
*7.C7 Beam and loading of (a) Prob. 7.16, (b) Prob. 7.185.

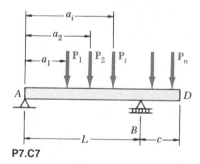

P7.C7

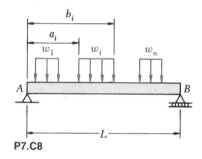

P7.C8

*7.C8 Beam and loading of (a) Prob. 7.66, (b) Prob. 7.182.

CHAPTER EIGHT

DEFLECTION OF BEAMS BY INTEGRATION

8.1. INTRODUCTION

In the preceding chapter we learned to design beams for strength. In this chapter and in the next, we shall be concerned with another aspect of the design of beams, namely, the determination of the *deflection* of prismatic beams under given loadings. Of particular interest is the determination of the *maximum deflection* of a beam under a given loading, since the design specifications of a beam will generally include a maximum allowable value for its deflection.

We saw in Sec. 4.5 that a prismatic beam subjected to pure bending is bent into an arc of circle and that, within the elastic range, the curvature of the neutral surface may be expressed as

$$\frac{1}{\rho} = \frac{M}{EI} \tag{4.21}$$

where M is the bending moment, E the modulus of elasticity, and I the moment of inertia of the cross section about its neutral axis.

When a beam is subjected to a transverse loading, Eq. (4.21) remains valid for any given transverse section, provided that Saint-Venant's principle applies. However, both the bending moment and the curvature of the neutral surface will vary from section to section. Denoting by x the distance of the section from the left end of the beam, we shall write

$$\frac{1}{\rho} = \frac{M(x)}{EI} \tag{8.1}$$

The knowledge of the curvature at various points of the beam will enable us to draw some general conclusions regarding the deformation of the beam under loading (Sec. 8.2).

To determine the slope and deflection of the beam at any given point, we shall first derive the following second-order linear differential equation, which governs the *elastic curve* characterizing the shape of the deformed beam (Sec. 8.3):

$$\frac{d^2y}{dx^2} = \frac{M(x)}{EI}$$

If the bending moment may be represented for all values of x by a single function $M(x)$, as in the case of the beams and loadings shown in Fig. 8.1, the slope $\theta = dy/dx$ and the deflection y at any point of the beam may be obtained through two successive integrations. The two constants of integration introduced in the process will be determined from the boundary conditions indicated in the figure.

However, if different analytical functions are required to represent the bending moment in various portions of the beam, different differential equations will also be required, leading to different functions defining the elastic curve in the various portions of the beam. In the case of the beam and loading of Fig. 8.2, for example, two differential equations are required, one for the portion of beam AD and the other for the portion DB. The first equation yields the functions θ_1 and y_1, and the second the functions θ_2 and y_2. Altogether, four constants of integration must be determined; two will be obtained by writing that the deflection is zero at A and B, and the other two by expressing that the portions of beam AD and DB have the same slope and the same deflection at D.

We shall observe in Sec. 8.4 that in the case of a beam supporting a distributed load $w(x)$, the elastic curve may be obtained directly from $w(x)$ through four successive integrations. The constants introduced in this process will be determined from the boundary values of V, M, θ, and y.

In Sec. 8.5, we shall discuss *statically indeterminate beams*, i.e., beams supported in such a way that the reactions at the supports involve four or more unknowns. Since only three equilibrium equations are available to determine these unknowns, the equilibrium equations must be supplemented by equations obtained from the boundary conditions imposed by the supports.

The method described earlier for the determination of the elastic curve when several functions are required to represent the bending moment M may be quite laborious, since it requires matching slopes and deflections at every transition point. We shall see in Sec. 8.6 that the use of *singularity functions* (previously discussed in Sec. 7.5) considerably simplifies the determination of θ and y at any point of the beam. An alternative method for the determination of θ and y, based on certain geometric properties of the elastic curve and involving the computation of areas and moments of areas under the bending-moment curve, will be discussed in Chap. 9.

The last part of the chapter (Secs. 8.7 and 8.8) is devoted to the *method of superposition*, which consists of determining separately, and then adding, the slope and deflection caused by the various loads applied to a beam. This procedure may be facilitated by the use of the table in Appendix D, which gives the slopes and deflections of beams for various loadings and types of support.

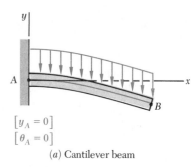

$$[y_A = 0]$$
$$[\theta_A = 0]$$

(*a*) Cantilever beam

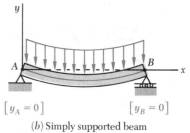

$$[y_A = 0] \qquad\qquad [y_B = 0]$$

(*b*) Simply supported beam

Fig. 8.1

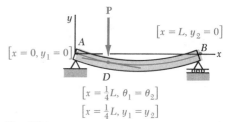

$$[x = L, y_2 = 0]$$
$$[x = 0, y_1 = 0]$$
$$[x = \tfrac{1}{4}L, \theta_1 = \theta_2]$$
$$[x = \tfrac{1}{4}L, y_1 = y_2]$$

Fig. 8.2

8.2. DEFORMATION OF A BEAM UNDER TRANSVERSE LOADING

At the beginning of this chapter, we recalled Eq. (4.21) of Sec. 4.5, which relates the curvature of the neutral surface and the bending moment in a beam in pure bending. We pointed out that this equation remains valid for any given transverse section of a beam subjected to a transverse loading, provided that Saint-Venant's principle applies. However, both the bending moment and the curvature of the neutral surface will vary from section to section. Denoting by x the distance of the section from the left end of the beam, we write

$$\frac{1}{\rho} = \frac{M(x)}{EI} \tag{8.1}$$

Consider, for example, a cantilever beam AB of length L subjected to a concentrated load $\mathbf{P}$ at its free end A (Fig. 8.3a). We have $M(x) = -Px$ and, substituting into (8.1),

$$\frac{1}{\rho} = -\frac{Px}{EI}$$

which shows that the curvature of the neutral surface varies linearly with x, from zero at A, where ρ_A itself is infinite, to $-PL/EI$ at B, where $|\rho_B| = EI/PL$ (Fig. 8.3b).

Consider now the overhanging beam AD of Fig. 8.4a, which supports two concentrated loads as shown. From the free-body diagram of the beam (Fig. 8.4b), we find that the reactions at the supports are $R_A = 1$ kN and $R_C = 5$ kN, respectively, and draw the corresponding bending-moment diagram (Fig. 8.5a). We note from the diagram that M, and thus the curvature of the beam, are zero at both ends of the beam, and also at

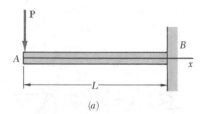

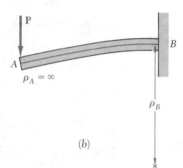

Fig. 8.3

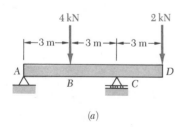

(a)

Fig. 8.4

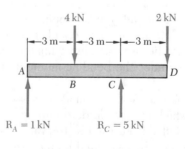

(b)

a point E located at $x = 4$ m.   Between A and E the bending moment is positive and the beam is concave upward; between E and D the bending moment is negative and the beam is concave downward (Fig. 8.5b).   We also note that the largest value of the curvature (i.e., the smallest value of the radius of curvature) occurs at the support C, where $|M|$ is maximum.

From the information obtained on its curvature, we may get a fairly good idea of the shape of the deformed beam.   However, the analysis and design of a beam usually require more precise information on the *deflection* and the *slope* of the beam at various points.   Of particular importance is the knowledge of the *maximum deflection* of the beam.   In the next section we shall use Eq. (8.1) to obtain a relation between the deflection y measured at a given point Q on the axis of the beam and the distance x of that point from some fixed origin (Fig. 8.6).   The relation obtained is the equation of the *elastic curve*, i.e., the equation of the curve into which the axis of the beam is transformed under the given loading (Fig. 8.6b).†

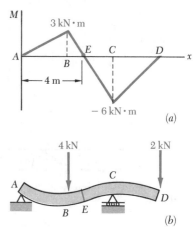

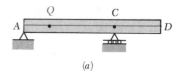

Fig. 8.5

8.3. EQUATION OF THE ELASTIC CURVE

We first recall from elementary calculus that the curvature of a plane curve at a point $Q(x,y)$ of the curve may be expressed as

$$\frac{1}{\rho} = \frac{\dfrac{d^2y}{dx^2}}{\left[1 + \left(\dfrac{dy}{dx}\right)^2\right]^{3/2}} \tag{8.2}$$

where dy/dx and d^2y/dx^2 are the first and second derivatives of the function $y(x)$ represented by that curve.   But, in the case of the elastic curve of a beam, the slope dy/dx is very small, and its square is negligible compared to unity.   We may write, therefore,

$$\frac{1}{\rho} = \frac{d^2y}{dx^2} \tag{8.3}$$

Substituting for $1/\rho$ from (8.3) into (8.1), we have

$$\frac{d^2y}{dx^2} = \frac{M(x)}{EI} \tag{8.4}$$

The equation obtained is a second-order linear differential equation; it is the governing differential equation for the elastic curve.

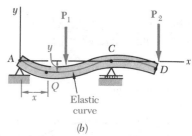

Fig. 8.6

†It should be noted that, in this chapter and the next, y represents a vertical displacement, while it was used in previous chapters to represent the distance of a given point in a transverse section from the neutral axis of that section.

The product EI is known as the *flexural rigidity* and, if it varies along the beam, as in the case of a beam of varying depth, we must express it as a function of x before proceeding to integrate Eq. (8.4). However, in the case of a prismatic beam, which is the case considered here, the flexural rigidity is constant. We may thus multiply both members of Eq. (8.4) by EI and integrate in x. We write

$$EI \frac{dy}{dx} = \int_0^x M(x)\ dx + C_1 \tag{8.5}$$

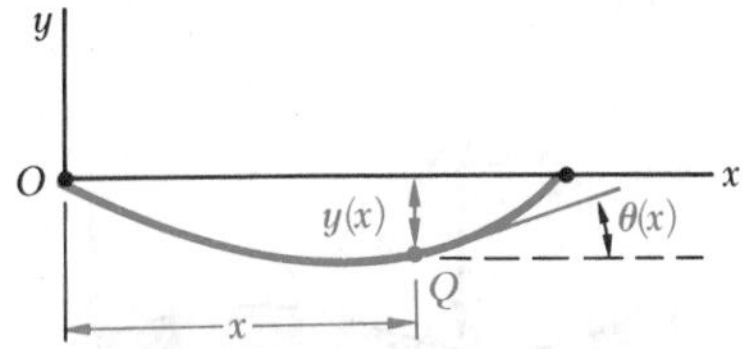

where C_1 is a constant of integration. Denoting by $\theta(x)$ the angle, measured in radians, that the tangent to the elastic curve at Q forms with the horizontal (Fig. 8.7), and recalling that this angle is very small, we have

$$\frac{dy}{dx} = \tan \theta \simeq \theta(x)$$

Thus, we may write Eq. (8.5) in the alternative form

$$EI\ \theta(x) = \int_0^x M(x)\ dx + C_1 \tag{8.5'}$$

Integrating both members of Eq. (8.5) in x, we have

$$EI\ y = \int_0^x \left[\int_0^x M(x)\ dx + C_1 \right] dx + C_2$$

$$EI\ y = \int_0^x dx \int_0^x M(x)\ dx + C_1 x + C_2 \tag{8.6}$$

where C_2 is a second constant, and where the first term in the right-hand member represents the function of x obtained by integrating twice in x the bending moment $M(x)$. If it were not for the fact that the constants C_1 and C_2 are as yet undetermined, Eq. (8.6) would define the deflection of the beam at any given point Q, and Eq. (8.5) or (8.5') would similarly define the slope of the beam at Q.

The constants C_1 and C_2 are determined from the *boundary conditions* or, more precisely, from the conditions imposed on the beam by its supports. Limiting our analysis in this section to *statically determinate beams*, i.e., to beams supported in such a way that the reactions at the supports may be obtained by the methods of statics, we note that only three types of beams need to be considered here (Fig. 8.8): (*a*) the *simply supported beam*, (*b*) the *overhanging beam*, and (*c*) the *cantilever beam*.

In the first two cases, the supports consist of a pin and bracket at A and of a roller at B, and require that the deflection be zero at each of these points. Letting first $x = x_A$, $y = y_A = 0$ in Eq. (8.6), and then $x = x_B$, $y = y_B = 0$ in the same equation, we obtain two equations which may be solved for C_1 and C_2. In the case of the cantilever beam (Fig. 8.8*c*), we note that both the deflection and the slope at A must be zero. Letting $x = x_A$, $y = y_A = 0$ in Eq. (8.6), and $x = x_A$, $\theta = \theta_A = 0$ in Eq. (8.5'), we obtain again two equations which may be solved for C_1 and C_2.

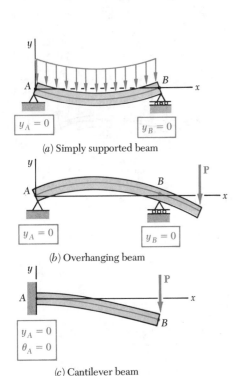

Fig. 8.7

(*a*) Simply supported beam

(*b*) Overhanging beam

(*c*) Cantilever beam

Fig. 8.8. Boundary conditions for statically determinate beams.

Example 8.01

The cantilever beam AB is of uniform cross section and carries a load **P** at its free end A (Fig. 8.9). Determine the equation of the elastic curve and the deflection and slope at A.

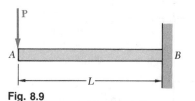

Fig. 8.9

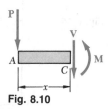

Fig. 8.10

Using the free-body diagram of the portion AC of the beam (Fig. 8.10), where C is located at a distance x from end A, we find

$$M = -Px \tag{8.7}$$

Substituting for M into Eq. (8.4) and multiplying both members by the constant EI, we write

$$EI\,\frac{d^2y}{dx^2} = -Px$$

Integrating in x, we obtain

$$EI\,\frac{dy}{dx} = -\tfrac{1}{2}Px^2 + C_1 \tag{8.8}$$

We now observe that at the fixed end B we have $x = L$ and $\theta = dy/dx = 0$ (Fig. 8.11). Substituting these values into (8.8) and solving for C_1, we have

$$C_1 = \tfrac{1}{2}PL^2$$

which we carry back into (8.8):

$$EI\,\frac{dy}{dx} = -\tfrac{1}{2}Px^2 + \tfrac{1}{2}PL^2 \tag{8.9}$$

Integrating both members of Eq. (8.9), we write

$$EI\,y = -\tfrac{1}{6}Px^3 + \tfrac{1}{2}PL^2x + C_2 \tag{8.10}$$

But, at B we have $x = L$, $y = 0$. Substituting into (8.10), we have

$$0 = -\tfrac{1}{6}PL^3 + \tfrac{1}{2}PL^3 + C_2$$
$$C_2 = -\tfrac{1}{3}PL^3$$

Carrying the value of C_2 back into Eq. (8.10), we obtain the equation of the elastic curve:

$$EI\,y = -\tfrac{1}{6}Px^3 + \tfrac{1}{2}PL^2x - \tfrac{1}{3}PL^3$$

or

$$y = \frac{P}{6EI}(-x^3 + 3L^2x - 2L^3) \tag{8.11}$$

The deflection and slope at A are obtained by letting $x = 0$ in Eqs. (8.11) and (8.9). We find

$$y_A = -\frac{PL^3}{3EI} \quad \text{and} \quad \theta_A = \left(\frac{dy}{dx}\right)_A = \frac{PL^2}{2EI}$$

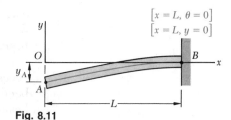

Fig. 8.11

Example 8.02

The simply supported prismatic beam AB carries a uniformly distributed load w per unit length (Fig. 8.12). Determine the equation of the elastic curve and the maximum deflection of the beam.

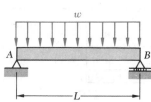

Fig. 8.12

Drawing the free-body diagram of the portion AD of the beam (Fig. 8.13) and taking moments about D, we find that

$$M = \tfrac{1}{2}wLx - \tfrac{1}{2}wx^2 \tag{8.12}$$

Substituting for M into Eq. (8.4) and multiplying both members of this equation by the constant EI, we write

$$EI\,\frac{d^2y}{dx^2} = -\frac{1}{2}wx^2 + \frac{1}{2}wLx \tag{8.13}$$

Integrating twice in x, we have

$$EI\,\frac{dy}{dx} = -\frac{1}{6}wx^3 + \frac{1}{4}wLx^2 + C_1 \tag{8.14}$$

$$EI\,y = -\frac{1}{24}wx^4 + \frac{1}{12}wLx^3 + C_1x + C_2 \tag{8.15}$$

Observing that $y = 0$ at both ends of the beam (Fig. 8.14), we first let $x = 0$ and $y = 0$ in Eq. (8.15) and obtain $C_2 = 0$. We then make $x = L$ and $y = 0$ in the same equation and write

$$0 = -\tfrac{1}{24}wL^4 + \tfrac{1}{12}wL^4 + C_1L$$

$$C_1 = -\tfrac{1}{24}wL^3$$

Carrying the values of C_1 and C_2 back into Eq. (8.15), we obtain the equation of the elastic curve:

$$EI\ y = -\tfrac{1}{24}wx^4 + \tfrac{1}{12}wLx^3 - \tfrac{1}{24}wL^3x$$

or

$$y = \frac{w}{24EI}(-x^4 + 2Lx^3 - L^3x) \qquad (8.16)$$

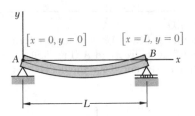

Fig. 8.14

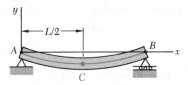

Fig. 8.15

Substituting into Eq. (8.14) the value obtained for C_1, we check that the slope of the beam is zero for $x = L/2$ and that the elastic curve has a minimum at the midpoint C of the beam (Fig. 8.15). Letting $x = L/2$ in Eq. (8.16), we have

$$y_C = \frac{w}{24EI}\left(-\frac{L^4}{16} + 2L\frac{L^3}{8} - L^3\frac{L}{2}\right) = -\frac{5wL^4}{384EI}$$

The maximum deflection or, more precisely, the maximum absolute value of the deflection, is thus

$$|y|_{max} = \frac{5wL^4}{384EI}$$

In each of the two examples considered so far, only one free-body diagram was required to determine the bending moment in the beam. As a result, a single function of x was used to represent M throughout the beam. This, however, is not generally the case. Concentrated loads, reactions at supports, or discontinuities in a distributed load will make it necessary to divide the beam into several portions, and to represent the bending moment by a different function $M(x)$ in each of these portions of beam. Each of the functions $M(x)$ will then lead to a different expression for the slope $\theta(x)$ and for the deflection $y(x)$. Since each of the expressions obtained for the deflection must contain two constants of integration, a large number of constants will have to be determined. As we shall see in the next example, the required additional boundary conditions may be obtained by observing that, while the shear and bending moment can be discontinuous at several points in a beam, the *deflection* and the *slope* of the beam *cannot be discontinuous* at any point.

Example 8.03

For the prismatic beam and the loading shown (Fig. 8.16), determine the slope and deflection at point D.

We must divide the beam into two portions, AD and DB, and determine the function $y(x)$ which defines the elastic curve for each of these portions.

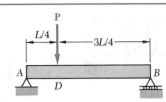

Fig. 8.16

1. From A to D ($x < L/4$). We draw the free-body diagram of a portion of beam AE of length $x < L/4$ (Fig. 8.17). Taking moments about E, we have

$$M_1 = \frac{3P}{4}x \qquad (8.17)$$

or, recalling Eq. (8.4),

$$EI\frac{d^2y_1}{dx^2} = \frac{3}{4}Px \qquad (8.18)$$

where $y_1(x)$ is the function which defines the elastic curve *for portion AD of the beam.* Integrating in x, we write

$$EI\,\theta_1 = EI\frac{dy_1}{dx} = \frac{3}{8}Px^2 + C_1 \qquad (8.19)$$

$$EI\,y_1 = \frac{1}{8}Px^3 + C_1x + C_2 \qquad (8.20)$$

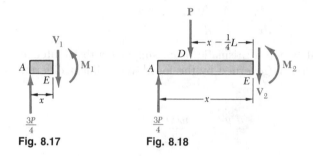

Fig. 8.17   **Fig. 8.18**

2. From D to B ($x > L/4$). We now draw the free-body diagram of a portion of beam AE of length $x > L/4$ (Fig. 8.18) and write

$$M_2 = \frac{3P}{4}x - P\left(x - \frac{L}{4}\right) \qquad (8.21)$$

or, recalling Eq. (8.4) and rearranging terms,

$$EI\frac{d^2y_2}{dx^2} = -\frac{1}{4}Px + \frac{1}{4}PL \qquad (8.22)$$

where $y_2(x)$ is the function which defines the elastic curve *for portion DB of the beam.* Integrating in x, we write

$$EI\,\theta_2 = EI\frac{dy_2}{dx} = -\frac{1}{8}Px^2 + \frac{1}{4}PLx + C_3 \qquad (8.23)$$

$$EI\,y_2 = -\frac{1}{24}Px^3 + \frac{1}{8}PLx^2 + C_3x + C_4 \qquad (8.24)$$

Determination of the Constants of Integration. The conditions which must be satisfied by the constants of integration have been summarized in Fig. 8.19. At the support A, where

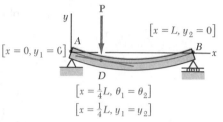

Fig. 8.19

the deflection is defined by Eq. (8.20), we must have $x = 0$ and $y_1 = 0$. At the support B, where the deflection is defined by Eq. (8.24), we must have $x = L$ and $y_2 = 0$. Also, the fact that there can be no sudden change in deflection or in slope at point D requires that $y_1 = y_2$ and $\theta_1 = \theta_2$ when $x = L/4$. We have therefore:

$$[x = 0,\ y_1 = 0],\ \text{Eq. (8.20):} \qquad 0 = C_2 \qquad (8.25)$$

$$[x = L,\ y_2 = 0],\ \text{Eq. (8.24):} \qquad 0 = \frac{1}{12}PL^3 + C_3L + C_4$$
$$(8.26)$$

$[x = L/4,\ \theta_1 = \theta_2]$, Eqs. (8.19) and (8.23):
$$\frac{3}{128}PL^2 + C_1 = \frac{7}{128}PL^2 + C_3 \quad (8.27)$$

$[x = L/4,\ y_1 = y_2]$, Eqs. (8.20) and (8.24):
$$\frac{PL^3}{512} + C_1\frac{L}{4} = \frac{11PL^3}{1536} + C_3\frac{L}{4} + C_4 \quad (8.28)$$

Solving these equations simultaneously, we find

$$C_1 = -\frac{7PL^2}{128},\ C_2 = 0,\ C_3 = -\frac{11PL^2}{128},\ C_4 = \frac{PL^3}{384}$$

Substituting for C_1 and C_2 into Eqs. (8.19) and (8.20), we write that for $x \leq L/4$,

$$EI\,\theta_1 = \frac{3}{8}Px^2 - \frac{7PL^2}{128} \qquad (8.29)$$

$$EI\,y_1 = \frac{1}{8}Px^3 - \frac{7PL^2}{128}x \qquad (8.30)$$

Letting $x = L/4$ in each of these equations, we find that the slope and deflection at point D are, respectively,

$$\theta_D = -\frac{PL^2}{32EI} \qquad \text{and} \qquad y_D = -\frac{3PL^3}{256EI}$$

We note that, since $\theta_D \neq 0$, the deflection at D is *not* the maximum deflection of the beam.

*8.4. DIRECT DETERMINATION OF THE ELASTIC CURVE FROM THE LOAD DISTRIBUTION

We saw in Sec. 8.3 that the equation of the elastic curve may be obtained by integrating twice the differential equation

$$\frac{d^2y}{dx^2} = \frac{M(x)}{EI} \tag{8.4}$$

where $M(x)$ is the bending moment in the beam. We now recall from Sec. 7.4 that, when a beam supports a distributed load $w(x)$, we have $dM/dx = V$ and $dV/dx = -w$ at any point of the beam. Differentiating both members of Eq. (8.4) with respect to x and assuming EI to be constant, we have therefore

$$\frac{d^3y}{dx^3} = \frac{1}{EI}\frac{dM}{dx} = \frac{V(x)}{EI} \tag{8.31}$$

and, differentiating again,

$$\frac{d^4y}{dx^4} = \frac{1}{EI}\frac{dV}{dx} = -\frac{w(x)}{EI}$$

We conclude that, when a prismatic beam supports a distributed load $w(x)$, its elastic curve is governed by the fourth-order linear differential equation

$$\frac{d^4y}{dx^4} = -\frac{w(x)}{EI} \tag{8.32}$$

Multiplying both members of Eq. (8.32) by the constant EI and integrating four times, we write

$$EI\frac{d^4y}{dx^4} = -w(x)$$

$$EI\frac{d^3y}{dx^3} = V(x) = -\int w(x)\,dx + C_1$$

$$EI\frac{d^2y}{dx^2} = M(x) = -\int dx \int w(x)\,dx + C_1x + C_2 \tag{8.33}$$

$$EI\frac{dy}{dx} = EI\,\theta\,(x) = -\int dx \int dx \int w(x)\,dx + \frac{1}{2}C_1x^2 + C_2x + C_3$$

$$EI\,y(x) = -\int dx \int dx \int dx \int w(x)\,dx + \frac{1}{6}C_1x^3 + \frac{1}{2}C_2x^2 + C_3x + C_4$$

The four constants of integration may be determined from the boundary conditions. These conditions include (*a*) the conditions imposed on the deflection or slope of the beam by its supports (cf. Sec. 8.3), and (*b*) the

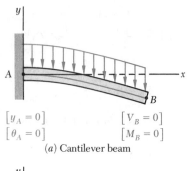

$$\left[\,y_A = 0\,\right] \qquad\qquad \left[\,V_B = 0\,\right]$$
$$\left[\,\theta_A = 0\,\right] \qquad\qquad \left[\,M_B = 0\,\right]$$

(*a*) Cantilever beam

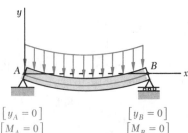

$$\left[\,y_A = 0\,\right] \qquad\qquad \left[\,y_B = 0\,\right]$$
$$\left[\,M_A = 0\,\right] \qquad\qquad \left[\,M_B = 0\,\right]$$

(*b*) Simply supported beam

Fig. 8.20 Boundary conditions for beams carrying a distributed load.

condition that V and M be zero at the free end of a cantilever beam, or that M be zero at both ends of a simply supported beam (cf. Sec. 7.4). This has been illustrated in Fig. 8.20.

The method presented here may be used effectively with cantilever or simply supported beams carrying a distributed load. In the case of overhanging beams, however, the reactions at the supports will cause discontinuities in the shear, i.e., in the third derivative of y, and different functions would be required to define the elastic curve over the entire beam.

Example 8.04

The simply supported prismatic beam AB carries a uniformly distributed load w per unit length (Fig. 8.21). Determine the equation of the elastic curve and the maximum deflection of the beam. (This is the same beam and loading as in Example 8.02.)

Since $w = $ constant, the first three of Eqs. (8.33) yield

$$EI \frac{d^4y}{dx^4} = -w$$

$$EI \frac{d^3y}{dx^3} = V(x) = -wx + C_1$$

$$EI \frac{d^2y}{dx^2} = M(x) = -\frac{1}{2}wx^2 + C_1x + C_2 \qquad (8.34)$$

Noting that the boundary conditions require that $M = 0$ at both ends of the beam (Fig. 8.22), we first let $x = 0$ and $M = 0$ in Eq. (8.34) and obtain $C_2 = 0$. We then make $x = L$ and $M = 0$ in the same equation and obtain $C_1 = \frac{1}{2}wL$.

Carrying the values of C_1 and C_2 back into Eq. (8.34), and integrating twice, we write

$$EI \frac{d^2y}{dx^2} = -\frac{1}{2}wx^2 + \frac{1}{2}wLx$$

$$EI \frac{dy}{dx} = -\frac{1}{6}wx^3 + \frac{1}{4}wLx^2 + C_3$$

$$EI\, y = -\frac{1}{24}wx^4 + \frac{1}{12}wLx^3 + C_3x + C_4 \qquad (8.35)$$

But the boundary conditions also require that $y = 0$ at both ends of the beam. Letting $x = 0$ and $y = 0$ in Eq. (8.35), we obtain $C_4 = 0$; letting $x = L$ and $y = 0$ in the same equation, we write

$$0 = -\tfrac{1}{24}wL^4 + \tfrac{1}{12}wL^4 + C_3L$$

$$C_3 = -\tfrac{1}{24}wL^3$$

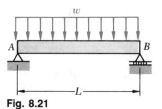

Fig. 8.21

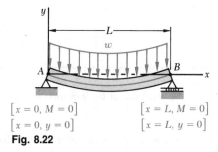

$$\left[x = 0,\, M = 0\right] \qquad \left[x = L,\, M = 0\right]$$
$$\left[x = 0,\, y = 0\right] \qquad \left[x = L,\, y = 0\right]$$

Fig. 8.22

Carrying the values of C_3 and C_4 back into Eq. (8.35) and dividing both members by EI, we obtain the equation of the elastic curve:

$$y = \frac{w}{24EI}(-x^4 + 2Lx^3 - L^3x) \qquad (8.36)$$

The value of the maximum deflection is obtained by making $x = L/2$ in Eq. (8.36). We have

$$|y|_{\max} = \frac{5wL^4}{384EI}$$

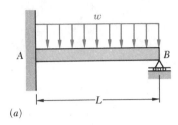

(a)

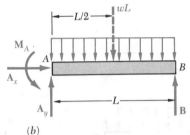

(b)

Fig. 8.23

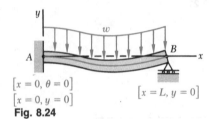

$[x = 0,\ \theta = 0]$

$[x = 0,\ y = 0]$ $[x = L,\ y = 0]$

Fig. 8.24

8.5. STATICALLY INDETERMINATE BEAMS

In the preceding sections, our analysis was limited to statically determinate beams. Consider now the prismatic beam AB (Fig. 8.23a), which has a fixed end at A and is supported by a roller at B. Drawing the free-body diagram of the beam (Fig. 8.23b), we note that the reactions involve four unknowns, while only three equilibrium equations are available, namely

$$\Sigma F_x = 0 \qquad \Sigma F_y = 0 \qquad \Sigma M_A = 0 \qquad (8.37)$$

Since only A_x may be determined from these equations, we conclude that the beam is *statically indeterminate*.

However, we recall from Chaps. 2 and 3 that, in a statically indeterminate problem, the reactions may be obtained by considering the *deformations* of the structure involved. We should, therefore, proceed with the computation of the slope and deformation along the beam. Following the method used in Sec. 8.3, we first express the bending moment $M(x)$ at any given point of AB in terms of the distance x from A, the given load, and the unknown reactions. Integrating in x, we obtain expressions for θ and y which contain two additional unknowns, namely the constants of integration C_1 and C_2. But altogether six equations are available to determine the reactions and the constants C_1 and C_2; they are the three equilibrium equations (8.37) and the three equations expressing that the boundary conditions are satisfied, i.e., that the slope and deflection at A are zero, and that the deflection at B is zero (Fig. 8.24). Thus, the reactions at the supports may be determined, and the equation of the elastic curve may be obtained.

Example 8.05

Determine the reactions at the supports for the prismatic beam of Fig. 8.23a.

Equilibrium Equations. From the free-body diagram of Fig. 8.23b we write

$\xrightarrow{+}\Sigma F_x = 0: \qquad A_x = 0$

$+\uparrow \Sigma F_y = 0: \qquad A_y + B - wL = 0 \qquad\qquad (8.38)$

$+\curvearrowleft\Sigma M_A = 0: \qquad M_A + BL - \tfrac{1}{2}wL^2 = 0$

Equation of Elastic Curve. Drawing the free-body diagram of a portion of beam AC (Fig. 8.25), we write

$+\curvearrowleft\Sigma M_C = 0: \qquad M + \tfrac{1}{2}wx^2 + M_A - A_y x = 0 \qquad (8.39)$

Solving Eq. (8.39) for M and carrying into Eq. (8.4), we write

$$EI\,\frac{d^2 y}{dx^2} = -\frac{1}{2}wx^2 + A_y x - M_A$$

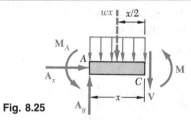

Fig. 8.25

Integrating in x, we have

$$EI\,\theta = EI\,\frac{dy}{dx} = -\frac{1}{6}wx^3 + \frac{1}{2}A_y x^2 - M_A x + C_1 \qquad (8.40)$$

$$EI\,y = -\frac{1}{24}wx^4 + \frac{1}{6}A_y x^3 - \frac{1}{2}M_A x^2 + C_1 x + C_2 \qquad (8.41)$$

Referring to the boundary conditions indicated in Fig. 8.24, we make $x = 0$, $\theta = 0$ in Eq. (8.40), $x = 0$, $y = 0$ in Eq. (8.41), and conclude that $C_1 = C_2 = 0$. Thus, we may rewrite Eq. (8.41) as follows:

$$EI\, y = -\tfrac{1}{24}wx^4 + \tfrac{1}{6}A_y x^3 - \tfrac{1}{2}M_A x^2 \qquad (8.42)$$

But the third boundary condition requires that $y = 0$ for $x = L$. Carrying these values into (8.42), we write

$$0 = -\tfrac{1}{24}wL^4 + \tfrac{1}{6}A_y L^3 - \tfrac{1}{2}M_A L^2$$

or

$$3M_A - A_y L + \tfrac{1}{4}wL^2 = 0 \qquad (8.43)$$

Solving this equation simultaneously with the three equilibrium equations (8.38), we obtain the reactions at the supports:

$$A_x = 0 \qquad A_y = \tfrac{5}{8}wL \qquad M_A = \tfrac{1}{8}wL^2 \qquad B = \tfrac{3}{8}wL$$

In the example we have just considered, there was one redundant reaction, i.e., there was one more reaction than could be determined from the equilibrium equations alone. The corresponding beam is said to be *statically indeterminate to the first degree*. Another example of a beam indeterminate to the first degree is provided in Sample Prob. 8.3. If the beam supports are such that two reactions are redundant (Fig. 8.26a), the beam is said to be *indeterminate to the second degree*. While there are now five unknown reactions (Fig. 8.26b), we find that four equations may be obtained from the boundary conditions (Fig. 8.26c). Thus, altogether seven equations are available to determine the five reactions and the two constants of integration.

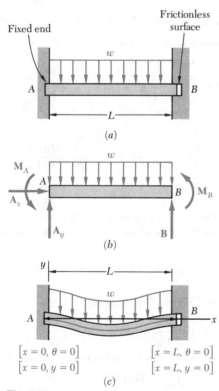

Fig. 8.26

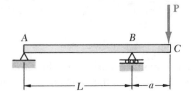

SAMPLE PROBLEM 8.1

The overhanging steel beam ABC carries a concentrated load $\mathbf{P}$ at end C. For portion AB of the beam, (a) derive the equation of the elastic curve, (b) determine the maximum deflection, (c) evaluate y_{max} for the following data:

$$W\ 360 \times 101 \qquad I = 301 \times 10^6\ \text{mm}^4 \qquad E = 200\ \text{GPa}$$
$$P = 250\ \text{kN} \qquad L = 5\ \text{m} \qquad a = 1.2\ \text{m}$$

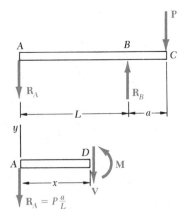

Free-Body Diagrams. Reactions: $\mathbf{R}_A = Pa/L \downarrow \qquad \mathbf{R}_B = P(1 + a/L) \uparrow$
Using the free-body diagram of the portion of beam AD of length x, we find

$$M = -P\frac{a}{L}x \qquad (0 < x < L)$$

Differential Equation of the Elastic Curve. We use Eq. (8.4) and write

$$EI\frac{d^2y}{dx^2} = -P\frac{a}{L}x$$

Noting that the flexural rigidity EI is constant, we integrate twice and find

$$EI\frac{dy}{dx} = -\frac{1}{2}P\frac{a}{L}x^2 + C_1 \tag{1}$$

$$EI\ y = -\frac{1}{6}P\frac{a}{L}x^3 + C_1 x + C_2 \tag{2}$$

Determination of Constants. For the boundary conditions shown, we have

$[x = 0, y = 0]$: From Eq. (2), we find $\quad C_2 = 0$
$[x = L, y = 0]$: Again using Eq. (2), we write

$$EI(0) = -\frac{1}{6}P\frac{a}{L}L^3 + C_1 L \qquad C_1 = +\frac{1}{6}PaL$$

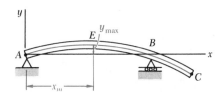

(a) **Equation of the Elastic Curve.** Substituting for C_1 and C_2 into Eqs. (1) and (2), we have

$$EI\frac{dy}{dx} = -\frac{1}{2}P\frac{a}{L}x^2 + \frac{1}{6}PaL \qquad \frac{dy}{dx} = \frac{PaL}{6EI}\left[1 - 3\left(\frac{x}{L}\right)^2\right] \tag{3}$$

$$EI\ y = -\frac{1}{6}P\frac{a}{L}x^3 + \frac{1}{6}PaLx \qquad y = \frac{PaL^2}{6EI}\left[\frac{x}{L} - \left(\frac{x}{L}\right)^3\right] \tag{4} \blacktriangleleft$$

(b) **Maximum Deflection in Portion AB.** The maximum deflection y_{max} occurs at point E where the slope of the elastic curve is zero. Setting $dy/dx = 0$ in Eq. (3), we write

$$0 = \frac{PaL}{6EI}\left[1 - 3\left(\frac{x_1}{L}\right)^2\right] \qquad x_1 = L/\sqrt{3} = 0.577\ L$$

We substitute $x_1/L = 0.577$ into Eq. (4) and have

$$y_{max} = \frac{PaL^2}{6EI}[(0.577) - (0.577)^3] \qquad y_{max} = 0.0642\frac{PaL^2}{EI} \blacktriangleleft$$

(c) **Evaluation of y_{max}.** For the data given, the value of y_{max} is

$$y_{max} = 0.0642\frac{(250\ \text{kN})(1.2\ \text{m})(5\ \text{m})^2}{(200\ \text{GPa})(301 \times 10^{-6}\ \text{m}^4)} \qquad y_{max} = 8.00\ \text{mm} \blacktriangleleft$$

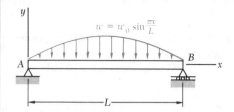

SAMPLE PROBLEM 8.2

For the beam and loading shown, determine (a) the equation of the elastic curve, (b) the slope at end A, (c) the maximum deflection.

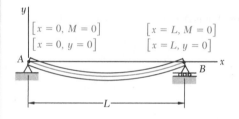

$[x = 0, M = 0]$ $[x = L, M = 0]$
$[x = 0, y = 0]$ $[x = L, y = 0]$

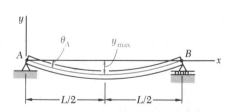

Differential Equation of the Elastic Curve. From Eq. (8.32),

$$EI\frac{d^4y}{dx^4} = -w(x) = -w_0 \sin\frac{\pi x}{L} \tag{1}$$

Integrate Eq. (1) twice:

$$EI\frac{d^3y}{dx^3} = V = +w_0\frac{L}{\pi}\cos\frac{\pi x}{L} + C_1 \tag{2}$$

$$EI\frac{d^2y}{dx^2} = M = +w_0\frac{L^2}{\pi^2}\sin\frac{\pi x}{L} + C_1x + C_2 \tag{3}$$

Boundary Conditions:

$[x = 0, M = 0]$: From Eq. (3), we find $C_2 = 0$
$[x = L, M = 0]$: Again using Eq. (3), we write

$$0 = w_0\frac{L^2}{\pi^2}\sin\pi + C_1L \qquad C_1 = 0$$

Thus:

$$EI\frac{d^2y}{dx^2} = +w_0\frac{L^2}{\pi^2}\sin\frac{\pi x}{L} \tag{4}$$

Integrate Eq. (4) twice:

$$EI\frac{dy}{dx} = EI\,\theta = -w_0\frac{L^3}{\pi^3}\cos\frac{\pi x}{L} + C_3 \tag{5}$$

$$EI\,y = -w_0\frac{L^4}{\pi^4}\sin\frac{\pi x}{L} + C_3x + C_4 \tag{6}$$

Boundary Conditions:

$[x = 0, y = 0]$: Using Eq. (6), we find $C_4 = 0$
$[x = L, y = 0]$: Again using Eq. (6), we find $C_3 = 0$

(a) **Equation of Elastic Curve** $\qquad EIy = -w_0\frac{L^4}{\pi^4}\sin\frac{\pi x}{L}$ ◄

(b) **Slope at End A.** For $x = 0$, we have

$$EI\,\theta_A = -w_0\frac{L^3}{\pi^3}\cos 0 \qquad\qquad \theta_A = \frac{w_0L^3}{\pi^3 EI} \text{◄}$$

(c) **Maximum Deflection.** For $x = \frac{1}{2}L$

$$EIy_{max} = -w_0\frac{L^4}{\pi^4}\sin\frac{\pi}{2} \qquad\qquad y_{max} = \frac{w_0L^4}{\pi^4 EI}\downarrow \text{◄}$$

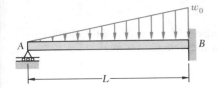

SAMPLE PROBLEM 8.3

For the uniform beam AB, (a) determine the reaction at A, (b) derive the equation of the elastic curve, (c) determine the slope at A. (Note that the beam is statically indeterminate to the first degree.)

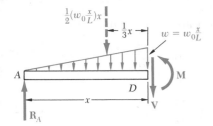

Bending Moment. Using the free body shown, we write

$$+\curvearrowleft \Sigma M_D = 0: \qquad R_A x - \frac{1}{2}\left(\frac{w_0 x^2}{L}\right)\frac{x}{3} - M = 0 \qquad M = R_A x - \frac{w_0 x^3}{6L}$$

Differential Equation of the Elastic Curve. We use Eq. (8.4) and write

$$EI\frac{d^2 y}{dx^2} = R_A x - \frac{w_0 x^3}{6L}$$

Noting that the flexural rigidity EI is constant, we integrate twice and find

$$EI\frac{dy}{dx} = EI\,\theta = \frac{1}{2}R_A x^2 - \frac{w_0 x^4}{24L} + C_1 \tag{1}$$

$$EI\,y = \frac{1}{6}R_A x^3 - \frac{w_0 x^5}{120L} + C_1 x + C_2 \tag{2}$$

Boundary Conditions. The three boundary conditions which must be satisfied are shown on the sketch

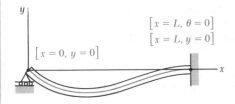

$$[x = 0, y = 0]: \qquad C_2 = 0 \tag{3}$$

$$[x = L, \theta = 0]: \qquad \frac{1}{2}R_A L^2 - \frac{w_0 L^3}{24} + C_1 = 0 \tag{4}$$

$$[x = L, y = 0]: \qquad \frac{1}{6}R_A L^3 - \frac{w_0 L^4}{120} + C_1 L + C_2 = 0 \tag{5}$$

(a) **Reaction at A.** Multiplying Eq. (4) by L, subtracting Eq. (5) member by member from the equation obtained, and noting that $C_2 = 0$, we have

$$\tfrac{1}{3}R_A L^3 - \tfrac{1}{30}w_0 L^4 = 0 \qquad\qquad R_A = \tfrac{1}{10}w_0 L \uparrow \quad \blacktriangleleft$$

We note that the reaction is independent of E and I. Substituting $R_A = \tfrac{1}{10}w_0 L$ into Eq. (4), we have

$$\tfrac{1}{2}(\tfrac{1}{10}w_0 L)L^2 - \tfrac{1}{24}w_0 L^3 + C_1 = 0 \qquad C_1 = -\tfrac{1}{120}w_0 L^3$$

(b) **Equation of the Elastic Curve.** Substituting for R_A, C_1, and C_2 into Eq. (2), we have

$$EI\,y = \frac{1}{6}\left(\frac{1}{10}w_0 L\right)x^3 - \frac{w_0 x^5}{120L} - \left(\frac{1}{120}w_0 L^3\right)x$$

$$y = \frac{w_0}{120EIL}\left(-x^5 + 2L^2 x^3 - L^4 x\right) \quad \blacktriangleleft$$

(c) **Slope at A.** Differentiating the above equation with respect to x, we write

$$\theta = \frac{dy}{dx} = \frac{w_0}{120EIL}\left(-5x^4 + 6L^2 x^2 - L^4\right)$$

Making $x = 0$, we have $\qquad \theta_A = -\dfrac{w_0 L^3}{120EI} \qquad\qquad \theta_A = \dfrac{w_0 L^3}{120EI} \swarrow \quad \blacktriangleleft$

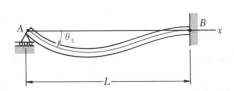

PROBLEMS

In the following problems assume that the flexural rigidity EI of each beam is constant.

8.1 through 8.6 For the loading shown, determine (*a*) the equation of the elastic curve for the cantilever beam AB, (*b*) the deflection at the free end, (*c*) the slope at the free end.

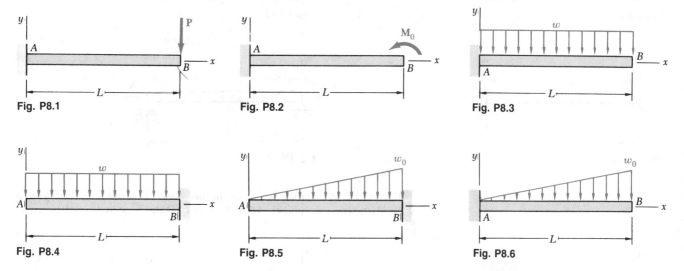

Fig. P8.1   Fig. P8.2   Fig. P8.3

Fig. P8.4   Fig. P8.5   Fig. P8.6

8.7 Solve parts *b* and *c* of Prob. 8.5, assuming that beam AB is a W 310 × 52 rolled-steel shape and that $w_0 = 45$ kN/m, $L = 4$ m, and $E = 200$ GPa.

8.8 Solve parts *b* and *c* of Prob. 8.4, assuming that beam AB is a W 150 × 24.0 rolled-steel shape and that $w = 9$ kN/m, $L = 2$ m, and $E = 200$ GPa.

8.9 and 8.10 The simple beam AB carries a distributed load which varies linearly as shown. Determine (*a*) the equation of the elastic curve, (*b*) the deflection at midspan, (*c*) the slope at end A. (*d*) Solve parts *b* and *c*, assuming that the beam AB is a W 360 × 79 rolled-steel shape and that $w_0 = 75$ kN/m, $L = 5$ m, and $E = 200$ GPa.

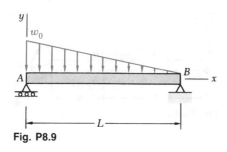

Fig. P8.9

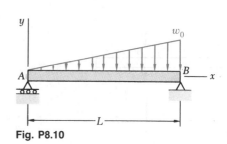

Fig. P8.10

8.11 through 8.14 For the beam and loading shown, determine (*a*) the equation of the elastic curve, (*b*) the slope at *A*, (*c*) the slope at *B*.

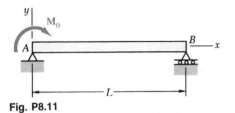

Fig. P8.11

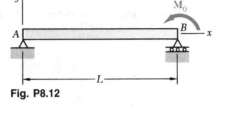

Fig. P8.12

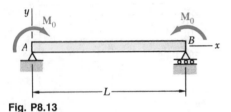

Fig. P8.13

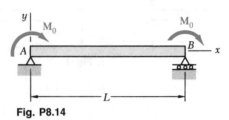

Fig. P8.14

8.15 and 8.16 For the beam and loading shown, determine (*a*) the equation of the elastic curve for portion *AB* of the beam, (*b*) the slope at *A*, (*c*) the slope at *B*.

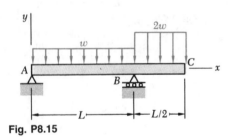

Fig. P8.15

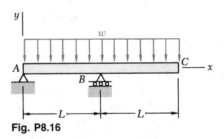

Fig. P8.16

8.17 and 8.18 For the cantilever beam and loading shown, determine (*a*) the equation of the elastic curve for portion *AB* of the beam, (*b*) the deflection at *B*, (*c*) the slope at *B*.

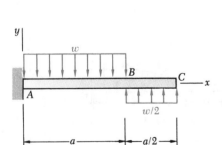

Fig. P8.17

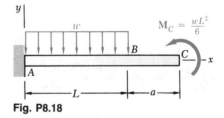

Fig. P8.18

8.19 Knowing that beam AB is a W 150 × 13.5 rolled-steel shape and that $P = 33$ kN, $L = 1.5$ m, and $E = 200$ GPa, determine (a) the slope at A, (b) the deflection at C.

8.20 Knowing that beam AB is a W 130 × 23.8 rolled-steel shape and that $P = 50$ kN, $L = 1.25$ m, and $E = 200$ GPa, determine (a) the slope at A, (b) the deflection at C.

8.21 Beam AE carries a distributed load $w = 4$ kN/m and two concentrated loads each of magnitude $P = 250$ N. Knowing that the beam is made of an aluminum rod of diameter $d = 36$ mm and that $L = 800$ mm, $a = 250$ mm, and $E = 72$ GPa, determine (a) the slope at B, (b) the deflection at the center C of the beam.

8.22 Beam AE carries a distributed load $w = 4.5$ kN/m and two concentrated loads each of magnitude $P = 700$ N. Knowing that the beam is made of a steel rod of diameter $d = 20$ mm and that $L = 500$ mm, $a = 150$ mm, and $E = 200$ GPa, determine (a) the slope at B, (b) the deflection at the center C of the beam.

8.23 Knowing that beam AE is an S 310 × 47.3 rolled-steel shape and that $w = 50$ kN/m, $a = 1.5$ m, and $E = 200$ GPa, determine (a) the slope at B, (b) the deflection at the center C of the beam.

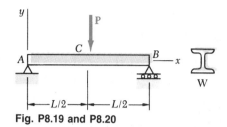

Fig. P8.19 and P8.20

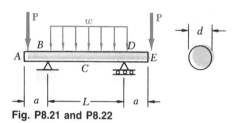

Fig. P8.21 and P8.22

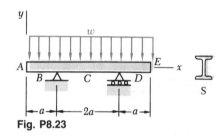

Fig. P8.23

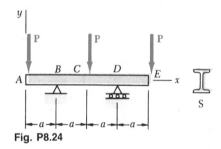

Fig. P8.24

8.24 Knowing that beam AE is an S 380 × 64 rolled-steel shape and that $P = 65$ kN, $a = 1.8$ m, and $E = 200$ GPa, determine (a) the slope at B, (b) the deflection at the center C of the beam.

8.25 and 8.26 For the beam and loading shown, determine (a) the equation of the elastic curve for portion BD, (b) the slope at B, (c) the deflection at the center C of the beam.

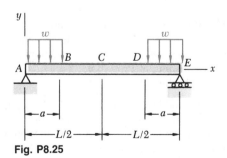

Fig. P8.25

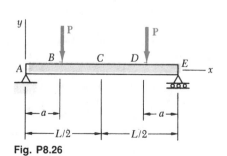

Fig. P8.26

8.27 For the beam and loading shown, knowing that $a = 2$ m and $w = 50$ kN/m, determine (*a*) the slope at support A, (*b*) the deflection at point C. Use $E = 200$ GPa.

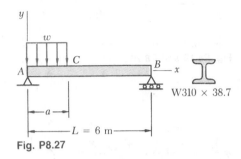

Fig. P8.27

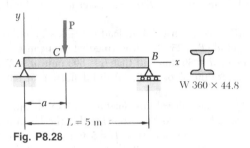

Fig. P8.28

8.28 For the beam and loading shown, knowing that $a = 1.5$ m and $P = 150$ kN, determine (*a*) the slope at support A, (*b*) the deflection at point C. Use $E = 200$ GPa.

8.29 (*a*) Determine the equation of the elastic curve for portion AB of the beam shown. (*b*) Knowing that the beam is a W 150×29.8 rolled-steel shape and that $w = 60$ kN/m, $L = 1$ m, and $E = 200$ GPa, determine the deflection at point A.

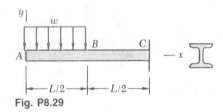

Fig. P8.29

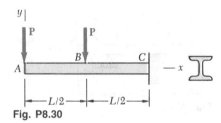

Fig. P8.30

8.30 (*a*) Determine the equation of the elastic curve for portion AB of the beam shown. (*b*) Knowing that the beam is a W 200×46.1 rolled-steel shape and that $P = 30$ kN, $L = 1.2$ m, and $E = 200$ GPa, determine the deflection at point A.

8.31 For the beam and loading of Example 8.03 (page 484), determine the magnitude and location of the maximum deflection.

8.32 Determine the magnitude and location of the maximum deflection of the beam AB.

8.33 and 8.34 For the beam and loading indicated, (*a*) express in terms of w_0, L, E, and I the magnitude and location of the maximum deflection, (*b*) calculate the value of the maximum deflection, using the given numerical data.

8.33 Beam and loading of Prob. 8.9.

8.34 Beam and loading of Prob. 8.10.

8.35 For the beam and loading of Prob. 8.15, determine the magnitude and location of the largest upward deflection in span AB.

8.36 For the beam and loading of Prob. 8.16, determine the magnitude and location of the largest upward deflection in span AB.

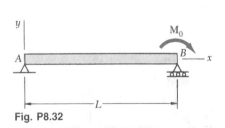

Fig. P8.32

*8.37 and 8.38   For the beam and loading shown, determine (a) the equation of the elastic curve, (b) the slope at end A, (c) the deflection at the midpoint of the span.

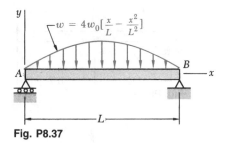

Fig. P8.37

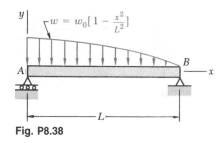

Fig. P8.38

*8.39 and 8.40   For the beam and loading shown, determine (a) the equation of the elastic curve, (b) the deflection at the free end.

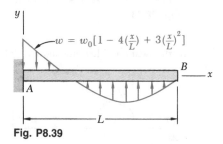

Fig. P8.39

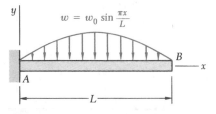

Fig. P8.40

8.41 through 8.44   For the beam and loading shown, determine the reaction at the roller support.

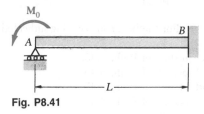

Fig. P8.41

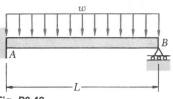

Fig. P8.42

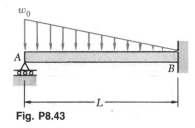

Fig. P8.43

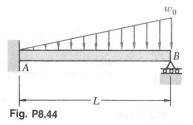

Fig. P8.44

8.45 through 8.48 Determine the reaction at the roller support and draw the bending-moment diagram for the beam and loading shown.

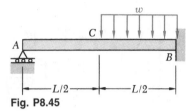

Fig. P8.45

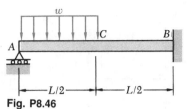

Fig. P8.46

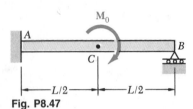

Fig. P8.47

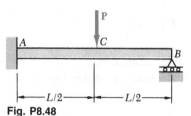

Fig. P8.48

8.49 and 8.50 For the beam and loading shown, determine the reaction at the roller support.

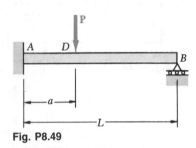

Fig. P8.49

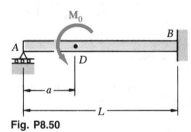

Fig. P8.50

8.51 through 8.54 For the beam and loading indicated, determine the deflection at point C.

    8.51 Beam and loading of Prob. 8.45.
    8.52 Beam and loading of Prob. 8.46.
    8.53 Beam and loading of Prob. 8.47.
    8.54 Beam and loading of Prob. 8.48.

8.55 and 8.56 Determine the reaction at A and draw the bending-moment diagram for the beam and loading shown.

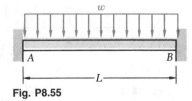

Fig. P8.55

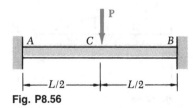

Fig. P8.56

8.57 and 8.58 Determine the reaction at *A* and draw the bending-moment diagram for the beam and loading shown.

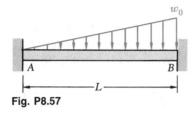

Fig. P8.57

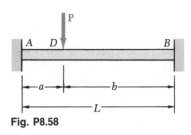

Fig. P8.58

*8.6. USING SINGULARITY FUNCTIONS TO DETERMINE THE SLOPE AND DEFLECTION OF A BEAM

Reviewing the work done so far in this chapter, we note that the integration method provides a convenient and effective way of determining the slope and deflection at any point of a prismatic beam, *as long as the bending moment may be represented by a single analytical function M(x)*.   However, when the loading of the beam is such that two different functions are needed to represent the bending moment over the entire length of the beam, as in Example 8.03 (Fig. 8.16), four constants of integration are required, and an equal number of equations, expressing continuity conditions at point *D*, as well as boundary conditions at the supports *A* and *B*, must be used to determine these constants.   If three or more functions were needed to represent the bending moment, additional constants and a corresponding number of additional equations would be required, resulting in rather lengthy computations.   We shall see in this section how these computations may be simplified through the use of the singularity functions discussed in Sec. 7.5.

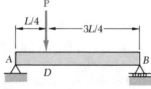

Fig. 8.16 (repeated)

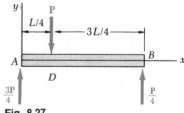

Fig. 8.27

Let us consider again the beam and loading of Example 8.03 (Fig. 8.16) and draw the free-body diagram of that beam (Fig. 8.27).   Using the appropriate singularity function, as explained in Sec. 7.5, to represent the contribution to the shear of the concentrated load **P**, we write

$$V(x) = \frac{3P}{4} - P\langle x - \tfrac{1}{4}L \rangle^0$$

Integrating in x and recalling from Sec. 7.5 that in the absence of any concentrated couple, the expression obtained for the bending moment will not contain any constant term, we have

$$M(x) = \frac{3P}{4} x - P\langle x - \tfrac{1}{4}L \rangle \qquad (8.44)$$

Substituting for $M(x)$ from (8.44) into Eq. (8.4), we write

$$EI\frac{d^2y}{dx^2} = \frac{3P}{4} x - P\langle x - \tfrac{1}{4}L \rangle \qquad (8.45)$$

and, integrating in x,

$$EI\,\theta = EI\frac{dy}{dx} = \frac{3}{8}Px^2 - \frac{1}{2}P\langle x - \tfrac{1}{4}L \rangle^2 + C_1 \qquad (8.46)$$

$$EI\,y = \frac{1}{8}Px^3 - \frac{1}{6}P\langle x - \tfrac{1}{4}L \rangle^3 + C_1x + C_2 \qquad (8.47)$$

The constants C_1 and C_2 may be determined from the boundary conditions shown in Fig. 8.28. Letting $x = 0$, $y = 0$ in Eq. (8.47), we have

$$0 = 0 - \frac{1}{6} P\langle - \tfrac{1}{4}L \rangle^3 + 0 + C_2$$

which reduces to $C_2 = 0$, since any bracket containing a negative quantity is equal to zero. Letting now $x = L$, $y = 0$, and $C_2 = 0$ in Eq. (8.47), we write

$$0 = \frac{1}{8} PL^3 - \frac{1}{6} P\langle \tfrac{3}{4}L \rangle^3 + C_1L$$

Since the quantity between brackets is positive, the brackets may be replaced by ordinary parentheses. Solving for C_1, we have

$$C_1 = - \frac{7PL^2}{128}$$

We check that the expressions obtained for the constants C_1 and C_2 are the same that were found earlier in Sec. 8.3. But the need for additional constants C_3 and C_4 has now been eliminated, and we do not have to write equations expressing that the slope and the deflection are continuous at point D.†

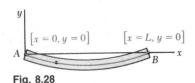

Fig. 8.28

$[x = 0, y = 0]$ $[x = L, y = 0]$

---

†The continuity conditions for the slope and deflection at D are "built-in" in Eqs. (8.46) and (8.47). Indeed, the difference between the expressions for the slope θ_1 in AD and the slope θ_2 in DB is represented by the term $-\tfrac{1}{2}P\langle x - \tfrac{1}{4}L \rangle^2$ in Eq. (8.46), and this term is equal to zero at D. Similarly, the difference between the expressions for the deflection y_1 in AD and the deflection y_2 in DB is represented by the term $-\tfrac{1}{6}P\langle x - \tfrac{1}{4}L \rangle^3$ in Eq. (8.47), and this term is also equal to zero at D.

Example 8.06

For the beam and loading shown (Fig. 8.29*a*) and using singularity functions, (*a*) express the slope and deflection as functions of the distance *x* from the support at *A*, (*b*) determine the deflection at the midpoint *D*. Use *E* = 200 GPa and *I* = 1.024 × 10^{-6} m^4.

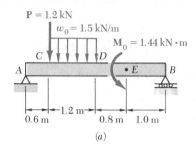

(*a*)

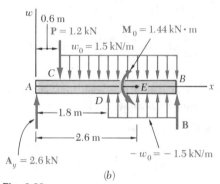

(*b*)

Fig. 8.29

(*a*) We note that the beam is loaded and supported in the same manner as the beam of Example 7.04. Referring to that example, we recall that the given distributed loading was replaced by the two equivalent open-ended loadings shown in Fig. 8.29*b* and that the following expressions were obtained for the shear and bending moment:

$$V(x) = -1.5\langle x - 0.6\rangle^1 + 1.5\langle x - 1.8\rangle^1 + 2.6 - 1.2\langle x - 0.6\rangle^0$$

$$M(x) = -0.75\langle x - 0.6\rangle^2 + 0.75\langle x - 1.8\rangle^2$$
$$+ 2.6x - 1.2\langle x - 0.6\rangle^1 - 1.44\langle x - 2.6\rangle^0$$

Integrating the last expression twice, we obtain

$$EI\theta = -0.25\langle x - 0.6\rangle^3 + 0.25\langle x - 1.8\rangle^3$$
$$+ 1.3x^2 - 0.6\langle x - 0.6\rangle^2 - 1.44\langle x - 2.6\rangle^1 + C_1 \quad (8.48)$$

$$EIy = -0.0625\langle x - 0.6\rangle^4 + 0.0625\langle x - 1.8\rangle^4$$
$$+ 0.4333x^3 - 0.2\langle x - 0.6\rangle^3 - 0.72\langle x - 2.6\rangle^2$$
$$+C_1x + C_2 \quad (8.49)$$

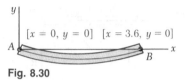

Fig. 8.30

The constants C_1 and C_2 may be determined from the boundary conditions shown in Fig. 8.30. Letting $x = 0$, $y = 0$ in Eq. (8.49) and noting that all the brackets contain negative quantities and, therefore, are equal to zero, we conclude that $C_2 = 0$. Letting now $x = 3.6$, $y = 0$, and $C_2 = 0$ in Eq. (8.49), we write

$$0 = -0.0625\langle 3.0\rangle^4 + 0.0625\langle 1.8\rangle^4$$
$$+ 0.4333\langle 3.6\rangle^3 - 0.2\langle 3.0\rangle^3 - 0.72\langle 1.0\rangle^2 + C_1\langle 3.6\rangle + 0$$

Since all the quantities between brackets are positive, the brackets may be replaced by ordinary parentheses. Solving for C_1, we find $C_1 = -2.692$.

(*b*) Substituting for C_1 and C_2 into Eq. (8.49) and making $x = x_D = 1.8$ m, we find that the deflection at point *D* is defined by the relation

$$EIy_D = -0.0625\langle 1.2\rangle^4 + 0.0625\langle 0\rangle^4$$
$$+0.4333(1.8)^3 - 0.2\langle 1.2\rangle^3 - 0.72\langle -0.8\rangle^2 - 2.692(1.8)$$

The last bracket contains a negative quantity and, therefore, is equal to zero. All the other brackets contain positive quantities and may be replaced by ordinary parentheses. We have

$$EIy_D = -0.0625(1.2)^4 + 0.0625(0)^4$$
$$+ 0.4333(1.8)^3 - 0.2(1.2)^3 - 0 - 2.692(1.8) = -2.794$$

Recalling the given numerical values of *E* and *I*, we write

$$(200 \text{ GPa})(1.024 \times 10^{-6} \text{ m}^4)y_D = -2.794 \text{ kN} \cdot \text{m}^3$$
$$y_D = -13.64 \times 10^{-3} \text{ m} = -13.64 \text{ mm}$$

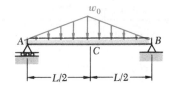

SAMPLE PROBLEM 8.4

For the prismatic beam and loading shown, determine (a) the equation of the elastic curve, (b) the slope at A, (c) the maximum deflection.

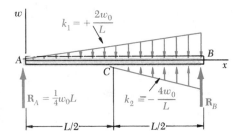

Bending Moment. The equation defining the bending moment of the beam was obtained in Sample Prob. 7.7. Using the modified loading diagram shown, we had [Eq. (3)]:

$$M(x) = -\frac{w_0}{3L}x^3 + \frac{2w_0}{3L}\langle x - \tfrac{1}{2}L\rangle^3 + \tfrac{1}{4}w_0 Lx$$

(a) **Equation of the Elastic Curve.** Using Eq. (8.4), we write

$$EI\frac{d^2y}{dx^2} = -\frac{w_0}{3L}x^3 + \frac{2w_0}{3L}\langle x - \tfrac{1}{2}L\rangle^3 + \tfrac{1}{4}w_0 Lx \qquad (1)$$

and, integrating twice in x,

$$EI\,\theta = -\frac{w_0}{12L}x^4 + \frac{w_0}{6L}\langle x - \tfrac{1}{2}L\rangle^4 + \frac{w_0 L}{8}x^2 + C_1 \qquad (2)$$

$$EI\,y = -\frac{w_0}{60L}x^5 + \frac{w_0}{30L}\langle x - \tfrac{1}{2}L\rangle^5 + \frac{w_0 L}{24}x^3 + C_1 x + C_2 \qquad (3)$$

Boundary Conditions.

$[x = 0, y = 0]$: Using Eq. (3) and noting that each bracket $\langle\ \rangle$ contains a negative quantity and, thus, is equal to zero, we find $C_2 = 0$.

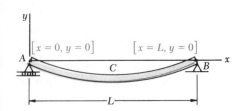

$[x = L, y = 0]$: Again using Eq. (3), we write

$$0 = -\frac{w_0 L^4}{60} + \frac{w_0}{30L}\left(\frac{L}{2}\right)^5 + \frac{w_0 L^4}{24} + C_1 L \qquad C_1 = -\frac{5}{192}w_0 L^3$$

Substituting C_1 and C_2 into Eqs. (2) and (3), we have

$$EI\,\theta = -\frac{w_0}{12L}x^4 + \frac{w_0}{6L}\langle x - \tfrac{1}{2}L\rangle^4 + \frac{w_0 L}{8}x^2 - \frac{5}{192}w_0 L^3 \qquad (4)$$

$$EI\,y = -\frac{w_0}{60L}x^5 + \frac{w_0}{30L}\langle x - \tfrac{1}{2}L\rangle^5 + \frac{w_0 L}{24}x^3 - \frac{5}{192}w_0 L^3 x \qquad (5) \blacktriangleleft$$

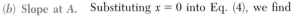

(b) **Slope at A.** Substituting $x = 0$ into Eq. (4), we find

$$EI\,\theta_A = -\frac{5}{192}w_0 L^3 \qquad\qquad \theta_A = \frac{5w_0 L^3}{192EI} \; \text{⦨} \quad \blacktriangleleft$$

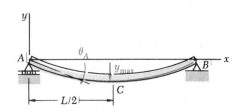

(c) **Maximum Deflection.** Because of the symmetry of the supports and loading, the maximum deflection occurs at point C, where $x = \tfrac{1}{2}L$. Substituting into Eq. (5), we obtain

$$EI\,y_{max} = w_0 L^4\left[-\frac{1}{60(32)} + 0 + \frac{1}{24(8)} - \frac{5}{192(2)}\right] = -\frac{w_0 L^4}{120EI}$$

$$y_{max} = \frac{w_0 L^4}{120EI}\;\downarrow \quad \blacktriangleleft$$

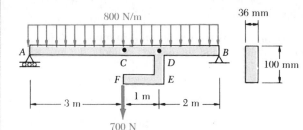

SAMPLE PROBLEM 8.5

The rigid bar DEF is welded at point D to the uniform steel beam AB. For the loading shown, determine (a) the equation of the elastic curve of the beam, (b) the deflection at the midpoint C of the beam. Use $E = 200$ GPa.

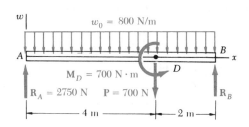

Bending Moment. The equation defining the bending moment of the beam was obtained in Sample Prob. 7.8. Using the modified loading diagram shown, we had [Eq. (3)]:

$$M(x) = [-0.4x^2 + 2.750x - 0.7\langle x - 4\rangle^1 - 0.7\langle x - 4\rangle^0]10^3 \text{ N} \cdot \text{m}$$

(a) **Equation of the Elastic Curve.** Using Eq. (8.4), we write

$$EI(d^2y/dx^2) = [-0.4x^2 + 2.75x - 0.7\langle x - 4\rangle^1 - 0.7\langle x - 4\rangle^0]10^3 \text{ N} \cdot \text{m} \quad (1)$$

and, integrating twice in x,

$$EI\,\theta = [-0.1333x^3 + 1.375x^2 - 0.350\langle x - 4\rangle^2 - 0.7\langle x - 4\rangle^1]10^3 + C_1 \text{ N} \cdot \text{m}^2 \quad (2)$$

$$EI\,y = [-0.0333x^4 + 0.458x^3 - 0.117\langle x - 4\rangle^3 - 0.350\langle x - 4\rangle^2]10^3$$
$$+ C_1 x + C_2 \text{ N} \cdot \text{m}^3 \quad (3)$$

Boundary Conditions.

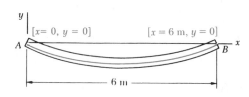

$[x = 0, y = 0]$: Using Eq. (3) and noting that each bracket $\langle\ \rangle$ contains a negative quantity and, thus, is equal to zero, we find $C_2 = 0$.

$[x = 6\text{ m}, y = 0]$: Again using Eq. (3) and noting that each bracket contains a positive quantity and, thus, may be replaced by a parenthesis, we write

$$0 = [-0.0333(6)^4 + 0.458(6)^3 - 0.117(2)^3 - 0.350(2)^2]10^3 + C_1(6)$$
$$C_1 = -8.906 \times 10^3$$

Substituting the values found for C_1 and C_2 into Eq. (3), we have

$$EI\,y = [-0.0333x^4 + 0.458x^3 - 0.117\langle x - 4\rangle^3 - 0.350\langle x - 4\rangle^2 - 8.906x]10^3 \quad (3') \blacktriangleleft$$

To determine EI, we recall that $E = 200$ GPa and compute

$$I = \tfrac{1}{12}bh^3 = \tfrac{1}{12}(0.036 \text{ m})(0.100 \text{ m})^3 = 3 \times 10^{-6} \text{ m}^4$$
$$EI = (200 \times 10^9 \text{ Pa})(3 \times 10^{-6} \text{ m}^4) = 600 \times 10^3 \text{ N} \cdot \text{m}^2$$

(b) **Deflection at Midpoint C.** Making $x = 3$ m in Eq. (3'), we write

$$EI\,y_C = [-0.0333(3)^4 + 0.458(3)^3 - 0.117\langle -1\rangle^3 - 0.350\langle -1\rangle^2 - 8.906(3)]10^3$$

Noting that each bracket is equal to zero and substituting for EI its numerical value, we have

$$(600 \times 10^3 \text{ N} \cdot \text{m}^2)y_C = -17.049 \times 10^3 \text{ N} \cdot \text{m}^3$$

and, solving for y_C:

$$y_C = -28.4 \text{ mm} \quad \blacktriangleleft$$

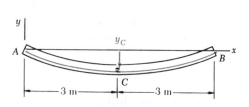

Note that the deflection obtained is *not* the maximum deflection.

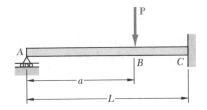

SAMPLE PROBLEM 8.6

For the uniform beam ABC, (a) express the reaction at A in terms of P, L, a, E, and I, (b) determine the reaction at A and the deflection under the load when $a = L/2$.

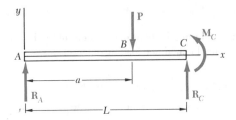

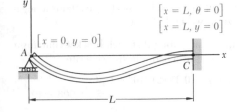

Reactions. For the given vertical load **P** the reactions are as shown. We note that they are statically indeterminate.

Shear and Bending Moment. Using a step function to represent the contribution of **P** to the shear, we write

$$V(x) = R_A - P\langle x - a \rangle^0$$

Integrating in x, we obtain the bending moment:

$$M(x) = R_A x - P\langle x - a \rangle^1$$

Equation of the Elastic Curve. Using Eq. (8.4), we write

$$EI \frac{d^2 y}{dx^2} = R_A x - P\langle x - a \rangle^1$$

Integrating twice in x,

$$EI \frac{dy}{dx} = EI\,\theta = \frac{1}{2} R_A x^2 - \frac{1}{2} P\langle x - a \rangle^2 + C_1$$

$$EI\,y = \frac{1}{6} R_A x^3 - \frac{1}{6} P\langle x - a \rangle^3 + C_1 x + C_2$$

Boundary Conditions. Noting that the bracket $\langle x - a \rangle$ is equal to zero for $x = 0$, and to $(L - a)$ for $x = L$, we write

$[x = 0,\ y = 0]$:	$C_2 = 0$	(1)
$[x = L,\ \theta = 0]$:	$\frac{1}{2} R_A L^2 - \frac{1}{2} P(L - a)^2 + C_1 = 0$	(2)
$[x = L,\ y = 0]$:	$\frac{1}{6} R_A L^3 - \frac{1}{6} P(L - a)^3 + C_1 L + C_2 = 0$	(3)

(a) **Reaction at A.** Multiplying Eq. (2) by L, subtracting Eq. (3) member by member from the equation obtained, and noting that $C_2 = 0$, we have

$$\frac{1}{3} R_A L^3 - \frac{1}{6} P(L - a)^2 [3L - (L - a)] = 0 \qquad R_A = P\left(1 - \frac{a}{L}\right)^2 \left(1 + \frac{a}{2L}\right) \uparrow \quad \blacktriangleleft$$

We note that the reaction is independent of E and I.

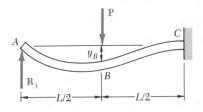

(b) **Reaction at A and Deflection at B when $a = \frac{1}{2}L$.** Making $a = \frac{1}{2}L$ in the expression obtained for R_A, we have

$$R_A = P(1 - \tfrac{1}{2})^2 (1 + \tfrac{1}{4}) = 5P/16 \qquad\qquad R_A = \tfrac{5}{16}P \uparrow \quad \blacktriangleleft$$

Substituting $a = L/2$ and $R_A = 5P/16$ into Eq. (2) and solving for C_1, we find $C_1 = -PL^2/32$. Making $x = L/2$, $C_1 = -PL^2/32$, and $C_2 = 0$ in the expression obtained for y, we have

$$y_B = -\frac{7PL^3}{768EI} \qquad\qquad y_B = \frac{7PL^3}{768EI} \downarrow \quad \blacktriangleleft$$

Note that the deflection obtained is *not* the maximum deflection.

PROBLEMS

Use singularity functions to solve the following problems and assume that the flexural rigidity EI of each beam is constant.

8.59 through 8.62 For the beam and loading shown, determine (a) the equation of the elastic curve, (b) the slope at end A, (c) the deflection at point C.

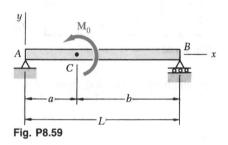

Fig. P8.59

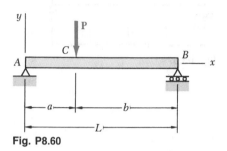

Fig. P8.60

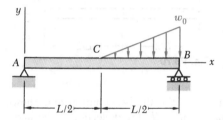

Fig. P8.61

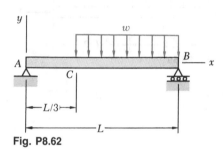

Fig. P8.62

8.63 through 8.66 For the beam and loading shown, determine (a) the equation of the elastic curve, (b) the deflection at the free end.

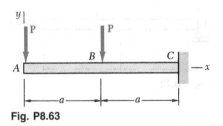

Fig. P8.63

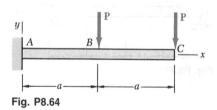

Fig. P8.64

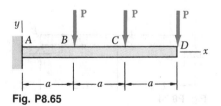

Fig. P8.65

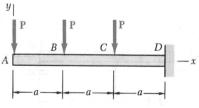

Fig. P8.66

8.67 and 8.68   For the beam and loading shown, determine the deflection at
(a) point B, (b) point C.

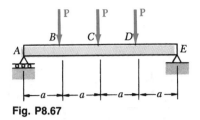

Fig. P8.67

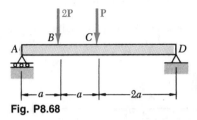

Fig. P8.68

8.69 and 8.70   For the beam and loading shown, determine (a) the equation
of the elastic curve, (b) the deflection at the free end.

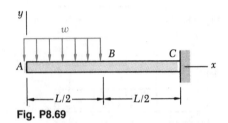

Fig. P8.69

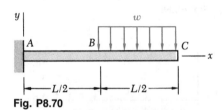

Fig. P8.70

8.71 thorough 8.74   For the beam and loading shown, determine (a) the
equation of the elastic curve, (b) the deflection at the midpoint C.

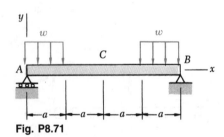

Fig. P8.71

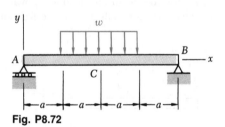

Fig. P8.72

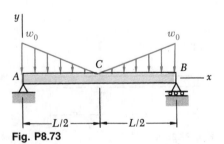

Fig. P8.73

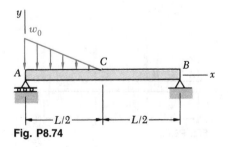

Fig. P8.74

8.75 and 8.76 For the beam and loading shown, determine (*a*) the equation of the elastic curve, (*b*) the deflection at point *B*, (*c*) the deflection at point *D*.

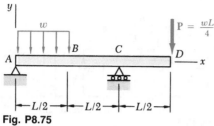

Fig. P8.75

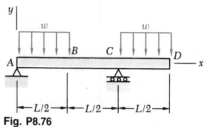

Fig. P8.76

8.77 For the beam and loading shown, determine (*a*) the slope at *B*, (*b*) the deflection at *A*. Use *E* = 200 GPa.

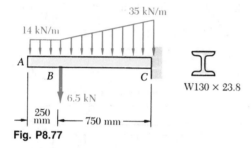

Fig. P8.77

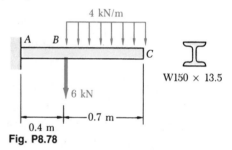

Fig. P8.78

8.78 For the beam and loading shown, determine (*a*) the slope at *B*, (*b*) the deflection at *C*. Use *E* = 200 GPa.

8.79 For the beam and loading shown, determine (*a*) the slope at end *A*, (*b*) the deflection at *B*. Use *E* = 200 GPa.

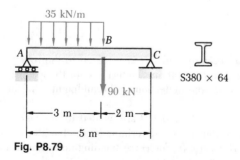

Fig. P8.79

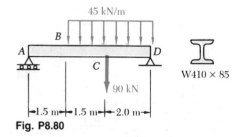

Fig. P8.80

8.80 For the beam and loading shown, determine (*a*) the slope at end *A*, (*b*) the deflection at *C*. Use *E* = 200 GPa.

8.81 Knowing that beam AD is made of a solid steel rod, determine (*a*) the slope at end A, (*b*) the deflection at point B.   Use $E = 200$ GPa.

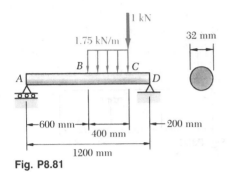

Fig. P8.81

8.82 For the wooden beam and loading shown, determine (*a*) the slope at end A, (*b*) the deflection at point C.   Use $E = 12$ GPa.

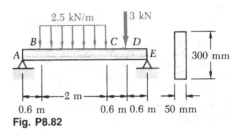

Fig. P8.82

8.83 Three rigid bars are welded to the steel rod AD as shown.   For the loading shown, determine the deflection (*a*) at point B, (*b*) at point C.   Use $E = 200$ GPa.

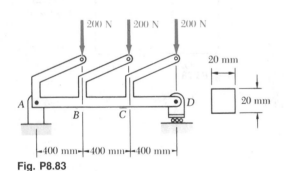

Fig. P8.83

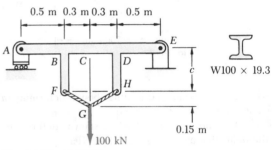

Fig. P8.84 and P8.85

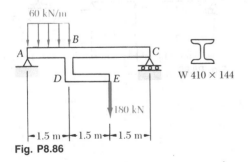

Fig. P8.86

8.84 The rigid bars BF and DH are welded to the rolled-steel beam AE as shown.   Knowing that $c = 150$ mm, determine for the loading shown (*a*) the deflection at point B, (*b*) the deflection at the midpoint C of the span.   Use $E = 200$ GPa.

8.85 The rigid bars BF and DH are welded to the rolled-steel beam AE as shown.   For the loading shown, determine (*a*) the length c of the bars for which the deflection at B is zero, (*b*) the corresponding deflection at the midpoint C of the span.   Use $E = 200$ GPa.

8.86 The rigid bar BDE is welded at point B to the rolled-steel beam AC. For the loading shown, determine (*a*) the slope at B, (*b*) the deflection at point B. Use $E = 200$ GPa.

8.87 The rigid bar *DEF* is welded at point *D* to the rolled-steel beam *AB*. For the loading shown, determine (*a*) the slope at *A*, (*b*) the deflection at the midpoint *C* of the beam. Use *E* = 200 GPa.

8.88 through 8.91† For the beam and loading indicated, write a computer program and use it to calculate the slope and deflection of the beam at intervals Δ*L* starting at point *A* and ending at the right-hand support.

8.88 Beam and loading of Prob. 8.80 with Δ*L* = 0.25 m.
8.89 Beam and loading of Prob. 8.79 with Δ*L* = 0.5 m.
8.90 Beam and loading of Prob. 8.82 with Δ*L* = 0.2 m.
8.91 Beam and loading of Prob. 8.81 with Δ*L* = 100 mm.

8.92 through 8.95 For the beam and loading shown, determine (*a*) the reaction at the roller support, (*b*) the deflection at point *B*.

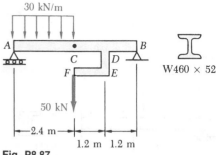

Fig. P8.87

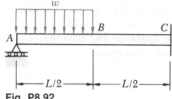

Fig. P8.92

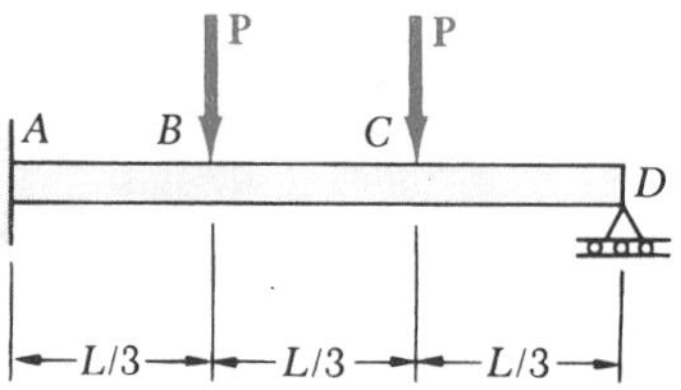

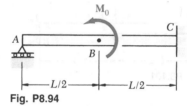

Fig. P8.94

Fig. P8.93

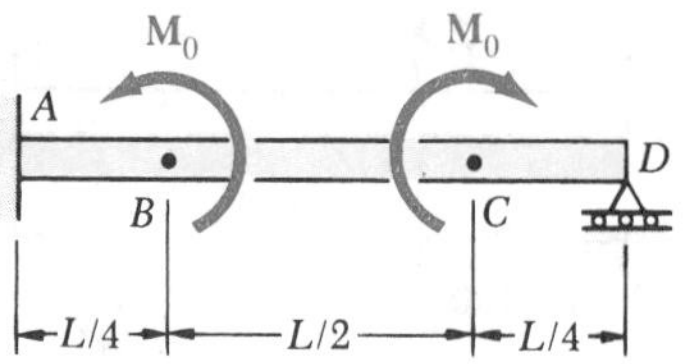

Fig. P8.95

8.96 For the beam and loading shown, determine (*a*) the reaction at *A*, (*b*) the deflection at point *B*. Use *E* = 200 GPa.

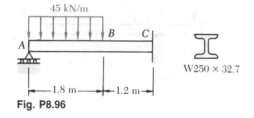

Fig. P8.96

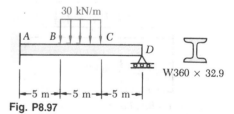

Fig. P8.97

8.97 For the beam and loading shown, **determine** (*a*) the reaction at *D*, (*b*) the deflection at point *B*. Use *E* = **200 GPa.**

†Problems 8.88 through 8.91 require the use of a computer.

8.98 For the beam and loading shown, determine (*a*) the reaction at *A*, (*b*) the deflection at point *B*. Use $E = 200$ GPa.

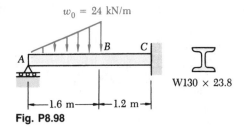

$w_0 = 24$ kN/m

W130 × 23.8

|—1.6 m—|—1.2 m—|

Fig. P8.98

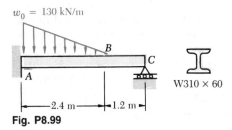

$w_0 = 130$ kN/m

W310 × 60

|——2.4 m——|—1.2 m—|

Fig. P8.99

8.99 For the beam and loading shown, determine (*a*) the reaction at *C*, (*b*) the deflection at point *B*. Use $E = 200$ GPa.

8.100 and 8.101 For the beam and loading shown, determine (*a*) the reaction at *A*, (*b*) the deflection at midspan.

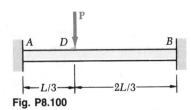

P

|—*L*/3—|———2*L*/3———|

Fig. P8.100

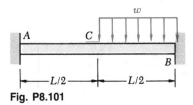

w

|———*L*/2———|———*L*/2———|

Fig. P8.101

8.102 and 8.103 For the beam and loading shown, determine (*a*) the reaction at *A*, (*b*) the bending moment at *D*.

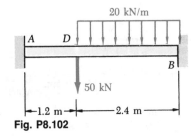

20 kN/m

50 kN

|—1.2 m—|———2.4 m———|

Fig. P8.102

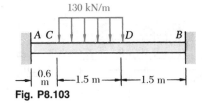

130 kN/m

|0.6 m| |—1.5 m—|—1.5 m—|

Fig. P8.103

8.104 through 8.107 For the beam and loading indicated, determine the magnitude and location of the largest downward deflection.

    8.104 Beam and loading of Prob. 8.79.
    8.105 Beam and loading of Prob. 8.80.
    8.106 Beam and loading of Prob. 8.81.
    8.107 Beam and loading of Prob. 8.82.

8.7. METHOD OF SUPERPOSITION

When a beam is subjected to several concentrated or distributed loads, it is often found convenient to compute separately the slope and deflection caused by each of the given loads. The slope and deflection due to the combined loads are then obtained by applying the principle of superposition (Sec. 2.12) and adding the values of the slope or deflection corresponding to the various loads.

Example 8.07

Determine the slope and deflection at D for the beam and loading shown (Fig. 8.31), knowing that the flexural rigidity of the beam is $EI = 100$ MN $\cdot$ m^2.

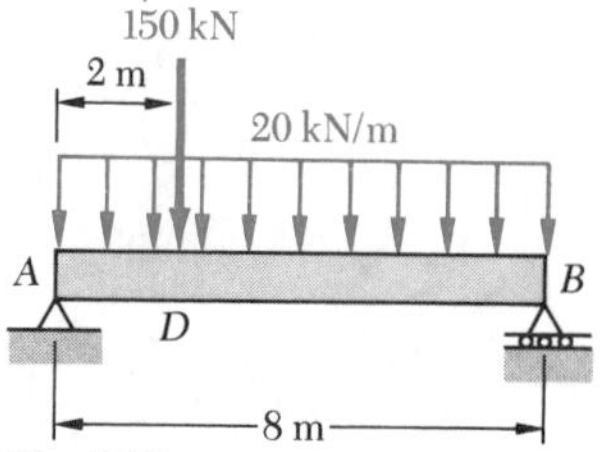

The slope and deflection at any point of the beam may be obtained by superposing the slopes and deflections caused respectively by the concentrated load and by the distributed load (Fig. 8.32).

Fig. 8.31

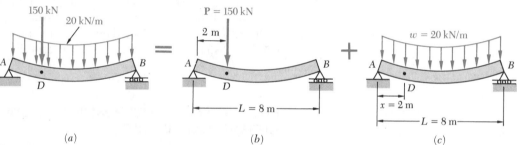

Fig. 8.32

(a)   *(b)*   *(c)*

Since the concentrated load in Fig. 8.32*b* is applied at quarter span, we may use the results obtained for the beam and loading of Example 8.03 and write

$$(\theta_D)_P = -\frac{PL^2}{32EI} = -\frac{(150 \times 10^3)(8)^2}{32(100 \times 10^6)} = -3 \times 10^{-3} \text{ rad}$$

$$(y_D)_P = -\frac{3PL^3}{256EI} = -\frac{3(150 \times 10^3)(8)^3}{256(100 \times 10^6)} = -9 \times 10^{-3} \text{ m}$$

$$= -9 \text{ mm}$$

On the other hand, recalling the equation of the elastic curve obtained for a uniformly distributed load in Example 8.02, we express the deflection in Fig. 8.32*c* as

$$y = \frac{w}{24EI}(-x^4 + 2Lx^3 - L^3x) \qquad (8.50)$$

and, differentiating with respect to x,

$$\theta = \frac{dy}{dx} = \frac{w}{24EI}(-4x^3 + 6Lx^2 - L^3) \qquad (8.51)$$

Making $w = 20$ kN/m, $x = 2$ m, and $L = 8$ m in Eqs. (8.51) and (8.50), we obtain

$$(\theta_D)_w = \frac{20 \times 10^3}{24(100 \times 10^6)}(-352) = -2.93 \times 10^{-3} \text{ rad}$$

$$(y_D)_w = \frac{20 \times 10^3}{24(100 \times 10^6)}(-912) = -7.60 \times 10^{-3} \text{ m}$$

$$= -7.60 \text{ mm}$$

Combining the slopes and deflections produced by the concentrated and the distributed loads, we have

$$\theta_D = (\theta_D)_P + (\theta_D)_w = -3 \times 10^{-3} - 2.93 \times 10^{-3}$$

$$= -5.93 \times 10^{-3} \text{ rad}$$

$$y_D = (y_D)_P + (y_D)_w = -9 \text{ mm} - 7.60 \text{ mm} = -16.60 \text{ mm}$$

To facilitate the task of practicing engineers, most structural and mechanical engineering handbooks include tables giving the deflections and slopes of beams for various loadings and types of support. Such a table will be found in Appendix D. We note that the slope and deflection of the beam of Fig. 8.31 could have been determined from that table.

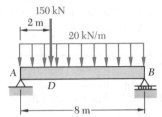

Fig. 8.31 (repeated)

Indeed, using the information given under cases 5 and 6, we could have expressed the deflection of the beam for any value $x \leq L/4$. Taking the derivative of the expression obtained in this way would have yielded the slope of the beam over the same interval. We also note that the slope at both ends of the beam may be obtained by simply adding the corresponding values given in the table. However, the maximum deflection of the beam of Fig. 8.31 *cannot* be obtained by adding the maximum deflections of cases 5 and 6, since these deflections occur at different points of the beam.†

8.8. APPLICATION OF SUPERPOSITION TO STATICALLY INDETERMINATE BEAMS

We shall often find it convenient to use the method of superposition to determine the reactions at the supports of a statically indeterminate beam. Considering first the case of a beam indeterminate to the first degree (cf. Sec. 8.5), we shall follow the approach described in Sec. 2.9. We designate one of the reactions as redundant and eliminate or modify accordingly the corresponding support. The redundant reaction is then treated as an unknown load which, together with the other loads, must produce deformations which are compatible with the original supports. The slope or deflection at the point where the support has been modified or eliminated is obtained by computing separately the deformations caused by the given loads and by the redundant reaction, and by superposing the results obtained. Once the reactions at the supports have been found, the slope and deflection may be determined in the usual way at any other point of the beam.

† An approximate value of the maximum deflection of the beam may be obtained by plotting the values of y corresponding to various values of x. The determination of the exact location and magnitude of the maximum deflection would require setting equal to zero the expression obtained for the slope of the beam and solving this equation for x.

Example 8.08

Determine the reactions at the supports for the prismatic beam and loading shown in Fig. 8.33. (This is the same beam and loading as in Example 8.05 of Sec. 8.5.)

We consider the reaction at B as redundant and release the beam from that support. The reaction $\mathbf{R}_B$ is now considered as an unknown load (Fig. 8.34a) and will be determined from the condition that the deflection of the beam at B must be zero. The solution is carried out by considering separately the deflection $(y_B)_w$ caused at B by the uniformly distributed load w (Fig. 8.34b) and the deflection $(y_B)_R$ produced at the same point by the redundant reaction $\mathbf{R}_B$ (Fig. 8.34c).

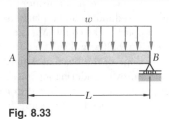

Fig. 8.33

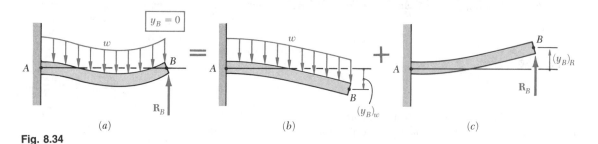

Fig. 8.34

From the table of Appendix D (cases 2 and 1), we find that

$$(y_B)_w = -\frac{wL^4}{8EI} \qquad (y_B)_R = +\frac{R_B L^3}{3EI}$$

Writing that the deflection at B is the sum of these two quantities and that it must be zero, we have

$$y_B = (y_B)_w + (y_B)_R = 0$$

$$y_B = -\frac{wL^4}{8EI} + \frac{R_B L^3}{3EI} = 0$$

and, solving for R_B,

$$R_B = \tfrac{3}{8}wL \qquad \mathbf{R}_B = \tfrac{3}{8}wL \uparrow$$

Drawing the free-body diagram of the beam (Fig. 8.35) and writing the corresponding equilibrium equations, we have

$+\uparrow \Sigma F_y = 0:\qquad R_A + R_B - wL = 0 \qquad\qquad (8.52)$
$$R_A = wL - R_B = wL - \tfrac{3}{8}wL = \tfrac{5}{8}wL$$
$$\mathbf{R}_A = \tfrac{5}{8}wL \uparrow$$

$+\curvearrowleft \Sigma M_A = 0:\qquad M_A + R_B L - (wL)(\tfrac{1}{2}L) = 0 \qquad (8.53)$
$$M_A = \tfrac{1}{2}wL^2 - R_B L = \tfrac{1}{2}wL^2 - \tfrac{3}{8}wL^2 = \tfrac{1}{8}wL^2$$
$$\mathbf{M}_A = \tfrac{1}{8}wL^2 \curvearrowright$$

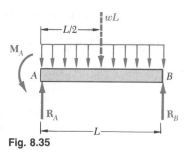

Fig. 8.35

Alternative Solution. We may consider the couple exerted at the fixed end A as redundant and replace the fixed end by a pin-and-bracket support. The couple $\mathbf{M}_A$ is now considered as an unknown load (Fig. 8.36a) and will be determined from the condition that the slope of the beam at A must be zero. The solution is carried out by considering separately the slope $(\theta_A)_w$ caused at A by the uniformity distributed load w (Fig. 8.36b) and the slope $(\theta_A)_M$ produced at the same point by the unknown couple $\mathbf{M}_A$ (Fig. 8.36c).

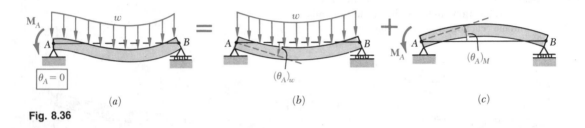

Fig. 8.36

Using the table of Appendix D (cases 6 and 7), and noting that in case 7, A and B must be interchanged, we find that

$$(\theta_A)_w = -\frac{wL^3}{24EI} \qquad (\theta_A)_M = \frac{M_A L}{3EI}$$

Writing that the slope at A is the sum of these two quantities and that it must be zero, we have

$$\theta_A = (\theta_A)_w + (\theta_A)_M = 0$$

$$\theta_A = -\frac{wL^3}{24EI} + \frac{M_A L}{3EI} = 0$$

and, solving for M_A,

$$M_A = \tfrac{1}{8}wL^2 \qquad \mathbf{M}_A = \tfrac{1}{8}wL^2 \; \rotatebox{90}{$\curvearrowright$}$$

The values of R_A and R_B may then be found from the equilibrium equations (8.52) and (8.53).

The beam considered in the preceding example was indeterminate to the first degree. In the case of a beam indeterminate to the second degree (cf. Sec. 8.5), two reactions must be designated as redundant, and the corresponding supports must be eliminated or modified accordingly. The redundant reactions are then treated as unknown loads which, simultaneously and together with the other loads, must produce deformations which are compatible with the original supports. (See Sample Prob. 8.9.)

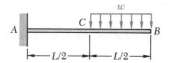

For the beam and loading shown, determine the slope and deflection at point B.

The given loading may be obtained by super-posing the loadings shown in the following "picture equation." The beam AB is, of course, the same in each part of the figure.

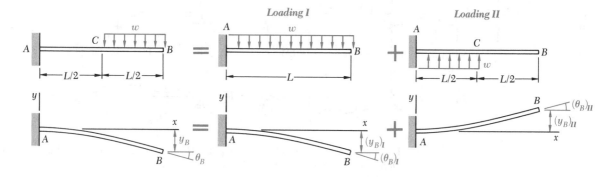

For each of the loadings I and II, we now determine the slope and deflection at B by using the table of *Beam Deflections and Slopes* in Appendix D.

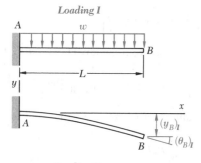

$$(\theta_B)_{\mathrm{I}} = -\frac{wL^3}{6EI} \qquad (y_B)_{\mathrm{I}} = -\frac{wL^4}{8EI}$$

$$(\theta_C)_{\mathrm{II}} = +\frac{w(L/2)^3}{6EI} = +\frac{wL^3}{48EI} \qquad (y_C)_{\mathrm{II}} = +\frac{w(L/2)^4}{8EI} = +\frac{wL^4}{128EI}$$

In portion CB, the bending moment for loading II is zero and thus the elastic curve is a straight line.

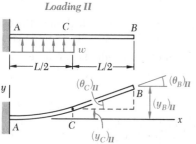

$$(\theta_B)_{\mathrm{II}} = (\theta_C)_{\mathrm{II}} = +\frac{wL^3}{48EI} \qquad (y_B)_{\mathrm{II}} = (y_C)_{\mathrm{II}} + (\theta_C)_{\mathrm{II}}\left(\frac{L}{2}\right)$$

$$= \frac{wL^4}{128EI} + \frac{wL^3}{48EI}\left(\frac{L}{2}\right) = +\frac{7wL^4}{384EI}$$

$$\theta_B = (\theta_B)_{\mathrm{I}} + (\theta_B)_{\mathrm{II}} = -\frac{wL^3}{6EI} + \frac{wL^3}{48EI} = -\frac{7wL^3}{48EI}$$

$$y_B = (y_B)_{\mathrm{I}} + (y_B)_{\mathrm{II}} = -\frac{wL^4}{8EI} + \frac{7wL^4}{384EI} = -\frac{41wL^4}{384EI}$$

515

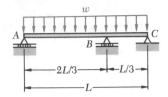

SAMPLE PROBLEM 8.8

For the uniform beam and loading shown, determine (a) the reaction at each support, (b) the slope at end A.

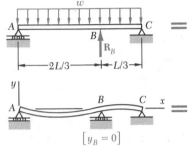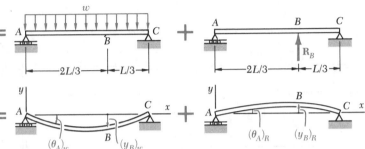

Principle of Superposition. The reaction $\mathbf{R}_B$ is designated as redundant and considered as an unknown load. The deflections due to the distributed load and to the reaction $\mathbf{R}_B$ are considered separately as shown below.

For each loading the deflection at point B is found by using the table of *Beam Deflections and Slopes* in Appendix D.

Distributed Loading. We use case 6, Appendix D

$$y = -\frac{w}{24EI}(x^4 - 2Lx^3 + L^3x)$$

At point B, $x = \frac{2}{3}L$:

$$(y_B)_w = -\frac{w}{24EI}\left[\left(\frac{2}{3}L\right)^4 - 2L\left(\frac{2}{3}L\right)^3 + L^3\left(\frac{2}{3}L\right)\right] = -0.01132\frac{wL^4}{EI}$$

Redundant Reaction Loading. From case 5, Appendix D, with $a = \frac{2}{3}L$ and $b = \frac{1}{3}L$, we have

$$(y_B)_R = -\frac{Pa^2b^2}{3EIL} = +\frac{R_B}{3EIL}\left(\frac{2}{3}L\right)^2\left(\frac{L}{3}\right)^2 = 0.01646\frac{R_BL^3}{EI}$$

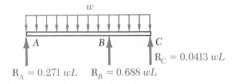

(a) Reactions at Supports. Recalling that $y_B = 0$, we write

$$y_B = (y_B)_w + (y_B)_R$$

$$0 = -0.01132\frac{wL^4}{EI} + 0.01646\frac{R_BL^3}{EI} \qquad\qquad \mathbf{R}_B = 0.688wL \uparrow \quad \blacktriangleleft$$

Since the reaction R_B is now known, we may use the methods of statics to determine the other reactions: $\qquad \mathbf{R}_A = 0.271wL \uparrow \qquad \mathbf{R}_C = 0.0413wL \uparrow \quad \blacktriangleleft$

(b) Slope at End A. Referring again to Appendix D, we have

Distributed Loading. $\quad (\theta_A)_w = -\dfrac{wL^3}{24EI} = -0.04167\dfrac{wL^3}{EI}$

Redundant Reaction Loading. For $P = -R_B = -0.688wL$ and $b = \frac{1}{3}L$

$$(\theta_A)_R = -\frac{Pb(L^2 - b^2)}{6EIL} = +\frac{0.688wL}{6EIL}\left(\frac{L}{3}\right)\left[L^2 - \left(\frac{L}{3}\right)^2\right] \qquad (\theta_A)_R = 0.03398\frac{wL^3}{EI}$$

Finally, $\qquad \theta_A = (\theta_A)_w + (\theta_A)_R$

$$\theta_A = -0.04167\frac{wL^3}{EI} + 0.03398\frac{wL^3}{EI} = -0.00769\frac{wL^3}{EI} \qquad \theta_A = 0.00769\frac{wL^3}{EI} \quad \blacktriangleleft$$

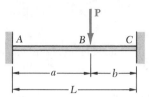

SAMPLE PROBLEM 8.9

For the beam and loading shown, determine the reaction at the fixed support C.

Principle of Superposition. Assuming the axial force in the beam to be zero, the beam ABC is indeterminate to the second degree and we choose two reaction components as redundant, namely, the vertical force $\mathbf{R}_C$ and the couple $\mathbf{M}_C$. The deformations caused by the given load $\mathbf{P}$, the force $\mathbf{R}_C$, and the couple $\mathbf{M}_C$ will be considered separately as shown.

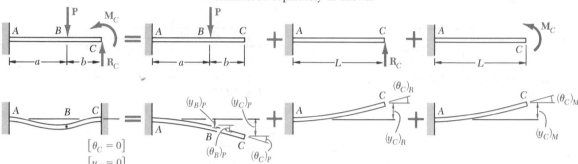

For each load, the slope and deflection at point C will be found by using the table of *Beam Deflections and Slopes* in Appendix D.

Load P. We note that, for this loading, portion BC of the beam is straight.

$$(\theta_C)_P = (\theta_B)_P = -\frac{Pa^2}{2EI} \qquad (y_C)_P = (y_B)_P + (\theta_B)_P b$$

$$= -\frac{Pa^3}{3EI} - \frac{Pa^2}{2EI}b = -\frac{Pa^2}{6EI}(2a + 3b)$$

Force R_C $\qquad (\theta_C)_R = +\dfrac{R_C L^2}{2EI} \qquad\qquad (y_C)_R = +\dfrac{R_C L^3}{3EI}$

Couple M_C $\qquad (\theta_C)_M = +\dfrac{M_C L}{EI} \qquad\qquad (y_C)_M = +\dfrac{M_C L^2}{2EI}$

Boundary Conditions. At end C the slope and deflection must be zero:

$[x = L,\ \theta_C = 0]$: $\qquad \theta_C = (\theta_C)_P + (\theta_C)_R + (\theta_C)_M$

$$0 = -\frac{Pa^2}{2EI} + \frac{R_C L^2}{2EI} + \frac{M_C L}{EI} \qquad (1)$$

$[x = L,\ y_C = 0]$: $\qquad y_C = (y_C)_P + (y_C)_R + (y_C)_M$

$$0 = -\frac{Pa^2}{6EI}(2a + 3b) + \frac{R_C L^3}{3EI} + \frac{M_C L^2}{2EI} \qquad (2)$$

Reaction Components at C. Solving simultaneously Eqs. (1) and (2), we find after reductions

$$R_C = +\frac{Pa^2}{L^3}(a + 3b) \qquad\qquad\qquad \mathbf{R_C = \frac{Pa^2}{L^3}(a + 3b)\uparrow} \quad \blacktriangleleft$$

$$M_C = -\frac{Pa^2 b}{L^2} \qquad\qquad\qquad\qquad \mathbf{M_C = \frac{Pa^2 b}{L^2}}\ \downarrow \quad \blacktriangleleft$$

Using the methods of statics, we may now determine the reaction at A.

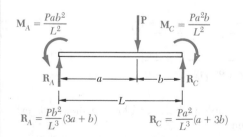

$$M_A = \frac{Pab^2}{L^2}$$

$$M_C = \frac{Pa^2 b}{L^2}$$

$$R_A = \frac{Pb^2}{L^3}(3a + b) \qquad\qquad R_C = \frac{Pa^2}{L^3}(a + 3b)$$

PROBLEMS

Use the method of superposition to solve the following problems and assume that the flexural rigidity EI of each beam is constant.

8.108 and 8.109 For the beam and loading shown, determine (a) the deflection at point C, (b) the slope at end A.

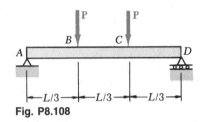

Fig. P8.108

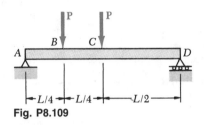

Fig. P8.109

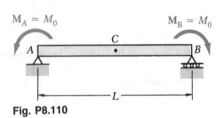

Fig. P8.110

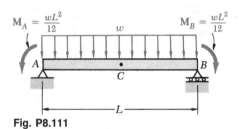

Fig. P8.111

8.110 and 8.111 For the beam and loading shown, determine (a) the deflection at the midpoint C, (b) the slope at end A.

8.112 through 8.115 For the cantilever beam and loading shown, determine the slope and deflection at the free end.

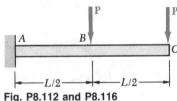

Fig. P8.112 and P8.116

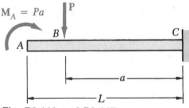

Fig. P8.113 and P8.117

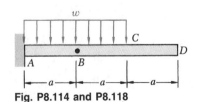

Fig. P8.114 and P8.118

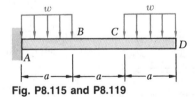

Fig. P8.115 and P8.119

8.116 through 8.119 For the cantilever beam and loading shown, determine the slope and deflection at point B.

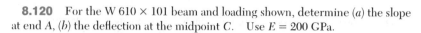

Fig. P8.120

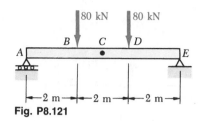

Fig. P8.121

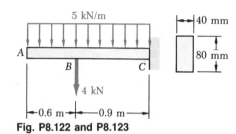

Fig. P8.122 and P8.123

8.120 For the W 610 × 101 beam and loading shown, determine (*a*) the slope at end *A*, (*b*) the deflection at the midpoint *C*. Use $E = 200$ GPa.

8.121 For the W 610 × 101 beam and loading shown, determine (*a*) the slope at end *A*, (*b*) the deflection at the midpoint *C*. Use $E = 200$ GPa.

8.122 For the cantilever beam and loading shown, determine the slope and deflection at end *A*. Use $E = 200$ GPa.

8.123 For the cantilever beam and loading shown, determine the slope and deflection at point *B*. Use $E = 200$ GPa.

8.124 For the cantilever beam and loading shown, determine the slope and deflection at end *C*. Use $E = 200$ GPa.

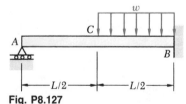

Fig. P8.124 and P8.125

8.125 For the cantilever beam and loading shown, determine the slope and deflection at point *B*. Use $E = 200$ GPa.

8.126 and 8.127 For the uniform beam shown, determine (*a*) the reaction at *A*, (*b*) the reaction at *B*.

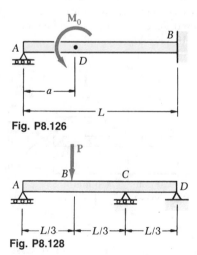

Fig. P8.126

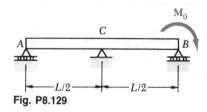

Fig. P8.127

Fig. P8.128

Fig. P8.129

8.128 and 8.129 For the uniform beam shown, determine the reaction at each of the three supports.

8.130 and 8.131 For the uniform beam shown, determine the reaction at B.

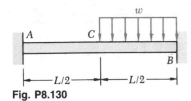

Fig. P8.130

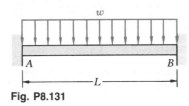

Fig. P8.131

8.132 Beam AC rests on beam DE as shown. Knowing that a W 200×19.3 rolled-steel shape is used for each beam, determine for the loading shown the deflection at point B. Use $E = 200$ GPa.

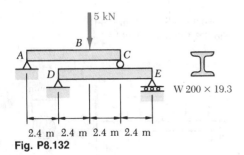

2.4 m 2.4 m 2.4 m 2.4 m

Fig. P8.132

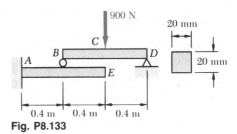

Fig. P8.133

8.133 Beam BD rests on the cantilever beam AE as shown. Knowing that a square rod of side 20 mm is used for each beam, determine for the loading shown the deflection (*a*) at point C, (*b*) at point E. Use $E = 200$ GPa.

8.134 Beam CE rests on beam AB as shown. Knowing that a W 250×49.1 rolled-steel shape is used for each beam, determine for the loading shown the deflection at point D. Use $E = 200$ GPa.

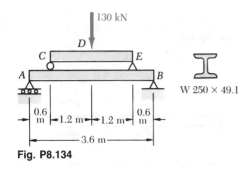

Fig. P8.134

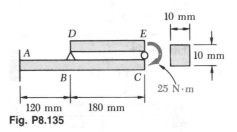

Fig. P8.135

8.135 Beam DE rests on the cantilever beam AC as shown. Knowing that a square rod of side 10 mm is used for each beam, determine the deflection at end C if the 25-N · m couple is applied (*a*) to end E of beam DE, (*b*) to end C of beam AC. Use $E = 200$ GPa.

8.136 For the loading shown, knowing that beams *AB* and *DE* have the same flexural rigidity, determine the reaction (*a*) at *A*, (*b*) at *D*.

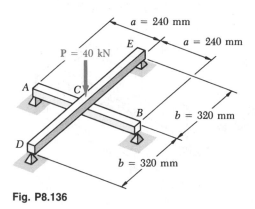

a = 240 mm

a = 240 mm

P = 40 kN

b = 320 mm

b = 320 mm

Fig. P8.136

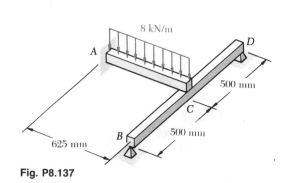

8 kN/m

500 mm

500 mm

625 mm

Fig. P8.137

8.137 For the loading shown, knowing that beams *AC* and *BD* have the same flexural rigidity, determine the reaction at *B*.

8.138 Before any load is applied, a gap $\delta_0 = 20$ mm exists between the W410 × 60 rolled-steel beam and the support at *C*.   Knowing that $E = 200$ GPa, determine the reaction at each support caused by a uniformly distributed load of 24 kN/m.

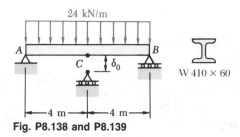

24 kN/m

δ_0

W 410 × 60

4 m   4 m

Fig. P8.138 and P8.139

8.139 Before the uniformly distributed load is applied, a gap δ_0 exists between the W 410 × 60 rolled-steel beam and the support at *C*.   Knowing that $E = 200$ GPa, determine the size of the gap for which each reaction will be equal to one-third of the total applied load.

8.140 Before the load **P** is applied, a gap $\delta_0 = 6$ mm exists between the S 75 × 8.5 rolled-steel beam and the suport at *B*.   Knowing that $E = 200$ GPa and $P = 12$ kN, determine (*a*) the reaction at *B*, (*b*) the deflection at *A*.

8.141 Before the load **P** is applied, a gap $\delta_0 = 6$ mm exists between the S 75 × 8.5 rolled-steel beam and the support at *B*.   Knowing that $E = 200$ GPa, determine the magnitude of **P** for which the deflection at *A* is 19 mm.

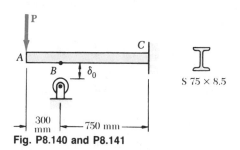

P

δ_0

S 75 × 8.5

300 mm   750 mm

Fig. P8.140 and P8.141

8.142 A 16-mm-diameter rod has been bent into the shape shown. Determine the deflection of end C caused by the 200-N force **P**. Use $E = 200$ GPa and $G = 80$ GPa.

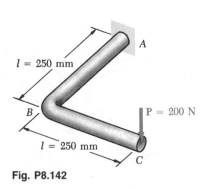

Fig. P8.142

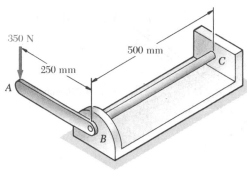

Fig. P8.143

8.143 A 22-mm-diameter rod BC is attached to the lever AB and to the fixed support at C. Lever AB has a uniform cross section 10 mm thick and 25 mm deep. For the loading shown, determine the deflection of point A. Use $E = 200$ GPa and $G = 77$ GPa.

REVIEW AND SUMMARY

This chapter was devoted to the determination by integration of slopes and deflections in prismatic beams under transverse loadings. Particular emphasis was placed on the computation of the maximum deflection of a beam under a given loading.

Deformation of a beam under transverse loading

We noted in Sec. 8.2 that Eq. (4.21) of Sec. 4.5, which relates the curvature $1/\rho$ of the neutral surface and the bending moment M in a prismatic beam in pure bending, may be applied to a beam under a transverse loading, but that both M and $1/\rho$ will vary from section to section. Denoting by x the distance from the left end of the beam, we wrote

$$\frac{1}{\rho} = \frac{M(x)}{EI} \qquad (8.1)$$

This equation enables us to determine the radius of curvature of the neutral surface for any value of x and to draw some general conclusions regarding the shape of the deformed beam.

In Sec. 8.3, we discussed how to obtain a relation between the deflection y of a beam, measured at a given point Q, and the distance x of that point from some fixed origin (Fig. 8.6b). Such a relation defines the *elastic curve* of a beam. Expressing the curvature $1/\rho$ in terms of the derivatives of the function $y(x)$ and substituting into (8.1), we obtained the following second-order linear differential equation:

$$\frac{d^2y}{dx^2} = \frac{M(x)}{EI} \tag{8.4}$$

Integrating this equation twice, we obtained the following expressions defining the slope $\theta(x) = dy/dx$ and the deflection $y(x)$, respectively:

$$EI\frac{dy}{dx} = \int_0^x M(x)\ dx + C_1 \tag{8.5}$$

$$EI\ y = \int_0^x dx \int_0^x M(x)\ dx + C_1x + C_2 \tag{8.6}$$

The product EI is known as the *flexural rigidity* of the beam; C_1 and C_2 are two constants of integration, which may be determined from the *boundary conditions* imposed on the beam by its supports (Fig. 8.8) [Example 8.01]. The maximum deflection may then be obtained by determining the value of x for which the slope is zero and the corresponding value of y [Example 8.02, Sample Prob. 8.1].

Elastic curve

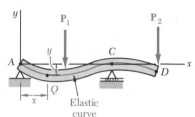

Fig. 8.6b

Boundary conditions

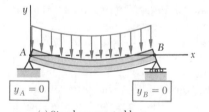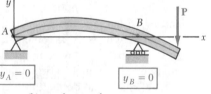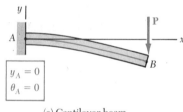

(a) Simply supported beam   (b) Overhanging beam

Fig. 8.8 Boundary conditions for statically determinate beams.

(c) Cantilever beam

When the loading is such that different analytical functions are required to represent the bending moment in various portions of the beam, then different differential equations are also required, leading to different functions representing the slope $\theta(x)$ and the deflection $y(x)$ in the various portions of the beam. In the case of the beam and loading considered in Example 8.03 (Fig. 8.19), two differential equations were required, one for the portion of beam AD and the other for the portion DB. The first equation yielded the functions θ_1 and y_1, and the second the functions θ_2 and y_2. Altogether, four constants of integration had to be determined; two were obtained by writing that the deflections at A and B were zero, and the other two by expressing that the portions of beam AD and DB had the same slope and the same deflection at D.

Elastic curve defined by different functions

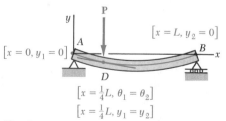

Fig. 8.19

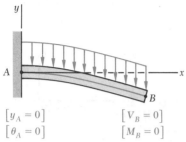

$[y_A = 0]$ $[V_B = 0]$

$[\theta_A = 0]$ $[M_B = 0]$

(a) Cantilever beam

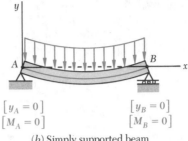

$[y_A = 0]$ $[y_B = 0]$

$[M_A = 0]$ $[M_B = 0]$

(b) Simply supported beam

Fig. 8.20 Boundary conditions for beams carrying a distributed load.

Direct determination of *y* from *w*

Statically indeterminate beams

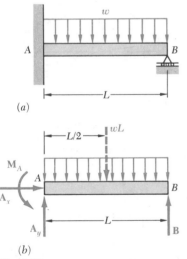

(a)

(b)

Fig. 8.23

We observed in Sec. 8.4 that in the case of a beam supporting a distributed load $w(x)$, the elastic curve may be determined directly from $w(x)$ through four successive integrations yielding V, M, θ, and y in that order. For the cantilever beam of Fig. 8.20a and the simply supported beam of Fig. 8.20b, the resulting four constants of integration may be determined from the four boundary conditions indicated in each part of the figure [Example 8.04, Sample Prob. 8.2].

In Sec. 8.5, we discussed *statically indeterminate beams*, i.e., beams supported in such a way that the reactions at the supports involved four or more unknowns. Since only three equilibrium equations are available to determine these unknowns, the equilibrium equations had to be supplemented by equations obtained from the boundary conditions imposed by the supports. In the case of the beam of Fig. 8.23, we noted that the reactions at the supports involved four unknowns, namely, M_A, A_x, A_y, and B. Such a beam is said to be *indeterminate to the first degree*. (If five unknowns were involved, the beam would be indeterminate to the *second degree*.) Expressing the bending moment $M(x)$ in terms of the four unknowns and integrating twice [Example 8.05], we determined the slope $\theta(x)$ and the deflection $y(x)$ in terms of the same unknowns and the constants of integration C_1 and C_2. The six unknowns involved in this computation were obtained by solving simultaneously the three equilibrium equations for the free body of Fig. 8.23b and the three equations expressing that $\theta = 0$, $y = 0$ for $x = 0$, and that $y = 0$ for $x = L$ (Fig. 8.24) [see also Sample Prob. 8.3].

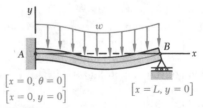

$[x = 0, \theta = 0]$

$[x = 0, y = 0]$ $[x = L, y = 0]$

Fig. 8.24

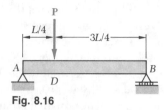

Fig. 8.16

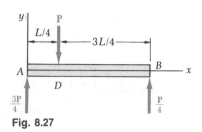

Fig. 8.27

The integration method provides an effective way for determining the slope and deflection at any point of a prismatic beam, as long as the bending moment M may be represented by a single analytical function. However, when several functions are required to represent M over the entire length of the beam, this method may become quite laborious, since it requires matching slopes and deflections at every transition point. We saw in Sec. 8.6 that the use of *singularity functions* (previously introduced in Sec. 7.5) considerably simplifies the determination of θ and y at any point of the beam. Considering again the beam of Example 8.03 (Fig. 8.16) and drawing its free-body diagram (Fig. 8.27), we expressed the shear at any point of the beam as

$$V(x) = \frac{3P}{4} - P\langle x - \tfrac{1}{4}L\rangle^0$$

where the step function $\langle x - \tfrac{1}{4}L\rangle^0$ is equal to zero when the quantity inside the brackets $\langle\ \rangle$ is negative, and equal to one otherwise. Integrating three times, we obtained successively

$$M(x) = \frac{3P}{4}x - P\langle x - \tfrac{1}{4}L\rangle \tag{8.44}$$

$$EI\ \theta = EI\frac{dy}{dx} = \tfrac{3}{8}Px^2 - \tfrac{1}{2}P\langle x - \tfrac{1}{4}L\rangle^2 + C_1 \tag{8.46}$$

$$EI\ y = \tfrac{1}{8}Px^3 - \tfrac{1}{6}P\langle x - \tfrac{1}{4}L\rangle^3 + C_1x + C_2 \tag{8.47}$$

where the brackets $\langle\ \rangle$ should be replaced by zero when the quantity inside is negative, and by ordinary parentheses otherwise. The constants C_1 and C_2 were determined from the boundary conditions shown in Fig. 8.28 [Example 8.06; Sample Probs. 8.4, 8.5, and 8.6].

The last part of the chapter was devoted to the *method of superposition*, which consists of determining separately, and then adding, the slope and deflection caused by the various loads applied to a beam [Sec. 8.7]. This procedure was facilitated by the use of the table of Appendix D, which gives the slopes and deflections of beams for various loadings and types of support [Example 8.07, Sample Prob. 8.7].

Use of singularity functions

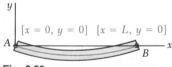

Fig. 8.28

Method of superposition

Statically indeterminate beams
by superposition

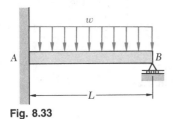

Fig. 8.33

The method of superposition may be used effectively with *statically indeterminate beams* [Sec. 8.8]. In the case of the beam of Example 8.08 (Fig. 8.33), which involves four unknown reactions and is thus indeterminate to the *first degree*, the reaction at B was considered as *redundant* and the beam was released from that support. Treating the reaction $\mathbf{R}_B$ as an unknown load and considering separately the deflections caused at B by the given distributed load and by $\mathbf{R}_B$, we wrote that the sum of these deflections was zero (Fig. 8.34). The equation obtained was solved for R_B [see also Sample Prob. 8.8]. In the case of a beam indeterminate to the *second degree*, i.e., with reactions at the supports involving five unknowns, two reactions must be designated as redundant, and the corresponding supports must be eliminated or modified accordingly [Sample Prob. 8.9].

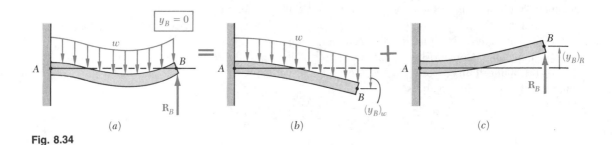

Fig. 8.34

REVIEW PROBLEMS

8.144 The uniform beam ABC supports a vertical 40-kN load at point C as shown. Using $E = 200$ GPa, determine (*a*) the slope at C, (*b*) the deflection at C.

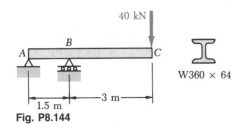

Fig. P8.144

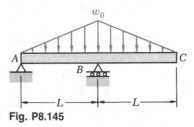

Fig. P8.145

8.145 For the beam and loading shown, determine (*a*) the slope at A, (*b*) the slope at B, (*c*) the deflection at C.

8.146 For the beam and loading shown, determine the deflection at (*a*) the midpoint *C* of span *AB*, (*b*) end *D*.   Use $E = 200$ GPa.

8.147 For the beam and loading shown, determine the magnitude and location of the largest deflection in span *AB*.   Use $E = 200$ GPa.

8.148 Three identical rods, each of length *L* and flexural rigidity *EI*, are welded to two cylinders *AB* and *CD* as shown.   Neglecting the deformation of the cylinders, determine the deflection at the midpoint of each rod caused by a vertical force **P** applied at the midpoint of rod *2*.

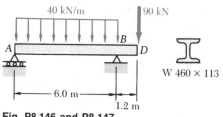

Fig. P8.146 and P8.147

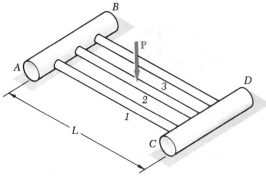

Fig. P8.148

8.149 For the cantilever beam and loading shown, determine the slope and deflection at point *B*.   Use $E = 200$ GPa.

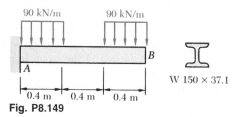

Fig. P8.149

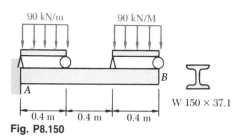

Fig. P8.150

8.150 The distributed loads shown are supported on short beams which rest on the cantilever beam *AB*.   Using $E = 200$ GPa, determine the slope and deflection at point *B* of the cantilever beam.

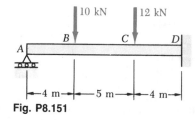

Fig. P8.151

8.151 For the beam and loading shown, determine (*a*) the reaction at *A*, (*b*) the reaction at *D*.

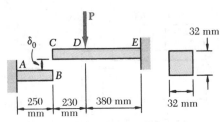

Fig. P8.152 and P8.153

8.152 Before the force **P** is applied, a gap $\delta_0 = 1.9$ mm exists between the ends of the cantilever bars AB and CDE.   Knowing that $E = 200$ GPa and $P = 3.5$ kN, determine (a) the reaction at A, (b) the reacton at E.

8.153 Before the force **P** is applied, a gap $\delta_0 = 1.9$ mm exists between the ends of the cantilever bars AB and CDE.   Knowing that $E = 200$ GPa, find the magnitude of the force **P** for which the deflection of point B is 0.6 mm.

8.154 For the beam and loading shown, determine the slope and deflection at B.   Use $E = 200$ GPa.

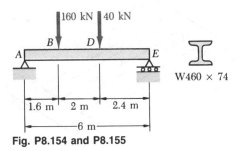

W460 × 74

Fig. P8.154 and P8.155

8.155 For the beam and loading shown, determine the magnitude and location of the largest deflection.   Use $E = 200$ GPa.

The following problems are designed to be solved with a computer.

8.C1 Several uniformly distributed loads and several concentrated loads may be applied to the cantilever beam AB.   Write a computer program which can be used to calculate the slope and deflection of the beam from $x = 0$ to $x = L$ at intervals ΔL.   Use this program to determine, at 0.1 m intervals, the slope and deflection for the beam and loading of (a) Prob. 8.149, (b) Prob. 8.150.   Use $E = 200$ GPa.

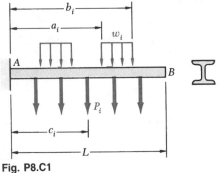

Fig. P8.C1

Fig. P8.C2

8.C2 The 5 m beam AB consists of a W 460 × 74 rolled-steel shape and supports a 45 kN/m distributed load as shown.   Write a computer program and use it to calculate for values of a from 0 to 5 m at 0.2 m intervals (a) the slope and deflection at D, (b) the lcoation and magnitude of the maximum deflection.   Use $E = 200$ GPa.

8.C3 The simple beam AB is of constant flexural rigidity EI and carries several concentrated loads as shown. Write a computer program which can be used to calculate the slope and deflection at points along the beam from $x = 0$ to $x = L$ at intervals ΔL and apply this program to the beam and loading of (a) Prob. 8.19 with $\Delta L = 6$ in., (b) Prob. 8.20 with $\Delta L = 0.125$ m, (c) Prob. 8.121 with $\Delta L = 0.5$ m.

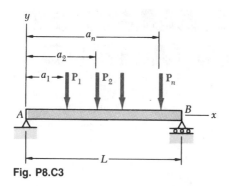

Fig. P8.C3

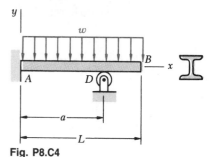

Fig. P8.C4

8.C4 The supports of beam AB consist of a fixed support at end A and a roller located at point D. Write a computer program which can be used to calculate the slope and deflection at the free end of the beam for values of a from 0 to L at intervals ΔL. Use this program to calculate values of θ_B and y_B for each of the following cases:

L	ΔL	w	E	Shape
(a) 10 m	1 m	8 kN/m	75 GPa	W 310 × 60
(b) 4 m	0.5 m	20 kN/M	200 GPa	W 250 × 58

CHAPTER NINE

DEFLECTION OF BEAMS BY MOMENT-AREA METHOD

*9.1. INTRODUCTION

In the preceding chapter, we used a mathematical method based on the integration of a differential equation to determine the deflection and slope of a beam at a given point. The bending moment was expressed as a function $M(x)$ of the distance x measured along the beam, and two successive integrations led to the functions $\theta(x)$ and $y(x)$ representing, respectively, the slope and deflection at any point of the beam.

In this chapter, we shall use certain geometric properties of the elastic curve to determine the deflection and slope of a beam at a given point. Instead of expressing the bending moment as a function $M(x)$ and integrating this function analytically, we shall draw the diagram representing the variation of M/EI over the length of the beam [Sec. 9.2]. We shall then derive two moment-area theorems. The *first moment-area theorem* will enable us to calculate the angle between the tangents to the beam at two points; the *second moment-area theorem* will be used to calculate the vertical distance from a point on the beam to a tangent at a second point.

In Sec. 9.3 the moment-area theorems will be used to determine the slope and deflection at selected points of cantilever beams and beams with symmetric loadings. In Sec. 9.4 we shall find that in many cases the areas and moments of areas defined by the M/EI diagram may be more easily determined if we draw the *bending-moment diagram by parts*. As we study the moment-area method, we shall observe that this method is particularly effective in the case of *beams of variable cross section*.

Beams with unsymmetric loadings and overhanging beams will be considered in Sec. 9.5. Since for an unsymmetric loading the maximum deflection does not occur at the center of a beam, we shall learn in Sec. 9.6 how to locate the point where the tangent is horizontal in order to determine the *maximum deflection*. Section 9.7 will be devoted to the solution of problems involving *statically indeterminate beams*.

530

*9.2. MOMENT-AREA THEOREMS

Consider a beam AB subjected to some arbitrary loading (Fig. 9.1a). We shall draw the diagram representing the variation along the beam of the quantity M/EI obtained by dividing the bending moment M by the flexural rigidity EI (Fig. 9.1b). We note that, except for a difference in the scales of ordinates, this diagram will be the same as the bending-moment diagram if the flexural rigidity of the beam is constant.

Recalling Eq. (8.4) of Sec. 8.3, and the fact that $dy/dx = \theta$, we write

$$\frac{d\theta}{dx} = \frac{d^2y}{dx^2} = \frac{M}{EI}$$

or

$$d\theta = \frac{M}{EI}dx \tag{9.1}†$$

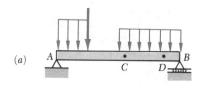

(a)

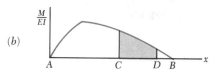

(b)

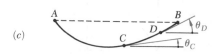

(c)

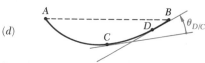

(d)

Fig. 9.1

Considering two arbitrary points C and D on the beam and integrating both members of Eq. (9.1) from C to D, we write

$$\int_{\theta_C}^{\theta_D} d\theta = \int_{x_C}^{x_D} \frac{M}{EI}dx$$

or

$$\theta_D - \theta_C = \int_{x_C}^{x_D} \frac{M}{EI}dx \tag{9.2}$$

where θ_C and θ_D denote the slope at C and D, respectively (Fig. 9.1c). But the right-hand member of Eq. (9.2) represents the area under the (M/EI) diagram between C and D, and the left-hand member the angle between the tangents to the elastic curve at C and D (Fig. 9.1d). Denoting this angle by $\theta_{D/C}$, we have

$$\theta_{D/C} = \text{area under } (M/EI) \text{ diagram between } C \text{ and } D \tag{9.3}$$

This is the *first moment-area theorem*.

We note that the angle $\theta_{D/C}$ and the area under the (M/EI) diagram have the same sign. In other words, a positive area (i.e., an area located above the x axis) corresponds to a counterclockwise rotation of the tangent to the elastic curve as we move from C to D, and a negative area corresponds to a clockwise rotation.

---

†This relation may also be derived without referring to the results obtained in Sec. 8.3, by noting that the angle $d\theta$ formed by the tangents to the elastic curve at P and P' is also the angle formed by the corresponding normals to that curve (Fig. 9.2). We thus have $d\theta = ds/\rho$ where ds is the length of the arc PP' and ρ the radius of curvature at P. Substituting for $1/\rho$ from Eq. (4.21), and noting that, since the slope at P is very small, ds is equal in first approximation to the horizontal distance dx between P and P', we write

$$d\theta = \frac{M}{EI}dx \tag{9.1}$$

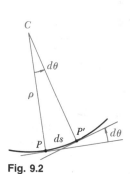

Fig. 9.2

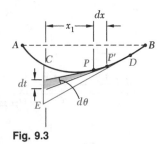

Fig. 9.3

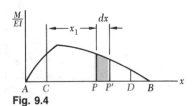

Fig. 9.4

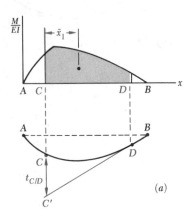

(a)

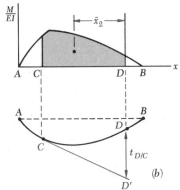

(b)

Fig. 9.5

Let us now consider two points P and P' located between C and D, and at a distance dx from each other (Fig. 9.3). The tangents to the elastic curve drawn at P and P' intercept a segment of length dt on the vertical through point C. Since the slope θ at P and the angle $d\theta$ formed by the tangents at P and P' are both small quantities, we may assume that dt is equal to the arc of circle of radius x_1 subtending the angle $d\theta$. We have, therefore,

$$dt = x_1 \, d\theta$$

or, substituting for $d\theta$ from Eq. (9.1),

$$dt = x_1 \frac{M}{EI} dx \tag{9.4}$$

We shall now integrate Eq. (9.4) from C to D. We note that, as point P describes the elastic curve from C to D, the tangent at P sweeps the vertical through C from C to E. The integral of the left-hand member is thus equal to the vertical distance from C to the tangent at D. This distance is denoted by $t_{C/D}$ and is called the *tangential deviation of C with respect to D*. We have, therefore,

$$t_{C/D} = \int_{x_C}^{x_D} x_1 \frac{M}{EI} dx \tag{9.5}$$

We now observe that $(M/EI) \, dx$ represents an element of area under the (M/EI) diagram, and $x_1(M/EI) \, dx$ the first moment of that element with respect to a vertical axis through C (Fig. 9.4). The right-hand member in Eq. (9.5), thus, represents the first moment with respect to that axis of the area located under the (M/EI) diagram between C and D.

We may, therefore, state the *second moment-area theorem* as follows: *The tangential deviation $t_{C/D}$ of C with respect to D is equal to the first moment with respect to a vertical axis through C of the area under the (M/EI) diagram between C and D.*

Recalling that the first moment of an area with respect to an axis is equal to the product of the area and of the distance from its centroid to that axis, we may also express the second moment-area theorem as follows:

$$\boxed{t_{C/D} = (\text{area between } C \text{ and } D) \, \bar{x}_1} \tag{9.6}$$

where the area refers to the area under the (M/EI) diagram, and where $\bar{x}_1$ is the distance from the centroid of the area to the vertical axis through C (Fig. 9.5a).

Care should be taken to distinguish between the tangential deviation of C with respect to D, denoted by $t_{C/D}$, and the tangential deviation of D with respect to C, which is denoted by $t_{D/C}$. The tangential deviation

$t_{D/C}$ represents the vertical distance from D to the tangent to the elastic curve at C, and is obtained by multiplying the area under the (M/EI) diagram by the distance $\bar{x}_2$ from its centroid to the vertical axis through D (Fig. 9.5b):

$$t_{D/C} = (\text{area between } C \text{ and } D)\ \bar{x}_2 \qquad (9.7)$$

We note that, if an area under the (M/EI) diagram is located above the x axis, its first moment with respect to a vertical axis will be positive; if it is located below the x axis, its first moment will be negative. As we may check from Fig. 9.5, a point with a *positive* tangential deviation is located *above* the corresponding tangent, while a point with a *negative* tangential deviation would be located *below* that tangent.

*9.3. APPLICATION TO CANTILEVER BEAMS AND BEAMS WITH SYMMETRIC LOADINGS

We recall that the first moment-area theorem derived in the preceding section defines the angle $\theta_{D/C}$ *between* the tangents at two points C and D of the elastic curve. Thus, the angle θ_D that the tangent at D forms with the horizontal, i.e., the slope at D, may be obtained only if the slope at C is known. Similarly, the second moment-area theorem defines the vertical distance of one point of the elastic curve from the tangent at another point. The tangential deviation $t_{D/C}$, therefore, will help us locate point D only if the tangent at C is known. We conclude that the two moment-area theorems may be applied effectively to the determination of slopes and deflections only if a certain *reference tangent* to the elastic curve has first been determined.

In the case of a *cantilever beam* (Fig. 9.6), the tangent to the elastic curve at the fixed end A is known and may be used as the reference tangent. Since $\theta_A = 0$, the slope of the beam at any point D is $\theta_D = \theta_{D/A}$ and may be obtained by the first moment-area theorem. On the other hand, the deflection y_D of point D is equal to the tangential deviation $t_{D/A}$ measured from the horizontal reference tangent at A and may be obtained by the second moment-area theorem.

In the case of a simply supported beam AB with a *symmetric loading* (Fig. 9.7a) or in the case of an overhanging symmetric beam with a symmetric loading (see Sample Prob. 9.2), the tangent at the center C of the beam must be horizontal by reason of symmetry and may be used as the reference tangent (Fig. 9.7b). Since $\theta_C = 0$, the slope at the support B is $\theta_B = \theta_{B/C}$ and may be obtained by the first moment-area theorem. We also note that $|y|_{\text{max}}$ is equal to the tangential deviation $t_{B/C}$ and may, therefore, be obtained by the second moment-area theorem. The slope at any other point D of the beam (Fig. 9.7c) is found in a similar fashion, and the deflection at D may be expressed as $y_D = t_{D/C} - t_{B/C}$.

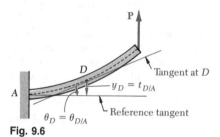

Fig. 9.6

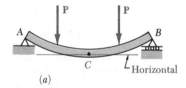

(a)

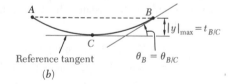

(b)

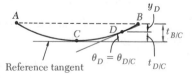

(c)

Fig. 9.7

Example 9.01

Determine the slope and deflection at end B of the prismatic cantilever beam AB when it is loaded as shown (Fig. 9.8), knowing that the flexural rigidity of the beam is $EI = 10$ MN $\cdot$ m^2.

We first draw the free-body diagram of the beam (Fig. 9.9a). Summing vertical components and moments about A, we find that the reaction at the fixed end A consists of a 50-kN upward vertical force $\mathbf{R}_A$ and a 60-kN $\cdot$ m counterclockwise couple $\mathbf{M}_A$. Next, we draw the bending-moment diagram (Fig. 9.9b) and determine from similar triangles the distance x_D from the end A to the point D of the beam where $M = 0$:

$$\frac{x_D}{60} = \frac{3 - x_D}{90} = \frac{3}{150} \qquad x_D = 1.2 \text{ m}$$

Dividing by the flexural rigidity EI the values obtained for M, we draw the (M/EI) diagram (Fig. 9.10) and compute the areas corresponding respectively to the segments AD and DB, assigning a positive sign to the area located above the x axis, and a negative sign to the area located below that axis. Using the first moment-area theorem, we write

$$\theta_{B/A} = \theta_B - \theta_A = \text{area from } A \text{ to } B = A_1 + A_2$$
$$= -\tfrac{1}{2}(1.2 \text{ m})(6 \times 10^{-3} \text{ m}^{-1})$$
$$\qquad\qquad + \tfrac{1}{2}(1.8 \text{ m})(9 \times 10^{-3} \text{ m}^{-1})$$
$$= -3.6 \times 10^{-3} + 8.1 \times 10^{-3}$$
$$= +4.5 \times 10^{-3} \text{ rad}$$

and, since $\theta_A = 0$,

$$\theta_B = +4.5 \times 10^{-3} \text{ rad}$$

Using now the second moment-area theorem, we write that the tangential deviation $t_{B/A}$ is equal to the first moment about a vertical axis through B of the total area between A and B. Expressing the moment of each partial area as the product of that area and of the distance from its centroid to the axis through B, we have

$$t_{B/A} = A_1(2.6 \text{ m}) + A_2(0.6 \text{ m})$$
$$= (-3.6 \times 10^{-3})(2.6 \text{ m}) + (8.1 \times 10^{-3})(0.6 \text{ m})$$
$$= -9.36 \text{ mm} + 4.86 \text{ mm} = -4.50 \text{ mm}$$

Since the reference tangent at A is horizontal, the deflection at B is equal to $t_{B/A}$ and we have

$$y_B = t_{B/A} = -4.50 \text{ mm}$$

The deflected beam has been sketched in Fig. 9.11.

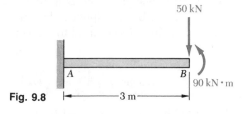

Fig. 9.8

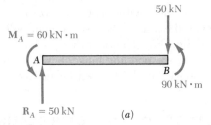

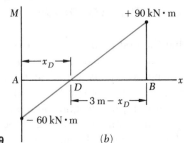

Fig. 9.9

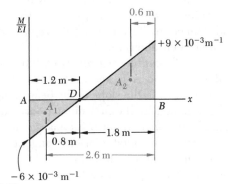

Fig. 9.10

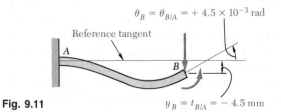

Fig. 9.11

*9.4. BENDING-MOMENT DIAGRAMS BY PARTS

In many applications the determination of the angle $\theta_{D/C}$ and of the tangential deviation $t_{D/C}$ is simplified if the effect of each load is evaluated independently. A separate (M/EI) diagram is drawn for each load, and the angle $\theta_{D/C}$ is obtained by adding algebraically the areas under the various diagrams. Similarly, the tangential deviation $t_{D/C}$ is obtained by adding the first moments of these areas about a vertical axis through D. A bending-moment or (M/EI) diagram plotted in this fashion is said to be *drawn by parts*.

When a bending-moment or (M/EI) diagram is drawn by parts, the various areas defined by the diagram consist of simple geometric shapes, such as rectangles, triangles, and parabolic spandrels. For convenience, the areas and centroids of these various shapes have been indicated in Fig. 9.12.

Shape		Area	c
Rectangle	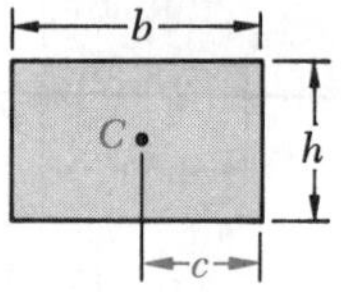	bh	$\dfrac{b}{2}$
Triangle	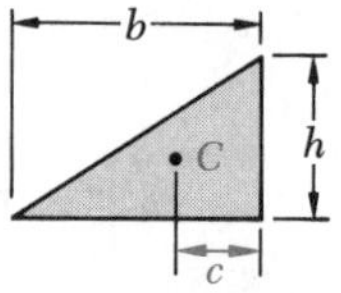	$\dfrac{bh}{2}$	$\dfrac{b}{3}$
Parabolic spandrel	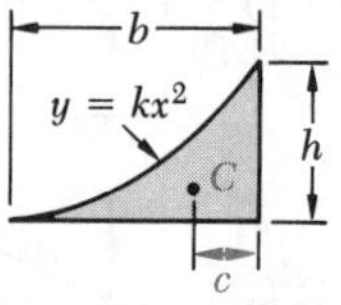 $y = kx^2$	$\dfrac{bh}{3}$	$\dfrac{b}{4}$
Cubic spandrel	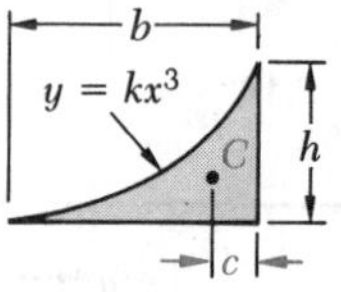 $y = kx^3$	$\dfrac{bh}{4}$	$\dfrac{b}{5}$
General spandrel	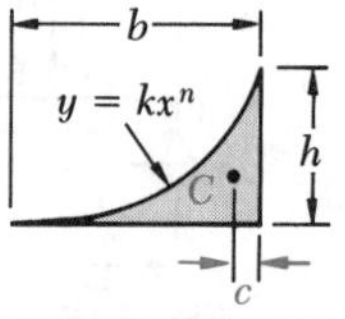 $y = kx^n$	$\dfrac{bh}{n+1}$	$\dfrac{b}{n+2}$

Fig. 9.12 Areas and centroids of common shapes.

Example 9.02

Determine the slope and deflection at end B of the prismatic beam of Example 9.01, drawing the bending-moment diagram by parts.

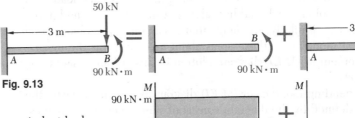

Fig. 9.13

We replace the given loading by the two equivalent loadings shown in Fig. 9.13, and draw the corresponding bending-moment and (M/EI) diagrams from right to left, starting at the free end B.

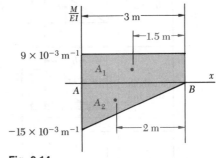

Applying the first moment-area theorem, and recalling that $\theta_A = 0$, we write

$$\theta_B = \theta_{B/A} = A_1 + A_2$$
$$= (9 \times 10^{-3} \text{ m}^{-1})(3 \text{ m}) - \tfrac{1}{2}(15 \times 10^{-3} \text{ m}^{-1})(3 \text{ m})$$
$$= 27 \times 10^{-3} - 22.5 \times 10^{-3} = 4.5 \times 10^{-3} \text{ rad}$$

Applying the second moment-area theorem, we compute the first moment of each area about a vertical axis through B and write

$$y_B = t_{B/A} = A_1(1.5 \text{ m}) + A_2(2 \text{ m})$$
$$= (27 \times 10^{-3})(1.5 \text{ m}) - (22.5 \times 10^{-3})(2 \text{ m})$$
$$= 40.5 \text{ mm} - 45 \text{ mm} = -4.5 \text{ mm}$$

It is convenient, in practice, to group into a single drawing the two portions of the (M/EI) diagram (Fig. 9.14).

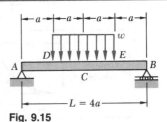

Fig. 9.14

Example 9.03

For the prismatic beam AB and the loading shown (Fig. 9.15), determine the slope at a support and the maximum deflection.

We first sketch the deflected beam (Fig. 9.16). Since the tangent at the center C of the beam is horizontal, it may be used as the reference tangent, and we have $|y|_{max} = t_{A/C}$. On

Fig. 9.15

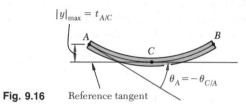

Fig. 9.16   Reference tangent

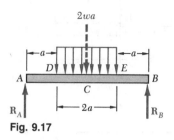

Fig. 9.17

the other hand, since $\theta_C = 0$, we write

$$\theta_{C/A} = \theta_C - \theta_A = -\theta_A \qquad \text{or} \qquad \theta_A = -\theta_{C/A}$$

From the free-body diagram of the beam (Fig. 9.17), we find that

$$R_A = R_B = wa$$

Next, we draw the shear and bending-moment diagrams for the portion AC of the beam. We draw these diagrams by parts, considering separately the effects of the reaction $\mathbf{R}_A$ and of the distributed load. However, for convenience, the two parts of each diagram have been plotted together (Fig. 9.18). We recall from Sec. 7.4 that, the distributed load being uniform, the corresponding parts of the shear and bending-moment diagrams will be, respectively, linear and parabolic. The area and centroid of the triangle and parabolic spandrel may be obtained by referring to Fig. 9.12. The areas of the triangle and spandrel are found to be, respectively,

$$A_1 = \frac{1}{2}(2a)\left(\frac{2wa^2}{EI}\right) = \frac{2wa^3}{EI}$$

and

$$A_2 = -\frac{1}{3}(a)\left(\frac{wa^2}{2EI}\right) = -\frac{wa^3}{6EI}$$

Applying the first moment-area theorem, we write

$$\theta_{C/A} = A_1 + A_2 = \frac{2wa^3}{EI} - \frac{wa^3}{6EI} = \frac{11wa^3}{6EI}$$

Recalling from Figs. 9.15 and 9.16 that $a = \frac{1}{4}L$ and $\theta_A = -\theta_{C/A}$, we have

$$\theta_A = -\frac{11wa^3}{6EI} = -\frac{11wL^3}{384EI}$$

Applying now the second moment-area theorem, we write

$$t_{A/C} = A_1\frac{4a}{3} + A_2\frac{7a}{4} = \left(\frac{2wa^3}{EI}\right)\frac{4a}{3} + \left(-\frac{wa^3}{6EI}\right)\frac{7a}{4} = \frac{19wa^4}{8EI}$$

and

$$|y|_{max} = t_{A/C} = \frac{19wa^4}{8EI} = \frac{19wL^4}{2048EI}$$

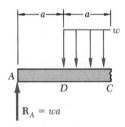

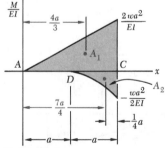

Fig. 9.18

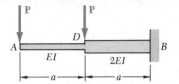

SAMPLE PROBLEM 9.1

The prismatic rods AD and DB are welded together to form the cantilever beam ADB. Knowing that the flexural rigidity is EI in portion AD of the beam and $2EI$ in portion DB, determine, for the loading shown, the slope and deflection at end A.

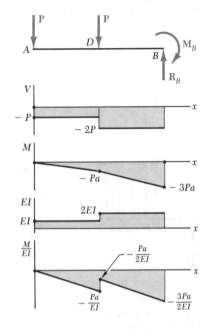

(M/EI) Diagram. We first draw the bending-moment diagram for the beam and then obtain the (M/EI) diagram by dividing the value of M at each point of the beam by the corresponding value of the flexural rigidity.

Reference Tangent. We choose the horizontal tangent at the fixed end B as the reference tangent. Since $\theta_B = 0$ and $y_B = 0$, we note that

$$\theta_A = -\theta_{B/A} \qquad y_A = t_{A/B}$$

Slope at A. Dividing the (M/EI) diagram into the three triangular portions shown, we write

$$A_1 = -\frac{1}{2}\frac{Pa}{EI}a = -\frac{Pa^2}{2EI}$$

$$A_2 = -\frac{1}{2}\frac{Pa}{2EI}a = -\frac{Pa^2}{4EI}$$

$$A_3 = -\frac{1}{2}\frac{3Pa}{2EI}a = -\frac{3Pa^2}{4EI}$$

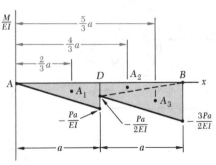

Using the first moment-area theorem, we have

$$\theta_{B/A} = A_1 + A_2 + A_3 = -\frac{Pa^2}{2EI} - \frac{Pa^2}{4EI} - \frac{3Pa^2}{4EI} = -\frac{3Pa^2}{2EI}$$

$$\theta_A = -\theta_{B/A} = +\frac{3Pa^2}{2EI} \qquad\qquad \theta_A = \frac{3Pa^2}{2EI} \measuredangle \ \blacktriangleleft$$

Deflection at A. Using the second moment-area theorem, we have

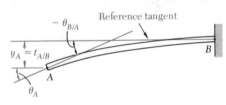

$$y_A = t_{A/B} = A_1\left(\frac{2}{3}a\right) + A_2\left(\frac{4}{3}a\right) + A_3\left(\frac{5}{3}a\right)$$

$$= \left(-\frac{Pa^2}{2EI}\right)\frac{2a}{3} + \left(-\frac{Pa^2}{4EI}\right)\frac{4a}{3} + \left(-\frac{3Pa^2}{4EI}\right)\frac{5a}{3}$$

$$y_A = -\frac{23Pa^3}{12EI} \qquad\qquad y_A = \frac{23Pa^3}{12EI}\downarrow \ \blacktriangleleft$$

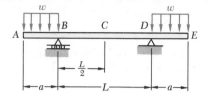

SAMPLE PROBLEM 9.2

For the prismatic beam and loading shown, determine the slope and deflection at end E.

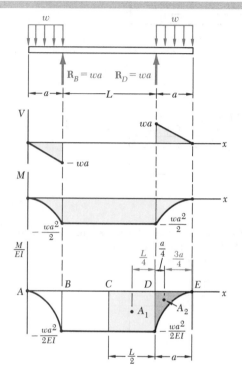

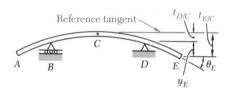

(M/EI) **Diagram.** From a free-body diagram of the beam, we determine the reactions and then draw the shear and bending-moment diagrams. Since the flexural rigidity of the beam is constant, we divide each value of M by EI and obtain the (M/EI) diagram shown.

Reference Tangent. Since the beam and its loading are symmetric with respect to the midpoint C, the tangent at C is horizontal and is used as the reference tangent. Referring to the sketch, we observe that, since $\theta_C = 0$,

$$\theta_E = \theta_C + \theta_{E/C} = \theta_{E/C} \tag{1}$$

$$y_E = t_{E/C} - t_{D/C} \tag{2}$$

Slope at E. Referring to the (M/EI) diagram and using the first moment-area theorem, we write

$$A_1 = -\frac{wa^2}{2EI}\left(\frac{L}{2}\right) = -\frac{wa^2 L}{4EI}$$

$$A_2 = -\frac{1}{3}\left(\frac{wa^2}{2EI}\right)(a) = -\frac{wa^3}{6EI}$$

Using Eq. (1), we have

$$\theta_E = \theta_{E/C} = A_1 + A_2 = -\frac{wa^2 L}{4EI} - \frac{wa^3}{6EI}$$

$$\theta_E = -\frac{wa^2}{12EI}(3L + 2a) \qquad \theta_E = \frac{wa^2}{12EI}(3L + 2a)\;\;\triangleleft$$

Deflection at E. Using the second moment-area theorem, we write

$$t_{D/C} = A_1\frac{L}{4} = \left(-\frac{wa^2 L}{4EI}\right)\frac{L}{4} = -\frac{wa^2 L^2}{16EI}$$

$$t_{E/C} = A_1\left(a + \frac{L}{4}\right) + A_2\left(\frac{3a}{4}\right)$$

$$= \left(-\frac{wa^2 L}{4EI}\right)\left(a + \frac{L}{4}\right) + \left(-\frac{wa^3}{6EI}\right)\left(\frac{3a}{4}\right)$$

$$= -\frac{wa^3 L}{4EI} - \frac{wa^2 L^2}{16EI} - \frac{wa^4}{8EI}$$

Using Eq. (2), we have

$$y_E = t_{E/C} - t_{D/C} = -\frac{wa^3 L}{4EI} - \frac{wa^4}{8EI}$$

$$y_E = -\frac{wa^3}{8EI}(2L + a) \qquad y_E = \frac{wa^3}{8EI}(2L + a)\downarrow\;\;\triangleleft$$

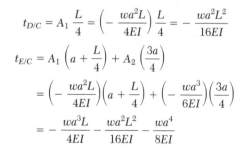

PROBLEMS

9.1 through 9.4   For the uniform cantilever beam and loading shown, determine (*a*) the slope at the free end, (*b*) the deflection at the free end.

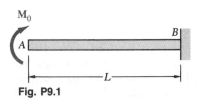

Fig. P9.1

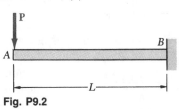

Fig. P9.2

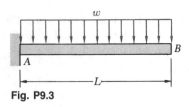

Fig. P9.3

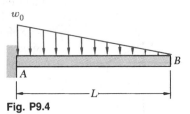

Fig. P9.4

9.5 through 9.8   For the uniform cantilever beam and loading shown, determine (*a*) the slope at the free end, (*b*) the deflection at the free end.

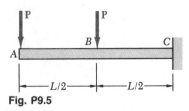

Fig. P9.5

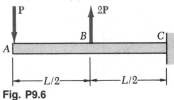

Fig. P9.6

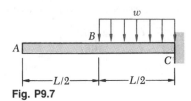

Fig. P9.7

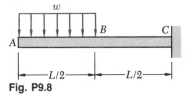

Fig. P9.8

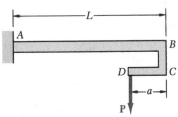

Fig. P9.9 and P9.10

9.9   The rigid bar *BCD* is welded at *B* to the uniform cantilever beam *AB*. Determine the slope and deflection at point *B* when $a = L/4$.

9.10   The rigid bar *BCD* is welded at *B* to the uniform cantilever beam *AB*. Determine the slope and deflection at point *B* when (*a*) $a = L/2$, (*b*) $a = 2L/3$.

9.11 For the cantilever beam and loading shown, determine (*a*) the slope at point *C*, (*b*) the deflection at point *C*. Use *E* = 200 GPa.

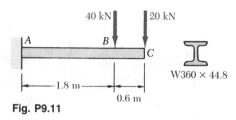

Fig. P9.11

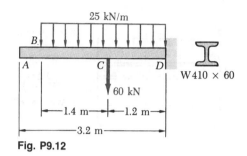

Fig. P9.12

9.12 For the cantilever beam and loading shown, determine (*a*) the slope at point *A*, (*b*) the deflection at point *A*. Use *E* = 200 GPa.

9.13 For the cantilever beam and loading shown, determine (*a*) the slope at point *A*, (*b*) the deflection at point *A*. Use *E* = 200 GPa.

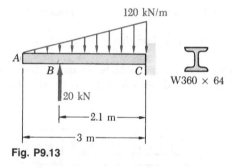

Fig. P9.13

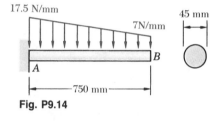

Fig. P9.14

9.14 For the cantilever beam and loading shown, determine (*a*) the slope at point *B*, (*b*) the deflection at point *B*. Use *E* = 200 GPa.

9.15 For the cantilever beam and loading shown, determine (*a*) the slope at point *C*, (*b*) the deflection at point *C*.

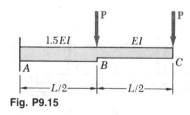

Fig. P9.15

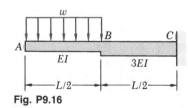

Fig. P9.16

9.16 For the cantilever beam and loading shown, determine (*a*) the slope at point *A*, (*b*) the deflection at point *A*.

9.17 Two cover plates are welded to the rolled-steel beam ABC as shown. Using $E = 200$ GPa, determine (a) the slope at end A, (b) the deflection at end A.

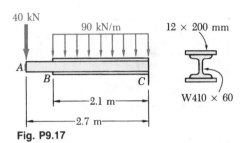

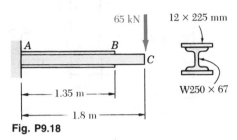

Fig. P9.17

Fig. P9.18

9.18 Two cover plates are welded to the rolled-steel beam ABC as shown. Using $E = 200$ GPa, determine (a) the slope at end C, (b) the deflection at end C.

9.19 Solve Prob. 9.18, assuming that an additional 65-kN load is applied at point B of the beam.

9.20 Solve Prob. 9.17, assuming that the 90-kN/m distributed load is removed.

9.21 through 9.24 For the prismatic beam and loading shown, determine (a) the slope at end A, (b) the deflection at the center C of the beam.

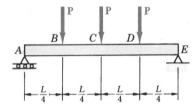

Fig. P9.21

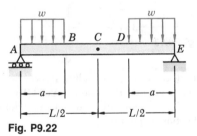

Fig. P9.22

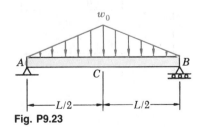

Fig. P9.23

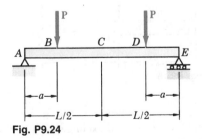

Fig. P9.24

9.25 For the beam and loading shown, determine (*a*) the slope at end *A*, (*b*) the deflection at the midpoint of the beam. Use *E* = 200 GPa.

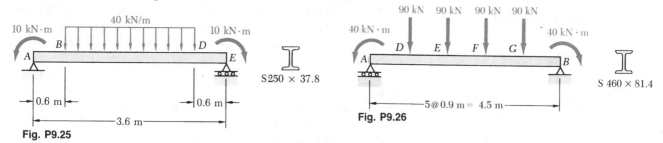

Fig. P9.25

Fig. P9.26

9.26 For the beam and loading shown, determine (*a*) the slope at end *A*, (*b*) the deflection at the midpoint of the beam. Use *E* = 200 GPa.

9.27 For the beam and loading shown, determine (*a*) the slope at end *A*, (*b*) the deflection at end *A*, (*c*) the deflection at the midpoint of the beam. Use *E* = 200 GPa.

9.28 For the beam and loading shown, determine (*a*) the slope at end *A*, (*b*) the deflection at end *A*, (*c*) the deflection at the midpoint of the beam. Use *E* = 200 GPa.

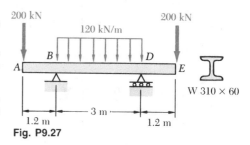

Fig. P9.27

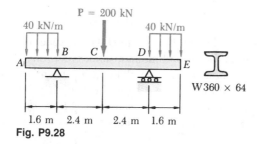

Fig. P9.28

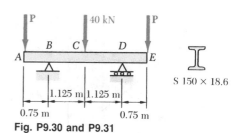

Fig. P9.30 and P9.31

9.29 For the beam and loading of Prob. 9.28, determine (*a*) the load **P** for which the deflection is zero at the midpoint *C* of the beam, (*b*) the corresponding deflection at end *A*. Use *E* = 200 GPa.

***9.30** For the beam and loading shown, determine (*a*) the value of *P* for which the deflection is zero at the midpoint *C* of the beam, (*b*) the range of values of *P* for which the absolute value of the deflection of the midpoint *C* of the beam is equal to or less than 2.5 mm. Use *E* = 200 GPa.

***9.31** For the beam and loading shown, determine (*a*) the value of *P* for which the deflection is zero at end *A* of the beam, (*b*) the range of values of *P* for which the absolute value of the deflection of end *A* of the beam is equal to or less than 2.5 mm. Use *E* = 200 GPa.

***9.32** A uniform rod *AE* is to be supported at two points *B* and *D*. Determine the distance *a* from the ends of the rod to the points of support if the downward deflections of points *A*, *C*, and *E* are to be equal.

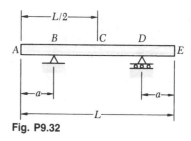

Fig. P9.32

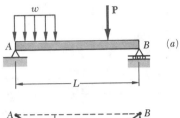

(a)

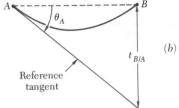

(b)

$t_{B/A}$

Reference
tangent

Fig. 9.19

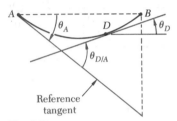

$\theta_{D/A}$

Reference
tangent

Fig. 9.20

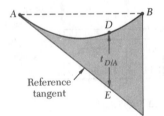

Reference
tangent

Fig. 9.21

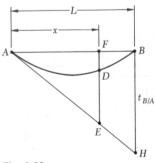

Fig. 9.23

*9.5. BEAMS WITH UNSYMMETRIC LOADINGS

We saw in Sec. 9.3 that, when a simply supported or overhanging beam carries a symmetric load, the tangent at the center C of the beam is horizontal and may be used as the reference tangent. When a simply supported or overhanging beam carries an unsymmetric load, it is generally not possible to determine by inspection the point of the beam where the tangent is horizontal. Other means must then be found for locating a reference tangent, i.e., a tangent of known slope to be used in applying either of the two moment-area theorems.

It is usually most convenient to select the reference tangent at one of the beam supports. Considering, for example, the tangent at the support A of the simply supported beam AB (Fig. 9.19a), we determine its slope by computing the tangential deviation $t_{B/A}$ of the support B with respect to A, and dividing $t_{B/A}$ by the distance L between the supports. Recalling that the tangential deviation of a point located above the tangent is positive, we write

$$\theta_A = -\frac{t_{B/A}}{L} \tag{9.8}$$

Once the slope of the reference tangent has been found, the slope θ_D of the beam at any point D (Fig. 9.20) may be determined by using the first moment-area theorem to obtain $\theta_{D/A}$, and then writing

$$\theta_D = \theta_A + \theta_{D/A} \tag{9.9}$$

The tangential deviation $t_{D/A}$ of D with respect to the support A may be obtained from the second moment-area theorem. We note that $t_{D/A}$ is equal to the segment ED (Fig. 9.21) and represents the vertical distance of D *from the reference tangent.* On the other hand, the deflection y_D of point D represents the vertical distance of D *from the horizontal line AB* (Fig. 9.22). Since y_D is equal in magnitude to the segment FD, it may be

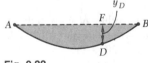

Fig. 9.22

expressed as the difference between EF and ED (Fig. 9.23). Observing from the similar triangles AFE and ABH that

$$\frac{EF}{x} = \frac{HB}{L} \qquad \text{or} \qquad EF = \frac{x}{L} t_{B/A}$$

and recalling the sign conventions used for deflections and tangential deviations, we write

$$y_D = ED - EF = t_{D/A} - \frac{x}{L} t_{B/A} \tag{9.10}$$

Example 9.04

For the prismatic beam and loading shown (Fig. 9.24), determine the slope and deflection at point D.

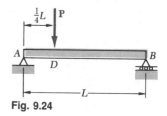

Fig. 9.24

Reference Tangent at Support A. We compute the reactions at the supports and draw the (M/EI) diagram (Fig. 9.25). We determine the tangential deviation $t_{B/A}$ of the support B with respect to the support A by applying the second moment-area theorem and computing the moments about a vertical axis through B of the areas A_1 and A_2. We have

$$A_1 = \frac{1}{2}\frac{L}{4}\frac{3PL}{16EI} = \frac{3PL^2}{128EI} \qquad A_2 = \frac{1}{2}\frac{3L}{4}\frac{3PL}{16EI} = \frac{9PL^2}{128EI}$$

$$t_{B/A} = A_1\left(\frac{L}{12} + \frac{3L}{4}\right) + A_2\left(\frac{L}{2}\right)$$

$$= \frac{3PL^2}{128EI}\frac{10L}{12} + \frac{9PL^2}{128EI}\frac{L}{2} = \frac{7PL^3}{128EI}$$

The slope of the reference tangent at A (Fig. 9.26) is

$$\theta_A = -\frac{t_{B/A}}{L} = -\frac{7PL^2}{128EI}$$

Slope at D. Applying the first moment-area theorem from A to D, we write

$$\theta_{D/A} = A_1 = \frac{3PL^2}{128EI}$$

Thus, the slope at D is

$$\theta_D = \theta_A + \theta_{D/A} = -\frac{7PL^2}{128EI} + \frac{3PL^2}{128EI} = -\frac{PL^2}{32EI}$$

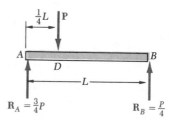

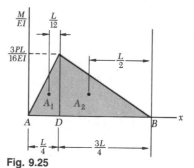

Fig. 9.25

Deflection at D. We first determine the tangential deviation $DE = t_{D/A}$ by computing the moment of the area A_1 about a vertical axis through D:

$$DE = t_{D/A} = A_1\left(\frac{L}{12}\right) = \frac{3PL^2}{128EI}\frac{L}{12} = \frac{PL^3}{512EI}$$

The deflection at D is equal to the difference between the segments DE and EF (Fig. 9.26). We have

$$y_D = DE - EF = t_{D/A} - \tfrac{1}{4}t_{B/A}$$

$$= \frac{PL^3}{512EI} - \frac{1}{4}\frac{7PL^3}{128EI}$$

$$y_D = -\frac{3PL^3}{256EI} = -0.01172PL^3/EI$$

Fig. 9.26

*9.6. MAXIMUM DEFLECTION

When a simply supported or overhanging beam carries an unsymmetric load, the maximum deflection generally does not occur at the center of the beam. To determine the maximum deflection of the beam, we should locate the point K of the beam where the tangent is horizontal, and compute the deflection at that point.

Our analysis must begin with the determination of a reference tangent at one of the supports. If support A is selected, the slope θ_A of the tangent at A is obtained by the method indicated in the preceding section, i.e., by computing the tangential deviation $t_{B/A}$ of support B with respect to A and dividing that quantity by the distance L between the two supports.

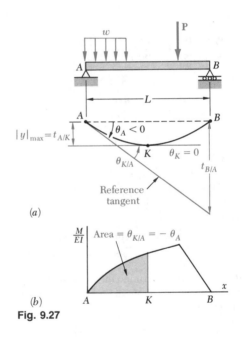

(a)

(b)

Fig. 9.27

Since the slope θ_K at point K is zero (Fig. 9.27a), we must have

$$\theta_{K/A} = \theta_K - \theta_A = 0 - \theta_A = -\theta_A$$

Recalling the first moment-area theorem, we conclude that point K may be determined by measuring under the (M/EI) diagram an area equal to $\theta_{K/A} = -\theta_A$ (Fig. 9.27b).

Observing that the maximum deflection $|y|_{max}$ is equal to the tangential deviation $t_{A/K}$ of support A with respect to K (Fig. 9.27a), we may obtain $|y|_{max}$ by computing the first moment with respect to the vertical axis through A of the area between A and K (Fig. 9.27b).

Example 9.05

Determine the maximum deflection for the beam of Example 9.04.

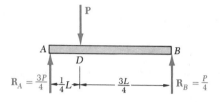

Determination of Point K Where Slope Is Zero. We recall from Example 9.04 that the slope at point D, where the load is applied, is negative. It follows that point K, where the slope is zero, is located between D and the support B (Fig. 9.28). Our computations, therefore, will be simplified if we relate the slope at K to the slope at B, rather than to the slope at A.

Since the slope at A has already been determined in Example 9.04, the slope at B may be obtained by writing

$$\theta_B = \theta_A + \theta_{B/A} = \theta_A + A_1 + A_2$$

$$\theta_B = -\frac{7PL^2}{128EI} + \frac{3PL^2}{128EI} + \frac{9PL^2}{128EI} = \frac{5PL^2}{128EI}$$

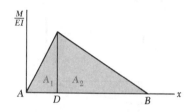

Observing that the bending moment at a distance u from end B is $M = \frac{1}{4}Pu$ (Fig. 9.29a), we express the area A' located between K and B under the (M/EI) diagram (Fig. 9.29b) as

$$A' = \frac{1}{2}\frac{Pu}{4EI}u = \frac{Pu^2}{8EI}$$

By the first moment-area theorem, we have

$$\theta_{B/K} = \theta_B - \theta_K = A'$$

and, since $\theta_K = 0$,

$$\theta_B = A'$$

Substituting the values obtained for θ_B and A', we write

$$\frac{5PL^2}{128EI} = \frac{Pu^2}{8EI}$$

and, solving for u,

$$u = \frac{\sqrt{5}}{4}L = 0.559L$$

Thus, the distance from the support A to point K is

$$AK = L - 0.559L = 0.441L$$

Maximum Deflection. The maximum deflection $|y|_{max}$ is equal to the tangential deviation $t_{B/K}$ and, thus, to the first moment of the area A' about a vertical axis through B (Fig. 9.29b). We write

$$|y|_{max} = t_{B/K} = A'\left(\frac{2u}{3}\right) = \frac{Pu^2}{8EI}\left(\frac{2u}{3}\right) = \frac{Pu^3}{12EI}$$

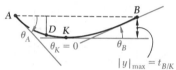

Fig. 9.28

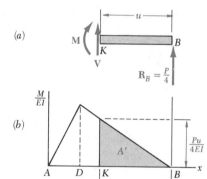

Fig. 9.29

Substituting the value obtained for u, we have

$$|y|_{max} = \frac{P}{12EI}\left(\frac{\sqrt{5}}{4}L\right)^3 = 0.01456PL^3/EI$$

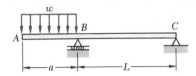

SAMPLE PROBLEM 9.3

For the beam and loading shown, (a) determine the deflection at end A, (b) evaluate y_A for the following data:

$$W\ 250 \times 49.1: I = 70.8 \times 10^6\ \text{mm}^4 \qquad E = 200\ \text{GPa}$$
$$a = 0.9\ \text{m} \qquad\qquad L = 1.65\ \text{m}$$
$$w = 200\ \text{kN/m}$$

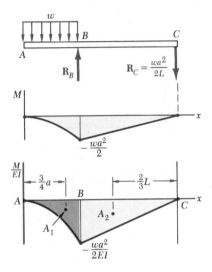

(M/EI) Diagram. We first draw the bending-moment diagram. Since the flexural rigidity EI is constant, we obtain the (M/EI) diagram shown which consists of a parabolic spandrel of area A_1 and a triangle of area A_2.

$$A_1 = \frac{1}{3}\left(-\frac{wa^2}{2EI}\right)a = -\frac{wa^3}{6EI}$$

$$A_2 = \frac{1}{2}\left(-\frac{wa^2}{2EI}\right)L = -\frac{wa^2 L}{4EI}$$

Reference Tangent at B. The reference tangent is drawn at point B as shown. Using the second moment-area theorem, we determine the tangential deviation of C with respect to B:

$$t_{C/B} = A_2\frac{2L}{3} = \left(-\frac{wa^2 L}{4EI}\right)\frac{2L}{3} = -\frac{wa^2 L^2}{6EI}$$

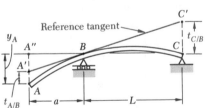

From the similar triangles $A''A'B$ and $CC'B$, we find

$$A''A' = t_{C/B}\left(\frac{a}{L}\right) = -\frac{wa^2 L^2}{6EI}\left(\frac{a}{L}\right) = -\frac{wa^3 L}{6EI}$$

Again using the second moment-area theorem, we write

$$t_{A/B} = A_1\frac{3a}{4} = \left(-\frac{wa^3}{6EI}\right)\frac{3a}{4} = -\frac{wa^4}{8EI}$$

(a) Deflection at End A

$$y_A = A''A' + t_{A/B} = -\frac{wa^3 L}{6EI} - \frac{wa^4}{8EI} = -\frac{wa^4}{8EI}\left(\frac{4}{3}\frac{L}{a}+1\right)$$

$$y_A = \frac{wa^4}{8EI}\left(1+\frac{4}{3}\frac{L}{a}\right)\downarrow \quad \blacktriangleleft$$

(b) Evaluation of y_A. Substituting the data given, we write

$$y_A = \frac{(200\ \text{N/mm})(900\ \text{mm})^4}{8(200\ \text{GPa})(70.8 \times 10^6\ \text{mm}^4)}\left(1+\frac{4}{3}\frac{1.65\ \text{m}}{0.9\ \text{m}}\right)$$

$$y_A = 3.99\ \text{mm}\downarrow \quad \blacktriangleleft$$

W 250 × 22.3

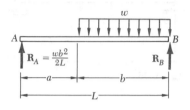

SAMPLE PROBLEM 9.4

For the beam and loading shown, determine the magnitude and location of the largest deflection. Use $E = 200$ GPa.

Reactions. Using the free-body diagram of the entire beam, we find

$$R_A = 16.81 \text{ kN} \uparrow \qquad R_B = 38.2 \text{ kN} \uparrow$$

(M/EI) Diagram. We draw the (M/EI) diagram by parts, considering separately the effects of the reaction $\mathbf{R}_A$ and of the distributed load. The areas of the triangle and of the spandrel are

$$A_1 = \frac{1}{2}\frac{R_A L}{EI} L = \frac{R_A L^2}{2EI} \qquad A_2 = \frac{1}{3}\left(-\frac{wb^2}{2EI}\right) b = -\frac{wb^3}{6EI}$$

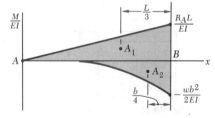

Reference Tangent. The tangent to the beam at support A is chosen as the reference tangent. Using the second moment-area theorem, we determine the tangential deviation $t_{B/A}$ of support B with respect to support A:

$$t_{B/A} = A_1 \frac{L}{3} + A_2 \frac{b}{4} = \left(\frac{R_A L^2}{2EI}\right)\frac{L}{3} + \left(-\frac{wb^3}{6EI}\right)\frac{b}{4} = \frac{R_A L^3}{6EI} - \frac{wb^4}{24EI}$$

Slope at A

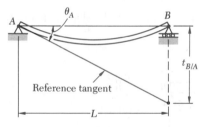

$$\theta_A = -\frac{t_{B/A}}{L} = -\left(\frac{R_A L^2}{6EI} - \frac{wb^4}{24EIL}\right) \tag{1}$$

Largest Deflection. The largest deflection occurs at point K, where the slope of the beam is zero. We write therefore

$$\theta_K = \theta_A + \theta_{K/A} = 0 \tag{2}$$

But

$$\theta_{K/A} = A_3 + A_4 = \frac{R_A x_m^2}{2EI} - \frac{w}{6EI}(x_m - a)^3 \tag{3}$$

We substitute for θ_A and $\theta_{K/A}$ from Eqs. (1) and (3) into Eq. (2):

$$-\left(\frac{R_A L^2}{6EI} - \frac{wb^4}{24EIL}\right) + \left[\frac{R_A x_m^2}{2EI} - \frac{w}{6EI}(x_m - a)^3\right] = 0$$

Substituting the numerical data, we have

$$-29.53\frac{10^3}{EI} + 8.405 x_m^2 \frac{10^3}{EI} - 4.167(x_m - 1.4)^3 \frac{10^3}{EI} = 0$$

Solving by trial and error for x_m, we find $x_m = 1.890$ m ◄
Computing the moments of A_3 and A_4 about a vertical axis through A, we have

$$|y|_m = t_{A/K} = A_3 \frac{2x_m}{3} + A_4\left[a + \frac{3}{4}(x_m - a)\right]$$

$$= \frac{R_A x_m^3}{3EI} - \frac{wa}{6EI}(x_m - a)^3 - \frac{w}{8EI}(x_m - a)^4$$

Using the given data, $R_A = 16.81$ kN, and $I = 28.7 \times 10^{-6}$ m^4, we find

$$y_m = 6.44 \text{ mm} \downarrow \text{ ◄}$$

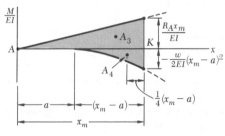

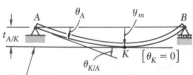

PROBLEMS

9.33 through 9.38   For the prismatic beam and loading shown, determine (*a*) the slope at point *A*, (*b*) the deflection at point *D*.

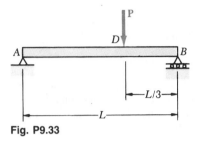

Fig. P9.33

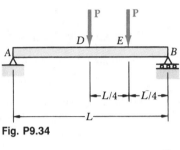

Fig. P9.34

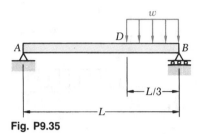

Fig. P9.35

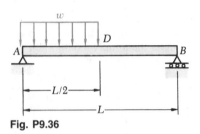

Fig. P9.36

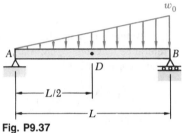

Fig. P9.37

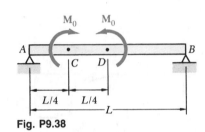

Fig. P9.38

9.39   For the beam and loading shown, determine (*a*) the slope at point *A*, (*b*) the deflection at point *C*.   Use $E = 200$ GPa.

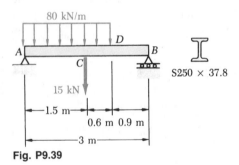

Fig. P9.39

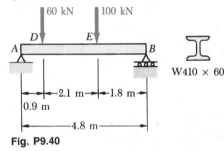

Fig. P9.40

9.40   For the beam and loading shown, determine (*a*) the slope at point *A*, (*b*) the deflection at point *E*.   Use $E = 200$ GPa.

9.41 and 9.42   For the beam and loading shown, determine (*a*) the slope at point *A*, (*b*) the deflection at point *E*.   Use $E = 200$ GPa.

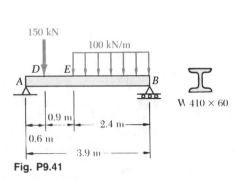

Fig. P9.41

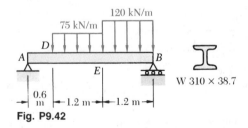

Fig. P9.42

9.43 The rigid bar DEC is welded at point C to the rolled-steel beam AB. For the loading shown, determine (a) the slope at point A, (b) the deflection at point C. Use $E = 200$ GPa.

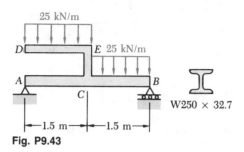

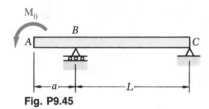

Fig. P9.43

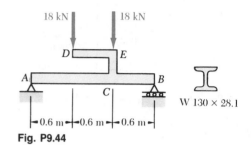

Fig. P9.44

9.44 The rigid bar DEC is welded at point C to the rolled-steel beam AB. For the loading shown, determine (a) the slope at point A, (b) the deflection at point C. Use $E = 200$ GPa.

9.45 For the beam and loading shown, determine (a) the slope at point A, (b) the deflection at point A.

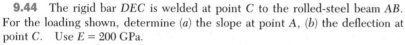

Fig. P9.45

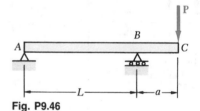

Fig. P9.46

9.46 For the beam and loading shown, determine (a) the slope at point C, (b) the deflection at point C.

9.47 For the beam and loading shown, determine (a) the slope at point A, (b) the slope at point B, (c) the deflection at point C.

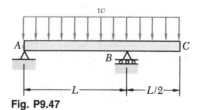

Fig. P9.47

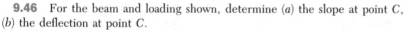

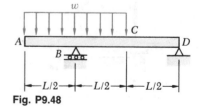

Fig. P9.48

9.48 For the beam and loading shown, determine the deflection at (a) point A, (b) point C.

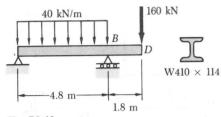

Fig. P9.49

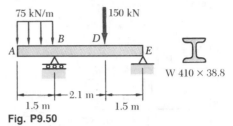

Fig. P9.50

9.49 For the beam and loading shown, determine (a) the slope at point B, (b) the deflection at point D.   Use $E = 200$ GPa.

9.50 For the beam and loading shown, determine the deflection at (a) point A, (b) point D.   Use $E = 200$ GPa.

9.51 Solve Prob. 9.50, assuming that the 150-kN load is replaced by a 45-kN load directed downward.

9.52 For the beam and loading shown, determine (a) the slope at point D, (b) the deflection at point E.   Use $E = 200$ GPa.

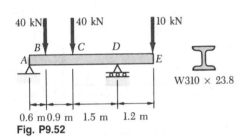

Fig. P9.52

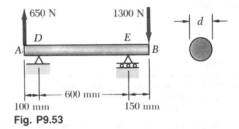

Fig. P9.53

9.53 Knowing that beam AB is made of a solid steel rod of diameter $d = 20$ mm, determine for the loading shown (a) the slope at point D, (b) the deflection at point A.   Use $E = 200$ GPa.

9.54 Solve Prob. 9.53, assuming that the 650-N force at A acts downward.

9.55 For the beam ACB the flexural rigidity is EI in portion AC and nEI in portion CB.   For the loading shown, determine the deflection of point C for (a) $n = \frac{1}{3}$, (b) $n = 1$, (c) $n = 3$.

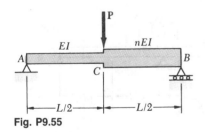

Fig. P9.55

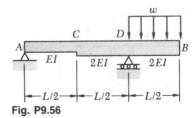

Fig. P9.56

9.56 For the beam and loading shown, determine (a) the slope at point D, (b) the deflection at point B.

9.57 For the beam and loading shown, determine (a) the slope at A, (b) the slope at B, (c) the deflection at the midpoint C.

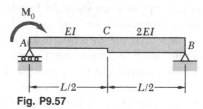

Fig. P9.57

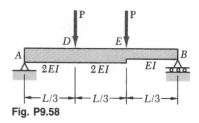

Fig. P9.58

9.58 For the beam and loading shown, determine the deflection (a) at point D, (b) at point E.

9.59 For the beam and loading shown, determine the magnitude and location of the maximum deflection.

9.60 through 9.66 For the beam and loading indicated, determine the magnitude and location of the largest downward deflection.

   9.60 Beam and loading of Prob. 9.33.
   9.61 Beam and loading of Prob. 9.37.
   9.62 Beam and loading of Prob. 9.36.
   9.63 Beam and loading of Prob. 9.39.
   9.64 Beam and loading of Prob. 9.40.
   9.65 Beam and loading of Prob. 9.41.
   9.66 Beam and loading of Prob. 9.42.

9.67 For the beam and loading of Prob. 9.49, determine the largest upward deflection in span AB.

9.68 For the beam and loading of Prob. 9.50, determine the largest downward deflection in span BE.

9.69 For the beam and loading of Prob. 9.53, determine the largest upward deflection in span DE.

9.70 For the beam and loading of Prob. 9.52, determine the largest downward deflection in span AD.

9.71 and 9.72 For the beam and loading indicated, determine the magnitude and location of the largest downward deflection.

   9.71 Beam and loading of Prob. 9.57.
   9.72 Beam and loading of Prob. 9.58.

9.73 For the beam and loading shown, determine (a) the value of a for which the slope at end A is zero, (b) the corresponding deflection at point C.

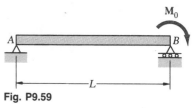

Fig. P9.59

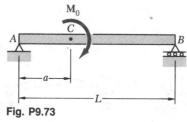

Fig. P9.73

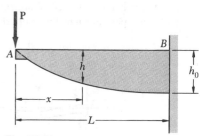

Fig. P9.74

*9.74 The constant-strength beam AB has a rectangular cross section of uniform width b and variable depth h. Express the deflection at end A in terms of P, L, and the flexural rigidity EI_0 at B. (*Hint:* Since the beam is of constant strength, Mc/I has a constant value along AB.)

*9.7. STATICALLY INDETERMINATE BEAMS

The reactions at the supports of a statically indeterminate beam may be determined by the moment-area method in much the same way that was described in Sec. 8.8. In the case of a beam indeterminate to the first degree, for example, we designate one of the reactions as redundant and eliminate or modify accordingly the corresponding support. The redundant reaction is then treated as an unknown load which, together with the other loads, must produce deformations which are compatible with the original supports. The compatibility condition is usually expressed by writing that the tangential deviation of one support with respect to another either is zero or has a predetermined value.

Two separate free-body diagrams of the beam are drawn. One shows *the given loads and the corresponding reactions* at the supports which have not been eliminated; the other shows *the redundant reaction and the corresponding reactions* at the same supports (see Example 9.06). An (*M/EI*) diagram is then drawn for each of the two loadings, and the desired tangential deviations are obtained by the second moment-area theorem. Superposing the results obtained, we express the required compatibility condition and determine the redundant reaction. The other reactions are obtained from the free-body diagram of the beam.

Once the reactions at the supports have been determined, the slope and deflection may be obtained by the moment-area method at any other point of the beam.

Example 9.06

Determine the reaction at the supports for the prismatic beam and loading shown (Fig. 9.30).

We consider the couple exerted at the fixed end A as redundant and replace the fixed end by a pin-and-bracket support. The couple $\mathbf{M}_A$ is now considered as an unknown load (Fig. 9.31*a*) and will be determined from the condition that the tangent to the beam at A must be horizontal. It follows that this tangent must pass through the support B and, thus, that the

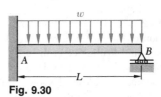

Fig. 9.30

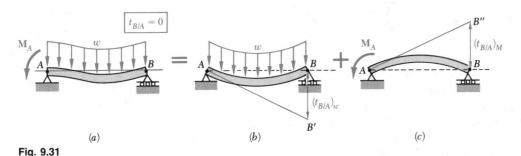

(a)   (b)   (c)

Fig. 9.31

tangential deviation $t_{B/A}$ of B with respect to A must be zero. The solution is carried out by computing separately the tangential deviation $(t_{B/A})_w$ caused by the uniformly distributed load w (Fig. 9.31b) and the tangential deviation $(t_{B/A})_M$ produced by the unknown couple $\mathbf{M}_A$ (Fig. 9.31c).

Considering first the free-body diagram of the beam under the known distributed load w (Fig. 9.32a), we determine the corresponding reactions at the supports A and B. We have

$$(\mathbf{R}_A)_1 = (\mathbf{R}_B)_1 = \tfrac{1}{2}wL \uparrow \qquad (9.11)$$

We may now draw the corresponding shear and (M/EI) diagrams (Figs. 9.32b and c). Observing that M/EI is represented by an arc of parabola, and recalling the formula, $A = \tfrac{2}{3}bh$, for the area under a parabola, we compute the first moment of this area about a vertical axis through B and write

$$(t_{B/A})_w = A_1\left(\frac{L}{2}\right) = \left(\frac{2}{3}L\frac{wL^2}{8EI}\right)\left(\frac{L}{2}\right) = \frac{wL^4}{24EI} \qquad (9.12)$$

Considering next the free-body diagram of the beam when it is subjected to the unknown couple $\mathbf{M}_A$ (Fig. 9.33a), we determine the corresponding reactions at A and B:

$$(\mathbf{R}_A)_2 = \frac{M_A}{L} \uparrow \qquad (\mathbf{R}_B)_2 = \frac{M_A}{L} \downarrow \qquad (9.13)$$

Drawing the corresponding (M/EI) diagram (Fig. 9.33b), we apply again the second moment-area theorem and write

$$(t_{B/A})_M = A_2\left(\frac{2L}{3}\right) = \left(-\frac{1}{2}L\frac{M_A}{EI}\right)\left(\frac{2L}{3}\right) = -\frac{M_A L^2}{3EI} \qquad (9.14)$$

Combining the results obtained in (9.12) and (9.14), and expressing that the resulting tangential deviation $t_{B/A}$ must be zero (Fig. 9.31), we have

$$t_{B/A} = (t_{B/A})_w + (t_{B/A})_M = 0$$

$$\frac{wL^4}{24EI} - \frac{M_A L^2}{3EI} = 0$$

and, solving for M_A,

$$M_A = +\tfrac{1}{8}wL^2 \qquad \mathbf{M}_A = \tfrac{1}{8}wL^2 \curvearrowright$$

Substituting for M_A into (9.13), and recalling (9.11), we obtain the values of R_A and R_B:

$$R_A = (R_A)_1 + (R_A)_2 = \tfrac{1}{2}wL + \tfrac{1}{8}wL = \tfrac{5}{8}wL$$
$$R_B = (R_B)_1 + (R_B)_2 = \tfrac{1}{2}wL - \tfrac{1}{8}wL = \tfrac{3}{8}wL$$

Alternative Solution. We may consider the reaction at B as redundant and release the beam from that support. The

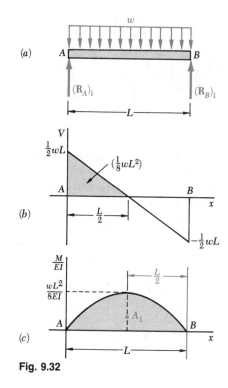

Fig. 9.32

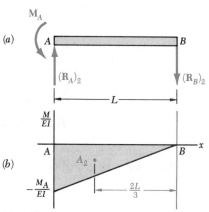

Fig. 9.33

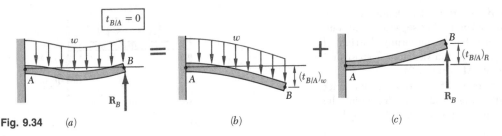

Fig. 9.34   (a)   (b)   (c)

reaction $\mathbf{R}_B$ is now considered as an unknown load (Fig. 9.34a) and will be determined from the condition that the end B of the beam must be on the same horizontal level as A. Since the tangent to the beam at A is horizontal, it follows that the tangential deviation $t_{B/A}$ of B with respect to A must be zero. The solution is carried out by computing separately the tangential deviation $(t_{B/A})_w$ caused by the uniformly distributed load w (Fig. 9.34b) and the tangential deviation $(t_{B/A})_R$ produced by the redundant reaction $\mathbf{R}_B$ (Fig. 9.34c).

First we consider the free-body diagram of the beam under the known distributed load w (Fig. 9.35a) and draw the corresponding (M/EI) diagram from right to left (Fig. 9.35b), thus avoiding the computation of the reaction at A. Observing that the area of the parabolic spandrel is

$$A_1 = \frac{1}{3}\left(-\frac{wL^2}{2EI}\right)L = -\frac{wL^3}{6EI}$$

and applying the second moment-area theorem, we write

$$(t_{B/A})_w = A_1\left(\frac{3L}{4}\right) = \left(-\frac{wL^3}{6EI}\right)\left(\frac{3L}{4}\right) = -\frac{wL^4}{8EI} \quad (9.15)$$

Considering next the free-body diagram of the beam when it is subjected to the redundant reaction $\mathbf{R}_B$ (Fig. 9.36a), we draw the (M/EI) diagram from right to left (Fig. 9.36b), avoiding again the computation of the reaction at A, and write

$$(t_{B/A})_R = A_2\left(\frac{2L}{3}\right) = \left(\frac{1}{2}\frac{R_B L}{EI}L\right)\left(\frac{2L}{3}\right) = \frac{R_B L^3}{3EI} \quad (9.16)$$

Combining the results obtained in (9.15) and (9.16), and expressing that the resulting tangential deviation $t_{B/A}$ must be zero (Fig. 9.34), we have

$$t_{B/A} = (t_{B/A})_w + (t_{B/A})_R = 0$$
$$-\frac{wL^4}{8EI} + \frac{R_B L^3}{3EI} = 0$$

and, solving for R_B,

$$R_B = \tfrac{3}{8}wL$$

The reaction at the fixed end A may then be obtained from the free-body diagram of the beam.

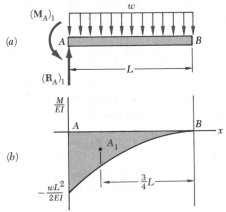

Fig. 9.35

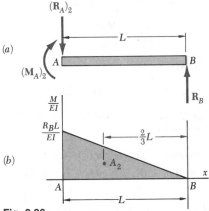

Fig. 9.36

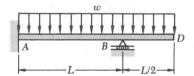

SAMPLE PROBLEM 9.5

For the prismatic beam and loading shown, determine the reactions at the supports.

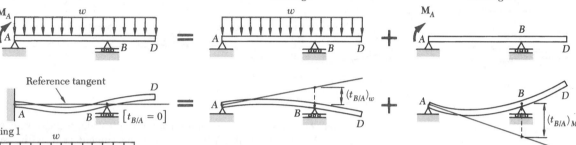

Solution. We note that the beam is indeterminate to the first degree and choose the couple M_A exerted by the fixed end A as redundant. We select the horizontal tangent at A as the reference tangent and note that the tangential

Loading 1   Loading 2

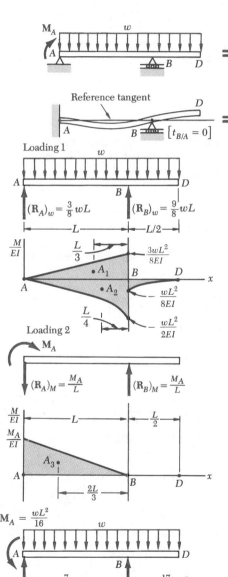

deviation $t_{B/A}$ of B with respect to A must be zero. The deformations caused by the distributed load w and the couple M_A (assumed clockwise) are considered separately.

For each loading we now determine the reactions, draw the (M/EI) diagram, and then find $t_{B/A}$.

Distributed Loading. (Loading 1)

$$(t_{B/A})_w = A_1 \frac{L}{3} + A_2 \frac{L}{4} = \left(\frac{1}{2} \frac{3wL^2}{8EI} L\right) \frac{L}{3} + \left(-\frac{1}{3} \frac{wL^2}{2EI} L\right) \frac{L}{4}$$

$$= \frac{wL^4}{EI}\left(\frac{1}{16} - \frac{1}{24}\right) = \frac{wL^4}{48EI}$$

Redundant Couple Loading. (Loading 2)

$$(t_{B/A})_M = A_3 \frac{2L}{3} = \left(\frac{1}{2} \frac{M_A}{EI} L\right) \frac{2L}{3} = \frac{M_A L^2}{3EI}$$

Couple M_A. Since the tangential deviation $t_{B/A}$ must be zero, we write

$$t_{B/A} = (t_{B/A})_w + (t_{B/A})_M = 0$$

$$\frac{wL^4}{48EI} + \frac{M_A L^2}{3EI} = 0 \qquad M_A = -\frac{wL^2}{16} \qquad M_A = \frac{wL^2}{16} \curvearrowright \quad \blacktriangleleft$$

Other Reactions. Substituting $M_A = -wL^2/16$ in the expressions indicated for $(R_A)_M$ and $(R_B)_M$ in the free-body diagram of the second loading, we have $(R_A)_M = (R_B)_M = -wL/16$. Combining the diagrams of both loadings, we obtain

$$+\uparrow R_A = (R_A)_w + (R_A)_M = \frac{3}{8} wL - \left(-\frac{wL}{16}\right) = \frac{7}{16} wL \qquad R_A = \frac{7}{16} wL \uparrow \quad \blacktriangleleft$$

$$+\uparrow R_B = (R_B)_w + (R_B)_M = \frac{9}{8} wL + \left(-\frac{wL}{16}\right) = \frac{17}{16} wL \qquad R_B = \frac{17}{16} wL \uparrow \quad \blacktriangleleft$$

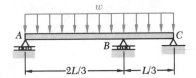

For the uniform beam and loading shown, determine the reaction at B.

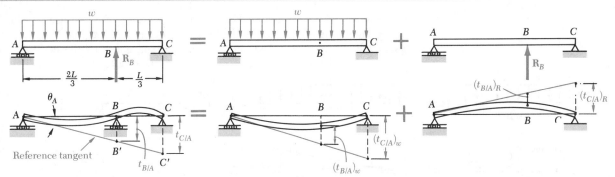

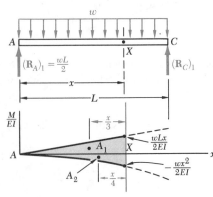

Reference tangent

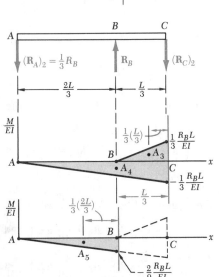

Solution. The beam is indeterminate to the first degree and we choose the reaction $\mathbf{R}_B$ as redundant. We next select the tangent at A as the reference tangent. From the similar triangles ABB' and ACC', we find that

$$\frac{t_{C/A}}{L} = \frac{t_{B/A}}{\frac{2}{3}L} \qquad t_{C/A} = \frac{3}{2}\,t_{B/A} \tag{1}$$

For each loading, we draw the (M/EI) diagram and then determine the tangential deviations of B and C with respect to A.

Distributed Loading. Considering the (M/EI) diagram from end A to an arbitrary point X, we write

$$(t_{X/A})_w = A_1\,\frac{x}{3} + A_2\,\frac{x}{4} = \left(\frac{1}{2}\,\frac{wLx}{2EI}\,x\right)\frac{x}{3} + \left(-\frac{1}{3}\,\frac{wx^2}{2EI}\,x\right)\frac{x}{4} = \frac{wx^3}{24EI}\,(2L - x)$$

Letting successively $x = L$ and $x = \frac{2}{3}L$, we have

$$(t_{C/A})_w = \frac{wL^4}{24EI} \qquad (t_{B/A})_w = \frac{4}{243}\,\frac{wL^4}{EI}$$

Redundant Reaction Loading

$$(t_{C/A})_R = A_3\,\frac{L}{9} + A_4\,\frac{L}{3} = \left(\frac{1}{2}\,\frac{R_B L}{3EI}\,\frac{L}{3}\right)\frac{L}{9} + \left(-\frac{1}{2}\,\frac{R_B L}{3EI}\,L\right)\frac{L}{3} = -\frac{4}{81}\,\frac{R_B L^3}{EI}$$

$$(t_{B/A})_R = A_5\,\frac{2L}{9} = \left[-\frac{1}{2}\,\frac{2R_B L}{9EI}\left(\frac{2L}{3}\right)\right]\frac{2L}{9} = -\frac{4}{243}\,\frac{R_B L^3}{EI}$$

Combined Loading. Adding the results obtained, we write

$$t_{C/A} = \frac{wL^4}{24EI} - \frac{4}{81}\,\frac{R_B L^3}{EI} \qquad t_{B/A} = \frac{4}{243}\,\frac{(wL^4 - R_B L^3)}{EI}$$

Reaction at B. Substituting for $t_{C/A}$ and $t_{B/A}$ into Eq. (1), we have

$$\left(\frac{wL^4}{24EI} - \frac{4}{81}\,\frac{R_B L^3}{EI}\right) = \frac{3}{2}\left[\frac{4}{243}\,\frac{(wL^4 - R_B L^3)}{EI}\right]$$

$$R_B = 0.6875wL \qquad\qquad \mathbf{R}_B = 0.688wL\uparrow \quad \blacktriangleleft$$

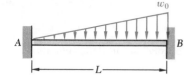

For the prismatic beam and loading shown, determine the reactions at the supports.

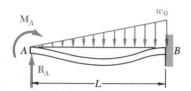

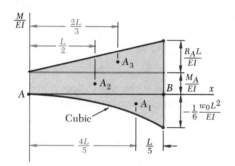

Solution. Assuming the axial force in the beam to be zero, we note that the beam is indeterminate to the second degree and we choose two reaction components as redundant, namely the vertical force $\mathbf{R}_A$ and the couple $\mathbf{M}_A$.

Reference Tangent. We select the tangent to the beam at support B as the reference tangent and note that both $\theta_{B/A}$ and $t_{A/B}$ are zero.†

(M/EI) Diagram. We draw the (M/EI) diagram by parts, considering separately the effects of the distributed load, the couple $\mathbf{M}_A$ and the force $\mathbf{R}_A$. We note that the diagram consists of three simple geometric shapes, namely, a triangle, a rectangle, and a cubic spandrel. Referring to Fig. 9.12, we write

$$A_1 = \frac{1}{4}\left(-\frac{w_0 L^2}{6EI}\right)L = -\frac{w_0 L^3}{24EI}$$

$$A_2 = \frac{M_A L}{EI} \qquad A_3 = \frac{R_A L^2}{2EI}$$

Moment-Area Theorems. Since $\theta_{B/A} = 0$, the area of the (M/EI) diagram between points A and B is equal to zero:

$$\theta_{B/A} = A_1 + A_2 + A_3 = 0$$

$$-\frac{w_0 L^3}{24EI} + \frac{M_A L}{EI} + \frac{R_A L^2}{2EI} = 0$$

$$M_A + \tfrac{1}{2}R_A L = \tfrac{1}{24}w_0 L^2 \qquad (1)$$

Since $t_{A/B} = 0$, the moment of the area of the (M/EI) diagram about a vertical axis through point A is zero.

$$t_{A/B} = A_1 \frac{4L}{5} + A_2 \frac{L}{2} + A_3 \frac{2L}{3} = 0$$

$$\left(-\frac{w_0 L^3}{24EI}\right)\frac{4L}{5} + \left(\frac{M_A}{EI}L\right)\frac{L}{2} + \left(\frac{R_A L^2}{2EI}\right)\frac{2L}{3} = 0$$

$$\tfrac{1}{2}M_A + \tfrac{1}{3}R_A L = \tfrac{1}{30}w_0 L^2 \qquad (2)$$

Reaction at A. Solving simultaneously Eqs. (1) and (2), we find

$$R_A = +\tfrac{3}{20}w_0 L \qquad\qquad R_A = \tfrac{3}{20}w_0 L \uparrow \blacktriangleleft$$

$$M_A = -\frac{w_0 L^2}{30} \qquad\qquad M_A = \frac{w_0 L^2}{30} \curvearrowright \blacktriangleleft$$

Reaction at B. Using the methods of statics, we find

$$R_B = \frac{7}{20}w_0 L \uparrow \qquad M_B = \frac{w_0 L^2}{20}\curvearrowleft \blacktriangleleft$$

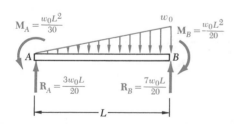

†*Note:* We observe that the tangential deviation of B with respect to A is also zero. Thus, in alternative solutions, we could use $\theta_{B/A} = 0$ and $t_{B/A} = 0$, or $t_{A/B} = 0$ and $t_{B/A} = 0$.

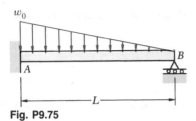

Fig. P9.75

PROBLEMS

9.75 through 9.78 For the beam and loading shown, determine the reaction at the roller support.

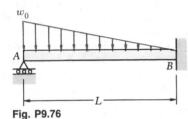

Fig. P9.76

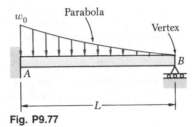

Fig. P9.77

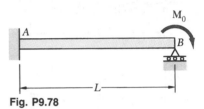

Fig. P9.78

9.79 through 9.82 Determine the reaction at the roller support and draw the bending-moment diagram for the beam and loading shown.

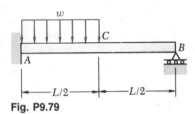

Fig. P9.79

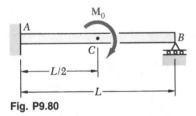

Fig. P9.80

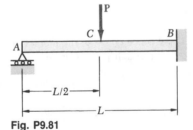

Fig. P9.81

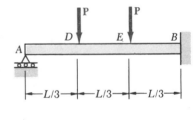

Fig. P9.82

9.83 and 9.84 Determine the reaction at the roller support and draw the bending-moment diagram for the beam and loading shown.

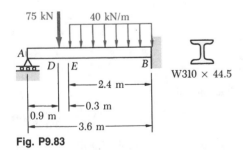

Fig. P9.83

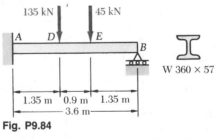

Fig. P9.84

9.85 Solve Prob. 9.84, assuming that the 135-kN load is replaced by a 100-kN load applied at the same point D of the beam.

9.86 Solve Prob. 9.83, assuming that the 75-kN load is removed.

9.87 through 9.90 For the beam and loading shown, determine the reaction at each support.

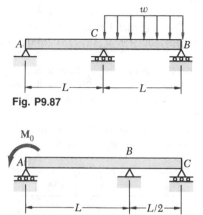

Fig. P9.87

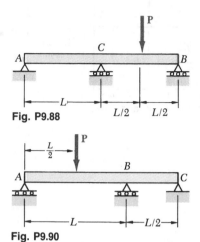

Fig. P9.88

Fig. P9.89

Fig. P9.90

9.91 through 9.94 Determine the reaction at A and draw the bending-moment diagram for the beam and loading shown.

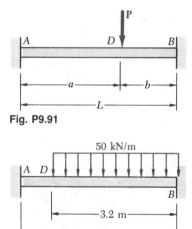

Fig. P9.91

Fig. P9.92

Fig. P9.93

Fig. P9.94

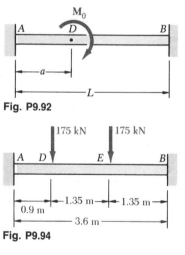

9.95 A hydraulic jack may be used to raise point B of the cantilever beam ABC. Knowing that after the 20-kN load is applied, point C is to have the same elevation as point A, determine (a) how much point B should be raised, (b) the reaction at B after point B has been raised and the 20-kN load has been applied. Use $E = 200$ GPa.

9.96 For the beam and loading of Prob. 9.95, determine (a) how much point B should be raised if the deflection at C is to be 10 mm downward as measured from the horizontal through A, (b) the reaction at B after point B has been raised and the 20-kN load has been applied.

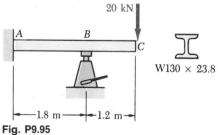

W130 × 23.8

Fig. P9.95

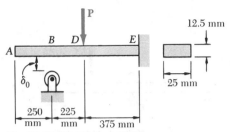

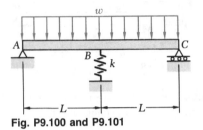

Fig. P9.97, P9.98, and P9.99

9.97 Before the load **P** is applied, a gap $\delta_0 = 5$ mm exists between the beam and the support at B. For $P = 350$ N, determine (*a*) the reaction at B, (*b*) the slope at B, (*c*) the deflection at A. Use $E = 200$ GPa.

9.98 Before the load **P** is applied, a gap $\delta_0 = 5$ mm exists between the beam and the support at B. Knowing that $E = 200$ GPa, determine (*a*) the magnitude P of the load for which the deflection at A is zero, (*b*) the corresponding reaction at B.

9.99 Before the load **P** is applied, a gap $\delta_0 = 5$ mm exists between the beam and the support at B. Knowing that $E = 200$ GPa, determine (*a*) the magnitude P of the load for which the slope at point B is zero, (*b*) the corresponding reaction at B.

9.100 For the beam and loading shown, determine the spring constant k for which the bending moment at B is $M_B = -wL^2/12$.

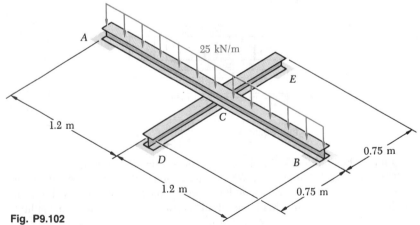

Fig. P9.100 and P9.101

9.101 For the beam and loading shown, determine the spring constant k for which the force in the spring is equal to one-third of the total load on the beam.

9.102 The support structure shown is made of two S 130 × 15 rolled-steel beams which are simply supported at each end and, when not loaded, touch at the midpoint C of each beam. For the distributed load shown, determine (*a*) the reaction at each support, (*b*) the deflection at the midpoint C. Use $E = 200$ GPa.

Fig. P9.102

9.103 Solve Prob. 9.102, assuming that the distributed load is replaced by a 40-kN load applied at the midpoint of beam ACB.

9.104 A solid 19-mm-diameter steel rod of length 1 m is attached at the midpoint C of the S 130×15 rolled-steel beam AE to help support the beam.   For $P = 40$ kN, determine (a) the tension in the rod, (b) the bending moment at points B and C of the beam, (c) the deflection at point C.   Use $E = 200$ GPa.

9.105 Solve Prob. 9.104, assuming that the 19-mm-diameter rod is replaced by a 12-mm-diameter rod of the same length.

9.106 Steel wires of 3.2-mm diameter are used to support the square steel rod ACB.   Knowing that each wire is initially taut, determine the additional tension in each wire when the two 2.5-kN forces are applied as shown.

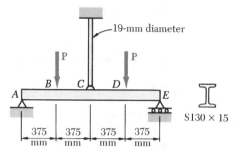

Fig. P9.104

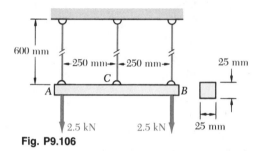

Fig. P9.106

9.107 Solve Prob. 9.106, assuming that a single 2.5-kN force is applied at point C.

REVIEW AND SUMMARY

This chapter was devoted to the determination of deflections and slopes of beams using the *moment-area method*.

In order to derive the *moment-area theorems* [Sec. 9.2], we first drew a diagram representing the variation along the beam of the quantity M/EI obtained by dividing the bending moment M by the flexural rigidity EI (Fig. 9.1).   We then derived the *first moment-area theorem*, which may be stated as follows: *The area under the (M/EI) diagram between two points is equal to the angle between the tangents to the elastic curve drawn at these points.*   Considering tangents at C and D, we wrote

$$\theta_{D/C} = \text{area under } (M/EI) \text{ diagram between } C \text{ and } D \tag{9.3}$$

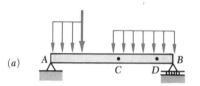

(a)

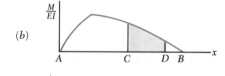

(b)

(c)

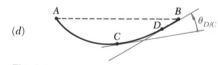

(d)

Fig. 9.1

First moment-area theorem

Again using the (M/EI) diagram and a sketch of the deflected beam (Fig. 9.5a), we drew a tangent at point D and considered the vertical distance $t_{C/D}$, which is called the *tangential deviation* of C with respect to D. We then derived the *second moment-area theorem*, which may be

Second moment-area theorem

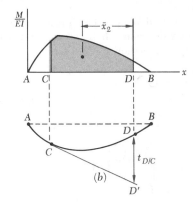

Fig. 9.5

stated as follows: *The tangential deviation $t_{C/D}$ of C with respect to D is equal to the first moment with respect to a vertical axis through C of the area under the (M/EI) diagram between C and D.* We were careful to distinguish between the tangential deviation of C with respect to D,

$$t_{C/D} = (\text{area between } C \text{ and } D)\, \bar{x}_1 \qquad (9.6)$$

and the tangential deviation of D with respect to C (Fig. 9.5b):

$$t_{D/C} = (\text{area between } C \text{ and } D)\, \bar{x}_2 \qquad (9.7)$$

In Sec. 9.3 we learned to determine the slope and deflection at points of *cantilever* beams and beams with *symmetric loadings*. For cantilever beams, the tangent at the fixed support is horizontal (Fig. 9.6); and for symmetrically loaded beams, the tangent is horizontal at the midpoint C of the beam (Fig. 9.7). Using the horizontal tangent as a *reference tangent*, we were able to determine slopes and deflections by using, respectively, the first and second moment-area theorems [Example 9.01, Sample Probs. 9.1 and 9.2]. We noted that to find a deflection which is not a tangential deviation (Fig. 9.7c), it is necessary to first determine which tangential deviations can be combined to obtain the desired deflection.

Cantilever Beams
Beams with symmetric loadings

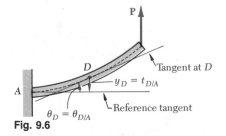

Fig. 9.6

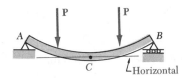

Fig. 9.7   (a)

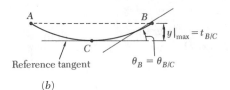

(b)

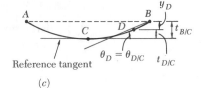

(c)

In many cases the application of the moment-area theorems is simplified if we consider the effect of each load separately [Sec. 9.4]. To do this we drew the (*M/EI*) *diagram by parts* by drawing a separate (*M/EI*) diagram for each load. The areas and the moments of areas under the several diagrams could then be added to determine slopes and tangential deviations for the original beam and loading [Examples 9.02 and 9.03].

In Sec. 9.5 we expanded the use of the moment-area method to cover beams with *unsymmetric loadings*. Observing that locating a horizontal tangent is usually not possible, we selected a reference tangent at one of the beam supports, since the slope of that tangent may be readily determined. For example, for the beam and loading shown in Fig. 9.19,

Bending-moment diagram by parts

Unsymmetric loadings

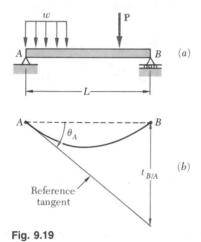

Fig. 9.19

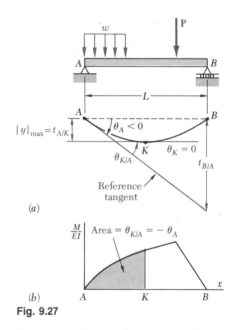

the slope of the tangent at *A* may be obtained by computing the tangential deviation $t_{B/A}$ and dividing it by the distance *L* between the supports *A* and *B*. Then, using both moment-area theorems and simple geometry, we could determine the slope and deflection at any point of the beam [Example 9.04, Sample Prob. 9.3].

The *maximum deflection* of an unsymmetrically loaded beam generally does not occur at midspan. The approach indicated in the preceding paragraph was used to determine point *K* where the maximum deflection occurs and the magnitude of that deflection [Sec. 9.6]. Observing that the slope at *K* is zero (Fig. 9.27), we concluded that $\theta_{K/A} = -\theta_A$. Recalling the first moment-area theorem, we determined the location of *K* by measuring under the (*M/EI*) diagram an area equal to $\theta_{K/A}$. The maximum deflection was then obtained by computing the tangential deviation $t_{A/K}$ [Example 9.05, Sample Probs. 9.3 and 9.4].

Maximum deflection

Statically indeterminate beams

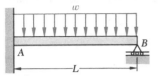

Fig. 9.30

In the last section of the chapter [Sec. 9.7] we considered the analysis of *statically indeterminate beams*. Since the reactions for the beam and loading shown in Fig. 9.30 cannot be determined by statics alone, we designated one of the reactions of the beam as redundant ($\mathbf{M}_A$ in Fig. 9.31a) and considered the redundant reaction as an unknown load. The tangential deviation of B with respect to A was considered separately for the distributed load (Fig. 9.31b) and for the redundant reaction (Fig. 9.31c). Expressing that under the combined action of the distributed load and of the couple $\mathbf{M}_A$ the tangential deviation of B with respect to A must be zero, we wrote

$$t_{B/A} = (t_{B/A})_w + (t_{B/A})_M = 0$$

From this expression we determined the magnitude of the redundant reaction $\mathbf{M}_A$ [Example 9.06, Sample Probs. 9.5, 9.6 and 9.7].

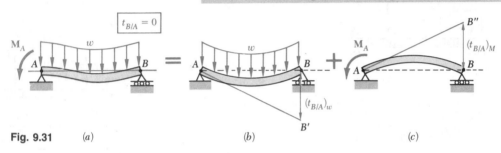

Fig. 9.31 (a) (b) (c)

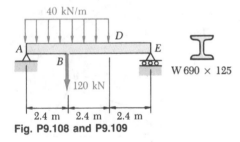

Fig. P9.108 and P9.109

W 690 × 125

REVIEW PROBLEMS

9.108 For the beam and loading shown, determine (a) the slope at end E, (b) the deflection at point B. Use $E = 200$ GPa.

9.109 For the beam and loading shown, determine the magnitude and location of the largest deflection. Use $E = 200$ GPa.

9.110 Two L-shaped bars are welded at points B and D to the rolled-steel beam AE. For the loading shown, determine (a) the slope at A, (b) the slope at B, (c) the deflection at B. Use $E = 200$ GPa.

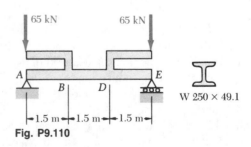

Fig. P9.110

9.111 For the beam and loading shown, determine (*a*) the bending moment at end *A*, (*b*) the deflection at the midpoint *C*.

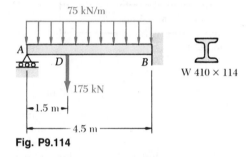

Fig. P9.111

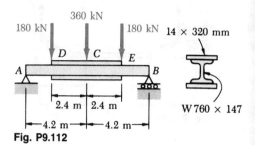

Fig. P9.112

9.112 Two cover plates are welded to a W760 × 147 rolled-steel beam as shown. Using $E = 200$ GPa, determine (*a*) the slope at end *A*, (*b*) the deflection at the midpoint *C*.

9.113 Solve Prob. 9.112, assuming that the two 180-kN loads are removed.

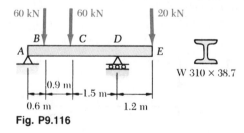

Fig. P9.114

9.114 For the beam and loading shown, determine (*a*) the reaction at *A*, (*b*) the deflection at *D*. Use $E = 200$ GPa.

9.115 Solve Prob. 9.114, assuming that the fixed support at *B* is replaced by a support consisting of a pin and bracket.

9.116 For the beam and loading shown, determine (*a*) the slope at *D*, (*b*) the deflection at *C*, (*c*) the deflection at *E*. Use $E = 200$ GPa.

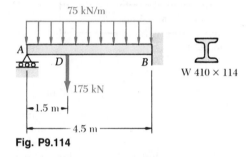

Fig. P9.116

9.117 Solve Prob. 9.116, assuming that the magnitude of the load at *E* is increased to 60 kN.

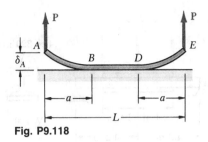

Fig. P9.118

9.118 Vertical forces **P** are applied at the ends A and E of a uniform steel rod which rests on a horizontal rigid surface. As the magnitude P of the applied forces is increased, portions AB and DE of the rod are lifted from the surface. Knowing that the rod is initially straight, and denoting by w its weight per unit length and by EI its flexural rigidity, determine the distance a and the deflection δ_A when (a) $P = wL/8$, (b) $P = wL/4$, (c) $P = 3wL/8$.

9.119 For the beam and loading shown, determine (a) the slope at A, (b) the deflection at E, (c) the deflection at F. Use $E = 200$ GPa.

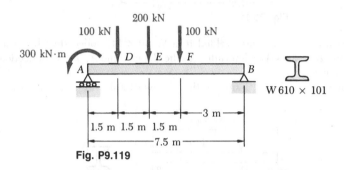

Fig. P9.119

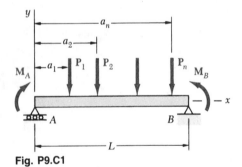

Fig. P9.C1

The following problems are designed to be solved with a computer.

9.C1 For the uniform beam and loading shown, write a computer program which can be used to calculate the slope and deflection of the beam from $x = 0$ to $x = L$ at intervals ΔL. Use this program to calculate the slope and deflection at each concentrated load for the beam of (a) Prob. 9.26, (b) Prob. 9.119

9.C2 Two 50-kN loads are maintained 1.5 m apart as they are moved slowly across beam AB. Write a computer program which can be used to calculate the deflection at the midpoint C of the beam for values of x from 0 to 4.8 m at 0.3 m intervals. Use $E = 200$ GPa.

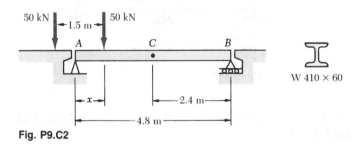

Fig. P9.C2

9.C3 The hydraulic jack shown may be used to raise or lower point B of the cantilever beam ABC. After the 25-kN load has been applied, the deflection of point C is to be 4 mm downward as measured from the horizontal through A. Write a computer program and use it to determine, for values of a from 0.6 m to 3 m at 0.3-m intervals, (a) the required deflection of point B, (b) the corresponding force applied by the jack at B. Use $E = 200$ GPa.

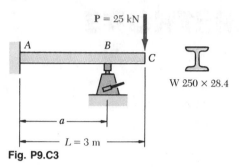

W 250 × 28.4

Fig. P9.C3

9.C4 Two brass rods, each of 6 × 6-mm cross section, are used to form the cantilever beams AB and CD. Before the load $P = 10$ N is applied, both of the beams are straight and are in contact at point C. Write a computer program and use it to determine, for values of a from 0 to 375 mm at 25-mm intervals, (a) the reaction at D, (b) the deflection at A, (c) the magnitude and location of the maximum stress. Use $E = 105$ GPa.

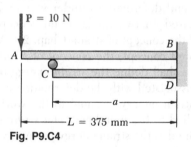

Fig. P9.C4

C H A P T E R   T E N

ENERGY METHODS

10.1. INTRODUCTION

In the previous chapters, we were concerned with the relations existing between forces and deformations under various loading conditions. Our analysis was based on two fundamental concepts, the concept of stress (Chap. 1) and the concept of strain (Chap. 2). We shall now introduce a third important concept, the concept of *strain energy*.

In Sec. 10.2 we shall define the *strain energy* of a member as the increase in energy associated with the deformation of the member. We shall see that the strain energy is equal to the work done by a slowly increasing load applied to the member. The *strain-energy density* of a material will be defined as the strain energy per unit volume and we shall see that it is equal to the area under the stress-strain curve of the material (Sec. 10.3). From the stress-strain diagram of a material we shall also define the *modulus of toughness* and *modulus of resilience* of the material.

In Sec. 10.4 we shall discuss the elastic strain energy associated with *normal stresses*, first in members under axial loading and then in members in bending. Later we shall consider the elastic strain energy associated with *shearing stresses* such as occur in torsional loadings of shafts and in transverse loadings of beams (Sec. 10.5). Strain energy for a *general state of stress* will be considered in Sec. 10.6, where we shall derive the *maximum-distortion-energy criterion* for yielding.

The effect of *impact loading* on members will be considered in Sec. 10.7. We shall learn to calculate both the *maximum stress* and the *maximum deflection* caused by a moving mass impacting on a member. Properties which increase the ability of a structure to withstand effectively impact loads will be discussed in Sec. 10.8.

In Sec. 10.9 we shall calculate the elastic strain energy of a member subjected to a *single concentrated load* and in Sec. 10.10 we shall learn to determine the deflection at the point of application of the single load.

In the last portion of the chapter we shall consider the strain energy of structures subjected to *several loads* (Sec. 10.11). *Castigliano's theorem* will be derived in Sec. 10.12 and used in Sec. 10.13 to determine the deflection at a given point of a structure subjected to several loads. In the last section we shall apply Castigliano's theorem to the analysis of indeterminate structures (Sec. 10.14).

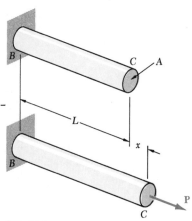

Fig. 10.1

10.2. STRAIN ENERGY

Consider a rod BC of length L and uniform cross-sectional area A, which is attached at B to a fixed support, and subjected at C to a *slowly increasing* axial load $\mathbf{P}$ (Fig. 10.1). As we noted in Sec. 2.2, by plotting the magnitude P of the load against the deformation x of the rod, we obtain a certain load-deformation diagram (Fig. 10.2) which is characteristic of the rod BC.

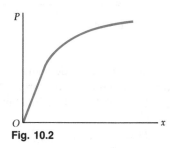

Fig. 10.2

Let us now consider the work dU done by the load $\mathbf{P}$ as the rod elongates by a small amount dx. This *elementary work* is equal to the product of the magnitude P of the load and of the small elongation dx. We write

$$dU = P\, dx \tag{10.1}$$

and note that the expression obtained is equal to the element of area of width dx located under the load-deformation diagram (Fig. 10.3). The *total work* U done by the load as the rod undergoes a deformation x_1 is thus

$$U = \int_0^{x_1} P\, dx$$

and is equal to the area under the load-deformation diagram between $x = 0$ and $x = x_1$.

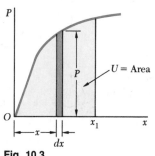

Fig. 10.3

The work done by the load **P** as it is slowly applied to the rod must result in the increase of some energy associated with the deformation of the rod. This energy is referred to as the *strain energy* of the rod. We have, by definition,

$$\text{Strain energy} = U = \int_0^{x_1} P \, dx \tag{10.2}$$

We recall that work and energy should be expressed in units obtained by multiplying units of length by units of force. Thus, if SI metric units are used, work and energy are expressed in N · m; this unit is called a *joule* (J).

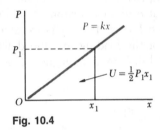

Fig. 10.4

In the case of a linear and elastic deformation, the portion of the load-deformation diagram involved may be represented by a straight line of equation $P = kx$ (Fig. 10.4). Substituting for P in Eq. (10.2), we have

$$U = \int_0^{x_1} kx \, dx = \tfrac{1}{2}kx_1^2$$

or

$$U = \tfrac{1}{2}P_1 x_1 \tag{10.3}$$

where P_1 is the value of the load corresponding to the deformation x_1.

The concept of strain energy is particularly useful in the determination of the effects of impact loadings on structures or machine components. Consider, for example, a body of mass m moving with a velocity v_0 which strikes the end B of a rod AB (Fig. 10.5a). Neglecting the inertia of the elements of the rod, and assuming no dissipation of energy during the impact, we find that the maximum strain energy U_m acquired by the rod (Fig. 10.5b) is equal to the original kinetic energy $T = \tfrac{1}{2}mv_0^2$ of the moving body. We may then determine the value P_m of the static load which would have produced the same strain energy in the rod, and obtain the value σ_m of the largest stress occurring in the rod by dividing P_m by the cross-sectional area of the rod.

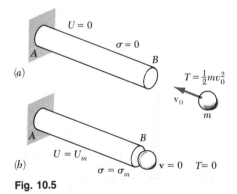

Fig. 10.5

10.3. STRAIN-ENERGY DENSITY

As we noted in Sec. 2.2, the load-deformation diagram for a rod BC depends upon the length L and the cross-sectional area A of the rod. The strain energy U defined by Eq. (10.2), therefore, will also depend upon the dimensions of the rod. In order to eliminate the effect of size from our discussion and direct our attention to the properties of the material, we shall consider the strain energy per unit volume. Dividing the strain energy U by the volume $V = AL$ of the rod (Fig. 10.1), and using Eq. (10.2), we have

$$\frac{U}{V} = \int_0^{x_1} \frac{P}{A} \frac{dx}{L}$$

Recalling that P/A represents the normal stress σ_x in the rod, and x/L the normal stain ϵ_x, we write

$$\frac{U}{V} = \int_0^{\epsilon_1} \sigma_x \, d\epsilon_x$$

where ϵ_1 denotes the value of the strain corresponding to the elongation x_1. The strain energy per unit volume, U/V, is referred to as the *strain-energy density* and will be denoted by the letter u. We have, therefore,

$$\text{Strain-energy density} = u = \int_0^{\epsilon_1} \sigma_x \, d\epsilon_x \qquad (10.4)$$

The strain-energy density u is expressed in units obtained by dividing units of energy by units of volume. Thus the strain-energy density is expressed in J/m^3 or its multiples kJ/m^3 and MJ/m^3.†

Referring to Fig. 10.6, we note that the strain-energy density u is equal to the area under the stress-strain curve, measured from $\epsilon_x = 0$ to $\epsilon_x = \epsilon_1$. If the material is unloaded, the stress returns to zero, but there is a permanent deformation represented by the strain ϵ_p, and only the portion of the strain energy per unit volume corresponding to the triangular area may be recovered. The remainder of the energy spent in deforming the material is dissipated in the form of heat.

The value of the strain-energy density obtained by setting $\epsilon_1 = \epsilon_R$ in Eq. (10.4), where ϵ_R is the strain at rupture, is known as the *modulus of toughness* of the material. It is equal to the area under the entire stress-strain diagram (Fig. 10.7) and represents the energy per unit volume required to cause the material to rupture. It is clear that the toughness of a material is related to its ductility as well as to its ultimate strength (Sec.

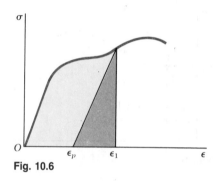

Fig. 10.6

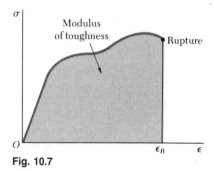

Fig. 10.7

† It may be noted that 1 J/m^3 and 1 Pa are both equal to 1 N/m^2. Thus, strain-energy density and stress are dimensionally equal and could be expressed in the same units.

2.3), and that the capacity of a structure to withstand an impact load depends upon the toughness of the material used.

If the stress σ_x remains within the proportional limit of the material, Hooke's law applies and we may write

$$\sigma_x = E\epsilon_x \tag{10.5}$$

Substituting for σ_x from (10.5) into (10.4), we have

$$u = \int_0^{\epsilon_1} E\epsilon_x \, d\epsilon_x = \frac{E\epsilon_1^2}{2} \tag{10.6}$$

or, using Eq. (10.5) to express ϵ_1 in terms of the corresponding stress σ_1,

$$u = \frac{\sigma_1^2}{2E} \tag{10.7}$$

The value u_Y of the strain-energy density obtained by setting $\sigma_1 = \sigma_Y$ in Eq. (10.7), where σ_Y is the yield strength, is called the *modulus of resilience* of the material. We have

$$u_Y = \frac{\sigma_Y^2}{2E} \tag{10.8}$$

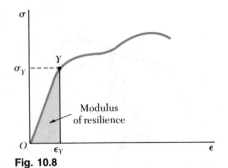

Fig. 10.8

The modulus of resilience is equal to the area under the straight-line portion OY of the stress-strain diagram (Fig. 10.8) and represents the energy per unit volume that the material may absorb without yielding. The capacity of a structure to withstand an impact load without being permanently deformed clearly depends upon the resilience of the material used.

Since the modulus of toughness and the modulus of resilience represent characteristic values of the strain-energy density of the material considered, they are both expressed in J/m^3 or its multiples.†

10.4. ELASTIC STRAIN ENERGY FOR NORMAL STRESSES

Since the rod considered in the preceding section was subjected to uniformly distributed stresses σ_x, the strain-energy density was constant throughout the rod and could be defined as the ratio U/V of the strain energy U and the volume V of the rod. In a structural element or machine part with a nonuniform stress distribution, the strain-energy density u may be defined by considering the strain energy of a small element of material of volume ΔV and writing

$$u = \lim_{\Delta V \to 0} \frac{\Delta U}{\Delta V}$$

† However, referring to the footnote on page 573, we note that the modulus of toughness and the modulus of resilience could be expressed in the same units as stress.

or

$$u = \frac{dU}{dV} \tag{10.9}$$

The expression obtained for u in Sec. 10.3 in terms of σ_x and ϵ_x remains valid, i.e., we still have

$$u = \int_0^{\epsilon_x} \sigma_x \, d\epsilon_x \tag{10.10}$$

but the stress σ_x, the strain ϵ_x, and the strain-energy density u will generally vary from point to point.

For values of σ_x within the proportional limit, we may set $\sigma_x = E\epsilon_x$ in Eq. (10.10) and write

$$u = \frac{1}{2}E\epsilon_x^2 = \frac{1}{2}\sigma_x\epsilon_x = \frac{1}{2}\frac{\sigma_x^2}{E} \tag{10.11}$$

The value of the strain energy U of a body subjected to uniaxial normal stresses may be obtained by substituting for u from Eq. (10.11) into Eq. (10.9) and integrating both members. We have

$$U = \int \frac{\sigma_x^2}{2E} dV \tag{10.12}$$

The expression obtained is valid only for elastic deformations and is referred to as the *elastic strain energy* of the body.

Strain Energy under Axial Loading. We recall from Sec. 2.16 that, when a rod is subjected to a centric axial loading, the normal stresses σ_x may be assumed uniformly distributed in any given transverse section. Denoting by A the area of the section located at a distance x from the end B of the rod (Fig. 10.9), and by P the internal force in that section, we write $\sigma_x = P/A$. Substituting for σ_x into Eq. (10.12), we have

$$U = \int \frac{P^2}{2EA^2} dV$$

or, setting $dV = A \, dx$,

$$U = \int_0^L \frac{P^2}{2AE} dx \tag{10.13}$$

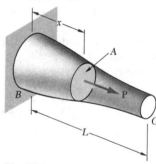

Fig. 10.9

In the case of a rod of uniform cross section subjected at its ends to equal and opposite forces of magnitude P (Fig. 10.10), Eq. (10.13) yields

$$U = \frac{P^2L}{2AE} \tag{10.14}$$

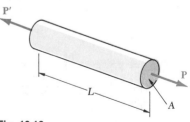

Fig. 10.10

Example 10.01

A rod consists of two portions BC and CD of the same material and same length, but of different cross sections (Fig. 10.11). Determine the strain energy of the rod when it is subjected to a centric axial load P, expressing the result in terms of P, L, E, the cross-sectional area A of portion CD, and the ratio n of the two diameters.

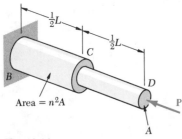

Fig. 10.11

We use Eq. (10.14) to compute the strain energy of each of the two portions, and add the expressions obtained:

$$U_n = \frac{P^2(\frac{1}{2}L)}{2AE} + \frac{P^2(\frac{1}{2}L)}{2(n^2A)E} = \frac{P^2L}{4AE}\left(1 + \frac{1}{n^2}\right)$$

or

$$U_n = \frac{1 + n^2}{2n^2} \frac{P^2L}{2AE} \qquad (10.15)$$

We check that, for $n = 1$, we have

$$U_1 = \frac{P^2L}{2AE}$$

which is the expression given in Eq. (10.14) for a rod of length L and uniform cross section of area A. We also note that, for $n > 1$, we have $U_n < U_1$; for example, when $n = 2$, we have $U_2 = (\frac{5}{8})U_1$. Since the maximum stress occurs in portion CD of the rod and is equal to $\sigma_{max} = P/A$, it follows that, for a given allowable stress, increasing the diameter of portion BC of the rod results in a *decrease* of the overall energy-absorbing capacity of the rod. Unnecessary changes in cross-sectional area should therefore be avoided in the design of members which may be subjected to loadings, such as impact loadings, where the energy-absorbing capacity of the member is critical.

Example 10.02

A load P is supported at B by two rods of the same material and of the same uniform cross section of area A (Fig. 10.12). Determine the strain energy of the system.

Denoting by F_{BC} and F_{BD}, respectively, the forces in members BC and BD, and recalling Eq. (10.14), we express the strain energy of the system as

$$U = \frac{F_{BC}^2(BC)}{2AE} + \frac{F_{BD}^2(BD)}{2AE} \qquad (10.16)$$

But we note from Fig. 10.12 that

$$BC = 0.6l \qquad BD = 0.8l$$

and from the free-body diagram of pin B and the corresponding force triangle (Fig. 10.13) that

$$F_{BC} = +0.6P \qquad F_{BD} = -0.8P$$

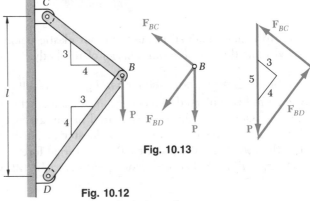

Fig. 10.13

Fig. 10.12

Substituting into Eq. (10.16), we have

$$U = \frac{P^2l[(0.6)^3 + (0.8)^3]}{2AE} = 0.364\frac{P^2l}{AE}$$

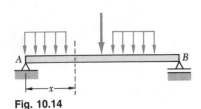

Fig. 10.14

Strain Energy in Bending. Consider a beam AB subjected to a given loading (Fig. 10.14), and let M be the bending moment at a distance x from end A. Neglecting for the time being the effect of shear, and taking into account only the normal stresses $\sigma_x = My/I$, we substitute this expression into Eq. (10.12) and write

$$U = \int \frac{\sigma_x^2}{2E}dV = \int \frac{M^2y^2}{2EI^2}dV$$

Setting $dV = dA\,dx$, where dA represents an element of the cross-sectional area, and recalling that $M^2/2EI^2$ is a function of x alone, we have

$$U = \int_0^L \frac{M^2}{2EI^2}(\int y^2\,dA)\,dx$$

Recalling that the integral within the parentheses represents the moment of inertia I of the cross section about its neutral axis, we write

$$U = \int_0^L \frac{M^2}{2EI}\,dx \qquad (10.17)$$

Example 10.03

Determine the strain energy of the prismatic cantilever beam AB (Fig. 10.15), taking into account only the effect of the normal stresses.

The bending moment at a distance x from end A is $M = -Px$. Substituting this expression into Eq. (10.17), we write

$$U = \int_0^L \frac{P^2x^2}{2EI}\,dx = \frac{P^2L^3}{6EI}$$

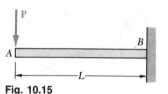

Fig. 10.15

10.5. ELASTIC STRAIN ENERGY FOR SHEARING STRESSES

When a material is subjected to plane shearing stresses τ_{xy}, the strain-energy density at a given point may be expressed as

$$u = \int_0^{\gamma_{xy}} \tau_{xy}\,d\gamma_{xy} \qquad (10.18)$$

where γ_{xy} is the shearing strain corresponding to τ_{xy} (Fig. 10.16a). We note that the strain-energy density u is equal to the area under the shearing-stress-strain diagram (Fig. 10.16b).

For values of τ_{xy} within the proportional limit, we have $\tau_{xy} = G\gamma_{xy}$, where G is the modulus of rigidity of the material. Substituting for τ_{xy} into Eq. (10.18) and performing the integration, we write

$$u = \frac{1}{2}G\gamma_{xy}^2 = \frac{1}{2}\tau_{xy}\gamma_{xy} = \frac{\tau_{xy}^2}{2G} \qquad (10.19)$$

The value of the strain energy U of a body subjected to plane shearing stresses may be obtained by recalling from Sec. 10.4 that

$$u = \frac{dU}{dV} \qquad (10.9)$$

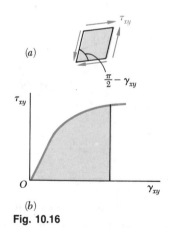

(a)

(b)

Fig. 10.16

Substituting for u from Eq. (10.19) into Eq. (10.9) and integrating both members, we have

$$U = \int \frac{\tau_{xy}^2}{2G} dV \qquad (10.20)$$

This expression defines the elastic strain associated with the shear deformations of the body. As the similar expression obtained in Sec. 10.4 for uniaxial normal stresses, it is valid only for elastic deformations.

Strain Energy in Torsion. Consider a shaft BC of length L subjected to one or several twisting couples. Denoting by J the polar moment of inertia of the cross section located at a distance x from B (Fig. 10.17), and by T the internal torque in that section, we recall that the shearing stresses in the section are $\tau_{xy} = T\rho/J$. Substituting for τ_{xy} into Eq. (10.20), we have

$$U = \int \frac{\tau_{xy}^2}{2G} dV = \int \frac{T^2 \rho^2}{2GJ^2} dV$$

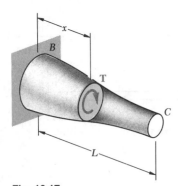

Fig. 10.17

Setting $dV = dA\, dx$, where dA represents an element of the cross-sectional area, and observing that $T^2/2GJ^2$ is a function of x alone, we write

$$U = \int_0^L \frac{T^2}{2GJ^2} (\int \rho^2\, dA)\, dx$$

Recalling that the integral within the parentheses represents the polar moment of inertia J of the cross section, we have

$$U = \int_0^L \frac{T^2}{2GJ} dx \qquad (10.21)$$

In the case of a shaft of uniform cross section subjected at its ends to equal and opposite couples of magnitude T (Fig. 10.18), Eq. (10.21) yields

$$U = \frac{T^2 L}{2GJ} \qquad (10.22)$$

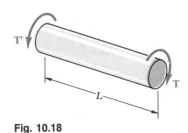

Fig. 10.18

Example 10.04

A circular shaft consists of two portions BC and CD of the same material and same length, but of different cross sections (Fig. 10.19). Determine the strain energy of the shaft when it is subjected to a twisting couple T at end D, expressing the result in terms of T, L, G, the polar moment of inertia J of the smaller cross section, and the ratio n of the two diameters.

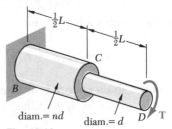

Fig. 10.19

We use Eq. (10.22) to compute the strain energy of each of the two portions of shaft, and add the expressions obtained. Noting that the polar moment of inertia of portion BC is equal to $n^4 J$, we write

$$U_n = \frac{T^2(\frac{1}{2}L)}{2GJ} + \frac{T^2(\frac{1}{2}L)}{2G(n^4 J)} = \frac{T^2 L}{4GJ}\left(1 + \frac{1}{n^4}\right)$$

or

$$U_n = \frac{1 + n^4}{2n^4} \frac{T^2 L}{2GJ} \qquad (10.23)$$

We check that, for $n = 1$, we have

$$U_1 = \frac{T^2 L}{2GJ}$$

which is the expression given in Eq. (10.22) for a shaft of length L and uniform cross section. We also note that, for $n > 1$, we have $U_n < U_1$; for example, when $n = 2$, we have $U_2 = (\frac{17}{32})U_1$. Since the maximum shearing stress occurs in the portion CD of the shaft and is proportional to the torque T, we note as we did earlier in the case of the axial loading of a rod that, for a given allowable stress, increasing the diameter of portion BC of the shaft results in a *decrease* of the overall energy-absorbing capacity of the shaft.

Strain Energy under Transverse Loading. In Sec. 10.4 we obtained an expression for the strain energy of a beam subjected to a transverse loading. However, in deriving that expression we took into account only the effect of the normal stresses due to bending and neglected the effect of the shearing stresses. We shall now take into account the effect of both types of stresses.

Example 10.05

Determine the strain energy of the rectangular cantilever beam AB (Fig. 10.20), taking into account the effect of both normal and shearing stresses.

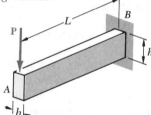

Fig. 10.20

We first recall from Example 10.03 that the strain energy due to the normal stresses σ_x is

$$U_\sigma = \frac{P^2 L^3}{6EI}$$

To determine the strain energy U_τ due to the shearing stresses τ_{xy}, we recall Eq. (5.13) of Sec. 5.6 and find that, for a beam with a rectangular cross section of width b and depth h,

$$\tau_{xy} = \frac{3}{2}\frac{V}{A}\left(1 - \frac{y^2}{c^2}\right) = \frac{3}{2}\frac{P}{bh}\left(1 - \frac{y^2}{c^2}\right)$$

Substituting for τ_{xy} into Eq. (10.20), we write

$$U_\tau = \frac{1}{2G}\left(\frac{3}{2}\frac{P}{bh}\right)^2 \int \left(1 - \frac{y^2}{c^2}\right)^2 dV$$

or, setting $dV = b\, dy\, dx$, and after reductions,

$$U_\tau = \frac{9P^2}{8Gbh^2}\int_{-c}^{c}\left(1 - 2\frac{y^2}{c^2} + \frac{y^4}{c^4}\right) dy \int_0^L dx$$

Performing the integrations, and recalling that $c = h/2$, we have

$$U_\tau = \frac{9P^2 L}{8Gbh^2}\left[y - \frac{2}{3}\frac{y^3}{c^2} + \frac{1}{5}\frac{y^5}{c^4}\right]_{-c}^{+c} = \frac{3P^2 L}{5Gbh} = \frac{3P^2 L}{5GA}$$

The total strain energy of the beam is thus

$$U = U_\sigma + U_\tau = \frac{P^2 L^3}{6EI} + \frac{3P^2 L}{5GA}$$

or, noting that $I/A = h^2/12$ and factoring the expression for U_σ,

$$U = \frac{P^2 L^3}{6EI}\left(1 + \frac{3Eh^2}{10GL^2}\right) = U_\sigma\left(1 + \frac{3Eh^2}{10GL^2}\right) \qquad (10.24)$$

Recalling from Sec. 2.14 that $G \geq E/3$, we conclude that the parenthesis in the expression obtained is less than $1 + 0.9(h/L)^2$ and, thus, that the relative error is less than $0.9(h/L)^2$ when the effect of shear is neglected. For a beam with a ratio h/L less than $\frac{1}{10}$, the percentage error is less than 0.9%. It is therefore customary in engineering practice to neglect the effect of shear in computing the strain energy of slender beams.

*10.6. STRAIN ENERGY FOR A GENERAL STATE OF STRESS

In the preceding sections, we determined the strain energy of a body in a state of uniaxial stress (Sec. 10.4) and in a state of plane shearing stress (Sec. 10.5). In the case of a body in a general state of stress characterized by the six stress components σ_x, σ_y, σ_z, τ_{xy}, τ_{yz}, and τ_{zx}, the strain-energy density may be obtained by adding the expressions given in Eqs. (10.10) and (10.18), as well as the four other expressions obtained through a permutation of the subscripts.

In the case of the elastic deformation of an isotropic body, each of the six stress-strain relations involved is linear, and the strain-energy density may be expressed as

$$u = \tfrac{1}{2}(\sigma_x\epsilon_x + \sigma_y\epsilon_y + \sigma_z\epsilon_z + \tau_{xy}\gamma_{xy} + \tau_{yz}\gamma_{yz} + \tau_{zx}\gamma_{zx}) \quad (10.25)$$

Recalling the relations (2.38) obtained in Sec. 2.14, and substituting for the strain components into (10.25), we have, for the most general state of stress at a given point of an elastic isotropic body,

$$u = \frac{1}{2E}[\sigma_x^2 + \sigma_y^2 + \sigma_z^2 - 2\nu(\sigma_x\sigma_y + \sigma_y\sigma_z + \sigma_z\sigma_x)]$$

$$+ \frac{1}{2G}(\tau_{xy}^2 + \tau_{yz}^2 + \tau_{zx}^2) \quad (10.26)$$

If the principal axes at the given point are used as coordinate axes, the shearing stresses become zero and Eq. (10.26) reduces to

$$u = \frac{1}{2E}[\sigma_a^2 + \sigma_b^2 + \sigma_c^2 - 2\nu(\sigma_a\sigma_b + \sigma_b\sigma_c + \sigma_c\sigma_a)] \quad (10.27)$$

where σ_a, σ_b, and σ_c are the principal stresses at the given point.

We now recall from Sec. 6.7 that one of the criteria used to predict whether a given state of stress will cause a ductile material to yield, namely, the maximum-distortion-energy criterion, is based on the determination of the energy per unit volume associated with the distortion, or change in shape, of that material. We shall, therefore, attempt to separate the strain-energy density u at a given point into two parts, a part u_v associated with a change in volume of the material at that point, and a part u_d associated with a distortion, or change in shape, of the material at the same point. We write

$$u = u_v + u_d \quad (10.28)$$

In order to determine u_v and u_d, we shall introduce the *average value* $\bar{\sigma}$ of the principal stresses at the point considered,

$$\bar{\sigma} = \frac{\sigma_a + \sigma_b + \sigma_c}{3} \quad (10.29)$$

and set

$$\sigma_a = \bar{\sigma} + \sigma_a' \qquad \sigma_b = \bar{\sigma} + \sigma_b' \qquad \sigma_c = \bar{\sigma} + \sigma_c' \quad (10.30)$$

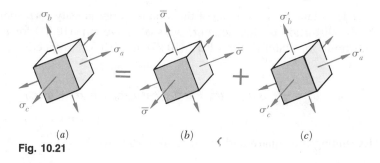

Fig. 10.21

Thus, the given state of stress (Fig. 10.21a) may be obtained by superposing the states of stress shown in Fig. 10.21b and c. We note that the state of stress described in Fig. 10.21b tends to change the volume of the element of material, but not its shape, since all the faces of the element are subjected to the same stress $\bar{\sigma}$. On the other hand, it follows from Eqs. (10.29) and (10.30) that

$$\sigma_a' + \sigma_b' + \sigma_c' = 0 \qquad (10.31)$$

which indicates that some of the stresses shown in Fig. 10.21c are tensile and others compressive. Thus, this state of stress tends to change the shape of the element. However, it does not tend to change its volume. Indeed, recalling Eq. (2.31) of Sec. 2.13, we note that the dilatation e (i.e., the change in volume per unit volume) caused by this state of stress is

$$e = \frac{1 - 2\nu}{E}(\sigma_a' + \sigma_b' + \sigma_c')$$

or $e = 0$, in view of Eq. (10.31). We conclude from these observations that the portion u_v of the strain-energy density must be associated with the state of stress shown in Fig. 10.21b, while the portion u_d must be associated with the state of stress shown in Fig. 10.21c.

It follows that the portion u_v of the strain-energy density corresponding to a change in volume of the element may be obtained by substituting $\bar{\sigma}$ for each of the principal stresses in Eq. (10.27). We have

$$u_v = \frac{1}{2E}[3\bar{\sigma}^2 - 2\nu(3\bar{\sigma}^2)] = \frac{3(1 - 2\nu)}{2E}\bar{\sigma}^2$$

or, recalling Eq. (10.29),

$$u_v = \frac{1 - 2\nu}{6E}(\sigma_a + \sigma_b + \sigma_c)^2 \qquad (10.32)$$

To obtain the portion u_d of the strain-energy density corresponding to the distortion of the element, we shall solve Eq. (10.28) for u_d and substitute for u and u_v from Eqs. (10.27) and (10.32), respectively. We write

$$u_d = u - u_v = \frac{1}{6E}[3(\sigma_a^2 + \sigma_b^2 + \sigma_c^2) - 6\nu(\sigma_a\sigma_b + \sigma_b\sigma_c + \sigma_c\sigma_a)$$
$$-(1 - 2\nu)(\sigma_a + \sigma_b + \sigma_c)^2]$$

Expanding the square and rearranging terms, we have

$$u_d = \frac{1 + \nu}{6E}[(\sigma_a^2 - 2\sigma_a\sigma_b + \sigma_b^2) + (\sigma_b^2 - 2\sigma_b\sigma_c + \sigma_c^2)$$
$$+(\sigma_c^2 - 2\sigma_c\sigma_a + \sigma_a^2)]$$

Noting that each of the parentheses inside the bracket is a perfect square, and recalling from Eq. (2.43) of Sec. 2.15 that the coefficient in front of the bracket is equal to $1/12G$, we obtain the following expression for the portion u_d of the strain-energy density, i.e., for the distortion energy per unit volume,

$$u_d = \frac{1}{12G}[(\sigma_a - \sigma_b)^2 + (\sigma_b - \sigma_c)^2 + (\sigma_c - \sigma_a)^2] \qquad (10.33)$$

In the case of *plane stress*, and assuming that the c axis is perpendicular to the plane of stress, we have $\sigma_c = 0$ and Eq. (10.33) reduces to

$$u_d = \frac{1}{6G}(\sigma_a^2 - \sigma_a\sigma_b + \sigma_b^2) \qquad (10.34)$$

Considering the particular case of a tensile-test specimen, we note that, at yield, we have $\sigma_a = \sigma_Y$, $\sigma_b = 0$, and thus $(u_d)_Y = \sigma_Y^2/6G$. The maximum-distortion-energy criterion for plane stress indicates that a given state of stress is safe as long as $u_d < (u_d)_Y$ or, substituting for u_d from Eq. (10.34), as long as

$$\sigma_a^2 - \sigma_a\sigma_b + \sigma_b^2 < \sigma_Y^2 \qquad (6.26)$$

which is the condition stated in Sec. 6.7 and represented graphically by the ellipse of Fig. 6.39. In the case of a general state of stress, the expression (10.33) obtained for u_d should be used. The maximum-distortion-energy criterion is then expressed by the condition

$$(\sigma_a - \sigma_b)^2 + (\sigma_b - \sigma_c)^2 + (\sigma_c - \sigma_a)^2 < 2\sigma_Y^2 \qquad (10.35)$$

which indicates that a given state of stress is safe if the point of coordinates σ_a, σ_b, σ_c is located within the surface defined by the equation

$$(\sigma_a - \sigma_b)^2 + (\sigma_b - \sigma_c)^2 + (\sigma_c - \sigma_a)^2 = 2\sigma_Y^2 \qquad (10.36)$$

One may verify that this surface is a circular cylinder of radius $\sqrt{2/3}\,\sigma_Y$ with an axis of symmetry forming equal angles with the three principal axes of stress.

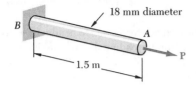

18 mm diameter

B

A

P

1.5 m

SAMPLE PROBLEM 10.1

During a routine manufacturing operation, rod AB must acquire an elastic strain energy of 12 J. Using $E = 200$ GPa, determine the required yield strength of the steel if the factor of safety with respect to permanent deformation is to be five.

Factor of Safety. Since a factor of safety of five is required, the rod should be designed for a strain energy of

$$U = 5(12 \text{ J}) = 60 \text{ J}$$

Strain-Energy Density. The volume of the rod is

$$V = AL = \frac{\pi}{4}(0.018 \text{ m})^2(1.5 \text{ m}) = 381.7 \times 10^{-6} \text{ m}^3$$

Since the rod is of uniform cross section, the required strain-energy density is

$$u = \frac{U}{V} = \frac{60 \text{ J}}{381.7 \times 10^{-6} \text{ m}^3} = 157.2 \times 10^3 \text{ J/m}^3$$

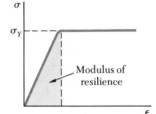

Modulus of resilience

Yield Strength. We recall that the modulus of resilience is equal to the strain-energy density when the maximum stress is equal to σ_Y. Using Eq. (10.8), we write

$$u = \frac{\sigma_Y^2}{2E}$$

$$157.2 \times 10^3 \text{ J/m}^3 = \frac{\sigma_Y^2}{2(200 \text{ GPa})} \qquad \sigma_Y = 251 \text{ MPa} \blacktriangleleft$$

$$\sigma_Y = 250.8 \text{ MPa}$$

Comment.

It is important to note that, since energy loads are not linearly related to the stresses they produce, factors of safety associated with energy loads should be applied to the energy loads and not to the stresses.

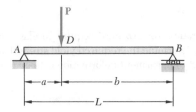

(a) Taking into account only the effect of normal stresses due to bending, determine the strain energy of the prismatic beam AB for the loading shown.　(b) Evaluate the strain energy, knowing that the beam is a W 250×67, $P = 180$ kN, $L = 3.6$ m, $a = 0.9$ m, $b = 2.7$ m, and $E = 200$ GPa.

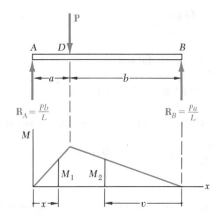

From A to D:

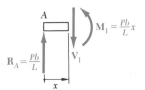

From B to D:

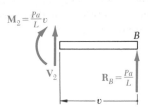

Bending Moment.　Using the free-body diagram of the entire beam, we determine the reactions

$$R_A = \frac{Pb}{L} \uparrow \qquad R_B = \frac{Pa}{L} \uparrow$$

For portion AD of the beam, the bending moment is

$$M_1 = \frac{Pb}{L}x$$

For portion DB, we note that the bending moment at a distance v from end B is

$$M_2 = \frac{Pa}{L}v$$

(a) Strain Energy.　Since strain energy is a scalar quantity, we add the strain energy of portion AD to that of portion DB to obtain the total strain energy of the beam.　Using Eq. (10.17), we write

$$U = U_{AD} + U_{DB}$$

$$= \int_0^a \frac{M_1^2}{2EI}\,dx + \int_0^b \frac{M_2^2}{2EI}\,dv$$

$$= \frac{1}{2EI}\int_0^a \left(\frac{Pb}{L}x\right)^2 dx + \frac{1}{2EI}\int_0^b \left(\frac{Pa}{L}v\right)^2 dv$$

$$= \frac{1}{2EI}\frac{P^2}{L^2}\left(\frac{b^2a^3}{3} + \frac{a^2b^3}{3}\right) = \frac{P^2a^2b^2}{6EIL^2}(a+b)$$

or, since $(a + b) = L$,　　　　　　　　　　　　　　　　$U = \dfrac{P^2a^2b^2}{6EIL}$ ◀

(b) Evaluation of the Strain Energy.　The moment of inertia of a W 250×67 rolled-steel shape is obtained from Appendix C and the given data is repeated here:

$$P = 180 \text{ kN} \qquad\qquad L = 3.6 \text{ m}$$
$$a = 0.9 \text{ m} \qquad\qquad b = 2.7 \text{ m}$$
$$E = 200 \text{ GPa} \qquad\qquad I = 103.2 \times 10^{-6} \text{ m}^4$$

Substituting into the expression for U, we have

$$U = \frac{(180 \text{ kN})^2(0.9 \text{ m})^2(2.7 \text{ m})^2}{6(200 \text{ GPa})(103.2 \times 10^{-6} \text{ m}^4)(3.6 \text{ m})}$$

$$U = 429 \text{ J} \quad ◀$$

PROBLEMS

10.1   Using $E = 200$ GPa, determine the modulus of resilience for each of the following grades of structural steel:

(*a*) ASTM A-36: $\sigma_Y = 250$ MPa
(*b*) ASTM A-441: $\sigma_Y = 320$ MPa
(*c*) ASTM A-514: $\sigma_Y = 690$ MPa

10.2   Determine the modulus of resilience for each of the following aluminum alloys:

(*a*) 1100-H14: $E = 70$ GPa, $\sigma_Y = 95$ MPa
(*b*) 2014-T6: $E = 75$ GPa, $\sigma_Y = 400$ MPa
(*c*) 7075-T6: $E = 72$ GPa, $\sigma_Y = 500$ MPa

10.3   Determine the modulus of resilience for each of the following alloys:

(*a*) Monel (cold-worked): $E = 180$ GPa, $\sigma_Y = 585$ MPa
(*b*) Monel (annealed): $E = 180$ GPa, $\sigma_Y = 220$ MPa
(*c*) Titanium alloy: $E = 114$ GPa, $\sigma_Y = 825$ MPa

10.4   Determine the modulus of resilience for each of the following metals:

(*a*) Stainless steel (cold-rolled): $E = 190$ GPa, $\sigma_Y = 520$ MPa
(*b*) Stainless steel (annealed): $E = 190$ GPa, $\sigma_Y = 260$ MPa
(*c*) Malleable cast iron: $E = 165$ GPa, $\sigma_Y = 230$ MPa

10.5   The stress-strain diagram shown has been drawn from data obtained during the tensile test of a specimen of structural steel. Using $E = 200$ GPa, (*a*) determine the modulus of resilience of the steel, (*b*) determine by approximate means the modulus of toughness of the steel.

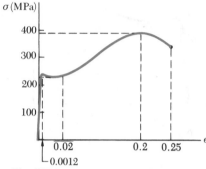

Fig. P10.5

10.6   The stress-strain diagram shown has been drawn from data obtained during the tensile test of an aluminum alloy.   Using $E = 72$ GPa, (*a*) determine the modulus of resilience of the alloy, (*b*) determine by approximate means the modulus of toughness of the alloy.

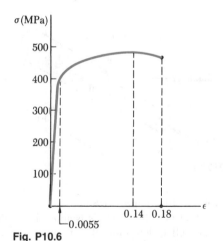

Fig. P10.6

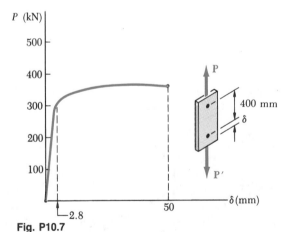

Fig. P10.7

10.7   The load-deformation diagram shown has been drawn from data obtained during the tensile test of a specimen of an aluminum alloy. Knowing that the cross-sectional area of the specimen was 600 mm² and that the deformation was measured using a 400-mm gage length, determine by approximate means (*a*) the modulus of resilience of the alloy, (*b*) the modulus of toughness of the alloy.

10.8 The load-deformation diagram shown has been drawn from data obtained during the tensile test of a 15-mm-diameter rod of structural steel. Knowing that the deformaton was measured using a 180-mm gage length, determine by approximate means (a) the modulus of resilience of the steel, (b) the modulus of toughness of the steel.

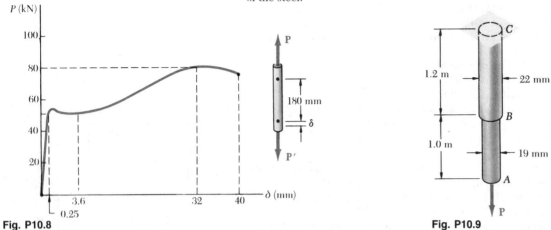

Fig. P10.8

Fig. P10.9

10.9 Using $E = 200$ GPa, determine (a) the strain energy of the steel rod ABC when $P = 65$ kN, (b) the corresponding strain-energy density in portions AB and BC of the rod.

10.10 Using $E = 75$ GPa, determine (a) the strain energy of the aluminum rod ABC when $a = 1.2$ m, (b) the corresponding strain-energy density in portions AB and BC of the rod.

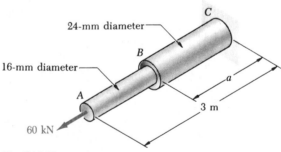

Fig. P10.10

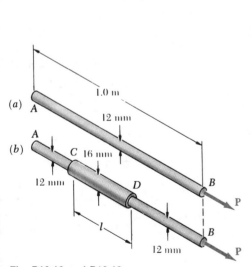

Fig. P10.12 and P10.13

10.11 Solve Prob. 10.10 when $a = 1.8$ m.

10.12 (a) A force $\mathbf{P}$ of magnitude 22 kN is applied to each of the steel rods shown. Using $E = 200$ GPa, determine the strain energy acquired by rod a. (b) Determine the required length l if the strain energy acquired by rod b is to be 90% of the strain energy acquired by rod a.

10.13 Each of the rods shown is made of a steel for which the yield strength is $\sigma_Y = 250$ MPa and the modulus of elasticity is $E = 200$ GPa. Knowing that $l = 375$ mm, determine the maximum strain energy which may be acquired by each rod without causing any permanent deformation.

10.14 Rod AB is made of an aluminum alloy for which the yield strength is $\sigma_Y = 300$ MPa and the modulus of elasticity is $E = 75$ GPa; rod BC is made of a grade of steel for which $\sigma_Y = 420$ MPa and $E = 200$ GPa. Determine the maximum strain energy which may be acquired by the composite rod ABC without causing any permanent deformation.

10.15 Solve Prob. 10.14, using the given material properties and assuming that rod ABC is made entirely of (a) steel, (b) aluminum.

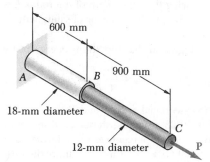

Fig. P10.14

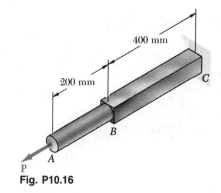

Fig. P10.16

10.16 A portion of a 22-mm-square steel bar has been machined to a 22-mm-diameter cylinder as shown. Knowing that the allowable normal stress of the steel is $\sigma_{all} = 150$ MPa and $E = 200$ GPa, determine, for the loading shown, the maximum strain energy which may be acquired by the bar ABC.

10.17 A uniform 10-mm-diameter rod AB of length $L = 3$ m is made of a high-strength stainless steel for which $\sigma_Y = 860$ MPa and $E = 193$ GPa. Knowing that a strain energy of 70 J must be acquired by the rod as the axial load $\mathbf{P}$ is applied, determine the factor of safety of the rod with respect to permanent deformation.

10.18 The 6-mm-diameter rod AB of length $L = 2.5$ m is made of an aluminum alloy for which $\sigma_Y = 320$ MPa and $E = 72$ GPa. Knowing that a strain energy of 10 J must be acquired by the rod as the axial load $\mathbf{P}$ is applied, determine the factor of safety of the rod with respect to permanent deformation.

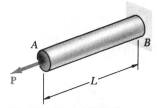

Fig. P10.17 and P10.18

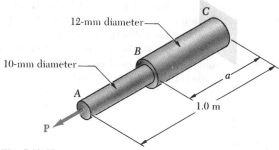

Fig. P10.19

10.19 The rod ABC is made of a steel for which $\sigma_Y = 345$ MPa and $E = 200$ GPa. Knowing that a strain energy of 4.5 J must be acquired by the rod as the axial load $\mathbf{P}$ is applied, determine the factor of safety of the rod with respect to permanent deformation when $a = 380$ mm.

10.20 Solve Prob. 10.19, assuming that $a = 635$ mm.

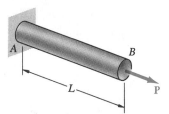

Fig. P10.21 and P10.22

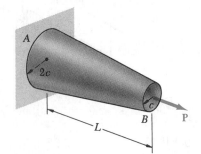

Fig. P10.23

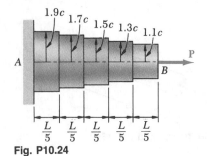

Fig. P10.24

10.21 Rod AB is made of a reinforcing steel for which $\sigma_Y = 275$ MPa and $E = 200$ GPa. A strain energy of 5 J must be acquired by the rod when the axial load **P** is applied. Knowing that the diameter of the rod is $d = 9$ mm, determine the length of the rod for which the factor of safety with respect to permanent deformation is six.

10.22 Rod AB is made of a high-strength steel for which $\sigma_Y = 750$ MPa and $E = 200$ GPa. A strain energy of 25 J must be acquired by the rod when the axial load **P** is applied. Knowing that the length of the rod is $L = 2.25$ m, determine the diameter of the rod for which the factor of safety with respect to permanent deformation is five.

10.23 Show by integration that the strain energy of the tapered rod AB is

$$U = \frac{1}{4}\frac{P^2 L}{EA_{\min}}$$

where $A_{\min}$ is the cross-sectional area at end B.

10.24 Solve Prob. 10.23, using the stepped rod shown as an approximation of the tapered rod. What is the percentage error in the answer obtained?

10.25 Using $E = 200$ GPa, determine by approximate means the maximum strain energy which may be acquired by the 240-mm portion AB of the control rod shown if the allowable normal stress is $\sigma_{\text{all}} = 175$ MPa.

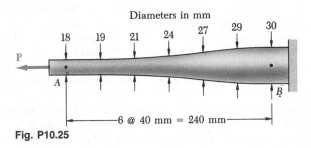

Fig. P10.25

10.26 Solve Prob. 10.25, assuming that where the diameters of the rod are 18 mm and 19 mm, the diameters are increased to 21 mm so that the left-hand third of the control rod is of a uniform 21-mm diameter.

10.27 through 10.30 In the truss shown, all members are made of the same material and have the uniform cross-sectional areas indicated. Determine the strain energy of the truss when the load **P** is applied.

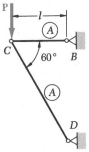

Fig. P10.27

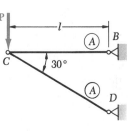

Fig. P10.28

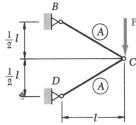

Fig. P10.29

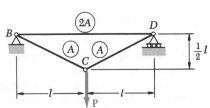

Fig. P10.30

10.31 Each member of the truss shown is made of steel and has the cross-sectional area shown.   Using $E = 200$ GPa, determine the strain energy of the truss for the loading shown.

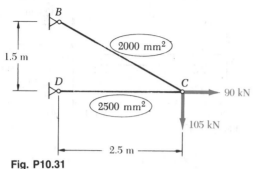

Fig. P10.31

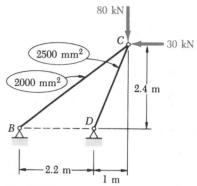

Fig. P10.32

10.32 Each member of the truss shown is made of aluminum and has the cross-sectional area shown.   Using $E = 72$ GPa, determine the strain energy of the truss for the loading shown.

10.33 Solve Prob. 10.32, assuming that the 30-kN load is removed.

10.34 Solve Prob. 10.31, assuming that the 90-kN load is removed.

10.35 through 10.38   Taking into account only the effect of normal stresses, determine the strain energy of the prismatic beam AB for the loading shown.

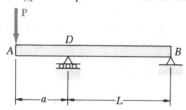

Fig. P10.35

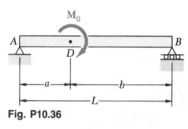

Fig. P10.36

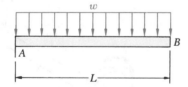

Fig. P10.37 and P10.39

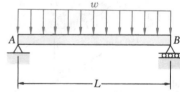

Fig. P10.38 and P10.40

10.39 Assuming that the prismatic beam AB has a rectangular cross section, show that for the given loading the maximum value of the strain-energy density in the beam is

$$u_m = 15\frac{U}{V}$$

where U is the strain energy of the beam and V is its volume.

10.40 Assuming that the prismatic beam AB has a rectangular cross section, show that for the given loading the maximum value of the strain-energy density in the beam is

$$u_m = \frac{45}{8}\frac{U}{V}$$

where U is the strain energy of the beam and V is its volume.

10.41 and 10.42 Using $E = 200$ GPa, determine the strain energy due to bending for the beam and loading shown. (Ignore the effect of shearing stresses.)

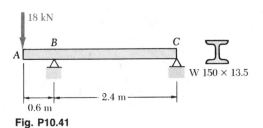

Fig. P10.41

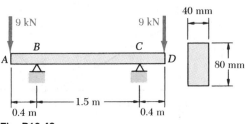

Fig. P10.42

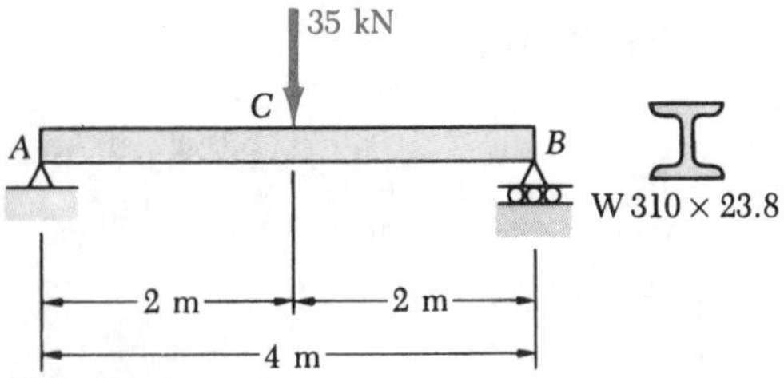

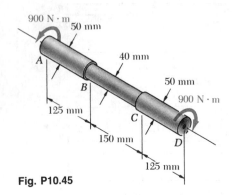

Fig. P10.43

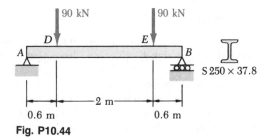

Fig. P10.44

10.43 and 10.44 Using $E = 200$ GPa, determine the strain energy due to bending for the beam and loading shown. (Ignore the effect of shearing stresses.)

10.45 The solid steel shaft shown is subjected to 900-N · m torques applied at ends A and D. Using $G = 77$ GPa, determine the strain energy of the shaft.

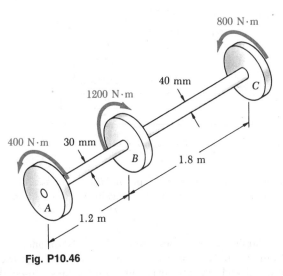

Fig. P10.45

Fig. P10.46

10.46 The torques shown are exerted on pulleys A, B, and C. Knowing that both shafts are solid and made of brass ($G = 39$ GPa), determine the strain energy of (a) shaft AB, (b) shaft BC.

10.47 The aluminum rod AB ($G = 26$ GPa) is bonded to the brass rod BD ($G = 39$ GPa). Knowing that portion CD of the brass rod is hollow and has an inner diameter of 40 mm, determine the total strain energy of the two rods.

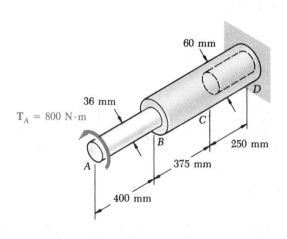

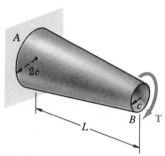

Fig. P10.47

Fig. P10.48

10.48 Two solid shafts are connected by the gears shown. Using $G = 77$ GPa, determine the strain energy of each shaft when a 1.1-kN · m torque **T** is applied at D.

10.49 Show by integration that the strain energy of the tapered rod AB is

$$U = \frac{7}{24} \frac{T^2 L}{GJ_{min}}$$

where J_{min} is the polar moment of inertia of the cross section of the rod at end B.

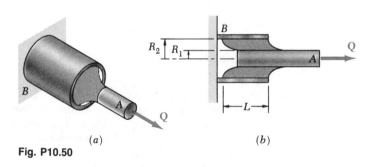

Fig. P10.49

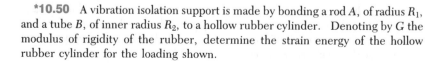

Fig. P10.50

***10.50** A vibration isolation support is made by bonding a rod A, of radius R_1, and a tube B, of inner radius R_2, to a hollow rubber cylinder. Denoting by G the modulus of rigidity of the rubber, determine the strain energy of the hollow rubber cylinder for the loading shown.

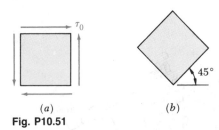

(*a*)   (*b*)

Fig. P10.51

10.51 For the state of stress shown in Fig. *a*, determine the stresses on an element oriented as shown in Fig. *b*. Compute the strain energy density for the given state of stress first by using Fig. *a* and then by using Fig. *b*. Equating the two results obtained, show that

$$G = \frac{E}{2(1 + \nu)}$$

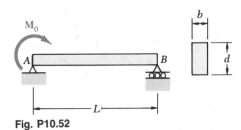

Fig. P10.52

10.52 Determine the strain energy of the prismatic beam *AB*, taking into account the effect of both normal and shearing stresses.

10.53 The state of stress shown occurs in a machine component which is made of a brass for which $\sigma_Y = 175$ MPa. Using the maximum-distortion-energy criterion, determine whether yield occurs when (*a*) $\sigma_z = +60$ MPa, (*b*) $\sigma_z = -60$ MPa.

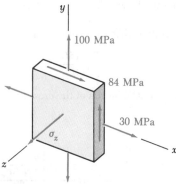

Fig. P10.53 and P10.54

10.54 The state of stress shown occurs in a machine component which is made of a brass for which $\sigma_Y = 175$ MPa. Using the maximum-distortion-energy criterion, determine the range of values of σ_z for which yield does not occur.

10.55 The state of stress shown occurs in a machine component which is made of a steel for which $\sigma_Y = 415$ MPa. Using the maximum-distortion-energy criterion, determine the factor of safety associated with the yield strength when (*a*) $\sigma_x = 80$ MPa, (*b*) $\sigma_x = 40$ MPa.

10.56 The state of stress shown occurs in a machine component which is made of a steel for which $\sigma_Y = 415$ MPa. Using the maximum-distortion-energy criterion, determine the range of values of σ_x for which the factor of safety associated with the yield strength is equal to or more than 2.5.

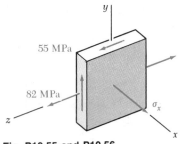

Fig. P10.55 and P10.56

10.7. IMPACT LOADING

Consider a rod BD of uniform cross section which is hit at its end B by a body of mass m moving with a velocity v_0 (Fig. 10.22a). As the rod deforms under the impact (Fig. 10.22b), stresses develop within the rod and reach a maximum value σ_m. After vibrating for a while, the rod will come to rest, and all stresses will disappear. Such a sequence of events is referred to as an *impact loading*.

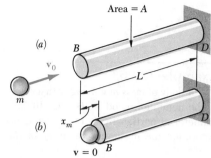

Fig. 10.22

In order to determine the maximum value σ_m of the stress occurring at a given point of a structure subjected to an impact loading, we shall make several simplifying assumptions.

First, we shall assume that the kinetic energy $T = \tfrac{1}{2}mv_0^2$ of the striking body is transferred entirely to the structure and, thus, that the strain energy U_m corresponding to the maximum deformation x_m is

$$U_m = \tfrac{1}{2}mv_0^2 \qquad (10.37)$$

This assumption leads to the following two specific requirements:

1. No energy should be dissipated during the impact.
2. The striking body should not bounce off the structure and retain part of its energy. This, in turn, necessitates that the inertia of the structure be negligible, compared to the inertia of the striking body.

In practice, neither of these requirements is satisfied, and only part of the kinetic energy of the striking body is actually transferred to the structure. Thus, assuming that all of the kinetic energy of the striking body is transferred to the structure leads to a conservative design of that structure.

We shall further assume that the stress-strain diagram obtained from a static test of the material is also valid under impact loading. Thus, for an elastic deformation of the structure, we may express the maximum value of the strain energy as

$$U_m = \int \frac{\sigma_m^2}{2E}\,dV \qquad (10.38)$$

In the case of the uniform rod of Fig. 10.22, the maximum stress σ_m has the same value throughout the rod, and we may write $U_m = \sigma_m^2 V/2E$. Solving for σ_m and substituting for U_m from Eq. (10.37), we write

$$\sigma_m = \sqrt{\frac{2U_m E}{V}} = \sqrt{\frac{mv_0^2 E}{V}} \qquad (10.39)$$

We note from the expression obtained that selecting a rod with a large volume V and a low modulus of elasticity E will result in a smaller value of the maximum stress σ_m for a given impact loading.

In most problems, the distribution of stresses in the structure is not uniform, and formula (10.39) does not apply. It is then convenient to determine the static load $\mathbf{P}_m$ which would produce the same strain energy as the impact loading, and compute from P_m the corresponding value σ_m of the largest stress occurring in the structure.

Example 10.06

A body of mass m moving with a velocity v_0 hits the end B of the nonuniform rod BCD (Fig. 10.23). Knowing that the diameter of portion BC is twice the diameter of portion CD, determine the maximum value σ_m of the stress in the rod.

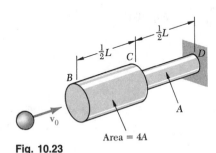

Area = 4A

Fig. 10.23

Making $n = 2$ in the expression (10.15) obtained in Example 10.01, we find that when rod BCD is subjected to a static load P_m, its strain energy is

$$U_m = \frac{5P_m^2 L}{16AE} \tag{10.40}$$

where A is the cross-sectional area of portion CD of the rod.

Solving Eq. (10.40) for P_m, we find that the static load which produces in the rod the same strain energy as the given impact loading is

$$P_m = \sqrt{\frac{16}{5} \frac{U_m AE}{L}}$$

where U_m is given by Eq. (10.37). The largest stress occurs in portion CD of the rod. Dividing P_m by the area A of that portion, we have

$$\sigma_m = \frac{P_m}{A} = \sqrt{\frac{16}{5} \frac{U_m E}{AL}} \tag{10.41}$$

or, substituting for U_m from Eq. (10.37),

$$\sigma_m = \sqrt{\frac{8}{5} \frac{mv_0^2 E}{AL}} = 1.265 \sqrt{\frac{mv_0^2 E}{AL}}$$

Comparing this value with the value obtained for σ_m in the case of the uniform rod of Fig. 10.22, and making $V = AL$ in Eq. (10.39), we note that the maximum stress in the rod of variable cross section is 26.5% larger than in the lighter uniform rod. Thus, as we observed earlier in our discussion of Example 10.01, increasing the diameter of portion BC of the rod results in a *decrease* of the energy-absorbing capacity of the rod.

Example 10.07

A block of weight W is dropped from a height h onto the free end of the cantilever beam AB (Fig. 10.24). Determine the maximum value of the stress in the beam.

As it falls through the distance h, the potential energy Wh of the block is transformed into kinetic energy. As a result of the impact, the kinetic energy in turn is transformed into strain energy. We have, therefore,†

$$U_m = Wh \tag{10.42}$$

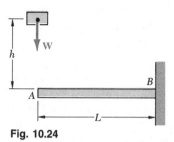

Fig. 10.24

---

†The total distance through which the block drops is actually $h + y_m$, where y_m is the maximum deflection of the end of the beam. Thus, a more accurate expression for U_m (see Sample Prob. 10.3) is
$$U_m = W(h + y_m) \tag{10.42'}$$
However, when $h \gg y_m$, we may neglect y_m and use Eq. (10.42).

Recalling the expression obtained for the strain energy of the cantilever beam AB in Example 10.03 and neglecting the effect of shear, we write

$$U_m = \frac{P_m^2 L^3}{6EI}$$

Solving this equation for P_m, we find that the static force which produces in the beam the same strain energy is

$$P_m = \sqrt{\frac{6U_m EI}{L^3}} \tag{10.43}$$

The maximum stress σ_m occurs at the fixed end B and is equal to

$$\sigma_m = \frac{|M|c}{I} = \frac{P_m L c}{I}$$

Substituting for P_m from (10.43), we write

$$\sigma_m = \sqrt{\frac{6U_m E}{L(I/c^2)}} \tag{10.44}$$

or, recalling (10.42),

$$\sigma_m = \sqrt{\frac{6WhE}{L(I/c^2)}}$$

10.8. DESIGN FOR IMPACT LOADS

We shall now compare the values obtained in the preceding section for the maximum stress σ_m (a) in the rod of uniform cross section of Fig. 10.22, (b) in the rod of variable cross section of Example 10.06, and (c) in the cantilever beam of Example 10.07, assuming that the latter has a circular cross section of radius c.

(a) We first recall from Eq. (10.39) that, if U_m denotes the amount of energy transferred to the rod as a result of the impact loading, the maximum stress in the rod of uniform cross section is

$$\sigma_m = \sqrt{\frac{2U_m E}{V}} \tag{10.45a}$$

where V is the volume of the rod.

(b) Considering next the rod of Example 10.06 and observing that the volume of the rod is

$$V = 4A(L/2) + A(L/2) = 5AL/2$$

we substitute $AL = 2V/5$ into Eq. (10.41) and write

$$\sigma_m = \sqrt{\frac{8U_m E}{V}} \tag{10.45b}$$

(c) Finally, recalling that $I = \frac{1}{4}\pi c^4$ for a beam of circular cross section, we note that

$$L(I/c^2) = L(\tfrac{1}{4}\pi c^4/c^2) = \tfrac{1}{4}(\pi c^2 L) = \tfrac{1}{4}V$$

where V denotes the volume of the beam. Substituting into Eq. (10.44), we express the maximum stress in the cantilever beam of Example 10.07 as

$$\sigma_m = \sqrt{\frac{24U_m E}{V}} \tag{10.45c}$$

We note that, in each case, the maximum stress σ_m is proportional to the square root of the modulus of elasticity of the material and inversely proportional to the square root of the volume of the member. Assuming all three members to have the same volume and to be of the same material, we also note that, for a given value of the absorbed energy, the uniform rod will experience the lowest maximum stress, and the cantilever beam the highest one.

This observation may be explained by the fact that, the distribution of stresses being uniform in case a, the strain energy will be uniformly distributed throughout the rod. In case b, on the other hand, the stresses in portion BC of the rod are only 25% as large as the stresses in portion CD. This uneven distribution of the stresses and of the strain energy results in a maximum stress σ_m twice as large as the corresponding stress in the uniform rod. Finally, in case c, where the cantilever beam is subjected to a transverse impact loading, the stresses vary linearly along the beam as well as across a transverse section. The very uneven resulting distribution of strain energy causes the maximum stress σ_m to be 3.46 times larger than if the same member had been loaded axially as in case a.

The properties noted in the three specific cases discussed in this section are quite general and may be observed in all types of structures and impact loadings. We thus conclude that a structure designed to withstand effectively an impact load should

1. Have a large volume
2. Be made of a material with a low modulus of elasticity and a high yield strength
3. Be shaped so that the stresses are distributed as evenly as possible throughout the structure

10.9. WORK AND ENERGY UNDER A SINGLE LOAD

When we first introduced the concept of strain energy at the beginning of this chapter, we considered the work done by an axial load $\mathbf{P}$ applied to the end of a rod of uniform cross section (Fig. 10.1). We defined the strain energy of the rod for an elongation x_1 as the work of the load $\mathbf{P}$ as it is slowly increased from 0 to the value P_1 corresponding to x_1. We wrote

$$\text{Strain energy} = U = \int_0^{x_1} P\, dx \qquad (10.2)$$

In the case of an elastic deformation, the work of the load $\mathbf{P}$, and thus the strain energy of the rod, were expressed as

$$U = \tfrac{1}{2}P_1 x_1 \qquad (10.3)$$

Later, in Secs. 10.4 and 10.5, we computed the strain energy of structural members under various loading conditions by determining the strain-energy density u at every point of the member and integrating u over the entire member.

However, when a structure or member is subjected to a *single concentrated load*, it is possible to use Eq. (10.3) to evaluate its elastic strain energy, provided, of course, that the relation between the load and the resulting deformation is known. For instance, in the case of the cantilever beam of Example 10.03 (Fig. 10.25), we write

$$U = \tfrac{1}{2}P_1 y_1$$

and, substituting for y_1 the value obtained from the table of *Beam Deflections and Slopes* of Appendix D,

$$U = \frac{1}{2}P_1\left(\frac{P_1 L^3}{3EI}\right) = \frac{P_1^2 L^3}{6EI} \tag{10.46}$$

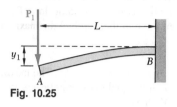

Fig. 10.25

A similar approach may be used to determine the strain energy of a structure or member subjected to a *single couple*. Recalling that the elementary work of a couple of moment M is $M\,d\theta$, where $d\theta$ is a small angle, we find, since M and θ are linearly related, that the elastic strain energy of a cantilever beam AB subjected to a single couple $\mathbf{M}_1$ at its end A (Fig. 10.26) may be expressed as

$$U = \int_0^{\theta_1} M\,d\theta = \tfrac{1}{2}M_1\theta_1 \tag{10.47}$$

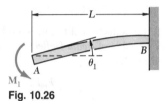

Fig. 10.26

where θ_1 is the slope of the beam at A. Substituting for θ_1 the value obtained from Appendix D, we write

$$U = \frac{1}{2}M_1\left(\frac{M_1 L}{EI}\right) = \frac{M_1^2 L}{2EI} \tag{10.48}$$

In a similar way, the elastic strain energy of a uniform circular shaft AB of length L subjected at its end B to a single torque $\mathbf{T}_1$ (Fig. 10.27) may be expressed as

$$U = \int_0^{\phi_1} T\,d\phi = \tfrac{1}{2}T_1\phi_1 \tag{10.49}$$

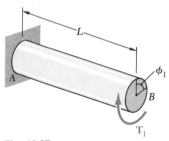

Fig. 10.27

Substituting for the angle of twist ϕ_1 from Eq. (3.16), we verify that

$$U = \frac{1}{2}T_1\left(\frac{T_1 L}{JG}\right) = \frac{T_1^2 L}{2JG}$$

as previously obtained in Sec. 10.5.

The method presented in this section may simplify the solution of many impact-loading problems.

Example 10.08

A block of mass m moving with a velocity $\mathbf{v}_0$ hits squarely the prismatic member AB at its midpoint C (Fig. 10.28). Determine (a) the equivalent static load P_m, (b) the maximum stress σ_m in the member, and (c) the maximum deflection x_m at point C.

(a) *Equivalent Static Load.* The maximum strain energy of the member is equal to the kinetic energy of the block before impact. We have

$$U_m = \tfrac{1}{2}mv_0^2 \qquad (10.50)$$

On the other hand, expressing U_m as the work of the equivalent horizontal static load as it is slowly applied at the midpoint C of the member, we write

$$U_m = \tfrac{1}{2}P_m x_m \qquad (10.51)$$

where x_m is the deflection of C corresponding to the static load P_m. From the table of *Beam Deflections and Slopes* of Appendix D, we find that

$$x_m = \frac{P_m L^3}{48EI} \qquad (10.52)$$

Substituting for x_m from (10.52) into (10.51), we write

$$U_m = \frac{1}{2}\frac{P_m^2 L^3}{48EI}$$

Solving for P_m and recalling Eq. (10.50), we find that the static load equivalent to the given impact loading is

$$P_m = \sqrt{\frac{96 U_m EI}{L^3}} = \sqrt{\frac{48 m v_0^2 EI}{L^3}} \qquad (10.53)$$

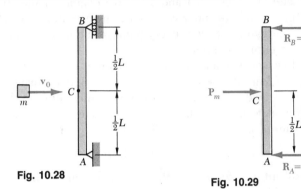

Fig. 10.28

Fig. 10.29

(b) *Maximum Stress.* Drawing the free-body diagram of the member (Fig. 10.29), we find that the maximum value of the bending moment occurs at C and is $M_{max} = P_m L/4$. The maximum stress, therefore, occurs in a transverse section through C and is equal to

$$\sigma_m = \frac{M_{max} c}{I} = \frac{P_m L c}{4I}$$

Substituting for P_m from (10.53), we write

$$\sigma_m = \sqrt{\frac{3 m v_0^2 EI}{L(I/c)^2}}$$

(c) *Maximum Deflection.* Substituting into Eq. (10.52) the expression obtained for P_m in (10.53), we have

$$x_m = \frac{L^3}{48EI}\sqrt{\frac{48 m v_0^2 EI}{L^3}} = \sqrt{\frac{m v_0^2 L^3}{48EI}}$$

10.10. DEFLECTION UNDER A SINGLE LOAD BY THE WORK-ENERGY METHOD

We saw in the preceding section that, if the deflection x_1 of a structure or member under a single concentrated load $\mathbf{P}_1$ is known, the corresponding strain energy U may be obtained by writing

$$U = \tfrac{1}{2}P_1 x_1 \qquad (10.3)$$

A similar expression may be used to obtain the strain energy of a structural member under a single couple $\mathbf{M}_1$:

$$U = \tfrac{1}{2}M_1 \theta_1 \qquad (10.47)$$

Conversely, if the strain energy U of a structure or member subjected to a single concentrated load $\mathbf{P}_1$ or couple $\mathbf{M}_1$ is known, Eq. (10.3) or (10.47) may be used to determine the corresponding deflection x_1 or angle θ_1. In order to determine the deflection under a single load applied to a structure consisting of several component parts, we may find it easier, rather than using one of the methods of Chaps. 8 and 9, to first compute the strain energy of the structure by integrating the strain-energy density over its various parts, as was done in Secs. 10.4 and 10.5, and then use either Eq. (10.3) or Eq. (10.47) to obtain the desired deflection. Similarly, the angle of twist ϕ_1 of a composite shaft may be obtained by integrating the strain-energy density over the various parts of the shaft and solving Eq. (10.49) for ϕ_1.

It should be kept in mind that the method presented in this section may be used *only if the given structure is subjected to a single concentrated load or couple*. As we shall see in Sec. 10.11, the strain energy of a structure subjected to several loads *cannot* be determined by computing the work of each load as if it were applied independently to the structure. We may also observe that, even if it were possible to compute the strain energy of the structure in this manner, only one equation would be available to determine the deflections corresponding to the various loads. In Secs. 10.12 and 10.13, we shall present another method based on the concept of strain energy, which may be used to determine the deflection or slope at a given point of a structure, even when that structure is subjected simultaneously to several concentrated loads, distributed loads, or couples.

Example 10.09

A load $\mathbf{P}$ is supported at B by two uniform rods of the same cross-sectional area A (Fig. 10.30). Determine the vertical deflection of point B.

The strain energy of the system under the given load was determined in Example 10.02. Equating the expression obtained for U to the work of the load, we write

$$U = 0.364 \frac{P^2 l}{AE} = \frac{1}{2} P y_B$$

and, solving for the vertical deflection of B,

$$y_B = 0.728 \frac{Pl}{AE}$$

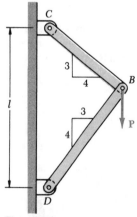

Fig. 10.30

Remark. We should note that, once the forces in the two rods have been obtained (see Example 10.02), the deformations $\delta_{B/C}$ and $\delta_{B/D}$ of the rods could be obtained by the method of Chap. 2. Determining the vertical deflection of point B from these deformations, however, would require a careful geometric analysis of the various displacements involved. The strain-energy method used here makes such an analysis unnecessary.

Example 10.10

Determine the deflection of end A of the cantilever beam AB (Fig. 10.31), taking into account the effect of (*a*) the normal stresses only, (*b*) both the normal and shearing stresses.

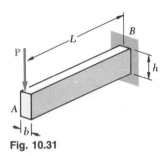

Fig. 10.31

(*a*) *Effect of Normal Stresses.* The work of the force **P** as it is slowly applied to A is

$$U = \tfrac{1}{2}Py_A$$

Substituting for U the expression obtained for the strain energy of the beam in Example 10.03, where only the effect of the normal stresses was considered, we write

$$\frac{P^2L^3}{6EI} = \frac{1}{2}Py_A$$

and, solving for y_A,

$$y_A = \frac{PL^3}{3EI}$$

(*b*) *Effect of Normal and Shearing Stresses.* We now substitute for U the expression (10.24) obtained in Example 10.05, where the effects of both the normal and shearing stresses were taken into account. We have

$$\frac{P^2L^3}{6EI}\left(1 + \frac{3Eh^2}{10GL^2}\right) = \frac{1}{2}Py_A$$

and, solving for y_A,

$$y_A = \frac{PL^3}{3EI}\left(1 + \frac{3Eh^2}{10GL^2}\right)$$

We note that the relative error when the effect of shear is neglected is the same that was obtained in Example 10.05, i.e., less than $0.9(h/L)^2$. As we indicated then, this is less than 0.9% for a beam with a ratio h/L less than $\tfrac{1}{10}$.

Example 10.11

A torque **T** is applied at the end D of shaft BCD (Fig. 10.32). Knowing that both portions of the shaft are of the same material and same length, but that the diameter of BC is twice the diameter of CD, determine the angle of twist for the entire shaft.

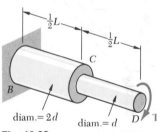

Fig. 10.32

The strain energy of a similar shaft was determined in Example 10.04 by breaking the shaft into its component parts BC and CD. Making $n = 2$ in Eq. (10.23), we have

$$U = \frac{17}{32}\frac{T^2L}{2GJ}$$

where G is the modulus of rigidity of the material, and J the polar moment of inertia of portion CD of the shaft. Setting U equal to the work of the torque as it is slowly applied to end D, and recalling Eq. (10.49), we write

$$\frac{17}{32}\frac{T^2L}{2GJ} = \frac{1}{2}T\phi_{D/B}$$

and, solving for the angle of twist $\phi_{D/B}$,

$$\phi_{D/B} = \frac{17TL}{32GJ}$$

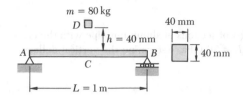

m = 80 kg

h = 40 mm

40 mm

40 mm

L = 1 m

SAMPLE PROBLEM 10.3

The block D of mass m is released from rest and falls a distance h before it strikes the midpoint C of the aluminum beam AB. Using $E = 70$ GPa, determine (*a*) the maximum deflection of point C, (*b*) the maximum stress which occurs in the beam.

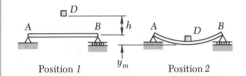

Position 1   Position 2

From Appendix D

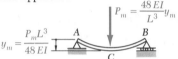

$$P_m = \frac{48\,EI}{L^3}y_m$$

$$y_m = \frac{P_m L^3}{48\,EI}$$

Principle of Work and Energy. Since the block is released from rest, we note that in position *1* both the kinetic energy and the strain energy are zero. In position *2*, where the maximum deflection y_m occurs, the kinetic energy is again zero. Referring to the table of *Beam Deflections and Slopes* of Appendix D, we find the expression for y_m shown. The strain energy of the beam in position *2* is

$$U_2 = \frac{1}{2}P_m y_m = \frac{1}{2}\frac{48EI}{L^3}y_m^2 \qquad U_2 = \frac{24EI}{L^3}y_m^2$$

We observe that the work done by the weight $\mathbf{W}$ of the block is $W(h + y_m)$. Equating the strain energy of the beam to the work done by $\mathbf{W}$, we have

$$\frac{24EI}{L^3}y_m^2 = W(h + y_m) \qquad (1)$$

(*a*) **Maximum Deflection of Point C.** From the given data we have

$$EI = (70 \times 10^9 \text{ Pa})\tfrac{1}{12}(0.04 \text{ m})^4 = 14.93 \times 10^3 \text{ N} \cdot \text{m}^2$$

$$L = 1 \text{ m} \qquad h = 0.040 \text{ m} \qquad W = mg = (80 \text{ kg})(9.81 \text{ m/s}^2) = 784.8 \text{ N}$$

Substituting into Eq. (1), we have

$$\frac{24(14.93 \times 10^3)}{(1)^3}y_m^2 - 784.8(0.040 + y_m) = 0$$

$$(358.3 \times 10^3)y_m^2 - 784.8 y_m - 31.39 = 0$$

Solving this quadratic equation, we find $\qquad\qquad y_m = 10.52 \text{ mm} \quad \blacktriangleleft$

(*b*) **Maximum Stress.** The value of P_m is

$$P_m = \frac{48EI}{L^3}y_m = \frac{48(14.93 \times 10^3 \text{ N} \cdot \text{m})}{(1 \text{ m})^3}(0.01052 \text{ m}) \qquad P_m = 7540 \text{ N}$$

Since $M_{\max} = \tfrac{1}{4}P_m L$, the maximum stress is

$$\sigma_m = \frac{M_{\max}c}{I} = \frac{(\tfrac{1}{4}P_m L)c}{I} = \frac{\tfrac{1}{4}(7540 \text{ N})(1 \text{ m})(0.020 \text{ m})}{\tfrac{1}{12}(0.040 \text{ m})^4}$$

$$\sigma_m = 176.7 \text{ MPa} \quad \blacktriangleleft$$

An approximation for the work done by the weight of the block may be obtained by omitting y_m from the expression for the work and from the right-hand member of Eq. (1), as was done in Example 10.07. If this approximation is used here, we find $y_m = 9.36$ mm; the error is 11.0%. However, if an 8-kg block is dropped from a height of 400 mm, producing the same value of Wh, omitting y_m from the right-hand member of Eq. (1) results in an error of only 1.2% A further discussion of this approximation is given in Prob. 10.78.

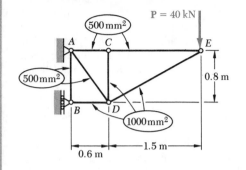

$P = 40$ kN

$500\,\text{mm}^2$

$500\,\text{mm}^2$

$1000\,\text{mm}^2$

0.8 m

1.5 m

0.6 m

SAMPLE PROBLEM 10.4

Members of the truss shown consist of sections of aluminum pipe with the cross-sectional areas indicated. Using $E = 70$ GPa, determine the vertical deflection of point E caused by the load $\mathbf{P}$.

Axial Forces in Truss Members. The reactions are found by using the free-body diagram of the entire truss. We then consider in sequence the equilibrium of joints, E, C, D, and B. At each joint we determine the forces indicated by dashed lines. At joint B, the equation $\Sigma F_x = 0$ provides a check of our computations.

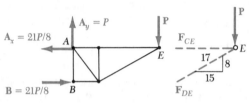

$A_y = P$

$A_x = 21P/8$

$B = 21P/8$

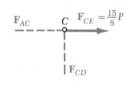

F_{CE}

F_{DE}

$\Sigma F_y = 0: F_{DE} = -\dfrac{17}{8}P$

$\Sigma F_x = 0: F_{CE} = +\dfrac{15}{8}P$

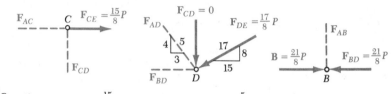

F_{AC} C $F_{CE} = \dfrac{15}{8}P$ F_{AD} $F_{CD} = 0$ $F_{DE} = \dfrac{17}{8}P$ F_{AB}

F_{CD} F_{BD} D $B = \dfrac{21}{8}P$ $F_{BD} = \dfrac{21}{8}P$ B

$\Sigma F_x = 0: F_{AC} = +\dfrac{15}{8}P$ $\Sigma F_y = 0: F_{AD} = +\dfrac{5}{4}P$ $\Sigma F_y = 0: F_{AB} = 0$

$\Sigma F_y = 0: F_{CD} = 0$ $\Sigma F_x = 0: F_{BD} = -\dfrac{21}{8}P$ $\Sigma F_x = 0: \text{(Checks)}$

Strain Energy. Noting that E is the same for all members, we express the strain energy of the truss as follows

$$U = \sum \frac{F_i^2 L_i}{2A_i E} = \frac{1}{2E} \sum \frac{F_i^2 L_i}{A_i} \tag{1}$$

where F_i is the force in a given member as indicated in the following table and where the summation is extended over all members of the truss.

Member	F_i	L_i, m	A_i, m^2	$\dfrac{F_i^2 L_i}{A_i}$
AB	0	0.8	500×10^{-6}	0
AC	$+15P/8$	0.6	500×10^{-6}	$4\,219P^2$
AD	$+5P/4$	1.0	500×10^{-6}	$3\,125P^2$
BD	$-21P/8$	0.6	1000×10^{-6}	$4\,134P^2$
CD	0	0.8	1000×10^{-6}	0
CE	$+15P/8$	1.5	500×10^{-6}	$10\,547P^2$
DE	$-17P/8$	1.7	1000×10^{-6}	$7\,677P^2$

$$\sum \frac{F_i^2 L_i}{A_i} = 29\,700P^2$$

Returning to Eq. (1), we have $U = (1/2E)(29.7 \times 10^3 P^2)$.

Principle of Work-Energy. We recall that the work done by the load $\mathbf{P}$ as it is gradually applied is $\frac{1}{2}Py_E$. Equating the work done by $\mathbf{P}$ to the strain energy U and recalling that $E = 70$ GPa and $P = 40$ kN, we have

$$\frac{1}{2}Py_E = U \qquad \frac{1}{2}Py_E = \frac{1}{2E}(29.7 \times 10^3 P^2)$$

or

$$y_E = \frac{1}{E}(29.7 \times 10^3 P) = \frac{(29.7 \times 10^3)(40 \times 10^3)}{70 \times 10^9}$$

$$y_E = 16.97 \times 10^{-3} \text{ m} \qquad\qquad y_E = 16.97 \text{ mm} \downarrow \quad \blacktriangleleft$$

PROBLEMS

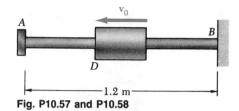

Fig. P10.57 and P10.58

10.57 A 5-kg collar D moves along the uniform rod AB and strikes a small plate attached to end A of the rod with a speed $v_0 = 6$ m/s. Using $E = 200$ GPa and knowing that the allowable stress in the rod is 250 MPa, determine the smallest diameter which may be used for the rod.

10.58 A 6-kg collar D has a speed $v_0 = 4.5$ m/s when it strikes a small plate attached to end A of the 20-mm-diameter rod AB. Using $E = 200$ GPa, determine (a) the equivalent static load for the rod, (b) the maximum stress in the rod, (c) the maximum deflection at end A.

10.59 Block E has a speed $v_0 = 4$ m/s when it strikes squarely the yoke BD which is attached to the 19-mm-diameter rods AB and CD. Knowing that the rods are made of a steel for which $\sigma_Y = 250$ MPa and $E = 200$ GPa, determine the weight of the block for which the factor of safety is four with respect to permanent deformation of the rods.

10.60 The 5.5-kg block E has a horizontal velocity v_0 when it strikes squarely the yoke BD which is attached to the 19-mm-diameter rods AB and CD. Knowing that the rods are made of a steel for which $\delta_Y = 250$ MPa and $E = 200$ GPa, determine the maximum allowable speed v_0 if the rods are not to be permanently deformed.

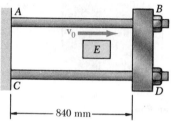

Fig. P10.59 and P10.60

10.61 Collar D is released from rest in the position shown and is stopped by the plate attached at end C of the vertical bronze rod ABC. Knowing that $E = 105$ GPa and $\sigma_{all} = 225$ MPa, determine the largest allowable mass of the collar.

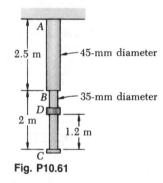

Fig. P10.61

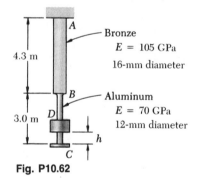

Fig. P10.62

10.62 The 9-kg collar D is released from rest when $h = 420$ mm and is stopped by a plate attached at end C of the vertical rod ABC. Determine (a) the maximum deflection at end C, (b) the equivalent static load, (c) the maximum stress which occurs in portion BC of the rod.

10.63 Solve Prob. 10.62, assuming that both portions of the rod ABC are made of bronze.

10.64 The 48-kg collar G is released from rest in the position shown and is stopped by the plate BDF which is attached to the 20-mm-diameter steel rod CD and to the 15-mm-diameter steel rods AB and EF. Knowing that for the grade of steel used $E = 200$ GPa and $\sigma_{all} = 180$ MPa, determine the largest allowable distance h.

10.65 Solve Prob. 10.64, assuming that the 20-mm-diameter steel rod CD is replaced by a 20-mm-diameter rod made of a grade of aluminum for which $E = 75$ GPa and $\sigma_{all} = 150$ MPa.

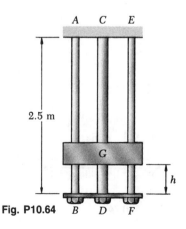

Fig. P10.64

10.66 A 12-kg block C moving horizontally with a velocity $\mathbf{v}_0$ hits post AB squarely as shown. Using $E = 200\ \text{GPa}$, determine the largest speed v_0 for which the maximum normal stress in the post does not exceed 125 MPa.

10.67 Solve Prob. 10.66, assuming that the post AB has been rotated 90° about its longitudinal axis.

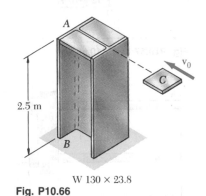

W 130 × 23.8

Fig. P10.66

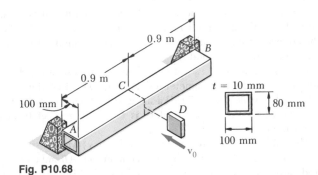

Fig. P10.68

10.68 An aluminum tube having the cross section shown is struck squarely in its midsection by a 6-kg block moving horizontally with a speed of 2 m/s. Using $E = 70\ \text{GPa}$, determine (a) the equivalent static load, (b) the maximum stress in the beam, (c) the maximum deflection at the midpoint C of the beam.

10.69 Solve Prob. 10.68, assuming that the tube has been replaced by a solid aluminum bar with the same outside dimensions as the tube.

10.70 The 20-kg block D is dropped from a height $h = 0.2$ m onto the steel beam AB. Knowing that $E = 200\ \text{GPa}$, determine (a) the maximum deflection at point E, (b) the maximum stress in the beam.

10.71 Solve Prob. 10.70, assuming that a W 100 × 19.3 rolled-steel shape is used for beam AB.

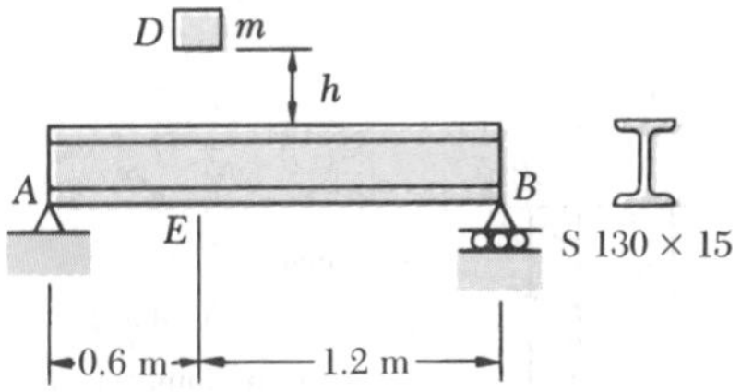

S 130 × 15

Fig. P10.70

10.72 and 10.73 The 3-kg block D is dropped from the position shown onto the end of a 16-mm-square steel bar. Knowing that $E = 200\ \text{GPa}$, determine (a) the maximum deflection at the free end of the bar, (b) the maximum bending moment in the bar, (c) the maximum normal stress in the bar.

10.74 The 2-kg block E is released from rest when $h = 40$ mm and strikes the 25-mm-square steel bar BD which is attached to 10-mm-diameter steel rods AB and CD. Knowing that $E = 200\ \text{GPa}$, determine the maximum deflection at the midpoint of the bar.

10.75 Solve Prob. 10.74, assuming that rods AB and CD are replaced by 3-mm-diameter steel cables.

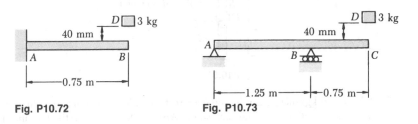

Fig. P10.72   **Fig. P10.73**

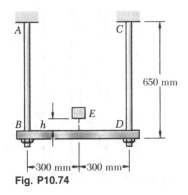

Fig. P10.74

10.76 A block of weight W is placed in contact with a beam at some given point D and released. Show that the resulting maximum deflection at point D is twice as large as the deflection due to a static load W applied at D.

10.77 A block of weight W is dropped from a height h onto the horizontal beam AB and hits it at point D. (a) Show that the maximum deflection y_m at point D may be expressed as

$$y_m = y_{st}\left(1 + \sqrt{1 + \frac{2h}{y_{st}}}\right)$$

where y_{st} represents the deflection at D caused by a static load W applied at that point and where the quantity in parentheses is referred to as the *impact factor*. (b) Compute the impact factor for the bar and impact loading of Prob. 10.72.

10.78 A block of weight W is dropped from a height h onto the horizontal beam AB and hits it at point D. (a) Denoting by y_m the exact value of the maximum deflection at D and by y'_m the value obtained by neglecting the effect of this deflection on the change in potential energy of the block, show that the absolute value of the relative error $(y'_m - y_m)/y_m$ never exceeds $y'_m/2h$. (b) Check the result obtained in part a by solving part a of Prob. 10.72 without taking y_m into account when determining the change in potential energy of the load, and comparing the answer obtained in this way with the exact answer to that problem.

10.79 and 10.80 Using the method of work and energy, determine the deflection at point D caused by the load $\mathbf{P}$.

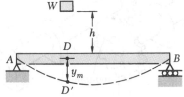

Fig. P10.77 and P10.78

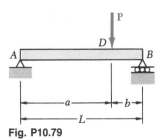

Fig. P10.79

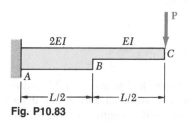

Fig. P10.80

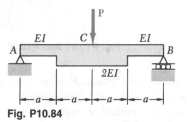

Fig. P10.81

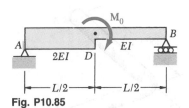

Fig. P10.82

10.81 and 10.82 Using the method of work and energy, determine the slope at point D caused by the couple $\mathbf{M}_0$.

10.83 and 10.84 Using the method of work and energy, determine the deflection at point C caused by the load $\mathbf{P}$.

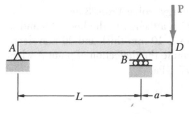

Fig. P10.83

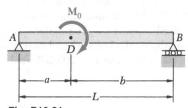

Fig. P10.84

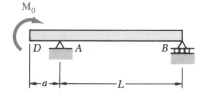

Fig. P10.85

10.85 Using the method of work and energy, determine the slope at point D caused by the couple $\mathbf{M}_0$.

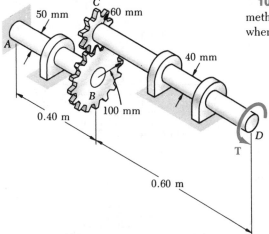

Fig. P10.86

10.86 Two solid steel shafts are connected by the gears shown. Using the method of work and energy, determine the angle through which end D rotates when $T = 820$ N · m. Use $G = 77$ GPa.

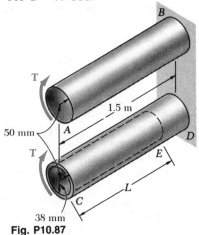

Fig. P10.87

10.87 Torques of the same magnitude T are applied to the steel shafts AB and CD. Using the method of work and energy, determine the length L of the hollow portion of shaft CD for which the angle of twist at C is equal to 1.25 times the angle of twist at A.

10.88 Using the method of work and energy, solve Prob. 3.34.

10.89 The 20-mm-diameter steel rod BC is attached to the lever AB and to the fixed support C. The uniform steel lever is 10 mm thick and 30 mm deep. Using the method of work and energy, determine the deflection of point A when $L = 600$ mm. Use $E = 200$ GPa and $G = 77$ GPa.

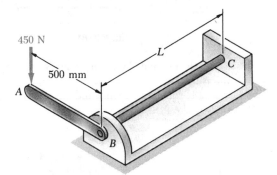

Fig. P10.89 and P10.90

Fig. P10.91

10.90 The 20-mm-diameter steel rod BC is attached to the lever AB and to the fixed support C. The uniform steel lever is 10 mm thick and 30 mm deep. Using the method of work and energy, determine the length L of rod BC for which the deflection of point A is 40 mm. Use $E = 200$ GPa and $G = 77$ GPa.

10.91 A 16-mm-diameter steel rod has been bent into the shape shown. Using the method of work and energy, determine the deflection of end C caused by the 200-N force $\mathbf{P}$. Use $E = 200$ GPa and $G = 77$ GPa.

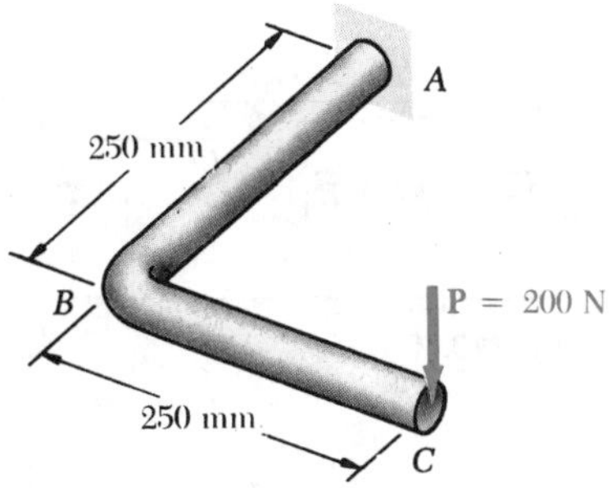

*10.92 The thin-walled hollow cylindrical member *AB* has a noncircular cross section of nonuniform thickness.   Using the expression for τ given in Eq. (3.53) of Sec. 3.13 and the expression for the strain-energy density given in Eq. (10.19) of Sec. 10.5, show that the angle of twist of member *AB* is

$$\phi = \frac{TL}{4\alpha^2 G}\oint \frac{ds}{t}$$

where *ds* is an element of the center line of the wall cross section and α is the area enclosed by that center line.

10.93 Each member of the truss shown has a uniform cross section of area *A*. Using the method of work and energy, determine the vertical deflection of the point of application of the load **P**.

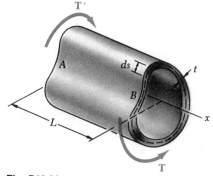

Fig. P10.92

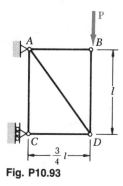

Fig. P10.93

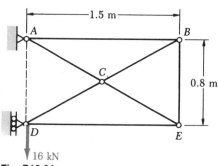

Fig. P10.94

10.94 Each member of the truss shown is made of steel and has a cross-sectional area of 400 mm². Using *E* = 200 GPa, determine the deflection of point *D* caused by the 16-kN load.

10.95 Each member of the truss shown is made of steel and has a cross-sectional area of 3500 mm². Using *E* = 200 GPa, determine the vertical deflection of point *B* caused by the 90-kN load.

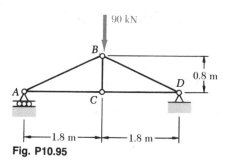

Fig. P10.95

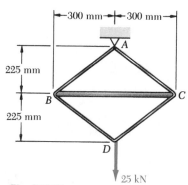

Fig. P10.96

10.96 In the assembly shown, member *BC* is a solid steel rod of 25-mm diameter and all other members are made of 12.5-mm-diameter steel rods.   Using *E* = 200 GPa, determine the deflection of point *D* caused by the 25-kN load.

10.97 through 10.99 For the truss and loading indicated, determine the vertical deflection of point *C*.

10.97 Truss and loading of Prob. 10.27.
10.98 Truss and loading of Prob. 10.28.
10.99 Truss and loading of Prob. 10.29.

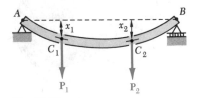

Fig. 10.33

*10.11. WORK AND ENERGY UNDER SEVERAL LOADS

In this section, we shall see how the strain energy of a structure subjected to several loads may be expressed in terms of the loads and the resulting deflections.

Consider an elastic beam AB subjected to two concentrated loads $\mathbf{P}_1$ and $\mathbf{P}_2$. The strain energy of the beam is equal to the work of $\mathbf{P}_1$ and $\mathbf{P}_2$ as they are slowly applied to the beam at C_1 and C_2, respectively (Fig. 10.33). However, in order to evaluate this work, we must first express the deflections x_1 and x_2 in terms of the loads $\mathbf{P}_1$ and $\mathbf{P}_2$.

Let us assume that only $\mathbf{P}_1$ is applied to the beam (Fig. 10.34). We note that both C_1 and C_2 are deflected and that their deflections are

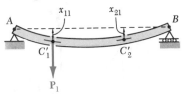

Fig. 10.34

proportional to the load P_1. Denoting these deflections by x_{11} and x_{21}, respectively, we write

$$x_{11} = \alpha_{11}P_1 \qquad x_{21} = \alpha_{21}P_1 \qquad (10.54)$$

where α_{11} and α_{21} are constants called *influence coefficients*. These constants represent the deflections of C_1 and C_2, respectively, when a unit load is applied at C_1 and are characteristics of the beam AB.

Let us now assume that only $\mathbf{P}_2$ is applied to the beam (Fig. 10.35).

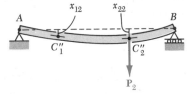

Fig. 10.35

Denoting by x_{12} and x_{22}, respectively, the resulting deflections of C_1 and C_2, we write

$$x_{12} = \alpha_{12}P_2 \qquad x_{22} = \alpha_{22}P_2 \qquad (10.55)$$

where α_{12} and α_{22} are the influence coefficients representing the deflections of C_1 and C_2, respectively, when a unit load is applied at C_2. Applying the principle of superposition, we express the deflections x_1 and x_2 of C_1 and C_2 when both loads are applied (Fig. 10.33) as

$$x_1 = x_{11} + x_{12} = \alpha_{11}P_1 + \alpha_{12}P_2 \qquad (10.56)$$
$$x_2 = x_{21} + x_{22} = \alpha_{21}P_1 + \alpha_{22}P_2 \qquad (10.57)$$

To compute the work done by P_1 and P_2, and thus the strain energy of the beam, we shall find it convenient to assume that P_1 is first applied slowly at C_1 (Fig. 10.36a). Recalling the first of Eqs. (10.54), we express

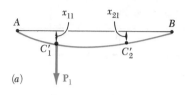

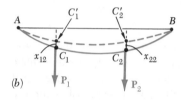

(a)

(b)

Fig. 10.36

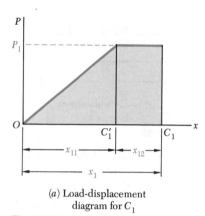

(a) Load-displacement diagram for C_1

Fig. 10.37

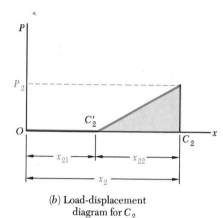

(b) Load-displacement diagram for C_2

the work of P_1 as

$$\tfrac{1}{2}P_1 x_{11} = \tfrac{1}{2}P_1(\alpha_{11}P_1) = \tfrac{1}{2}\alpha_{11}P_1^2 \qquad (10.58)$$

and note that P_2 does no work while C_2 moves through x_{21}, since it has not yet been applied to the beam.

Now we slowly apply P_2 at C_2 (Fig. 10.36b); recalling the second of Eqs. (10.55), we express the work of P_2 as

$$\tfrac{1}{2}P_2 x_{22} = \tfrac{1}{2}P_2(\alpha_{22}P_2) = \tfrac{1}{2}\alpha_{22}P_2^2 \qquad (10.59)$$

But, as P_2 is slowly applied at C_2, the point of application of P_1 moves through x_{12} from C_1' to C_1, and the load P_1 does work. Since P_1 is *fully applied* during this displacement (Fig. 10.37), its work is equal to $P_1 x_{12}$ or, recalling the first of Eqs. (10.55),

$$P_1 x_{12} = P_1(\alpha_{12}P_2) = \alpha_{12}P_1P_2 \qquad (10.60)$$

Adding the expressions obtained in (10.58), (10.59), and (10.60), we express the strain energy of the beam under the loads P_1 and P_2 as

$$U = \tfrac{1}{2}(\alpha_{11}P_1^2 + 2\alpha_{12}P_1P_2 + \alpha_{22}P_2^2) \qquad (10.61)$$

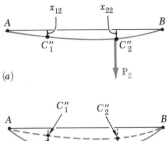

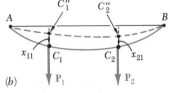

(a)

(b)

Fig. 10.38

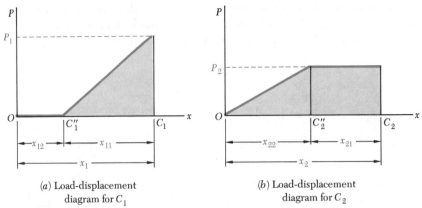

(a) Load-displacement
diagram for C_1

(b) Load-displacement
diagram for C_2

Fig. 10.39

If the load $\mathbf{P}_2$ had first been applied to the beam (Fig. 10.38a), and then the load $\mathbf{P}_1$ (Fig. 10.38b), the work done by each load would have been as shown in Fig. 10.39. Calculations similar to those we have just carried out would lead to the following alternative expression for the strain energy of the beam:

$$U = \tfrac{1}{2}(\alpha_{22}P_2^2 + 2\alpha_{21}P_2P_1 + \alpha_{11}P_1^2) \tag{10.62}$$

Equating the right-hand members of Eqs. (10.61) and (10.62), we find that $\alpha_{12} = \alpha_{21}$, and thus conclude that the deflection produced at C_1 by a unit load applied at C_2 is equal to the deflection produced at C_2 by a unit load applied at C_1. This is known as *Maxwell's reciprocal theorem*, after the British physicist James Clerk Maxwell (1831–1879).

While we are now able to express the strain energy U of a structure subjected to several loads as a function of these loads, we cannot use the method of Sec. 10.10 to determine the deflection of such a structure. Indeed, computing the strain energy U by integrating the strain-energy density u over the structure and substituting the expression obtained into (10.61) would yield only one equation, which clearly could not be solved for the various coefficients α.

*10.12. CASTIGLIANO'S THEOREM

We recall the expression obtained in the preceding section for the strain energy of an elastic structure subjected to two loads $\mathbf{P}_1$ and $\mathbf{P}_2$:

$$U = \tfrac{1}{2}(\alpha_{11}P_1^2 + 2\alpha_{12}P_1P_2 + \alpha_{22}P_2^2) \tag{10.61}$$

where α_{11}, α_{12} and α_{22} are the influence coefficients associated with the points of application C_1 and C_2 of the two loads. Differentiating both members of Eq. (10.61) with respect to P_1 and recalling Eq. (10.56), we write

$$\frac{\partial U}{\partial P_1} = \alpha_{11}P_1 + \alpha_{12}P_2 = x_1 \tag{10.63}$$

Differentiating both members of Eq. (10.61) with respect to P_2, recalling Eq. (10.57), and keeping in mind that $\alpha_{12} = \alpha_{21}$, we have

$$\frac{\partial U}{\partial P_2} = \alpha_{12}P_1 + \alpha_{22}P_2 = x_2 \tag{10.64}$$

More generally, if an elastic structure is subjected to n loads $\mathbf{P}_1$, $\mathbf{P}_2$, . . . , $\mathbf{P}_n$, the deflection x_j of the point of application of $\mathbf{P}_j$, measured along the line of action of $\mathbf{P}_j$, may be expressed as the partial derivative of the strain energy of the structure with respect to the load $\mathbf{P}_j$. We write

$$x_j = \frac{\partial U}{\partial P_j} \tag{10.65}$$

This is *Castigliano's theorem,* named after the Italian engineer Alberto Castigliano (1847–1884) who first stated it.†

Recalling that the work of a couple $\mathbf{M}$ is $\frac{1}{2}M\theta$, where θ is the angle of rotation at the point where the couple is slowly applied, we note that Castigliano's theorem may be used to determine the slope of a beam at the point of application of a couple $\mathbf{M}_j$. We have

$$\theta_j = \frac{\partial U}{\partial M_j} \tag{10.68}$$

Similarly, the angle of twist ϕ_j in a section of a shaft where a torque $\mathbf{T}_j$ is slowly applied is obtained by differentiating the strain energy of the shaft with respect to T_j:

$$\phi_j = \frac{\partial U}{\partial T_j} \tag{10.69}$$

---

†In the case of an elastic structure subjected to n loads $\mathbf{P}_1$, $\mathbf{P}_2$, . . . , $\mathbf{P}_n$, the deflection of the point of application of $\mathbf{P}_j$, measured along the line of action of $\mathbf{P}_j$, may be expressed as

$$x_j = \sum_k \alpha_{jk}P_k \tag{10.66}$$

and the strain energy of the structure is found to be

$$U = \tfrac{1}{2}\sum_i \sum_k \alpha_{ik}P_iP_k \tag{10.67}$$

Differentiating U with respect to P_j, and observing that P_j is found in terms corresponding to either $i = j$ or $k = j$, we write

$$\frac{\partial U}{\partial P_j} = \frac{1}{2}\sum_k \alpha_{jk}P_k + \frac{1}{2}\sum_i \alpha_{ij}P_i$$

or, since $\alpha_{ij} = \alpha_{ji}$, $\quad \dfrac{\partial U}{\partial P_j} = \dfrac{1}{2}\sum_k \alpha_{jk}P_k + \dfrac{1}{2}\sum_i \alpha_{ji}P_i = \sum_k \alpha_{jk}P_k$

Recalling Eq. (10.66), we verify that $\quad x_j = \dfrac{\partial U}{\partial P_j} \tag{10.65}$

*10.13. DEFLECTIONS BY CASTIGLIANO'S THEOREM

We saw in the preceding section that the deflection x_j of a structure at the point of application of a load $\mathbf{P}_j$ may be determined by computing the partial derivative $\partial U/\partial P_j$ of the strain energy U of the structure. As we recall from Secs. 10.4 and 10.5, the strain energy U is obtained by integrating or summing over the structure the strain energy of each element of the structure. We shall find that the calculation by Castigliano's theorem of the deflection x_j is simplified if the differentiation with respect to the load P_j is carried out before the integration or summation.

In the case of a beam, for example, we recall from Sec. 10.4 that

$$U = \int_0^L \frac{M^2}{2EI}\,dx \tag{10.17}$$

and determine the deflection x_j of the point of application of the load $\mathbf{P}_j$ by writing

$$x_j = \frac{\partial U}{\partial P_j} = \int_0^L \frac{M}{EI}\frac{\partial M}{\partial P_j}\,dx \tag{10.70}$$

In the case of a truss consisting of n uniform members of length L_i, cross-sectional area A_i, and internal force F_i, we recall Eq. (10.14) and express the strain energy U of the truss as

$$U = \sum_{i=1}^{n} \frac{F_i^2 L_i}{2A_i E} \tag{10.71}$$

The deflection x_j of the point of application of the load $\mathbf{P}_j$ is obtained by differentiating with respect to P_j each term of sum. We write

$$x_j = \frac{\partial U}{\partial P_j} = \sum_{i=1}^{n} \frac{F_i L_i}{A_i E}\frac{\partial F_i}{\partial P_j} \tag{10.72}$$

Example 10.12

The cantilever beam AB supports a uniformly distributed load w and a concentrated load $\mathbf{P}$ as shown (Fig. 10.40). Knowing that $L = 2$ m, $w = 4$ kN/m, $P = 6$ kN, and $EI = 5$ MN $\cdot$ m^2, determine the deflection at A.

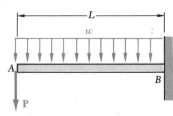

Fig. 10.40

The deflection y_A of the point A where the load $\mathbf{P}$ is applied is obtained from Eq. (10.70). Since $\mathbf{P}$ is vertical and directed downward, y_A represents a vertical deflection and is positive downward. We have

$$y_A = \frac{\partial U}{\partial P} = \int_0^L \frac{M}{EI}\frac{\partial M}{\partial P}\,dx \tag{10.73}$$

The bending moment M at a distance x from A is

$$M = -(Px + \tfrac{1}{2}wx^2) \tag{10.74}$$

and its derivative with respect to P is

$$\frac{\partial M}{\partial P} = -x$$

Substituting for M and $\partial M/\partial P$ into Eq. (10.73), we write

$$y_A = \frac{1}{EI}\int_0^L \left(Px^2 + \frac{1}{2}wx^3\right)dx$$

$$y_A = \frac{1}{EI}\left(\frac{PL^3}{3} + \frac{wL^4}{8}\right) \tag{10.75}$$

Substituting the given data, we have

$$y_A = \frac{1}{5 \times 10^6 \text{ N} \cdot \text{m}^2}\left[\frac{(6 \times 10^3 \text{ N})(2 \text{ m})^3}{3} + \frac{(4 \times 10^3 \text{ N/m})(2 \text{ m})^4}{8}\right]$$

$$y_A = 4.8 \times 10^{-3} \text{ m} \qquad y_A = 4.8 \text{ mm} \downarrow$$

We note that the computation of the partial derivative $\partial M/\partial P$ *could not have been carried out* if the numerical value of P had been substituted for P in the expression (10.74) for the bending moment.

We may observe that the deflection x_j of a structure at a given point C_j can be obtained by the direct application of Castigliano's theorem only if a load $\mathbf{P}_j$ happens to be applied at C_j in the direction in which x_j is to be determined. When no load is applied at C_j, or when a load is applied in a direction other than the desired one, we may still obtain the deflection x_j by Castigliano's theorem if we use the following procedure: We apply a fictitious or "dummy" load $\mathbf{Q}_j$ at C_j in the direction in which the deflection x_j is to be determined and use Castigliano's theorem to obtain the deflection

$$x_j = \frac{\partial U}{\partial Q_j} \tag{10.76}$$

due to $\mathbf{Q}_j$ and the actual loads. Making $Q_j = 0$ in Eq. (10.76) yields the deflection at C_j in the desired direction under the given loading.

The slope θ_j of a beam at a point C_j may be determined in a similar manner by applying a fictitious couple $\mathbf{M}_j$ at C_j, computing the partial derivative $\partial U/\partial M_j$, and making $M_j = 0$ in the expression obtained.

Example 10.13

The cantilever beam AB supports a uniformly distributed load w (Fig. 10.41). Determine the deflection and slope at A.

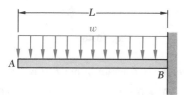

Fig. 10.41

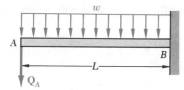

Fig. 10.42

Deflection at A. We apply a dummy downward load $\mathbf{Q}_A$ at A (Fig. 10.42) and write

$$y_A = \frac{\partial U}{\partial Q_A} = \int_0^L \frac{M}{EI}\frac{\partial M}{\partial Q_A}dx \tag{10.77}$$

The bending moment M at a distance x from A is

$$M = -Q_A x - \tfrac{1}{2}wx^2 \tag{10.78}$$

and its derivative with respect to Q_A is

$$\frac{\partial M}{\partial Q_A} = -x \tag{10.79}$$

Substituting for M and $\partial M/\partial Q_A$ from (10.78) and (10.79) into (10.77), and making $Q_A = 0$, we obtain the deflection at A for the given loading:

$$y_A = \frac{1}{EI}\int_0^L (-\tfrac{1}{2}wx^2)(-x)\,dx = +\frac{wL^4}{8EI}$$

Since the dummy load was directed downward, the positive sign indicates that

$$y_A = \frac{wL^4}{8EI} \downarrow$$

Slope at A. We apply a dummy counterclockwise couple M_A at A (Fig. 10.43) and write

$$\theta_A = \frac{\partial U}{\partial M_A}$$

Recalling Eq. (10.17), we have

$$\theta_A = \frac{\partial}{\partial M_A} \int_0^L \frac{M^2}{2EI}\, dx = \int_0^L \frac{M}{EI} \frac{\partial M}{\partial M_A}\, dx \qquad (10.80)$$

The bending moment M at a distance x from A is

$$M = -M_A - \tfrac{1}{2}wx^2 \qquad (10.81)$$

and its derivative with respect to M_A is

$$\frac{\partial M}{\partial M_A} = -1 \qquad (10.82)$$

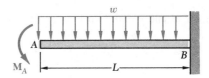

Fig. 10.43 M_A

Substituting for M and $\partial M/\partial M_A$ from (10.81) and (10.82) into (10.80), and making $M_A = 0$, we obtain the slope at A for the given loading:

$$\theta_A = \frac{1}{EI} \int_0^L \left(-\tfrac{1}{2}wx^2\right)(-1)\, dx = +\frac{wL^3}{6EI}$$

Since the dummy couple was counterclockwise, the positive sign indicates that the angle θ_A is also counterclockwise:

$$\theta_A = \frac{wL^3}{6EI} \;\measuredangle$$

Example 10.14

A load $\mathbf{P}$ is supported at B by two rods of the same material and of the same cross-sectional area A (Fig. 10.44). Determine the horizontal and vertical deflection of point B.

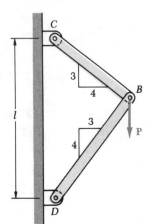

Fig. 10.44

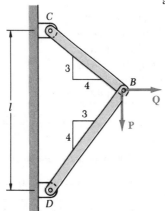

Fig. 10.45

We apply a dummy horizontal load $\mathbf{Q}$ at B (Fig. 10.45). From Castigliano's theorem we have

$$x_B = \frac{\partial U}{\partial Q} \qquad y_B = \frac{\partial U}{\partial P}$$

Recalling from Sec. 10.4 the expression (10.14) for the strain energy of a rod, we write

$$U = \frac{F_{BC}^2(BC)}{2AE} + \frac{F_{BD}^2(BD)}{2AE}$$

where F_{BC} and F_{BD} represent the forces in BC and BD, respectively. We have, therefore,

$$x_B = \frac{\partial U}{\partial Q} = \frac{F_{BC}(BC)}{AE} \frac{\partial F_{BC}}{\partial Q} + \frac{F_{BD}(BD)}{AE} \frac{\partial F_{BD}}{\partial Q} \qquad (10.83)$$

and

$$y_B = \frac{\partial U}{\partial P} = \frac{F_{BC}(BC)}{AE} \frac{\partial F_{BC}}{\partial P} + \frac{F_{BD}(BD)}{AE} \frac{\partial F_{BD}}{\partial P} \qquad (10.84)$$

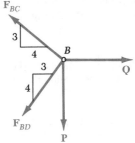

Fig. 10.46

From the free-body diagram of pin B (Fig. 10.46), we obtain

$$F_{BC} = 0.6P + 0.8Q \qquad F_{BD} = -0.8P + 0.6Q \qquad (10.85)$$

Differentiating these expressions with respect to Q and P, we write

$$\frac{\partial F_{BC}}{\partial Q} = 0.8 \qquad \frac{\partial F_{BD}}{\partial Q} = 0.6$$

$$\frac{\partial F_{BC}}{\partial P} = 0.6 \qquad \frac{\partial F_{BD}}{\partial P} = -0.8$$

$$(10.86)$$

Substituting from (10.85) and (10.86) into both (10.83) and (10.84), making $Q = 0$, and noting that $BC = 0.6l$ and $BD = 0.8l$, we obtain the horizontal and vertical deflections of point B under the given load $\mathbf{P}$:

$$x_B = \frac{(0.6P)(0.6l)}{AE}(0.8) + \frac{(-0.8P)(0.8l)}{AE}(0.6) = -0.096\frac{Pl}{AE}$$

$$y_B = \frac{(0.6P)(0.6l)}{AE}(0.6) + \frac{(-0.8P)(0.8l)}{AE}(-0.8) = +0.728\frac{Pl}{AE}$$

Referring to the directions of the loads $\mathbf{Q}$ and $\mathbf{P}$, we conclude that

$$x_B = 0.096\frac{Pl}{AE}\leftarrow \qquad y_B = 0.728\frac{Pl}{AE}\downarrow$$

We check that the expression obtained for the vertical deflection of B is the same that was found in Example 10.09.

*10.14. STATICALLY INDETERMINATE STRUCTURES

The reactions at the supports of a statically indeterminate elastic structure may be determined by Castigliano's theorem. In the case of a structure indeterminate to the first degree, for example, we designate one of the reactions as redundant and eliminate or modify accordingly the corresponding support. The redundant reaction is then treated as an unknown load which, together with the other loads, must produce deformations which are compatible with the original supports. We first calculate the strain energy U of the structure due to the combined action of the given loads and the redundant reaction. Observing that the partial derivative of U with respect to the redundant reaction represents the deflection (or slope) at the support which has been eliminated or modified, we then set this derivative equal to zero and solve the equation obtained for the redundant reaction.† The remaining reactions may be obtained from the equations of statics.

†This is in the case of a rigid support allowing no deflection. For other types of support, the partial derivative of U should be set equal to the allowed deflection.

Example 10.15
Determine the reactions at the supports for the prismatic beam and loading shown (Fig. 10.47).

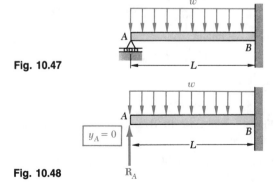

Fig. 10.47

Fig. 10.48

The beam is statically indeterminate to the first degree. We consider the reaction at A as redundant and release the beam from that support. The reaction $\mathbf{R}_A$ is now considered as an unknown load (Fig. 10.48) and will be determined from the condition that the deflection y_A at A must be zero. By Castigliano's theorem $y_A = \partial U/\partial R_A$, where U is the strain energy of the beam under the distributed load and the redundant reaction. Recalling Eq. (10.70), we write

$$y_A = \frac{\partial U}{\partial R_A} = \int_0^L \frac{M}{EI}\frac{\partial M}{\partial R_A}\,dx \qquad (10.87)$$

We now express the bending moment M for the loading of Fig. 10.48.

The bending moment at a distance x from A is

$$M = R_A x - \tfrac{1}{2}wx^2 \qquad (10.88)$$

and its derivative with respect to R_A is

$$\frac{\partial M}{\partial R_A} = x \qquad (10.89)$$

Substituting for M and $\partial M/\partial R_A$ from (10.88) and (10.89) into (10.87), we write

$$y_A = \frac{1}{EI} \int_0^L \left(R_A x^2 - \frac{1}{2}wx^3 \right) dx = \frac{1}{EI} \left(\frac{R_A L^3}{3} - \frac{wL^4}{8} \right)$$

Setting $y_A = 0$ and solving for R_A, we have

$$R_A = \tfrac{3}{8}wL \qquad R_A = \tfrac{3}{8}wL \uparrow$$

From the conditions of equilibrium for the beam, we find that the reaction at B consists of the following force and couple:

$$\mathbf{R}_B = \tfrac{5}{8}wL \uparrow \qquad \mathbf{M}_B = \tfrac{1}{8}wL^2 \downarrow$$

Example 10.16

A load $\mathbf{P}$ is supported at B by three rods of the same material and the same cross-sectional area A (Fig. 10.49). Determine the force in each rod.

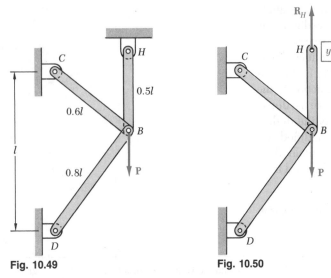

Fig. 10.49   **Fig. 10.50**

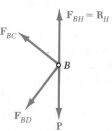

Fig. 10.51

Then, from the free-body diagram of pin B (Fig. 10.51), we obtain

$$F_{BC} = 0.6P - 0.6R_H \qquad F_{BD} = 0.8R_H - 0.8P \qquad (10.92)$$

Differentiating with respect to R_H the force in each rod, we write

$$\frac{\partial F_{BC}}{\partial R_H} = -0.6 \qquad \frac{\partial F_{BD}}{\partial R_H} = 0.8 \qquad \frac{\partial F_{BH}}{\partial R_H} = 1 \qquad (10.93)$$

Substituting from (10.91), (10.92), and (10.93) into (10.90), and noting that the lengths BC, BD, and BH are respectively equal to $0.6l$, $0.8l$, and $0.5l$, we write

$$y_H = \frac{1}{AE} [(0.6P - 0.6R_H)(0.6l)(-0.6)$$
$$+ (0.8R_H - 0.8P)(0.8l)(0.8) + R_H(0.5l)(1)]$$

Setting $y_H = 0$, we obtain

$$1.228R_H - 0.728P = 0$$

and, solving for R_H,

$$R_H = 0.593P$$

Carrying this value into Eqs. (10.91) and (10.92), we obtain the forces in the three rods:

$$F_{BC} = +0.244P \qquad F_{BD} = -0.326P \qquad F_{BH} = +0.593P$$

The structure is statically indeterminate to the first degree. We consider the reaction at H as redundant and release rod BH from its support at H. The reaction $\mathbf{R}_H$ is now considered as an unknown load (Fig. 10.50) and will be determined from the condition that the deflection y_H of point H must be zero. By Castigliano's theorem $y_H = \partial U/\partial R_H$, where U is the strain energy of the three-rod system under the load $\mathbf{P}$ and the redundant reaction $\mathbf{R}_H$. Recalling Eq. (10.72), we write

$$y_H = \frac{F_{BC}(BC)}{AE}\frac{\partial F_{BC}}{\partial R_H} + \frac{F_{BD}(BD)}{AE}\frac{\partial F_{BD}}{\partial R_H} + \frac{F_{BH}(BH)}{AE}\frac{\partial F_{BH}}{\partial R_H} \qquad (10.90)$$

We note that the force in rod BH is equal to R_H and write

$$F_{BH} = R_H \qquad (10.91)$$

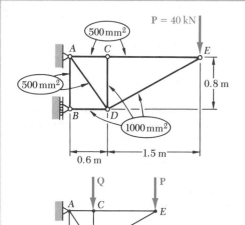

$P = 40$ kN

500 mm²

A C

500 mm²

0.8 m

B D

1000 mm²

1.5 m

0.6 m

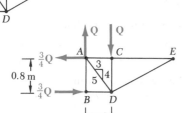

Q   P

A C

E

B D

$\frac{3}{4}Q$

0.8 m

$\frac{3}{4}Q$

Q   Q

A   C   E

3

5 4

B   D

0.6 m

SAMPLE PROBLEM 10.5

For the truss and loading of Sample Prob. 10.4, determine the vertical deflection of joint C.

Castigliano's Theorem. Since no vertical load is applied at joint C, we introduce the dummy load $\mathbf{Q}$ as shown. Using Castigliano's theorem, we have, since $E = $ constant,

$$y_C = \sum \left(\frac{F_i L_i}{A_i E}\right)\frac{\partial F_i}{\partial Q} = \frac{1}{E}\sum\left(\frac{F_i L_i}{A_i}\right)\frac{\partial F_i}{\partial Q} \qquad (1)$$

where F_i is the force in a given member under the combined loading of $\mathbf{P}$ and $\mathbf{Q}$.

Force in Members. Considering in sequence the equilibrium of joints E, C, B, and D, we determine the force in each member caused by the load $\mathbf{Q}$.

Joint E: $F_{CE} = F_{DE} = 0$

Joint C: $F_{AC} = 0$; $F_{CD} = -Q$

Joint B: $F_{AB} = 0$; $F_{BD} = -\frac{3}{4}Q$

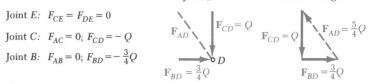

Joint D Force triangle

F_{AD} $F_{CD} = Q$ $F_{CD} = Q$ $F_{AD} = \frac{5}{4}Q$

$F_{BD} = \frac{3}{4}Q$ D $F_{BD} = \frac{3}{4}Q$

The force in each member caused by the load $\mathbf{P}$ was previously found in Sample Prob. 10.4. The total force in each member under the combined action of $\mathbf{Q}$ and $\mathbf{P}$ is shown in the following table. Forming $\partial F_i/\partial Q$ for each member, we then compute $(F_i L_i/A_i)(\partial F_i/\partial Q)$ as indicated in the table.

Member	F_i	$\partial F_i/\partial \mathbf{Q}$	L_i, m	A_i, m²	$\left(\dfrac{F_i L_i}{A_i}\right)\dfrac{\partial F_i}{\partial \mathbf{Q}}$
AB	0	0	0.8	500×10^{-6}	0
AC	$+15P/8$	0	0.6	500×10^{-6}	0
AD	$+5P/4 + 5Q/4$	$\frac{5}{4}$	1.0	500×10^{-6}	$+3125P + 3125Q$
BD	$-21P/8 - 3Q/4$	$-\frac{3}{4}$	0.6	1000×10^{-6}	$+1181P + 338Q$
CD	$-Q$	-1	0.8	1000×10^{-6}	$+ 800Q$
CE	$+15P/8$	0	1.5	500×10^{-6}	0
DE	$-17P/8$	0	1.7	1000×10^{-6}	0

$$\sum\left(\frac{F_i L_i}{A_i}\right)\frac{\partial F_i}{\partial Q} = 4306P + 4263Q$$

Deflection of C. Substituting into Eq. (1), we have

$$y_C = \frac{1}{E}\sum\left(\frac{F_i L_i}{A_i}\right)\frac{\partial F_i}{\partial Q} = \frac{1}{E}(4306P + 4263Q)$$

Since the load $\mathbf{Q}$ is not part of the original loading, we set $Q = 0$. Substituting the given data, $P = 40$ kN and $E = 70$ GPa, we find

$$y_C = \frac{4306(40 \times 10^3 \text{ N})}{70 \times 10^9 \text{ Pa}} = 2.46 \times 10^{-3} \text{ m} \qquad y_C = 2.46 \text{ mm} \downarrow \quad \blacktriangleleft$$

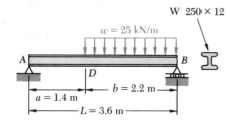

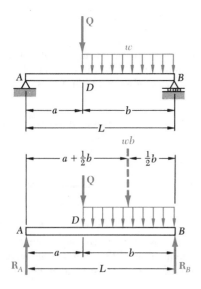

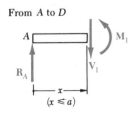

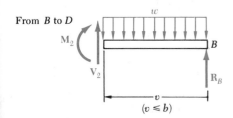

From A to D

$(x \leqslant a)$

From B to D

$(v \leqslant b)$

SAMPLE PROBLEM 10.6

For the beam and loading shown, determine the deflection at point D. Use $E = 200$ GPa.

Castigliano's Theorem. Since the given loading does not include a vertical load at point D, we introduce the dummy load $\mathbf{Q}$ as shown. Using Castigliano's theorem and noting that the flexural rigidity EI is constant, we write

$$y_D = \int \frac{M}{EI}\left(\frac{\partial M}{\partial Q}\right) dx = \frac{1}{EI} \int M\left(\frac{\partial M}{\partial Q}\right) dx \tag{1}$$

We shall perform the integration separately for portions AD and DB of the beam.

Reactions. Using the free-body diagram of the entire beam, we find

$$\mathbf{R}_A = \frac{wb^2}{2L} + Q\frac{b}{L} \uparrow \qquad \mathbf{R}_B = \frac{wb(a + \frac{1}{2}b)}{L} + Q\frac{a}{L} \uparrow$$

Portion AD of Beam. Using the free body shown, we find

$$M_1 = R_A x = \left(\frac{wb^2}{2L} + Q\frac{b}{L}\right)x \qquad \frac{\partial M_1}{\partial Q} = +\frac{bx}{L}$$

Substituting into Eq. (1) and integrating from A to D,

$$\frac{1}{EI} \int M_1 \frac{\partial M_1}{\partial Q} dx = \frac{1}{EI} \int_0^a R_A x\left(\frac{bx}{L}\right) dx = \frac{R_A a^3 b}{3EIL}$$

We substitute for R_A and then set $Q = 0$, since the load $\mathbf{Q}$ is not part of the given loading.

$$\frac{1}{EI} \int M_1 \frac{\partial M_1}{\partial Q} dx = \frac{wa^3 b^3}{6EIL^2} \tag{2}$$

Portion DB of Beam. Using the free body shown, we find that the bending moment at a distance v from end B is

$$M_2 = R_B v - \frac{wv^2}{2} = \left[\frac{wb(a + \frac{1}{2}b)}{L} + Q\frac{a}{L}\right]v - \frac{wv^2}{2} \qquad \frac{\partial M_2}{\partial Q} = +\frac{av}{L}$$

Substituting into Eq. (1) and integrating from point B where $v = 0$, to point D where $v = b$, we write

$$\frac{1}{EI} \int M_2 \frac{\partial M_2}{\partial Q} dv = \frac{1}{EI} \int_0^b \left(R_B v - \frac{wv^2}{2}\right)\left(\frac{av}{L}\right) dv = \frac{R_B ab^3}{3EIL} - \frac{wab^4}{8EIL}$$

Substituting for R_B and setting $Q = 0$,

$$\frac{1}{EI} \int M_2 \frac{\partial M_2}{\partial Q} dv = \left[\frac{wb(a + \frac{1}{2}b)}{L}\right]\frac{ab^3}{3EIL} - \frac{wab^4}{8EIL} = \frac{5a^2 b^4 + ab^5}{24EIL^2} w \tag{3}$$

Deflection at Point D. Combining the results given in Eqs. (2) and (3) and recalling Eq. (1), we have

$$y_D = \frac{wab^3}{24EIL^2}(4a^2 + 5ab + b^2) = \frac{wab^3}{24EIL^2}(4a + b)(a + b) = \frac{wab^3}{24EIL}(4a + b)$$

Referring to Appendix C, we find that, for a W 250×22, $I = 28.7 \times 10^{-6}$ m^4. Substituting the given data, we have

$$y_D = 5.86 \times 10^3 \text{ m} \qquad y_D = 5.86 \text{ mm} \downarrow$$

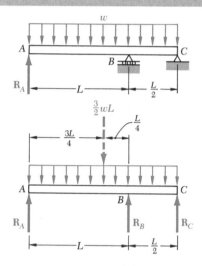

SAMPLE PROBLEM 10.7

For the uniform beam and loading shown, determine the reactions at the supports.

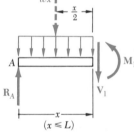

From A to B

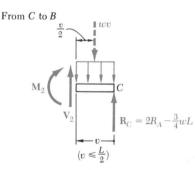

From C to B

Castigliano's Theorem. The beam is indeterminate to the first degree and we choose the reaction $\mathbf{R}_A$ as redundant. Using Castigliano's theorem, we shall determine the deflection at A due to the combined action of $\mathbf{R}_A$ and the distributed load. Since the flexural rigidity EI is constant, we write

$$y_A = \int \frac{M}{EI}\left(\frac{\partial M}{\partial R_A}\right) dx = \frac{1}{EI} \int M \frac{\partial M}{\partial R_A} dx \qquad (1)$$

The integration will be performed separately for portions AB and BC of the beam. Finally, $\mathbf{R}_A$ is obtained by setting y_A equal to zero.

Free Body: Entire Beam. We express the reactions at B and C in terms of R_A and the distributed load

$$R_B = \tfrac{9}{4}wL - 3R_A \qquad R_C = 2R_A - \tfrac{3}{4}wL \qquad (2)$$

Portion AB of Beam. Using the free-body diagram shown, we find the bending moment M_1 and compute its partial derivative with respect to R_A.

$$M_1 = R_A x - \frac{wx^2}{2} \qquad \frac{\partial M_1}{\partial R_A} = x$$

Substituting into Eq. (1) and integrating from A to B, we have

$$\frac{1}{EI}\int M_1 \frac{\partial M}{\partial R_A} dx = \frac{1}{EI}\int_0^L \left(R_A x^2 - \frac{wx^3}{2}\right) dx = \frac{1}{EI}\left(\frac{R_A L^3}{3} - \frac{wL^4}{8}\right) \qquad (3)$$

Portion BC of Beam. We have

$$M_2 = \left(2R_A - \frac{3}{4}wL\right)v - \frac{wv^2}{2} \qquad \frac{\partial M_2}{\partial R_A} = 2v$$

Substituting into Eq. (1) and integrating from C, where $v = 0$, to B, where $v = \frac{1}{2}L$, we have

$$\frac{1}{EI}\int M_2 \frac{\partial M_2}{\partial R_A} dv = \frac{1}{EI}\int_0^{L/2}\left(4R_A v^2 - \frac{3}{2}wLv^2 - wv^3\right) dv$$

$$= \frac{1}{EI}\left(\frac{R_A L^3}{6} - \frac{wL^4}{16} - \frac{wL^4}{64}\right) = \frac{1}{EI}\left(\frac{R_A L^3}{6} - \frac{5wL^4}{64}\right) \qquad (4)$$

Reaction at A. Adding the expressions obtained in (3) and (4), we determine y_A and set it equal to zero

$$y_A = \frac{1}{EI}\left(\frac{R_A L^3}{3} - \frac{wL^4}{8}\right) + \frac{1}{EI}\left(\frac{R_A L^3}{6} - \frac{5wL^4}{64}\right) = 0$$

Solving for R_A, $\qquad R_A = \dfrac{13}{32}wL \qquad\qquad\qquad R_A = \dfrac{13}{32}wL \uparrow$ ◀

Reactions at B and C. Substituting for R_A into Eqs. (2), we obtain

$$R_B = \frac{33}{32}wL \uparrow \qquad R_C = \frac{wL}{16} \uparrow \quad ◀$$

619

PROBLEMS

10.100 through 10.102 Using the information provided in Appendix D, compute the work of the loads as they are applied to the beam (*a*) if the load **P** is applied first, (*b*) if the couple **M$_0$** is applied first.

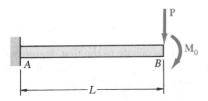

Fig. P10.100

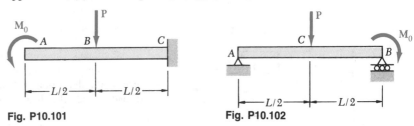

Fig. P10.101

Fig. P10.102

10.103 through 10.105 For the beam and loading shown, (*a*) compute the work of the loads as they are applied successively to the beam, using the information provided in Appendix D, (*b*) compute the strain energy of the beam by the method of Sec. 10.4 and show that it is equal to the work obtained in part *a*.

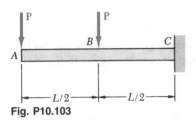

Fig. P10.103

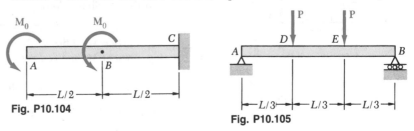

Fig. P10.104

Fig. P10.105

10.106 and 10.107 For the prismatic beam shown, determine the deflection at point *D*.

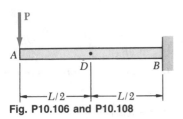

Fig. P10.106 and P10.108

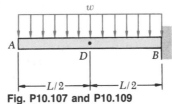

Fig. P10.107 and P10.109

10.108 and 10.109 For the prismatic beam shown, determine the slope at point *D*.
10.110 and 10.111 For the prismatic beam shown, determine the deflection at point *D*.

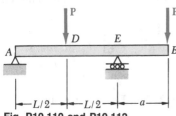

Fig. P10.110 and P10.112

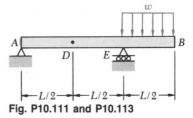

Fig. P10.111 and P10.113

10.112 and 10.113 For the prismatic beam shown, determine the slope at point *D*.

10.114   For the prismatic beam shown, determine the slope at end A.

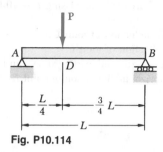

Fig. P10.114

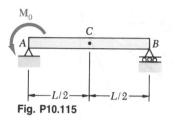

Fig. P10.115

10.115   For the prismatic beam shown, determine (a) the slope at end B, (b) the deflection at the midpoint C.

10.116   For the beam and loading shown, determine (a) the slope at point C, (b) the deflections at point C.   Use $E = 200$ GPa.

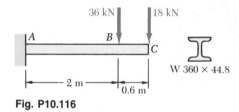

Fig. P10.116

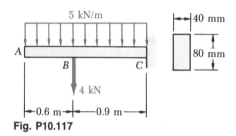

Fig. P10.117

10.117   For the beam and loading shown, determine (a) the slope at point B, (b) the deflection at point B.   Use $E = 200$ GPa.

10.118   For the beam and loading of Prob. 10.44, determine the deflection at point D.

10.119   For the beam and loading of Prob. 10.42, determine the deflection at point A.

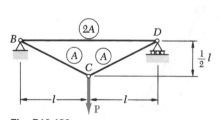

Fig. P10.120

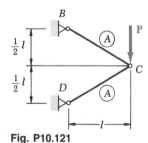

Fig. P10.121

10.120 and 10.121   For the truss and loading shown, determine the horizontal and vertical deflection of joint C.

10.122 and 10.123   Each member of the truss shown is made of steel and has the cross-sectional area shown.   Using $E = 200$ GPa, determine the deflection indicated.

   10.122   Vertical deflection of joint C.
   10.123   Horizontal deflection of joint C.

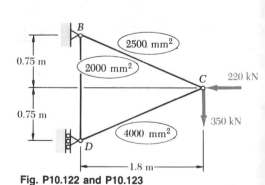

Fig. P10.122 and P10.123

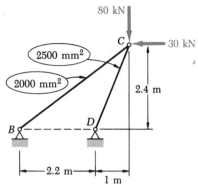

Fig. P10.124 and P10.125

10.124 and 10.125 Each member of the truss shown is made of aluminum and has the cross-sectional area shown. Using $E = 70$ GPa, determine the deflection indicated.

    10.124 Horizontal deflection of joint C.

    10.125 Vertical deflection of joint C.

    *10.126 For the uniform rod and loading shown and using Castigliano's theorem, determine (a) the horizontal deflection of point B, (b) the vertical deflection of point B.

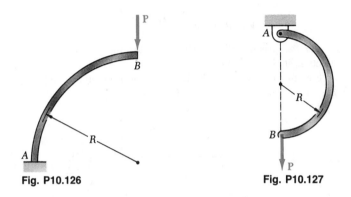

Fig. P10.126   Fig. P10.127

    *10.127 For the uniform rod and loading shown, and using Castigliano's theorem, determine the deflection of point B.

    *10.128 Three rods, each of the same flexural rigidity EI, are welded to form the frame $ABCD$. For the loading shown, determine (a) the deflection of point D, (b) the angle formed by the frame at point D and the vertical.

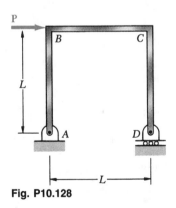

Fig. P10.128

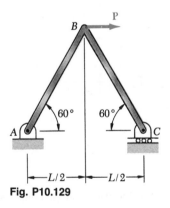

Fig. P10.129

    *10.129 Two rods of the same flexural rigidity EI are welded at B. For the loading shown, determine (a) the horizontal deflection of point B, (b) the vertical deflection of point B.

10.130 through 10.133 Determine the reaction at the roller support and draw the bending-moment diagram for the beam and loading shown.

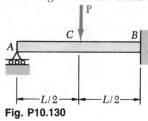

Fig. P10.130

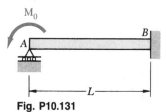

Fig. P10.131

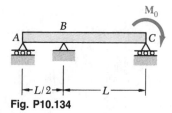

Fig. P10.132

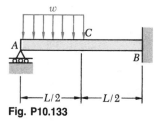

Fig. P10.133

10.134 and 10.135 For the uniform beam and loading shown, determine the reaction at each support.

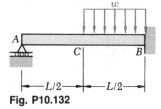

Fig. P10.134

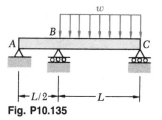

Fig. P10.135

10.136 and 10.137 Three members of the same material and same cross-sectional area are used to support the load **P**. Determine the force in each member when $\phi = 30°$.

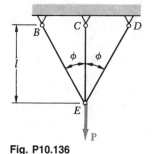

Fig. P10.136

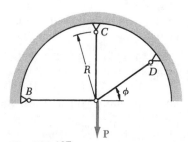

Fig. P10.137

10.138 Several members of the same material and same cross-sectional area are used to support the load **P**. Determine the force in member *CF*.

10.139 Solve Prob. 10.138, assuming that the load **P** is moved along its line of action and applied at point *C*.

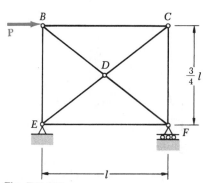

Fig. P10.138

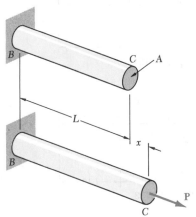

Fig. 10.1

Strain energy

Strain-energy density

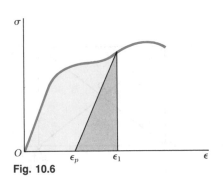

Fig. 10.6

Modulus of toughness

REVIEW AND SUMMARY

This chapter was devoted to the study of strain energy and to the ways in which it may be used to determine the stresses and deformations in structures subjected to both static and impact loadings.

In Sec. 10.2 we considered a uniform rod subjected to a slowly increasing axial load P (Fig. 10.1). We noted that the area under the load-

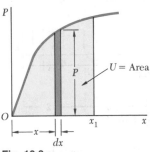

Fig. 10.3

deformation diagram (Fig. 10.3) represents the work done by P. This work is equal to the *strain energy* of the rod associated with the deformation caused by the load P:

$$\text{Strain energy} = U = \int_0^{x_1} P \, dx \qquad (10.2)$$

Since the stress is uniform throughout the rod, we were able to divide the strain energy by the volume of the rod and obtain the strain energy per unit volume, which we defined as the *strain-energy density* of the material [Sec. 10.3]. We found that

$$\text{Strain-energy density} = u = \int_0^{\epsilon_1} \sigma_x \, d\epsilon_x \qquad (10.4)$$

and noted that the strain-energy density is equal to the area under the stress-strain diagram of the material (Fig. 10.6). As we saw in Sec. 10.4, Eq. (10.4) remains valid when the stresses are not uniformly distributed, but the strain-energy density will then vary from point to point. If the material is unloaded, there is a permanent strain ϵ_p and only the strain-energy density corresponding to the triangular area is recovered, the remainder of the energy having been dissipated in the form of heat during the deformation of the material.

The area under the entire stress-strain diagram was defined as the *modulus of toughness* and is a measure of the total energy which can be acquired by the material.

If the normal stress σ remains within the proportional limit of the material, we may express the strain-energy density u as

$$u = \frac{\sigma^2}{2E}$$

The area under the stress-strain curve from zero strain to the strain ϵ_Y at yield (Fig. 10.8) is referred to as the *modulus of resilience* of the material and represents the energy per unit volume that the material may absorb without yielding. We wrote

$$u_Y = \frac{\sigma_Y^2}{2E} \qquad (10.8)$$

In Sec. 10.4 we considered the strain energy associated with *normal stresses*. We saw that if a rod of length L and *variable cross-sectional area A* is subjected at its end to a centric axial load **P**, the strain energy of the rod is

$$U = \int_0^L \frac{P^2}{2AE}\,dx \qquad (10.13)$$

If the rod is of *uniform cross section* of area A, the strain energy is

$$U = \frac{P^2 L}{2AE} \qquad (10.14)$$

We saw that for a beam subjected to transverse loads (Fig. 10.14) the strain energy associated with the normal stresses is

$$U = \int_0^L \frac{M^2}{2EI}\,dx \qquad (10.17)$$

where M is the bending moment and EI the flexural rigidity of the beam.

The strain energy associated with *shearing stresses* was considered in Sec. 10.5. We found that the strain-energy density for a material in pure shear is

$$u = \frac{\tau_{xy}^2}{2G} \qquad (10.19)$$

where τ_{xy} is the shearing stress and G the modulus of rigidity of the material.

For a shaft of length L and uniform cross section subjected at its ends to couples of magnitude T (Fig. 10.18) the strain energy was found to be

$$U = \frac{T^2 L}{2GJ} \qquad (10.22)$$

where J is the polar moment of inertia of the cross-sectional area of the shaft.

Modulus of resilience

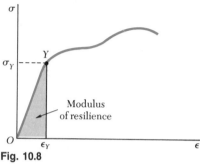

Fig. 10.8

Strain energy under axial load

Strain energy due to bending

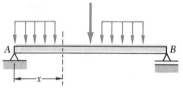

Fig. 10.14

Strain energy due to shearing stresses

Strain energy due to torsion

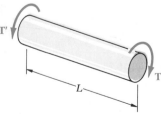

Fig. 10.18

General state of stress

In Sec. 10.6 we considered the strain energy of an elastic isotropic material under a general state of stress and expressed the strain-energy density at a given point in terms of the principal stresses σ_a, σ_b, and σ_c at that point:

$$u = \frac{1}{2E}[\sigma_a^2 + \sigma_b^2 + \sigma_c^2 - 2\nu(\sigma_a\sigma_b + \sigma_b\sigma_c + \sigma_c\sigma_a)] \qquad (10.27)$$

The strain-energy density at a given point was divided into two parts: u_v, associated with a change in volume of the material at that point, and u_d, associated with a distortion of the material at the same point. We wrote $u = u_v + u_d$, where

$$u_v = \frac{1 - 2\nu}{6E}(\sigma_a + \sigma_b + \sigma_c)^2 \qquad (10.32)$$

and

$$u_d = \frac{1}{12G}[(\sigma_a - \sigma_b)^2 + (\sigma_b - \sigma_c)^2 + (\sigma_c - \sigma_a)^2] \qquad (10.33)$$

Using the expression obtained for u_d, we derived the maximum-distortion-energy criterion, which was used in Sec. 6.7 to predict whether a ductile material would yield under a given state of plane stress.

Impact loading

Equivalent static load

In Sec. 10.7 we considered the *impact loading* of an elastic structure being hit by a mass moving with a given velocity. We assumed that the kinetic energy of the mass is transferred entirely to the structure and defined the *equivalent static load* as the load which would cause the same deformations and stresses as are caused by the impact loading.

After discussing several examples, we noted that a structure designed to withstand effectively an impact load should be shaped in such a way that stresses are evenly distributed throughout the structure, and that the material used should have a low modulus of elasticity and a high yield strength [Sec. 10.8].

Members subjected to a single load

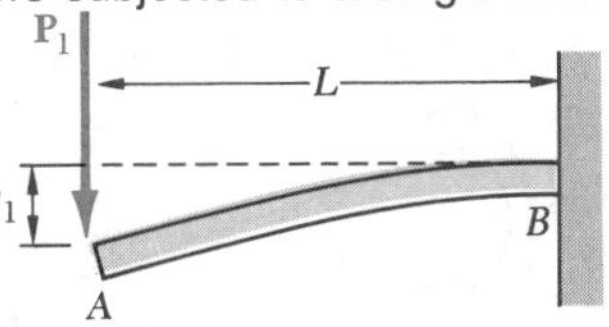

Fig. 10.25

The strain energy of structural members subjected to a *single load* was considered in Sec. 10.9. In the case of the beam and loading of Fig. 10.25 we found that the strain energy of the beam is

$$U = \frac{P_1^2 L^3}{6EI} \qquad (10.46)$$

Observing that the work done by the load $\mathbf{P}$ is equal to $\frac{1}{2}P_1 y_1$, we equated the work of the load and the strain energy of the beam and determined the deflection y_1 at the point of application of the load [Sec. 10.10 and Example 10.10].

The method just described is of limited value, since it is restricted to structures subjected to a single concentrated load and to the determination of the deflection at the point of application of that load. In the remaining sections of the chapter, we presented a more general method, which may be used to determine deflections at various points of structures subjected to several loads.

In Sec. 10.11 we discussed the strain energy of a structure subjected to several loads, and in Sec. 10.12 introduced *Castigliano's theorem*, which states that the deflection x_j, of the point of application of a load $\mathbf{P}_j$ measured along the line of action of $\mathbf{P}_j$ is equal to the partial derivative of the strain energy of the structure with respect to the load $\mathbf{P}_j$. We wrote

Castigliano's theorem

$$x_j = \frac{\partial U}{\partial P_j} \tag{10.65}$$

We also found that we could use Castigliano's theorem to determine the *slope* of a beam at the point of application of a couple $\mathbf{M}_j$ by writing

$$\theta_j = \frac{\partial U}{\partial M_j} \tag{10.68}$$

and the *angle of twist* in a section of a shaft where a torque $\mathbf{T}_j$ is applied by writing

$$\phi_j = \frac{\partial U}{\partial T_j} \tag{10.69}$$

In Sec. 10.13, Castigliano's theorem was applied to the determination of deflections and slopes at various points of a given structure. The use of "dummy" loads enabled us to include points where no actual load was applied. We also observed that the calculation of a deflection x_j was simplified if the differentiation with respect to the load P_j was carried out before the integration. In the case of a beam, recalling Eq. (10.17), we wrote

$$x_j = \frac{\partial U}{\partial P_j} = \int_0^L \frac{M}{EI} \frac{\partial M}{\partial P_j} \, dx \tag{10.70}$$

Similarly, for a truss consisting of n members, the deflection x_j at the point of application of the load $\mathbf{P}_j$ was found by writing

$$x_j = \frac{\partial U}{\partial P_j} = \sum_{i=1}^{n} \frac{F_i L_i}{A_i E} \frac{\partial F_i}{\partial P_j} \tag{10.72}$$

The chapter concluded [Sec. 10.14] with the application of Castigliano's theorem to the analysis of *statically indeterminate structures* [Sample Prob. 10.7, Examples 10.15 and 10.16].

Indeterminate structures

REVIEW PROBLEMS

10.140 A 70 kg diver jumps from a height of 600 mm onto end A of a diving board having the uniform cross section shown. Assuming the diver's legs remain rigid and using $E = 12$ GPa for the board, determine (a) the maximum deflection of point A, (b) the maximum bending stress in the board, (c) the equivalent static load.

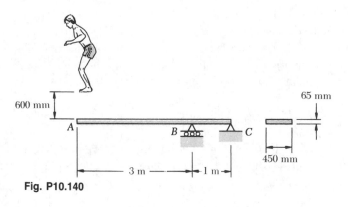

Fig. P10.140

10.141 Solve Prob. 10.140, assuming that portion AB of the diving board is 3.15 m long and that portion BC is 0.85 m long.

10.142 The steel bar ABC has a cross section of 20×20 mm, and is subjected to a load $P = 250$ N. Using $E = 200$ GPa, determine the vertical deflection of point C.

10.143 For the bar and loading of Prob. 10.142, determine the horizontal deflection of point C.

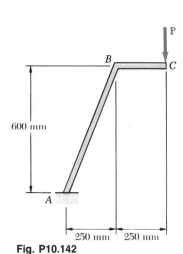

Fig. P10.142

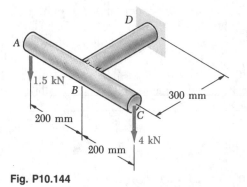

Fig. P10.144

10.144 Two solid steel rods, each of 56-mm diameter, are welded together to form the bracket shown. Using $E = 200$ GPa and $G = 77$ GPa, determine the vertical deflection of point C.

10.145 For the bracket and loading of Prob. 10.144, determine the vertical deflection of point A.

10.146 Two cover plates are welded to a rolled-steel beam as shown. Using $E = 200$ GPa, determine the slope and deflection at point A.

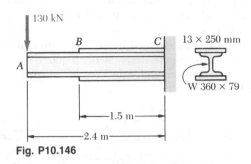

130 kN

B C 13×250 mm

A

W 360 × 79

1.5 m

2.4 m

Fig. P10.146

10.147 Solve Prob. 10.146, assuming that the length BC of the cover plates is reduced to 1.0 m.

10.148 Bars AB and CD are made of steel and have the cross-sectional areas shown. Assuming that lever BCE is rigid and using $E = 200$ GPa, determine the deflection of point E.

10.149 Bars AB and CD have cross-sectional areas as shown, and the lever BCE has a square cross section of side 24 mm. Knowing that all parts are made of steel and using $E = 200$ GPa, determine the deflection of point E.

10.150 The assembly ABC is made of a steel for which $E = 200$ GPa and $\sigma_Y = 320$ MPa. Knowing that a strain energy of 5 J must be acquired by the assembly as the axial load $\mathbf{P}$ is applied, determine the factor of safety with respect to permanent deformation when (a) $x = 300$ mm, (b) $x = 600$ mm.

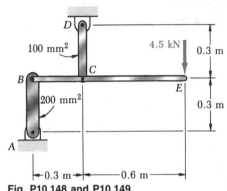

D

100 mm²   4.5 kN   0.3 m

B C E

200 mm²   0.3 m

A

0.3 m   0.6 m

Fig. P10.148 and P10.149

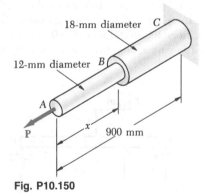

18-mm diameter C

12-mm diameter B

A

P x 900 mm

Fig. P10.150

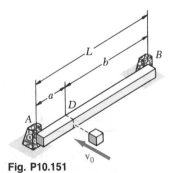

L

b B

a

A D

v_0

Fig. P10.151

10.151 The simply supported beam AB is struck squarely at D by a block of mass m moving horizontally with a velocity $\mathbf{v}_0$. Show that the resulting maximum normal stress σ_m in the beam due to bending is independent of the location of point D.

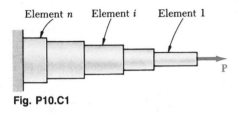

Fig. P10.C1

The following problems are designed to be solved with a computer.

10.C1 A rod consisting of n elements, each of which is homogeneous and of uniform cross section, is subjected to a load **P** applied at its free end. The length of element i is denoted by L_i, and its diameter by d_i. (a) Denoting by E the modulus of elasticity of the material used in the rod, write a computer program which can be used to determine the strain energy acquired by the rod and the deformation measured at the free end. (b) Use this program to solve Probs. 2.15a, 10.9a, and 10.10a.

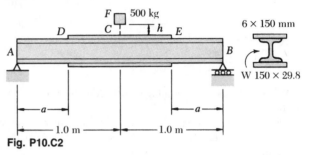

Fig. P10.C2

10.C2 Two 6×150-mm cover plates are welded to a W 150×29.8 rolled-steel beam as shown. The 500-kg block F is to be dropped from a height $h = 30$ mm onto the beam. (a) Write a computer program and use it to calculate the maximum normal stress on transverse sections just to the left of D and at the center of the beam for values of a from 0 to 1 m at 100 mm intervals. (b) From the values considered in part a, select the distance a for which the maximum normal stress is as small as possible. Use $E = 200$ GPa.

10.C3 The 12-kg block D is dropped from a height h onto the free end of the steel bar AB. For the steel used, $\sigma_{all} = 105$ MPa and $E = 200$ GPa. (a) Write a computer program and use it to calculate the maximum allowable height h for values of the length L from 100 to 900 mm at 100-mm intervals. (b) From the values of L considered in part a, select the length corresponding to the largest allowable height.

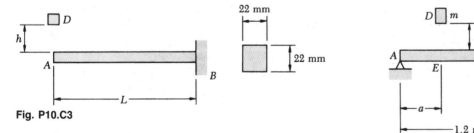

Fig. P10.C3

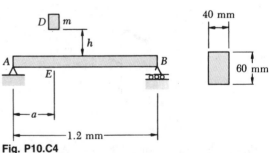

Fig. P10.C4

10.C4 The block D of mass $m = 10$ kg is dropped from a height $h = 450$ mn. onto the aluminum beam AB. Knowing that $E = 70$ GPa, write a computer program and use it to calculate the maximum deflection at point E and the maximum normal stress in the beam for values of a from 100 to 600 mm at 100-mm intervals.

CHAPTER ELEVEN

COLUMNS

11.1. INTRODUCTION

In the preceding chapters, we had two primary concerns: (1) the strength of the structure, i.e., its ability to support a specified load without experiencing excessive stress; (2) the ability of the structure to support a specified load without undergoing unacceptable deformations. In this chapter, we shall be concerned with the *stability* of the structure, i.e., its ability to support a given load without experiencing a sudden change in its configuration. Our discussion will relate chiefly to columns, i.e., to the analysis and design of vertical prismatic members supporting axial loads.

Before considering the stability of columns, we shall analyze in Sec. 11.2 the stability of a simplified model consisting of two rigid rods connected by a pin and a spring and supporting a load **P**. We shall note that if its equilibrium is disturbed, the system will return to its original equilibrium position as long as P does not exceed a certain value P_{cr}, called the *critical load*. However, if $P > P_{cr}$, the system will move away from its original position and settle in a new position of equilibrium. In the first case, the system is said to be *stable*, and in the second case, it is said to be *unstable*.

In Sec. 11.3, we shall begin our study of the *stability of elastic columns* by considering a pin-ended column subjected to a centric axial load. *Euler's formula* for the critical load of the column will be derived and from that formula we shall determine the corresponding critical normal stress in the column. By applying a factor of safety to the critical load, we shall be able to determine the allowable load which may be applied to a pin-ended column.

631

In Sec. 11.4, we shall expand our study to the analysis of the stability of columns with different end conditions. We shall simplify these analyses by learning how to determine the *effective length* of a column, i.e., the length of a pin-ended column having the same critical load.

In Sec. 11.5, we shall consider columns supporting eccentric axial loads; these column have transverse deflections for all magnitudes of the load. We shall derive an expression for the maximum deflection under a given load and use it to determine the maximum normal stress in the column. Finally, the *secant formula* which relates the average and maximum stresses in a column will be developed.

In the first sections of the chapter, each column is initially assumed to be a straight homogeneous prism. In the last part of the chapter, we shall consider real columns which are designed and analyzed using empirical formulas set forth by professional organizations. In Sec. 11.6, formulas will be presented for the allowable stress in columns made of steel, aluminum, or timber and subjected to a centric axial load. In the last section of the chapter (Sec. 11.7), we shall consider the design of columns under an eccentric axial load.

11.2. STABILITY OF STRUCTURES

Suppose we are to design a column *AB* of length *L* to support a given load **P** (Fig. 11.1). The column will be pin-connected at both ends and we shall assume that **P** is a centric axial load. If the cross-sectional area *A* of the column is selected so that the value $\sigma = P/A$ of the stress on a transverse section is less than the allowable stress σ_{all} for the material used, and if the deformation $\delta = PL/AE$ falls within the given specifications, we might conclude that the column has been properly designed. However, it may happen that, as the load is applied, the column will *buckle;* instead of remaining straight, it will suddenly become sharply curved (Fig. 11.2). Clearly, a column which buckles under the specified load is not properly designed.

Before getting into the actual discussion of the stability of elastic columns, we shall try to get some insight on the problem by considering a simplified model consisting of two rigid rods *AC* and *BC* connected at *C* by a pin and a torsional spring of constant *K* (Fig. 11.3).

If the two rods and the two forces **P** and **P'** are perfectly aligned, the system will remain in the position of equilibrium shown in Fig. 11.4a as long as it is not disturbed. But suppose that we move *C* slightly to the right, so that each rod now forms a small angle $\Delta\theta$ with the vertical (Fig. 11.4b). Will the system return to its original equilibrium position, or will it move further away from that position? In the first case, the system is said to be *stable,* and in the second case, it is said to be *unstable*.

To determine whether the two-rod system is stable or unstable, we shall consider the forces acting on rod *AC* (Fig. 11.5). These forces con-

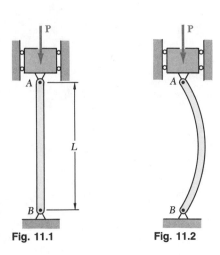

Fig. 11.1   **Fig. 11.2**

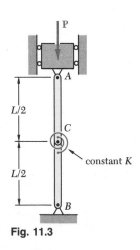

Fig. 11.3

sist of two couples, namely the couple formed by **P** and **P′**, of moment $P(L/2) \sin \Delta\theta$, which tends to move the rod away from the vertical, and the couple **M** exerted by the spring, which tends to bring the rod back into its original vertical position. Since the angle of deflection of the spring is $2\,\Delta\theta$, the moment of the couple **M** is $M = K(2\,\Delta\theta)$. If the moment of the second couple is larger than the moment of the first couple, the system tends to return to its original equilibrium position; the system is stable. If the moment of the first couple is larger than the moment of the second couple, the system tends to move away from its original equilibrium position; the system is unstable. The value of the load for which the two couples balance each other is called the *critical load* and is denoted by P_{cr}. We have

$$P_{cr}(L/2) \sin \Delta\theta = K(2\,\Delta\theta) \tag{11.1}$$

or, since $\sin \Delta\theta \approx \Delta\theta$,

$$P_{cr} = 4K/L \tag{11.2}$$

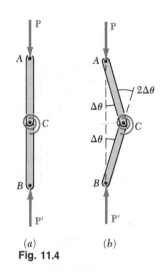

(a)   (b)

Fig. 11.4

Clearly, the system is stable for $P < P_{cr}$, that is, for values of the load smaller than the critical value, and unstable for $P > P_{cr}$.

Let us assume that a load $P > P_{cr}$ has been applied to the two rods of Fig. 11.3 and that the system has been disturbed. Since $P > P_{cr}$, the system will move further away from the vertical and, after some oscillations, will settle into a new equilibrium position (Fig. 11.6a). Considering the equilibrium of the free body AC (Fig. 11.6b), we obtain an equation similar to Eq. (11.1), but involving the finite angle θ, namely

$$P(L/2) \sin \theta = K(2\theta)$$

or

$$\frac{PL}{4K} = \frac{\theta}{\sin \theta} \tag{11.3}$$

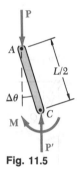

Fig. 11.5

The value of θ corresponding to the equilibrium position represented in Fig. 11.6 is obtained by solving Eq. (11.3) by trial and error. But we observe that, for any positive value of θ, we have $\sin \theta < \theta$. Thus, Eq. (11.3) yields a value of θ different from zero only when the left-hand member of the equation is larger than one. Recalling Eq. (11.2), we note that this is indeed the case here, since we have assumed $P > P_{cr}$. But, if we had assumed $P < P_{cr}$, the second equilibrium position shown in Fig. 11.6 would not exist and the only possible equilibrium position would be the position corresponding to $\theta = 0$. We thus check that, for $P < P_{cr}$, the position $\theta = 0$ must be stable.

This observation applies to structures and mechanical systems in general, and will be used in the next section, where we shall discuss the stability of elastic columns.

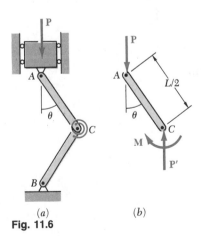

(a)   (b)

Fig. 11.6

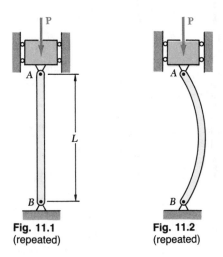

Fig. 11.1
(repeated)

Fig. 11.2
(repeated)

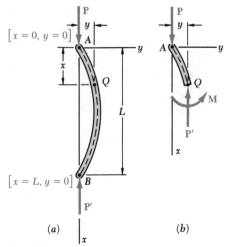

Fig. 11.7

11.3. EULER'S FORMULA FOR PIN-ENDED COLUMNS

Returning to the column AB considered in the preceding section (Fig. 11.1), we propose to determine the critical value of the load $\mathbf{P}$, i.e., the value P_{cr} of the load for which the position shown in Fig. 11.1 ceases to be stable. If $P > P_{cr}$, the slightest misalignment or disturbance will cause the column to buckle, i.e., to assume a curved shape as shown in Fig. 11.2.

Our approach will be to determine the conditions under which the configuration of Fig. 11.2 is possible. Since a column may be considered as a beam placed in a vertical position and subjected to an axial load, we shall proceed as in Chap. 8 and denote by x the distance from end A of the column to a given point Q of its elastic curve, and by y the deflection of that point (Fig. 11.7a). It follows that the x axis will be vertical and directed downward, and the y axis horizontal and directed to the right. Considering the equilibrium of the free body AQ (Fig. 11.7b), we find that the bending moment at Q is $M = -Py$. Substituting this value for M in Eq. (8.4) of Sec. 8.3, we write

$$\frac{d^2y}{dx^2} = \frac{M}{EI} = -\frac{P}{EI}y \tag{11.4}$$

or, transposing the last term,

$$\frac{d^2y}{dx^2} + \frac{P}{EI}y = 0 \tag{11.5}$$

This equation is a linear, homogeneous differential equation of the second order with constant coefficients. Setting

$$p^2 = \frac{P}{EI} \tag{11.6}$$

we may write Eq. (11.5) in the form

$$\frac{d^2y}{dx^2} + p^2y = 0 \tag{11.7}$$

which is the same as that of the differential equation for simple harmonic motion except, of course, that the independent variable is now the distance x instead of the time t. The general solution of Eq. (11.7) is

$$y = A \sin px + B \cos px \tag{11.8}$$

as we may easily check by computing d^2y/dx^2 and substituting for y and d^2y/dx^2 into (11.7).

Recalling the boundary conditions which must be satisfied at ends A and B of the column (Fig. 11.7a), we first make $x = 0$, $y = 0$ in Eq. (11.8) and find that $B = 0$. Substituting next $x = L$, $y = 0$, we obtain

$$A \sin pL = 0 \tag{11.9}$$

This equation is satisfied either if $A = 0$, or if $\sin pL = 0$. If the first of these conditions is satisfied, Eq. (11.8) reduces to $y = 0$ and the column is straight (Fig. 11.1). For the second condition to be satisfied, we must have $pL = n\pi$ or, substituting for p from (11.6) and solving for P,

$$P = \frac{n^2\pi^2 EI}{L^2} \tag{11.10}$$

The smallest of the values of P defined by Eq. (11.10) is that corresponding to $n = 1$. We thus have

$$P_{cr} = \frac{\pi^2 EI}{L^2} \tag{11.11}$$

The expression obtained is known as *Euler's formula,* after the Swiss mathematician Leonhard Euler (1707–1783). Substituting this expression for P in Eq. (11.6) and the value obtained for p into Eq. (11.8), and recalling that $B = 0$, we write

$$y = A \sin\frac{\pi x}{L} \tag{11.12}$$

which is the equation of the elastic curve after the column has buckled (Fig. 11.2). We note that the value of the maximum deflection, $y_m = A$, is indeterminate. This is due to the fact that the differential equation (11.5) is a linearized approximation of the actual governing differential equation for the elastic curve.†

If $P < P_{cr}$, the condition $\sin pL = 0$ cannot be satisfied, and the solution given by Eq. (11.12) does not exist. We must then have $A = 0$, and the only possible configuration for the column is a straight one. Thus, for $P < P_{cr}$ the straight configuration of Fig. 11.1 is stable.

In the case of a column with a circular or square cross section, the moment of inertia I of the cross section is the same about any centroidal axis, and the column is as likely to buckle in one plane as another, except for the restraints which may be imposed by the end connections. For other shapes of cross section, the critical load should be computed by making $I = I_{min}$ in Eq. (11.11); if buckling occurs, it will take place in a plane perpendicular to the corresponding principal axis of inertia.

The value of the stress corresponding to the critical load is called the *critical stress* and is denoted by σ_{cr}. Recalling Eq. (11.11) and setting $I = Ar^2$, where A is the cross-sectional area and r its radius of gyration, we have

$$\sigma_{cr} = \frac{P_{cr}}{A} = \frac{\pi^2 EAr^2}{AL^2}$$

† We recall that the equation $d^2y/dx^2 = M/EI$ was obtained in Sec. 8.3 by assuming that the slope dy/dx of the beam could be neglected and that the exact expression given in Eq. (8.3) for the curvature of the beam could be replaced by $1/\rho = d^2y/dx^2$.

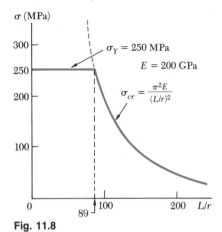

Fig. 11.8

or

$$\sigma_{cr} = \frac{\pi^2 E}{(L/r)^2} \tag{11.13}$$

The quantity L/r is called the *slenderness ratio* of the column. It is clear, in view of the remark of the preceding paragraph, that the minimum value of the radius of gyration r should be used in computing the slenderness ratio and the critical stress in a column.

Equation (11.13) shows that the critical stress is proportional to the modulus of elasticity of the material, and inversely proportional to the square of the slenderness ratio of the column. The plot of σ_{cr} versus L/r is shown in Fig. 11.8 for structural steel, assuming $E = 200$ GPa and $\sigma_Y = 250$ MPa. We should keep in mind that no factor of safety has been used in plotting σ_{cr}. We also note that, if the value obtained for σ_{cr} from Eq. (11.13) or from the curve of Fig. 11.8 is larger than the yield strength σ_Y, this value is of no interest to us, since the column will yield in compression and cease to be elastic before it has a chance to buckle.

Our analysis of the behavior of a column has been based so far on the assumption of a perfectly aligned centric load. In practice, this is seldom the case, and in Sec. 11.5 we shall take into account the effect of the eccentricity of the loading. This approach will lead to a smoother transition from the buckling failure of long, slender columns to the compression failure of short, stubby columns. It will also provide us with a more realistic view of the relation between the slenderness ratio of a column and the load which causes it to fail.

Example 11.01

A 2-m-long pin-ended column of square cross section is to be made of Douglas fir. Assuming $E = 13$ GPa, $\sigma_{all} = 12$ MPa for compression parallel to the grain, and using a factor of safety of 2.5 in computing Euler's critical load for buckling, determine the size of the cross section if the column is to safely support (*a*) a 100-kN load, (*b*) a 200-kN load.

(*a*) *For the 100-kN Load.* Using the given factor of safety, we make

$$P_{cr} = 2.5(100 \text{ kN}) = 250 \text{ kN} \qquad L = 2 \text{ m} \qquad E = 13 \text{ GPa}$$

in Euler's formula (11.11) and solve for I. We have

$$I = \frac{P_{cr}L^2}{\pi^2 E} = \frac{(250 \times 10^3 \text{ N})(2 \text{ m})^2}{\pi^2(13 \times 10^9 \text{ Pa})} = 7.794 \times 10^{-6} \text{ m}^4$$

Recalling that, for a square of side a, we have $I = a^4/12$, we write

$$\frac{a^4}{12} = 7.794 \times 10^{-6} \text{ m}^4 \qquad a = 98.3 \text{ mm} \approx 100 \text{ mm}$$

We check the value of the normal stress in the column:

$$\sigma = \frac{P}{A} = \frac{100 \text{ kN}}{(0.100 \text{ m})^2} = 10 \text{ MPa}$$

Since σ is smaller than the allowable stress, a 100×100-mm cross section is acceptable.

(*b*) *For the 200-kN Load.* Solving again Eq. (11.11) for I, but making now $P_{cr} = 2.5(200) = 500$ kN, we have

$$I = 15.588 \times 10^{-6} \text{ m}^4, \quad \frac{a^4}{12} = 15.588 \times 10^{-6}, \quad a = 116.95 \text{ mm}$$

The value of the normal stress is

$$\sigma = \frac{P}{A} = \frac{200 \text{ kN}}{(0.11695 \text{ m})^2} = 14.62 \text{ MPa}$$

Since this value is larger than the allowable stress, the dimension obtained is not acceptable, and we must select the cross section on the basis of its resistance to compression. We write

$$A = \frac{P}{\sigma_{all}} = \frac{200 \text{ kN}}{12 \text{ MPa}} = 16.67 \times 10^{-3} \text{ m}^2$$

$$a^2 = 16.67 \times 10^{-3} \text{ m}^2 \qquad a = 129.1 \text{ mm}$$

A 130×130-mm cross section is acceptable.

11.4. EXTENSION OF EULER'S FORMULA TO COLUMNS WITH OTHER END CONDITIONS

Euler's formula (11.11) was derived in the preceding section for a column which was pin-connected at both ends. We shall now see how the critical load P_{cr} may be determined for columns with different end conditions.

In the case of a column with one free end A supporting a load $\mathbf{P}$ and one fixed end B (Fig. 11.9a), we may observe that the column will behave as the upper half of a pin-connected column (Fig. 11.9b). The critical

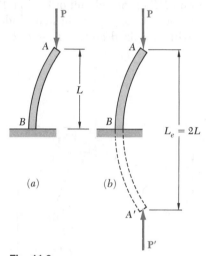

Fig. 11.9

load for the column of Fig. 11.9a is thus the same as for the pin-ended column of Fig. 11.9b and may be obtained from Euler's formula (11.11) by using a column length equal to twice the actual length L of the given column. We say that the *effective length* L_e of the column of Fig. 11.9 is equal to $2L$ and substitute $L_e = 2L$ in Euler's formula:

$$P_{cr} = \frac{\pi^2 EI}{L_e^2} \tag{11.11'}$$

The critical stress is found in a similar way from the formula

$$\sigma_{cr} = \frac{\pi^2 E}{(L_e/r)^2} \tag{11.13'}$$

The quantity L_e/r is referred to as the *effective slenderness ratio* of the column and, in the case considered here, is equal to $2L/r$.

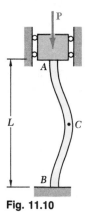

Fig. 11.10

Consider next a column with two fixed ends A and B supporting a load **P** (Fig. 11.10). The symmetry of the supports and of the loading about a horizontal axis through the midpoint C requires that the shear at C and the horizontal components of the reactions at A and B be zero (Fig. 11.11). It follows that the restraints imposed upon the upper half AC of the column by the support at A and by the lower half CB are identical (Fig. 11.12). Portion AC must thus be symmetric about its midpoint D, and this point must be a point of inflection, where the bending moment is zero. A similar reasoning shows that the bending moment at the midpoint E of the lower half of the column must also be zero (Fig. 11.13a). Since the bending moment at the ends of a pin-ended column is zero, it follows that the portion DE of the column of Fig. 11.13a must behave as a pin-ended column (Fig. 11.13b). We thus conclude that the effective length of a column with two fixed ends is $L_e = L/2$.

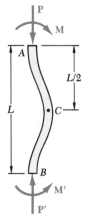

Fig. 11.11

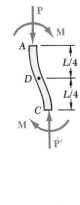

Fig. 11.12

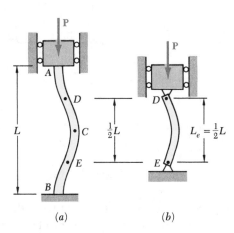

(a)   (b)

Fig. 11.13

In the case of a column with one fixed end B and one pin-connected end A supporting a load **P** (Fig. 11.14), we must write and solve the differential equation of the elastic curve to determine the effective length of the column. From the free-body diagram of the entire column (Fig. 11.15), we first note that a transverse force **V** is exerted at end A, in addition to the axial load **P**, and that **V** is statically indeterminate. Considering now the free-body diagram of a portion AQ of the column (Fig. 11.16), we find that the bending moment at Q is

$$M = -Py - Vx$$

Substituting this value into Eq. (8.4) of Sec. 8.3, we write

$$\frac{d^2y}{dx^2} = \frac{M}{EI} = -\frac{P}{EI}y - \frac{V}{EI}x$$

Transposing the term containing y and setting

$$p^2 = \frac{P}{EI} \qquad (11.6)$$

as we did in Sec. 11.3, we write

$$\frac{d^2y}{dx^2} + p^2y = -\frac{V}{EI}x \qquad (11.14)$$

This equation is a linear, nonhomogeneous differential equation of the second order with constant coefficients. Observing that the left-hand member of Eqs. (11.7) and (11.14) are identical, we conclude that the general solution of Eq. (11.14) may be obtained by adding a particular solution of Eq. (11.14) to the solution (11.8) obtained for Eq. (11.7). Such a particular solution is easily seen to be

$$y = -\frac{V}{p^2EI}x$$

or, recalling (11.6),

$$y = -\frac{V}{P}x \qquad (11.15)$$

Adding the solutions (11.8) and (11.15), we write the general solution of Eq. (11.14) as

$$y = A \sin px + B \cos px - \frac{V}{P}x \qquad (11.16)$$

The constants A and B, and the magnitude V of the unknown transverse force **V** are obtained from the boundary conditions indicated in Fig. 11.15. Making first $x = 0$, $y = 0$ in Eq. (11.16), we find that $B = 0$. Making next $x = L$, $y = 0$, we obtain

$$A \sin pL = \frac{V}{P}L \qquad (11.17)$$

Finally, computing

$$\frac{dy}{dx} = Ap \cos px - \frac{V}{P}$$

and making $x = L$, $dy/dx = 0$, we have

$$Ap \cos pL = \frac{V}{P} \qquad (11.18)$$

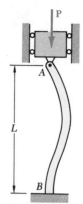

Fig. 11.14

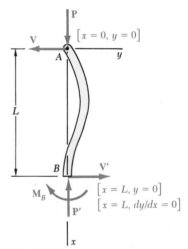

Fig. 11.15

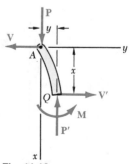

Fig. 11.16

Dividing (11.17) by (11.18) member by member, we conclude that a solution of the form (11.16) can exist only if

$$\tan pL = pL \tag{11.19}$$

Solving this equation by trial and error, we find that the smallest value of pL which satisfies (11.19) is

$$pL = 4.4934 \tag{11.20}$$

Carrying the value of p defined by Eq. (11.20) into Eq. (11.6) and solving for P, we obtain the critical load for the column of Fig. 11.14:

$$P_{cr} = \frac{20.19EI}{L^2} \tag{11.21}$$

The effective length of the column is obtained by equating the right-hand members of Eqs. (11.11′) and (11.21):

$$\frac{\pi^2 EI}{L_e^2} = \frac{20.19EI}{L^2}$$

Solving for L_e, we find that the effective length of a column with one fixed end and one pin-connected end is $L_e = 0.699L \approx 0.7L$.

The effective lengths corresponding to the various end conditions considered in this section are shown in Fig. 11.17.

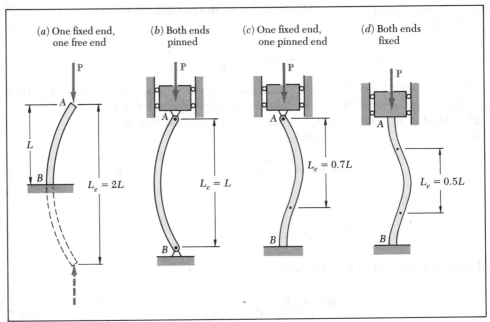

Fig. 11.17 Effective length of column for various end conditions.

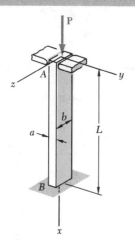

SAMPLE PROBLEM 11.1

An aluminum column of length L and rectangular cross section has a fixed end B and supports a centric load at A. Two smooth and rounded fixed plates restrain end A from moving in one of the vertical planes of symmetry of the column, but allow it to move in the other plane. (a) Determine the ratio a/b of the two sides of the cross section corresponding to the most efficient design against buckling. (b) Design the most efficient cross section for the column, knowing that $L = 500$ mm, $E = 70$ GPa, $P = 20$ kN, and that a factor of safety of 2.5 is required.

Buckling in xy Plane. Referring to Fig. 11.17, we note that the effective length of the column with respect to buckling in this plane is $L_e = 0.7L$. The radius of gyration r_z of the cross section is obtained by writing

$$I_z = \tfrac{1}{12}ba^3 \qquad A = ab$$

and, since $I_z = Ar_z^2$, $\qquad r_z^2 = \dfrac{I_z}{A} = \dfrac{\frac{1}{12}ba^3}{ab} = \dfrac{a^2}{12} \qquad r_z = a/\sqrt{12}$

The effective slenderness ratio of the column with respect to buckling in the xy plane is

$$\frac{L_e}{r_z} = \frac{0.7L}{a/\sqrt{12}} \tag{1}$$

Buckling in xz Plane. The effective length of the column with respect to buckling in this plane is $L_e = 2L$, and the corresponding radius of gyration is $r_y = b/\sqrt{12}$. Thus,

$$\frac{L_e}{r_y} = \frac{2L}{b/\sqrt{12}} \tag{2}$$

(a) Most Efficient Design. The most efficient design is that for which the critical stresses corresponding to the two possible modes of buckling are equal. Referring to Eq. (11.13′), we note that this will be the case if the two values obtained above for the effective slenderness ratio are equal. We write

$$\frac{0.7L}{a/\sqrt{12}} = \frac{2L}{b/\sqrt{12}}$$

and, solving for the ratio a/b, $\qquad \dfrac{a}{b} = \dfrac{0.7}{2} \qquad\qquad \dfrac{a}{b} = 0.35 \blacktriangleleft$

(b) Design for Given Data. Since $F.S. = 2.5$ is required,

$$P_{cr} = (F.S.)P = (2.5)(20 \text{ kN}) = 50 \text{ kN}$$

Using $a = 0.35b$, we have $A = ab = 0.35b^2$ and

$$\sigma_{cr} = \frac{P_{cr}}{A} = \frac{50 \times 10^3 \text{ N}}{0.35b^2}$$

Making $L = 0.500$ m in Eq. (2), we have $L_e/r_y = 3.464/b$. Substituting for L_e/r and σ_{cr} into Eq. (11.13′), we write

$$\sigma_{cr} = \frac{\pi^2 E}{(L_e/r)^2} \qquad \frac{50 \times 10^3 \text{ N}}{0.35b^2} = \frac{\pi^2(70 \times 10^9 \text{ Pa})}{(3.464/b)^2}$$

$$b = 39.7 \text{ mm} \qquad a = 0.35b = 13.9 \text{ mm} \blacktriangleleft$$

PROBLEMS

11.1   Knowing that the spring at A is of constant k and that the bar AB is rigid, determine the critical load P_{cr}.

11.2   Knowing that the torsional spring at B is of constant K and that the bar AB is rigid, determine the critical load P_{cr}.

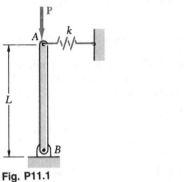

Fig. P11.1

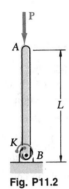

Fig. P11.2

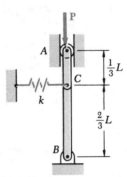

Fig. P11.3

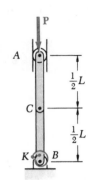

Fig. P11.4

11.3   Two rigid bars AC and BC are connected as shown to a spring of constant k.   Knowing that the spring can act in either tension or compression, determine the critical load P_{cr} for the system.

11.4   Two rigid bars AC and BC are connected by a pin at C as shown. Knowing that the torsional spring at B is of constant K, determine the critical load P_{cr} for the system.

11.5   The rigid bar AD is attached to two springs of constant k and is in equilibrium in the position shown.   Knowing that the equal and opposite loads P and P' *remain vertical*, determine the magnitude P_{cr} of the critical loading for the system.   Each spring can act in either tension or compression.

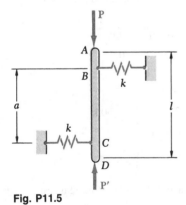

Fig. P11.5

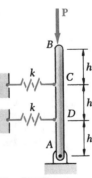

Fig. P11.6

11.6   The rigid rod AB is attached to a hinge at A and to two springs, each of constant $k = 250$ N/mm.   If $h = 0.75$ m, determine the critical load.   Each spring can act in either tension or compression.

11.7 If $m = 150$ kg, $h = 500$ mm, and the constant of each spring is $k = 2.5$ kN/m, determine the range of values of the distance d for which the equilibrium of the rigid rod AB is stable in the position shown. Each spring can act in either tension or compression.

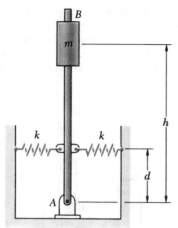

Fig. P11.7

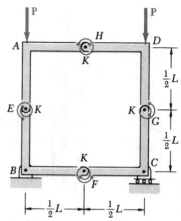

Fig. P11.8

11.8 A frame consists of four L-shaped members connected by four torsional springs each of constant K. Knowing that equal forces **P** are applied at points A and D as shown, determine the critical load P_{cr} for the system.

11.9 The steel rod BC is attached to the rigid bar AB and to the fixed support at C. Knowing that $G = 77$ GPa, determine the diameter d of rod BC for which the critical load P_{cr} of the system is 350 N.

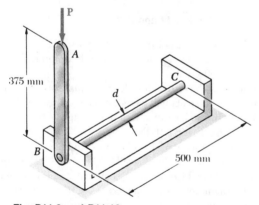

Fig. P11.9 and P11.10

11.10 The steel rod BC is attached to the rigid bar AB and to the fixed support at C. Knowing that $G = 77$ GPa, determine the critical load P_{cr} for the system when $d = 12$ mm.

11.11 Determine the critical load of a wooden stick which is 0.9 m long and has a 5×30 mm rectangular cross section. Use $E = 12$ GPa.

11.12 Determine the critical load of a wooden meter stick which has a 4×28-mm rectangular cross section. Use $E = 12$ GPa.

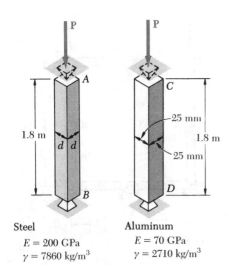

Steel
$E = 200$ GPa
$\gamma = 7860$ kg/m^3

Aluminum
$E = 70$ GPa
$\gamma = 2710$ kg/m^3

Fig. P11.15 and P11.16

11.13 Determine the critical load of a round wooden dowel which is 0.9 m long and has a diameter of (*a*) 10 mm, (*b*) 15 mm. Use $E = 12$ GPa.

11.14 Determine the critical load of an aluminum tube which is 1.5 m long and has a 16-mm outer diameter and a 1.25-mm wall thickness. Use $E = 70$ GPa.

11.15 Determine (*a*) the critical load for the aluminum strut, (*b*) the dimension *d* for which the steel strut will have the same critical load. (*c*) Express the weight of the aluminum strut as a percentage of the weight of the steel strut.

11.16 Determine the dimension *d* so that the aluminum and steel struts will have the same weight and compute the critical load for each strut.

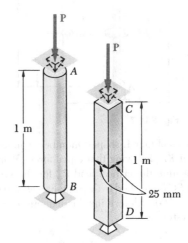

Fig. P11.17 and P11.18

11.17 Determine the radius of the round strut so that the round and square struts will have the same cross-sectional area and compute the critical load for each strut. Use $E = 200$ GPa.

11.18 Determine (*a*) the critical load for the square strut, (*b*) the radius of the round strut for which both struts will have the same critical load. (*c*) Express the cross-sectional area of the square strut as a percentage of the cross-sectional area of the round strut. Use $E = 200$ GPa.

11.19 (*a*) Determine the critical load of a 0.9-m hollow aluminum rod which has a 50-mm outer diameter and a 6-mm wall thickness. (*b*) Determine the critical load of a redesigned rod in which the cross-sectional area is reduced by 50% while the length and outer diameter are unchanged. Use $E = 70$ GPa.

11.20 The design of a hollow steel rod requires a cross-sectional area of 1000 mm^2. Knowing that the effective length of the rod is 2 m, determine the critical load when the ratio of the outer diameter to the inner diameter is (*a*) 2.00, (*b*) 1.50. Use $E = 200$ GPa.

11.21 A compression member of 2-m effective length consists of a solid 40-mm-diameter brass rod. To reduce the weight of the member by 25%, the solid rod is replaced by a hollow rod of the cross section shown. Determine (*a*) the percent reduction in the critical load, (*b*) the value of the critical load for the hollow rod. Use $E = 105$ GPa.

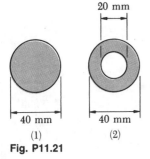

20 mm

40 mm
(1)

40 mm
(2)

Fig. P11.21

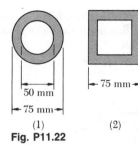

Fig. P11.22

(1)   (2)

11.22 Two brass rods used as compression members, each of 1.8-m effective length, have the cross sections shown. (*a*) Determine the wall thickness of the hollow square rod for which the rods have the same cross-sectional area. (*b*) Using $E = 100$ GPa, determine the critical load of each rod.

11.23 and 11.24 A column of effective length L can be made by securely nailing together identical planks in each of the arrangements shown. For the thickness of planks indicated, determine the ratio of the critical load using arrangement *a* to the critical load using arrangement *b*.

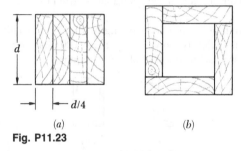

(*a*)   (*b*)
Fig. P11.23

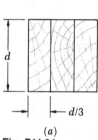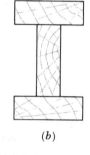

(*a*)   (*b*)
Fig. P11.24

11.25 Supports A and B of the pin-ended column shown are a fixed distance L apart. Knowing that at a temperature T_0 the force in the column is zero and that buckling occurs when the temperature is $T_1 = T_0 + \Delta T$, express ΔT in terms of b, L, and the coefficient of thermal expansion α.

11.26 A pin-ended steel column of 100×100-mm cross section is supported as described in Prob. 11.25. Knowing that initially the force in the column is zero, determine the required length L if buckling is to occur for a temperature rise $\Delta T = 40°C$. Use $E = 200$ GPa and $\alpha = 11.7 \times 10^{-6}/°C$.

11.27 and 11.28 A compression member of 3-m effective length is made by welding together two L $75 \times 50 \times 6$ steel angles as shown. Using $E = 200$ GPa, determine the allowable centric load for the member if a factor of safety of 3 is required.

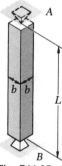

Fig. P11.25

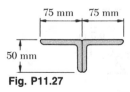

Fig. P11.27

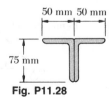

Fig. P11.28

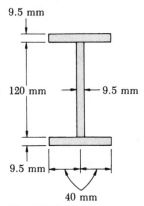

Fig. P11.29

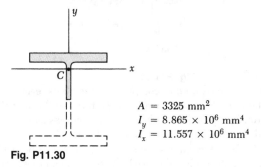

Fig. P11.30

11.29 A column of 6-m effective length is to be made from three plates as shown. Using $E = 200$ GPa, determine the factor of safety with respect to buckling for a centric load of 16 kN.

$A = 3325$ mm^2
$I_y = 8.865 \times 10^6$ mm^4
$I_x = 11.557 \times 10^6$ mm^4

11.30 A column of 2.5-m effective length is made of half of a W 200 × 52 rolled-steel shape. Using the geometric properties of the cross section shown, determine the factor of safety if the allowable centric load is 1150 kN. Use $E = 200$ GPa.

11.31 A column of 3.5-m effective length is made by welding together two 89 × 64 × 6.4-mm angles as shown. Using $E = 200$ GPa, determine the allowable centric load if a factor of safety of 2.8 is required.

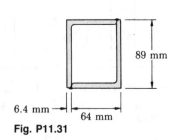

Fig. P11.31

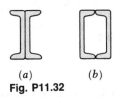

(a)   (b)
Fig. P11.32

11.32 A column of 4-m effective length is to be made by welding together two C 150 × 12 rolled-steel channels. Using $E = 200$ GPa, determine for each arrangement shown the allowable centric load if a factor of safety of 3.2 is required.

11.33 and 11.34 Knowing that a factor of safety of 2.6 is required, determine the largest load **P** which may be applied to the structure shown. Use $E = 200$ GPa and consider only buckling in the plane of the structure.

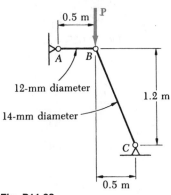

Fig. P11.33

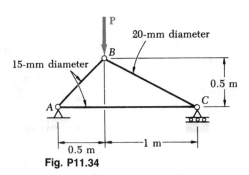

Fig. P11.34

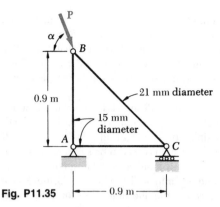

Fig. P11.35

11.35 Knowing that a factor of safety of 2.5 is required, determine the largest load **P** which may be applied to the structure when $\alpha = 75°$. Use $E = 200$ GPa and consider only buckling in the plane of the structure.

11.36 For the structure in Prob. 11.35, determine (*a*) the angle α for which the factor of safety with respect to buckling is maximum, (*b*) the corresponding factor of safety for $P = 7$ kN.

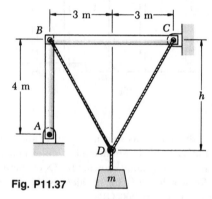

Fig. P11.37

11.37 The pin-ended members AB and BC consist of sections of aluminum pipe of 120-mm outer diameter and 10-mm wall thickness. Knowing that a factor of safety of 3.5 is required, determine the mass m of the largest block which may be supported by the cable arrangement shown when $h = 4$ m. Use $E = 70$ GPa and consider only buckling in the plane of the structure.

11.38 For the structure of Prob. 11.37, determine (*a*) the distance h for which the factor of safety with respect to buckling is maximum, (*b*) the corresponding factor of safety for $m = 7500$ kg.

11.39 Members AB and CD are 27-mm-diameter steel rods, and members BC and AD are 21-mm-diameter steel rods. When the turnbuckle is tightened, the diagonal member AC is put in tension. Knowing that a factor of safety with respect to buckling of 3.2 is required, determine the largest allowable tension in AC when $h = 2.4$ m. Use $E = 200$ GPa and consider only buckling in the plane of the structure.

11.40 For the structure in Prob. 11.39, determine (*a*) the distance h for which the factor of safety with respect to buckling is the same for each of the four compression members, (*b*) the corresponding allowable tension in member AC if the factor of safety is 2.8.

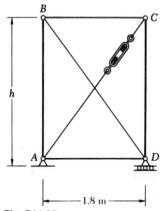

Fig. P11.39

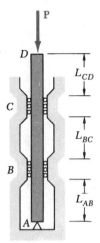

Fig. P11.41

11.41 A 25-mm-diameter aluminum strut is maintained in the position shown by a pin support at A and by sets of rollers at B and C which prevent rotation of the strut in the plane of the figure. Knowing that $L_{AB} = 0.9$ m, $L_{BC} = 1.2$ m, and $L_{CD} = 0.3$ m, determine the allowable load $\mathbf{P}$ using a factor of safety with respect to buckling of 3.2. Consider only buckling in the plane of the figure and use $E = 77$ GPa.

11.42 For the strut of Prob. 11.41, knowing that $L_{AB} = 0.9$ m, determine (a) the largest lengths of L_{BC} and L_{CD} which may be used if the allowable load $\mathbf{P}$ is to be as large as possible, (b) the magnitude of the corresponding allowable load.

11.43 A rigid block of mass m can be supported by two columns in each of the four ways shown. Each column consists of an aluminum tube which has a 44-mm outer diameter and a 4-mm wall thickness. Using $E = 70$ GPa and a factor of safety of 2.8, determine the allowable load for each support condition.

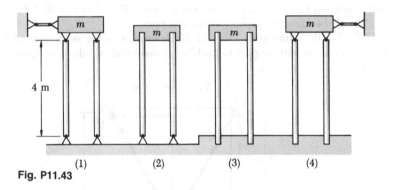

Fig. P11.43

11.44 Each of the five struts shown consists of a solid steel rod. (a) Knowing that the strut of Fig. (1) is of 20-mm diameter, determine the factor of safety with respect to buckling for the loading shown. (b) Determine the diameter of each of the other struts for which the factor of safety is the same as the factor of safety obtained in part a. Use $E = 200$ GPa.

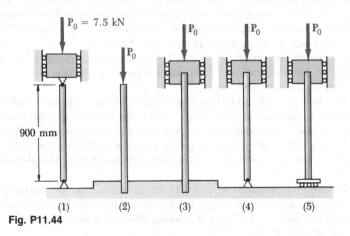

Fig. P11.44

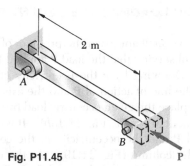

Fig. P11.45

11.45 The uniform aluminum bar AB has a 20 × 36-mm rectangular cross section and is supported by pins and brackets as shown. Each end of the bar may rotate freely about a horizontal axis through the pin, but rotation about a vertical axis is prevented by the brackets. Using $E = 70$ GPa, determine the allowable centric load **P** if a factor of safety of 2.5 is required.

11.46 Column ACB has a uniform rectangular cross section and is braced in the xz plane at its midpoint C. (*a*) Determine the ratio b/d for which the factor of safety is the same with respect to buckling in the xz and yz planes. (*b*) Using the ratio found in part *a*, design the cross section of the column for which the factor of safety will be 3.0 when $P = 4400$ N, $L = 1$ m, and $E = 200$ GPa.

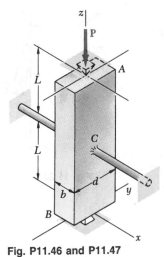

Fig. P11.46 and P11.47

11.47 The steel column ACB has a uniform rectangular cross section with $b = 12$ mm and $d = 22$ mm. The column is braced in the xz plane at its midpoint C and carries a centric load **P** of magnitude 3800 N. Knowing that a factor of safety of 3.2 is required, determine the largest allowable length L. Use $E = 200$ GPa.

11.48 Column AB carries a centric load **P** of magnitude 65 kN. Cables BC and BD are taut and prevent motion of point B in the xz plane. Using Euler's formula and a factor of safety of 2.2, and neglecting the tension in the cables, determine the maximum allowable length L. Use $E = 200$ GPa.

11.49 A W 200 × 22.5 rolled-steel shape is used with the support and cable arrangement shown in Prob. 11.48. Knowing that $L = 7.2$ m, determine the allowable centric load **P** if a factor of safety of 2.2 is required. Use $E = 200$ GPa.

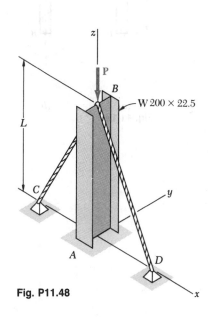

Fig. P11.48

*11.5. ECCENTRIC LOADING; THE SECANT FORMULA

In this section we shall approach the problem of column buckling in a different way, by observing that the load **P** applied to a column is never perfectly centric. Denoting by e the ecentricity of the load, i.e., the distance between the line of action of **P** and the axis of the column (Fig. 11.18a), we shall replace the given eccentric load by a centric force **P** and a couple $\mathbf{M}_A$ of moment $M_A = Pe$ (Fig. 11.18b). It is clear that, no matter how small the load **P** and the eccentricity e, the couple $\mathbf{M}_A$ will cause some bending of the column (Fig. 11.19). As the eccentric load is increased, both the couple $\mathbf{M}_A$ and the axial force **P** increase, and both cause the column to bend further. Viewed in this way, the problem of buckling is not a question of determining how long the column can remain straight and stable under an increasing load, but rather how much the column can be permitted to bend under the increasing load, if the allowable stress is not to be exceeded and if the deflection y_{max} is not to become excessive.

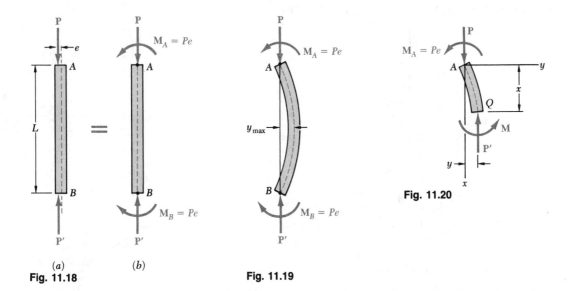

(a)   (b)
Fig. 11.18

Fig. 11.19

Fig. 11.20

We shall first write and solve the differential equation of the elastic curve, proceeding in the same manner as we did earlier in Secs. 11.3 and 11.4. Drawing the free-body diagram of a portion AQ of the column and choosing the coordinate axes as shown (Fig. 11.20), we find that the bending moment at Q is

$$M = -Py - M_A = -Py - Pe \qquad (11.22)$$

Substituting the value of M into Eq. (8.4) of Sec. 8.3, we write

$$\frac{d^2y}{dx^2} = \frac{M}{EI} = -\frac{P}{EI}y - \frac{Pe}{EI}$$

Transposing the term containing y and setting

$$p^2 = \frac{P}{EI} \tag{11.6}$$

as done earlier, we write

$$\frac{d^2y}{dx^2} + p^2y = -p^2e \tag{11.23}$$

Since the left-hand member of this equation is the same as that of Eq. (11.7), which was solved in Sec. 11.3, we write the general solution of Eq. (11.23) as

$$y = A \sin px + B \cos px - e \tag{11.24}$$

where the last term is a particular solution of Eq. (11.23).

The constants A and B are obtained from the boundary conditions shown in Fig. 11.21. Making first $x = 0$, $y = 0$ in Eq. (11.24), we have

$$B = e$$

Making next $x = L$, $y = 0$, we write

$$A \sin pL = e(1 - \cos pL) \tag{11.25}$$

Recalling that

$$\sin pL = 2 \sin\frac{pL}{2} \cos\frac{pL}{2}$$

and

$$1 - \cos pL = 2 \sin^2 \frac{pL}{2}$$

and substituting into Eq. (11.25), we obtain, after reductions,

$$A = e \tan\frac{pL}{2}$$

Substituting for A and B into Eq. (11.24), we write the equation of the elastic curve:

$$y = e\left(\tan\frac{pL}{2} \sin px + \cos px - 1\right) \tag{11.26}$$

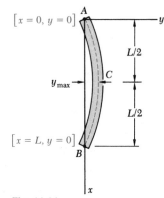

Fig. 11.21

The value of the maximum deflection is obtained by setting $x = L/2$ in Eq. (11.26). We have

$$y_{\max} = e\left(\tan \frac{pL}{2} \sin \frac{pL}{2} + \cos \frac{pL}{2} - 1\right)$$

$$= e\left(\frac{\sin^2 \dfrac{pL}{2} + \cos^2 \dfrac{pL}{2}}{\cos \dfrac{pL}{2}} - 1\right)$$

$$y_{\max} = e\left(\sec \frac{pL}{2} - 1\right) \tag{11.27}$$

Recalling Eq. (11.6), we write

$$y_{\max} = e\left[\sec\left(\sqrt{\frac{P}{EI}}\,\frac{L}{2}\right) - 1\right] \tag{11.28}$$

We note from the expression obtained that $y_{\max}$ becomes infinite when

$$\sqrt{\frac{P}{EI}}\,\frac{L}{2} = \frac{\pi}{2} \tag{11.29}$$

While the deflection does not actually become infinite, it nevertheless becomes unacceptably large, and P should not be allowed to reach the critical value which satisfies Eq. (11.29). Solving (11.29) for P, we have

$$P_{cr} = \frac{\pi^2 EI}{L^2} \tag{11.30}$$

which is the value that we obtained in Sec. 11.3 for a column under a centric load. Solving (11.30) for EI and substituting into (11.28), we may express the maximum deflection in the alternative form

$$y_{\max} = e\left(\sec \frac{\pi}{2}\sqrt{\frac{P}{P_{cr}}} - 1\right) \tag{11.31}$$

The maximum stress $\sigma_{\max}$ occurs in the section of the column where the bending moment is maximum, i.e., in the transverse section through the midpoint C, and may be obtained by adding the normal stresses due, respectively, to the axial force and the bending couple exerted on that section (cf. Sec. 4.13). We have

$$\sigma_{\max} = \frac{P}{A} + \frac{M_{\max}c}{I} \tag{11.32}$$

From the free-body diagram of the portion AC of the column (Fig. 11.22), we find that

$$M_{max} = Py_{max} + M_A = P(y_{max} + e)$$

Substituting this value into (11.32) and recalling that $I = Ar^2$, we write

$$\sigma_{max} = \frac{P}{A}\left[1 + \frac{(y_{max} + e)c}{r^2}\right] \qquad (11.33)$$

Substituting for y_{max} the value obtained in (11.28), we write

$$\sigma_{max} = \frac{P}{A}\left[1 + \frac{ec}{r^2}\sec\left(\sqrt{\frac{P}{EI}}\frac{L}{2}\right)\right] \qquad (11.34)$$

An alternative form for σ_{max} may be obtained by substituting for y_{max} from (11.31) into (11.33). We have

$$\sigma_{max} = \frac{P}{A}\left(1 + \frac{ec}{r^2}\sec\frac{\pi}{2}\sqrt{\frac{P}{P_{cr}}}\right) \qquad (11.35)$$

The equation obtained may be used with any end conditions, as long as the appropriate value is used for the critical load (cf. Sec. 11.4).

We note that, since σ_{max} does not vary linearly with the load P, the principle of superposition does not apply to the determination of the stress due to the simultaneous application of several loads; the resultant load must first be computed, and then Eq. (11.34) or Eq. (11.35) may be used to determine the corresponding stress. For the same reason, any given factor of safety should be applied to the load, and not to the stress.

Making $I = Ar^2$ in Eq. (11.34) and solving for the ratio P/A in front of the bracket, we write

$$\frac{P}{A} = \frac{\sigma_{max}}{1 + \frac{ec}{r^2}\sec\left(\frac{1}{2}\sqrt{\frac{P}{EA}}\frac{L_e}{r}\right)} \qquad (11.36)$$

where the effective length is used to make the formula applicable to various end conditions. This formula is referred to as *the secant formula;* it defines the force per unit area, P/A, which causes a specified maximum stress σ_{max} in a column of given effective slenderness ratio, L_e/r, for a given value of the ratio ec/r^2, where e is the eccentricity of the applied load. We note that, since P/A appears in both members, it is necessary to solve a transcendental equation by trial and error to obtain the value of P/A corresponding to a given column and loading condition.

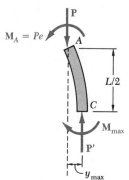

Fig. 11.22

Equation (11.36) was used to draw the curves shown in Fig. 11.23 for a steel column, assuming the values of E and σ_Y shown in the figure. These curves make it possible to determine the load per unit area P/A, which causes the column to yield for given values of the ratios L_e/r and ec/r^2.

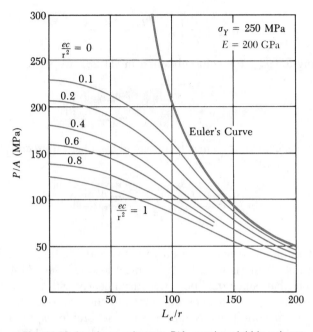

Fig. 11.23 Load per unit area, P/A, causing yield in column.

We note that, for small values of L_e/r, the secant is almost equal to 1 in Eq. (11.36), and P/A may be assumed equal to

$$\frac{P}{A} = \frac{\sigma_{\max}}{1 + \dfrac{ec}{r^2}} \tag{11.37}$$

a value which could be obtained by neglecting the effect of the lateral deflection of the column and using the method of Sec. 4.13. On the other hand, we note from Fig. 11.23 that, for large values of L_e/r, the curves corresponding to the various values of the ratio ec/r^2 get very close to Euler's curve defined by Eq. (11.13′), and thus that the effect of the eccentricity of the loading on the value of P/A becomes negligible. The secant formula is chiefly useful for intermediate values of L_e/r. However, to use it effectively, we should know the value of the eccentricity e of the loading, and this quantity, unfortunately, is seldom known with any degree of accuracy.

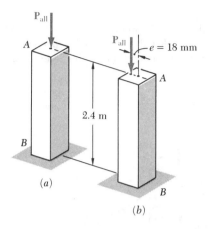

(a)

2.4 m

A

B

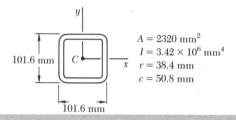

(b)

SAMPLE PROBLEM 11.2

The uniform column AB consists of a 2.4-m section of structural tubing having the cross section shown. (a) Using Euler's formula and a factor of safety of two, determine the allowable centric load for the column and the corresponding normal stress.   (b) Assuming that the allowable load, found in part a, is applied as shown at a point 18 mm from the geometric axis of the column, determine the horizontal deflection of the top of the column and the maximum normal stress in the column.   Use $E = 200$ GPa.

$A = 2320$ mm^2
$I = 3.42 \times 10^6$ mm^4
$r = 38.4$ mm
$c = 50.8$ mm

101.6 mm

101.6 mm

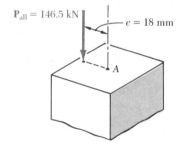

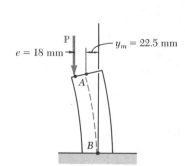

Effective Length.   Since the column has one end fixed and one end free, its effective length is

$$L_e = 2(2.4 \text{ m}) = 4.8 \text{ m}$$

Critical Load.   Using Euler's formula, we write

$$P_{cr} = \frac{\pi^2 EI}{L_e^2} = \frac{\pi^2 (200 \text{ GPa})(3.42 \times 10^{-6} \text{ m}^4)}{(4.8 \text{ m})^2} \qquad P_{cr} = 293 \text{ kN}$$

(a) Allowable Load and Stress.   For a factor of safety of 2, we find

$$P_{all} = \frac{P_{cr}}{F.S.} = \frac{293 \text{ kN}}{2} \qquad\qquad P_{all} = 146.5 \text{ kN} \blacktriangleleft$$

and

$$\sigma = \frac{P_{all}}{A} = \frac{146.5 \text{ kN}}{2320 \times 10^{-6} \text{ m}^2} \qquad\qquad \sigma = 63.1 \text{ MPa} \blacktriangleleft$$

(b) Eccentric Load.   We observe that column AB and its loading are identical to the upper half of the column of Fig. 11.18 which was used in the derivation of the secant formulas; we conclude that the formulas of Sec. 11.5 apply directly to the case considered here.   Recalling that $P_{all}/P_{cr} = \frac{1}{2}$ and using Eq. (11.31), we compute the horizontal deflection of point A:

$$y_m = e\left[\sec\left(\frac{\pi}{2}\sqrt{\frac{P}{P_{cr}}}\right) - 1\right] = (18 \text{ mm})\left[\sec\left(\frac{\pi}{2\sqrt{2}}\right) - 1\right]$$

$$= (18 \text{ mm})[2.252 - 1] \qquad\qquad y_m = 22.5 \text{ mm} \blacktriangleleft$$

The maximum normal stress is obtained from Eq. (11.35):

$$\sigma_m = \frac{P}{A}\left[1 + \frac{ec}{r^2}\sec\left(\frac{\pi}{2}\sqrt{\frac{P}{P_{cr}}}\right)\right]$$

$$= \frac{146.5 \text{ kN}}{2320 \times 10^{-6} \text{ m}^2}\left[1 + \frac{(18 \text{ mm})(50.8 \text{ mm})}{(38.4 \text{ mm})^2}\sec\left(\frac{\pi}{2\sqrt{2}}\right)\right]$$

$$= (63.15 \text{ MPa})[1 + 0.620(2.252)] \qquad\qquad \sigma_m = 151.3 \text{ MPa} \blacktriangleleft$$

PROBLEMS

11.50 An axial load **P** is applied to the 35-mm-diameter steel rod *AB* as shown. When $P = 90$ kN, it is observed that the horizontal deflection of the midpoint *C* is 0.75 mm. Using $E = 200$ GPa, determine (*a*) the eccentricity *e* of the load, (*b*) the maximum stress in the rod.

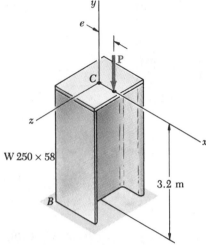

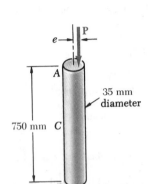

Fig. P11.50

Fig. P11.51

11.51 The axial load **P** is applied at a point located on the *x* axis at a distance *e* from the geometric axis of the W 250 × 58 rolled-steel column *BC*. When $P = 350$ kN, it is observed that the horizontal deflection of the top of the column is 5 mm. Using $E = 200$ GPa, determine (*a*) the eccentricity *e* of the load, (*b*) the maximum stress in the column.

11.52 An axial load of magnitude $P = 15$ kN is applied at a point *D* which is 4 mm from the geometric axis of the square aluminum bar *BC*. Using $E = 70$ GPa, determine (*a*) the horizontal deflection of end *C*, (*b*) the maximum stress in the bar.

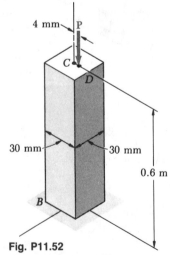

Fig. P11.52

11.53 The line of action of an axial load **P** of magnitude 265 kN is parallel to the geometric axis of the column AB and intersects the x axis at $x = 20$ mm. Using $E = 200$ GPa, determine (a) the deflection of the midpoint C of the column, (b) the maximum stress in the column.

11.54 The line of action of an axial load **P** is parallel to the geometric axis of the column AB and intersects the x axis at $x = 20$ mm. Using $E = 200$ GPa, determine (a) the load **P** for which the horizontal deflection of the midpoint C of the column is 12 mm, (b) the corresponding maximum stress in the column.

11.55 An axial load **P** is applied at a point located on the x axis at a distance $e = 12$ mm from the geometric axis of the W 250×58 rolled-steel column BC. Using $E = 200$ GPa, determine (a) the load **P** for which the horizontal deflection of the top of the column is 15 mm, (b) the corresponding maximum stress in the column.

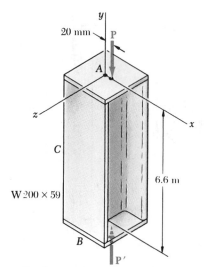

Fig. P11.53 and P11.54

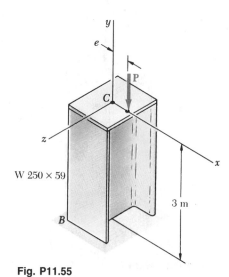

Fig. P11.55

11.56 The steel bar AB has a 10×10-mm square cross section and is held by pins which are a fixed distance apart and are located at a distance $e = 0.9$ mm from the geometric axis of the bar. Knowing that at temperature T_0 the pins are in contact with the bar and that the force in the bar is zero, determine the increase in temperature for which the bar will just make contact with point C if $d = 0.3$ mm. Use $E = 200$ GPa and the coefficient of thermal expansion $\alpha = 11.7 \times 10^{-6}/°C$.

11.57 For the bar of Prob. 11.56, determine the required distance d for which the bar will just make contact with point C when the temperature increases by 60°C.

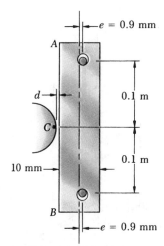

Fig. P11.56 and P11.57

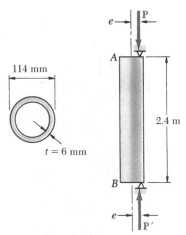

114 mm

$t = 6$ mm

2.4 m

Fig. P11.58 and P11.59

11.58 A pipe having the cross section shown is used as a 2.4 m column. For the grade of steel used $\sigma_Y = 250$ MPa and $E = 200$ GPa. Knowing that a factor of safety of 2.5 with respect to permanent deformation is required, determine the allowable load **P** when the eccentricity e is (*a*) 12 mm, (*b*) 6 mm. (*Hint*: Since the factor of safety must be applied to the load **P**, not to the stress, use Fig. 11.23 to determine P_Y.)

11.59 Solve Prob. 11.58, assuming that the length of the column is increased to 3.5 m.

11.60 Axial loads of magnitude P are applied parallel to the geometric axis of the column *AB* and intersect the x axis at a distance $e = 12$ mm from the geometric axis. For the grade of steel used, $\sigma_Y = 250$ MPa and $E = 200$ GPa. Knowing that a factor of safety of 2.5 with respect to permanent deformation is required, determine (*a*) the magnitude P of the allowable load when the length L is 4.25 m, (*b*) the ratio of the load found in part *a* to the magnitude of the allowable centric load for the column. (See hint of Prob. 11.58.)

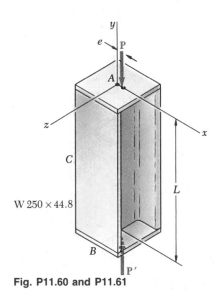

W 250 × 44.8

Fig. P11.60 and P11.61

11.61 Solve Prob. 11.60, assuming that the length of the column is reduced to 2.75 m.

11.62 A 250-kN axial load **P** is applied to the W 200 × 35.9 rolled-steel column *BC* which is free at its top *C* and fixed at its base *B*. Knowing that the eccentricity of the load is $e = 6$ mm, determine the largest permissible length L if the allowable stress in the column is 80 MPa. Use $E = 200$ GPa.

11.63 A 100-kN axial load **P** is applied to the W 150 × 18 rolled-steel column *BC* which is free at its top *C* and fixed at its base *B*. Knowing that the eccentricity of the load is $e = 6$ mm, determine the largest permissible length L if the allowable stress in the column is 80 MPa. Use $E = 200$ GPa.

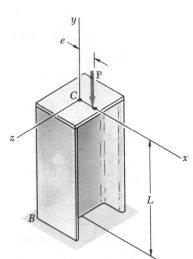

Fig. P11.62 and P11.63

11.64 Axial loads of magnitude $P = 90$ kN are applied parallel to the geometric axis of the W 200×22.5 rolled-steel column AB and intersect the x axis at a distance e from the geometric axis. Knowing that $\sigma_{all} = 80$ MPa and $E = 200$ GPa, determine the largest permissible length L when (a) $e = 6$ mm, (b) $e = 12$ mm.

11.65 Axial loads of magnitude $P = 600$ kN are applied parallel to the geometric axis of the W 250×80 rolled-steel column AB and intersect the x axis at a distance e from the geometric axis. Knowing that $\sigma_{all} = 80$ MPa and $E = 200$ GPa, determine the largest permissible length L when (a) $e = 6$ mm, (b) $e = 12$ mm.

11.66 A 60-kN axial load is to be applied with an eccentricity $e = 8$ mm to the circular steel rod BC which is free at its top C and fixed at its base B. Knowing that the rods available for use have diameters in increments of 4 mm from 40 mm to 80 mm, determine the lightest rod which may be used if $\sigma_{all} = 120$ MPa. Use $E = 200$ GPa.

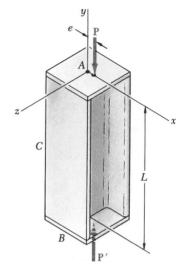

Fig. P11.64 and P11.65

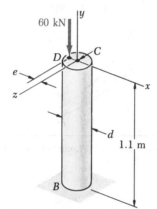

Fig. P11.66 and P11.67

11.67 Solve Prob. 11.66, assuming that the 60-kN axial load will be applied to the rod with an eccentricity $e = \frac{1}{2}d$.

11.68 An axial load P is applied at a point located on the x axis at a distance $e = 15$ mm from the geometric axis of the W 200×41.7 rolled-steel column BC. Knowing that the column is free at its top C and fixed at its base B and that $\sigma_Y = 250$ MPa and $E = 200$ GPa, determine the allowable load P if a factor of safety of 2.5 with respect to yield is required. (*Hint:* Since the factor of safety must be applied to the load P, not to the stresses, use Fig. 11.23 to determine P_Y.)

11.69 An axial load P of magnitude 220 kN is applied at a point located on the x axis at a distance $e = 6$ mm from the geometric axis of the W 200×41.7 rolled-steel column BC. Knowing that the column is free at its top C and fixed at its base B and that $\sigma_Y = 250$ MPa and $E = 200$ GPa, determine the factor of safety with respect to yield. (See hint of Prob. 11.68.)

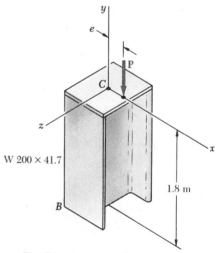

Fig. P11.68 and P11.69

11.6. DESIGN OF COLUMNS UNDER A CENTRIC LOAD

In the preceding sections, we have determined the critical load of a column by using Euler's formula, and we have investigated the deformations and stresses in eccentrically loaded columns by using the secant formula. In each case we assumed that all stresses remained below the proportional limit and that the column was initially a straight homogeneous prism. Real columns fall short of such an idealization, and in practice the design of columns is based on empirical formulas which reflect the results of numerous laboratory tests.

Over the last century, very many steel columns have been tested by applying to them a centric axial load and increasing the load until failure occurred. The results of such tests are represented in Fig. 11.24 where,

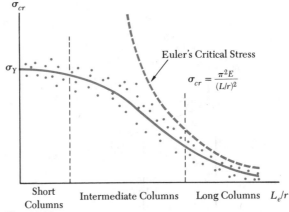

Fig. 11.24

for each of many tests, a point has been plotted with its ordinate equal to the normal stress σ_{cr} at failure, and its abscissa equal to the corresponding value of the effective slenderness ratio, L_e/r. Although there is considerable scatter in the test results, regions corresponding to three types of failure may be observed. For long columns, where L_e/r is large, failure is closely predicted by Euler's formula, and the value of σ_{cr} is observed to depend on the modulus of elasticity E of the steel used, but not on its yield strength σ_Y. For very short columns and compression blocks, failure occurs essentially as a result of yield, and we have $\sigma_{cr} \approx \sigma_Y$. Columns of intermediate length comprise those cases where failure is dependent on both σ_Y and E. In this range, column failure is an extremely complex phenomenon, and test data have been used extensively to guide the development of specifications and design formulas.

Empirical formulas which express an allowable stress or critical stress in terms of the effective slenderness ratio were first introduced over a century ago, and since then have undergone a continuous process of refinement and improvement. Typical empirical formulas used to approxi-

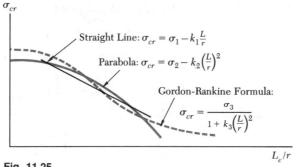

Fig. 11.25

mate test data are shown in Fig. 11.25. Since a single formula is usually not adequate for all values of L_e/r, different formulas, each with a definite range of applicability, have been developed for various materials. In each case we must check that the formula we propose to use is applicable for the value of L_e/r for the column involved. Furthermore, we must determine whether the formula provides the value of the critical stress for the column, in which case this value must be divided by the appropriate factor of safety, or whether it provides directly the allowable stress.

Specific formulas for the design of columns under centric loading will now be considered for three different materials.

Structural Steel. The formulas most widely used for the design of steel columns under a centric load are found in the specifications of the American Institute of Steel Construction.[†] As we shall see, a parabolic expression is used to predict σ_{all} for columns of short and intermediate lengths, and an Euler-type relation is used for long columns. These relations are developed in two steps:

1. First a curve representing the variation of σ_{cr} with L/r is obtained (Fig. 11.26). It is important to note that this curve does not incorporate any factor of safety. The portion AB of this curve is an arc of parabola defined by an equation of the form

$$\sigma_{\text{cr}} = \sigma_0 - k\left(\frac{L}{r}\right)^2 \tag{11.38}$$

while portion BE is part of Euler's curve DBE defined by the equation

$$\sigma_{\text{cr}} = \frac{\pi^2 E}{(L/r)^2} \tag{11.39}$$

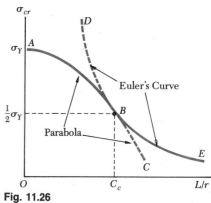

Fig. 11.26

We note that, since $\sigma_{\text{cr}} = \sigma_Y$ for $L/r = 0$, the constant σ_0 in Eq. (11.38) must be equal to σ_Y. On the other hand, it is assumed in the AISC specifications, as well as in others,[‡] that at point B where the parabola joins Euler's curve, the critical stress is equal to half of the yield stress.

[†]*Manual of Steel Construction*, 9th ed., American Institute of Steel Construction, New York, 1989.

[‡]E.g., American Association of State Highways Officials.

Denoting by C_c the value of L/r at that point, Eq. (11.38) yields therefore

$$\tfrac{1}{2}\sigma_Y = \sigma_Y - kC_c^2$$

and $k = \sigma_Y/2C_c^2$. Substituting for σ_0 and k into Eq. (11.38), we find that

$$\text{for } L/r \leq C_c: \qquad \sigma_{\text{cr}} = \sigma_Y\left[1 - \frac{(L/r)^2}{2C_c^2}\right] \qquad (11.40)$$

and, recalling (11.39), that

$$\text{for } L/r \geq C_c: \qquad \sigma_{\text{cr}} = \frac{\pi^2 E}{(L/r)^2} \qquad (11.41)$$

Making $\sigma_{\text{cr}} = \tfrac{1}{2}\sigma_Y$ and $L/r = C_c$ in Eq. (11.41), we find that

$$C_c^2 = \frac{2\pi^2 E}{\sigma_Y} \qquad (11.42)$$

2. A factor of safety must be introduced to obtain the final AISC design formulas defining σ_{all} as a function of L/r. For $L/r \geq C_c$, that is, for long columns, a constant factor of safety of 1.92 is used. Dividing the value obtained in Eq. (11.41) for σ_{cr} by this factor of safety, and noting that the AISC specifications state that L/r preferably should not exceed 200, we write†

$$\frac{L}{r} \geq C_c: \qquad \sigma_{\text{all}} = \frac{\sigma_{\text{cr}}}{F.S.} = \frac{\pi^2 E}{1.92(L/r)^2} \qquad (11.43)$$

For short and intermediate-length columns, the following formula is used to determine the factor of safety:

$$F.S. = \frac{5}{3} + \frac{3}{8}\frac{L/r}{C_c} - \frac{1}{8}\left(\frac{L/r}{C_c}\right)^3 \qquad (11.44)$$

Dividing the expression obtained in (11.40) for σ_{cr} by this factor of safety, we write

$$\frac{L}{r} < C_c: \qquad \sigma_{\text{all}} = \frac{\sigma_{\text{cr}}}{F.S.} = \frac{\sigma_Y}{F.S.}\left[1 - \frac{1}{2}\left(\frac{L/r}{C_c}\right)^2\right] \qquad (11.45)$$

The formulas obtained may be used with any consistent set of units.

We observe that, by using Eqs. (11.42), (11.43), (11.44), and (11.45), we may determine the allowable axial stress for a given grade of steel and any given allowable value of L/r. The procedure is to first compute C_c by substituting the given value of σ_Y into Eq (11.42). For values of L/r larger than C_c, we then use Eq. (11.43) to determine σ_{all}, and for values of L/r smaller than C_c, we determine σ_{all} from Eqs. (11.44) and (11.45). For the convenience of the designer, values of the allowable stress have been tabulated in the AISC Manual of Steel Construction for different grades of steel and for all values of L/r from 1 to 200. The variation of σ_{all} with L/r is shown in Fig. 11.27 for three different grades of structural steel.

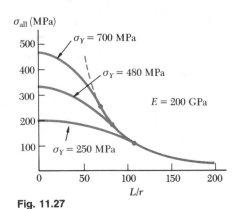

Fig. 11.27

†The AISC actually specifies $\sigma_{\text{all}} = 12\pi^2 E/23(L/r)^2$.

Example 11.02

Determine the longest unsupported length L for which the S 100×11.5 rolled-steel compression member AB can safely carry the centric load shown (Fig. 11.28). Assume $\sigma_Y = 290$ MPa and $E = 200$ GPa.

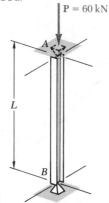

Fig. 11.28

From Appendix C we find that, for a S 100×11.5 shape,

$$A = 1452 \text{ mm}^2 \qquad r_x = 41.6 \text{ mm} \qquad r_y = 14.75 \text{ mm}$$

If the 60-kN load is to be safely supported, we must have

$$\sigma_{all} = \frac{P}{A} = \frac{60 \times 10^3 \text{ N}}{1452 \times 10^{-6} \text{ m}^2} = 41.3 \times 10^6 \text{ Pa}$$

On the other hand, for the given yield strength, Eq. (11.42) gives

$$C_c^2 = \frac{2\pi^2 E}{\sigma_Y} = \frac{2\pi^2(200 \times 10^9)}{290 \times 10^6} = 13.61 \times 10^3 \qquad C_c = 116.7$$

Assuming $L/r \geq C_c$, we use Eq. (11.43) and write

$$\sigma_{all} = \frac{\pi^2 E}{1.92(L/r)^2} = \frac{\pi^2(200 \times 10^9 \text{ Pa})}{1.92(L/r)^2} = \frac{1.028 \times 10^{12} \text{ Pa}}{(L/r)^2}$$

Equating this expression to the required value of σ_{all}, we write

$$\frac{1.028 \times 10^{12} \text{ Pa}}{(L/r)^2} = 41.3 \times 10^6 \text{ Pa} \qquad L/r = 157.8$$

Our assumption that $L/r \geq C_c$ was correct. Choosing the smaller of the two radii of gyration, we have

$$\frac{L}{r_y} = \frac{L}{14.75 \times 10^{-3} \text{ m}} = 157.8 \qquad L = 2.33 \text{ m}$$

Aluminum. Many aluminum alloys are available for use in structural and machine construction. For each alloy the specifications of the Aluminum Association† provide three formulas for the allowable stress in columns under centric loading. The variation of σ_{all} with L/r defined by these formulas is shown in Fig. 11.29. We note that for short columns σ_{all} is constant, for intermediate columns a linear relation between σ_{all} and L/r is used, and for long columns an Euler-type formula is used. Specific formulas for use in the design of buildings and similar structures are given below for two commonly used alloys.

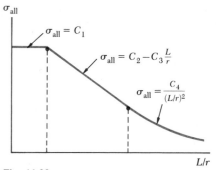

Fig. 11.29

Alloy 6061-T6:

$$L/r \leq 9.5: \qquad \sigma_{all} = 131 \text{ MPa} \tag{11.46}$$

$$9.5 < L/r < 66: \qquad \sigma_{all} = [139 - 0.868(L/r)] \text{ MPa} \tag{11.47}$$

$$L/r \geq 66: \qquad \sigma_{all} = \frac{351 \times 10^3 \text{ MPa}}{(L/r)^2} \tag{11.48}$$

Alloy 2014-T6 (Alclad):

$$L/r \leq 12: \qquad \sigma_{all} = 193 \text{ MPa} \tag{11.49}$$

$$12 < L/r < 55: \qquad \sigma_{all} = [212 - 1.585(L/r)] \text{ MPa} \tag{11.50}$$

$$L/r \geq 55: \qquad \sigma_{all} = \frac{372 \times 10^3 \text{ MPa}}{(L/r)^2} \tag{11.51}$$

†*Specifications for Aluminum Structures*, Aluminum Association, Inc., Washington D.C., 1986.

Timber. For the design of timber columns the specifications of the American Institute of Timber Construction† provide formulas for the allowable stress in short, intermediate, and long columns under centric loading. For a column with a *rectangular* cross section of sides b and d, where $d < b$, the variation of σ_{all} with L/d is shown in Fig. 11.30.

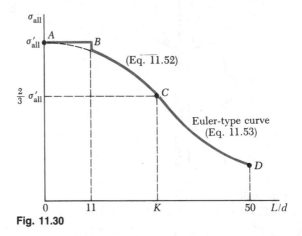

Fig. 11.30

For *short columns* σ_{all} is constant and equal to the allowable stress for compression parallel to the grain σ'_{all}. At $L/d = 11$, which is the demarcation between short and intermediate columns, the AITC specifications introduce at point B a small discontinuity as shown in Fig. 11.30.

For *intermediate columns* the variation of σ_{all} with L/d is defined by the following equation:

$$\sigma_{all} = \sigma'_{all}\left[1 - \frac{1}{3}\left(\frac{L/d}{K}\right)^4\right] \tag{11.52}$$

where K denotes the value of L/d at point C.

For *long columns* the value of σ_{all} is predicted by Euler's formula with a factor of safety of 2.74. Recalling that the radius of gyration of a rectangular area of side d is $r = d/\sqrt{12}$, we write

$$\sigma_{all} = \frac{\pi^2 E}{2.74(L/r)^2} = \frac{\pi^2 E}{2.74(12L^2/d^2)}$$

$$\sigma_{all} = \frac{0.3E}{(L/d)^2} \tag{11.53}$$

Columns in which L/d exceeds 50 are not permitted by the AITC specifications.

As shown in Fig. 11.30, the AITC specifications require that at point C, where the curve BC joins Euler's curve CD, the allowable stress is equal to two-thirds of the allowable stress σ'_{all} parallel to the grain. Mak-

†*Timber Construction Manual*, third edition, American Institute of Timber Construction, John Wiley & Sons, New York, 1985

ing $L/d = K$ and $\sigma_{all} = \frac{2}{3}\sigma'_{all}$ in Eq. (11.53), we write

$$\frac{2}{3}\sigma'_{all} = \frac{0.3E}{K^2}$$

Solving for K, we find†

$$K = 0.671\sqrt{\frac{E}{\sigma'_{all}}} \qquad (11.54)$$

---

† In the foregoing discussion we were concerned with columns having a rectangular cross section, since these columns are the most commonly used. A more general presentation, applicable to columns of arbitrary cross section, would lead to the following formulas, where σ_{all} is expressed in terms of the slenderness ratio L/r of the column:

$$0 < \frac{L}{r} < 38: \qquad \sigma_{all} = \sigma'_{all} \qquad (11.55)$$

$$38 < \frac{L}{r} < K': \qquad \sigma_{all} = \sigma'_{all}\left[1 - \frac{1}{3}\left(\frac{L/r}{K'}\right)^4\right] \qquad (11.56)$$

$$K' < \frac{L}{r} < 173: \qquad \sigma_{all} = \frac{\pi^2 E}{2.74(L/r)^2} \qquad (11.57)$$

where

$$K' = 2.324\sqrt{\frac{E}{\sigma'_{all}}} \qquad (11.58)$$

Example 11.03

Knowing that the effective length of column AB (Fig. 11.31) is 4 m, and that it must safely carry a 150-kN load, design the column using a square cross section. The timber to be used is a grade of Douglas fir for which $E = 12$ GPa and $\sigma'_{all} = 9$ MPa for compression parallel to the grain.

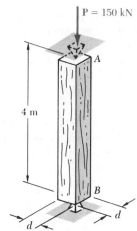

$P = 150$ kN

4 m

A

B

d

Fig. 11.31   d

We first compute the value of K corresponding to the given values of E and σ'_{all}. Using Eq. (11.54), we write

$$K = 0.671\sqrt{\frac{E}{\sigma'_{all}}} = 0.671\sqrt{\frac{12 \times 10^3 \text{ MPa}}{9 \text{ MPa}}} = 24.5$$

Since the side d of the square cross section of the column is not known, a value of L/d must be assumed; we shall *assume* $L/d > K$ and use Eq. (11.53). We write

$$\sigma_{all} = \frac{P}{A} = \frac{0.3E}{(L/d)^2}$$

Substituting the given data.

$$P = 150 \times 10^3 \text{ N} \qquad E = 12 \times 10^3 \text{ MPa}$$
$$L = 4 \text{ m} = 4000 \text{ mm}$$

we write

$$\frac{150 \times 10^3 \text{ N}}{d^2} = \frac{0.3(12 \times 10^3 \text{ MPa})}{4000 \text{ mm}}$$

Solving for d, we have

$$d^4 = 667 \times 10^6 \qquad d = 160.7 \text{ mm}$$

For $d = 160.7$ mm, we write

$$\frac{L}{d} = \frac{4000}{160.7} = 24.9 > K$$

Our assumption is correct.

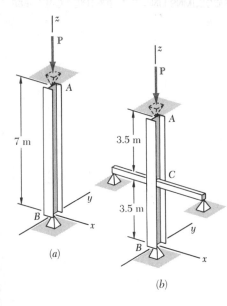

(a)

(b)

SAMPLE PROBLEM 11.3

Column AB consists of a W 250×58 rolled-steel shape made of a grade of steel for which $\sigma_Y = 260$ MPa and $E = 200$ GPa. Determine the centric load **P** which may be applied (a) if the effective length is 7 m in all directions, (b) if bracing is provided to prevent the movement of the midpoint C in the xz plane. (Assume that the movement of point C in the yz plane is not affected by the bracing.)

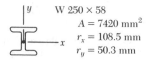

W 250×58

$A = 7420$ mm²

$r_x = 108.5$ mm

$r_y = 50.3$ mm

Solution. We first compute the value of C_c corresponding to the given yield strength $\sigma_Y = 260$ MPa.

$$C_c^2 = \frac{2\pi^2 E}{\sigma_Y} = \frac{2\pi^2(200\ \text{GPa})}{260\ \text{MPa}} = 15.18 \times 10^3 \qquad C_c = 123.2$$

(a) **Effective Length = 7 m.** Since $r_y < r_x$, buckling will take place in the xz plane. For $L = 7$ m and $r = r_y = 50.3$ mm the slenderness ratio is

$$\frac{L}{r_y} = \frac{7\ \text{m}}{0.0503\ \text{m}} = 139.2$$

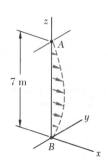

Since $L/r > C_c$, we use Eq. (11.43), with $E = 200$ GPa

$$\sigma_{\text{all}} = \frac{\pi^2 E}{1.92(L/r)^2} = \frac{\pi^2(200\ \text{GPa})}{1.92(139.2)^2} = 53.06\ \text{MPa}$$

$$P_{\text{all}} = \sigma_{\text{all}} A = (53.06\ \text{MPa})(7420 \times 10^{-6}\ \text{m}^2) \qquad P_{\text{all}} = 394\ \text{kN} \quad \blacktriangleleft$$

(b) **Bracing at Midpoint C.** Since bracing prevents movement of point C in the xz plane but not in the yz plane, we must compute the slenderness ratio corresponding to buckling in each plane and determine which is larger.

xz **Plane:** Effective Length = 3.5 m, $r = r_y = 50.3$ mm

$\qquad\quad\ L/r = (3.5\ \text{m})/(0.0503\ \text{m}) = 69.6$

yz **Plane:** Effective Length = 7 m, $r = r_x = 108.5$ mm

$\qquad\quad\ L/r = (7\ \text{m})/(0.1085\ \text{m}) = 64.5$

Since the larger slenderness ratio corresponds to a smaller allowable load, we choose $L/r = 69.6$. For $L/r < C_c$, the column is of intermediate length and we use Eqs. (11.44) and (11.45). We first compute the factor of safety, then the allowable stress, and finally the allowable load **P**.

$$F.S. = \frac{5}{3} + \frac{3}{8}\left(\frac{L/r}{C_c}\right) - \frac{1}{8}\left(\frac{L/r}{C_c}\right)^3 = \frac{5}{3} + \frac{3}{8}\left(\frac{69.6}{123.2}\right) - \frac{1}{8}\left(\frac{69.6}{123.2}\right)^3 = 1.86$$

$$\sigma_{\text{all}} = \frac{\sigma_Y}{F.S.}\left[1 - \frac{1}{2}\left(\frac{L/r}{C_c}\right)^2\right] = \frac{260\ \text{MPa}}{1.86}\left[1 - \frac{1}{2}\left(\frac{69.6}{123.2}\right)^2\right] = 117.5\ \text{MPa}$$

$$P_{\text{all}} = \sigma_{\text{all}} A = (117.5\ \text{MPa})(7420 \times 10^{-6}\ \text{m}^2) \qquad P_{\text{all}} = 872\ \text{kN} \quad \blacktriangleleft$$

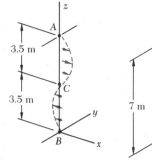

Buckling in xz plane   Buckling in yz plane

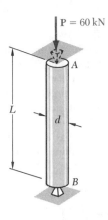

P = 60 kN

A

L

d

B

c

d

SAMPLE PROBLEM 11.4

Using the aluminum alloy 2014-T6, determine the smallest diameter rod which may be used to support the centric load $P = 60$ kN if (a) $L = 750$ mm, (b) $L = 300$ mm.

Solution. For the cross section of a solid circular rod, we have

$$I = \frac{\pi}{4}c^4 \qquad A = \pi c^2 \qquad r = \sqrt{\frac{I}{A}} = \sqrt{\frac{\pi c^4/4}{\pi c^2}} = \frac{c}{2}$$

(a) **Length of 750 mm.** Since the diameter of the rod is not known, a value of L/r must be assumed; we shall *assume* that $L/r > 55$ and use Eq. (11.51). For the centric load **P**, we have $\sigma = P/A$ and write

$$\frac{P}{A} = \sigma_{\text{all}} = \frac{372 \times 10^3 \text{ MPa}}{(L/r)^2}$$

$$\frac{60 \times 10^3 \text{ N}}{\pi c^2} = \frac{372 \times 10^9 \text{ Pa}}{\left(\dfrac{0.750 \text{ m}}{c/2}\right)^2}$$

$$c^4 = 115.5 \times 10^{-9} \text{ m}^4 \qquad c = 18.44 \text{ mm}$$

For $c = 18.44$ mm, the slenderness ratio is

$$\frac{L}{r} = \frac{L}{c/2} = \frac{750 \text{ mm}}{(18.44 \text{ mm})/2} = 81.3 > 55$$

Our assumption is correct, and for $L = 750$ mm, the required diameter is

$$d = 2c = 2(18.44 \text{ mm}) \qquad\qquad d = 36.9 \text{ mm} \quad \blacktriangleleft$$

(b) **Length of 300 mm.** We again *assume* that $L/r > 55$. Using Eq. (11.51), and following the procedure used in part a, we find that $c = 11.66$ mm and $L/r = 51.5$. Since L/r is less than 55, our assumption is wrong; we now assume that $12 < L/r < 55$ and use Eq. (11.50') for the design of this rod.

$$\frac{P}{A} = \sigma_{\text{all}} = \left[212 - 1.585\left(\frac{L}{r}\right)\right]\text{MPa}$$

$$\frac{60 \times 10^3 \text{ N}}{\pi c^2} = \left[212 - 1.585\left(\frac{0.3 \text{ m}}{c/2}\right)\right]10^6 \text{ Pa}$$

$$c = 12.00 \text{ mm}$$

For $c = 12.00$ mm, the slenderness ratio is

$$\frac{L}{r} = \frac{L}{c/2} = \frac{300 \text{ mm}}{(12.00 \text{ mm})/2} = 50$$

Our second assumption that $12 < L/r < 55$ is correct. For $L = 300$ mm, the required diameter is

$$d = 2c = 2(12.00 \text{ mm}) \qquad\qquad d = 24.0 \text{ mm} \quad \blacktriangleleft$$

PROBLEMS

11.70 Determine the allowable centric load for a column of 6-m effective length which is made of the following rolled-steel shape: (*a*) W 200 × 35.9, (*b*) W 200 × 86. Use σ_Y = 250 MPa and E = 200 GPa.

11.71 Determine the allowable centric load for a column of 7-m effective length which is made of the following rolled-steel shape: (*a*) W 250 × 49.1, (*b*) W 310 × 143. Use σ_Y = 250 MPa and E = 200 GPa.

11.72 A column with the cross section shown has a 4-m effective length. Knowing that σ_Y = 250 MPa and E = 200 GPa, determine the largest centric load which may be applied to the column.

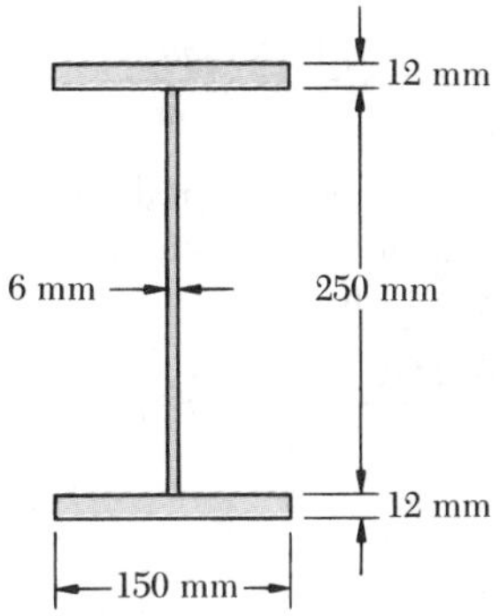

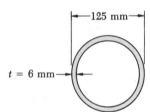

6 mm   250 mm
12 mm
12 mm
150 mm

Fig. P11.72

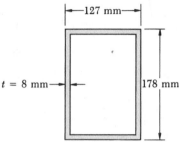

127 mm

t = 8 mm   178 mm

Fig. P11.73

11.73 A rectangular structural tube having the cross section shown is used as a column of 5-m effective length. Knowing that σ_Y = 250 MPa and E = 200 GPa, determine the largest centric load which may be applied to the column.

11.74 A steel pipe having the cross section shown is used as a column. Knowing that σ_Y = 320 MPa and E = 200 GPa, determine the allowable centric load if the effective length of the column is (*a*) 6 m, (*b*) 4 m.

11.75 A steel pipe having the cross section shown is used as a column. Knowing that σ_Y = 250 MPa and E = 200 GPa, determine the allowable centric load if the effective length of the column is (*a*) 6 m, (*b*) 4 m.

11.76 A square tube with the cross section shown is used as a column. Knowing that σ_Y = 250 MPa and E = 200 GPa, determine the allowable centric load if the effective length of the column is (*a*) 9.6 m, (*b*) 7.2 m.

125 mm

t = 6 mm

Fig. P11.74 and P11.75

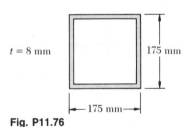

t = 8 mm   175 mm

175 mm

Fig. P11.76

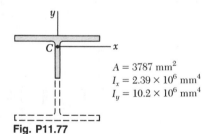

y
C x

$A = 3787 \text{ mm}^2$
$I_x = 2.39 \times 10^6 \text{ mm}^4$
$I_y = 10.2 \times 10^6 \text{ mm}^4$

Fig. P11.77

11.77 A column is made of half of a W 200 × 59 rolled-steel shape, the cross section of which has the geometric properties shown. Knowing that σ_Y = 250 MPa and E = 200 GPa, determine the allowable centric load if the effective length of the column is (*a*) 3.6 m, (*b*) 2.25 m.

11.78 A W 200 × 59 rolled-steel shape is used to form a column of 5-m effective length. Knowing that $E = 200$ GPa, determine the allowable centric load if the yield strength of the grade of steel used is (*a*) $\sigma_Y = 250$ MPa, (*b*) $\sigma_Y = 345$ MPa.

11.79 Solve Prob. 11.78, assuming that the effective length of the column is changed to 7 m.

11.80 A W 250 × 67 rolled-steel shape is used to form a column of 5-m effective length. Knowing that $E = 200$ GPa, determine the allowable centric load if the yield strength of the grade of steel used is (*a*) $\sigma_Y = 250$ MPa, (*b*) $\sigma_Y = 345$ MPa.

11.81 Solve Prob. 11.80, assuming that the effective length of the column is changed to 7.5 m.

11.82 and 11.83 Four planks, each of 38 × 190-mm cross section, are securely nailed together as shown to form a column. Knowing that for the grade of southern pine used $E = 12$ GPa and that the allowable stress for compression parallel to the grain is 10 MPa, determine the allowable centric load if the effective length of the column is (*a*) 7 m, (*b*) 3 m.

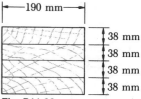
Fig. P11.82

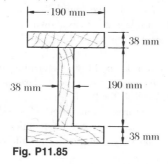
Fig. P11.83

11.84 and 11.85 Three planks, each of 38 × 190-mm cross section, are securely nailed together as shown to form a column. Knowing that for the grade of southern pine used $E = 12$ GPa and that the allowable stress for compression parallel to the grain is 10 MPa, determine the allowable centric load if the effective length of the column is (*a*) 2.4 m, (*b*) 4.4 m.

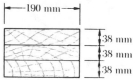
Fig. P11.84

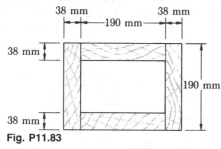
Fig. P11.85

11.86 Four planks, each of 38 × 140-mm cross section, are securely nailed together as shown to form a column. Knowing that for the grade of southern pine used $E = 12$ GPa and that the allowable stress for compression parallel to the grain is 10 MPa, determine the allowable centric load for the column if its effective length is (*a*) 3.2 m, (*b*) 1.6 m.

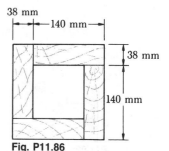
Fig. P11.86

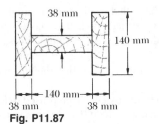

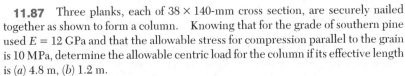

Fig. P11.87

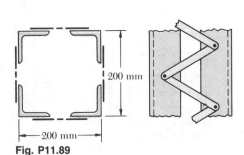

Fig. P11.88

11.87 Three planks, each of 38 × 140-mm cross section, are securely nailed together as shown to form a column. Knowing that for the grade of southern pine used $E = 12$ GPa and that the allowable stress for compression parallel to the grain is 10 MPa, determine the allowable centric load for the column if its effective length is (*a*) 4.8 m, (*b*) 1.2 m.

11.88 A column of 6-m effective length is obtained by connecting two 180 × 14.6 steel channels as shown. Knowing that $\sigma_Y = 250$ MPa and $E = 200$ GPa, determine the allowable centric load for the column.

11.89 A column of 5.4-m effective length is obtained by connecting four 76 × 76 × 9.5-mm steel angles with lacing bars as shown. Knowing that $\sigma_Y = 250$ MPa and $E = 200$ GPa, determine the allowable centric load for the column.

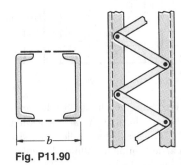

Fig. P11.90

200 mm

200 mm

Fig. P11.89

11.90 A column of 7-m effective length is obtained by connecting two C 250 × 37 steel channels with lacing bars as shown. Knowing that $\sigma_Y = 250$ MPa and $E = 200$ GPa, determine the allowable centric load for the column when $b = 180$ mm.

11.91 For the column of Prob. 11.90, determine (*a*) the smallest distance b for which the allowable normal stress σ_{all} is as large as possible, (*b*) the corresponding allowable centric load for the column.

11.92 and 11.93 A compression member of 3-m effective length is obtained by bolting together two L 102 × 76 × 6.4 steel angles as shown. Knowing that $\sigma_Y = 250$ MPa and $E = 200$ GPa, determine the allowable centric load for the compression member.

Fig. P11.92

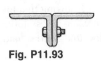

Fig. P11.93

Fig. P11.94

Fig. P11.95

11.94 and 11.95 A compression member of 9-m effective length is obtained by welding two 10-mm-thick steel plates to a W 250 × 80 rolled-steel shape as shown. Knowing that $\sigma_Y = 320$ MPa and $E = 200$ GPa, determine the allowable centric load for the compression member.

11.96 A 16-mm-diameter rod made of the aluminum alloy 6061-T6 is used as a compression strut. Determine the allowable centric load for the strut if its effective length is (a) 120 mm, (b) 240 mm.

11.97 Bar AB is free at its end A and fixed at its base B. Determine the allowable centric load **P** if the aluminum alloy used is (a) 6061-T6, (b) 2014-T6.

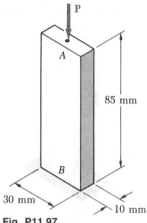

Fig. P11.97

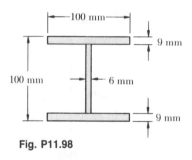

Fig. P11.98

11.98 A compression member has the cross section shown and an effective length of 1.2 m. Knowing that the aluminum alloy used is 2014-T6, determine the allowable centric load.

11.99 Solve Prob. 11.98, assuming that the effective length of the compression member is increased to 1.6 m.

11.100 A compression member has the cross section shown and an effective length of 2 m. Knowing that the aluminum alloy used is 2014-T6, determine the allowable centric load.

11.101 Solve Prob. 11.100, assuming that the effective length of the compression member is increased to 2.75 m.

11.102 A column of 4.5-m effective length must carry a centric load of 900 kN. Knowing that $\sigma_Y = 350$ MPa and $E = 200$ GPa, select the wide-flange shape of 250-mm nominal depth which should be used.

11.103 A column of 7.8-m effective length must carry a centric load of 1100 kN. Knowing that $\sigma_Y = 350$ MPa and $E = 200$ GPa, select the wide-flange shape of 300-mm nominal depth which should be used.

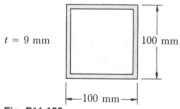

Fig. P11.100

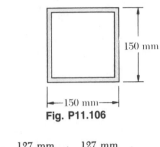

Fig. P11.106

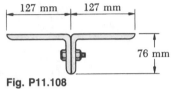

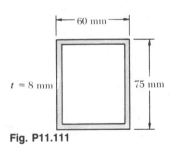

Fig. P11.108

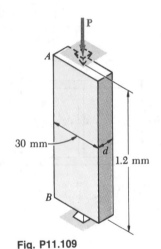

Fig. P11.109

11.104 A column of 4.4-m effective length must carry a centric load of 1150 kN. Knowing that $\sigma_Y = 350$ MPa and $E = 200$ GPa, select the wide-flange shape of 360-mm nominal depth which should be used.

11.105 A column of 7.5-m effective length must carry a centric load of 220 kN. Knowing that $\sigma_Y = 250$ MPa and $E = 200$ GPa, select the wide-flange shape of 200-mm nominal depth which should be used.

11.106 A square structural tube having the cross section shown is used as a column of 7.8-m effective length to carry a centric load of 290 kN. Knowing that the tubes available for use are made with wall thicknesses in increments of 1.5 mm from 6.5 mm to 20 mm, determine the lightest tube which may be used. Use $\sigma_Y = 250$ MPa and $E = 200$ GPa.

11.107 Solve Prob. 11.106, assuming that the effective length of the column is decreased to 6 m.

11.108 Two 127×76-mm angles are bolted together as shown for use as a column of 3-m effective length to carry a centric load of 165 kN. Knowing that the angles available for use are made with thicknesses of 6.4 mm, 9.5 mm, and 12.7 mm, determine the lightest angles which may be used. Use $\sigma_Y = 250$ MPa and $E = 200$ GPa.

11.109 A centric load **P** must be supported by the steel bar AB. Knowing that $\sigma_Y = 250$ MPa and $E = 200$ GPa, determine the smallest dimension d of the cross section which may be used when (a) $P = 60$ kN, (b) $P = 30$ kN.

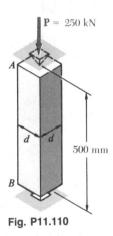

Fig. P11.110

11.110 For a rod made of the aluminum alloy 2014-T6, select the smallest square cross section which may be used if the rod is to carry a 250 kN centric load.

11.111 A structural tube with the cross section shown is used as a column to carry a 150-kN centric load. Determine the largest allowable effective length of the tube if the aluminum alloy used is (a) 6061-T6, (b) 2014-T6.

11.112 An aluminum tube of 75-mm outer diameter and 2-m effective length must carry a centric load of 95 kN. Knowing that the tubes available for use are made of alloy 6061-T6 and have wall thicknesses in increments of 2 mm up to 12 mm, determine the lightest tube which may be used.

Fig. P11.111

11.113 Using the aluminum alloy 2014-T6, determine the largest allowable length of the aluminum bar *AB* for a centric load **P** of magnitude (*a*) 150 kN, (*b*) 90 kN, (*c*) 25 kN.

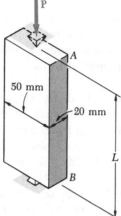

Fig. P11.113

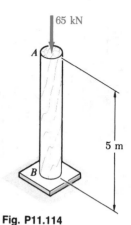

Fig. P11.114

11.114 A 65-kN centric load is applied to a 5-m pole which is free at its top *A* and fixed at its base *B*. Using a timber for which $E = 11$ GPa and the allowable stress for compression parallel to the grain is 9 MPa, determine the smallest diameter pole which may be used.

11.115 A column of 3-m effective length is to be made by securely nailing together boards of 24×100-mm cross section. Knowing that $E = 11$ GPa and the allowable stress for compression parallel to the grain is 9 MPa, determine the number of boards which must be used to support the centric load shown when (*a*) $P = 30$ kN, (*b*) $P = 40$ kN.

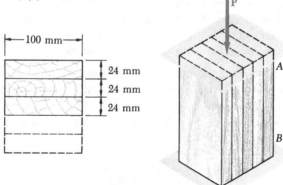

Fig. P11.115

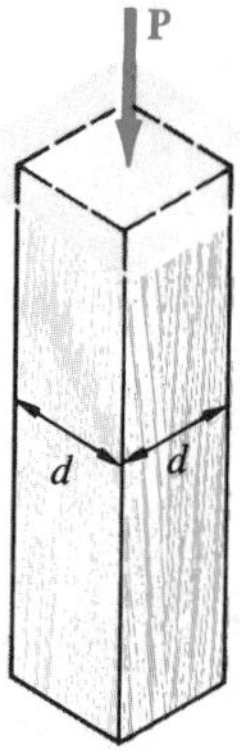

Fig. P11.116

11.116 A 200-kN centric load is applied to a timber column of 4.8-m effective length. Using a grade of southern pine for which $E = 12$ GPa and the allowable stress for compression parallel to the grain is 10 MPa, determine the smallest square cross section which may be used.

11.117 Solve Prob. 11.116, assuming that the effective length of the column is (*a*) 3.6 m, (*b*) 1.2 m.

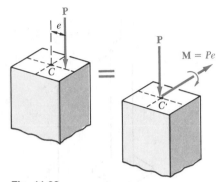

Fig. 11.32

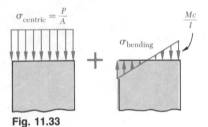

Fig. 11.33

11.7. DESIGN OF COLUMNS UNDER AN ECCENTRIC LOAD

In this section, we shall consider the design of columns subjected to an eccentric load. We shall see how the empirical formulas developed in the preceding section for columns under a centric load may be modified and used when the load **P** applied to the column has an eccentricity e which is known.

We first recall from Sec. 4.13 that an eccentric axial load **P** applied in a plane of symmetry of the column may be replaced by an equivalent system consisting of a centric load **P** and a couple **M** of moment $M = Pe$, where e is the distance from the line of action of the load to the longitudinal axis of the column (Fig. 11.32). The normal stresses exerted on a transverse section of the column may then be obtained by superposing the stresses due, respectively, to the centric load **P** and to the couple **M** (Fig. 11.33), provided that the section considered is not too close to either end of the column, and as long as the stresses involved do not exceed the proportional limit of the material. The normal stresses due to the eccentric load **P** may thus be expressed as

$$\sigma = \sigma_{\text{centric}} + \sigma_{\text{bending}} \qquad (11.59)$$

Recalling the results obtained in Sec. 4.13, we find that the maximum compressive stress in the column is

$$\sigma_{\text{max}} = \frac{P}{A} + \frac{Mc}{I} \qquad (11.60)$$

In a properly designed column, the maximum stress defined by Eq. (11.60) should not exceed the allowable stress for the column. Two alternative approaches may be used to satisfy this requirement, namely, the *allowable-stress method* and the *interaction method*.

a. Allowable-Stress Method. This method is based on the assumption that the allowable stress for an eccentrically loaded column is the same as if the column were centrically loaded. We must have, therefore, $\sigma_{\text{max}} \le \sigma_{\text{all}}$, where σ_{all} is the allowable stress under a centric load, or substituting for σ_{max} from Eq. (11.60),

$$\frac{P}{A} + \frac{Mc}{I} \le \sigma_{\text{all}} \qquad (11.61)$$

The allowable stress is obtained from the formulas of Sec. 11.6 which, for a given material, express σ_{all} as a function of the slenderness ratio of the column. The major engineering codes require that the largest value of the slenderness ratio of the column be used to determine the allowable stress, whether or not this value corresponds to the actual plane of bending. This requirement sometimes results in an overly conservative design.

Example 11.04

A column of 125-mm-square cross section and 3.6-m length is made of a grade of Douglas fir for which $E = 12$ GPa and $\sigma'_{all} = 9$ MPa for compression parallel to the grain. Using the allowable-stress method, determine the maximum load P which may safely be supported with an eccentricity of 50 mm.

We first compute the value of K corresponding to the given values of E and σ'_{all}. Using Eq. (11.54), we write

$$K = 0.671 \sqrt{\frac{E}{\sigma'_{all}}} = 0.671 \sqrt{\frac{12 \times 10^3 \text{ MPa}}{9 \text{ MPa}}} = 24.5$$

We next compute $L/d = (3600 \text{ mm})/(125 \text{ mm}) = 28.8$.

Since $L/d > K$, we use Eq. (11.53) to determine the allowable stress for a timber column of rectangular cross section subjected to a centric load. We have

$$\sigma_{all} = \frac{0.3E}{(L/d)^2} = \frac{0.3(12 \times 10^3 \text{ MPa})}{(28.8)^2} = 4.34 \text{ MPa}$$

We may now use Eq. (11.61) to determine the maximum permissible load P. Substituting the given data

$$A = (125 \text{ mm})^2 = 15\,625 \text{ mm}^2$$

$$I = \tfrac{1}{12}(125 \text{ mm})^4 = 20.3 \times 10^6 \text{ mm}^4$$

$$c = 62.5 \text{ mm} \qquad M = Pe = P(50 \text{ mm})$$

into Eq. (11.61), we write

$$\frac{P}{15\,625 \text{ mm}^2} + \frac{P(50 \text{ mm})(62.5 \text{ mm})}{20.3 \times 10^6 \text{ mm}^4} \leq 4.34 \text{ MPa}$$

$$P \leq 19.9 \text{ kN}$$

The maximum load which may be safely applied is $P = 19.9$ kN.

b. Interaction Method. We recall that the allowable stress for a column subjected to a centric load (Fig. 11.34a) is generally smaller than the allowable stress for a column in pure bending (Fig. 11.34b), since the former takes into account the possibility of buckling. Therefore, when we use the allowable-stress method to design an eccentrically loaded column and write that the sum of the stresses due to the centric load **P** and the bending couple **M** (Fig. 11.34c) must not exceed the allowable stress for a centrically loaded column, the resulting design is generally overly conservative. An improved method of design may be developed by rewriting Eq. (11.61) in the form

$$\frac{P/A}{\sigma_{all}} + \frac{Mc/I}{\sigma_{all}} \leq 1 \qquad (11.62)$$

and substituting for σ_{all} in the first and second terms the values of the allowable stress which correspond, respectively, to the centric loading of Fig. 11.34a and to the pure bending of Fig. 11.34b. We have

$$\frac{P/A}{(\sigma_{all})_{centric}} + \frac{Mc/I}{(\sigma_{all})_{bending}} \leq 1 \qquad (11.63)$$

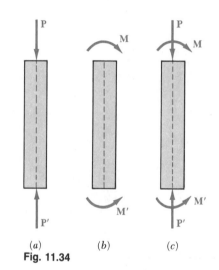

(a)   (b)   (c)

Fig. 11.34

The type of formula obtained is known as an *interaction formula*.

We note that, when $M = 0$, the use of this formula results in the design of a centrically loaded column by the method of Sec. 11.6. On the other hand, when $P = 0$, the use of the formula results in the design of a beam in pure bending by the method of Chap. 4. When P and M are both different from zero, the interaction formula results in a design which takes into account the capacity of the member to resist bending as well as axial loading. In all cases, $(\sigma_{all})_{centric}$ will be determined by using the largest slenderness ratio of the column, regardless of the plane in which bending takes place.[†]

When the eccentric load **P** is not applied in a plane of symmetry of the column, it causes bending about both of the principal axes of the cross section. We recall from Sec. 4.15 that the load **P** may then be replaced by a centric load **P** and two couples represented by the couple vectors $\mathbf{M}_x$

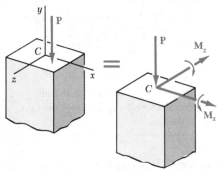

Fig. 11.35

and $\mathbf{M}_z$ shown in Fig. 11.35. The interaction formula to be used in this case is

$$\frac{P/A}{(\sigma_{all})_{centric}} + \frac{|M_x|z_{max}/I_x}{(\sigma_{all})_{bending}} + \frac{|M_z|x_{max}/I_z}{(\sigma_{all})_{bending}} \le 1 \qquad (11.64)$$

[†]This procedure is required by all major codes for the design of steel, aluminum, and timber compression members. In addition, many specifications call for the use of an additional factor in the second term of Eq. (11.63); this factor takes into account the additional stresses resulting from the deflection of the column due to bending.

Example 11.05

Use the interaction method to determine the maximum load P which may be safely supported by the column of Example 11.04 with an eccentricity $e = 50$ mm.

The value of $(\sigma_{all})_{centric}$ has already been determined in Example 11.04. On the other hand, the value of $(\sigma_{all})_{bending}$ is equal to the specified allowable stress for compression parallel to the grain. We have, therefore,

$$(\sigma_{all})_{centric} = 4.34 \text{ MPa} \qquad (\sigma_{all})_{bending} = 9 \text{ MPa}$$

Substituting these values into Eq. (11.63), we write

$$\frac{P/A}{4.34 \text{ MPa}} + \frac{Mc/I}{9 \text{ MPa}} \le 1$$

or, recalling the numerical data obtained in Example 11.04,

$$\frac{P/15\,625}{4.34} + \frac{P(50)(62.5)/20.3 \times 10^6}{9} \le 1$$

$$P \le 31.4 \text{ kN}$$

The maximum load which may be safely applied is thus $P = 31.4$ kN.

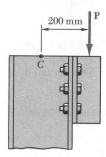

SAMPLE PROBLEM 11.5

Using the allowable-stress method, determine the largest load **P** which may be safely carried by a W 310 × 74 steel column of 4.5-m effective length. Use $E = 200$ GPa and $\sigma_Y = 250$ MPa.

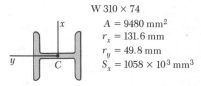

W 310 × 74
$A = 9480 \text{ mm}^2$
$r_x = 131.6$ mm
$r_y = 49.8$ mm
$S_x = 1058 \times 10^3 \text{ mm}^3$

Solution. The largest slenderness ratio of the column is $L/r_y = (4.5 \text{ m})/(0.0498 \text{ m}) = 90.4$. Using Eq. (11.42) with $E = 200$ GPa and $\sigma_Y = 250$ MPa, we find that $C_c = 125.7$. Since $L/r < C_c$, we use Eqs. (11.44) and (11.45) and find $F.S. = 1.890$ and

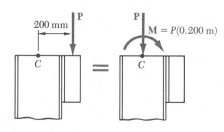

$$\sigma_{all} = (\sigma_{all})_{centric} = 98.1 \text{ MPa}$$

For the given column and loading, we have

$$\frac{P}{A} = \frac{P}{9.48 \times 10^{-3} \text{ m}^2} \qquad \frac{Mc}{I} = \frac{M}{S} = \frac{P(0.200 \text{ m})}{1.058 \times 10^{-3} \text{ m}^3}$$

Substituting into Eq. (11.61), we write

$$\frac{P}{A} + \frac{Mc}{I} \leq \sigma_{all}$$

$$\frac{P}{9.48 \times 10^{-3} \text{ m}^2} + \frac{P(0.200 \text{ m})}{1.058 \times 10^{-3} \text{ m}^3} \leq 98.1 \text{ MPa} \qquad P \leq 333 \text{ kN}$$

The largest allowable load **P** is thus

$$\mathbf{P} = 333 \text{ kN} \downarrow \quad \blacktriangleleft$$

SAMPLE PROBLEM 11.6

Using the interaction method, solve Sample Prob. 11.5. Assume $(\sigma_{all})_{bending} = 150$ MPa.

Solution. Using Eq. (11.63), we write

$$\frac{P/A}{(\sigma_{all})_{centric}} + \frac{Mc/I}{(\sigma_{all})_{bending}} \leq 1$$

Substituting the given allowable bending stress and the allowable centric stress found in Sample Prob. 11.5, as well as the other given data, we have

$$\frac{P/(9.48 \times 10^{-3} \text{ m}^2)}{98.1 \times 10^6 \text{ Pa}} + \frac{P(0.200 \text{ m})/(1.058 \times 10^{-3} \text{ m}^3)}{150 \times 10^6 \text{ Pa}} \leq 1$$

$$P \leq 428 \text{ kN}$$

The largest allowable load **P** is thus

$$\mathbf{P} = 428 \text{ kN} \downarrow \quad \blacktriangleleft$$

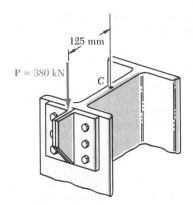

$P = 380$ kN

125 mm

C

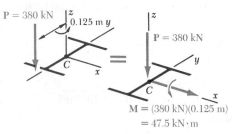

$P = 380$ kN

0.125 m y

C

x

$=$

z

$P = 380$ kN

y

C

x

$M = (380$ kN$)(0.125$ m$)$
$= 47.5$ kN·m

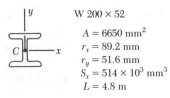

W 200 × 52

$A = 6650$ mm^2
$r_x = 89.2$ mm
$r_y = 51.6$ mm
$S_x = 514 \times 10^3$ mm^3
$L = 4.8$ m

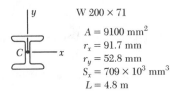

W 200 × 71

$A = 9100$ mm^2
$r_x = 91.7$ mm
$r_y = 52.8$ mm
$S_x = 709 \times 10^3$ mm^3
$L = 4.8$ m

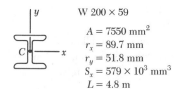

W 200 × 59

$A = 7550$ mm^2
$r_x = 89.7$ mm
$r_y = 51.8$ mm
$S_x = 579 \times 10^3$ mm^3
$L = 4.8$ m

SAMPLE PROBLEM 11.7

A steel column having an effective length of 4.8-m is loaded eccentrically as shown. Using the interaction method, select the wide-flange shape of 200-mm nominal depth which should be used. Assume $E = 200$ GPa and $\sigma_Y = 260$ MPa, and use an allowable stress in bending of 150 MPa.

Solution. In order to assist us in selecting a trial section, we use the allowable-stress method with $\sigma_{all} = 150$ MPa and write

$$\sigma_{all} = \frac{P}{A} + \frac{Mc}{I_x} = \frac{P}{A} + \frac{Mc}{Ar_x^2} \qquad (1)$$

From Appendix C we observe for shapes of 200-mm nominal depth that $c \approx 100$ mm and $r_x \approx 90$ mm. Substituting into Eq. (1), we have

$$150 \text{ MPa} = \frac{380 \text{ kN}}{A} + \frac{(47.5 \text{ kN·m})(0.100 \text{ m})}{A(0.090 \text{ m})^2} \qquad A \approx 6440 \text{ mm}^2$$

We select for a first trial shape: W 200 × 52.

Trial 1: W 200 × 52. The allowable stresses are:

Allowable Bending Stress: (see data)) $\qquad (\sigma_{all})_{bending} = 150$ MPa

Allowable Axial Stress: The largest slenderness ratio is $L/r_y = (4.8 \text{ m})/(0.0516 \text{ m}) = 93.0$. For $E = 200$ GPa and $\sigma_Y = 260$ MPa, Eq. (11.42) yields $C_c = 123.2$ and we note that $L/r < C_c$. Using Eqs. (11.44) and (11.45) we find

$$(\sigma_{all})_{centric} = 98.1 \text{ MPa}$$

For the W 200 × 52 trial shape, we have

$$\frac{P}{A} = \frac{380 \text{ kN}}{6650 \times 10^{-6} \text{ m}^2} = 57.1 \text{ MPa} \qquad \frac{Mc}{I} = \frac{M}{S_x} = \frac{47.5 \text{ kN·m}}{514 \times 10^{-6} \text{ m}^3} = 92.4 \text{ MPa}$$

With this data we find that the left-hand member of Eq. (11.63) is

$$\frac{P/A}{(\sigma_{all})_{centric}} + \frac{Mc/I}{(\sigma_{all})_{bending}} = \frac{57.1 \text{ MPa}}{98.1 \text{ MPa}} + \frac{92.4 \text{ MPa}}{150 \text{ MPa}} = 1.198$$

Since $1.198 > 1.000$, the requirement expressed by the interaction formula is not satisfied; we must select a larger trial shape.

Trial 2: W 200 × 71. Following the procedure used in trial 1, we write

$$\frac{L}{r_y} = \frac{4.8 \text{ m}}{0.0528 \text{ m}} = 90.9 \qquad (\sigma_{all})_{centric} = 99.9 \text{ MPa}$$

$$\frac{P}{A} = \frac{380 \text{ kN}}{9100 \times 10^{-6} \text{ m}^2} = 41.8 \text{ MPa} \qquad \frac{Mc}{I} = \frac{M}{S_x} = \frac{47.5 \text{ kN·m}}{709 \times 10^{-6} \text{ m}^3} = 67.0 \text{ MPa}$$

Substituting into Eq. (11.63):

$$\frac{P/A}{(\sigma_{all})_{centric}} + \frac{Mc/I}{(\sigma_{all})_{bending}} = \frac{41.8 \text{ MPa}}{99.9 \text{ MPa}} + \frac{67.0 \text{ MPa}}{150 \text{ MPa}} = 0.865 < 1.000$$

The W 200 × 71 shape is satisfactory but may be unnecessarily large.

Trial 3: W 200 × 59. Following again the same procedure, we find that the interaction formula is not satisfied.

Selection of Shape. The shape to be used is $\qquad$ W 200 × 71 ◀

PROBLEMS

11.118 A 220-mm-diameter timber pole is free at its top A and fixed at its base B. For the grade of timber used $E = 12$ GPa and the allowable stress for compression parallel to the grain is 10 MPa. Using the allowable-stress method, determine the largest eccentric load $\mathbf{P}$ which may be applied as shown.

11.119 Solve Prob. 11.118, using the interaction method.

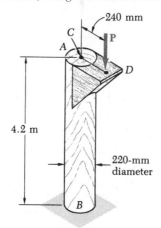

Fig. P11.118

11.120 A timber column of 120×180-mm cross section has an effective length of 2.5 m. For the grade of timber used $E = 12$ GPa and the allowable stress for compression parallel to the grain is 10 MPa. Using the allowable-stress method, determine the largest eccentric load $\mathbf{P}$ which may be applied when (a) $e = 12$ mm, (b) $e = 24$ mm.

11.121 Solve Prob. 11.120 when (a) $e = 18$ mm, (b) $e = 36$ mm.

11.122 A uniform column of 5.5-m effective length consists of a section of steel tubing having the cross section shown. Using the interaction method, determine the allowable load $\mathbf{P}$ when the eccentricity is (a) $e = 0$, (b) $e = 30$ mm. Assume $E = 200$ GPa, $\sigma_Y = 250$ MPa, and an allowable stress in bending of 150 MPa.

11.123 Solve Prob. 11.122, using the allowable-stress method.

11.124 A steel compression member of 2.7-m effective length supports an eccentric load as shown. Using the allowable-stress method, determine the allowable load $\mathbf{P}$. Assume $E = 200$ GPa and $\sigma_Y = 250$ MPa.

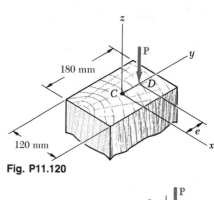

Fig. P11.120

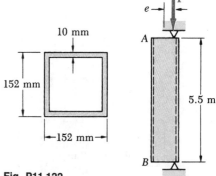

Fig. P11.122

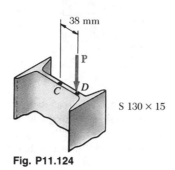

Fig. P11.124

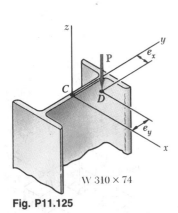

Fig. P11.125

11.125 A column of 6-m effective length consists of a W 310 × 74 rolled-steel shape. For the grade of steel used σ_Y = 350 MPa, σ_{all} = 200 MPa in bending, and E = 200 GPa. Using the interaction method, determine the allowable load **P** if the eccentricity is (a) e_x = 25 mm, e_y = 0, (b) e_x = 0, e_y = 25 mm.

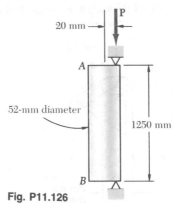

Fig. P11.126

11.126 An eccentric load is applied at a point 20 mm from the geometric axis of a 52-mm-diameter rod made of the aluminum alloy 6061-T6 for which the allowable stress in bending is 150 MPa. Using the interaction method, determine the allowable load **P**.

11.127 Solve Prob. 11.126 using the allowable-stress method, assuming that the aluminum alloy used is 2014-T6.

11.128 An eccentric load **P** of magnitude 160 kN is applied as shown to the tube AB which is made of the aluminum alloy 6061-T6. Using the interaction method and an allowable stress in bending of 150 MPa, determine the largest allowable eccentricity e.

11.129 Solve Prob. 11.128, assuming that the magnitude P of the eccentric load is increased to 240 kN.

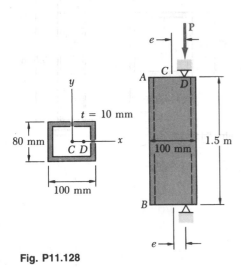

Fig. P11.128

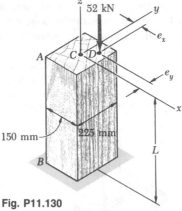

Fig. P11.130

11.130 A timber column AB is free at its top A and fixed at its base B and supports a 52-kN load as shown. For the grade of timber used E = 12 GPa and the allowable stress for compression parallel to the grain is 10 MPa. Using the allowable-stress method, determine the largest length L which may be used when (a) e_x = 25 mm, e_y = 0, (b) e_x = 0, e_y = 25 mm.

11.131 Solve Prob. 11.130 when $e_x = e_y$ = 25 mm.

11.132 Two L $102 \times 76 \times 9.5$ steel angles are welded together to form the column AB. An axial load **P** of magnitude 60 kN is applied to point D. Using the allowable-stress method, determine the largest allowable length L. Assume $E = 200$ GPa and $\sigma_Y = 250$ MPa.

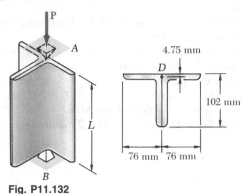

Fig. P11.132

11.133 Solve Prob. 11.132, using the interaction method, for a load of magnitude $P = 80$ kN and an allowable stress in bending of 150 MPa.

11.134 A 15-kN load is applied with an eccentricity $e = 0.2d$ to a square timber column having an effective length of 3.5 m. For the grade of Douglas fir used $E = 12.5$ GPa and the allowable stress for compression parallel to the grain is 7.5 MPa. Using the allowable-stress method, determine the smallest allowable dimension d.

11.135 A 45-kN load is applied with an eccentricity of 40 mm to the timber post AB. For the grade of timber used $E = 12.5$ GPa and the allowable stress for compression parallel to the grain is 7.5 MPa. Using the interaction method, determine the smallest allowable diameter of the post.

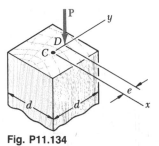

Fig. P11.134

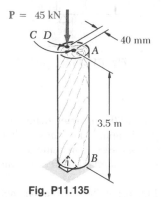

Fig. P11.135

11.136 Solve Prob. 11.135 using the allowable-stress method.

11.137 A square timber column of 4.5-m effective length supports a 65-kN load with eccentricities $e = 45$ mm as shown. For the grade of southern pine used $E = 11$ GPa and the allowable stress for compression parallel to the grain is 9 MPa. Using the interaction method, determine the smallest allowable dimension d.

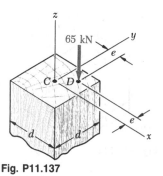

Fig. P11.137

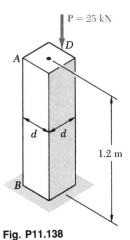

$P = 25$ kN

A

D

d d

1.2 m

B

Fig. P11.138

11.138 A 25-kN vertical load **P** is applied at the midpoint of one edge of the square cross section of the steel compression member AB, which is free at its top A and fixed at its base B. Knowing that for the grade of steel used $E = 200$ GPa and $\sigma_Y = 250$ MPa and using the allowable-stress method, determine the smallest allowable dimension d.

11.139 Solve Prob. 11.138, assuming that the vertical load **P** is applied at the corner of the square cross section.

11.140 A rectangular steel tube is to carry a load of 60 kN having an eccentricity $e = 25$ mm. The tubes available for use have wall thicknesses in increments of 1.5 mm from 6 mm to 12 mm. Using the interaction method, determine the lightest tube which may be used. Assume $E = 200$ GPa and $\sigma_Y = 250$ MPa, and an allowable stress in bending of 150 MPa.

11.141 Solve Prob. 11.140, using the allowable-stress method.

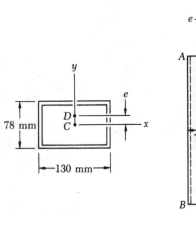

y

e

D
C x

78 mm

—130 mm—

Fig. P11.140

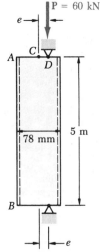

$P = 60$ kN

e

A   C
  D

78 mm   5 m

B

e

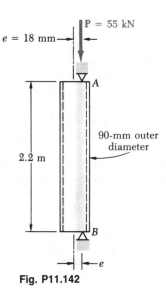

$P = 55$ kN

$e = 18$ mm

A

90-mm outer diameter

2.2 m

B

e

Fig. P11.142

11.142 An aluminum tube of 90-mm outer diameter is to carry a load of 55 kN having an eccentricity $e = 18$ mm. The tubes available for use are made of alloy 6061-T6 and have wall thicknesses in increments of 3 mm from 6 mm to 15 mm. Using the allowable-stress method, determine the lightest tube which may be used.

11.143 Solve Prob. 11.142, using the interaction method, for a load **P** = 75 kN and an allowable stress in bending of 150 MPa.

11.144 A compression member of rectangular cross section has an effective length of 0.9 m and is made of the aluminum alloy 2014-T6 for which the allowable stress in bending is 165 MPa. Using the interaction method, determine the smallest dimension d of the cross section which may be used when $e = 10$ mm.

11.145 Solve Prob. 11.144, assuming that $e = 5$ mm.

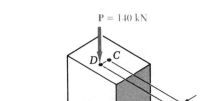

$P = 140$ kN

D C

e

60 mm d

Fig. P11.144

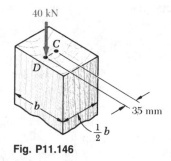

Fig. P11.146

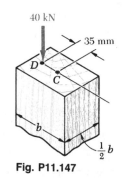

Fig. P11.147

11.146 and 11.147 A timber column of rectangular cross section has an 2.5-m effective length and must support a 40-kN load as shown. Using the allowable-stress method, design the column so that one side of its cross section will be twice the other side. Assume $E = 12$ GPa and an allowable stress of 10 MPa for compression parallel to the grain.

11.148 A steel column of 6.6-m effective length must support an eccentric load of 160 kN at a point D located on the x axis as shown. Using the interaction method, select the wide-flange shape of 310 mm nominal depth which should be used. Assume $E = 200$ GPa, $\sigma_Y = 250$ MPa, and $\sigma_{all} = 150$ MPa in bending.

11.149 Solve Prob. 11.148, selecting a wide-flange shape of 250 mm nominal depth.

11.150 A steel column of 7.4-m effective length must carry a load of 245 kN with an eccentricity of 62 mm as shown. Using the interaction method, select the wide-flange shape of 250-mm nominal depth which should be used. Assume $E = 200$ GPa, $\sigma_Y = 250$ MPa, and $\sigma_{all} = 150$ MPa in bending.

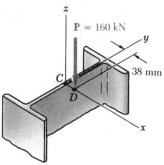

Fig. P11.148

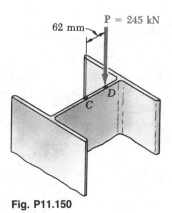

Fig. P11.150

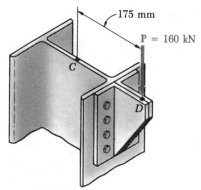

Fig. P11.151

11.151 A steel column of 5-m effective length must carry a load of 160 kN with an eccentricity of 175 mm as shown. Using the interaction method, select the wide-flange shape of 200-mm nominal depth which should be used. Assume $E = 200$ GPa, $\sigma_Y = 250$ MPa, and $\sigma_{all} = 150$ MPa in bending.

Critical load

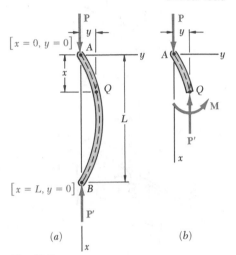

$[x = 0, y = 0]$

$[x = L, y = 0]$

(a)

(b)

Fig. 11.7

Euler's formula

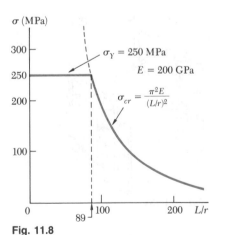

σ (MPa)

$\sigma_Y = 250$ MPa

$E = 200$ GPa

$\sigma_{cr} = \dfrac{\pi^2 E}{(L/r)^2}$

Fig. 11.8

Slenderness ratio

REVIEW AND SUMMARY

This chapter was devoted to the design and analysis of columns, i.e., prismatic members supporting axial loads. In order to gain insight into the behavior of columns, we first considered in Sec. 11.2 the equilibrium of a simple model and found that for values of the load P exceeding a certain value P_{cr}, called the *critical load*, two equilibrium positions of the model were possible: the original position with zero transverse deflections and a second position involving deflections which could be quite large. This led us to conclude that the first equilibrium position was unstable for $P > P_{cr}$, and stable for $P < P_{cr}$, since in the latter case it was the only possible equilibrium position.

In Sec. 11.3, we considered a pin-ended column of length L and of constant flexural rigidity EI subjected to an axial centric load P. Assuming that the column had buckled (Fig. 11.7), we noted that the bending moment at point Q was equal to $-Py$ and wrote

$$\frac{d^2y}{dx^2} = \frac{M}{EI} = -\frac{P}{EI}y \qquad (11.4)$$

Solving this differential equation, subject to the boundary conditions corresponding to a pin-ended column, we determined the smallest load P for which buckling can take place. This load, known as the *critical load* and denoted by P_{cr}, is given by *Euler's formula*:

$$P_{cr} = \frac{\pi^2 EI}{L^2} \qquad (11.11)$$

where L is the length of the column. For this load or any larger load, the equilibrium of the column is unstable and transverse deflections will occur.

Denoting the cross-sectional area of the column by A and its radius of gyration by r, we determined the critical stress σ_{cr} corresponding to the critical load P_{cr}:

$$\sigma_{cr} = \frac{\pi^2 E}{(L/r)^2} \qquad (11.13)$$

The quantity L/r is called the *slenderness ratio* and we plotted σ_{cr} as a function of L/r (Fig. 11.8). Since our analysis was based on stresses remaining below the yield strength of the material, we noted that the column would fail by yielding when $\sigma_{cr} > \sigma_Y$.

In Sec. 11.4, we discussed the critical load of columns with various end conditions and wrote

$$P_{cr} = \frac{\pi^2 EI}{L_e^2} \qquad (11.11')$$

where L_e is the *effective length* of the column, i.e., the length of an equivalent pin-ended column. The effective lengths of several columns with various end conditions were calculated and shown in Fig. 11.17 on page 640.

Effective length

In Sec. 11.5, we considered columns supporting an *eccentric axial load*. For a pin-ended column subjected to a load P applied with an eccentricity e, we replaced the load by a centric axial load and a couple of moment $M_A = Pe$ (Figs. 11.18a and 11.19) and derived the following expression for the maximum transverse deflection:

$$y_{max} = e \left[\sec \left(\sqrt{\frac{P}{EI}} \frac{L}{2} \right) - 1 \right] \qquad (11.28)$$

Eccentric axial load. Secant formula.

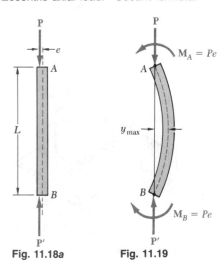

Fig. 11.18a   Fig. 11.19

We then determined the maximum stress in the column, and from the expression obtained for that stress, we derived the *secant formula*:

$$\frac{P}{A} = \frac{\sigma_{max}}{1 + \frac{ec}{r^2} \sec \left(\frac{1}{2} \sqrt{\frac{P}{EA}} \frac{L_e}{r} \right)} \qquad (11.36)$$

This equation may be solved for the force per unit area, P/A, which causes a specified maximum stress σ_{max} in a pin-ended column or any other column of effective slenderness ratio L_e/r.

In the first part of the chapter we considered each column as a straight homogeneous prism. Since imperfections exist in all real columns, the *design of real columns* is done by using empirical formulas based on laboratory tests and set forth in specifications codes issued by professional organizations. In Sec. 11.6, we discussed the design of *centrically loaded columns* made of steel, aluminum, or timber. For each material, the design of the column was based on formulas expressing the allowable stress as a function of the slenderness ratio L/r of the column.

Design of real columns

Centrically loaded columns

In the last section of the chapter [Sec. 11.7], we studied two methods used for the design of columns under an *eccentric* load. The first method was the *allowable-stress method*, a conservative method in which it is

Eccentrically loaded columns

Allowable-stress method

assumed that the allowable stress is the same as if the column were centrically loaded. The allowable-stress method requires that the following inequality be satisfied:

$$\frac{P}{A} + \frac{Mc}{I} \leq \sigma_{\text{all}} \qquad (11.61)$$

Interaction method

The second method was the *interaction method,* a method used in most modern specifications. In this method the allowable stress for a centrically loaded column is used for the portion of the total stress due to the axial load and the allowable stress in bending for the stress due to bending. Thus, the inequality to be satisfied is

$$\frac{P/A}{(\sigma_{\text{all}})_{\text{centric}}} + \frac{Mc/I}{(\sigma_{\text{all}})_{\text{bending}}} \leq 1 \qquad (11.63)$$

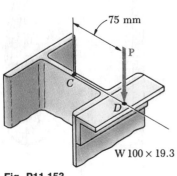

Fig. P11.153

REVIEW PROBLEMS

11.152 A timber column has a 200 × 200-mm square cross section and a 4-m effective length. For the grade of southern pine used $E = 12$ MPa and the allowable stress for compression parallel to the grain is 7 MPa. Using the interaction method, determine the largest allowable eccentricity of a 100-kN axial load.

11.153 A steel compression member of 3.5-m effective length supports an eccentric load as shown. Using the interaction method, determine the allowable load **P**. Assume $E = 200$ GPa, $\sigma_Y = 250$ MPa, and $\sigma_{\text{all}} = 150$ MPa in bending.

11.154 Knowing that a factor of safety of 2.8 is required and using Euler's formula, determine the largest load **P** which may be applied to the structure shown when $\theta = 30°$. Use $E = 200$ GPa and consider only buckling in the plane of the structure.

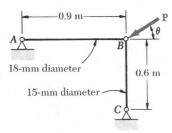

Fig. P11.154 and P11.155

11.155 (*a*) Considering only buckling in the plane of the structure shown and using Euler's formula, determine the value of θ between 0 and 90° for which the allowable magnitude of the load **P** is maximum. (*b*) Determine the corresponding maximum value of P knowing that a factor of safety of 3.2 is required. Use $E = 200$ GPa.

11.156 A W 150 × 13.5 rolled-steel shape is used to form a column of 3-m effective length. Using $E = 200$ GPa and Euler's formula, determine the allowable centric axial load if a factor of safety of 2.6 is required.

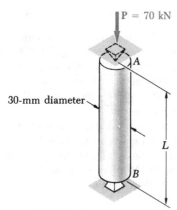

Fig. P11.157

11.157 Determine the length L of the aluminum rod AB for which the allowable centric load is 70 kN if the aluminum alloy used is (a) 6061-T6, (b) 2014-T6. Use the Aluminum Association design formulas.

11.158 An axial load $\mathbf{P}$ of magnitude 700 kN is applied at a point on the x axis at a distance $e = 6$ mm from the geometric axis of the W 200×46.1 rolled-steel column BC. Using $E = 200$ GPa, determine (a) the horizontal deflection of end C, (b) the maximum stress in the column.

11.159 An axial load $\mathbf{P}$ is applied at a point on the x axis at a distance $e = 6$ mm from the geometric axis of the W 200×46.1 rolled-steel column BC. Using the allowable-stress method, determine the allowable load $\mathbf{P}$. Assume $E = 200$ GPa, $\sigma_Y = 250$ MPa, and $\sigma_{\text{all}} = 150$ MPa in bending.

11.160 Solve Prob. 11.159, using the interaction method.

11.161 The uniform aluminum bar AB has a 20×36-mm rectangular cross section and is supported as shown. Using $E = 70$ GPa and using Euler's formula, determine the allowable centric load $\mathbf{P}$ if a factor of safety of 2.5 is required.

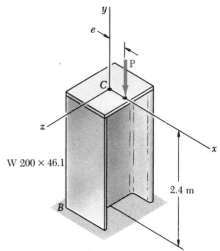

Fig. P11.158 and P11.159

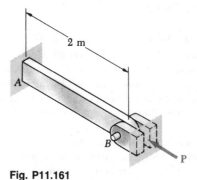

Fig. P11.161

11.162 A column of 4-m effective length must carry a centric load of 800 kN. Knowing that $\sigma_Y = 250$ MPa and $E = 200$ GPa and using AISC design formulas, select the wide-flange shape of 250-mm nominal depth which should be used.

11.163 Solve part b of Sample Prob. 11.2, assuming that the axial load $\mathbf{P}$ is applied with an eccentricity $e = 20$ mm at a point located on one of the diagonals of the cross section of the square structural tube.

The following problems are designed to be solved with a computer.

11.C1 A solid steel rod having an effective length of 0.6 m is to be used as a compression strut which will carry a centric load **P**. For the grade of steel to be used $E = 200$ GPa and $\sigma_Y = 250$ MPa. Knowing that a factor of safety of 3 is required and using Euler's formula, write a computer program and use it to calculate the allowable centric load P_{all} for values of the diameter of the rod from 2.5 mm to 40 mm at 2.5 mm intervals.

11.C2 An axial load **P** is applied at a point located on the x axis at a distance $e = 12$ mm from the geometric axis of the W 200×52 rolled-steel column BC. Using $E = 200$ GPa, write a computer program and use it to calculate for values of P from 200 to 750 kN at 50-kN intervals (a) the horizontal deflection of end C, (b) the maximum stress in the column.

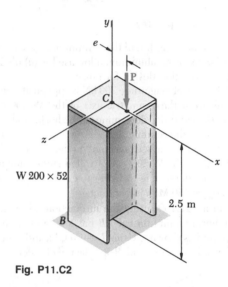

Fig. P11.C2

11.C3 A column of effective length L is made from a rolled-steel shape and carries a centric axial load **P**. The yield strength for the grade of steel used is denoted by σ_Y, the modulus of elasticity by E, the cross-sectional area of the selected shape by A, and its smallest radius of gyration by r. Using the AISC design formulas, write a computer program which can be used with any consistent set of units to determine the allowable load **P**. Use this program to solve (a) Prob. 11.70, (b) Prob. 11.71.

11.C4 A column of effective length L is made from a rolled-steel shape and is loaded eccentrically as shown. The yield strength of the grade of steel used is denoted by σ_Y, the allowable stress in bending by σ_{all}, the modulus of elasticity by E, the cross-sectional area of the selected shape by A, and its smallest radius of gyration by r. Write a computer program which can be used with any consistent set of units to determine the allowable load **P**, using either the allowable-stress method or the interaction method. Use this program to solve (a) Prob. 11.153, (b) Prob. 11.159, (c) Prob. 11.160.

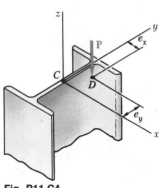

Fig. P11.C4

APPENDICES

Appendix A. Moments of Areas   690

Appendix B. Typical Properties of Selected
             Materials Used in Engineering   700

Appendix C. Properties of Rolled-Steel Shapes†   704

Appendix D. Beam Deflections and Slopes   716

†Courtesy of the American Institute of Steel Construction, Chicago, Illinois.

APPENDIX A

MOMENTS OF AREAS

A.1. FIRST MOMENT OF AN AREA; CENTROID OF AN AREA

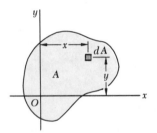

Fig. A.1

Consider an area A located in the xy plane (Fig. A.1). Denoting by x and y the coordinates of an element of area dA, we define the *first moment of the area A with respect to the x axis* as the integral

$$Q_x = \int_A y \, dA \tag{A.1}$$

Similarly, the *first moment of the area A with respect to the y axis* is defined as the integral

$$Q_y = \int_A x \, dA \tag{A.2}$$

We note that each of these integrals may be positive, negative, or zero, depending on the position of the coordinate axes. If SI units are used, the first moments Q_x and Q_y are expressed in m^3 or mm^3; if U.S. customary units are used, they are expressed in ft^3 or in^3.

The *centroid of the area A* is defined as the point C of coordinates $\bar{x}$ and $\bar{y}$ (Fig. A.2), which satisfy the relations

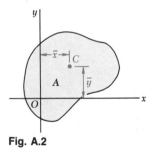

Fig. A.2

$$\int_A x \, dA = A\bar{x} \qquad \int_A y \, dA = A\bar{y} \tag{A.3}$$

Comparing Eqs. (A.1) and (A.2) with Eqs. (A.3), we note that the first moments of the area A may be expressed as the products of the area and of the coordinates of its centroid:

$$Q_x = A\bar{y} \qquad Q_y = A\bar{x} \tag{A.4}$$

When an area possesses an *axis of symmetry*, the first moment of the area with respect to that axis is zero. Indeed, considering the area A of Fig. A.3, which is symmetric with respect to the y axis, we observe that to every element of area dA of abscissa x corresponds an element of area dA' of abscissa $-x$. It follows that the integral in Eq. (A.2) is zero and, thus, that $Q_y = 0$. It also follows from the first of the relations (A.3) that $\bar{x} = 0$. Thus, if an area A possesses an axis of symmetry, its centroid C is located on that axis.

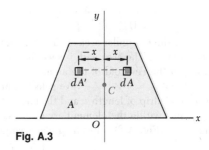

Fig. A.3

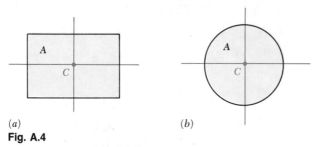

(a)   (b)

Fig. A.4

Since a rectangle possesses two axes of symmetry (Fig. A.4a), the centroid C of a rectangular area coincides with its geometric center. Similarly, the centroid of a circular area coincides with the center of the circle (Fig. A.4b).

When an area possesses a *center of symmetry* O, the first moment of the area about any axis through O is zero. Indeed, considering the area A of Fig. A.5, we observe that to every element of area dA of coordinates x and y corresponds an element of area dA' of coordinates $-x$ and $-y$. It follows that the integrals in Eqs. (A.1) and (A.2) are both zero, and that $Q_x = Q_y = 0$. It also follows from Eqs. (A.3) that $\bar{x} = \bar{y} = 0$, that is, the centroid of the area coincides with its center of symmetry.

When the centroid C of an area can be located by symmetry, the first moment of that area with respect to any given axis may be readily obtained from Eqs. (A.4). For example, in the case of the rectangular area of Fig. A.6, we have

$$Q_x = A\bar{y} = (bh)(\tfrac{1}{2}h) = \tfrac{1}{2}bh^2$$

and

$$Q_y = A\bar{x} = (bh)(\tfrac{1}{2}b) = \tfrac{1}{2}b^2h$$

In most cases, however, it is necessary to perform the integrations indicated in Eqs. (A.1) through (A.3) to determine the first moments and the centroid of a given area. While each of the integrals involved is actually a double integral, it is possible in many applications to select elements of area dA in the shape of thin horizontal or vertical strips, and thus to reduce the computations to integrations in a single variable. This is illustrated in Example A.01. Centroids of common geometric shapes are indicated in a table inside the back cover of this book.

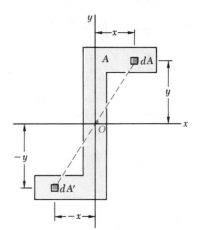

Fig. A.5

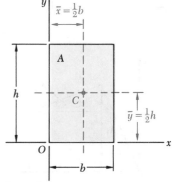

Fig. A.6

Example A.01

For the triangular area of Fig. A.7, determine (a) the first moment Q_x of the area with respect to the x axis, (b) the ordinate $\bar{y}$ of the centroid of the area.

(a) *First Moment* Q_x. We select as an element of area a horizontal strip of length u and thickness dy, and note that all the points within the element are at the same distance y from the x axis (Fig. A.8). From similar triangles, we have

$$\frac{u}{b} = \frac{h - y}{h} \qquad u = b\frac{h - y}{h}$$

and

$$dA = u \, dy = b\frac{h - y}{h} \, dy$$

The first moment of the area with respect to the x axis is

$$Q_x = \int_A y \, dA = \int_0^h yb\frac{h - y}{h} \, dy = \frac{b}{h}\int_0^h (hy - y^2) \, dy$$

$$= \frac{b}{h}\left[h\frac{y^2}{2} - \frac{y^3}{3}\right]_0^h \qquad Q_x = \tfrac{1}{6}bh^2$$

(b) *Ordinate of Centroid.* Recalling the first of Eqs. (A.4) and observing that $A = \tfrac{1}{2}bh$, we have

$$Q_x = A\bar{y} \qquad \tfrac{1}{6}bh^2 = (\tfrac{1}{2}bh)\bar{y}$$

$$\bar{y} = \tfrac{1}{3}h$$

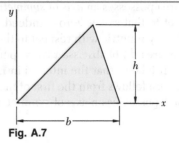

Fig. A.7

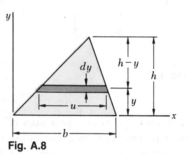

Fig. A.8

A.2. DETERMINATION OF THE FIRST MOMENT AND CENTROID OF A COMPOSITE AREA

Consider an area A, such as the trapezoidal area shown in Fig. A.9, which may be divided into simple geometric shapes. As we saw in the preceding section, the first moment Q_x of the area with respect to the x axis is represented by the integral $\int y \, dA$, which extends over the entire area A. Dividing A into its component parts A_1, A_2, A_3, we write

$$Q_x = \int_A y \, dA = \int_{A_1} y \, dA + \int_{A_2} y \, dA + \int_{A_3} y \, dA$$

or, recalling the second of Eqs. (A.3),

$$Q_x = A_1\bar{y}_1 + A_2\bar{y}_2 + A_3\bar{y}_3$$

where $\bar{y}_1$, $\bar{y}_2$, and $\bar{y}_3$ represent the ordinates of the centroids of the component areas. Extending this result to an arbitrary number of component

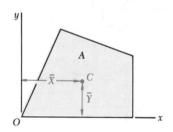

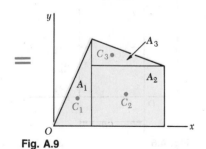

Fig. A.9

areas, and noting that a similar expression may be obtained for Q_y, we write

$$Q_x = \sum_i A_i \bar{y}_i \qquad Q_y = \sum_i A_i \bar{x}_i \qquad \text{(A.5)}$$

To obtain the coordinates $\bar{X}$ and $\bar{Y}$ of the centroid C of the composite area A, we substitute $Q_x = A\bar{Y}$ and $Q_y = A\bar{X}$ into Eqs. (A.5). We have

$$A\bar{Y} = \sum_i A_i \bar{y}_i \qquad A\bar{X} = \sum_i A_i \bar{x}_i$$

Solving for $\bar{X}$ and $\bar{Y}$ and recalling that the area A is the sum of the component areas A_i, we write

$$\bar{X} = \frac{\sum_i A_i \bar{x}_i}{\sum_i A_i} \qquad \bar{Y} = \frac{\sum_i A_i \bar{y}_i}{\sum_i A_i} \qquad \text{(A.6)}$$

Example A.02

Locate the centroid C of the area A shown in Fig. A.10.

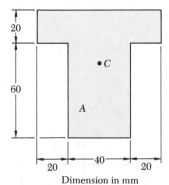

Dimension in mm

Fig. A.10

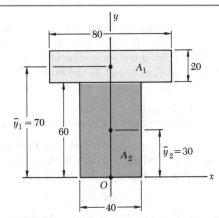

Fig. A.11   Dimensions in mm

	Area, mm²	$\bar{y}_i$, mm	$A_i\bar{y}_i$, mm³
A_1	$(20)(80) = 1600$	70	112×10^3
A_2	$(40)(60) = 2400$	30	72×10^3
	$\sum_i A_i = 4000$		$\sum_i A_i\bar{y}_i = 184 \times 10^3$

Selecting the coordinate axes shown in Fig. A.11, we note that the centroid C must be located on the y axis, since this axis is an axis of symmetry; thus, $\bar{X} = 0$.

Dividing A into its component parts A_1 and A_2, we use the second of Eqs. (A.6) to determine the ordinate $\bar{Y}$ of the centroid. The actual computation is best carried out in tabular form.

$$\bar{Y} = \frac{\sum_i A_i\bar{y}_i}{\sum_i A_i} = \frac{184 \times 10^3 \text{ mm}^3}{4 \times 10^3 \text{ mm}^2} = 46 \text{ mm}$$

Example A.03

Referring to the area A of Example A.02, we consider the horizontal x' axis through its centroid C. (Such an axis is called a *centroidal axis*.) Denoting by A' the portion of A located above that axis (Fig. A.12), determine the first moment of A' with respect to the x' axis.

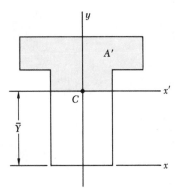

Fig. A.12

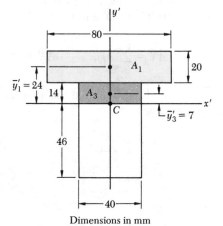

Dimensions in mm

Fig. A.13

Solution. We divide the area A' into its components A_1 and A_3 (Fig. A.13). Recalling from Example A.02 that C is located 46 mm above the lower edge of A, we determine the ordinates $\bar{y}_1'$ and $\bar{y}_3'$ of A_1 and A_3 and express the first moment $Q_{x'}'$ of A' with respect to x' as follows:

$$Q_{x'}' = A_1\bar{y}_1' + A_3\bar{y}_3'$$
$$= (20 \times 80)(24) + (14 \times 40)(7) = 42.3 \times 10^3 \text{ mm}^3$$

Alternative Solution. We first note that since the centroid C of A is located on the x' axis, the first moment $Q_{x'}$ of the *entire area* A with respect to that axis is zero:

$$Q_{x'} = A\bar{y}' = A(0) = 0$$

Denoting by A'' the portion of A located below the x' axis and by $Q_{x'}''$ its first moment with respect to that axis, we have therefore

$$Q_{x'} = Q_{x'}' + Q_{x'}'' = 0 \quad \text{or} \quad Q_{x'}' = -Q_{x'}''$$

which shows that the first moments of A' and A'' have the same magnitude and opposite signs. Referring to Fig. A.14, we write

$$Q_{x'}'' = A_4\bar{y}_4' = (40 \times 46)(-23) = -42.3 \times 10^3 \text{ mm}^3$$

and

$$Q_{x'}' = -Q_{x'}'' = +42.3 \times 10^3 \text{ mm}^3$$

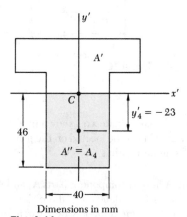

Dimensions in mm

Fig. A.14

A.3. SECOND MOMENT, OR MOMENT OF INERTIA, OF AN AREA; RADIUS OF GYRATION

Consider again an area A located in the xy plane (Fig. A.1) and the element of area dA of coordinates x and y. The *second moment*, or *moment of inertia*, of the area A with respect to the x axis, and the second moment, or moment of inertia, of A with respect to the y axis are defined, respectively, as

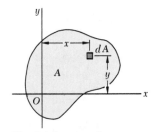

$$I_x = \int_A y^2 \, dA \qquad I_y = \int_A x^2 \, dA \tag{A.7}$$

Fig. A.1 (repeated)

These integrals are referred to as *rectangular moments of inertia*, since they are computed from the rectangular coordinates of the element dA. While each integral is actually a double integral, it is possible in many applications to select elements of area dA in the shape of thin horizontal or vertical strips, and thus reduce the computations to integrations in a single variable. This is illustrated in Example A.04.

We now define the *polar moment of inertia* of the area A with respect to point O (Fig. A.15) as the integral

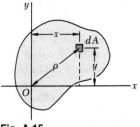

$$J_O = \int_A \rho^2 \, dA \tag{A.8}$$

Fig. A.15

where ρ is the distance from O to the element dA. While this integral is again a double integral, it is possible in the case of a circular area to select elements of area dA in the shape of thin circular rings, and thus reduce the computation of J_O to a single integration (see Example A.05).

We note from Eqs. (A.7) and (A.8) that the moments of inertia of an area are positive quantities. If SI units are used, moments of inertia are expressed in m^4 or mm^4.

An important relation may be established between the polar moment of inertia J_O of a given area and the rectangular moments of inertia I_x and I_y of the same area. Noting that $\rho^2 = x^2 + y^2$, we write

$$J_O = \int_A \rho^2 \, dA = \int_A (x^2 + y^2) \, dA = \int_A y^2 \, dA + \int_A x^2 \, dA$$

or

$$J_O = I_x + I_y \tag{A.9}$$

The *radius of gyration* of an area A with respect to the x axis is defined as the quantity r_x, which satisfies the relation

$$I_x = r_x^2 A \tag{A.10}$$

where I_x is the moment of inertia of A with respect to the x axis.   Solving Eq. (A.10) for r_x, we have

$$r_x = \sqrt{\frac{I_x}{A}} \tag{A.11}$$

In a similar way, we may define the radii of gyration with respect to the y axis and the origin O.   We write

$$I_y = r_y^2 A \qquad r_y = \sqrt{\frac{I_y}{A}} \tag{A.12}$$

$$J_O = r_O^2 A \qquad r_O = \sqrt{\frac{J_O}{A}} \tag{A.13}$$

Substituting for J_O, I_x, and I_y in terms of the corresponding radii of gyration in Eq. (A.9), we observe that

$$r_O^2 = r_x^2 + r_y^2 \tag{A.14}$$

Example A.04

For the rectangular area of Fig. A.16, determine (a) the moment of inertia I_x of the area with respect to the centroidal x axis, (b) the corresponding radius of gyration r_x.

(a) *Moment of Inertia I_x.*   We select as an element of area a horizontal strip of length b and thickness dy (Fig. A.17). Since all the points within the strip are at the same distance y from the x axis, the moment of inertia of the strip with respect to that axis is

$$dI_x = y^2\, dA = y^2(b\, dy)$$

Integrating from $y = -h/2$ to $y = +h/2$, we write

$$I_x = \int_A y^2\, dA = \int_{-h/2}^{+h/2} y^2(b\, dy) = \tfrac{1}{3}b[y^3]_{-h/2}^{+h/2}$$

$$= \tfrac{1}{3}b\left(\frac{h^3}{8} + \frac{h^3}{8}\right)$$

or

$$I_x = \tfrac{1}{12}bh^3$$

(b) *Radius of Gyration r_x.*   From Eq. (A.10), we have

$$I_x = r_x^2 A \qquad \tfrac{1}{12}bh^3 = r_x^2(bh)$$

and, solving for r_x,

$$r_x = h/\sqrt{12}$$

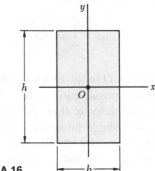

Fig. A.16

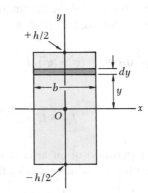

Fig. A.17

Example A.05

For the circular area of Fig. A.18, determine (a) the polar moment of inertia J_O, (b) the rectangular moments of inertia I_x and I_y.

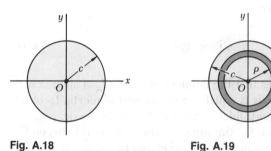

Fig. A.18   **Fig. A.19**

(a) *Polar Moment of Inertia.* We select as an element of area a ring of radius ρ and thickness $d\rho$ (Fig. A.19). Since all the points within the ring are at the same distance ρ from the origin O, the polar moment of inertia of the ring is

$$dJ_O = \rho^2 \, dA = \rho^2 (2\pi\rho \, d\rho)$$

Integrating in ρ from 0 to c, we write

$$J_O = \int_A \rho^2 \, dA = \int_0^c \rho^2 (2\pi\rho \, d\rho) = 2\pi \int_0^c \rho^3 \, d\rho$$

$$J_O = \tfrac{1}{2}\pi c^4$$

(b) *Rectangular Moments of Inertia.* Because of the symmetry of the circular area, we have $I_x = I_y$. Recalling Eq. (A.9), we write

$$J_O = I_x + I_y = 2I_x \qquad \tfrac{1}{2}\pi c^4 = 2I_x$$

and, thus,

$$I_x = I_y = \tfrac{1}{4}\pi c^4$$

The results obtained in the preceding two examples, and the moments of inertia of other common geometric shapes, are listed in a table inside the back cover of this book.

A.4. PARALLEL-AXIS THEOREM

Consider the moment of inertia I_x of an area A with respect to an arbitrary x axis (Fig. A.20). Denoting by y the distance from an element of area dA to that axis, we recall from Sec. A.3 that

$$I_x = \int_A y^2 \, dA$$

Let us now draw the *centroidal x' axis*, i.e., the axis parallel to the x axis which passes through the centroid C of the area. Denoting by y' the distance from the element dA to that axis, we write $y = y' + d$, where d is the distance between the two axes. Substituting for y in the integral representing I_x, we write

$$I_x = \int_A y^2 dA = \int_A (y' + d)^2 dA$$

$$I_x = \int_A y'^2 \, dA + 2d \int_A y' \, dA + d^2 \int_A dA \qquad (A.15)$$

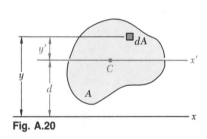

Fig. A.20

The first integral in Eq. (A.15) represents the moment of inertia $\bar{I}_{x'}$ of the area with respect to the centroidal x' axis. The second integral represents the first moment $Q_{x'}$ of the area with respect to the x' axis and is equal to zero, since the centroid C of the area is located on that axis. Indeed, we recall from Sec. A.1 that

$$Q_{x'} = A\bar{y}' = A(0) = 0$$

Finally, we observe that the last integral in Eq. (A.15) is equal to the total area A. We have, therefore,

$$I_x = \bar{I}_{x'} + Ad^2 \qquad (A.16)$$

This formula expresses that the moment of inertia I_x of an area with respect to an arbitrary x axis is equal to the moment of inertia $\bar{I}_{x'}$ of the area with respect to the centroidal x' axis parallel to the x axis, *plus* the product Ad^2 of the area A and of the square of the distance d between the two axes. This result is known as the *parallel-axis theorem*. It makes it possible to determine the moment of inertia of an area with respect to a given axis, when its moment of inertia with respect to a centroidal axis of the same direction is known. Conversely, it makes it possible to determine the moment of inertia $\bar{I}_{x'}$ of an area A with respect to a centroidal axis x', when the moment of inertia I_x of A with respect to a parallel axis is known, by *subtracting* from I_x the product Ad^2. We should note that the parallel-axis theorem may be used *only if one of the two axes involved is a centroidal axis*.

A similar formula may be derived, which relates the polar moment of inertia J_O of an area with respect to an arbitrary point O and the polar moment of inertia $\bar{J}_C$ of the same area with respect to its centroid C. Denoting by d the distance between O and C, we write

$$J_O = \bar{J}_C + Ad^2 \qquad (A.17)$$

A.5. DETERMINATION OF THE MOMENT OF INERTIA OF A COMPOSITE AREA

Consider a composite area A made of several component parts A_1, A_2, and so forth. Since the integral representing the moment of inertia of A may be subdivided into integrals extending over A_1, A_2, and so forth, the moment of inertia of A with respect to a given axis will be obtained by adding the moments of inertia of the areas A_1, A_2, and so forth, with respect to the same axis. The moment of inertia of an area made of several of the common shapes shown in the table inside the back cover of this book may thus be obtained from the formulas given in that table. Before adding the moments of inertia of the component areas, however, the parallel-axis theorem should be used to transfer each moment of inertia to the desired axis. This is shown in Example A.06.

Example A.06

Determine the moment of inertia $\bar{I}_x$ of the area shown with respect to the centroidal x axis (Fig. A.21).

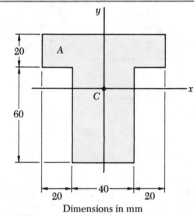

Location of Centroid. The centroid C of the area must first be located. However, this has already been done in Example A.02 for the given area. We recall from that example that C is located 46 mm above the lower edge of the area A.

Computation of Moment of Inertia. We divide the area A into the two rectangular areas A_1 and A_2 (Fig. A.22), and compute the moment of inertia of each area with respect to the x axis.

Fig. A.21

Rectangular Area A_1. To obtain the moment of inertia $(I_x)_1$ of A_1 with respect to the x axis, we first compute the moment of inertia of A, *with respect to its own centroidal axis x'*. Recalling the formula derived in part a of Example A.04 for the centroidal moment of inertia of a rectangular area, we have

$$(\bar{I}_{x'})_1 = \tfrac{1}{12}bh^3 = \tfrac{1}{12}(80 \text{ mm})(20 \text{ mm})^3 = 53.3 \times 10^3 \text{ mm}^4$$

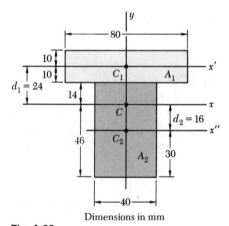

Using the parallel-axis theorem, we transfer the moment of inertia of A_1 from its centroidal axis x' to the parallel axis x:

$$(I_x)_1 = (\bar{I}_{x'})_1 + A_1 d_1^2 = 53.3 \times 10^3 + (80 \times 20)(24)^2$$
$$= 975 \times 10^3 \text{ mm}^4$$

Rectangular Area A_2. Computing the moment of inertia of A_2 with respect to its centroidal axis x'', and using the parallel-axis theorem to transfer it to the x axis, we have

Fig. A.22

$$(\bar{I}_{x''})_2 = \tfrac{1}{12}bh^3 = \tfrac{1}{12}(40)(60)^3 = 720 \times 10^3 \text{ mm}^4$$
$$(I_x)_2 = (\bar{I}_{x''})_2 + A_2 d_2^2 = 720 \times 10^3 + (40 \times 60)(16)^2$$
$$= 1334 \times 10^3 \text{ mm}^4$$

Entire Area A. Adding the values computed for the moments of inertia of A_1 and A_2 with respect to the x axis, we obtain the moment of inertia $\bar{I}_x$ of the entire area:

$$\bar{I}_x = (I_x)_1 + (I_x)_2 = 975 \times 10^3 + 1334 \times 10^3$$
$$\bar{I}_x = 2.31 \times 10^6 \text{ mm}^4$$

Appendix B. Typical Properties of Selected Materials Used in Engineering[1,5]
(U.S. Customary Units)

Material	Specific Weight, lb/in³	Ultimate Strength			Yield Strength[3]		Modulus of Elasticity, 10⁶ psi	Modulus of Rigidity, 10⁶ psi	Coefficient of Thermal Expansion, 10⁻⁶/°F	Ductility, Percent Elongation in 2 in.
		Tension, ksi	Compression,[2] ksi	Shear, ksi	Tension, ksi	Shear, ksi				
STEEL:										
Structural (ASTM-A36)	0.284	58			36	21	29	11.2	6.5	23
High-strength-low-alloy										
ASTM-A242	0.284	70			50	30	29	11.2	6.5	22
ASTM-A441	0.284	67			46		29	11.2	6.5	21
ASTM-A572	0.284	60			42		29	11.2	6.5	24
Quenched & tempered										
ASTM A-514	0.284	110			100	55	29	11.2	6.5	18
Stainless, AISI 302										
Cold-rolled	0.286	125			75		28	10.8	9.6	12
Annealed	0.286	95			38	22	28	10.8	9.6	50
Reinforcing Steel										
Medium strength	0.283	70			40		29	11	6.5	
High strength	0.283	90			60		29	11	6.5	
CAST IRON:										
Gray Cast Iron										
4.5% C, ASTM A-48	0.260	25	95	35			10	4.1	6.7	0.5
Malleable Cast Iron										
2% C, 1% Si, ASTM A-47	0.264	50	90	48	33		24	9.3	6.7	10
ALUMINUM:										
Alloy 1100-H14 (99% Al)	0.098	16		10	14	8	10.1	3.7	13.1	9
Alloy 2014-T6	0.101	66		40	58	33	10.9	3.9	12.8	13
Alloy 2024-T4	0.101	68		41	47		10.6		12.9	19
Alloy 5456-H116	0.095	46		27	33	19	10.4		13.3	16
Alloy 6061-T6	0.098	38		24	35	20	10.1	3.7	13.1	17
Alloy 7075-T6	0.101	83		48	73		10.4	4	13.1	11
COPPER										
Oxygen-free copper (99.9% Cu)										
Annealed	0.322	32		22	10		17	6.4	9.4	45
Hard-drawn	0.322	57		29	53		17	6.4	9.4	4
Yellow Brass (65% Cu, 35% Zn)										
Cold-rolled	0.306	74		43	60	36	15	5.6	11.6	8
Annealed	0.306	46		32	15	9	15	5.6	11.6	65
Red Brass (85% Cu, 15% Zn)										
Cold-rolled	0.316	85		46	63		17	6.4	10.4	3
Annealed	0.316	39		31	10		17	6.4	10.4	48
Tin bronze (88 Cu, 8Sn, 4Zn)	0.318	45			21		14		10	30
Manganese bronze (63 Cu, 25 Zn, 6 Al, 3 Mn, 3 Fe)	0.302	95			48		15		12	20
Aluminum bronze 81 Cu, 4 Ni, 4 Fe, 11 Al)	0.301	90	130		40		16	6.1	9	6

(Table continued on page 702)

Material	Density, kg/m³	Ultimate Strength			Yield Strength[3]		Modulus of Elasticity, GPa	Modulus of Rigidity, GPa	Coefficient of Thermal Expansion, 10^{-6}/°C	Ductility, Percent Elongation in 50 mm
		Tension, MPa	Compression,[2] MPa	Shear, MPa	Tension, MPa	Shear, MPa				
STEEL:										
Structural (ASTM-A36)	7860	400			250	145	200	77	11.7	23
High-strength-low-alloy										
ASTM-A242	7860	480			345	205	200	77	11.7	22
ASTM-A441	7860	460			320		200	77	11.7	21
ASTM-A572	7860	415			290		200	77	11.7	24
Quenched & tempered										
ASTM A-514	7860	760			690	380	200	77	11.7	18
Stainless, AISI 302										
Cold-rolled	7920	860			520		190	75	17.3	12
Annealed	7920	655			260	150	190	75	17.3	50
Reinforcing Steel										
Medium strength	7860	480			275		200	77	11.7	
High strength	7860	620			415		200	77	11.7	
CAST IRON:										
Gray Cast Iron										
4.5% C, ASTM A-48	7200	170	655	240			69	28	12.1	0.5
Malleable Cast Iron										
2% C, 1% Si, ASTM A-47	7300	345	620	330	230		165	65	12.1	10
ALUMINUM:										
Alloy 1100-H14 (99% Al)	2710	110		70	100	55	70	26	23.6	9
Alloy 2014-T6	2800	455		275	400	230	75	27	23.0	13
Alloy 2024-T4	2800	470		280	325		73		23.2	19
Alloy 5456-H116	2630	320		185	230	130	72		23.9	16
Alloy 6061-T6	2710	260		165	240	140	70	26	23.6	17
Alloy 7075-T6	2800	570		330	500		72	28	23.6	11
COPPER										
Oxygen-free copper										
(99.9% Cu)										
Annealed	8910	220		150	70		120	44	16.9	45
Hard-drawn	8910	390		200	265		120	44	16.9	4
Yellow Brass										
(65% Cu, 35% Zn)										
Cold-rolled	8470	510		300	410	250	105	39	20.9	8
Annealed	8470	320		220	100	60	105	39	20.9	65
Red Brass										
(85% Cu, 15% Zn)										
Cold-rolled	8740	585		320	435		120	44	18.7	3
Annealed	8740	270		210	70		120	44	18.7	48
Tin bronze	8800	310			145		95		18.0	30
(88 Cu, 8Sn, 4Zn)										
Manganese bronze	8360	655			330		105		21.6	20
(63 Cu, 25 Zn, 6 Al, 3 Mn, 3 Fe)										
Aluminum bronze	8330	620	900		275		110	42	16.2	6
81 Cu, 4 Ni, 4 Fe, 11 Al)										

(Table continued on page 703)

Appendix B. Typical Properties of Selected Materials Used in Engineering[1,5]
(U.S. Customary Units)

Continued from page 700

Material	Specific Weight, lb/in³	Ultimate Strength			Yield Strength[3]		Modulus of Elasticity, 10⁶ psi	Modulus of Rigidity, 10⁶ psi	Coefficient of Thermal Expansion, 10⁻⁶/°F	Ductility, Percent Elongation in 2 in.
		Tension, ksi	Compression,[2] ksi	Shear, ksi	Tension, ksi	Shear, ksi				
MAGNESIUM ALLOYS										
Alloy AZ80 (Forging)	0.065	50		23	36		6.5	2.4	14	6
Alloy AZ31 (Extrusion)	0.064	37		19	29		6.5	2.4	14	12
TITANIUM										
Alloy (6% Al, 4% V)	0.161	130			120		16.5		5.3	10
MONEL ALLOY 400(Ni-Cu)										
Cold-worked	0.319	98			85	50	26		7.7	22
Annealed	0.319	80			32	18	26		7.7	46
CUPRONICKEL										
(90% Cu, 10% Ni)										
Annealed	0.323	53			16		20	7.5	9.5	35
Cold-worked	0.323	85			79		20	7.5	9.5	3
TIMBER, air dry:										
Douglas-fir	0.017	15	7.2	1.1			1.9	.1	Varies	
Spruce, Sitka	0.015	8.6	5.6	1.1			1.5	.07	1.7 to 2.5	
Shortleaf pine	0.018		7.3	1.4			1.7			
Western white pine	0.014		5.0	1.0			1.5			
Ponderosa pine	0.015	8.4	5.3	1.1			1.3			
White oak	0.025		7.4	2.0			1.8			
Red oak	0.024		6.8	1.8			1.8			
Western hemlock	0.016	13	7.2	1.3			1.6			
Shagbark hickory	0.026		9.2	2.4			2.2			
Redwood	0.015	9.4	6.1	0.9			1.3			
CONCRETE										
Medium strength	0.084		4.0				3.6		5.5	
High strength	0.084		6.0				4.5		5.5	
PLASTICS										
Nylon, type 6/6, (molding compound)	0.0412	11	14		6.5		0.4		80	50
Polycarbonate	0.0433	9.5	12.5		9		0.35		68	110
Polyester, PBT (thermoplastic)	0.0484	8	11		8		0.35		75	150
Polyester elastomer	0.0433	6.5		5.5			0.03			500
Polystyrene	0.0374	8	13		8		0.45		70	2
Vinyl, rigid PVC	0.0520	6	10		6.5		0.45		75	40
Rubber	0.033	2							90	600
Granite (Avg. values)	0.100	3	35	5			10	4	4	
Marble (Avg. values)	0.100	2	18	4			8	3	6	
Sandstone (Avg. values)	0.083	1	12	2			6	2	5	
Glass, 98% silica	0.079		7				9.6	4.1	44	

[1] Properties of metals vary widely as a result of variations in composition, heat treatment, and mechanical working.

[2] For ductile metals the compression strength is generally assumed to be equal to the tension strength.

[3] Offset of 0.2 percent.

[4] Timber properties are for loading parallel to the grain.

[5] See also *Mark's Mechanical Engineering Handbook,* 9th ed., McGraw-Hill, New York, 1987; *Annual Book of ASTM,* American Society for Testing Materials, Philadelphia, Pa.; *Metals Handbook,* American Society for Metals, Metals Park, Ohio; and *Aluminum Construction Manual,* The Aluminum Association, Washington, DC.

Continued from page 701

Material	Density, kg/m³	Ultimate Strength Tension, MPa	Ultimate Strength Compression,[2] MPa	Ultimate Strength Shear, MPa	Yield Strength[3] Tension, MPa	Yield Strength[3] Shear, MPa	Modulus of Elasticity, GPa	Modulus of Rigidity, GPa	Coefficient of Thermal Expansion, $10^{-6}/°C$	Ductility, Percent Elongation in 50 mm
MAGNESIUM ALLOYS										
Alloy AZ80 (Forging)	1800	345		160	250		45	16	25.2	6
Alloy AZ31 (Extrusion)	1770	255		130	200		45	16	25.2	12
TITANIUM										
Alloy (6% Al, 4% V)	4730	900			830		115		9.5	10
MONEL ALLOY 400(Ni-Cu)										
Cold-worked	8830	675			585	345	180		13.9	22
Annealed	8830	550			220	125	180		13.9	46
CUPRONICKEL										
(90% Cu, 10% Ni)										
Annealed	8940	365			110		140	52	17.1	35
Cold-worked	8940	585			545		140	52	17.1	3
TIMBER, air dry:										
Douglas-fir	470	100	50	7.6			13	0.7	Varies	
Spruce, Sitka	415	60	39	7.6			10	0.5	3.0 to 4.5	
Shortleaf pine	500		50	9.7			12			
Western white pine	390		34	7.0			10			
Ponderosa pine	415	55	36	7.6			9			
White oak	690		51	13.8			12			
Red oak	660		47	12.4			12			
Western hemlock	440	90	50	10.0			11			
Shagbark hickory	720		63	16.5			15			
Redwood	415	65	42	6.2			9			
CONCRETE										
Medium strength	2320		28				25		9.9	
High strength	2320		40				30		9.9	
PLASTICS										
Nylon, type 6/6, (molding compound)	1140	75	95		45		2.8		144	50
Polycarbonate	1200	65	85		35		2.4		122	110
Polyester, PBT (thermoplastic)	1340	55	75		55		2.4		135	150
Polyester elastomer	1200	45		40			0.2			500
Polystyrene	1030	55	90		55		3.1		125	2
Vinyl, rigid PVC	1440	40	70		45		3.1		135	40
Rubber	910	15							162	600
Granite (Avg. values)	2770	20	240	35			70	4	7.2	
Marble (Avg. values)	2770	15	125	28			55	3	10.8	
Sandstone (Avg. values)	2300	7	85	14			40	2	9.0	
Glass, 98% silica	2190		50				65	4.1	80	

[1] Properties of metals vary widely as a result of variations in composition, heat treatment, and mechanical working.

[2] For ductile metals the compression strength is generally assumed to be equal to the tension strength.

[3] Offset of 0.2 percent.

[4] Timber properties are for loading parallel to the grain.

[5] See also *Mark's Mechanical Engineering Handbook,* 9th ed., McGraw-Hill, New York, 1987; *Annual Book of ASTM,* American Society for Testing Materials, Philadelphia, Pa.; *Metals Handbook,* American Society for Metals, Metals Park, Ohio; and *Aluminum Construction Manual,* The Aluminum Association, Washington, DC.

Appendix C. Properties of Rolled-Steel Shapes
(U.S. Customary Units)

W Shapes
(Wide-Flange Shapes)

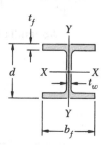

Designation†	Area A, in^2	Depth d, in.	Flange Width b_f, in.	Flange Thickness t_f, in.	Web Thickness t_w, in.	Axis X-X I_x, in^4	S_x, in^3	r_x, in.	Axis Y-Y I_y, in^4	S_y, in^3	r_y, in.
W36 × 300	88.3	36.74	16.655	1.680	0.945	20300	1110	15.2	1300	156	3.83
135	39.7	35.55	11.950	0.790	0.600	7800	439	14.0	225	37.7	2.38
W33 × 201	59.1	33.68	15.745	1.150	0.715	11500	684	14.0	749	95.2	3.56
118	34.7	32.86	11.480	0.740	0.550	5900	359	13.0	187	32.6	2.32
W30 × 173	50.8	30.44	14.985	1.065	0.655	8200	539	12.7	598	79.8	3.43
99	29.1	29.65	10.450	0.670	0.520	3990	269	11.7	128	24.5	2.10
W27 × 146	42.9	27.38	13.965	0.975	0.605	5630	411	11.4	443	63.5	3.21
84	24.8	26.71	9.960	0.640	0.460	2850	213	10.7	106	21.2	2.07
W24 × 104	30.6	24.06	12.750	0.750	0.500	3100	258	10.1	259	40.7	2.91
68	20.1	23.73	8.965	0.585	0.415	1830	154	9.55	70.4	15.7	1.87
W21 × 101	29.8	21.36	12.290	0.800	0.500	2420	227	9.02	248	40.3	2.89
62	18.3	20.99	8.240	0.615	0.400	1330	127	8.54	57.5	13.9	1.77
44	13.0	20.66	6.500	0.450	0.350	843	81.6	8.06	20.7	6.36	1.26
W18 × 106	31.1	18.73	11.200	0.940	0.590	1910	204	7.84	220	39.4	2.66
76	22.3	18.21	11.035	0.680	0.425	1330	146	7.73	152	27.6	2.61
50	14.7	17.99	7.495	0.570	0.355	800	88.9	7.38	40.1	10.7	1.65
35	10.3	17.70	6.000	0.425	0.300	510	57.6	7.04	15.3	5.12	1.22
W16 × 77	22.6	16.52	10.295	0.760	0.455	1110	134	7.00	138	26.9	2.47
57	16.8	16.43	7.120	0.715	0.430	758	92.2	6.72	43.1	12.1	1.60
40	11.8	16.01	6.995	0.505	0.305	518	64.7	6.63	28.9	8.25	1.57
31	9.12	15.88	5.525	0.440	0.275	375	47.2	6.41	12.4	4.49	1.17
26	7.68	15.69	5.500	0.345	0.250	301	38.4	6.26	9.59	3.49	1.12
W14 × 370	109	17.92	16.475	2.660	1.655	5440	607	7.07	1990	241	4.27
145	42.7	14.78	15.500	1.090	0.680	1710	232	6.33	677	87.3	3.98
82	24.1	14.31	10.130	0.855	0.510	882	123	6.05	148	29.3	2.48
68	20.0	14.04	10.035	0.720	0.415	723	103	6.01	121	24.2	2.46
53	15.6	13.92	8.060	0.660	0.370	541	77.8	5.89	57.7	14.3	1.92
43	12.6	13.66	7.995	0.530	0.305	428	62.7	5.82	45.2	11.3	1.89
38	11.2	14.10	6.770	0.515	0.310	385	54.6	5.88	26.7	7.88	1.55
30	8.85	13.84	6.730	0.385	0.270	291	42.0	5.73	19.6	5.82	1.49
26	7.69	13.91	5.025	0.420	0.255	245	35.3	5.65	8.91	3.54	1.08
22	6.49	13.74	5.000	0.335	0.230	199	29.0	5.54	7.00	2.80	1.04

†A wide-flange shape is designated by the letter W followed by the nominal depth in inches and the weight in pounds per foot.

(Table continued on page 706)

Appendix C. Properties of Rolled-Steel Shapes
(SI Units)

W Shapes
(Wide-Flange Shapes)

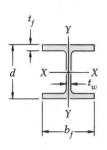

Designation†	Area A, mm²	Depth d, mm	Flange Width b_f, mm	Flange Thickness t_f, mm	Web Thickness t_w, mm	Axis X-X I_x 10^6 mm⁴	S_x 10^3 mm³	r_x mm	Axis Y-Y I_y 10^6 mm⁴	S_y 10^3 mm³	r_y mm
W920 × 446	57000	933	423	42.7	24.0	8450	18110	386	541	2560	97.3
201	25600	903	304	20.1	15.2	3250	7200	356	93.7	616	60.5
W840 × 299	38100	855	400	29.2	18.2	4790	11200	356	312	1560	90.4
176	22400	835	292	18.8	14.0	2460	5890	330	77.8	533	58.9
W760 × 257	32800	773	381	27.1	16.6	3410	8820	323	249	1307	87.1
147	18800	753	265	17.0	13.2	1660	4410	297	53.3	402	53.3
W690 × 217	27700	695	355	24.8	15.4	2340	6730	290	184.4	1039	81.5
125	16000	678	253	16.3	11.7	1186	3500	272	44.1	349	52.6
W610 × 155	19700	611	324	19.0	12.7	1290	4220	256	107.8	665	73.9
101	13000	603	228	14.9	10.5	762	2530	243	29.3	257	47.5
W530 × 150	19200	543	312	20.3	12.7	1007	3710	229	103.2	662	73.4
92	11800	533	209	15.6	10.2	554	2080	217	23.9	229	45.0
66	8390	525	165	11.4	8.9	351	1337	205	8.62	104.5	32.0
W460 × 158	20100	476	284	23.9	15.0	795	3340	199.1	91.6	645	67.6
113	14400	463	280	17.3	10.8	554	2390	196.3	63.3	452	66.3
74	9480	457	190	14.5	9.0	333	1457	187.5	16.69	175.7	41.9
52	6650	450	152	10.8	7.6	212	942	178.8	6.37	83.8	31.0
W410 × 114	14600	420	261	19.3	11.6	462	2200	177.8	57.4	440	62.7
85	10800	417	181	18.2	10.9	316	1516	170.7	17.94	198.2	40.6
60	7610	407	178	12.8	7.7	216	1061	168.4	12.03	135.2	39.9
46.1	5880	403	140	11.2	7.0	156.1	775	162.8	5.16	73.7	29.7
38.8	4950	399	140	8.8	6.4	125.3	628	159.0	3.99	57.0	28.4
W360 × 551	70300	455	418	67.6	42.0	2260	9930	179.6	828	3960	108.5
216	27500	375	394	27.7	17.3	712	3800	160.8	282	1431	101.1
122	15500	363	257	21.7	13.0	367	2020	153.7	61.6	479	63.0
101	12900	357	255	18.3	10.5	301	1686	152.7	50.4	395	62.5
79	10100	354	205	16.8	9.4	225	1271	149.6	24.0	234	48.8
64	8130	347	203	13.5	7.7	178.1	1027	147.8	18.81	185.3	48.0
57	7230	358	172	13.1	7.9	160.2	895	149.4	11.11	129.2	39.4
44.8	5710	352	171	9.8	6.9	121.1	688	145.5	8.16	95.4	37.8
39.0	4960	353	128	10.7	6.5	102.0	578	143.5	3.71	58.0	27.4
32.9	4190	349	127	8.5	5.8	82.8	474	140.7	2.91	45.8	26.4

†A wide-flange shape is designated by the letter W followed by the nominal depth in millimeters and the mass in kilograms per meter.

(Table continued on page 707)

Appendix C. Properties of Rolled-Steel Shapes

(U.S. Customary Units)

Continued from page 704

W Shapes
(Wide-Flange Shapes)

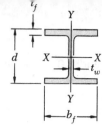

Designation†	Area A, in²	Depth d, in.	Flange Width b_f, in.	Flange Thickness t_f, in.	Web Thickness t_w, in.	Axis X-X I_x, in⁴	Axis X-X S_x, in³	Axis X-X r_x, in.	Axis Y-Y I_y, in⁴	Axis Y-Y S_y, in³	Axis Y-Y r_y, in.
W12 × 96	28.2	12.71	12.160	0.900	0.550	833	131	5.44	270	44.4	3.09
72	21.1	12.25	12.040	0.670	0.430	597	97.4	5.31	195	32.4	3.04
50	14.7	12.19	8.080	0.640	0.370	394	64.7	5.18	56.3	13.9	1.96
40	11.8	11.94	8.005	0.515	0.295	310	51.9	5.13	44.1	11.0	1.93
35	10.3	12.50	6.560	0.520	0.300	285	45.6	5.25	24.5	7.47	1.54
30	8.79	12.34	6.520	0.440	0.260	238	38.6	5.21	20.3	6.24	1.52
26	7.65	12.22	6.490	0.380	0.230	204	33.4	5.17	17.3	5.34	1.51
22	6.48	12.31	4.030	0.425	0.260	156	25.4	4.91	4.66	2.31	0.848
16	4.71	11.99	3.990	0.265	0.220	103	17.1	4.67	2.82	1.41	0.773
W10 × 112	32.9	11.36	10.415	1.250	0.755	716	126	4.66	236	45.3	2.68
68	20.0	10.40	10.130	0.770	0.470	394	75.7	4.44	134	26.4	2.59
54	15.8	10.09	10.030	0.615	0.370	303	60.0	4.37	103	20.6	2.56
45	13.3	10.10	8.020	0.620	0.350	248	49.1	4.33	53.4	13.3	2.01
39	11.5	9.92	7.985	0.530	0.315	209	42.1	4.27	45.0	11.3	1.98
33	9.71	9.73	7.960	0.435	0.290	170	35.0	4.19	36.6	9.20	1.94
30	8.84	10.47	5.810	0.510	0.300	170	32.4	4.38	16.7	5.75	1.37
22	6.49	10.17	5.750	0.360	0.240	118	23.2	4.27	11.4	3.97	1.33
19	5.62	10.24	4.020	0.395	0.250	96.3	18.8	4.14	4.29	2.14	0.874
15	4.41	9.99	4.000	0.270	0.230	68.9	13.8	3.95	2.89	1.45	0.810
W8 × 58	17.1	8.75	8.220	0.810	0.510	228	52.0	3.65	75.1	18.3	2.10
48	14.1	8.50	8.110	0.685	0.400	184	43.3	3.61	60.9	15.0	2.08
40	11.7	8.25	8.070	0.560	0.360	146	35.5	3.53	49.1	12.2	2.04
35	10.3	8.12	8.020	0.495	0.310	127	31.2	3.51	42.6	10.6	2.03
31	9.13	8.00	7.995	0.435	0.285	110	27.5	3.47	37.1	9.27	2.02
28	8.25	8.06	6.535	0.465	0.285	98.0	24.3	3.45	21.7	6.63	1.62
24	7.08	7.93	6.495	0.400	0.245	82.8	20.9	3.42	18.3	5.63	1.61
21	6.16	8.28	5.270	0.400	0.250	75.3	18.2	3.49	9.77	3.71	1.26
18	5.26	8.14	5.250	0.330	0.230	61.9	15.2	3.43	7.97	3.04	1.23
15	4.44	8.11	4.015	0.315	0.245	48.0	11.8	3.29	3.41	1.70	0.876
13	3.84	7.99	4.000	0.255	0.230	39.6	9.91	3.21	2.73	1.37	0.843
W6 × 25	7.34	6.38	6.080	0.455	0.320	53.4	16.7	2.70	17.1	5.61	1.52
20	5.87	6.20	6.020	0.365	0.260	41.4	13.4	2.66	13.3	4.41	1.50
16	4.74	6.28	4.030	0.405	0.260	32.1	10.2	2.60	4.43	2.20	0.967
12	3.55	6.03	4.000	0.280	0.230	22.1	7.31	2.49	2.99	1.50	0.918
9	2.68	5.90	3.940	0.215	0.170	16.4	5.56	2.47	2.20	1.11	0.905
W5 × 19	5.54	5.15	5.030	0.430	0.270	26.2	10.2	2.17	9.13	3.63	1.28
16	4.68	5.01	5.000	0.360	0.240	21.3	8.51	2.13	7.51	3.00	1.27
W4 × 13	3.83	4.16	4.060	0.345	0.280	11.3	5.46	1.72	3.86	1.90	1.00

†A wide-flange shape is designated by the letter W followed by the nominal depth in inches and the weight in pounds per foot.

Appendix C. Properties of Rolled-Steel Shapes

(SI Units)

Continued from page 705

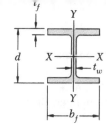

W Shapes
(Wide-Flange Shapes)

Designation†	Area A, mm²	Depth d, mm	Flange Width b_f, mm	Flange Thickness t_f, mm	Web Thickness t_w, mm	Axis X-X I_x 10⁶ mm⁴	Axis X-X S_x 10³ mm³	Axis X-X r_x mm	Axis Y-Y I_y 10⁶ mm⁴	Axis Y-Y S_y 10³ mm³	Axis Y-Y r_y mm
W310 × 143	18200	323	309	22.9	14.0	347	2150	138.2	112.4	728	78.5
107	13600	311	306	17.0	10.9	248	1595	134.9	81.2	531	77.2
74	9480	310	205	16.3	9.4	164.0	1058	131.6	23.4	228	49.8
60	7610	303	203	13.1	7.5	129.0	851	130.3	18.36	180.9	49.0
52	6650	317	167	13.2	7.6	118.6	748	133.4	10.20	122.2	39.1
44.5	5670	313	166	11.2	6.6	99.1	633	132.3	8.45	101.8	38.6
38.7	4940	310	165	9.7	5.8	84.9	548	131.3	7.20	87.3	38.4
32.7	4180	313	102	10.8	6.6	64.9	415	124.7	1.940	38.0	21.5
23.8	3040	305	101	6.7	5.6	42.9	281	118.6	1.174	23.2	19.63
W250 × 167	21200	289	265	31.8	19.2	298.0	2060	118.4	98.2	741	68.1
101	12900	264	257	19.6	11.9	164.0	1242	112.8	55.8	434	65.8
80	10200	256	255	15.6	9.4	126.1	985	111.0	42.8	336	65.0
67	8580	257	204	15.7	8.9	103.2	803	110.0	22.2	218	51.1
58	7420	252	203	13.5	8.0	87.0	690	108.5	18.73	184.5	50.3
49.1	6260	247	202	11.0	7.4	70.8	573	106.4	15.23	150.8	49.3
44.8	5700	266	148	13.0	7.6	70.8	532	111.3	6.95	93.9	34.8
32.7	4190	258	146	9.1	6.1	49.1	381	108.5	4.75	65.1	33.8
28.4	3630	260	102	10.0	6.4	40.1	308	105.2	1.796	35.2	22.2
22.3	2850	254	102	6.9	5.8	28.7	226	100.3	1.203	23.6	20.6
W200 × 86	11000	222	209	20.6	13.0	94.9	855	92.7	31.3	300	53.3
71	9100	216	206	17.4	10.2	76.6	709	91.7	25.3	246	52.8
59	7550	210	205	14.2	9.1	60.8	579	89.7	20.4	199.0	51.8
52	6650	206	204	12.6	7.9	52.9	514	89.2	17.73	173.8	51.6
46.1	5890	203	203	11.0	7.2	45.8	451	88.1	15.44	152.1	51.3
41.7	5320	205	166	11.8	7.2	40.8	398	87.6	9.03	108.8	41.1
35.9	4570	201	165	10.2	6.2	34.5	343	86.9	7.62	92.4	40.9
31.3	3970	210	134	10.2	6.4	31.3	298	88.6	4.07	60.7	32.0
26.6	3390	207	133	8.4	5.8	25.8	249	87.1	3.32	49.9	31.2
22.5	2860	206	102	8.0	6.2	20.0	194.2	83.6	1.419	27.8	22.3
19.3	2480	203	102	6.5	5.8	16.48	162.4	81.5	1.136	22.3	21.4
W150 × 37.1	4740	162	154	11.6	8.1	22.2	274	68.6	7.12	92.5	38.6
29.8	3790	157	153	9.3	6.6	17.23	219	67.6	5.54	72.4	38.1
24.0	3060	160	102	10.3	6.6	13.36	167.0	66.0	1.844	36.2	24.6
18.0	2290	153	102	7.1	5.8	9.20	120.3	63.2	1.245	24.4	23.3
13.5	1730	150	100	5.5	4.3	6.83	91.1	62.7	0.916	18.32	23.0
W130 × 28.1	3590	131	128	10.9	6.9	10.91	166.6	55.1	3.80	59.4	32.5
23.8	3040	127	127	9.1	6.1	8.87	139.7	54.1	3.13	49.3	32.3
W100 × 19.3	2470	106	103	8.8	7.1	4.70	88.7	43.7	1.607	31.2	25.4

†A wide-flange shape is designated by the letter W followed by the nominal depth in millimeters and the mass in kilograms per meter.

Appendix C. Properties of Rolled-Steel Shapes
(U.S. Customary Units)

S Shapes
(American Standard Shapes)

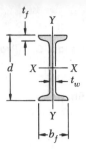

Designation†	Area A, in^2	Depth d, in.	Flange Width b_f, in.	Flange Thickness t_f, in.	Web Thickness t_w, in.	Axis X-X I_x, in^4	Axis X-X S_x, in^3	Axis X-X r_x, in.	Axis Y-Y I_y, in^4	Axis Y-Y S_y, in^3	Axis Y-Y r_y, in.
S24 × 100	29.4	24.00	7.247	0.871	0.747	2390	199	9.01	47.8	13.2	1.27
× 90	26.5	24.00	7.124	0.871	0.624	2250	187	9.22	44.9	12.6	1.30
× 79.9	23.5	24.00	7.001	0.871	0.501	2110	175	9.47	42.3	12.1	1.34
S20 × 95	27.9	20.00	7.200	0.916	0.800	1610	161	7.60	49.7	13.8	1.33
× 85	25.0	20.00	7.053	0.916	0.653	1520	152	7.79	46.2	13.1	1.36
75	22.1	20.00	6.391	0.789	0.641	1280	128	7.60	29.6	9.28	1.16
65.4	19.2	20.00	6.250	0.789	0.500	1180	118	7.84	27.4	8.77	1.19
S18 × 70	20.6	18.00	6.251	0.691	0.711	926	103	6.71	24.1	7.72	1.08
54.7	16.1	18.00	6.001	0.691	0.461	804	89.4	7.07	20.8	6.94	1.14
S15 × 50	14.7	15.00	5.640	0.622	0.550	486	64.8	5.75	15.7	5.57	1.03
42.9	12.6	15.00	5.501	0.622	0.411	447	59.6	5.95	14.4	5.23	1.07
S12 × 50	14.7	12.00	5.477	0.659	0.687	305	50.8	4.55	15.7	5.74	1.03
40.8	12.0	12.00	5.252	0.659	0.462	272	45.4	4.77	13.6	5.16	1.06
35	10.3	12.00	5.078	0.544	0.428	229	38.2	4.72	9.87	3.89	0.980
31.8	9.35	12.00	5.000	0.544	0.350	218	36.4	4.83	9.36	3.74	1.00
S10 × 35	10.3	10.00	4.944	0.491	0.594	147	29.4	3.78	8.36	3.38	0.901
25.4	7.46	10.00	4.661	0.491	0.311	124	24.7	4.07	6.79	2.91	0.954
S8 × 23	6.77	8.00	4.171	0.425	0.441	64.9	16.2	3.10	4.31	2.07	0.798
18.4	5.41	8.00	4.001	0.425	0.271	57.6	14.4	3.26	3.73	1.86	0.831
S7 × 20	5.88	7.00	3.860	0.392	0.450	42.4	12.1	2.69	3.17	1.64	0.734
.15.3	4.50	7.00	3.662	0.392	0.252	36.7	10.5	2.86	2.64	1.44	0.766
S6 × 17.25	5.07	6.00	3.565	0.359	0.465	26.3	8.77	2.28	2.31	1.30	0.675
12.5	3.67	6.00	3.332	0.359	0.232	22.1	7.37	2.45	1.82	1.09	0.705
S5 × 14.75	4.34	5.00	3.284	0.326	0.494	15.2	6.09	1.87	1.67	1.01	0.620
10	2.94	5.00	3.004	0.326	0.214	12.3	4.92	2.05	1.22	0.809	0.643
S4 × 9.5	2.79	4.00	2.796	0.293	0.326	6.79	3.39	1.56	0.903	0.646	0.569
7.7	2.26	4.00	2.663	0.293	0.193	6.08	3.04	1.64	0.764	0.574	0.581
S3 × 7.5	2.21	3.00	2.509	0.260	0.349	2.93	1.95	1.15	0.586	0.468	0.516
5.7	1.67	3.00	2.330	0.260	0.170	2.52	1.68	1.23	0.455	0.390	0.522

†An American Standard Beam is designated by the letter S followed by the nominal depth in inches and the weight in pounds per foot.

Appendix C. Properties of Rolled-Steel Shapes
(SI Units)

S Shapes
(American Standard Shapes)

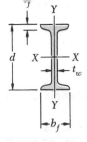

Designation†	Area A, mm²	Depth d, mm	Flange Width b_f, mm	Flange Thickness t_f, mm	Web Thickness t_w, mm	Axis X-X I_x 10⁶ mm⁴	Axis X-X S_x 10³ mm³	Axis X-X r_x mm	Axis Y-Y I_y 10⁶ mm⁴	Axis Y-Y S_y 10³ mm³	Axis Y-Y r_y mm
S610 × 149	18970	610	184	22.1	19.0	995	3260	229	19.90	216	32.3
134	17100	610	181	22.1	15.8	937	3070	234	18.69	207	33.0
118.9	15160	610	178	22.1	12.7	878	2880	241	17.61	197.9	34.0
S510 × 141	18000	508	183	23.3	20.3	670	2640	193.0	20.69	226.	33.8
127	16130	508	179	23.3	16.6	633	2490	197.9	19.23	215.	34.5
112	14260	508	162	20.1	16.3	533	2100	193.0	12.32	152.1	29.5
97.3	12390	508	159	20.1	12.7	491	1933	199.1	11.40	143.4	30.2
S460 × 104	13290	457	159	17.6	18.1	385	1685	170.4	10.03	126.2	27.4
81.4	10390	457	152	17.6	11.7	335	1466	179.6	8.66	113.9	29.0
S380 × 74	9480	381	143	15.8	14.0	202	1060	146.1	6.53	91.3	26.2
64	8130	381	140	15.8	10.4	186.1	977	151.1	5.99	85.6	27.2
S310 × 74	9480	305	139	16.8	17.4	127.0	833	115.6	6.53	94.0	26.2
60.7	7740	305	133	16.8	11.7	113.2	742	121.2	5.66	85.1	26.9
52	6640	305	129	13.8	10.9	95.3	625	119.9	4.11	63.7	24.9
47.3	6032	305	127	13.8	8.9	90.7	595	122.7	3.90	61.4	25.4
S250 × 52	6640	254	126	12.5	15.1	61.2	482	96.0	3.48	55.2	22.9
37.8	4806	254	118	12.5	7.9	51.6	406	103.4	2.83	48.0	24.2
S200 × 34	4368	203	106	10.8	11.2	27.0	266	78.7	1.794	33.8	20.3
27.4	3484	203	102	10.8	6.9	24.0	236	82.8	1.553	30.4	21.1
S180 × 30	3794	178	97	10.0	11.4	17.65	198.3	68.3	1.319	27.2	18.64
22.8	2890	178	92	10.0	6.4	15.28	171.7	72.6	1.099	23.9	19.45
S150 × 25.7	3271	152	90	9.1	11.8	10.95	144.1	57.9	0.961	21.4	17.15
18.6	2362	152	84	9.1	5.8	9.20	121.1	62.2	0.758	18.05	17.91
S130 × 22.0	2800	127	83	8.3	12.5	6.33	99.7	47.5	0.695	16.75	15.75
15	1884	127	76	8.3	5.3	5.12	80.6	52.1	0.508	13.37	16.33
S100 × 14.1	1800	102	70	7.4	8.3	2.83	55.5	39.6	0.376	10.74	14.45
11.5	1452	102	67	7.4	4.8	2.53	49.6	41.6	0.318	9.49	14.75
S75 × 11.2	1426	76	63	6.6	8.9	1.22	32.1	29.2	0.244	7.75	13.11
8.5	1077	76	59	6.6	4.3	1.05	27.6	31.3	0.189	6.41	13.26

†An American Standard Beam is designated by the letter S followed by the nominal depth in millimeters and the mass in kilograms per meter.

Appendix C. Properties of Rolled-Steel Shapes
(U.S. Customary Units)

C Shapes
(American Standard Channels)

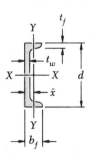

Designation†	Area A, in²	Depth d, in.	Flange		Web Thickness t_w, in.	Axis X-X			Axis Y-Y			
			Width b_f, in.	Thickness t_f, in.		I_x, in⁴	S_x, in³	r_x, in.	I_y, in⁴	S_y, in³	r_y, in.	$\bar{x}$, in.
C15 × 50	14.7	15.00	3.716	0.650	0.716	404	53.8	5.24	11.0	3.78	0.867	0.799
40	11.8	15.00	3.520	0.650	0.520	349	46.5	5.44	9.23	3.36	0.886	0.778
33.9	9.96	15.00	3.400	0.650	0.400	315	42.0	5.62	8.13	3.11	0.904	0.787
C12 × 30	8.82	12.00	3.170	0.501	0.510	162	27.0	4.29	5.14	2.06	0.763	0.674
25	7.35	12.00	3.047	0.501	0.387	144	24.1	4.43	4.47	1.88	0.780	0.674
20.7	6.09	12.00	2.942	0.501	0.282	129	21.5	4.61	3.88	1.73	0.799	0.698
C10 × 30	8.82	10.00	3.033	0.436	0.673	103	20.7	3.42	3.94	1.65	0.669	0.649
25	7.35	10.00	2.886	0.436	0.526	91.2	18.2	3.52	3.36	1.48	0.676	0.617
20	5.88	10.00	2.739	0.436	0.379	78.9	15.8	3.66	2.81	1.32	0.691	0.606
15.3	4.49	10.00	2.600	0.436	0.240	67.4	13.5	3.87	2.28	1.16	0.713	0.634
C9 × 20	5.88	9.00	2.648	0.413	0.448	60.9	13.5	3.22	2.42	1.17	0.642	0.583
15	4.41	9.00	2.485	0.413	0.285	51.0	11.3	3.40	1.93	1.01	0.661	0.586
13.4	3.94	9.00	2.433	0.413	0.233	47.9	10.6	3.48	1.76	0.962	0.668	0.601
C8 × 18.75	5.51	8.00	2.527	0.390	0.487	44.0	11.0	2.82	1.98	1.01	0.599	0.565
13.75	4.04	8.00	2.343	0.390	0.303	36.1	9.03	2.99	1.53	0.853	0.615	0.553
11.5	3.38	8.00	2.260	0.390	0.220	32.6	8.14	3.11	1.32	0.781	0.625	0.571
C7 × 14.75	4.33	7.00	2.299	0.366	0.419	27.2	7.78	2.51	1.38	0.779	0.564	0.532
12.25	3.60	7.00	2.194	0.366	0.314	24.2	6.93	2.60	1.17	0.702	0.571	0.525
9.8	2.87	7.00	2.090	0.366	0.210	21.3	6.08	2.72	0.968	0.625	0.581	0.541
C6 × 13	3.83	6.00	2.157	0.343	0.437	17.4	5.80	2.13	1.05	0.642	0.525	0.514
10.5	3.09	6.00	2.034	0.343	0.314	15.2	5.06	2.22	0.865	0.564	0.529	0.500
8.2	2.40	6.00	1.920	0.343	0.200	13.1	4.38	2.34	0.692	0.492	0.537	0.512
C5 × 9	2.64	5.00	1.885	0.320	0.325	8.90	3.56	1.83	0.632	0.449	0.489	0.478
6.7	1.97	5.00	1.750	0.320	0.190	7.49	3.00	1.95	0.478	0.378	0.493	0.484
C4 × 7.25	2.13	4.00	1.721	0.296	0.321	4.59	2.29	1.47	0.432	0.343	0.450	0.459
5.4	1.59	4.00	1.584	0.296	0.184	3.85	1.93	1.56	0.319	0.283	0.449	0.458
C3 × 6	1.76	3.00	1.596	0.273	0.356	2.07	1.38	1.08	0.305	0.268	0.416	0.455
5	1.47	3.00	1.498	0.273	0.258	1.85	1.24	1.12	0.247	0.233	0.410	0.438
4.1	1.21	3.00	1.410	0.273	0.170	1.66	1.10	1.17	0.197	0.202	0.404	0.437

†An American Standard Channel is designated by the letter C followed by the nominal depth in inches and the weight in pounds per foot.

Appendix C.   Properties of Rolled-Steel Shapes
(SI Units)

C Shapes
(American Standard Channels)

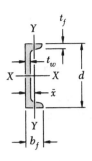

Designation†	Area A, mm²	Depth d, mm	Flange Width b_f, mm	Flange Thickness t_f, mm	Web Thickness t_w, mm	Axis X-X I_x 10⁶ mm⁴	Axis X-X S_x 10³ mm³	Axis X-X r_x mm	Axis Y-Y I_y 10⁶ mm⁴	Axis Y-Y S_y 10³ mm³	Axis Y-Y r_y mm	Axis Y-Y $\bar{x}$ mm
C380 × 74	9480	381	94	16.5	18.2	168.2	883	133.1	4.58	62.1	22.0	20.3
60	7610	381	89	16.5	13.2	145.3	763	138.2	3.84	55.5	22.5	19.76
50.4	6426	381	86	16.5	10.2	131.1	688	142.7	3.38	51.2	23.0	19.99
C310 × 45	5690	305	80	12.7	13.0	67.4	442	109.0	2.14	34.0	19.38	17.12
37	4742	305	77	12.7	9.8	59.9	393	112.5	1.861	31.1	19.81	17.12
30.8	3929	305	74	12.7	7.2	53.7	352	117.1	1.615	28.7	20.29	17.73
C250 × 45	5690	254	76	11.1	17.1	42.9	338	86.9	1.640	27.6	16.99	16.48
37	4742	254	73	11.1	13.4	38.0	299	89.4	1.399	24.4	17.17	15.67
30	3794	254	69	11.1	9.6	32.8	258	93.0	1.170	21.8	17.55	15.39
22.8	2897	254	65	11.1	6.1	28.1	221	98.3	0.949	18.29	18.11	16.10
C230 × 30	3794	229	67	10.5	11.4	25.4	222	81.8	1.007	19.29	16.31	14.81
22	2845	229	63	10.5	7.2	21.2	185.2	86.4	0.803	16.69	16.79	14.88
19.9	2542	229	61	10.5	5.9	19.94	174.2	88.4	0.733	16.03	16.97	15.27
C200 × 27.9	3555	203	64	9.9	12.4	18.31	180.4	71.6	0.824	16.60	15.21	14.35
20.5	2606	203	59	9.9	7.7	15.03	148.1	75.9	0.637	14.17	15.62	14.05
17.1	2181	203	57	9.9	5.6	13.57	133.7	79.0	0.549	12.92	15.88	14.50
C180 × 22.0	2794	178	58	9.3	10.6	11.32	127.2	63.8	0.574	12.90	14.33	13.51
18.2	2323	178	55	9.3	8.0	10.07	113.2	66.0	0.487	11.69	14.50	13.34
14.6	1852	178	53	9.3	5.3	8.86	99.6	69.1	0.403	10.26	14.76	13.74
C150 × 19.3	2471	152	54	8.7	11.1	7.24	95.3	54.1	0.437	10.67	13.34	13.06
15.6	1994	152	51	8.7	8.0	6.33	83.3	56.4	0.360	9.40	13.44	12.70
12.2	1548	152	48	8.7	5.1	5.45	71.7	59.4	0.288	8.23	13.64	13.00
C130 × 13.4	1703	127	47	8.1	8.3	3.70	58.3	46.5	0.263	7.54	12.42	12.14
10.0	1271	127	44	8.1	4.8	3.12	49.1	49.5	0.199	6.28	12.52	12.29
C100 × 10.8	1374	102	43	7.5	8.2	1.911	37.5	37.3	0.180	5.74	11.43	11.66
8.0	1026	102	40	7.5	4.7	1.602	31.4	39.6	0.133	4.69	11.40	11.63
C75 × 8.9	1135	76	40	6.9	9.0	0.862	22.7	27.4	0.127	4.47	10.57	11.56
7.4	948	76	37	6.9	6.6	0.770	20.3	28.4	0.103	3.98	10.41	11.13
6.1	781	76	35	6.9	4.3	0.691	18.18	29.7	0.082	3.43	10.26	11.10

†An American Standard Channel is designated by the letter C followed by the nominal depth in millimeters and the mass in kilograms per meter.

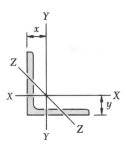

Appendix C. Properties of Rolled-Steel Shapes

(U.S. Customary Units)

Angles

Equal Legs

Size and Thickness, in.	Weight per Foot, lb/ft	Area, in²	Axis X-X and Axis Y-Y				Axis Z-Z r, in.
			I, in⁴	S, in³	r, in.	x or y, in.	
L8 × 8 × 1	51.0	15.0	89.0	15.8	2.44	2.37	1.56
³⁄₄	38.9	11.4	69.7	12.2	2.47	2.28	1.58
¹⁄₂	26.4	7.75	48.6	8.36	2.50	2.19	1.59
L6 × 6 × 1	37.4	11.0	35.5	8.57	1.80	1.86	1.17
³⁄₄	28.7	8.44	28.2	6.66	1.83	1.78	1.17
⁵⁄₈	24.2	7.11	24.2	5.66	1.84	1.73	1.18
¹⁄₂	19.6	5.75	19.9	4.61	1.86	1.68	1.18
³⁄₈	14.9	4.36	15.4	3.53	1.88	1.64	1.19
L5 × 5 × ³⁄₄	23.6	6.94	15.7	4.53	1.51	1.52	0.975
⁵⁄₈	20.0	5.86	13.6	3.86	1.52	1.48	0.978
¹⁄₂	16.2	4.75	11.3	3.16	1.54	1.43	0.983
³⁄₈	12.3	3.61	8.74	2.42	1.56	1.39	0.990
L4 × 4 × ³⁄₄	18.5	5.44	7.67	2.81	1.19	1.27	0.778
⁵⁄₈	15.7	4.61	6.66	2.40	1.20	1.23	0.779
¹⁄₂	12.8	3.75	5.56	1.97	1.22	1.18	0.782
³⁄₈	9.8	2.86	4.36	1.52	1.23	1.14	0.788
¹⁄₄	6.6	1.94	3.04	1.05	1.25	1.09	0.795
L3½ × 3½ × ¹⁄₂	11.1	3.25	3.64	1.49	1.06	1.06	0.683
³⁄₈	8.5	2.48	2.87	1.15	1.07	1.01	0.687
¹⁄₄	5.8	1.69	2.01	0.794	1.09	0.968	0.694
L3 × 3 × ¹⁄₂	9.4	2.75	2.22	1.07	0.898	0.932	0.584
³⁄₈	7.2	2.11	1.76	0.833	0.913	0.888	0.587
¹⁄₄	4.9	1.44	1.24	0.577	0.930	0.842	0.592
L2½ × 2½ × ¹⁄₂	7.7	2.25	1.23	0.724	0.739	0.806	0.487
³⁄₈	5.9	1.73	0.984	0.566	0.753	0.762	0.487
¹⁄₄	4.1	1.19	0.703	0.394	0.769	0.717	0.491
³⁄₁₆	3.07	0.902	0.547	0.303	0.778	0.694	0.495
L2 × 2 × ³⁄₈	4.7	1.36	0.479	0.351	0.594	0.636	0.389
¹⁄₄	3.19	0.938	0.348	0.247	0.609	0.592	0.391
¹⁄₈	1.65	0.484	0.190	0.131	0.626	0.546	0.398

Appendix C. **Properties of Rolled-Steel Shapes**
                        (SI Units)

Angles
Equal Legs

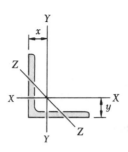

| Size and Thickness, mm | Mass per Meter, kg/m | Area, mm² | Axis X-X and Axis Y-Y | | | | Axis Z-Z |
			I 10^6 mm⁴	S 10^3 mm³	r mm	x or y mm	r mm
L203 × 203 × 25.4	75.9	9680	37.0	259	61.8	60.2	39.6
19.0	57.9	7360	29.0	200	62.8	57.9	40.1
12.7	39.3	5000	20.2	137.0	63.6	55.6	40.4
L152 × 152 × 25.4	55.7	7100	14.78	140.4	45.6	47.2	29.7
19.0	42.7	5445	11.74	109.1	46.4	45.2	29.7
15.9	36.0	4590	10.07	92.8	46.8	43.9	30.0
12.7	29.2	3710	8.28	75.5	47.2	42.7	30.0
9.5	22.2	2800	6.41	57.8	47.8	41.7	30.2
L127 × 127 × 19.0	35.1	4480	6.53	74.2	38.2	38.6	24.8
15.9	29.8	3780	5.66	63.3	38.7	37.6	24.8
12.7	24.1	3070	4.70	51.8	39.2	36.3	25.0
9.5	18.3	2330	3.64	39.7	39.5	35.3	25.1
L102 × 102 × 19.0	27.5	3510	3.19	46.0	30.1	32.3	19.76
15.9	23.4	2970	2.77	39.3	30.5	31.2	19.79
12.7	19.0	2420	2.31	32.3	30.9	30.0	19.86
9.5	14.6	1845	1.815	24.9	31.4	29.0	20.0
6.4	9.8	1252	1.265	17.21	31.8	27.7	20.2
L89 × 89 × 12.7	16.5	2100	1.515	24.4	26.9	26.9	17.35
9.5	12.6	1600	1.195	18.85	27.3	25.7	17.45
6.4	8.6	1090	0.837	13.01	27.7	24.6	17.63
L76 × 76 × 12.7	14.0	1774	0.924	17.53	22.8	23.7	14.83
9.5	10.7	1361	0.733	13.65	23.2	22.6	14.91
6.4	7.3	929	0.516	9.46	23.6	21.4	15.04
L64 × 64 × 12.7	11.4	1452	0.512	11.86	18.78	20.5	12.37
9.5	8.7	1116	0.410	9.28	19.17	19.35	12.37
6.4	6.1	768	0.293	6.46	19.53	18.21	12.47
4.8	4.6	581	0.228	4.97	19.81	17.63	12.57
L51 × 51 × 9.5	7.0	877	0.1994	5.75	15.08	16.15	9.88
6.4	4.7	605	0.1448	4.05	15.47	15.04	9.93
3.2	2.4	312	0.0791	2.15	15.92	13.87	10.11

Appendix C. **Properties of Rolled-Steel Shapes**
(U.S. Customary Units)

Angles
Unequal Legs

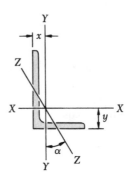

Size and Thickness, in.	Weight per Foot, lb/ft.	Area, in²	Axis X-X				Axis Y-Y				Axis Z-Z	
			I_x, in⁴	S_x, in³	r_x, in.	y, in.	I_y, in⁴	S_y, in³	r_y, in.	x, in.	r_z, in.	tan α
L8 × 6 × 1	44.2	13.0	80.8	15.1	2.49	2.65	38.8	8.92	1.73	1.65	1.28	0.543
¾	33.8	9.94	63.4	11.7	2.53	2.56	30.7	6.92	1.76	1.56	1.29	0.551
½	23.0	6.75	44.3	8.02	2.56	2.47	21.7	4.79	1.79	1.47	1.30	0.558
L6 × 4 × ¾	23.6	6.94	24.5	6.25	1.88	2.08	8.68	2.97	1.12	1.08	0.860	0.428
½	16.2	4.75	17.4	4.33	1.91	1.99	6.27	2.08	1.15	0.987	0.870	0.440
⅜	12.3	3.61	13.5	3.32	1.93	1.94	4.90	1.60	1.17	0.941	0.877	0.446
L5 × 3 × ½	12.8	3.75	9.45	2.91	1.59	1.75	2.58	1.15	0.829	0.750	0.648	0.357
⅜	9.8	2.86	7.37	2.24	1.61	1.70	2.04	0.888	0.845	0.704	0.654	0.364
¼	6.6	1.94	5.11	1.53	1.62	1.66	1.44	0.614	0.861	0.657	0.663	0.371
L4 × 3 × ½	11.1	3.25	5.05	1.89	1.25	1.33	2.42	1.12	0.864	0.827	0.639	0.543
⅜	8.5	2.48	3.96	1.46	1.26	1.28	1.92	0.866	0.879	0.782	0.644	0.551
¼	5.8	1.69	2.77	1.00	1.28	1.24	1.36	0.599	0.896	0.736	0.651	0.558
L3½ × 2½ × ½	9.4	2.75	3.24	1.41	1.09	1.20	1.36	0.760	0.704	0.705	0.534	0.486
⅜	7.2	2.11	2.56	1.09	1.10	1.16	1.09	0.592	0.719	0.660	0.537	0.496
¼	4.9	1.44	1.80	0.755	1.12	1.11	0.777	0.412	0.735	0.614	0.544	0.506
L3 × 2 × ½	7.7	2.25	1.92	1.00	0.924	1.08	0.672	0.474	0.546	0.583	0.428	0.414
⅜	5.9	1.73	1.53	0.781	0.940	1.04	0.543	0.371	0.559	0.539	0.430	0.428
¼	4.1	1.19	1.09	0.542	0.957	0.993	0.392	0.260	0.574	0.493	0.435	0.440
L2½ × 2 × ⅜	5.3	1.55	0.912	0.547	0.768	0.831	0.514	0.363	0.577	0.581	0.420	0.614
¼	3.62	1.06	0.654	0.381	0.784	0.787	0.372	0.254	0.592	0.537	0.424	0.626

Appendix C.   Properties of Rolled-Steel Shapes
(SI Units)

Angles
Unequal Legs

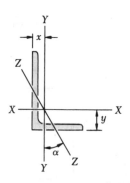

Size and Thickness, mm	Mass per Meter kg/m	Area mm²	Axis X-X				Axis Y-Y				Axis Z-Z	
			I_x 10⁶ mm⁴	S_x 10³ mm³	r_x mm	y mm	I_y 10⁶ mm⁴	S_y 10³ mm³	r_y mm	x mm	r_z mm	tan α
L203 × 152 × 25.4	65.5	8390	33.6	247	63.3	67.3	16.15	146.2	43.9	41.9	32.5	0.543
19.0	50.1	6410	26.4	192	64.2	65.0	12.78	113.4	44.7	39.6	32.8	0.551
12.7	34.1	4350	18.44	131	65.1	62.7	9.03	78.5	45.6	37.3	33.0	0.558
L152 × 102 × 19.0	35.0	4480	10.20	102.4	47.7	52.8	3.61	48.7	28.4	27.4	21.8	0.428
12.7	24.0	3060	7.24	71.0	48.6	50.5	2.61	34.1	29.2	25.1	22.1	0.440
9.5	18.2	2330	5.62	54.4	49.1	49.3	2.04	26.2	29.6	23.9	22.3	0.446
L127 × 76 × 12.7	19.0	2420	3.93	47.7	40.3	44.5	1.074	18.85	21.1	19.05	16.46	0.357
9.5	14.5	1845	3.07	36.7	40.8	43.2	0.849	14.55	21.5	17.88	16.61	0.364
6.4	9.8	1252	2.13	25.1	41.2	42.2	0.599	10.06	21.9	16.69	16.84	0.371
L102 × 76 × 12.7	16.4	2100	2.10	31.0	31.6	33.8	1.007	18.35	21.9	21.0	16.23	0.543
9.5	12.6	1600	1.648	23.9	32.1	32.5	0.799	14.19	22.3	19.86	16.36	0.551
6.4	8.6	1090	1.153	16.39	32.5	31.5	0.566	9.82	22.8	18.69	16.54	0.558
L89 × 64 × 12.7	13.9	1774	1.349	23.1	27.6	30.5	0.566	12.45	17.88	17.91	13.56	0.486
9.5	10.7	1361	1.066	17.86	28.0	29.5	0.454	9.70	18.26	16.76	13.64	0.496
6.4	7.3	929	0.749	12.37	28.4	28.2	0.323	6.75	18.65	15.60	13.82	0.506
L76 × 51 × 12.7	11.5	1452	0.799	16.39	23.5	27.4	0.280	7.77	13.89	14.81	10.87	0.414
9.5	8.8	1116	0.637	12.80	23.9	26.4	0.226	6.08	14.20	13.69	10.92	0.428
6.4	6.1	768	0.454	8.88	24.3	25.2	0.1632	4.26	14.58	12.52	11.05	0.440
L64 × 51 × 9.5	7.9	1000	0.380	8.96	19.51	21.1	0.214	5.95	14.66	14.76	10.67	0.614
6.4	5.4	684	0.272	6.24	19.94	20.0	0.1548	4.16	15.04	13.64	10.77	0.626

Beam and Loading	Elastic Curve	Maximum Deflection	Slope at End	Equation of Elastic Curve
1		$-\dfrac{PL^3}{3EI}$	$-\dfrac{PL^2}{2EI}$	$y = \dfrac{P}{6EI}(x^3 - 3Lx^2)$
2		$-\dfrac{wL^4}{8EI}$	$-\dfrac{wL^3}{6EI}$	$y = -\dfrac{w}{24EI}(x^4 - 4Lx^3 + 6L^2x^2)$
3		$-\dfrac{ML^2}{2EI}$	$-\dfrac{ML}{EI}$	$y = -\dfrac{M}{2EI}x^2$
4		$-\dfrac{PL^3}{48EI}$	$\pm\dfrac{PL^2}{16EI}$	For $x \leq \frac{1}{2}L$: $y = \dfrac{P}{48EI}(4x^3 - 3L^2x)$
5		For $a > b$: $-\dfrac{Pb(L^2 - b^2)^{3/2}}{9\sqrt{3}EIL}$ at $x_m = \sqrt{\dfrac{L^2 - b^2}{3}}$	$\theta_A = -\dfrac{Pb(L^2 - b^2)}{6EIL}$ $\theta_B = +\dfrac{Pa(L^2 - a^2)}{6EIL}$	For $x < a$: $y = \dfrac{Pb}{6EIL}[x^3 - (L^2 - b^2)x]$ For $x = a$: $y = -\dfrac{Pa^2b^2}{3EIL}$
6		$-\dfrac{5wL^4}{384EI}$	$\pm\dfrac{wL^3}{24EI}$	$y = -\dfrac{w}{24EI}(x^4 - 2Lx^3 + L^3x)$
7		$\dfrac{ML^2}{9\sqrt{3}EI}$	$\theta_A = +\dfrac{ML}{6EI}$ $\theta_B = -\dfrac{ML}{3EI}$	$y = -\dfrac{M}{6EIL}(x^3 - L^2x)$

INDEX

Allowable load, 25
Allowable stress, 24
Allowable-stress method, 674–675
Aluminum columns, design of, 663
American standard beam, 192
American standard channels (*see* C shapes)
American standard shapes, properties of, 708–709
Angle of rotation, relative, 132
Angle of twist:
  for circular shaft: in elastic range, 132
    in plastic range, 153
  for rectangular bar, 164
  for thin-walled hollow shaft, 168
Angle shapes, 325, 327
  properties of, 712–715
Anticlastic curvature, 194
Areas of common shapes, 535, *inside back cover*
Axial loading, 5, 40
  centric, 6
    elastic deformations under, 52
    Mohr's circle for, 357
    of columns, 526–640, 660–665
    plastic deformations under, 95–97
    strain due to, 90–92
    stress due to, 23, 90–92
  eccentric, 6
    general case of, 247, 249
    in plane of symmetry, 233–235
    of column, 650–654, 674–676

Bar with rectangular cross section, torsion of, 164
Bauschinger effect, 49

Beams, 278
  curved (*see* Curved members)
  deflection of (*see* Deflection of beams)
  design of, 410–463
  made of several materials, 204
  nonprismatic, 458–459
  normal stresses in, 191, 410
  of constant strength, 458–459
  principal stresses in, 443–446
  reinforced-concrete, 207
  shearing stresses in, 284–288, 411
  slope of (*see* Slope of beams)
  (*See also specific types of beams*)
Bearing stress, 9, 12
Bending, 183
  deformations in, 187–192
  in plane of symmetry, 183–225
  of curved members, 257–263
  of members made of several materials, 204
  plastic deformations in (*see* Plastic deformations, for beams in pure bending)
  stress in (*see* Normal stress, in bending)
  unsymmetric, 242–246
Bending moment, 184
  sign convention for, 184, 413
  ultimate, 219
Bending-moment diagram, 413–415
  by parts, 535–537
Biaxial stress, 87, 367
Bolts, stress in, 8
Boundary conditions:
  for beams carrying a distributed load, 486
  for statically determinate beams, 482

Boundary conditions (*Cont.*):
  for statically indeterminate beams, 488, 489
Box beam, 296
Bracing, 26
Breaking strength, 44
Brittle material, 43, 123
  fracture criteria for, 369–371
Buckling, 26, 300
  local, 372
Building codes, 26
Bulk modulus, 78

Cantilever beam, 482
  analysis of, 533–534
Castigliano, Alberto, 611
Castigliano's theorem, 610–611
  analysis of statically indeterminate structures by, 615–616
  deflections by, 612–615
Centric load *or* centric loading (*see* Axial loading, centric)
Centroid:
  of a cross section, 190
  of an area, 690–694
Centroidal axis:
  of an area, 694–697
  principal, 243
Centroids of common shapes, tables of, 535, *inside back cover*
Changes in temperature, 64
Channel section, 320–324
  shear center of, 322
  shearing stresses in, 323
Channel shapes (*see* C shapes)
Circular hole, stress distribution near, 93

Circular shafts:
  deformations in, 117–120
  elastic torsion formula for, 121
  made of an elastoplastic material, 151–156
  of variable cross section, 132
  plastic deformations in, 150–156
  residual stresses in, 154–156
  statically indeterminate, 133
  strains in, 119
  stress-concentration factors for, 146
  stresses in, 116, 117–123
Clebsch, A., 436
Clevis, 10
Coefficient of thermal expansion, 64
  of selected materials, 700–703
Columns, 631–676
  centric loading of, 631–640
  design of: under a centric load, 660–665
    under an eccentric load, 674–676
  eccentric loading of, 650–654
Combined loadings, stresses under, 307–309
Common shapes, tables of areas and centroids of, 535 *inside back cover*
Components of stress, 20
Composite beams, 204
Compression, modulus of, 78
Computer programming of singularity functions, 437
Concentrated load, 278
  stresses under, 463
Concrete, design specifications for, 26
Connections, 11
Constant strength, beams of, 458–459
Coulomb, Charles Augustin de, 370
Coulomb's criterion, 370
Coupling, flange, 145
Cover plate, 458
Cracks, 46, 371
Creep, 48
Crippling, 446
Critical load, 633, 635
Critical stress, 635
C shapes, properties of, 710–711
Curvature:
  anticlastic, 194
  of composite member, 206
  of neutral surface, 192

Curvature (*Cont.*):
  of transverse section of beam, 194
Curved beams (*see* Curved members)
Curved members:
  bending of, 257–263
  neutral surface in, 260
  stresses in, 262
Cycles, loading, 50
Cyclic loading, 26
Cylindrical pressure vessels, 377

Deflection of beams:
  by Castigliano's theorem, 612–615
  by energy method, 598–599, 612–615
  by integration, 478–514
  by moment-area method, 530–556
  by superposition, 478–514
  table of, 716
Deformations:
  in bending, 187–192
  in circular shafts, 117–120
  in transverse cross section of beam, 194
  permanent (*see* Permanent deformations)
  plastic (*see* Plastic deformations)
  under axial loading, 51
Density of selected materials, 701, 703
Design, 4
  for impact loads, 595–596
  of beams, 410–463
  of columns: under a centric load, 660–665
    under an eccentric load, 674–676
  of prismatic beams, 446–448
  of transmission shafts, 143–144, 460–461
Design load, 25
Design specifications, 26
Deviation, tangential, 532
Displacement, relative, 53
Distributed forces, 3
Distributed load, 278
  stress under, 462
Double shear, 8
Ductile material, 43, 123
  yield criteria for, 367–369

Ductility, 45
  of selected materials, 700–703
Dynamic loading, 26
  (*See also* Impact loading)

Eccentric load *or* eccentric loading (*see* Axial loading, eccentric)
Effective length of column, 637
  for various end conditions, 640
Effective slenderness ratio, 637
Elastic core:
  in beam, 220
  in circular shaft, 152
Elastic curve, equation of, 481–485, 486–487
Elastic flexure formulas, 191
Elasticity, modulus of, 47
  of selected materials, 700–703
Elastic limit, 48
Elastic section modulus, 191
Elastic strain energy (*see* Strain energy)
Elastic torque, maximum, 152
Elastic torsion formulas, 121
Elastic zone (*see* Elastic core)
Elastoplastic material, 95
  axial loading of member made of, 95–100
  bending of member made of, 220–225
  torsion of shaft made of, 151–156
Elementary work, 571
Elongation, percent, 45
  of selected materials, 700–703
End conditions for columns, 640
Endurance limit, 51
Energy-absorbing capacity, 576, 594, 596
Energy density, 573
Energy loading (*see* Impact loading)
Energy methods, 570–616
  analysis of statically indeterminate structures by, 615–616
  deflections by, 598–599, 612–615
Equation of elastic curve, 481–485, 486–487
Equivalent static load, 594, 598
Euler, Leonhard, 635
Euler's formula, 635

Factor of safety, 25
  for steel columns, 25
Failure criteria, 367–371
Fatigue, 25, 50
Fatigue limit, 51
Fillet, stress distribution near:
  in circular shaft, 146
  in flat bar: in bending, 208
    under axial loading, 93
First moment:
  of area, 690–694
  of cross section, 190
Flange coupling, 145
Flange of S- or W-beam, stresses in,
    287, 297
Flat bars, stress-concentration factors
    for:
  in bending, 208
  under axial loading, 94
Flexural rigidity, 482
Flexural stress, 191
Flexure formulas, 191
Fluctuating loading, 50
Forces, distributed, 3
Fracture criteria for brittle materials,
    369–371
Fracture mechanics, 371
Free-body diagram, 2
Frequency, 144

Gage length, 42
Generalized Hooke's law (*see* Hooke's
    law)
Gigapascal, 3
Groove, stress distribution near, 208
Gyration, radius of (*see* Radius, of
    gyration)

Hexagon of Tresca, 368
Highway bridges, design specifications
    for, 26
Hooke, Robert, 47
Hooke's law, 47
  for general state of stress, 81
  for multiaxial loading, 77
  for shearing stress and strain, 81
Hoop stress, 377
Horizontal shear, 282, 294

Horsepower, 144
Hydrostatic pressure, 78

Impact factor, 605
Impact loading, 572, 593–595
  design for, 595–596
Impulsive loading, 26
  (*See also* Impact loading)
Indeterminate (*see entries beginning
    with term:* Statically
    indeterminate)
Inertia, moment of (*see* Moment of
    inertia)
Influence coefficient, 608
Integration method for deflection of
    beams, 478–514
Interaction formula, 675
Interaction method, 675
Internal force, 52
Internal torque, 122
Isotropic material, 74, 81

Joule (unit), 572

Kern, 256
Keyway, 145
Kilopascal, 3

Lateral strain, 74
Limit:
  elastic, 48
  endurance, 51
  fatigue, 51
  proportional, 47
Load, 24–25
  concentrated (*see* Concentrated load)
  critical, 633, 635
  distributed (*see* Distributed load)
  equivalent static, 594, 598
  (*See also* Loading)
Load-deformation diagram, 40
Loading(s):
  axial (*see* Axial loading)
  centric (*see* Axial loading, centric)
  combined, 307–309
  cyclic, 26

Loading(s) (*Cont.*):
  dynamic, 26
    (*See also* impact, *below*)
  eccentric (*see* Axial loading,
    eccentric)
  energy (*see* impact, *below*)
  fluctuating, 50
  impact, 572, 593–595
  impulsive, 26
    (*See also* impact, *above*)
  multiaxial, 75,77
  repeated, 50
  reverse, 50
  transverse, 277–326
Loading cycles, 50
Local buckling, 446
Longitudinal stress in cylindrical
    pressure vessels, 377, 378
Lower yield point, 45

Macaulay, W. H., 436
Macaulay's brackets (*see* Singularity
    functions)
Macroscopic cracks, 371
Material:
  brittle, 43, 123
    fracture criteria for a, 369–371
  ductile, 43, 123
    yield criteria for, 367–369
  elastoplastic (*see* Elastoplastic
    material)
Materials, table of typical properties
    of, 700–703
Maximum deflection of beam:
  by integration, 483, 490
  by moment-area method, 546–547,
    549
Maximum-distortion-energy criterion:
  for general state of stress, 582
  for plane stress, 368
Maximum elastic moment, 220
Maximum elastic torque, 152
Maximum in-plane shearing strain, 388
Maximum in-plane shearing stress,
    346, 365
Maximum-normal-strain criterion, 370
Maximum-normal-stress criterion, 369,
    370
Maximum shearing strain, 388, 390

Maximum shearing stress, 346, 364
Maximum-shearing-stress criterion, 367
Maxwell, James Clerk, 610
Maxwell's reciprocal theorem, 610
Megapascal, 3
Members, secondary, 26
Membrane analogy, 164
Microscopic cracks, 46, 371
Modulus:
　bulk, 78
　of compression, 78
　of elasticity, 47
　　for selected materials, 700–703
　of resilience, 574
　of rigidity, 81
　　for selected materials, 700–703
　of rupture: in bending, 219
　　in torsion, 151
　of toughness, 573
　section: elastic, 191
　　plastic, 222
　shear, 81
　Young's, 47
Mohr, Otto, 353, 370
Mohr's circle:
　for centric axial loading, 357
　for moments and products of inertia,
　　251
　for plane strain, 387–388, 391
　for plane stress, 353–357, 364–365,
　　391–392
　for stresses: in cylindrical pressure
　　vessels, 378
　　in spherical pressure vessels, 379
　for three-dimensional strain,
　　390–392
　for three-dimensional stress,
　　364–365
　for torsional loading, 357
Mohr's criterion, 370–371
Moment:
　bending (*see* Bending moment)
　maximum elastic, 220
　plastic, 221
Moment-area method for deflection of
　beams, 530–556
Moment-area theorems, 530–533
Moment of inertia:
　of area, 695–706

Moment of inertia (*Cont.*):
　of beam section, 191
　of composite area, 698–699
　of cross section of shaft, 121
　polar, 121, 695
　rectangular, 695
Moments of areas, 690–699
Moments of inertia of common shapes,
　inside back cover
Multiaxial loading, 75
　generalized Hooke's law for, 77
Multiaxial stress, 75

Necking, 44
Neutral axis:
　for eccentric axial loading in plane of
　　symmetry, 234
　for general case of eccentric axial
　　loading, 248
　for symmetric bending, 188
　　in elastic range, 190
　　in plastic range, 224
　for unsymmetric bending, 245
Neutral surface:
　of curved beam, 258
　of prismatic beam, 188
　radius of curvature of, 192, 221, 261
Noncircular members:
　torsion of solid, 163–165
　torsion of thin-walled hollow,
　　166–168
Nonprismatic beams, 458–459
Normal strain:
　in bending, 189
　under axial loading, 40–41, 91
Normal stress, 5
　due to centric axial loading, 23,
　　90–92
　due to combined loading, 307–309
　due to eccentric axial loading, 234,
　　247
　due to transverse loading, 280
　in bending: of curved members, 262
　　of prismatic members, 186
　　in elastic range, 190, 193
　　in plastic range, 218
　in torsion, 123
　maximum, in beams, 410

Offset method, 45
Overhanging beam, 482

Parallel-axis theorem, 697–698
Pascal (unit), 3
Percent elongation, 45
　of selected materials, 700–703
Percent reduction in area, 45
Permanent deformations, 48
　in bending, 225
　in torsion, 155
　under axial loading, 95
Permanent set, 48, 97
Pin-ended columns, 634–636
Pins, stress in, 8
Plane strain, 87
　transformation of 384–388
Plane stress, 87, 340, 391
　transformation of, 342–357
Plastic deformations, 48
　for beams in pure bending, 218–225
　　made of an elastoplastic material:
　　　and of nonrectangular cross
　　　　section, 222–223
　　　and of rectangular cross section,
　　　　220–222
　　with a horizontal and a vertical
　　　plane of symmetry, 218–223
　　with a single vertical plane of
　　　symmetry, 224
　for beams under transverse loading,
　　298–299
　for circular shafts, 150–156
　under axial loading, 95–97
Plastic hinge, 298
Plastic moment, 221
Plastic section modulus, 222
Plastic torque, 153
Plastic zone:
　in circular shaft, 152
　in member in pure bending, 220
　in member under transverse loading,
　　298–299
Poisson, Siméon Denis, 74
Poisson's ratio, 74
Polar moment of inertia:
　of area, 579
　of cross section of shaft, 121

Power, 143
Pressure, hydrostatic, 78
Pressure vessels, stresses in thin-walled, 377–379
Principal axes of strain, 388, 390
Principal centroidal axes, 243
Principal planes of stress, 345, 363
Principal strains, 388, 390
Principal stresses, 345, 363
    in a beam, 369–372
Principle of superposition, 76
Prismatic beams, design of, 446–448
Properties:
    of rolled-steel shapes, 704–715
    of selected materials, 700–703
Proportional limit, 47
Pure bending, 184
    deformations in, 187–192
    stresses in (see Normal stress, in bending)

Radius:
    of curvature of neutral surface for prismatic beam, 192, 221
    of gyration: of area, 695–697
        of cross section of column, 635
    of neutral surface for curved beam, 261
Reciprocal theorem, 610
Reduction in area, percent, 45
Redundant reaction:
    for a beam, 512, 555
    under axial loading, 61
Reference tangent, 533, 544
Reinforced-concrete beams, 207
Relations:
    among E, ν, and G, 82–84
    between load and shear, 422–423
    between shear and bending moment, 423–425
Relative angle of rotation, 132
Relative displacement, 53
Repeated loadings, 50
Residual stresses, 25
    in bending, 224–225
    in torsion, 154–156
    under axial loading 99–100
Resilience, modulus of 574

Reverse loading, 50
Rigidity:
    flexural, 482
    modulus of, 81
        for selected materials, 700–703
Rivets, stresses in, 8
Rolled-steel shapes, properties of, 704–715
Rosette, strain, 393
Rotation, relative angle of, 132
Rupture, modulus of:
    in bending, 219
    in torsion, 151

Safety, factor of, 25
    for steel columns, 662
Saint-Venant, Adhémar Barré de, 92
Saint-Venant's criterion, 370
Saint-Venant's principle, 92
S-beam, 192
    (See also S shapes)
Secant formula, 653
Secondary members, 26
Second moment of an area, 695–699
Section modulus:
    elastic, 191
    plastic, 222
Shafts:
    circular (see Circular shafts)
    design of, 143, 460–461
    noncircular (see Noncircular members)
    transmission (see Transmission shafts)
Shape factor, 222
Shear, 7, 413
    double, 8
    horizontal, 282, 294
    single, 8
    vertical, 284
Shear center, 297, 321
    of angle shape, 325
    of channel shape, 322
    of Z shape, 326
Shear diagram, 413–415
Shear flow:
    in thin-walled hollow shafts, 167
    under transverse loading, 282
        in thin-walled members, 297

Shearing force, 278
Shearing strain, 80
    in circular shafts, 119
Shearing stress, 7, 279
    due to axial loading, 20, 23
    due to combined loading, 307–309
    due to transverse loading: of angle shape, 327
        of beam, 284–288
        of channel shape, 323
        of narrow rectangular beam, 286–288
        of S- and W-beams, 286
        of thin-walled member, 296–297, 320–326
        of Z shape, 326
    in torsion: of circular shafts, 116, 120–123, 151–154
        of rectangular bar, 164
        of thin-walled hollow shaft, 167
    maximum, 346, 364
    maximum, in beam, 411
    maximum in-plane, 346, 365
Shearing stress and strain, Hooke's law for, 81
Shearing-stress-strain diagram, 150
Shear modulus, 81
σ-n curve, 51
Sign convention:
    for bending moment, 184, 413
    for shear, 413
    for shearing strain, 80
    for stress, 3, 21
Simple structures, analysis of, 9
Simply supported beam, 482
Single shear, 8
Singularity functions:
    computer programming of, 437
    definition of, 434
    for shear and bending moment in beams, 432–457
    for slope and deflection of beams, 499–510
SI units, table of, *inside front cover*
Slenderness ratio, 636
    effective, 637
Slip, 48
Slope of beams:
    by Castigliano's theorem, 613

Slope of beams (*Cont.*):
  by integration method, 482, 486, 500
  by moment-area method, 531, 544
Specific weight of selected materials, 700, 702
Spherical pressure vessels, 379
S shapes, properties of, 708–709
Stability:
  of columns, 631–676
  of structures, 632–633
Statically determinate beams, boundary conditions for, 482
Statically indeterminate beams:
  analysis of: by Castigliano's theorem, 615–616
    by integration, 402–403
    by moment-area method, 554–556
    by superposition, 512–514
  boundary conditions for, 488, 489
Statically indeterminate forces, 39
Statically indeterminate problems, 60
Statically indeterminate shafts, 133
Steel beams, 192
  design of, 446–448
  properties of, 704–715
Steel columns, design of, 661–663
Steel, design specifications for, 26
Step function, 434
Stiffeners, 446
Stiffness, 48
Strain:
  lateral, 74
  measurement of, 393
  normal (*see* Normal strain)
  principal, 388, 390
  principal axes of, 388, 390
  shearing (*see* Shearing strain)
  thermal, 64
  transformation of, 331–339
  true, 46
Strain energy, 571
  for general state of stress, 580–582
  for normal stresses, 574–577
    in bending, 576–577
    under axial loading, 575–576
  for shearing stresses, 577–579
    in torsion, 578–579
    under transverse loading, 579

Strain-energy density:
  for general state of stress, 580–582
  for normal stresses, 573–574
  for shearing stresses, 577
Strain-hardening, 45
Strain rosette, 393
Strength:
  breaking, 44
  ultimate, 24, 44
    of selected materials, 700–703
  yield, 44
    of selected materials, 700–703
Stress, 3
  allowable, 24
  bearing, 9, 12
  biaxial, 87, 367
  components of, 20
  critical, 635
  due to axial loading (*see under* Normal stress)
  due to combined loading, 307–309
  due to transverse loading (*see* Shearing stress, due to transverse loading)
  flexural, 191
  general state of, 362–363
  hoop, 377
  in bending (*see* Normal stress, in bending)
  in bolts, 8
  in pins, 8
  in rivets, 8
  in thin-walled pressure vessels, 377–379
  in torsion (*see* Shearing stress, in torsion)
  longitudinal, 377, 378
  multiaxial, 75
  normal (*see* Normal stress)
  principal (*see* Principal stresses)
  principal planes of, 345, 363
  residual (*see* Residual stresses)
  shearing (*see* Shearing stress)
  sign convention for 3, 21
  transformation of, 339–366
  true, 46
  ultimate, 24–25
  under an applied load in a beam, 461–463

Stress (*Cont.*):
  uniform distribution of, 6
  yield (*see* Yield strength)
Stress-concentration factors:
  for circular shafts, 146
  for flat bars: in bending, 208
    under axial loading, 94
Stress concentrations:
  for flat bars: in bending, 208
    under axial loading, 93–94
  in circular shafts, 145
Stress-strain diagram, 42
Stress trajectory, 446
Structures, analysis of, 9
Superposition method, 61
  for deflection of beams, 426–429
  for determination of stresses, 307–309
Superposition principle, 76
Symbols, xix
Symmetric loading, deflection of beams with, 533–534

Tangential deviation, 440
Temperature changes, 64
Tensile test, 42
Testing machine, 43
  torsion, 130
Thermal expansion, coefficient of, 64
  of selected materials, 700–703
Thermal strain, 64
Thin-walled hollow shafts, 166–168
Thin-walled members:
  under nonsymmetric transverse loading, 320–326
  under symmetric transverse loading, 296–297
Thin-walled pressure vessels, 377–379
Three-dimensional analysis:
  of strain, 390–392
  of stress, 362–366
Timber columns, design of, 416
Timber, design specifications for, 26
Torque, 114
  internal, 122
  maximum elastic, 152
  plastic, 153
  ultimate, 151

Torsion:
　of circular shafts, 114–156
　of noncircular members, 163–165
　of rectangular bars, 164
　of thin-walled hollow shafts, 166–168
Torsional loading, Mohr's circle for, 357
Torsion formulas, 121
Torsion testing machine, 130
Toughness, modulus of, 573
Transformation:
　of plane strain, 384–388
　of plane stress, 342, 357
　of three-dimensional strain, 390–392
　of three-dimensional stress, 362–366
Transformed section:
　of beam made of several materials, 205
　of reinforced-concrete beam, 207
Transmission shafts, 114
　design of, 143–144, 384–385
Transverse cross section of beam:
　curvature of, 194
　deformations in, 194
Transverse loading, 277, 326
Tresca, Henri Edouard, 368
Tresca's criterion, 368
Tresca's hexagon, 368

True strain, 46
True stress, 46
Twist, angle of (see Angle of twist)
Twisting of thin-walled members under transverse loading, 320
Two-force members, 9

Ultimate bending moment, 219
Ultimate load, 24–25
Ultimate strength, 24, 44
　of selected materials, 700–703
Ultimate stress, 24–25
Ultimate torque, 151
Unsymmetric bending, 242–246
Unsymmetric loading:
　deflection of beam under, 544–546
　of thin-walled members, 320–326
Upper yield point, 45

Variable cross section:
　beams of, 458–459
　shafts of, 132
Vertical shear, 284
von Mises, Richard, 368
von Mises criterion, 368

Warping, 163
Watt (unit), 144

W-beam, 192
　(See also W shapes)
Web of S- or W-beam, stresses in, 286–287
Wide-flange beam, 192
Wide-flange shapes (see W shapes)
Work:
　of a couple, 597
　of a load, 571–572, 596–597
　of several loads, 608–610
Work-energy method for deflection under a single load, 598–599
Working load, 25
W shapes, properties of, 708–709

Yield criteria for ductile materials, 367–369
Yield or yielding, 43
Yield point, 45
Yield strength, 45
　of selected materials, 700–703
Yield stress (see Yield strength)
Young, Thomas, 47
Young's modulus, 47

Z shape, 326

ANSWERS TO EVEN-NUMBERED PROBLEMS†

CHAPTER 1

1.2	$\sigma_{AB} = 42.4$ MPa; $\sigma_{BC} = -35.7$ MPa.
1.4	73.9 kN.
1.6	23.4 MPa.
1.8	2145 kN.
1.10	33.1 kN.
1.12	6.67 MPa.
1.14	(*a*) 28.8 MPa. (*b*) -48.8 MPa.
1.16	(*a*) 30.5 MPa. (*b*) -24.9 MPa.
1.18	2000 mm².
1.20	158.4 MPa.
1.22	-2.24 MPa.
1.24	195.3 MPa.
1.26	42.3 mm.
1.28	43.4 mm.
1.30	(*a*) 126.7 MPa. (*b*) 139.3 MPa. (*c*) 185.7 MPa.
1.32	(*a*) 39.6 MPa. (*b*) 74.7 MPa. (*c*) 37.4 MPa.
1.34	(*a*) 3.44 kN ⦜60°. (*b*) 73.3 MPa. (*c*) 34.5 MPa.
1.36	$\sigma = 319$ kPa; $\tau = 149$ kPa.
1.38	16.58 kN. (*b*) 242 kPa.
1.40	1246 kN.
1.42	(*a*) 3.00. (*b*) 31.3 mm.
1.44	33.7 mm.
1.46	3.40.
1.48	21.2 mm.
1.50	1.683 kN ↓.
1.52	1.142 kN ↓.
1.54	(*a*) 4 MPa. (*b*) $b = 577$ mm.
1.56	(*a*) 4.67, fails in shear. (*b*) 3.19, fails in shear. (*c*) 2.52, fails in tension.
1.58	3.11.
1.60	(*a*) 3.31. (*b*) 26.7 MPa. (*c*) $b = 37.1$ mm, $c = 88.4$ mm.
1.62	(*a*) 39.7 MPa. (*b*) 34.7 MPa. (*c*) 31.2 MPa.
1.64	(*a*) 44 mm. (*b*) 180 mm.

CHAPTER 2

2.2	(*a*) 6.91 mm. (*b*) 160.0 MPa.
2.4	(*a*) 0.564 mm. (*b*) 1.43%.
2.6	(*a*) 55.0 m. (*b*) 3.46.
2.8	0.641 mm.
2.10	(*a*) -0.1549 mm. (*b*) 0.1019 mm ↓.
2.12	(*a*) 104.8 kN. (*b*) 0.267 mm.
2.14	(*a*) $+0.794$ mm. (*b*) $+0.484$ mm.
2.16	(*a*) $\sigma_t = +67.9$ MPa; $\sigma_r = -55.6$ MPa. (*b*) $\delta_t = +0.2425$ mm; $\delta_r = -0.1325$ mm.
2.18	(*a*) $+0.155$ mm. (*b*) 0.231 mm. (*c*) 124 kPa.
2.20	44.0 kN.
2.22	4.54 kN.
2.24	(*a*) $+1.222$ mm. (*b*) $+1.910$ mm.
2.26	4.41 kN.
2.28	$\rho g h^2 / 4E$ ↓.
2.30	$Ph/\pi abE$ ↓.
2.34	(*a*) -57.1 MPa. (*b*) -85.7 MPa.
2.36	(*a*) $\sigma_s = -124.6$ MPa; $\sigma_a = -43.6$ MPa. (*b*) -0.156 mm.
2.38	(*a*) $\mathbf{R}_A = 62.8$ kN ←; $\mathbf{R}_E = 37.2$ kN ←. (*b*) 46.3 μm →.
2.40	(*a*) $\mathbf{R}_A = 62.5$ kN →; $\mathbf{R}_D = 32.5$ kN ←. (*b*) $+0.453$ mm.
2.42	(*a*) $\mathbf{R}_A = 80.1$ kN →; $\mathbf{R}_D = 50.1$ kN ←. (*b*) $+0.391$ mm.

† Partial answers are provided or referenced for *all* problems designed to be solved with a computer.

2.44 (a) $\mathbf{R}_A = 78.0$ kN $\leftarrow$; $\mathbf{R}_E = 22.0$ kN $\leftarrow$.
(b) 0.1228 mm $\rightarrow$.

2.46 (a) $\Delta T_A = \Delta T_D = 230$ N;
$\Delta T_B = \Delta T_C = 770$ N. (b) 0.1839 mm.

2.48 (a) $F_{BC} = 4.17$ kN T; $F_{DE} = 1.667$ kN C.
(b) 57.9 μm $\rightarrow$.

2.50 (a) $T_{AD} = 8.32$ kN; $T_{CE} = 11.09$ kN.
(b) 1.655 mm $\uparrow$.

2.52 $T_A = P/12$, $T_B = P/3$, $T_D = 7P/12$.

2.54 -47.0 MPa.

2.56 $\sigma_s = -13.13$ kPa; $\sigma_c = +383$ kPa.

2.58 (a) $\sigma_{AB} = -202$ MPa; $\sigma_{BC} = -72.6$ MPa.
(b) 106.0 μm $\uparrow$.

2.60 (a) $\sigma_{AB} = -58.9$ MPa; $\sigma_{BC} = -132.6$ MPa.
(b) 0.218 mm.

2.62 (a) $\sigma_{AC} = -101.3$ MPa.
$\sigma_{CE} = -180.2$ MPa. (b) $+128.8$ μm.

2.64 (a) 107°C. (b) $+0.178$ mm.

2.66 (a) 0.205 mm. (b) -9.55 μm.

2.68 0.3989.

2.70 (a) $+13.61$ μm. (b) $+3.22$ μm.
(c) $+11.90$ μm.

2.72 (a) $+0.1234$ mm. (b) -14.6 μm.

2.74 $\epsilon' = \frac{1}{2}(1 - \nu)\epsilon_x$.

2.78 (a) $e = 409 \times 10^{-6}$, $\Delta V = 19.29$ mm³.
(b) $e = -290 \times 10^{-6}$, $\Delta V = -5.67 \times 10^{-6}$ m³.

2.80 (a) -34.3 μm. (b) -25.7 μm.
(c) -1163 mm³.

2.82 17.8 MPa.

2.84 1.190 mm $\downarrow$.

2.86 $a = 42.9$ mm; $b = 160.7$ mm.

2.88 16.46 kN.

2.92 20.3 kN.

2.94 (a) 121 MPa. (b) 137 MPa. (c) 162 MPa.

2.96 (a) 97.3 kN; 0.834 mm. (b) 180.0 kN;
1.714 mm.

2.98 (a) 86.5 kN. (b) 4 mm.

2.100 (a) 201 kN; 0.5 mm. (b) 201 kN; 3.5 mm.

2.102 176.7 kN; 3.84 mm.

2.104 (b) $\delta_m = 0.318$ mm $\downarrow$; $\sigma_{AC} = 250$ MPa;
$\sigma_{CB} = -318$ MPa; $\delta_P = 0.0342$ mm.

2.106 (a) 316 kN; 250 MPa; 414 MPa; 0.1476 mm.
(b) 426 kN; 250 MPa; 690 MPa; 0.603 mm.

2.108 $\sigma_{AC} = \sigma_{CB} = -34.2$ MPa.

2.110 (a) -65.6 MPa; 98.4 MPa.
(b) -176.0 MPa; 264 MPa.

2.112 (a) 0.767 mm; $\sigma_{AD} = 250$ MPa;
$\sigma_{BE} = 153.5$ MPa.
(b) 0.1802 mm; $\sigma_{AD} = -10.81$ MPa;
$\sigma_{BE} = 36.0$ MPa.

2.114 (a) 204 kN; $\sigma_{BD} = -250$ MPa,
$\sigma_{CE} = +233$ MPa. (b) $+0.0377$ mm;
$\sigma_{BD} = +17.44$ MPa, $\sigma_{CE} = +10.47$ MPa.

2.116 (a) -250 MPa. (b) $+54.2$ MPa.

2.118 (a) $\sigma_{AC} = -150$ MPa; $\sigma_{CB} = -250$ MPa.
(b) 106.9 μm $\rightarrow$.

2.120 20.8 kN.

2.122 (a) 8.55 kN. (b) 1.815 mm $\leftarrow$.

2.124 (a) 40 μm $\rightarrow$. (b) -40 MPa.

2.126 (a) -187.8 MPa; 0.031 mm $\leftarrow$.
(b) -250 MPa; 0.092 mm $\leftarrow$.

2.128 $\Delta T_A = \Delta T_B = 397$ N, $\Delta T_C = 105.3$ N.

2.130 (a) 82.5 MPa. (b) 72.5 MPa.

CHAPTER 3

3.2 (a) 5.17 kN · m. (b) 87.2 MPa.

3.4 (a) 6.63 kN · m. (b) 9.56 kN · m.

3.6 (a) 17.9 MPa. (b) 28.5 MPa. (c) 34.8 MPa.

3.8 (a) CD. (b) 85.9 MPa.

3.10 477 N · m.

3.12 (a) 2.40 kN · m. (b) 53.5 mm.

3.14 $\tau_{AB} = 47.7$ MPa; $\tau_{CD} = 35.4$ MPa.

3.16 1.10 kN · m.

3.18 (a) 55.0 MPa. (b) 45.3 MPa. (c) 47.7 MPa.

3.20 (a) 920 N · m; fails in shear.
(b) 522 N · m; fails in tension.

3.22 (a) $T/w = (\tau_m/2\rho g)[(c_1^2 + c_2^2)/c_2]$.
(b) $T/w = (T/w)_0[1 + (c_1/c_2)^2]$.

3.24 (a) 4.26°. (b) 4.32°.

3.26 48.4 MPa.

3.28 (a) 1.51°). (b) 3.45°).

3.30 (a) 0.556°). (b) 0.247°).

3.32 (a) 1.43 kN · m. (b) 1.02°.

3.34 5.25°.

3.36 (a) 410 N · m. (b) 4.00°.

3.38 31.7 mm.

3.40 (a) 22.5 mm. (b) 29.0 mm.

3.42 17.47 MPa. (b) 27.6 MPa. (c) 2.05°.

3.44 (a) 1.22 kN · m. (b) 2.75°.

3.46 (*a*) $\mathbf{M}_A = 1090$ N $\cdot$ m $)$;
$\mathbf{M}_C = 310$ N $\cdot$ m $)$. (*b*) 47.4 MPa.
(*c*) 28.8 MPa.

3.48 (*a*) $\mathbf{M}_A = 11.38$ kN $\cdot$ m $)$;
$\mathbf{M}_C = 1.12$ kN $\cdot$ m $)$. (*b*) 29.7 MPa.
(*c*) 13.53 MPa.

3.50 (*a*) 23.8 MPa. (*b*) 35.4 MPa.

3.52 4.12 kN $\cdot$ m.

3.54 (*a*) 82.5 MPa. (*b*) 0.281°.

3.58 (*a*) 60.8 MPa. (*b*) 30.4 MPa.

3.60 (*a*) 20.1 mm. (*b*) 15.94 mm.

3.62 (*a*) 18.76 kW. (*b*) 24.2 MPa.

3.64 (*a*) 65.5 MPa. (*b*) 4.06°.

3.66 (*a*) 31.1 MPa. (*b*) 51.9 MPa.

3.68 22.1 mm.

3.70 50.0 kW.

3.72 (*a*) 37.2 MPa. (*b*) 22.6 MPa.

3.74 1.7 mm.

3.76 (*a*) 120.7 MPa; 37.5 mm.
(*b*) 145 MPa, 23.7 mm.

3.78 (*a*) 993 N $\cdot$ m. (*b*) 1015 N $\cdot$ m.

3.80 (*a*) 2.59°. (*b*) 5.46°.

3.82 (*a*) 976 N $\cdot$ m. (*b*) 8.63 mm.

3.84 (*a*) 5.96 kN $\cdot$ m; 17.98°.
(*b*) 7.31 kN $\cdot$ m; 27.0°.

3.86 (*a*) 13.8 mm. (*b*) 79.2 mm.

3.90 (*a*) 551 N $\cdot$ m. (*b*) 6.67°.

3.92 (*a*) 3.40 kN $\cdot$ m. (*b*) 129 MPa.

3.94 (*a*) 3.41 kN $\cdot$ m. (*b*) 128.7 MPa.

3.96 (*a*) 65.9 MPa. (*b*) 44.8 MPa.

3.98 36.4 MPa, at surface of shaft.

3.100 9.42°.

3.102 1.18°.

3.104 (*a*) 26.4 MPa, at inner surface. (*b*) 0.706°.

3.106 $T_1 = 531$ N $\cdot$ m; $T_Y = 769$ N $\cdot$ m.

3.108 (*a*) 40.1 MPa; 0.653°.
(*b*) 50.9 MPa; 0.917°.

3.110 (*a*) 1.41 kN $\cdot$ m; 3.03°.
(*b*) 1.77 kN $\cdot$ m; 2.22°.

3.112 (*a*) 29.8 mm. (*b*) 30.4 mm. (*c*) 27.6 mm.

3.114 (*a*) 307 mm. (*b*) 227 mm. (*c*) 344 mm.

3.116 $\tau_A/\tau_B = 0.737$.

3.118 $\phi_A/\phi_B = 1.198$.

3.120 (*a*) 900 N $\cdot$ m. (*b*) 8.79°.

3.122 (*a*) 39.7 MPa. (*b*) 24.2 MPa. (*c*) 4.72°.

3.124 $\tau_a = 45.0$ MPa; $\tau_b = 22.5$ MPa.

3.126 5.44 kN $\cdot$ m.

3.128 (*a*) $T = T_0(1 - e/t)$. (*b*) 10%, 50%, 90%.

3.130 1.70°.

3.132 $\tau_2/\tau_1 = 3c/t°3$, $\phi_2/\phi_1 = 3(c/t)^2$.

3.134 (*a*) 30.2 MPa. (*b*) 61.1 MPa. (*c*) 39.8 MPa.

3.136 6.18°.

3.138 (*a*) 953 rpm. (*b*) 1647 rpm.

3.140 (*a*) 101.4 MPa. (*b*) 40.6 MPa.

3.142 (*a*) 66.0 MPa. (*b*) 6.06°.

3.144 (*a*) 2.36 kN $\cdot$ m. (*b*) 150 MPa. (*c*) 17.86°.

CHAPTER 4

4.2 (*a*) -46.0 MPa. (*b*) 29.1 MPa.

4.4 129.6 kN $\cdot$ m.

4.6 11.01 kN $\cdot$ m.

4.8 73.2 MPa; -102.4 MPa.

4.10 166 MPa; -65.6 MPa.

4.12 (*a*) 33.6 kN. (*b*) 5.00 kN.

4.14 39.6 kN.

4.16 4.29 kN $\cdot$ m.

4.18 177.8 kN $\cdot$ m.

4.20 (*a*) 83.0 MPa; 90.4 m.
(*b*) 138.3 MPa; 32.5 m.

4.22 449 MPa.

4.24 (*a*) 0.602 mm. (*b*) 0.203 N $\cdot$ m.

4.26 (*a*) $\sqrt{2}$. (*b*) $\sqrt{3}$.

4.28 0.950.

4.30 (*a*) 46.4 m. (*b*) 154.7 m. (*c*) 0.0444°.

4.32 (*a*) 0.064 mm. (*b*) 0.0257 mm.

4.36 2.22 kN $\cdot$ m.

4.38 67.2 kN $\cdot$ m.

4.40 (*a*) 66.2 MPa. (*b*) -112.4 mPa.

4.42 (*a*) -14.11 MPa. (*b*) 121.5 mPa.

4.44 20.1 kN $\cdot$ m.

4.46 79.1 kN $\cdot$ m.

4.48 (*a*) 156.8 MPa. (*b*) -11.23 MPa.

4.50 (*a*) 1.674×10^{-3} m². (*b*) 90.8 kN $\cdot$ m.

4.52 39.8 m.

4.54 176 m.

4.56 (*a*) 80.2 MPa. (*b*) 60.1 MPa.
(*c*) 13.37 MPa. (*d*) 52.4 m.

4.58 (*a*) 39.3 MPa. (*b*) 120.0 MPa. (*c*) 46.7 m.

4.62 (*a*) 96 MPa. (*b*) 137 MPa.

4.64 (*a*) 310 N · m. (*b*) 390 N · m.

4.66 (*a*) 53 MPa. (*b*) 95 MPa.

4.68 (*a*) 1.26 kN · m. (*b*) 1.42 kN · m.

4.70 (*a*) 2.40 kN · m. (*b*) 3.41 kN · m.

4.72 (*a*) 14.63 mm. (*b*) 5.85 m.

4.74 (*a*) 19.44 mm. (*b*) 7.78 m.

4.76 (*a*) 33.2 kN · m; 34.5 m.
   (*b*) 43.8 kN · m; 17.24 m.

4.78 (*a*) 20.6 kN · m; 29.3 m.
   (*b*) 26.9 kN · m; 16.00 m.

4.80 (*a*) 46.4 kN · m. (*b*) 1.364.

4.82 (*a*) 28.5 kN · m. (*b*) 1.385.

4.84 12.16 kN · m.

4.86 6.09 kN · m.

4.88 759 N · m.

4.90 (*a*) −92.3 MPa. (*b*) 105.4 MPa.

4.92 (*a*) −73.3 MPa. (*b*) 115.6 MPa.

4.94 (*a*) 92.3 MPa. (*b*) $y = 0$, $y = \pm 37.9$ mm.
   (*c*) 50.6 m.

4.96 (*a*) 92.1 MPa. (*b*) $y = 0$, $y = \pm 33.7$ mm.
   (*c*) 55.7 m.

4.98 (*a*) 0.707 ρ_Y. (*b*) 6.09 ρ_Y.

4.100 (*a*) 292 MPa. (*b*) 7.01 mm.

4.102 (*a*) 5.54 m. (*b*) 11.40 kN · m.

4.104 (*a*) 310 MPa. (*b*) 1.12 kN · m.

4.108 (*a*) −1.415 MPa. (*b*) −4.28 MPa.
   (*c*) −7.07 MPa.

4.110 (*a*) 34.0 MPa. (*b*) 36.3 MPa.

4.112 (*a*) 0.926 MPa. (*b*) −14.81 MPa.

4.114 (*a*) 5 mm. (*b*) 4.77 MPa.

4.116 2.62 kN.

4.118 134.3 kN ≤ *P* ≤ 230 kN.

4.120 (*a*) 47.6 MPa. (*b*) −49.4 MPa.
   (*c*) 9.80 mm below top edge.

4.122 (*a*) −3.50 MPa. (*b*) −5.75 MPa.
   (*c*) −4.67 MPa.
   (*d*) −8.96 MPa. (*e*) −7.00 MPa.

4.124 (*a*) 91.1 MPa. (*b*) −64.5 MPa.
   (*c*) 5.89 mm below *A*.

4.126 106.1 kN.

4.128 34.1 kN ↓ ; 40.2 kN ↑ .

4.130 (*a*) 990 kN. (*b*) 54.5 mm.

4.132 **P** = 337 kN ↓ ; **Q** = 338 kN ↓ .

4.134 (*a*) 90 mm. (*b*) 40 MPa.

4.136 (*a*) −1.151 MPa. (*b*) 7.31 MPa. (*c*) 66.6°↘.

4.138 (*a*) 115.2 MPa. (*b*) −76.4 MPa. (*c*) 79.8°↗.

4.140 (*a*) −29.3 MPa. (*b*) −144.8 MPa.
   (*c*) 41.5°↘.

4.142 (*a*) 66.3°↗. (*b*) At *D*: 40.4 MPa.

4.144 (*a*) 22.9°↗. (*b*) At *E*: 92.0 MPa.

4.146 (*a*) 11.3°↘. (*b*) At *D*: 110.6 MPa.

4.148 103.3 MPa.

4.150 −17.81 MPa.

4.152 32.1 MPa.

4.154 (*a*) 9.23 MPa. (*b*) −11.92 MPa.
   (*c*) 69.8 mm from *A*.

4.156 (*a*) $\sigma_A = 0.261$ MPa, $\sigma_B = -1.826$ MPa.
   (*b*) 12.5 mm from *A* and 19.17 mm from *D*.

4.158 31.2 mm.

4.160 51.6 kN.

4.162 3.18 kN · m.

4.164 3.04 kN · m.

4.172 (*a*) 91.4 MPa. (*b*) 85.7 MPa.

4.174 $\sigma_A = -97.4$ MPa; $\sigma_B = 44.2$ MPa.

4.176 106.1 MPa; −87.5 MPa.

4.178 (*a*) −16.90 MPa. (*b*) −50.3 MPa.

4.180 4.37 kN.

4.182 (*a*) −104.1 MPa. (*b*) 92.3 MPa.

4.184 (*a*) −73.7 MPa. (*b*) 56.7 MPa.

4.186 (*a*) 22.5 MPa. (*b*) −16.60 MPa.
   (*c*) −3.56 MPa.

4.188 (*a*) 104.2 MPa. (*b*) −68.4 MPa.

4.190 +56.4 MPa.

4.192 (*a*) −58.4 MPa. (*b*) 59.5 MPa.

4.200 187.9 MPa.

4.202 (*a*) 432 kN.
   (*b*) $y = 1.600$ mm, $z = 6.00$ mm.

4.204 (*a*) 37.1°↗. (*b*) 167.2 MPa; −125.0 MPa.

4.206 9.86 kN · m.

4.208 (*a*) $\sigma_A = -3M_1/bt^2$, $\sigma_B = 6M_1/bt^2$,
   $\sigma_C = -6M_1/bt^2$, $\sigma_D = 3M_1/bt^2$. (*b*) $\frac{1}{3}\rho_1$.

4.210 (*a*) 3.31 kN · m. (*b*) 50.0 m.

CHAPTER 5

5.2 1303 N.

5.4 738 N.

5.6 193 kN.

5.8 54.8 kN.

5.10 (*a*) 142 MPa. (*b*) 82.3 MPa.
   (*c*) 95.5 MPa.

5.12   (*a*) 140.9 MPa. (*b*) 13.01 MPa.
        (*c*) 17.63 MPa.

5.14   (*a*) 146.6 MPa. (*b*) 4.38 MPa.

5.16   (*a*) 131 MPa. (*b*) 19.7 MPa.

5.18   210 kN.

5.20   7.67 kN.

5.22   (*a*) 8.75 MPa. (*b*) 15.0 MPa.

5.24   32.7 MPa.

5.26   327 N.

5.28   (*a*) 417 kPa. (*b*) 520 kPa.

5.30   146 kN/m.

5.32   (*a*) 46.6 MPa. (*b*) 43.8 MPa.

5.34   (*a*) 4.55 MPa. (*b*) 3.93 MPa.

5.36   (*a*) 7.75 MPa. (*b*) 3.41 MPa. (*c*) 27.9 MPa,
        (*d*) 58.0 MPa.

5.38   (*a*) 75.0 MPa. (*b*) 58.0 MPa.
        (*c*) 15.13 MPa.

5.40   83.3 MPa.

5.42   9.57 mm.

5.44   (*a*) 20.7 MPa. (*b*) 18.9 MPa.

5.46   (*a*) 23.2 MPa. (*b*) 35.2 MPa.

5.48   (*a*) 40.9 MPa. (*b*) 49.6 MPa.
        (*c*) 57.8 MPa. (*d*) 58.7 MPa.

5.50   20.2 mm.

5.54   (*a*) 20.7 MPa. (*b*) 7.25 MPa.

5.56   (*a*) 10.90 MPa. (*b*) 18.26 MPa.

5.58   (*a*) $\sigma = +81.7$ MPa; $\tau = 0$.
        (*b*) $\sigma = +4.60$ MPa; $\tau = 5.79$ MPa.
        (*c*) $\sigma = -72.5$ MPa, $\tau = 0$.

5.60   (*a*) $\sigma = -129.7$ MPa; $\tau = 0$.
        (*b*) $\sigma = -8.94$ MPa; $\tau = 6.71$ MPa.
        (*c*) $\sigma = +111.8$ MPa; $\tau = 0$.

5.62   (*a*) 51.7 kN.
        (*b*) $\sigma = -3.02$ MPa; $\tau = 390$ kPa.

5.64   (*a*) $\sigma = -212$ MPa; $\tau = 0$.
        (*b*) $\sigma = -24.0$ MPa; $\tau = 17.44$ MPa.

5.66   (*a*) $\sigma = +31.3$ MPa; $\tau = 20.1$ MPa.
        (*b*) $\sigma = -16.8$ MPa; $\tau = 20.1$ MPa.

5.68   (*a*) $\sigma = +79.6$ MPa; $\tau = 7.96$ MPa.
        (*b*) $\sigma = 0$; $\tau = 13.26$ MPa.

5.70 $\sigma = -57.9$ MPa; $\tau = 4.17$ MPa.

5.72 $\sigma = -48.9$ MPa; $\tau = 17.8$ MPa.

5.74   (*a*) $\sigma = +18.3$ MPa; $\tau = 7.50$ MPa.
        (*b*) $\sigma = +45.8$ MPa; $\tau = 5.62$ MPa.
        (*c*) $\sigma = +73.3$ MPa; $\tau = 0$.

5.76   (*a*) $\sigma = +18.39$ MPa; $\tau = 391$ kPa.

        (*b*) $\sigma = +21.3$ MPa; $\tau = 293$ kPa.
        (*c*) $\sigma = +24.1$ MPa; $\tau = 0$.

5.78   (*a*) $\sigma = -7.11$ MPa; $\tau = 0$.
        (*b*) $\sigma = -5.45$ MPa; $\tau = 133.3$ kPa.
        (*c*) $\sigma = -3.79$ MPa; $\tau = 0$.

5.80   (*a*) $\sigma = -115.9$ MPa; $\tau = 0$.
        (*b*) $\sigma = -39.4$ MPa; $\tau = 2.62$ MPa.

5.82   (*a*) $\sigma = +86.5$ MPa; $\tau = 0$.
        (*b*) $\sigma = +57.0$ MPa; $\tau = 3.32$ MPa.

5.84   (*a*) $\sigma = +110.8$ MPa; $\tau = 0$.
        (*b*) $\sigma = +56.6$ MPa; $\tau = 4.06$ MPa.

5.86   (*a*) -129.6 MPa. (*b*) 827 kPa.

5.88   (*a*) -13.32 MPa. (*b*) 17.87 MPa.

5.90   (*a*) $+42.4$ MPa. (*b*) $+70.9$ MPa.

5.92   (*a*) $+7.50$ MPa. (*b*) $+11.25$ MPa.
        (*c*) 56.3°; 13.52 MPa.

5.94   (*a*) 97.1 MPa. (*b*) 85.5 MPa. (*c*) 30.9 MPa.

5.96   (*a*) 22.2 MPa. (*b*) 76.9 MPa.

5.98   118.5 MPa.

5.100 $e = (b^2 - a^2)/2(a + b + \frac{1}{6}h)$.

5.102   (*a*) $e = 20.8$ mm.
        (*b*) $\tau = 49.6$ kN at *B* in flange *AB*;
        $\tau = 57.9$ kN at midpoint of web *BD*.

5.104   (*a*) $e = 23.2$ mm. (*b*) $\tau = 12.36$ MPa at *B* in
        flange *BD*; $\tau = 25.2$ MPa at midpoint of web
        AG.

5.106   (*a*) $e = 9.12$ mm.
        (*b*) $\tau = 50.6$ MPa at *A* in flange *AB*;
        $\tau = 88.6$ MPa at midpoint of web *AH*.

5.108 $e = 14.6$ mm.

5.110 $c = 62.1$ mm.

5.112 $e = 0.1443a$.

5.114 $e = 2a$.

5.116   40 mm.

5.118 $e = 55.3$ mm.

5.120   (*a*) 144.6 N · m. (*b*) 65.9 MPa.

5.122   (*a*) 22.9 N · m. (*b*) 34.3 MPa.

5.126   (*a*) Vertical leg: $y = 2a/5$, $\tau = 27P/20at$;
        horizontal leg: $z = 2a/3$, $\tau = P/4at$.

5.128   (*a*) $z = 0.337a$; $\tau = 1.012P/at$.
        (*b*) $y = 0.495a$; $\tau = 0.0221P/at$.

5.130 $\tau = (P/ta^3)(a^2 - 2ay + 0.75y^2)$.

5.136 $\tau = (P/ta^3)(0.429a^2 + 1.288ay - 1.717y^2)$.

5.138 $\tau = (P/ta^3)(-0.215a^2 + 0.644x^2)$.

5.140   (*a*) 126 mm. (*b*) 254 kPa.

5.142   39.5 MPa.

5.144 (a) $\sigma = +108.0$ MPa; $\tau = 0$.
(b) $\sigma = -11.9$ MPa; $\tau = 39.9$ MPa.
(c) $\sigma = -131.8$ MPa; $\tau = 0$.

5.146 (a) 2.22 kN. (b) 973 kPa.

5.148 (a) $\sigma = +20.4$ MPa; $\tau = 14.34$ MPa.
(b) $\sigma = -21.5$ MPa; $\tau = 19.98$ MPa.

5.150 (a) $e = 50.6$ mm. (b) $\tau = 0.791$ MPa at B in segment BD; $\tau = 2.61$ MPa at midpoint of web DF.

CHAPTER 6

6.2 $\sigma = -6.07$ MPa; $\tau = 24.9$ MPa$\angle 50°$.

6.4 $\sigma = -62.0$ MPa; $\tau = 22.0$ MPa$\angle 60°$.

6.6 (a) $-26.6°$; $+63.4°$.
(b) $\sigma_{max} = +190.0$ MPa ($\theta = -26.6°$); $\sigma_{min} = -10.00$ MPa.

6.8 (a) $-30.9°$; $+59.1°$.
(b) $\sigma_{max} = 193.8$ MPa ($\theta = -30.9°$); $\sigma_{min} = -123.8$ MPa.

6.10 (a) $+18.4°$; $+108.4°$. (b) 100.0 MPa.
(c) $+90.0$ MPa.

6.12 (a) $+14.1°$; $+104.1°$. (b) 158.8 MPa.
(c) 35.0 MPa.

6.14 (a) -37.8 MPa; 52.8 MPa; 55.5 MPa.
(b) -29.1 MPa; 44.1 MPa; -61.6 MPa.

6.16 (a) -107.0 MPa; -23.0 MPa; $+7.83$ MPa.
(b) -58.0 MPa; -72.0 MPa; -42.1 MPa.

6.18 (a) 1.214 MPa. (b) -0.882 MPa.

6.20 $\sigma = -35.7$ MPa; $\tau = -3.52$ MPa.

6.22 (a) $\sigma_{max} = 2.06$ MPa ($\theta = 19.9°$); $\sigma_{min} = -15.70$ MPa ($\theta = 70.1°$).
(b) 8.88 MPa.

6.24 (a) $\sigma_{max} = 1.757$ MPa; $\sigma_{min} = -19.18$ MPa; $\tau_{max} = 10.47$ MPa.
(b) $\sigma_{max} = 28.0$ MPa; $\sigma_{min} = -1.009$ MPa; $\tau_{max} = 14.50$ MPa.

6.26 (a) 109.2 MPa, 45.0°); 45.0°).
(b) 48.2 MPa; 34.4°); 55.6°).

6.28 (a) $\sigma_{max} = 108.0$ MPa; $\sigma_{min} = -5.76$ MPa; $\tau_{max} = 56.9$ MPa.
(b) $\sigma_{max} = 31.6$ MPa; $\sigma_{min} = -31.6$ MPa; $\tau_{max} = 31.6$ MPa.

6.48 $\sigma_{max} = 168.6$ MPa ($\theta_p = 33.8°$); $\sigma_{min} = 6.42$ MPa ($\theta = 123.8°$).

6.50 $\sigma_{max} = 1.732\tau_0$ ($\theta_p = 90°$); $\sigma_{min} = -1.732\tau_0$ ($\theta_p = 0$).

6.52 $\sigma_{max} = 1.732\sigma_0$ ($\theta_p = 45°$); $\sigma_{min} = -1.732\sigma_0$ ($\theta_p = 45°$).

6.54 -120 MPa $\leq \tau_{xy} \leq 120$ MPa.

6.56 $-148.6° \leq \theta \leq 5.52°$; $31.4° \leq \theta \leq 185.5°$.

6.58 (a) 56.3°) and 33.7°). (b) -10 MPa.
(c) 65 MPa.

6.60 (a) 90.8 MPa ($\theta_p = 24.6°$); -0.82 MPa ($\theta_p = 65.4°$).
(b) 45.8 MPa.

6.64 (a) 69.0 MPa. (b) 104.0 MPa.

6.66 (a) 78 MPa. (b) 90 MPa.

6.68 (a) 111 MPa. (b) 78 MPa. (c) 66 MPa.

6.70 (a) 91 MPa. (b) 91 MPa. (c) 108 MPa.

6.72 (a) 40 MPa. (b) 72 MPa.

6.74 12 MPa and 56 MPa.

6.76 (a) Yes. (b) No. (c) Yes.

6.78 (a) No. (b) No. (c) Yes.

6.80 (a) 1.772. (b) 1.875.

6.82 (a) 1.565. (b) 1.726.

6.84 708 N · m.

6.86 299 kN.

6.88 No rupture.

6.90 Rupture will occur.

6.92 42.4 MPa.

6.94 351 N · m.

6.96 $\sigma_{max} = 49.8$ MPa; $\tau_{max} = 2.94$ MPa.

6.98 3.49.

6.100 $\sigma_{max} = 89.0$ MPa; $\tau_{max} = 44.5$ MPa.

6.102 13.59 m.

6.104 (a) 139.1 MPa. (b) 45.2 MPa.

6.106 4.51 MPa.

6.108 (a) 419 kPa. (b) 558 kPa.

6.110 (a) -15.23 MPa. (b) 22.6 MPa.

6.112 (a) $\sigma_{max} = 77.4$ MPa; $\tau_{max} = 38.7$ MPa (out of plane).
(b) $\sigma_{max} = 73.1$ MPa; $\tau_{max} = 51.9$ MPa (in plane).

6.114 (a) 37.7 MPa. (b) 1.810 MPa.

6.116 $\epsilon_{x'} = +357\,\mu$; $\epsilon_{y'} = -157.1\,\mu$; $\gamma_{x'y'} = -613\,\mu$.

6.118 $\epsilon_{x'} = +93.6\,\mu$; $\epsilon_{y'} = -13.58\,\mu$; $\gamma_{x'y'} = -641\,\mu$.

6.120 See Prob. 6.116.

6.122 See Prob. 6.118.

6.124 (a) $+410\,\mu$ at $31.0°$); $+70\,\mu$ at $59.0°$).
(b) $340\,\mu$. (c) $410\,\mu$.

6.126 (a) $+325\,\mu$ at $8.1°$); $+75\,\mu$ at $98.1°$).
(b) $250\,\mu$. (c) $325\,\mu$.

6.128 (a) $\theta p = 50.2°$); $\epsilon_a = 192.0\,\mu$;
$\epsilon_b = -52.0\,\mu$; $\epsilon_c = -60.0\,\mu$. (b) $244\,\mu$.
(c) $252\,\mu$.

6.130 (a) $\theta_p = 56.3°$); $\epsilon_a = 330\,\mu$;
$\epsilon_b = 70.0\,\mu$; $\epsilon_c = -200\,\mu$.
(b) $260\,\mu$. (c) $530\,\mu$.

6.132 (a) $\theta_p = 30.0°$); $\epsilon_a = 250\,\mu$; $\epsilon_b = 100\,\mu$.
(b) $150\,\mu$.

6.134 (a) $127.5\,\mu$.
(b) $\epsilon_a = 220\,\mu$; $\epsilon_b = -150\,\mu$; $\gamma_{\max} = 370\,\mu$.

6.136 (a) $\epsilon_a = 250\,\mu$; $\epsilon_b = 100\,\mu$; $\epsilon_c = -150\,\mu$.
(b) $400\,\mu$.

6.138 (a) $\epsilon_a = 220\,\mu$; $\epsilon_b = -150\,\mu$; $\epsilon_c = -30\,\mu$.
(b) $370\,\mu$.

6.140 $\theta_p = 31.0°$); $\sigma_{\max} = 21.6$ MPa;
$\sigma_{\min} = 2.61$ MPa.

6.142 $\theta_p = 31.1°$); $\epsilon_a = 1720\,\mu$; $\epsilon_b = -626\,\mu$;
$\epsilon_c = -466\,\mu$.

6.144 12.30 kN $\cdot$ m.

6.146 80.0 kN.

6.148 (b) $\epsilon_{\max} = 750\,\mu$; $\epsilon_{\min} = -250\,\mu$;
$\gamma_{\max} = 1000\,\mu$.

6.150 (a) $59.8°$); $30.2°$).
(b) 39.5 MPa; -116.7 MPa.
(c) 78.1 MPa (in plane).

6.152 (a) $18.4°$); $108.4°$).
(b) 82.5 MPa; 7.50 MPa.
(c) 41.2 MPa (out of plane).

6.154 (a) 775 kPa. (b) -2.69 MPa.

6.156 800 mm.

6.158 (a) $\epsilon_a = 120\,\mu$; $\epsilon_b = 420\,\mu$; $\epsilon_c = -180\,\mu$.
(b) $\sigma_{\max} = 96.0$ MPa; $\sigma_{\min} = 48.0$ MPa.
(c) 2.46 MPa.

6.160 (a) 1.500. (b) 1.688.

CHAPTER 7

7.2 $M_B = Pab/L$.
7.4 $M_A = -\frac{1}{2}wL^2$.
7.6 $M_B = M_C = Pa$.
7.8 From A to B: $V = Pb/L$, $M = Pbx/L$;

From B to C: $V = -Pa/L$,
$M = Pa(L - x)/L$.

7.10 $V = w(L - x)$; $M = -\frac{1}{2}w(L - x)^2$.

7.12 From A to B: $V = P$, $M = Px$;
From B to C: $V = 0$, $M = Pa$;
From C to D: $V = -P$, $M = P(L - x)$.

7.14 (a) 600 N. (b) 150 N $\cdot$ m.
7.16 (a) 42 kN. (b) 27 kN $\cdot$ m
7.18 (a) 62.5 kN. (b) 47.6 kN $\cdot$ m.
7.20 Just to the left of F: $V = -3$ kN,
$M = -2.4$ kN $\cdot$ m.
7.22 (a) 50 kN. (b) 32.9 kN $\cdot$ m.
7.24 (a) 6.56 MPa. (b) 375 kPa.
7.26 (a) 111.0 MPa. (b) 33.9 MPa.
7.28 108 MPa.
7.30 17.2 MPa.
7.32 (a) 865 mm. (b) 104.2 MPa.
7.34 (a) 0.420 m. (b) 60.9 MPa.
7.36 (a) $a = (\sqrt{2} - 1)(L/2)$. (b) 10.57 MPa.
7.48 (a) 54 kN. (b) 29.4 kN $\cdot$ m.
7.50 (a) 100.8 kN. (b) 133.1 kN $\cdot$ m.
7.52 $V = (w_0L/6)[1 - 3(x/L)^2]$;
$M = (w_0L^2/6)[x/L - (x/L)^3]$;
$M_{\max} = 0.0642w_0L^2$.
7.54 $V = (w_0L/3)[2 - 3x/L + (x/L)^3]$;
$M = -(w_0L^2/12)[3 - 8x/L + 6(x/L)^2 - (x/L)^4]$;
$|M|_{\max} = \frac{1}{4}w_0L^2$.
7.60 (a) 81.2 kN $\cdot$ m. (b) 148.2 MPa.
7.62 (a) 135.0 kN $\cdot$ m. (b) 158.6 MPa.
7.64 (a) 68.3 MPa. (b) 31.6 MPa.
7.66 (a) 127.3 MPa. (b) 29.4 MPa.
7.68 (a) $V_F = -1800$ N, $M_F = 540$ kN $\cdot$ m.
(b) 62.5 MPa.
7.70 (a) $V = wa - w\langle x - a \rangle + w\langle x - 3a \rangle$,
$M = wax - \frac{1}{2}w\langle x - a \rangle^2 + \frac{1}{2}w\langle x - 3a \rangle^2$.
(b) $\frac{3}{2}wa^2$.
7.72 (a) $V = 3w_0a - w_0x - w_0\langle x - a \rangle +$
$w_0\langle x - 3a \rangle$, $M = 3w_0ax - \frac{1}{2}w_0x^2 -$
$\frac{1}{2}w_0\langle x - a \rangle^2 + \frac{1}{2}w_0\langle x - 3a \rangle^2$. (b) $\frac{7}{2}wa^2$.
7.74 (a) $V = -(w_0/L)(Lx - x^2 + \langle x - \frac{1}{2}L \rangle^2)$,
$M = -(w_0/6L)(3Lx^2 - 2x^3 + 2\langle x - \frac{1}{2}L \rangle^3)$.
(b) $5w_0L^2/24$.
7.76 (a) $V = -1500x + 3000\langle x - 0.8 \rangle^0 +$
$3000\langle x - 3.2 \rangle^0$, $M = -750x^2 +$
$3000\langle x - 0.8 \rangle^1 + 3000\langle x - 3.2 \rangle^1$.
(b) 600 N $\cdot$ m.

7.78 (a) $V = -12 + 39\langle x - 0.9\rangle^0 -$
$24\langle x - 2.1\rangle^0 - 24\langle x - 3.3\rangle^0$,
$M = -12x + 39\langle x - 0.9\rangle^1 -$
$24\langle x - 2.1\rangle^1 - 24\langle x - 33\rangle^1$. (b) 25.2 kN · m.

7.80 (a) $V = 180 - 50x - 100\langle x - 0.8\rangle^0$,
$M = 180x - 25x^2 - 100\langle x - 0.8\rangle^1$.
(b) $x_m = 1.6$ m; $M_{max} = 144$ kN · m;
$\sigma_{max} = 152.9$ MPa.

7.82 (a) $V = 2.73 - 0.722x^2 + 0.722\langle x - 1.8\rangle^2$,
$M = 2.73x - 0.241x^3 + 0.241\langle x - 1.8\rangle^3$.
(b) $x_m = 1.95$ m; $M_{max} = 3.54$ kN · m;
$\sigma_{max} = 5.90$ MPa.

7.86 (b) x_m 1039 mm; $M_{max} = 1.465$ kN · m;
$\sigma_{max} = 3.82$ MPa.

7.88 (a) $V_C = 4.05$ kN, $M_C = 4.02$ kN · m.
(b) $x_m = 0.670$ m; $M_{max} = 4.57$ kN · m.

7.90 $d = 189.7$ mm.

7.92 $d = 311$ mm.

7.94 $b = 156.3$ mm.

7.96 $d = 349$ mm.

7.98 W 530×66.

7.100 W 360×32.9.

7.102 W 530×66.

7.104 W 530×92.

7.106 299.3 mm.

7.108 9 mm.

7.110 (a) W 410×60. (b) W 530×66.
(c) W 360×44.8.

7.112 S 510×97.3.

7.114 (a) $w_A = 11.46$ kN/m, $w_B = 6.88$ kN/m.
(b) 24.0 mm.

7.116 91.9 kN/m.

7.118 15.06 kN.

7.120 (a) 2026 N · m. (b) 1.14 m.

7.122 (a) 2.74 m. (b) W 530×66.

7.124 $b = 60.0$ m, $x = 0.342$ m.

7.128 (a) 58.4 MPa. (b) 108.2 MPa.
(c) 107.6 MPa.

7.130 (a) 130.4 MPa. (b) 160.9 MPa.
(c) 97.6 MPa.

7.132 $h = h_0(x/L)^{1/2}$.

7.134 $h = 2h_0[(x/L) - (x/L)^2]^{1/2}$.

7.136 For $x \leq \frac{1}{2}L$: $h = h_0(2x/L)^{1/2}$;
For $x > \frac{1}{2}L$: $h = h_0[2(L - x)/L]^{1/2}$.

7.138 (a) $b = b_0 x/L$. (b) $b = b_0(x/L)^2$.

7.140 $d = 1.587d_0[(x/L) - (x/L)^2]^{1/3}$.

7.142 (a) 151.6 MPa. (b) 128.6 MPa.

7.144 (a) 292 mm. (b) 4.175 m.

7.146 194.6 kN.

7.148 300.5 kN.

7.150 (a) 281.7 kN. (b) 3.17 m.

7.152 (a) $x = \frac{1}{2}Lh_0/h_1$.
(b) $\sigma_m = 3wL^2/[4bh_0(2h_1 - h_0)]$.

7.154 (a) $x = 375$ mm. (b) 28.5 kN · m.

7.156 39.0 mm.

7.158 36.5 mm.

7.160 (a) 37.7 mm. (b) 30.6 mm.

7.162 42.6 mm.

7.164 45.9 mm.

7.166 31.5 mm.

7.168 $d_{BC} = 21.7$ mm; $d_{CD} = 33.4$ mm.

7.170 (a) 54.1 MPa. (b) 38.1 MPa.

7.174 (a) 73.8 MPa. (b) 94.9 MPa.

7.176 $\tau_{max} = 75.6$ MPa; $\sigma_{max} = 137.4$ MPa.

7.178 87.8 kN/m.

7.180 $b = 59.0$ mm.

7.182 (a) 1.434 m. (b) 14.0 MPa.

7.186 56.0 mm.

7.188 $d = 323$ mm.

CHAPTER 8

8.2 (a) $y = M_0x^2/2EI$. (b) $M_0L^2/2EI \uparrow$.
(c) $M_0L/EI \measuredangle$.

8.4 (a) $y = -(w/24EI)(x^4 - 4L^3x + 3L^1)$.
(b) $wL^4/8EI \downarrow$. (c) $wL^3/6EI \measuredangle$.

8.6 (a) $y = (w_0/120EIL)(-x^5 + 10L^2x^3 - 20L^3x^2)$. (b) $11w_0L^4/120EI \downarrow$.
(c) $w_0L^3/8EI \measuredangle$.

8.8 (b) 6.74 mm $\downarrow$.
(c) 4.49×10^{-3} rad $\measuredangle$.

8.10 (a) $y = (w_0/360EIL)(-3x^5 + 10L^2x^3 - 7L^4x)$. (b) $5w_0L^4/768EI \downarrow$.
(c) $7w_0L^3/360EI \measuredangle$. (d) 6.78 mm $\downarrow$;
4.05×10^{-3} rad $\measuredangle$.

8.12 (a) $y = (M_0/6EIL)(x^3 - L^2x)$.
(b) $M_0L/6EI \measuredangle$. (c) $M_0L/3EI \measuredangle$.

8.14 (a) $y = (M_0/6EIL)(-2x^3 + 3Lx^2 - L^2x)$.
(b) and (c) $M_0L/6EI \measuredangle$.

8.16 (a) $y = (w/24EI)(-x^4 + L^3x)$.
(b) $wL^3/24EI\angle$. (c) $wL^3/8EI\diagdown$.

8.18 (a) $y = (w/24EI)(-x^4 + 4Lx^3 - 4L^2x^2)$.
(b) $wL^4/24EI \downarrow$. (c) 0.

8.20 (a) 2.75×10^{-3} rad $\diagdown$. (b) 1.147 mm $\downarrow$.

8.22 (a) 1.790×10^{-3} rad $\angle$. (b) 0.242 mm $\downarrow$.

8.24 (a) 4.24×10^{-3} rad $\angle$. (b) 3.39 mm $\uparrow$.

8.26 (a) $y = (Pa/6EI)(3x^2 - 3Lx + a^2)$.
(b) $(Pa/2EI)(L - 2a)\diagdown$.
(c) $(Pa/24EI)(3L^2 - 4a^2) \downarrow$.

8.28 (a) 9.21×10^{-3} rad $\diagdown$. (b) 11.38 mm $\downarrow$.

8.30 (a) $y = (P/48EI)(-8x^3 + 30L^2x - 21L^3)$.
(b) 2.48 mm $\downarrow$.

8.32 $0.0642M_0L^2/EI \uparrow$ at $x = 0.577L$.

8.34 (a) $0.00652w_0L^4/EI \downarrow$ at $x = 0.519L$.
(b) 6.79 mm $\downarrow$.

8.36 $0.01969wL^4/EI \uparrow$ at $x = 0.630L$.

8.38 (a) $y = (w_0/360EIL^2)(x^6 - 15L^2x^4 + 25L^3x^3 - 11L^5x)$. (b) $11w_0L^3/360EI \diagdown$.
(c) $0.00916w_0L^4/EI \downarrow$.

8.40 (a) $y = -\dfrac{w_0L^4}{\pi^4EI}\left[\sin\dfrac{\pi x}{L} - \dfrac{\pi^3}{6}\left(\dfrac{x}{L}\right)^3 + \dfrac{\pi^3}{2}\left(\dfrac{x}{L}\right)^2 - \pi\left(\dfrac{x}{L}\right)\right]$.
(b) $[(\pi^2 - 3)/3\pi^3]w_0L^4/EI \downarrow$.

8.42 $R_B = 3wL/8 \uparrow$.

8.44 $R_B = 11w_0L/40 \uparrow$.

8.46 $R_A = 41wL/128 \uparrow$; $M_C = +9wL^2/256$; $M_B = -7wL^2/128$.

8.48 $R_B = 5P/16 \uparrow$; $M_A = -3PL/16$; $M_C = +5PL/32$.

8.50 $R_A = 3M_0(L^2 - a^2)/2L^3 \uparrow$.

8.52 $3.09 \times 10^{-3}wL^4/EI \downarrow$.

8.54 $7PL^3/768EI \downarrow$.

8.56 $R_A = P/2 \uparrow$, $M_A = PL/8\rangle$;
$M_A = M_B = -PL/8$; $M_C = +PL/8$.

8.58 $R_A = Pb^2(3a + b)/L^2 \uparrow$, $M_A = Pab^2/L^2\rangle$;
$M_A = -Pab^2/L^2$; $M_B = -Pa^2b/L^2$;
$M_D = +2Pa^2b^2/L^3$.

8.60 (a) $y = (Pb/6EIL)[x^3 - (L^2 - b^2)x - (L/b)\langle x - a\rangle^3]$. (b) $(Pb/6EIL)(L^2 - b^2)\diagdown$.
(c) $Pa^2b^2/3EIL \downarrow$.

8.62 (a) $y = (w/1944EI)[72Lx^3 - 81\langle x - L/3\rangle^4 - 56L^3x]$. (b) $7wL^3243EI \diagdown$.

8.64 (c) $2wL^4/243EI \downarrow$.
(a) $y = (P/6EI)[-9ax^2 + 2x^3 - \langle x - a\rangle^3]$.
(b) $7Pa^3/2EI \downarrow$.

8.66 (a) $y = (P/6EI)[-x^3 - \langle x - a\rangle^3 - \langle x - 2a\rangle^3 + 42a^2x - 90a^3]$.
(b) $15Pa^3/EI \downarrow$.

8.68 (a) $29Pa^3/12EI \downarrow$. (b) $19Pa^3/6EI \downarrow$.

8.70 (a) $y = (w/48EI)[-9L^2x^2 + 4Lx^3 - 2\langle x - \frac{1}{2}L\rangle^4]$. (b) $41wL^4/384EI \downarrow$.

8.72 (a) $y = (w/24EI)[4ax^3 - \langle x - a\rangle^4 + \langle x - 3a\rangle^4 - 44a^3x]$. (b) $19wa^4/8EI \downarrow$.

8.74 (a) $y = (w_0/5760EIL)[96x^5 - 96\langle x - \frac{1}{2}L\rangle^5 - 240Lx^4 + 200L^2x^3 - 53L^4x]$.
(b) $3w_0L^4/1280EI \downarrow$.

8.76 (a) $y = (w/384EI)[-16x^4 + 16Lx^3 + 16\langle x - \frac{1}{2}L\rangle^4 - 16\langle x - L\rangle^4 + 48L\langle x - L\rangle^3 - L^3x]$. (b) $wL^4/768EI \uparrow$.
(c) $5wL^4/256EI \downarrow$.

8.78 (a) 0.802×10^{-3} rad $\diagdown$. (b) 0.844 mm $\downarrow$.

8.80 (a) 4.74×10^{-3} rad $\diagdown$. (b) 7.87 mm.

8.82 (a) 3.90×10^{-3} rad $\diagdown$. (b) 4.06 mm $\downarrow$.

8.84 (a) 3.18 mm $\downarrow$. (b) 4.29 mm $\downarrow$.

8.86 (a) 2.65×10^{-3} rad $\diagdown$. (b) 5.20 mm $\downarrow$.

8.92 (a) $41wL/128 \uparrow$. (b) $19wL^4/6144EI \downarrow$.

8.94 (a) $9M_0/8L \uparrow$. (b) $M_0L^2/128EI \downarrow$.

8.96 (a) 46.7 kN $\uparrow$. (b) 1.187 mm $\downarrow$.

8.98 (a) 8.95 kN $\uparrow$. (b) 1.677 mm $\downarrow$.

8.100 (a) $R_A = 20P/27 \uparrow$, $M_A = 4PL/27\rangle$.
(b) $5PL^3/1296EI \downarrow$.

8.102 (a) $R_A = 51.3$ kN $\uparrow$, $M_A = 39.5$ kN $\cdot$ m$\rangle$.
(b) $+22.0$ kN $\cdot$ m.

8.104 10.97 mm; 2.52 m from A.

8.106 4.23 mm; 0.699 m from A.

8.108 (a) $5PL^3/162EI \downarrow$. (b) $PL^2/9EI \diagdown$.

8.110 (a) $M_0L^2/8EI \uparrow$. (b) $M_0L/2EI \angle$.

8.112 $5PL^2/8EI \diagdown$; $7PL^3/16EI \downarrow$.

8.114 $4wa^3/3EI \diagdown$; $10wa^4/3EI \downarrow$.

8.116 $PL^2/2EI \diagdown$; $7PL^3/48EI \downarrow$.

8.118 $7wa^3/6EI \diagdown$; $17wa^4/24EI \downarrow$.

8.120 (a) 5.83×10^{-3} rad $\diagdown$. (b) 11.22 mm $\downarrow$.

8.122 12.99×10^{-3} rad $\angle$; 14.97 mm $\downarrow$.

8.124 11.37×10^{-3} rad $\diagdown$; 8.24 mm $\downarrow$.

8.126 (a) $R_A = (3M_0/2L^2)(L^2 - a^2) \uparrow$.
(b) $R_B = -R_A$, $M_B = (M_0/2L^2)(L^2 - 3a^2)\rangle$.

8.128 $R_A = 3P/8 \uparrow$; $R_C = 7P/8 \uparrow$; $R_D = P/4 \downarrow$.

8.130 $R_B = 13wL/32 \uparrow$, $M_B = 11wL^2/192\rangle$.

8.132 13.46 mm $\downarrow$.

8.134 4.96 mm $\downarrow$.

8.136 (a) 14.07 kN $\uparrow$. (b) 5.93 kN $\uparrow$.

8.138 $\mathbf{R}_A = \mathbf{R}_B = 76.5$ kN $\uparrow$; $\mathbf{R}_C = 39.9$ kN $\downarrow$.

8.140 (a) 10.24 kN $\uparrow$. (b) 11.08 mm $\downarrow$.

8.142 9.31 mm $\downarrow$.

8.144 (a) 6.74×10^{-3} rad $\searrow$. (b) 15.16 mm $\downarrow$.

8.146 (a) 3.90 mm $\downarrow$. (b) 1.092 mm $\uparrow$.

8.148 $y_1 = y_3 = PL^3/192EI \downarrow$; $y_2 = PL^3/96EI \downarrow$.

8.150 4.54×10^{-3} rad $\searrow$; 3.89 mm $\downarrow$.

8.152 (a) $\mathbf{R}_A = 1.100$ kN $\uparrow$, $\mathbf{M}_A = 275$ Nm $\rangle$.
(b) $\mathbf{R}_B = 2.40$ kN $\uparrow$, $\mathbf{M}_B = 659$ Nm $\rangle$.

8.154 3.53×10^{-3} rad $\searrow$; 8.39 mm $\downarrow$.

CHAPTER 9

9.2 (a) $PL^2/2EI \measuredangle$. (b) $PL^3/3EI \downarrow$.

9.4 (a) $w_0 L^3/24EI \searrow$. (b) $w_0 L^4/30EI \downarrow$.

9.6 (a) $PL^2/4EI \measuredangle$. (b) $PL^3/4EI \downarrow$.

9.8 (a) $7wL^3/48EI \measuredangle$. (b) $41wL^4/384EI \downarrow$.

9.10 (a) 0; $PL^3/12EI \downarrow$. (b) $PL^2/6EI \measuredangle$; 0.

9.12 (a) 2.70×10^{-3} rad $\measuredangle$. (b) 7.12 mm $\downarrow$.

9.14 (a) 16.8×10^{-3} rad $\searrow$. (b) 9.63 mm. $\downarrow$.

9.16 (a) $wL^3 16EI \measuredangle$. (b) $47wL^4/1152EI \downarrow$.

9.18 (a) 2.77×10^{-3} rad $\searrow$. (b) 3.19 mm $\downarrow$.

9.20 (a) 1.791×10^{-3} rad $\measuredangle$. (b) 3.11 mm $\downarrow$.

9.22 (a) $wa^2(3L - 2a)/12EI \searrow$.
(b) $wa^2(3L^2 - 2a^2)/48EI \downarrow$.

9.24 (a) $Pa(L - a)2EI \searrow$.
(b) $Pa(3L^2 - 4a^2)/24EI \downarrow$.

9.26 (a) 4.10×10^{-3} rad $\searrow$. (b) 6.20 mm $\downarrow$.

9.28 (a) 3.87×10^{-3} rad $\searrow$. (b) 6.50 mm $\uparrow$.
(c) 8.80 mm $\downarrow$.

9.30 (a) 20 kN. (b) 10.3 kN $\leq P \leq$ 29.7 kN.

9.32 (a) $0.223L$.

9.34 (a) $39PL^2/384EI \searrow$. (b) $9PL^3/256EI \downarrow$.

9.36 (a) $3wL^3/128EI \searrow$. (b) $5wL^4 768EI \downarrow$.

9.38 (a) $5M_0L/32EI \searrow$. (b) $3M_0L^2/64EI \downarrow$.

9.40 (a) 4.34×10^{-3} rad $\searrow$. (b) 6.17 mm $\downarrow$.

9.42 (a) 5.20×10^{-3} rad $\searrow$. (b) 5.01 mm $\downarrow$.

9.44 (a) 3.63×10^{-3} rad $\searrow$. (b) 1.99 mm $\downarrow$.

9.46 (a) $Pa(2L + 3a)/6EI \searrow$.
(b) $Pa^2(L + a)/3EI \downarrow$.

9.48 (a) $13wL^4/768EI \downarrow$. (b) $wL^4/768EI \uparrow$.

9.50 (a) 1.28 mm $\downarrow$. (b) 2.995 mm $\downarrow$.

9.52 (a) 2.57×10^{-3} rad $\measuredangle$. (b) 2.41 mm $\uparrow$.

9.54 (a) 20.7×10^{-3} rad $\measuredangle$. (b) 2.21 mm $\downarrow$.

9.56 (a) $3wL^3/128EI \measuredangle$. (b) $wL^4/64EI \downarrow$.

9.58 (a) $0.01749PL^3/EI \downarrow$. (b) $0.01955PL^3/EI \downarrow$.

9.60 (a) $17.92 \times 10^{-3}PL^3/EI$, at $0.544L$ from A.

9.62 6.56×10^{-3} wL^4/EI, at $0.460L$ from A.

9.64 6.59 mm, at 2.45 m from A.

9.66 5.19 mm, at 1.55 m from A.

9.68 2.996 mm, at 2.08 m from A.

9.70 3.35 mm, at 1.387 m from A.

9.72 24.0×10^{-3} PL^3/EI, at $0.537L$ from A.

9.74 $y_A = 2PL^3/3EI_0 \downarrow$.

9.76 $0.275w_0L \uparrow$.

9.78 $3M_0/2L \uparrow$.

9.80 $\mathbf{R}_B = 9M_0/8L \uparrow$; $M_A = +M_0/8$.

9.82 $\mathbf{R}_A = 2P/3 \uparrow$; $M_B = -PL/3$.

9.84 $\mathbf{R}_B = 45.8$ kN $\uparrow$; $M_{\max} = 62.5$ kN $\cdot$ m at D;
$M_{\min} = -119$ kN $\cdot$ m at A.

9.86 $\mathbf{R}_A = 17.78$ kN $\uparrow$;
$M_{\max} = 25.3$ kN $\cdot$ m at 1.644 m from A;
$M_{\min} = -51.2$ kN $\cdot$ m at B.

9.88 $\mathbf{R}_A = 3P/32 \downarrow$; $\mathbf{R}_B = 13P/32 \uparrow$;
$\mathbf{R}_C = 11P/16 \uparrow$.

9.90 $\mathbf{R}_A = 3P/8 \uparrow$; $\mathbf{R}_B = 7P/8 \uparrow$; $\mathbf{R}_C = P/4 \downarrow$.

9.92 $\mathbf{R}_A = 6M_0a(L - a)/L^3 \downarrow$,
$\mathbf{M}_A = M_0(L - a)(L - 3a)/L^2 \rangle$;
$M_B = +M_0a(3a - 2L)/L^2$.

9.94 $\mathbf{R}_A = 203$ kN $\uparrow$, $\mathbf{M}_A = 144$ kN $\cdot$ m $\rangle$;
$M_E = +76.6$ kN $\cdot$ m.

9.96 (a) 1.900 mm. (b) 41.7 kN $\uparrow$.

9.98 (a) 2007 N. (b) 875 N $\uparrow$.

9.100 $k = 84EI/L^3$.

9.102 (a) $\mathbf{R}_A = \mathbf{R}_B = 14.93$ kN $\uparrow$;
$\mathbf{R}_D = \mathbf{R}_E = 15.07$ kN $\uparrow$. (b) 2.07 mm $\downarrow$.

9.104 (a) 43.8 kN. (b) $M_B = +6.79$ kN $\cdot$ m;
$M_C = -1.42$ kN $\cdot$ m. (c) 0.770 mm $\downarrow$.

9.106 $\Delta T_A = \Delta T_B = 1.81$ kN; $\Delta T_C = 1.37$ kN.

9.108 (a) 3.11×10^{-3} rad $\measuredangle$. (b) 7.07 mm $\downarrow$.

9.110 (a) 5.16×10^{-3} rad $\searrow$. (b) 0.
(c) 5.16 mm $\downarrow$.

9.112 (a) 5.24×10^{-3} rad $\searrow$. (b) 12.90 mm. $\downarrow$.

9.114 (a) 217 kN $\uparrow$. (b) 3.29 mm $\downarrow$.

9.116 (a) 1.59×10^{-3} rad $\measuredangle$. (b) 2.32 mm $\downarrow$.
(c) 1.23 mm $\uparrow$.

9.118 (a) $L/4$, $wL^4/6144EI$. (b) $L/2$, $wL^4 384EI$.
(c) $L/2$, $wL^4/128EI$.

CHAPTER 10

10.2	(a) 64.5 kJ/m^3. (b) 1067 kJ/m^3. (c) 1736 kJ/m^3.
10.4	(a) 712 kJ/m^3. (b) 177.9 kJ/m^3. (c) 160.3 kJ/m^3.
10.6	(a) 1.09 MJ/m^3. (b) 83 MJ/m^3.
10.8	(a) 196 kJ/m^3. (b) 83 MJ/m^3.
10.10	(a) 279 J. (b) $u_{AB} = 594$ kJ/m^3. (c) $u_{BC} = 117.3$ kJ/m^3.
10.12	(a) 10.70 J. (b) 0.229 m.
10.14	80.4 J.
10.16	10.86 J.
10.18	5.03.
10.20	4.19.
10.22	7.09 mm.
10.24	$U = 0.2486 PL^2/EA_{min}$; -0.576 percent.
10.26	4.90 J.
10.28	$3.81 P^2 l/AE$.
10.30	$1.898 P^2 l/AE$.
10.32	59.8 J.
10.34	228 J.
11.36	$M_0^2(a^3 + b^3)/6EIL^2$.
10.38	$w^2 L^5/240EI$.
10.42	36.1 J.
10.44	339 J.
10.46	(a) 31.0 J. (b) 58.8 J.
10.48	$U_{AB} = 6.91$ J; $U_{CD} = 7.68$ J.
10.50	$U = (Q^2/4\pi\,GL)\ln(R_2/R_1)$.
10.52	$U = \dfrac{M_0^2 L}{6EI}\left(1 + \dfrac{3Eh^2}{10GL^2}\right)$.
10.54	11.04 MPa $\le \sigma_z \le 141.0$ MPa.
10.56	-74.9 MPa $\le \sigma_x \le 156.9$ MPa.
10.58	(a) 79.8 kN. (b) 254 MPa. (c) 1.523 mm.
10.60	5.20 m/s.
10.62	(a) 6.57 mm. (b) 11.28 kN. (c) 99.8 MPa.
10.64	285 mm.
10.66	3.45 m/s.
10.68	(a) 7.54 kN. (b) 41.3 MPa. (c) 3.18 mm.
10.70	(a) 2.71 mm. (b) 143.6 MPa.
10.72	(a) 21.6 mm. (b) 125.9 N · m. (c) 184.4 MPa.
10.74	1.057 mm.
10.78	(b) $y'_m = 17.40$ mm; $\lvert(y'_m - y_m)/y_m\rvert = 0.1946$; $y'_m/2h = 0.218$.

10.80	$Pa^2(a + L)/3EI\downarrow$.
10.82	$M_0(L + 3a)/3EI\ \text{↘}$.
10.84	$\tfrac{3}{4}Pa^3/EI\downarrow$.
10.86	2.56°.
10.90	385 mm.
10.94	10.68 mm $\downarrow$.
10.96	1.333 mm $\downarrow$.
10.98	$7.62 Pl/AE\downarrow$.
10.100	(a) and (b) $(P^2L^2 + 3M_0PL + 3M_0^2)L/6EI$.
10.102	(a) and (b) $(P^2L^2 - 6M_0PL + 16M_0^2)L/96EI$.
10.104	$5M_0^2 L/4EI$.
10.106	$5PL^3/48EI\downarrow$.
10.108	$3PL^2/8EI\angle$.
10.110	$(PL^3/48EI)/[1 - 3(a/L)]\downarrow$.
10.112	$PaL/24EI\angle$.
10.114	$7PL^2/128EI\ \text{↘}$.
10.116	(a) 5.48×10^{-3} rad ↘. (b) 10.10 mm $\downarrow$.
10.118	3.77 mm $\downarrow$.
10.120	$Pl/2AE\leftarrow$; $3.80 Pl/AE\downarrow$.
10.122	3.94 mm $\downarrow$.
10.124	1.906 mm $\rightarrow$.
10.126	(a) $PR^3/2EI\rightarrow$. (b) $\pi PR^3/4EI\downarrow$.
10.128	(a) $5PL^3/6EI\rightarrow$. (b) $PL^2/6EI$.
10.130	$\mathbf{R}_A = 5P/16\uparrow$; $M_B = -3PL/16$.
10.132	$\mathbf{R}_A = 7wL/128\uparrow$; $M_B = -9wL/128$.
10.134	$\mathbf{R}_A = 2M_0/3L\uparrow$; $\mathbf{R}_B = 2M_0/L\downarrow$; $\mathbf{R}_C = 4M_0/3L\uparrow$.
10.136	$F_{BE} = F_{DE} = 0.326P$; $F_{CE} = 0.435P$.
10.138	$-0.375P$.
10.140	(a) 357 mm $\downarrow$. (b) 34.8 MPa. (c) 3680 N.
10.142	8.74 mm $\downarrow$.
10.144	1.027 mm $\downarrow$.
10.146	4.79×10^{-3} rad $\angle$; 7.09 mm $\downarrow$.
10.148	0.743 mm $\downarrow$.
10.150	(a) 3.28. (b) 4.25.

CHAPTER 11

11.2	K/L.
11.4	K/L.
11.6	313 kN.
11.8	$8K/L$.
11.10	836 N.
11.12	17.69 N.
11.14	487 N.

11.16 14.7 mm; (aluminum) 6.94 kN, (steel) 2.36 kN.

11.18 (*a*) 64.3 kN. (*b*) 14.27 mm. (*c*) 97.7%.

11.20 (*a*) 65.4 kN. (*b*) 102.1 kN.

11.22 (*a*) 9.34 mm. (*b*) $P_1 = 380$ kN, $P_2 = 548$ kN.

11.24 1.421.

11.26 4.19 m.

11.28 37.2 kN.

11.30 2.43.

11.32 (*a*) 42.4 kN. (*b*) 168.4 kN.

11.34 4.00 kN.

11.36 (*a*) 60.1°. (*b*) 2.35.

11.38 (*a*) 6.75 m. (*b*) 6.19.

11.40 (*a*) 2.52 m. (*b*) 3.57 kN.

11.42 (*a*) $L_{BC} = 1.26$ m, $L_{CD} = 0.315$ m. (*b*) 19.5 kN.

11.44 (*a*) 2.55. (*b*) $d_2 = 28.3$ mm, $d_3 = 14.14$ mm, $d_4 = 16.73$ mm, $d_5 = 20.0$ mm.

11.46 (*a*) 0.500. (*b*) $b = 14.15$ mm, $d = 28.3$ mm.

11.48 6.32 m.

11.50 (*a*) 1.13 mm. (*b*) 133.7 MPa.

11.52 (*a*) 4.32 mm. (*b*) 44.4 MPa.

11.54 (*a*) 301 kN. (*b*) 88.7 MPa.

11.56 37.2°C.

11.58 (*a*) 122 kN. (*b*) 147 kN.

11.60 (*a*) 182 kN. (*b*) 0.6.

11.62 2.16 m.

11.64 (*a*) 4.13 m. (*b*) 2.29 m.

11.66 $d = 52$ mm.

11.68 240 kN.

11.70 (*a*) 218 kN. (*b*) 861 kN.

11.72 92.6 kN.

11.74 (*a*) 113.7 kN. (*b*) 239 kN.

11.76 (*a*) 278 kN. (*b*) 454 kN.

11.78 (*a*) 701 kN. (*b*) 807 kN.

11.80 (*a*) 787 kN. (*b*) 900 kN.

11.82 (*a*) 49.0 kN. (*b*) 239 kN.

11.84 (*a*) 168 kN. (*b*) 52.7 kN.

11.86 (*a*) 198.4 kN. (*b*) 213 kN.

11.88 186 kN.

11.90 919 kN.

11.92 192.4 kN.

11.94 894 kN.

11.96 (*a*) 22.7 kN. (*b*) 17.5 kN.

11.98 316 kN.

11.100 416 kN.

11.102 W 250×67.

11.104 W 360×79.

11.106 9.5-mm thickness.

11.108 9.5-mm thickness.

11.110 $d = 41.4$ mm.

11.112 10-mm thickness.

11.114 241 mm.

11.116 $d = 191$ mm.

11.118 7.24 kN.

11.120 (*a*) 121 kN. (*b*) 94 kN.

11.122 (*a*) 537 kN. (*b*) 376 kN.

11.124 37.5 kN.

11.126 45.3 kN.

11.128 36.6 mm.

11.130 (*a*) 2.56 m. (*b*) 2.81 m.

11.132 5.11 m.

11.134 101.9 mm.

11.136 164.4 mm.

11.138 50.9 mm.

11.140 6-mm thickness.

11.142 9-mm thickness.

11.144 46.5 mm.

11.146 $b = 203$ mm.

11.148 W 310×52.

11.150 W 250×58.

11.152 56.7 mm.

11.154 5.18 kN.

11.156 77.3 kN.

11.158 (*a*) 8.45 mm. (*b*) 185 MPa.

11.160 489 kN.

11.162 W 250×67.

Centroids of Common Shapes of Areas and Lines

Shape		$\bar{x}$	$\bar{y}$	Area
Triangular area			$\dfrac{h}{3}$	$\dfrac{bh}{2}$
Quarter-circular area		$\dfrac{4r}{3\pi}$	$\dfrac{4r}{3\pi}$	$\dfrac{\pi r^2}{4}$
Semicircular area		0	$\dfrac{4r}{3\pi}$	$\dfrac{\pi r^2}{2}$
Semiparabolic area		$\dfrac{3a}{8}$	$\dfrac{3h}{5}$	$\dfrac{2ah}{3}$
Parabolic area		0	$\dfrac{3h}{5}$	$\dfrac{4ah}{3}$
Parabolic spandrel		$\dfrac{3a}{4}$	$\dfrac{3h}{10}$	$\dfrac{ah}{3}$
Circular sector		$\dfrac{2r\sin\alpha}{3\alpha}$	0	αr^2
Quarter-circular arc		$\dfrac{2r}{\pi}$	$\dfrac{2r}{\pi}$	$\dfrac{\pi r}{2}$
Semicircular arc		0	$\dfrac{2r}{\pi}$	πr
Arc of circle		$\dfrac{r\sin\alpha}{\alpha}$	0	$2\alpha r$

Moments of Inertia of Common Geometric Shapes

Rectangle		$\bar{I}_{x'} = \frac{1}{12}bh^3$ $\bar{I}_{y'} = \frac{1}{12}b^3h$ $I_x = \frac{1}{3}bh^3$ $I_y = \frac{1}{3}b^3h$ $J_C = \frac{1}{12}bh(b^2 + h^2)$
Triangle		$\bar{I}_{x'} = \frac{1}{36}bh^3$ $I_x = \frac{1}{12}bh^3$
Circle		$\bar{I}_x = \bar{I}_y = \frac{1}{4}\pi r^4$ $J_O = \frac{1}{2}\pi r^4$
Semicircle		$I_x = I_y = \frac{1}{8}\pi r^4$ $J_O = \frac{1}{4}\pi r^4$
Quarter circle		$I_x = I_y = \frac{1}{16}\pi r^4$ $J_O = \frac{1}{8}\pi r^4$
Ellipse		$\bar{I}_x = \frac{1}{4}\pi ab^3$ $\bar{I}_y = \frac{1}{4}\pi a^3b$ $J_O = \frac{1}{4}\pi ab(a^2 + b^2)$